ANALYTICAL HEAT DIFFUSION THEORY

ANALYTICAL HEAT DIFFUSION THEORY

A. V. LUIKOV

HEAT AND MASS TRANSFER INSTITUTE
MINSK, BSSR, USSR

Edited by James P. Hartnett

DEPARTMENT OF ENERGY ENGINEERING
UNIVERSITY OF ILLINOIS, CHICAGO, ILLINOIS

1968

ACADEMIC PRESS

A Subsidiary of Harcourt Brace Jovanovich, Publishers

New York London Toronto Sydney San Francisco

First published in the Russian language under the title
TEORIYA TEPLOPROVODNOSTI
2nd Edition, Izd. "Vysshaya Shkola," Moscow, 1967

ACADEMIC PRESS, INC.
111 Fifth Avenue, New York, New York 10003

United Kingdom Edition published by
ACADEMIC PRESS, INC. (LONDON) LTD.
24/28 Oval Road, London NW1

LIBRARY OF CONGRESS CATALOG CARD NUMBER: 67-23167

PRINTED IN THE UNITED STATES OF AMERICA

80 81 82 9 8 7 6 5 4 3 2

EDITOR'S PREFACE

The editor is pleased to have played a role in introducing this textbook by Academician A. V. Luikov, member of the BSSR Academy of Sciences and Director of the Heat and Mass Transfer Laboratory in Minsk, BSSR, USSR.

This work is a revised edition of an earlier book by Academician Luikov which was widely used throughout the Soviet Union and the surrounding socialist countries. The presentation is unique in that it not only treats heat conduction problems by the classical methods such as separation of variables, but, in addition, it emphasizes the advantages of the transform method, particularly in obtaining short time solutions of many transient problems. In such cases, the long time solution may be obtained from the classical approach, and by interpolation, a very good estimate is obtained for intermediate times. The text is also noteworthy in that it covers a wide variety of geometrical shapes and treats boundary conditions of constant surface temperature, and constant surface heat flux, as well as the technically important case of a convective boundary condition.

The level of the book is advanced undergraduate or graduate. In addition to its value as a textbook, the availability of many technically important results in the form of tables and curves should make the book a valuable asset to the practicing engineer. The editor is convinced that the work will be well received by the English reading audience.

James P. Hartnett

October, 1968

CONTENTS

INTRODUCTION

The study of heat transfer is one of the most important fields of modern science and is of great practical significance in industrial power engineering, chemical technology, civil engineering, light industry, and other branches of technology. For example, design of thermal apparatus, design of walls of buildings undergoing changing thermal effects (heat insulation of buildings, furnaces, pipe lines), heating of machines, thermal stresses in bridges, and many other problems involve the solution of unsteady heat conduction problems. Investigation of the kinetics of sorption, drying, combustion, and other processes of chemical technology depends on the solution of diffusion problems, similar to those of transient heat conduction. Some aspects of unsteady heat transfer acquire special importance in rocket and propulsion engineering, where thermal apparatus operate under transient conditions. Thus the analytical heat conduction theory is widely applicable to the solution of various engineering problems.

The present work differs considerably from earlier monographs of the author, the last being published in 1952. While in the book " Heat Conduction and Diffusion " [68] all the problems were solved by contour integration, in the books published in 1948 and 1952 [69, 71] operational calculus and the Laplace transformation were used as the main solution methods.

In the present monograph, numerous problems of the heat conduction theory are solved by different methods: the method of separation of variables (a classical method for the solution of such problems), the integral Laplace, Fourier, and Hankel transformations, as well as operational calculus. For asymptotic solutions, the contour integration should be applied. Thus, the present monograph is a summary of all those previously published by the present author.

The book is a text for students and it considers in detail the solution of unsteady-state heat conduction problems of basic bodies (semi-infinite body, infinite plate, solid cylinder, sphere, hollow cylinder) by various methods (separation of variables, the operational Hankel and Fourier integral transformations).

Thus, the reader who becomes acquainted with peculiarities of every approach can choose for his work the most realistic method of solution with regard to conservation of time, labor, and accuracy for application to design problems.

The solutions are presented in generalized variables and obtained by the methods of the similarity theory. Much numerical information in the form of tables and text figures drawn by the author and excellent nomograms taken from Schneider's work [102] is given, allowing rapid engineering calculation which will prompt wider application of the solution to engineering practice. In addition, the solution of the most important problems are presented in two forms, one of which is convenient for calculation with small Fourier numbers, and the other with large ones.

The experience of the author in teaching heat conduction theory at various institutions has shown the necessity of presenting the whole process of solution in detail with main manipulations and calculations, and to consider the problems according to their difficulty.

In Chapters 4–6, detailed solutions are therefore illustrated by a large number of numerical examples and the problems are classified according to the mode of interaction of a body with the surroundings but not according to the geometry of bodies. This approach has been more effective from a pedagogical viewpoint.

Great attention is paid to the solution of problems with boundary conditions of the fourth kind. Such solutions are necessary for realistic studies in the field of unsteady convective heat transfer. A special chapter (Chapter 13) is devoted to the solution of problems with variable thermal coefficients.

In Chapter 14, a short description is given of application of the Laplace, Fourier, and Hankel transforms to the solution of unsteady-state heat conduction problems.

Readers who are interested in more profound problems of the heat conduction theory (asymptotic approximations, etc.) are referred to Chapter 15, where a brief description of the theory of analytical functions and their application to the solution of heat conduction problems is presented.

Reference material in the form of formulas and tables is given in the Appendixes.

The book is designed mainly for students and engineers who have the usual background in mathematics. The author aims to educate them in the solution of problems of unsteady-state heat transfer encountered in engineering design.

The author expresses his deep gratitude to Professor J. P. Hartnett for the interest shown in this manuscript and the work done by him in editing the English translation. He also wishes to thank Mrs. T. Kortneva, G. Malyavskaya, and E. Bogacheva from the Heat and Mass Transfer Institute of the BSSR Academy of Sciences for the English translation of this book.

CHAPTER

1

PHYSICAL FUNDAMENTALS OF HEAT TRANSFER

In the present chapter the main principles of the phenomenological heat conduction theory are given.

When heat is transferred from one part of a body to another, or from one body to another which is in contact with it, the process is usually referred to as heat conduction. In the phenomenological heat conduction theory, the molecular structure of a substance is neglected; the substance is considered as a continuous medium (continuum), but not as a combination of separate discrete particles. Such a model of the substance may be adopted when solving problems on heat propagation, provided that differential volumes are large compared with molecule sizes and distances between them. In all the following calculations and examples, the body is assumed isotropic and uniform.

1.1 Temperature Field

Any physical phenomenon, including the heat transfer process, occurs in time and space. Analytical investigation into heat conduction reduces, therefore, to the study of space-time variations of the main physical quantity (temperature) peculiar to a certain process, i.e., to the solution of the equation

$$t = f(x, y, z, \tau), \tag{1.1.1}$$

where x, y, z are Cartesian space coordinates and τ is time.

The instantaneous values of temperature at all points of the space of interest is called a temperature field. Since temperature is a scalar quantity, then so is a temperature field.

A distinction is made between a steady and transient temperature field. A transient temperature field is one in which the temperature varies not only in space but also with time; in other words, "temperature is a function of space and time" (the unsteady state). Equation (1.1.1) is a mathematical representation of a transient temperature field.

A steady temperature field is one in which the temperature at any point never varies with time, i.e., it is a function of the space coordinates solely (the steady state):

$$t = \Phi(x, y, z), \qquad \partial t/\partial \tau = 0. \tag{1.1.2}$$

In some problems, a transient temperature field becomes asymptotically steady when $\tau \to \infty$.

A temperature field governed by Eq. (1.1.1) or (1.1.2) is spatial (three-dimensional) since t is a function of three coordinates. When temperature is a function of two coordinates, the field is then two-dimensional:

$$t = F(x, y, \tau), \qquad \partial t/\partial z = 0.$$

If temperature is a function of one space coordinate alone, then the field is one-dimensional:

$$t = \varphi(x, \tau), \qquad \partial t/\partial y = \partial t/\partial z = 0.$$

The field of an infinite plate (i.e., the width and the length are infinitely large compared with its thickness) affords an example of a one-dimensional temperature field, the heat flow being normal to the plate surface.

If the points of the field with equal temperatures are connected, an isothermal surface results. Intersection of the isothermal surface with a plane

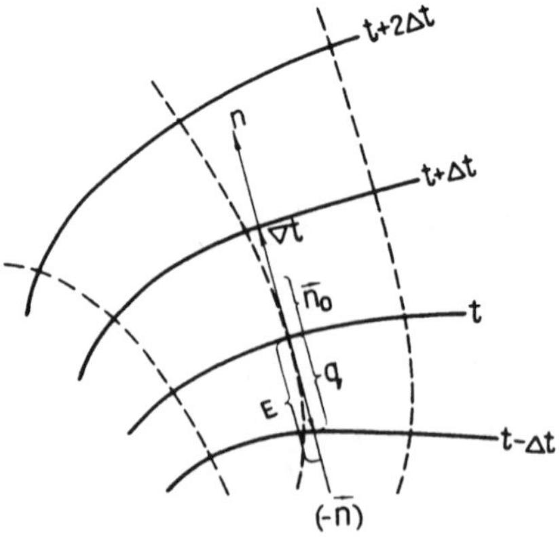

Fig. 1.1 Isotherms of temperature field. (Letters with arrows correspond to boldface type in the text.)

yields a family of isotherms on the plane surface (lines corresponding to equal temperature). Isothermal surfaces and isothermal lines never intersect inside the field when it is continuous. In Fig. 1.1, several isothermal lines are shown, each differing from its neighbor by the amount Δt.

By definition, the temperature does not change along the isotherm, but in any other direction it may vary. Temperature change per unit length is a maximum along the normal to the isothermal surface.

The temperature increase along the normal to the isothermal surface is characterized by a temperature gradient (grad t). A temperature gradient is a vector along the normal to the isothermal surface in the direction of the increasing temperature, i.e.,

$$\text{grad}\, t = \mathbf{n}_0(\partial t/\partial n), \tag{1.1.3}$$

where $\mathbf{n}_0$ denotes a unit vector,[1] along the normal in the direction of the temperature change (see Fig. 1.1) and $\partial t/\partial n$ is the temperature derivative along the normal (n) to the isothermal surface. A temperature gradient is therefore the first temperature derivative along the normal to the isothermal surface. The gradient is also denoted by ∇. Gradient components along the Cartesian coordinates are identical to the appropriate partial derivaties, so that

$$\text{grad}\, t = \nabla t = \mathbf{i}\frac{\partial t}{\partial x} + \mathbf{j}\frac{\partial t}{\partial y} + \mathbf{k}\frac{\partial t}{\partial z} \tag{1.1.4}$$

where $\mathbf{i}$, $\mathbf{j}$, $\mathbf{k}$ are mutual orthogonal vectors of a unit length along the coordinate axes. This relation is possible because any vector may be represented as a vectorial sum of three components along the coordinate axes.

The concept of a temperature-field intensity may be introduced:

$$\mathbf{E} = -\,\text{grad}\, t. \tag{1.1.5}$$

The vector $\mathbf{E}$ is referred to as a vector of the temperature field intensity.

1.2 The Fundamental Fourier Heat Conduction Law

The necessary condition for heat conduction is the existence of a temperature gradient. Experience shows that heat is transferred by conduction in the direction normal to the isothermal surface from a higher temperature level to a lower one.

[1] Vectors are shown by boldface type and their scalar values by italic type.

The quantity of heat transferred per unit time per unit area of the isothermal surface is referred to as a heat flux; the appropriate vector is obtained by the relation

$$\mathbf{q} = (-\mathbf{n}_0)\frac{dQ}{d\tau}\frac{1}{S}, \tag{1.2.1}$$

where $dQ/d\tau$ is quantity of heat transferred per unit time, or the heat-flow rate, S is the isothermal surface area, and $(-\mathbf{n}_0)$ is a unit vector along the normal to the area S in the direction of the decreasing temperature.

The vector $\mathbf{q}$ is therefore called a heat flux vector, the direction of which is opposite to that of the temperature gradient (both vectors follow the normal to the isothermal surface, but their directions are opposite to each other).

The projection of the vector $\mathbf{q}$ on any arbitrary direction l is also the vector $\mathbf{q}_l$, the scalar quantity of which is $q \cos(n, l)$.

The lines which coincide with the direction of the vector $\mathbf{q}$ are referred to as heat-flow lines. These are perpendicular to the isothermal surfaces at the intersection points. A tangent to the heat-flow lines taken in the opposite direction yields the temperature gradient direction (Fig. 1.1).

The fundamental heat-conduction law may be formulated as follows: the heat flux is proportional to the temperature field intensity, or the heat flux is proportional to the temperature gradient, i.e.,

$$\mathbf{q} = \lambda\mathbf{E} = -\lambda \operatorname{grad} t = -\lambda \nabla t = -\lambda\mathbf{n}_0(\partial t/\partial n), \tag{1.2.2}$$

where λ is the proportionality factor called thermal conductivity.

To reveal the physical significance of the thermal conductivity, we shall write the basic relation (1.2.2) for a steady one-dimensional temperature field for the situation where the temperature depends only on one coordinate which is normal to the isothermal surfaces. The scalar quantity of the heat-flux vector is

$$q = -\lambda\frac{dt}{dx} \qquad \left(\frac{\partial t}{\partial \tau} = \frac{\partial t}{\partial y} = \frac{\partial t}{\partial z} = 0\right). \tag{1.2.3}$$

If the temperature gradient is a constant value ($dt/dx = \text{const}$), which means the temperature variation with x follows the linear law, then it may be written

$$\frac{dt}{dx} = \frac{t_2 - t_1}{x_2 - x_1} = \text{const}. \tag{1.2.4}$$

Hence the heat-flow rate $dQ/d\tau$ is also a constant value

$$dQ/d\tau = Q/\tau = \text{const}, \tag{1.2.5}$$

where Q is the quantity of heat flowing in the time τ.

It follows from Eqs. (1.2.1)–(1.2.5) that

$$\frac{Q}{S\tau} = -\lambda \frac{t_2 - t_1}{x_2 - x_1} = \lambda \frac{t_1 - t_2}{x_2 - x_1}, \tag{1.2.6}$$

since $t_1 > t_2$ and $x_2 > x_1$.

Thus the thermal conductivity is equal to the heat flowing per unit time and per unit surface when the temperature difference per unit length of the normal is 1 degree. Thermal conductivity has dimensions of kcal/m hr °C or W/m °C. Thermal conductivity is a physical property of a body characterizing its ability to transfer heat. The physical significance of the thermal conductivity and its dependence on basic properties of a body may be better understood when we consider the heat transfer mechanism in a body in a specific state.

The relation $\lambda/(x_2 - x_1) = \lambda/\Delta x$ (kcal/m² hr °C or W/m² °C) is called thermal conductance of a certain portion of a body and the inverse value $\Delta x/\lambda$ (m² hr °C/kcal) (m² °C/W) is the thermal resistance of this portion of a body. The magnitude of the thermal conductivity of materials varies over wide ranges: from 0.0074 kcal/m hr °C (0.0086 W/m °C) for tetrachloro-methane at 100°C to 358 kcal/m hr °C (416 W/m °C) for silver at 0°C. Thermal conductivity depends on chemical composition, physical structure, and state of the material.

Heat conduction in gases and vapors depends mainly on molecular transfer of the kinetic energy of molecular movement, and therefore conductivity values (λ) are naturally small for gases and vapors.

In liquids a mechanism of heat transfer by conduction is similar to that of sound propagation (propagation of longitudinal waves). Thermal conductivities of liquids are therefore higher than those of gases. Molecular structure of crystals favors heat transfer.

In metals, heat transfer by conduction depends mainly on energy transfer by free electrons. Difference in thermal conductivities of various nonuniform materials are caused by the voidage effect. For fibrous materials, anisotropy is a destruction of uniformity which results in different thermal conductivity in various directions. Thermal conductivity depends on temperature, which, for many metals, decreases with increase in temperature following a linear law.

Thermal conductivity of gases increases with temperature but it is practically independent of pressure except for very high (> 2000 atm) and very low ($<$ 10 mm Hg) pressure levels. For gas mixtures thermal conductivity λ may be determined only experimentally since the additive law is not valid.[2]

Liquids have thermal conductivity values of 0.08 to 0.6 kcal/m hr°C. For the majority of liquids, thermal conductivity decreases with increasing temperature, with water and glycerine being exceptions to this rule. Thermal conductivity of buildings and heat insulating materials varies from 0.02 to 2.5 kcal/m hr °C, increasing approximately linearly with temperature. Materials with low thermal conductivity ($\lambda \leq 0.2$ kcal/m hr °C) are usually referred to as heat insulating materials.

For engineering calculations, the thermal conductivity of gases may be assumed to change with temperature according to a linear law

$$\beta_t = d\lambda/dt = \text{const}.$$

For a certain temperature range, $\Delta t = t_2 - t_1$, thermal conductivity may be assumed constant and equal to the arithmetic mean value of λ at t_2 and t_1.

The Fourier heat conduction law (1.2.2) may be written in another form. Expressing the internal energy per unit volume of a body by U_v, the scalar value of the temperature gradient may be written

$$\frac{\partial t}{\partial n} = \left(\frac{\partial t}{\partial U_v}\right)_v \frac{\partial U_v}{\partial n} = \frac{1}{C_v}\frac{\partial U_v}{\partial n}, \tag{1.2.7}$$

where C_v is the volume heat capacity at a constant volume (kcal/m³ °C)

$$C_v = \left(\frac{\partial U_v}{\partial t}\right)_v = c_v\gamma \tag{1.2.8}$$

where c_v is the specific heat at a constant volume (kcal/kg °C) and γ is the density of a body (kg/m³). Consequently, the heat-conduction equation will be of the form

$$\mathbf{q} = -\mathbf{n}_0\lambda\frac{\partial t}{\partial n} = -a_v \nabla U_v, \tag{1.2.9}$$

where a_v is the thermal diffusivity at a constant volume of a body (v = const)

$$a_v = \lambda/C_v = \lambda/c_v\gamma. \tag{1.2.10}$$

[2] *Editor's note*: Empirical methods for calculating the thermal conductivity of binary mixtures are given in the treatise by J. O. Hirschfelder, C. F. Curtis, and R. B. Bird "Molecular Theory of Gases and Liquids," Wiley, New York, 1954.

According to Eq. (1.2.9) *the heat flux is directly proportional to the gradient of the internal energy of a body per unit volume. The proportionality factor equals the thermal diffusivity when the body volume is constant.*

For a steady one-dimensional heat flow ($q =$ const, $\partial U_v/\partial n =$ const), Eq. (1.2.9) may be written as

$$\frac{Q}{S\tau} = a_v \frac{U_{v1} - U_{v2}}{x_2 - x_1}. \tag{1.2.11}$$

Thus, *the coefficient a_v is equal to the heat flowing per unit time and per unit surface area when the difference of the volume concentration of internal energy per unit length of the normal is 1 kcal/m³.*

Consequently, thermal diffusivity has the dimension of

$$[a_v] = \frac{\text{kcal m}^3\text{ m}}{\text{m}^2\text{ hr kcal}} = \frac{\text{m}^2}{\text{hr}}.$$

Its physical significance is that it characterizes the molecular transfer of the internal energy of a body.

With constant pressure ($p =$ const) the thermal diffusivity a_p (m²/hr), is obtained from the relation

$$a_p = \frac{\lambda}{c_p\gamma} = \frac{\lambda}{C_p} \tag{1.2.12}$$

where c_p and C_p are specific and pressure constant heat capacities of a body, respectively,

$$C_p = c_p\gamma \left(\frac{\partial H_v}{\partial t}\right)_p, \tag{1.2.13}$$

where H_v is the enthalpy per unit volume (kcal/m³).

The physical significance of the thermal diffusivity a_p is that it characterizes the molecular enthalpy transfer within a body.

The heat conduction law may be written as

$$\mathbf{q} = -\mathbf{n}_0\lambda \left(\frac{\partial t}{\partial H_v}\right)_p \frac{\partial H_v}{\partial n} = -a_p \nabla H_v. \tag{1.2.14}$$

Thus thermal diffusivity is a diffusion coefficient of internal energy (a_v) or enthalpy (a_p) depending on conditions of body interaction with the surrounding medium ($v =$ const or $p =$ const).

For solids, the specific heat capacity at a constant volume c_v differs only slightly from the specific heat capacity c_p at a constant pressure. It may therefore be assumed that $c_v = c_p = c$. In the analytical heat conduction theory of solids, the thermal diffusivities are assumed to be one and the same, independent of the conditions of body conjunction with the surrounding medium, i.e.,

$$a = a_p = a_v = \lambda/c\gamma. \qquad (1.2.15)$$

We now return to basic relation (1.2.2). The scalar value of the heat flux vector is

$$q = -\lambda(\partial t/\partial n). \qquad (1.2.16)$$

Components of the vector **q** along the coordinate axes x, y, z are designated by $\mathbf{q}_x$, $\mathbf{q}_y$, $\mathbf{q}_z$, scalar values of which are

$$q_x = q\cos(n, x) = -\lambda(\partial t/\partial n)\cos(n, x) = -\lambda(\partial t/\partial x), \qquad (1.2.17)$$

$$q_y = -\lambda(\partial t/\partial y), \qquad (1.2.18)$$

$$q_z = -\lambda(\partial t/\partial z), \qquad (1.2.19)$$

respectively.

The amount of heat flowing through the elementary area dS_l making an angle of ψ with the isothermal surface (or with a plane tangent to the isothermal surface) can be found from

$$q_l = q\cos\psi = \frac{dQ}{d\tau}\frac{1}{dS_n}\cos\psi = \frac{dQ}{d\tau}\frac{1}{dS_l} \qquad (1.2.20)$$

since (see Fig. 1.2)

$$dS_n = dS_l\cos\psi$$

is a projection of the area dS_l on the isothermal surface.

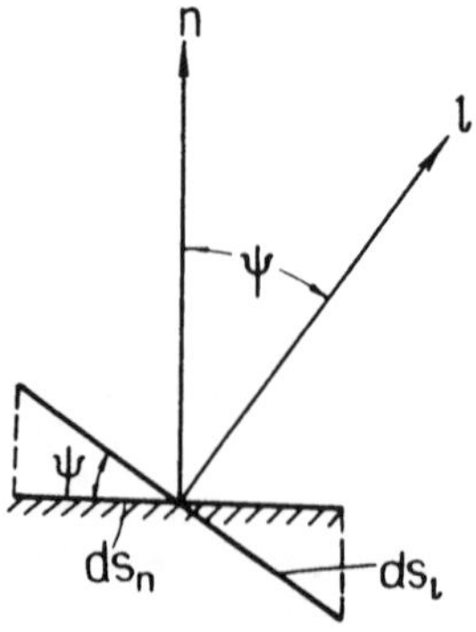

Fig. 1.2 Heat flux through elementary area.

Equation (1.2.20) yields

$$dQ = q_l\, dS_l\, d\tau = q(dS_l \cos\psi)\, d\tau = q\, dS_n\, d\tau. \qquad (1.2.21)$$

Equation (1.2.21) considers transfer of heat dQ in two ways: (1) heat flow through the area dS_l and along the normal (l) to it; and (2) heat flow through the projection of the area dS_l on the isothermal surface along the normal (n) to the latter.

The quantity of heat Q which flows through the surface S of a finite size in the time τ is

$$Q = -\lambda \int_0^\tau \int_{(S)} \partial t/\partial l\, dS_l\, d\tau. \qquad (1.2.22)$$

Thus, to determine the quantity of heat which flows through a surface of a solid, the temperature field inside the body must be known. The determination of the temperature field is the main problem of mathematical heat conduction theory.

1.3 Heat Distribution in the High Rate Processes

In the phenomenological heat conduction theory, the heat distribution velocity is assumed infinitely large. This assumption is confirmed by calculations of temperature fields in various bodies at the usual conditions encountered in practice. However, in rarefied media in transient heat-transfer processes of high rates, it should be taken into account that the heat distribution velocity is not infinitely high, but has a certain finite, although very great, value w_r.

Vernotte [125], followed by C. Cattaneo, was the first to pay attention to this. The present author independently suggested a hypothesis on finite heat and mass distribution for studying heat and moisture transfer in capillary-porous bodies [65].

This velocity will be

$$w_r = (\lambda/c\gamma\tau_r)^{1/2}, \qquad (1.3.1)$$

where τ_r is a time constant or relaxation time. For nitrogen, w_r is about 150 m/sec, and $\tau_r \approx 10^{-9}$ sec; for metals it is less, say for aluminum, $\tau_r \approx 10^{-11}$ sec. Experimental measurements of τ_r is impossible because of the limitations of existing measuring techniques. But for gases in supersonic flows, the effect of the finite heat distribution velocity w_r on heat transfer becomes more noticeable. In this case, the law of heat distribution will be of the form

$$\mathbf{q} = -\lambda \nabla t - \tau_r(\partial \mathbf{q}/\partial \tau). \tag{1.3.2}$$

For a steady heat flux ($\mathbf{q}$ = const, $\partial \mathbf{q}/\partial \tau = 0$), Eq. (1.3.2) is identical with Eq. (1.2.2). For nonstationary processes of high rate, the second term of Eq. (1.3.2) may become comparable with the first.

Equation (1.3.2) is similar to the viscous flow equation for non-Newtonian fluids (viscoelastic fluids). We shall consider this in more detail.

A century ago, Maxwell pointed out the similarities between mechanical properties of fluids and solid bodies, using the concept of relaxation. Relaxation is a phenomenon of progressive diffusion of elastic shear stress with constant values of the prescribed strain, i.e., constant dissipation of elastic energy stored in a body under strain by conversion of it into heat. Relaxation phenomena as well as diffusion processes are inseparably linked with random thermal molecular motion.

If the relaxation time is very large compared with the usual observation time, the fluid then behaves like a solid. If relaxation time is very short, a body exhibits properties of a viscous fluid. Between limiting states (perfectly elastic solids and viscous (Newtonian) fluids) there exists a continuous series of transitions which form the variety of real bodies of intermediate nature. For two visco-elastic fluids (non-Newtonian), the shear stress p depends on the shear strain ε. Close to the body surface this relation may be written as follows:

$$p = \eta \frac{d\varepsilon}{d\tau} - \frac{\eta}{G} \frac{dp}{d\tau}, \tag{1.3.3}$$

where η is a viscosity, G is the shear modulus, and $d\varepsilon/d\tau$ is the shear strain propagation velocity. η/G equals the relaxation time τ_r ($\tau_r = \eta/G$). We shall denote the strain propagation velocity through $\dot{\varepsilon}$ ($\dot{\varepsilon} = d\varepsilon/d\tau$). This results in

$$p = \eta\dot{\varepsilon} - \tau_r(dp/d\tau). \tag{1.3.4}$$

If a relaxation time is short ($\tau_r \to 0$), then from Eq. (1.3.4) we obtain the Newtonian equation of a viscous flow for a laminar plane-parallel flow[3]

[3] Equation (1.3.5) is an approximate one which is valid for a particular case of a parallel plane flow. In a general case, the shear stress

$$p_{xy} = -\eta\left(\frac{\partial w_x}{\partial y} + \frac{\partial w_y}{\partial x}\right).$$

For a parallel plane flow, $\partial w_y/\partial x \ll \partial w_x/\partial y$, i.e., $p_{xy} = -\eta(\partial w_x/\partial y)$.

$$p = \eta\dot{\varepsilon} = -\eta(\partial w_x/\partial y) \tag{1.3.5}$$

since

$$\dot{\varepsilon} = \frac{d\varepsilon}{d\tau} = -\frac{d}{d\tau}\left(\frac{dl_x}{dy}\right) = -\frac{\partial w_x}{dy}, \tag{1.3.6}$$

where y is the normal to the x direction and w_x is the fluid velocity ($w_x = dl_x/d\tau$). If the viscosity η is very large ($\eta \to \infty$), then from Eq. (1.3.3) the equation of Hook's classical law is obtained:

$$0 = \frac{d\varepsilon}{d\tau} - \frac{1}{G}\frac{dp}{d\tau}$$

i.e.,

$$p = G\varepsilon_1 = \gamma w_c^2\varepsilon, \tag{1.3.7}$$

where w_c is the shear strain propagation velocity (a velocity of propagation of transverse waves)

$$w_c = (G/\gamma)^{1/2}. \tag{1.3.8}$$

Similarly, from Eq. (1.3.2) we obtain the following limiting cases. Equation (1.3.2) may be written in a scalar form as follows

$$-\frac{\partial t}{\partial n} = \frac{q}{\lambda} + \frac{1}{c\gamma w_r^2}\frac{\partial q}{\partial \tau}. \tag{1.3.9}$$

If thermal conductivity is low ($\lambda \to 0$), and the velocity w_r is high ($w_r \to \infty$), then from Eq. (1.3.9) the classical equation of the Fourier law is obtained

$$q = -\lambda\frac{\partial t}{\partial n}. \tag{1.3.10}$$

If the thermal conductivity is high ($\lambda \to \infty$) or the volume heat capacity $c\gamma$ is small ($c\gamma \to 0$), we obtain from Eq. (1.3.9) after some manipulations

$$q = c\gamma(w_r^2/w_t)\,\Delta t, \tag{1.3.11}$$

where w_t is an isotherm propagation velocity which is

$$w_t = dn/d\tau \tag{1.3.12}$$

Explanation of the concept of an isotherm propagation velocity will be given below.

Let an equation of the isothermal surface be

$$t(x, y, z, \tau) = \text{const}. \tag{1.3.13}$$

The total differential of this surface is

$$\frac{\partial t}{\partial \tau}\, d\tau + \frac{\partial t}{\partial n}\, dn = 0. \tag{1.3.14}$$

This equation may be written as

$$\frac{\partial t}{\partial \tau} + \frac{\partial t}{\partial n}\frac{dn}{d\tau} = \frac{\partial t}{\partial \tau} + w_t \frac{\partial t}{\partial n} = 0. \tag{1.3.15}$$

The time derivative of the normal to the isothermal surface $dn/d\tau$ is the propagation velocity of the isothermal surface.

One can see from Eq. (1.3.11) that the heat flux is directly proportional to the temperature displacement, Δt, the volume heat capacity of a body, $c\gamma$, the square of the heat distribution velocity, w_r^2, and inversely proportional to the velocity of isothermal surface propagation, w_t.

1.4 Heat Distribution Equation in Liquid and Gas Mixtures

In gas mixtures and in solutions, heat transfer by conduction is combined with mass transfer. When there is a temperature gradient in such systems, thermal diffusion (the Soret effect) occurs, and mass diffusion causes heat transfer which is referred to as diffusion heat conduction or diffusion thermo (the Dufour effect). For a binary gas mixture, say of the density ϱ (kg/m^3) the heat flux is

$$\mathbf{q} = -\lambda \nabla t - D\varrho Q^* \nabla \varrho_{10} \tag{1.4.1}$$

where D is the interdiffusion coefficient (m^2/hr) ϱ_{10} is the relative mass fraction of the first component ($\varrho_{10} = \varrho_1/\varrho$), and ϱ_1 is the volume concentration of the first component

$$\varrho_1 + \varrho_2 = \varrho; \qquad \varrho_{10} + \varrho_{20} = 1. \tag{1.4.2}$$

Consequently, $\nabla \varrho_{10} = -\nabla \varrho_{20}$. In Eq. (1.4.1), heat transfer due to mass diffusion (i.e., diffusion thermo) is neglected.

The specific heat of isothermal transfer is Q^* (kcal/kg) which is equal to the heat amount transferred by unit mass under isothermal conditions. Q^* may be expressed through the chemical potential of the first component μ_1 of the mixture and the thermal diffusion coefficient k_t

$$Q^* = \frac{\varrho k_T}{\varrho_{20}} \left(\frac{(\partial \mu_1)}{\partial \varrho_{10}}\right)_{\varrho, T} \tag{1.4.3}$$

Thus the heat flux $\mathbf{q}$ depends not only on the temperature gradient ∇t but also on the concentration gradient $\nabla \varrho_{10}$.

The mass flux $\mathbf{j}_1$ for the first component is

$$\mathbf{j}_1 = - D\varrho(\nabla \varrho_{10} + (k_T/T) \nabla t). \quad (1.4.4)$$

Thus, the mass flux $\mathbf{j}_1$ depends on the concentration gradient $\nabla \varrho_{10}$ and the temperature gradient ∇t.

Transfer of heat and dissolved matter in solutions is described by similar equations (1.4.1) and (1.4.4). In this case the quantity $\varrho k_T/T$ is referred to as the Soret coefficient σ ($\sigma = \varrho k_T/T$). The heat transfer process is therefore inseparably linked with mass transfer and is a complex heat and mass transfer process.

Heat transfer combined with mass transfer is treated by the irreversible thermodynamics. A flux of some substance $\mathbf{j}_i$ (energy, mass, electricity, etc.) is caused by the action of all the thermodynamic forces $\mathbf{X}_k$ ($k = 1, 2, 3, ..., n$)

$$\mathbf{j}_i = \sum_{k=1}^{n} L_{ik}\mathbf{X}_k, \qquad i = 1, 2, 3, ..., n. \quad (1.4.5)$$

Relation (1.4.5) is a system of linear Onsager relations, which is a basic relation of the irreversible thermodynamics theory. The quantities L_{ik} are referred to as kinetic coefficients for which the reciprocal equation holds:

$$L_{ik} = L_{ki}. \quad (1.4.6)$$

Thermodynamics forces $\mathbf{X}_i$ and fluxes $\mathbf{j}_i$ should satisfy the basic formula of the irreversible thermodynamics theory,

$$T\frac{dS}{d\tau} = \sum_i \mathbf{j}_i\mathbf{X}_i, \quad (1.4.7)$$

where S is the entropy of the system considered. Using the Gibbs equation

$$T\,dS = dU + p\,dV - \sum_{k=1}^{n} \mu_k\,dM_k \quad (1.4.8)$$

(where U is the intrinsic energy, μ_k is the chemical potential, V is the volume and M is the mass), and differential equations of energy and mass transfer, one may define the thermodynamic forces $\mathbf{X}_i$ by formula (1.4.7).

For instance, in case of transfer of internal energy and mass in a gas mixture, the thermodynamic forces will be

$$\mathbf{X}_u = -\frac{1}{T}\nabla T; \qquad \mathbf{X}_{mk} = \mathbf{F}_k - T\nabla\frac{\mu_k}{T} \tag{1.4.9}$$

where $\mathbf{F}_k$ is the external force.

Accounting for the relation $\sum_{k=1}^{n}\mathbf{j}_k = 0$, fluxes of energy $\mathbf{j}_u$ and mass $\mathbf{j}_{mi}$ of the ith component will be

$$\mathbf{j}_u = L_{uu}\frac{1}{T}\nabla T - \sum_{k=1}^{n} L_{uk}\left[T\nabla\left(\frac{\mu_k - \mu_r}{T}\right) - (\mathbf{F}_k - \mathbf{F}_n)\right], \tag{1.4.10}$$

$$\mathbf{j}_{mi} = -L_{iu}\frac{1}{T}\nabla T - \sum_{k=1}^{n} L_{ik}\left[T\nabla\left(\frac{\mu_k - \mu_n}{T}\right) - (\mathbf{F}_k - \mathbf{F}_n)\right], \tag{1.4.11}$$

where L_{uu}, L_{uk}, L_{iu}, L_{ik} are the kinetic Onsager coefficients.

The fluxes of energy $\mathbf{j}_u$ and heat $\mathbf{j}_q$ are related by

$$\mathbf{j}_q = \mathbf{j}_u - \sum_{k=1}^{n} h_k\,\mathbf{j}_k, \tag{1.4.12}$$

where h_k is the specific enthalpy of the kth component.

In this case the thermodynamic force of heat $\mathbf{X}_q$ and mass $\mathbf{X}_{mk}$ transfer will be

$$\mathbf{X}_q = \mathbf{X}_u = -\frac{1}{T}\nabla T; \qquad \mathbf{X}_{mk} = -(\nabla\mu_k)_T. \tag{1.4.13}$$

Then, in the absence of the external forces field ($\mathbf{F}_k = 0$) we shall have

$$\mathbf{j}_q = -L_{qq}\frac{1}{T}\nabla T - \sum_{k=1}^{n-1} L_{qk}[\nabla(\mu_k - \mu_n)]_T, \tag{1.4.14}$$

$$\mathbf{j}_{mi} = -L_{iq}\frac{1}{T}\nabla T - \sum_{k=1}^{n-1} L_{ik}[\nabla(\mu_k - \mu_n)]_T. \tag{1.4.15}$$

The kinetic coefficients L_{qq}, L_{qk}, L_{iq}, L_{ik} are expressed through heat and mass transfer coefficients. For a binary gas mixture ($n = 1, 2$) with constant pressure ($p = \text{const}$) according to the Gibbs–Duhem equation,

$$\varrho_{10}\,d\mu_1 + \varrho_{20}\,d\mu_2 = 0, \tag{1.4.16}$$

the gradient $\nabla(\mu_k - \mu_n)_T$ will be

$$\nabla(\mu_k - \mu_n)_T = \frac{1}{\varrho_{20}}\left(\frac{\partial\mu_1}{\partial\varrho_{10}}\right)_{p,T}\nabla\varrho_{10}. \tag{1.4.17}$$

This then yields

$$\mathbf{j}_q = -L_{qq}\frac{1}{T}\nabla T - L_{q1}\frac{\mu_1'}{\varrho_{20}}\nabla\varrho_{10} \tag{1.4.18}$$

$$\mathbf{j}_i = L_{1q}\frac{1}{T}\nabla T - L_{11}\frac{\mu_1'}{\varrho_{20}}\nabla\varrho_{10} \tag{1.4.19}$$

where μ_1' denotes $(\partial\mu_1/\partial\varrho_{10})_{p,T}$.

Comparing Eqs. (1.4.18) and (1.4.19) with Eqs. (1.4.1) and (1.4.4) and assuming $\mathbf{j}_q \equiv \mathbf{q}$, we find

$$\lambda = \frac{L_{qq}}{T};\qquad D = \frac{L_{11}\mu_1'}{\varrho_2};\qquad Q^* = \frac{L_{q1}}{L_{11}};\qquad k_T = \frac{L_{1q}\varrho_{20}}{L_{11}\mu_1'}. \tag{1.4.20}$$

Molecular heat transfer may be considered similarly in combination with any other substance transfer in more complex systems.

1.5 Differential Heat Conduction Equation

To solve problems involving the temperature field determination, one should obtain a differential heat conduction equation. A differential equation is a mathematical relation (expressed by a differential equation) between physical quantities characterizing the phenomenon considered, these quantities being functions of space and time. Such an equation characterizes the physical process at any point of a body at any moment.

A differential heat conduction equation provides a relation between temperature, time, and coordinates of an elementary volume.

A differential equation will be derived by a simplified method. A one-dimensional temperature field is assumed (heat propagates in only one direction; say, in the direction of the x axis). Thermal coefficients are assumed to be independent of spatial coordinates and time.

From a uniform and isotropic infinite plate, we single out an elementary parallelepiped of the volume $dx\ dy\ dz$ (Fig. 1.3). The heat amount flowing

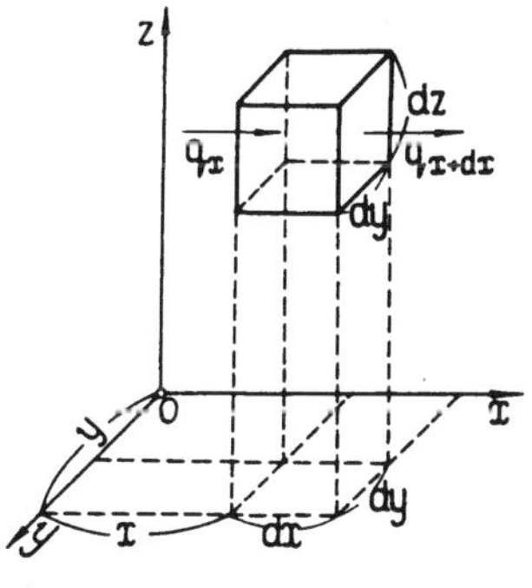

Fig. 1.3 Heat flux through elementary volume.

in through the left side, $dy\,dz$, into the parallelepiped per unit time is $q_x\,dy\,dz$, and the heat amount flowing out through the opposite side per unit time is $q_{x+dx}\,dy\,dz$.

If $q_x > q_{x+dx}$, then the elementary parallelepiped will be heated. Then the difference between these flows according to the law of energy conservation is equal to the heat accumulated in this elementary parallelepiped,[4] i.e.,

$$q_x\,dy\,dz - q_{x+dx}\,dy\,dz = c\gamma \frac{\partial t}{\partial \tau}\,dx\,dy\,dz. \tag{1.5.1}$$

The quantity q_{x+dx} is an unknown function of x. If it is expanded in Taylor's series and only the first two terms of the series are retained, it may be written

$$q_{x+dx} \approx q_x + \frac{\partial q_x}{\partial x}\,dx.$$

Then from Eq. (1.5.1) it follows that

$$-\frac{\partial q_x}{\partial x}\,dx\,dy\,dz = c\gamma \frac{\partial t}{\partial \tau}\,dx\,dy\,dz.$$

Using the heat conduction equation $q_x = -\lambda(\partial t/\partial x)$, we obtain

$$c\gamma \frac{\partial t}{\partial \tau} = \lambda \frac{\partial^2 t}{\partial x^2}$$

or

$$\frac{\partial t}{\partial \tau} = a \frac{\partial^2 t}{\partial x^2}. \tag{1.5.2}$$

Expression (1.5.2) is a differential heat conduction equation for a one-dimensional heat flow. If heat propagates along the normal to the isothermal surfaces, then the vector $\mathbf{q}$ may be expanded in three components along the coordinate axes. The heat stored in the elementary volume will be equal to the sum

$$-\left(\frac{\partial q_x}{\partial x} + \frac{\partial q_y}{\partial y} + \frac{\partial q_z}{\partial z}\right) dx\,dy\,dz.$$

[4] The heat accumulated is calculated from the elementary relation $\Delta \dot{Q} = cM\,\Delta\dot{\vartheta} = c\gamma V\,\Delta\dot{\vartheta}$, where $\Delta\dot{\vartheta}$ is a temperature increment in a body per unit time with the mass M and the volume V; c is a specific heat.

Then the differential equation can be written

$$\frac{\partial t}{\partial \tau} = a\left(\frac{\partial^2 t}{\partial x^2} + \frac{\partial^2 t}{\partial y^2} + \frac{\partial^2 t}{\partial z^2}\right) = a\,\nabla^2 t\,, \tag{1.5.3}$$

where

$$\nabla^2 = \frac{\partial^2}{\partial x^2} + \frac{\partial^2}{\partial y^2} + \frac{\partial^2}{\partial z^2}$$

is a Laplace operator. (Laplace operators expressed in spherical and cylindrical coordinates are represented in Appendix.)

Inside a body, there are sometimes heat sources, which may be positive or negative. Moisture evaporation inside the material with heating is an example of a negative source. Let the specific strength (quantity of heat absorbed or released per unit time per unit volume) of these sources be w (kcal/m^3 hr). Then the heat amount generated in the elementary volume per unit time will be $w\,dx\,dy\,dz$; this amount should be subtracted from the amount of the stored heat for the equality (1.5.1) to hold. The resulting differential heat conduction equation with heat sources will be of the form

$$\frac{\partial t}{\partial \tau} = a\,\nabla^2 t + \frac{w}{c\gamma}. \tag{1.5.4}$$

Differential equation (1.5.3) may be derived by a more general method using the Ostrogradsky–Gauss transformation. We have a medium in which the volume V bounded by the surface S will be considered. Heat is transferred in this medium by conduction. According to Eq. (1.2.2), the heat amount which flows through the surface S per unit time is

$$\int_{(S)} \lambda \operatorname{grad} t\, d\mathbf{S} = \int_{(S)} \lambda \mathbf{n}_0 \cdot \operatorname{grad} t\, dS.$$

Here the integral is taken over the whole surface S. When the body is free of heat sources, this heat flow causes changes in the internal energy of the medium in the given volume per unit time by the value

$$\frac{\partial}{\partial \tau}\int_{(V)} c\gamma t\, dv = \int_{(V)} c\gamma\,\frac{\partial t}{\partial \tau}\, dv.$$

Here the integral is taken over the whole volume V.

Following the law of energy conservation, change in the internal energy of the medium in the volume V is equal to the heat loss through the surface S bounding the given volume V, i.e.,

$$\int_{(V)} c\gamma\,\frac{\partial t}{\partial \tau}\, dv = \int_{(S)} \mathbf{n}_0\,\lambda \operatorname{grad} t\, dS. \tag{1.5.5}$$

Using the Ostrogradsky–Gauss transformation we obtain

$$\int_{(S)} \mathbf{n}_0 \lambda \operatorname{grad} t \, dS = \int_{(V)} \operatorname{div}(\lambda \operatorname{grad} t) \, dv.$$

Then Eq. (1.5.5) can be written

$$\int_{(V)} c\gamma \frac{\partial t}{\partial \tau} \, dv \int_{(V)} \operatorname{div}(\lambda \operatorname{grad} t) \, dv. \tag{1.5.6}$$

Hence, we obtain

$$c\gamma \frac{\partial t}{\partial \tau} = \operatorname{div}(\lambda \operatorname{grad} t). \tag{1.5.7}$$

If the thermal conductivity is independent of temperature, then from Eq. (1.5.7) the differential heat conduction equation is obtained as

$$\frac{\partial t}{\partial \tau} = a \operatorname{div}(\operatorname{grad} t) = a \nabla^2 t. \tag{1.5.8}$$

For a one-dimensional symmetrical temperature field, $\nabla^2 t$ is a function of one space coordinate. This will be explained by an example of an infinite circular cylinder. If the axis of such a cylinder coincides with the coordinate z, then the temperature at any place of the cylinder depends only on the coordinates x and y. When the cylinder is cooled or heated uniformly at any point located at the distance r from the cylindrical axis, the temperature at a certain moment will be the same. Thus, isothermal surfaces are cylindrical surfaces which are located coaxially with the cylinder. The radial coordinate r (radius-vector) and the coordinates x and y are related by

$$r^2 = x^2 + y^2. \tag{1.5.9}$$

Then the differential heat conduction equation

$$\frac{\partial t}{\partial \tau} = a \left(\frac{\partial^2 t}{\partial x^2} + \frac{\partial^2 t}{\partial y^2} \right) \tag{1.5.10}$$

may be transformed for an infinite cylinder as

$$\frac{\partial t}{\partial x} = \frac{\partial t}{\partial r} \frac{dr}{\partial x} = \frac{\partial t}{\partial r} \frac{x}{(x^2 + y^2)^{1/2}} = \frac{\partial t}{\partial r} \frac{x}{r} \tag{1.5.11}$$

$$\frac{\partial t}{\partial y} = \frac{\partial t}{\partial r} \frac{\partial r}{\partial y} = \frac{\partial t}{\partial r} \frac{y}{r}. \tag{1.5.12}$$

Differentiating Eq. (1.5.11) with respect to x and Eq. (1.5.12) with respect to y, we obtain

$$\frac{\partial^2 t}{\partial x^2} = \frac{\partial^2 t}{\partial r^2}\frac{x^2}{r^2} + \frac{\partial t}{\partial r}\frac{y^2}{r^3}, \tag{1.5.13}$$

$$\frac{\partial^2 t}{\partial y^2} = \frac{\partial^2 t}{\partial r^2}\frac{y^2}{r^2} + \frac{\partial t}{\partial r}\frac{x^2}{r^3}. \tag{1.5.14}$$

Adding Eq. (1.5.13) to Eq. (1.5.14) and taking into account equation (1.5.9), we obtain for the heat conduction equation (1.5.10)

$$\frac{\partial t}{\partial \tau} = a\left(\frac{\partial^2 t}{\partial r^2} + \frac{1}{r}\frac{\partial t}{\partial r}\right). \tag{1.5.15}$$

In a general case when temperature depends on all three coordinates (x, y, z), the differential heat conduction equation of a finite cylinder is of the form

$$\frac{\partial t}{\partial \tau} = a\left(\frac{\partial^2 t}{\partial r^2} + \frac{1}{r}\frac{\partial t}{\partial r} + \frac{1}{r^2}\frac{\partial^2 t}{\partial \theta^2} + \frac{\partial^2 t}{\partial z^2}\right), \tag{1.5.16}$$

where θ is a polar angle (an angle between the radius-vector r and the axis x). The differential heat conduction equation in spherical coordinates is given in Appendix III.

To conclude Section 1.5, we will consider the physical significanee of thermal diffusivity. As already shown, the coefficient a is a diffusion coefficient of internal energy (a_v) or enthalpy (a_p), depending on the interaction of the system with the surrounding medium. The coefficient a may, however, have another physical significance.

The author has already shown that for the isothermal surface $t(x, y, z, \tau) = \text{const}$, the following relation holds

$$\frac{\partial t}{\partial \tau} + w_t \nabla t = 0. \tag{1.5.17}$$

Then from differential heat conduction equation (1.5.8) we have

$$a = -w_t \frac{\nabla t}{\nabla^2 t}. \tag{1.5.18}$$

The quantity $\nabla t/\nabla^2 t$ is a ratio of the first differential parameter to the second, the parameters being invariant with respect to the linear group of the transformation. Consequently, this relation will be any value with a dimension m^{-1}. Denoting this quantity by A ($A = -\nabla t/\nabla^2 t$), we shall have

$$a = Aw_t. \tag{1.5.19}$$

Thus, *thermal diffusivity is proportional to the velocity of isothermal surface propagation. The quantity which is inverse to the thermal diffusivity* $1/a$ *characterizes inertia properties of the body with respect to the temperature field propagation.*

One of the substances having great thermal inertia is water, the thermal diffusivity of which at a temperature of 90°C and pressure of 1 atm is 0.0005 m^2/hr ($1/a = 2000$ hr/m^2). Gases have low thermal inertia, for instance, air under the same conditions shows a diffusivity of 0.0925 m^2/hr ($1/a = 10.8$ hr/m^2).

The value of the thermal diffusivity depends on temperature and, for porous bodies, on density and moisture content. Thermal conductivity, thermal diffusivity, and heat capacity may be therefore considered constant only as an approximation.

1.6 Hyperbolic Heat Conduction Equation

In Section 1.3, heat transfer in unsteady processes of high rate was shown to be governed by the generalized Fourier law (Eq. 1.3.2). In this case the differential heat conduction equation will be of another form.

From the heat-balance equation for a one-dimensional temperature field we have

$$-\frac{\partial q_x}{\partial x} = c\gamma \frac{\partial t}{\partial \tau}. \tag{1.6.1}$$

Replacement of q_x by the appropriate expression from Eq. (1.3.2) yields

$$q_x = -\lambda \frac{\partial t}{\partial x} - \tau_r \frac{\partial q_x}{\partial \tau} \tag{1.6.2}$$

Assuming λ and τ_r to be constant, we shall have

$$\lambda \frac{\partial^2 t}{\partial x^2} + \tau_r \frac{\partial^2 q_x}{\partial x \, \partial \tau} = c\gamma \frac{\partial t}{\partial \tau}. \tag{1.6.3}$$

Differentiation of Eq. (1.6.1) with respect to τ gives

$$\frac{\partial^2 q_x}{\partial x \, \partial \tau} = -c\gamma \frac{\partial^2 t}{\partial \tau^2}. \tag{1.6.4}$$

It follows that the differential equation (1.6.3) may be written

$$\frac{\partial t}{\partial \tau} + \tau_r \frac{\partial^2 t}{\partial \tau^2} = a \frac{\partial^2 t}{\partial x^2}. \tag{1.6.5}$$

For a three-dimensional temperature field, the differential heat conduction equation may be similarly written as[5]

$$\frac{\partial t}{\partial \tau} + \tau_r \frac{\partial^2 t}{\partial \tau^2} = a \nabla^2 t. \tag{1.6.6}$$

In the subsequent portion of this section, Eq. (1.6.6) will be analyzed. This equation may be written as

$$c\gamma \frac{\partial t}{\partial \tau} + \frac{\lambda \dfrac{\partial^2 t}{\partial \tau^2}}{w_r^2} = \lambda \nabla^2 t, \tag{1.6.7}$$

since according to relation (1.3.1), $w_r^2 = a/\tau_r$. For low gas pressures $c\gamma$ is small ($c\gamma \to 0$) and the mean free molecular path which determines w_r increases considerably. The first term of Eq. (1.6.7) may therefore be neglected. We then have a differential heat distribution equation which is the same as the hyperbolic wave equation

$$\frac{\partial^2 t}{\partial \tau^2} = w_r^2 \nabla^2 t. \tag{1.6.8}$$

The generalized heat conduction equation (1.6.6) is therefore called a hyperbolic heat conduction equation. For the heat transfer distribution, the additional term in the heat conduction equation $(a/w_r^2)(\partial^2 t/\partial \tau^2)$, accounting for a finite heat distribution velocity, is small, so that in ordinary engineering calculations it may be neglected.

Calculation of moisture transfer in porous bodies is quite different. Until now, a diffusion equation of a parabolic type has been usually used for the solution of moisture transport in porous bodies. However, as it has been shown in the author's works, in these cases the hyperbolic parabolic equation of moisture diffusion in capillary-porous bodies should be used.

[5] The possibility of extension of Eq. (1.6.5) to the three-dimensional case is considered by the author [66].

1.7 A System of Differential Heat and Mass Transfer Equations

A system of differential equations of heat and mass transfer in a gas mixture flow is derived following the laws of energy and mass conservation.

A differential equation corresponding to the law of mass conservation of the kth component may be written as follows

$$\frac{\partial \varrho_k}{\partial \tau} = - \operatorname{div} \varrho_k \mathbf{w}_k + I_k \tag{1.7.1}$$

where ϱ_k is the volumetric concentration of the kth component and $\mathbf{w}_k$ is the velocity of its motion which is related to the velocity $\mathbf{w}$ of the center of mass of the mixture by

$$\mathbf{w} = \frac{1}{\varrho} \sum_k \varrho_k \mathbf{w}_k . \tag{1.7.2}$$

The source I_k of the kth component mass is a result of phase or chemical conversions. Summing up equation (1.7.2) over all the components of the mixture, we obtain from Eq. (1.7.1) the ordinary continuity equation

$$\partial \varrho / \partial \tau = - \operatorname{div} \varrho \mathbf{w} \tag{1.7.3}$$

since the sum of all the sources and sinks in the volume considered of the mixture is $\Sigma_k I_k = 0$.

Equation (1.7.3) may be written in another form. Since

$$\operatorname{div} \varrho \mathbf{w} = \mathbf{w} \nabla \varrho + \varrho \operatorname{div} \mathbf{w} ,$$

we shall have

$$d\varrho / d\tau = - \varrho \operatorname{div} \mathbf{w} \tag{1.7.4}$$

where $d\varrho/d\tau$ is the total or substantial derivative

$$\frac{d\varrho}{d\tau} = \frac{\partial \varrho}{\partial \tau} + \mathbf{w} \nabla \varrho . \tag{1.7.5}$$

A diffusion mass flux $\mathbf{j}_k$ of the kth component is

$$\mathbf{j}_k = \varrho_k (\mathbf{w}_k - \mathbf{w}) \tag{1.7.6}$$

Summing up equation (1.7.6) over all the components and accounting for Eq. (1.7.2) we obtain

$$\sum_k \mathbf{j}_k = 0. \tag{1.7.7}$$

The value of $\varrho_k \mathbf{w}_k$ will be found from Eq. (1.7.6) and substitution of it into Eq. (1.7.1) yields

$$(\partial \varrho_k / \partial \tau) + \operatorname{div} \varrho_k \mathbf{w} = \quad \operatorname{div} \mathbf{j}_k + I_k. \qquad (1.7.8)$$

If the dimensionless concentration ($\varrho_{k0} = \varrho_k / \varrho$) is denoted through ϱ_{k0}, then Eq. (1.7.8) may be written

$$\varrho\,(d\varrho_{k0}/d\tau) = -\operatorname{div} \mathbf{j}_k + I_k. \qquad (1.7.9)$$

If $\mathbf{j}_k$ is replaced by corresponding equation (1.5.4), Eq. (1.7.9) for a binary gas mixture ($k = 1, 2$) will be of the form

$$\varrho(d\varrho_{10}/d\tau) = \operatorname{div}[D\varrho(\nabla \varrho_{10} + (k_T/T)\nabla t)] + I_1. \qquad (1.7.10)$$

Equation (1.7.10) is a differential mass-transfer equation.
A differential energy-transfer equation will be

$$\varrho(dh/d\tau) = -\operatorname{div} \mathbf{j}_u. \qquad (1.7.11)$$

If $\mathbf{j}_u$ is replaced by Eq. (1.5.12), then for a binary gas mixture

$$\varrho \frac{d}{\partial \tau}(h_1 \varrho_{10} + h_2 \varrho_{20}) = -\operatorname{div} \mathbf{q} - \operatorname{div}(h_1 - h_2)\,\mathbf{j}_1. \qquad (1.7.12)$$

If the specific heat capacity at constant pressure is denoted by c_{pk} ($c_{pk} = dh_k/dt$) and the volumetric heat capacity at constant pressure by $c_p\varrho$, then

$$c_p\varrho = c_{p1}\varrho_1 + c_{p2}\varrho_2. \qquad (1.7.13)$$

Usage of equation (1.5.1) results in

$$c_p\varrho(dt/d\tau) = \operatorname{div}(\lambda \nabla t) + \operatorname{div}(D\varrho Q^* \nabla \varrho_{10}) + (h_1 - h_2)I_1 - (c_{p1} - c_{p2})\,\mathbf{j}_1 \nabla t. \qquad (1.7.14)$$

The left-hand side of the differential equation (1.7.14) represents an enthalpy change of a gas mixture with time ($c_p\varrho(\partial t/\partial \tau)$ and enthalpy transfer by a mixture flow ($c_p\varrho \mathbf{w} \nabla t$). The first term of the right-hand side describes the heat transfer by conduction, the second term represents heat transfer due to diffusion-heat conduction (the Dufour effect), and the third term is a heat source or sink caused by phase or chemical conversion and the last term in Eq. (1.7.14) describes enthalpy transfer by diffusion.

The difference ($c_{p1} - c_{p2}$) is usually small, and the last term may be therefore neglected. It should, however, be noted that in transpiration cooling when the injected gas is hydrogen or helium, neglect of this term leads to substantial error. If all the coefficients are assumed to be constant, then for a fixed gas mixture ($\mathbf{w} = 0$, $d/d\tau = \partial/\partial\tau$), the following system of differential equations will be obtained from Eqs. (1.7.10) and (1.7.14):

$$\frac{\partial \varrho_{10}}{\partial \tau} = D\,\nabla^2 \varrho_{10} + D\,\frac{k_T}{T}\,\nabla^2 t + \frac{I_1}{\varrho}, \tag{1.7.15}$$

$$\frac{\partial t}{\partial \tau} = a\,\nabla^2 t + \frac{DQ^*}{c_p}\,\nabla^2 \varrho_{10} + \frac{(h_1 - h_2)}{c_p \varrho}\,I_1. \tag{1.7.16}$$

A similar system of differential heat and mass transfer equations describes heat and mass transfer in solutions.

If thermal diffusion is neglected ($k_T = 0$, $Q^* = 0$), then from Eq. (1.7.16) the differential Fourier heat conduction equation with a heat source will be obtained. The differential mass-diffusion equation (1.7.15) is similar to the differential Fourier heat conduction equation. All the solutions obtained for nonstationary conduction problems may therefore be applied to the estimation of nonstationary mass diffusion.

1.8 End Conditions

The differential heat conduction equation relates time and space variations of the body temperature; it governs the heat transfer process in a body. To determine the temperature field inside a body at any moment, i.e., to solve a differential equation, one should know the temperature distribution at the initial moment (initial condition) the geometry of the body, and the law of interaction between the surrounding medium and the body surface (boundary condition).

A combination of an initial and a boundary condition is referred to as an end condition. The initial condition is called a time end condition and the boundary condition a space end condition.

The initial condition describes the temperature distribution inside a body at the initial moment, i.e.,

$$t(x, y, z, 0) = f(x, y, z). \tag{1.8.1}$$

In a number of problems, where $f(k, y, z)$ is a known function, a uniform temperature distribution is assumed at the initial moment; then

$$t(x, y, z, 0) = t_0 = \text{const.} \tag{1.8.2}$$

The boundary condition may be prescribed in several ways:

(a) *In a boundary condition of the first kind, the temperature distribution over the body surface at any moment is prescribed, i.e.,*

$$t_s(\tau) = f(\tau) \tag{1.8.3}$$

where $t_s(\tau)$ is the temperature at the body surface.

A particular example is $t_s(\tau) = t_a = \text{const}$, i.e., the surface temperature is constant during the whole heat transfer process. This may be achieved by special temperature control or with special conditions of heat transfer between the surroundings and the body surface (see the boundary condition of the third kind).

(b) *A boundary condition of the second kind occurs when heat flux for every point of the body surface is given as a function of time, i.e.,*

$$q_s(\tau) = f(\tau). \tag{1.8.4}$$

The simplest case of the boundary condition of the second kind is that of a constant heat flux

$$q_s(\tau) = q_c = \text{const.} \tag{1.8.5}$$

This case of heat transfer takes place, for example, with heating bodies in high temperature furnaces where heat is mainly transferred by radiation according to the Stefan–Boltzmann law when the body temperature is considerably lower than that of the radiating surfaces (see Chapter 5).

(c) *Usually a boundary condition of the third kind is characteristic of the law of convective heat transfer between the solid surface and the surrounding medium.* In this case, the heat quantity transferred per unit time per unit surface area to the surrounding medium with the temperature t_a in the process of cooling ($t_s > t_a$) is directly proportional to the temperature difference between the surface and the surrounding medium, i.e.,

$$q_s = \alpha(t_s - t_a), \tag{1.8.6}$$

where α is the proportionality factor called a heat transfer coefficient (kcal/m^2 hr °C or W/m^2 °C). For the process of body heating a similar relation may be written, with t_s and t_a interchanged.

The heat transfer coefficient is estimated as the heat amount released (or received) per unit surface area per unit time, the per unit temperature difference between the body surface and the surrounding medium. Relation (1.8.6) may be obtained from the Fourier heat conduction law assuming heat transfer from the body to the fluid adjacent to it follows the Fourier law

$$\mathbf{q}_s = -\lambda_f\left(\frac{\partial t_f}{\partial n}\right)_s \mathbf{n}_0 = \frac{-\lambda_f(t_s - t_a)\mathbf{n}_0}{\delta} = \alpha(t_s - t_a)(-\mathbf{n}_0), \quad (1.8.7)$$

where λ_f is the thermal conducitivity of the fluid and δ is a conventional thickness of the boundary layer.

Thus, the heat flux vector $\mathbf{q}_s$ follows the normal to the isothermal surface $(-n)$, its scalar value being equal to q_s (see Fig. 1.1).

If we denote λ_f/δ by α ($\alpha = \lambda_f/\delta$), we shall obtain relation (1.8.6).

The conventional thickness of the boundary layer δ depends on the fluid velocity and its physical properties. The heat transfer coefficient therefore depends on the fluid velocity and its temperature, and in general, varies along the body surface. The heat transfer coefficient may be approximately assumed constant, independent of the temperature, and uniform over the whole surface.

In unsteady heat transfer processes between the body surface and the fluid flow, a conventional thickness of a boundary layer will depend not only on the fluid velocity and its physical properties, but also on thermal properties of a body and will continuously change with time: $\delta = f(\tau)$. However, for particular problems of unsteady-state heat transfer,

$$q_s = \alpha(t_s - t_a) = \text{const}, \quad (1.8.6a)$$

the relation may be taken as a boundary condition and as a predicting formula to the first approximation, but not as a relation of the convective-transfer law.

Boundary conditions of the third kind may also be used for radiant heating or cooling. Following the Stefan–Boltzmann law, the radiant heat flux between two surfaces is

$$q_s(\tau) = \sigma^*[T_s^4(\tau) - T_a^4] \quad (1.8.8)$$

where σ^* is a reduced emissivity (i.e., the product of the Stefan-Boltzmann constant and the emissivity) and T_a is the absolute surface temperature of the heat-receiving medium. With small temperature difference $(T_s - T_a)$, the relation may be approximately written as

$$q_s(\tau) = \sigma^*\{[T_s^2(\tau) + T_a^2][T_s(\tau) + T_a]\}\,[T_s(\tau) - T_a] = \alpha(t)[t_s(\tau) - t_a], \quad (1.8.9)$$

where $\alpha(t)$ is the radiant heat transfer coefficient measured in the same units as a convective heat transfer coefficient

$$\alpha(t) = [T_s(\tau) + T_a][T_s^2(\tau) + T_a^2]\sigma^* = \sigma^* b(t). \quad (1.8.10)$$

The value of the coefficient $b(t)$ is shown in Table 1.1.

TABLE 1.1. COEFFICIENT $b(t) \times 10^8$ AS A FUNCTION OF T_s AND T_a

Medium temperature t_a (°C)	Surface temperature, t_s (°C)														Medium temperature, T_a (°K)
	−273	−200	−100	0	100	200	300	400	500	600	700	800	900	1000	
−273	0														0
−200	0.0039	0.0156													73
−100	0.0518	0.0867	0.2070												173
0	0.2034	0.2764	0.465	0.814											273
100	0.519	0.644	0.923	1.380	2.076										373
200	1.058	1.251	1.630	2.225	3.070	4.733									473
300	1.882	2.155	2.673	3.408	4.422	5.77	7.53								573
400	3.050	3.418	4.08	4.99	6.19	7.75	9.73	12.19							673
500	4.62	5.10	5.94	7.03	8.44	10.23	12.46	15.19	18.48						773
600	6.66	7.24	8.28	9.59	11.30	13.27	15.77	18.70	22.38	26.61					873
700	9.21	9.96	11.19	12.72	14.62	16.92	19.71	23.04	26.96	31.55	36.84				973
800	12.35	13.26	14.72	16.50	18.66	21.26	24.36	28.01	32.29	37.29	42.93	49.5			1073
900	16.14	17.21	18.92	20.97	23.42	26.33	29.76	33.76	38.41	43.75	49.85	56.8	64.6		1173
1000	20.45	21.88	23.87	26.21	28.96	32.20	35.98	40.35	45.38	51.1	57.6	65.0	73.3	82.6	1273
	Surface temperature, T_s (°K)														
	0	073	173	273	373	473	573	673	773	873	973	1073	1173	1273	

Relation (1.8.9) is the expression of the Newton law for cooling or heating a body. Although relation (1.8.9) is similar to expression (1.8.6.) for the convective heat transfer law $q_s = \text{const}$, its physical significance is quite different. The radiant heat transfer coefficient $\alpha(t)$ depends on the temperature (see Table 1.1), as well as on the properties of body surfaces involved in radiant heat transfer. If the temperature $t_s(\tau)$ changes negligibly, then the coefficient $\alpha(t)$ may be approximately assumed constant. This is due to the restriction that the temperature difference $(t_s - t_a)$ is small and thereby t_s may be written instead of t_a. Here, the convective contribution to heat flux may be assumed equal to $\alpha_k \, \Delta t$ where α_k is a convective heat transfer coefficient. In this case, in the relation

$$q_s(\tau) = \alpha[t_s(\tau) - t_a], \tag{1.8.11}$$

the coefficient α is the total heat-transfer coefficient

$$\alpha = \alpha_k + \alpha(t). \tag{1.8.12}$$

It should, however, be remembered that in unsteady-state heat transfer, formula (1.8.10) describes the radiant heat transfer law to the first approximation and the contribution of convective heat transfer should be sufficiently small such that the time change of the coefficient α_k and its dependence on thermal properties of the body may be neglected.

In the subsequent sections we shall refer to the unsteady-state heat transfer, the mechanism of which is described by relation (1.8.11) as the Newton law of heat transfer.

According to the energy-conservation law, the heat quantity $q_s(\tau)$ transferred from the body surface equals that transferred to the body surface from the inside per unit time per unit surface by heat conduction

$$q_s(\tau) = \alpha[t_s(\tau) - t_a(\tau)] = -\lambda(\partial t/\partial n)_s, \tag{1.8.13}$$

where for generality of the problem statement, the temperature t_a is assumed variable and the heat transfer coefficient $\alpha(t)$ is approximately taken as constant $[\alpha(t) = \alpha = \text{const}]$.

The boundary condition is usually written as

$$\lambda\left(\frac{\partial t}{\partial n}\right)_s + \alpha[t_s(\tau) - t_a(\tau)] = 0. \tag{1.8.14}$$

From a boundary condition of the third kind, the simplest form of a boundary condition of the first kind may be obtained as a particular case

$t_s(\tau) = t_a = \text{const}$. If the ratio α/λ approaches infinity (the heat transfer coefficient is very large, $\alpha \to \infty$, or thermal conductivity is small, $\lambda \to 0$), then

$$t_s(\tau) - t_a(\tau) = \lim_{\alpha/\lambda \to \infty} \left\{ \frac{1}{(\alpha/\lambda)} \left(\frac{\partial t}{\partial n} \right)_s \right\} = 0, \tag{1.8.15}$$

hence

$$t_s(\tau) = t_a = \text{const},$$

i.e., the surface temperature is the same during the whole heat transfer process and equal to that of the surrounding medium.

(d) *A boundary condition of the fourth kind corresponds to heat transfer between a solid body and the surrounding medium (convective transfer of a body with a liquid) or to the case of heat transfer of solid bodies in contact when the temperature of the interface surfaces is equal.* For the solid body exposed to a fluid medium, the heat transfer from the fluid adjacent to the body surface (a laminar boundary layer or laminar sublayer) follows the heat conduction law (molecular heat transfer), i.e., heat transfer corresponds to the boundary condition of the fourth kind

$$t_s(\tau) = [t_a(\tau)]_s. \tag{1.8.16}$$

Besides the equality of temperatures the equality of heat fluxes is also valid:

$$-\lambda_a(\partial t_a/\partial n)_s = -\lambda(\partial t/\partial n)_s. \tag{1.8.17}$$

Thus, for convective heat transfer between a solid and liquid in case of a steady-state temperature field ($\partial t/\partial \tau = 0$), the boundary condition of the third kind (relation (1.8.14)) is used; in case of the unsteady-state temperature field ($\partial t/\partial \tau \neq 0$) it is necessary to apply boundary conditions of the fourth kind (relations (1.8.16, 1.8.17)). With unsteady-state radiative heat transfer it is necessary to apply the second-kind boundary conditions (relation (1.8.4)). At small temperature differences $[t_s(\tau) - t_a(\tau)]$ it is possible to use the Newton law, i.e., the boundary conditions of the third kind. The quantity α will represent the coefficient of radiative heat transfer $[\alpha = \alpha(t) = \text{const}]$.

The four kinds of boundary conditions are shown graphically in Fig. 1.4. A scalar value of the heat-flux vector is proportional to the absolute value of the temperature gradient which numerically equals the slope of the tangent to the temperature distribution curve along the normal to the isothermal surface, i.e.,

$$(\partial t/\partial n)_s = \tan \psi_s.$$

In Fig. 1.4 are shown four elements of the surface ΔS with the normal n to it (the normal is positive when it is directed outward). The temperature is shown on the ordinate.

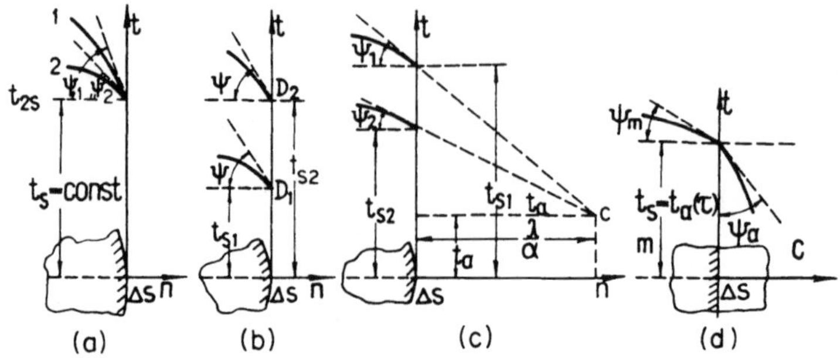

Fig. 1.4. Various means of prescription of conditions on the surface. (a) $\tan\psi = \text{var}$, $t_s = \text{const}$; (b) $\tan\psi = \text{const}$, $t_s = \text{var}$; (c) $\tan\psi = \text{var}$, $t_s = \text{var}$; (d) $\tan\psi_m/\tan\psi_a = \text{const}$, $t_s = \text{var}$.

The boundary condition of the first kind occurs when $t_s(\tau)$ is prescribed; in the simplest case $t_s(\tau) = \text{const}$. A slope of the tangent to the temperature curve at the body surface should be found, as well as the heat amount released by the surface (see Fig. 1.4a).

Problems involving boundary conditions of the second kind are quite different; a slope of the tangent to the temperature curve at the body surface (see Fig. 1.4b) is given; the surface temperature is to be found.

In problems with boundary conditions of the third kind, the temperature and the slope of the tangent to the temperature curve are variable but on the external normal the point C is prescribed through which all the tangents to the temperature curve should pass (see Fig. 1.4c). It follows from condition (1.8.14) that

$$\tan\psi_s = \left(\frac{\partial t}{\partial n}\right)_s = \frac{t_s(\tau) - t_a}{(\lambda/\alpha)}. \tag{1.8.18}$$

A slope of the tangent to the temperature curve at the surface equals the ratio of the value $t_s(\tau) - t_a$ to the length λ/α of the corresponding retangular triangle. The length λ/α is a constant value, and the temperature difference $t_s(\tau) - t_a$ changes continuously during the heat transfer process and this change is directly proportional to $\tan\psi_s$. Hence, the leading point C becomes invariable.

In problems with boundary conditions of the fourth kind, the ratio of the slope of the tangents to the temperature curves in a body and in the medium on their interface (see Fig. 1.4d)

$$\tan \psi_m / \tan \psi_a = \lambda_a / \lambda = \text{const} \tag{1.8.19}$$

account for a perfect thermal contact (tangents near the interface cross at one and the same point).

The differential equations together with initial and boundary conditions determine the problem completely, i.e., if we know the geometry of a body and the initial and boundary conditions, we can solve the differential equation and thus find the temperature distribution function at any moment. Thus as a result of the solution, we find the function

$$t(x, y, z, \tau) = f(x, y, z, \tau). \tag{1.8.20}$$

The function $f(x, y, z, \tau)$ should satisfy the differential equation (when it is introduced into the heat conduction differential equation instead of t, it should become an identity) and the initial and boundary conditions.

From the uniqueness theorem (see Appendix II), if some function $t(x, y, z, \tau)$ satisfies the differential heat conduction equation and initial and boundary conditions, it is a unique solution of the problem.

1.9 Methods for Calculating the Heat Flow

In the process of heating and cooling, a body receives or releases a definite quantity of heat. There are three methods for determining the heat flow in the heat transfer process:

a. The First Calculation Method. Crossing the surface element dS during the time $d\tau$, the amount of heat transfer is equal to

$$-\lambda(\partial t/\partial n)\, dS\, d\tau. \tag{1.9.1}$$

To determine the amount of heat ΔQ received by a body for the time $\Delta\tau = \tau_2 - \tau_1$, relation (1.9.1) should be integrated over the whole surface S and time interval $\Delta\tau$:

$$\Delta Q = -\int_{\tau_1}^{\tau_2} \int_{(S)} \lambda(\partial t/\partial n)\, dS\, d\tau. \tag{1.9.2}$$

Frequently, the temperature and the temperature gradient are independent of the surface position S; the calculation formula (1.9.2) then simplifies to

$$\Delta Q = Q_2 - Q_1 = -\lambda S \int_{\tau_1}^{\tau_2} (\partial t/\partial n)\, d\tau. \tag{1.9.3}$$

b. The Second Calculation Method. The surface element dS during the time $d\tau$ receives from the surrounding medium the amount of heat

$$\alpha(t_a - t_s)\, dS\, d\tau. \tag{1.9.4}$$

To determine the total amount ΔQ received by the whole surface of a body, Eq. (1.9.4) is integrated over the whole surface and the time interval $\Delta\tau = \tau_2 - \tau_1$.

If a surface temperature is constant and the coefficient α is independent of the temperature, then we shall have

$$\Delta Q = Q_2 - Q_1 = \alpha S \int_{\tau_1}^{\tau_2} [t_a - t_s(\tau)]\, d\tau. \tag{1.9.5}$$

c. The Third Calculation Method. The volume element $dv = dx\, dy\, dz$ is heated for the time $\Delta\tau = \tau_2 - \tau_1$ in the temperature range from t_1 to t_2; it receives the heat amount equal to

$$c\gamma(t_2 - t_1)\, dv. \tag{1.9.6}$$

The total heat amount ΔQ which was supplied for heating for the time $\Delta\tau$ will be found after integration over the whole volume

$$\begin{aligned} \Delta Q = Q_2 - Q_1 &= c\gamma \int_{(V)} (t_2 - t_1)\, dv \\ &= c\gamma V \cdot (1/V) \int_{(V)} (t_2 - t_1)\, dv. \end{aligned} \tag{1.9.7}$$

Denoting an average (integral) temperature over the whole body volume through $\bar{t}$, i.e.,

$$\bar{t} = (1/V) \int_{(V)} t\, dv,$$

we may write

$$\Delta Q = Q_2 - Q_1 = c\gamma V(\bar{t}_2 - \bar{t}_1) \tag{1.9.8}$$

since in the heating process $\bar{t}_2 > \bar{t}_1$.

The amount of heat $(Q - Q_0)$ supplied during the time τ from the beginning of the process $(\tau_1 = 0)$ will be

$$Q - Q_0 = c\gamma V(\bar{t} - \bar{t}_0), \tag{1.9.9}$$

where $\bar{t}_0$ is the mean (integral) temperature. If the initial temperature is equal at all the points, i.e., when $\bar{t}_0 = t_0 = \text{const}$, the specific heat flow rate will be

$$\Delta Q_v = c\gamma(\bar{t} - t_0). \tag{1.9.10}$$

Thus the main problem in this calculation method is to find $\bar{t}(\tau)$. Consider some examples.

Example 1. We have a plate $2R$ in thickness. The width ($2h$) and length ($2l$) of the plate are considerably greater than its thickness, so that the temperature gradient along the length and over the width of the plate is zero (a one-dimensional case). Then temperature at any point of the plate will be dependent on x and τ, i.e., $t(x, \tau)$.

If the coordinate origin is in the center of the plate, then the mean integral temperature will be

$$\begin{aligned}\bar{t}(\tau) &= \frac{1}{V}\int_{(V)} t(x, \tau)\, dv \\ &= \frac{1}{2R\,2l\,2h}\int_{-R}^{R}\int_{-l}^{l}\int_{-h}^{h} t(x, \tau)\, dx\, dy\, dz \\ &= \frac{1}{2R}\int_{-R}^{R} t(x\ \tau)\, dx = \frac{1}{R}\int_{0}^{R} t(x, \tau)\, dx.\end{aligned} \tag{1.9.11}$$

Example 2. Let the temperature of a spherical body (a sphere) be a function of r and τ (a symmetrical problem), i.e., $t(r, \tau)$. Then the mean temperature (over the volume) $\bar{t}(\tau)$ will be

$$\begin{aligned}\bar{t}(\tau) &= \frac{3}{4\pi R^3}\int_{0}^{R}\int_{\theta=0}^{2\pi}\int_{\psi=0}^{\pi} t(r, \tau) r^2 \sin\theta\, d\theta\, d\psi\, dr \\ &= \frac{3}{R^3}\int_{0}^{R} r^2 t(r\ \tau)\, dr,\end{aligned} \tag{1.9.12}$$

since the volume element is

$$dv = r^2 \sin\theta\, d\theta\, d\psi\, dr.$$

Example 3. Consider a cylinder with the radius R and the length l. The length of the cylinder is considerably greater than the diameter ($l \gg 2R$), therefore the cylinder may be considered infinite and the temperature inside it is a function of r and τ, i.e., $t(r, \tau)$. Then the mean cylinder temperature is

$$\begin{aligned}\bar{t}(\tau) &= \frac{1}{\pi R^2 l}\int_{0}^{R}\int_{0}^{2\pi}\int_{0}^{l} t(r, \tau) r\, dr\, d\theta\, dz \\ &= \frac{2}{R^2}\int_{0}^{R} r t(r, \tau)\, dr,\end{aligned} \tag{1.9.13}$$

since $dv = r\, dr\, d\theta\, dz$.

In the subsequent problems the mean temperature $\bar{t}$ will be determined simultaneously with the temperature t. If the value of $\bar{t}$ is known, we may determine the heat content Q and the heat spent for heating $(Q - Q_0)$ or the heat loss $(Q_0 - Q)$ when cooling.

In the cases when determination of $\bar{t}$ is difficult, the heat transfer rate may be calculated by formulas (1.9.3) and (1.9.5).

CHAPTER

2

THEORY OF GENERALIZED VARIABLES

Introduction

Due to the development of computing techniques, many problems of mathematical physics are being solved as concrete numerical relations. A great number of variables are being introduced to specify and to bring a problem nearer to a real process. It is, however, very difficult to systematize such calculation results and to find implicit relationships between variables. It is therefore highly valuable to use methods of similarity theory, which at present may be referred to as the theory of generalized variables [15].

The essence of this theory is that on the basis of general physical concepts, a set of relationships is proved not to be an intrinsic property of the studied problems caused by their physical nature. Indeed, the effect of different factors represented by various quantities becomes apparent not separately, but in common. It is therefore not necessary to consider these individual quantities, but rather their sets or complexes of a definite physical significance. The similarity-theory methods make it possible by analyzing differential equations and boundary conditions to determine these complexes which may be called generalized variables.

For each problem there exists a set of characteristic variables, in terms of which the problem should be investigated. A transition from the ordinary physical quantities to those of the complex kind, which are formed from the same quantities but in definite dimensionless combinations, gives important advantages. First, the number of variables decreases. Internal relations characterizing a process are more distinctly expressed in these quantities.

In addition, the new variables also possess another advantage. A prescribed value of a complex of quantities may be obtained as a result of an infinite number of various combinations of the physical quantities making up the complex. Consequently, not just a single set but an infinite set of initial physical quantities satisfies the fixed values of the new variables. This means that an infinite number of various cases, not just a single case, are investigated when solving a problem in new variables. Thus, the new variables are by their essence generalized. The replacement of ordinary variables by the generalized ones is the main problem of the similiarity theory or the theory of generalized variables.

It is quite obvious that the similarity theory may be used more successfully when it is impossible to integrate a differential equation and to define the relation between variables in explicit form. If it is possible to solve the differential equation then in principle, there is no need to use the similarity theory. However, even in this case the similarity theory has certain use. If solutions are presented as dimensionless variables, then the number of variables will be essentially reduced. Moreover, a solution in such a form allows the establishment of internal relations between variables and parameters entering the dimensionless groups, which thereby reveal the physical significance of the solutions in more detail.

2.1 Dimensionless Quantities

Differential equations describing a process reflect our concepts of the physical mechanism of a process. For instance, the differential heat conduction equation is a particular case of the energy conservation law and is reduced to the statement that a change in internal energy of an elementary volume of a body is equal to the heat amount which is transferred between the elementary volume and the whole volume of a body. The basic equations of the problem represent a special form of the general physical laws. The development of the process is defined by relations between separate terms of equations. Such relations should be introduced as characteristic variables for the problem in question. In many cases, each term of the equation, however, represents complex differential expressions including the main variables of the problem. Hence, it is necessary to establish rules of a transition from differential expressions to those in finite form.

The similarity theory gives a general method to directly transform expressions having differential operators into the simplest algebraic ones. The essence of this method is that a real process is replaced by a simple conven-

tional scheme where all the differential operators preserve their constant value in space and time.

As an example, consider the operation of the mth fold differentiation with respect to the independent variable x. Then, the function $z = d^m y/dx^m$ will result in an effect of this operator on the function $y = f(x)$. The effect of x and y on z is, however, of interest to us. Therefore, we assume $d^m y/dx^m =$ const. It is obvious that to satisfy this condition it is necessary to assume $y = Ax^m$, hence $A = y/x^m$ where x and y are any values of the variables obtained from the relation $y = f(x)$. Let any values of the variables x and y be known or given according to the condition (x_0, y_0). These prescribed values of x_0 and y_0 are referred to as parametric values. Therefore, $A \sim y_0/x_0{}^m$. Hence, $z \sim y_0/x_0{}^m$.

Thus, the law of constructing z in terms of x and y lies in the fact that z is determined as a quantity proportional to y_0 in the first power and inversely proportional to x_0 in the mth power. Only this relation is important when settling a question about the structure of a complex corresponding to the operator z. Parameters x_0 and y_0 are given according to the specific condition.

Thus, it is not difficult to pass from the derivatives to the simplest algebraic expressions. We illustrate this transition by the example of heat transfer of a heated plate into the surrounding medium. If heat is transferred following the Newton law described by the third boundary condition, we have

$$\alpha(t_s - t_a) = -\lambda(\partial t/\partial x)_s, \tag{2.1.1}$$

where t_s and t_a are temperatures of a wall and medium, respectively; α is the heat transfer coefficient; λ is the heat conduction coefficient of wall material; and the subscript s applies to plate surfaces.

Now assume $\partial t/\partial x =$ const. In this case this assumption is not a replacement of a real process by the conventional scheme as it is valid at a stationary temperature distribution. Thus, a transition from a general case to the particular one of a stationary temperature distribution is performed. In this case the temperature t becomes a function of only one variable, the x coordinate. Under these conditions we have

$$dt/dx = -\delta t/l, \tag{2.1.2}$$

where δt is the temperature drop over the plate thickness l.

If the temperature difference Δt is introduced and defined by

$$\Delta t = t_s - t_a, \tag{2.1.3}$$

the Eq. (2.1.1) assumes the form

$$\alpha \, \Delta t = \lambda \frac{\delta t}{l}$$

or

$$\delta t / \Delta t = (\alpha / \lambda) \, l. \tag{2.1.4}$$

So from properties which are determined from the basic Eq. (2.1.1), the ratio between the temperature drop and the temperature difference is defined immediately by the expression $(\alpha/\lambda)l$.

In general, with a variable temperature gradient ($\partial t/\partial x =$ var), the temperature distribution is obviously different from linear and this simple result loses its validity because now it is impossible to represent the derivative $(\partial t/\partial x)_s$ as $-\,\delta t/l$. However, assuming

$$(\partial t/\partial x)_s = -\, \varepsilon(\delta t/l), \tag{2.1.5}$$

the factor ε (degree of distortion) will depend only on the configuration of the temperature distribution curve. For all similar temperature distributions, ε would be constant.

Now Eq. (2.1.4) may be rewritten

$$\frac{\delta t}{\Delta t} = \frac{1}{\varepsilon} \frac{\alpha}{\lambda} l. \tag{2.1.6}$$

Thus, in general, to determine the relation $\delta t/\Delta t$, it is necessary to know ε. But it has already been established that for similar distributions ε is constant. Therefore, if ε is determined for some similar distributions, it is also valid for all similar distributions. Simple considerations, not given here, show that similar distributions are those which correspond to identical values of $(\alpha/\lambda)l$. But this means that the factor ε is a single-valued function of the value $(\alpha/\lambda)l$ and consequently, Eq. (2.1.6) may be given as

$$\frac{\delta t}{\Delta t} = F\left(\frac{\alpha}{\lambda} l\right). \tag{2.1.7}$$

This result is important because the solution is presented as a function of one argument, although it is quite obvious that the temperature distribution is caused by the effect of three parameters, α, l, and λ.

The real meaning of this remarkable fact is that, in accordance with our notion of the physical nature of a process (expressed in the basic equation), each of these parameters alone is not of essential importance but rather their definite combination $(\alpha/\lambda)l$.

The expression $(\alpha/\lambda)l$ represents a typical example of a generalized variable or a complex which applies to the problem being considered. *In the similarity theory, such variables are usually named as the similarity criteria and designated by the first two letters of the name of the scientist who made a significant contribution to the development of this field of science.* The expression $(\alpha/\lambda)l$ is referred to as the Biot criterion.

$$(\alpha/\lambda)l = \mathrm{Bi}. \tag{2.1.8}$$

Hence, Eq. (2.1.7) may be rewritten as

$$\delta t/\Delta t = F(\mathrm{Bi}). \tag{2.1.9}$$

A transition from initial variables α, l, λ to a new variable Bi, results not only in a decrease in the number of arguments; in parallel to this, the analysis itself changes. Actually, when α, l, λ are taken as basic quantities, some particular case for each set of the given values of these parameters is obtainable. In contrast to this, any particular case does not uniquely satisfy the given value of the Biot number, because this value may be realized by the infinite number of various combinations of quantities α, l, λ.

Thus, *when the value of the Biot number is fixed, not one particular phenomenon but an infinite number of various phenomena are determined. Hence, owing to a new concept, the particular case (corresponding to the given value of an argument) is not a single phenomenon but an infinite number of analogous phenomena. In this sense it may be said that new variables are generalized ones and, consequently, the whole analysis acquires a generalized character.*

The physical significance of Eq. (2.1.9) lies in the fact that the temperature distributions are similar between themselves, for which the relation $\delta t/\Delta t$ has the same value, corresponding to the given values of the Biot number. However, the solution is not completely determined because a nonstationary process is being investigated. It is necessary to clarify how to determine those time instants for which similar distributions are obtainable.

It is quite obvious that processes in different systems will not develop together, because the rate of evolution of a temperature field depends both on λ and l. Hence, there appears a problem on the rules of determining mutually appropriate time instants. To solve this problem we turn to the basic heat conduction equation, which establishes a relation between the rate of evolution of temperature in time and the temperature distribution in space.

For a one-dimensional problem we have

$$\frac{\partial t}{\partial \tau} = a \frac{\partial^2 t}{\partial x^2}, \tag{2.1.10}$$

where a is the thermal diffusivity of a plate material.

If, in this case, our method of transition to the conventional scheme with constant values of derivatives is assumed, then the derivative $\partial t/\partial \tau$ may be replaced by $\delta t_\tau/\tau$ and the derivative $\partial^2 t/\partial x^2$ by $\delta t_l/l^2$ (subscripts l and τ denote changes in temperature for time τ and along the length l, respectively). Hence, for the scheme under consideration

$$\delta t_\tau/\delta t_l = a\tau/l^2. \tag{2.1.11}$$

If at any moment, for example, at the start of a process, the temperature distributions are similar, this similarity will be preserved if and only if the relation between space and time changes remains constant. Whence it follows that for all systems the expression $a\tau/l^2$ should be constant.

For a given initial temperature distribution, any subsequent distribution depends on the duration of the process τ, the thermal diffusivity a, and the system size l. The definite combination $a\tau/l^2$ of these quantities is of great importance. *It is obvious that the expression $a\tau/l^2$ represents a generalized variable which is usually referred to as the Fourier criterion or number*

$$a\tau/l^2 = \mathrm{Fo}. \tag{2.1.12}$$

Hence, it follows that *the Fourier number has the meaning of generalized time. It may therefore be called as the homochronous number* (*homochronity means uniformity in time*; *if for two systems the relation l^2/a is constant, then for them homochronity obviously transforms into synchronity*).

Gukhman in his recent work [45] distinguishes between similarity criteria and similarity numbers. *Similarity criteria are those complexes which as a whole consist of parameters given according to the condition of the problem. The Biot criterion is a typical example. The complex $a\tau/l^2$ is not a criterion but a generalized variable or the Fourier number.* However, if according to the problem condition some characteristic time is given, for example, a period of oscillation of a surrounding medium temperature τ_0, then the analogous complex $a\tau_0/l^2$ will be called the Fourier criterion $a\tau_0/l^2 = \mathrm{Fo}'$. In this case the complex $a\tau/l^2$ may be presented as the product of the Fourier criterion $a\tau_0/l^2$ by the dimensionless variable τ/τ_0 of a parametric type, i.e.,

$$\mathrm{Fo} = \frac{a\tau}{l^2} = \mathrm{Fo}'\left(\frac{\tau}{\tau_0}\right) \tag{2.1.13}$$

In this case[1] all the similarity criteria are dimensionless parameters and the generalized variables of complex type represent numbers.

Returning to our specific problem, we have

$$\delta t/\Delta t = \varphi(\mathrm{Bi}, \mathrm{Fo}). \tag{2.1.14}$$

Note that this solution determining the temperature conditions on the plate surface could be changed so that it would be valid for any point inside the plate. The location of points inside a plate will be fixed by the relation x/l. Then, designating the local temperature, measured relative to that of the surrounding medium through t and some characteristic temperature, given in the problem condition as t_0, we shall have

$$t/t_0 = \Phi(\mathrm{Bi}, \mathrm{Fo}, x/l). \tag{2.1.15}$$

One can see that besides the Biot criterion and Fourier number the number of arguments is affected by the ratio x/l. This ratio expresses one of the conditions of a problem: "For a point, located at the x distance from the plate surface l thick, determine ..., etc." Such relations immediately introduced on the basis of the problem condition are referred to as variables of a parametric type. Obviously, the ratio t/t_0 in the left-hand side of the equation also represents a parametric variable, as it also satisfies a definite part of the problem condition "... to determine a temperature t if the initial temperature is equal to t_0."

Thus, *dimensionless quantities, similarity criteria, and dimensionless variables are distinguished in the theory of generalized variables. Similarity criteria, consisting of constant dimensionless parameters of a problem, may be of two kinds. The similarity criteria of parametric type represent the ratio of the parameters of the same dimensions which are given according to the problem condition (ratio of length to height, or width to a parallelepiped height, etc.). The complex-type criteria combine the different kind of parameters (Biot criteria, Fourier number, etc.).*

Dimensionless variables represent a ratio of a variable to a constant parameter or their combination. Two types of dimensionless variables are therefore distinguished. The ratio of a local variable to a parameter of the same dimension (x/l, t/t_0, etc.) is the simplest type of a variable.

[1] The group $a\tau/l^2$ may be designated as a criterion or a number depending on the nature of τ. When τ is a current time the group is a dimensionless number. When τ is a known fixed quantity (for example, the period of temperature oscillation) the group is referred to as a criterion.

If a parameter corresponding to a certain variable is not given, then the complex, consisting of a variable and some different parameters, is constructed, for example $a\tau/l^2$ *where* l^2/a *is a complex of different parameters having the dimension of time. Such variables of the complex type will be named as numbers (Fourier number* $\mathrm{Fo} = a\tau/l^2$*).*

As an example, it is possible to give the solution of a problem of plate heating in a medium with a constant heat source W expressed in kcal/m³ hr.

In our notations, the solution of this problem has the form (see Sections 9.2–9.4)

$$\frac{t - t_0}{t_a - t_0} = f\left(\mathrm{Bi}, \frac{x}{l}\, \mathrm{Fo}, \mathrm{Po}\right) \tag{2.1.16}$$

where t_a is the medium temperature, t_0 is the initial temperature of a body

$$\mathrm{Po} = \frac{Wl^2}{\lambda(t_a - t_0)}. \tag{2.1.17}$$

If t_a and t_0 are given, then the complex Po is a parameter and at the same time it is a similarity criterion (the Pomerantsev criterion) and the quantity $(t - t_0)/(t_a - t_0)$ is a dimensionless variable.

However, if the stationary problem ($\mathrm{Fo} \to \infty$) is considered, the initial temperature t_0 does not enter the solution of a problem and the medium temperature t_a may serve as the datum for the reading of the body temperature $(t - t_a)$. Then the Pomerantsev criterion loses its physical significance. In this case the quantity

$$\frac{t - t_a}{(Wl^2/\lambda)} = f\left(\mathrm{Bi}, \frac{x}{l}\right) \tag{2.1.18}$$

is a relative variable.

The complex of the different paramenters Wl^2/λ has the dimension of temperature and serves as the given temperature parameter.

The difference in the Biot and Nusselt numbers may be given as a second example. The Biot number, which plays an essential role in the determination of the temperature field of a solid, represents the ratio of thermal resistance of the wall l/λ to the convective resistance of heat transfer on the surface $1/\alpha$, both resistances being specified according to the conditions of the problem. Thus the Biot number is a parameter, i.e., the similarity criterion.

In contrast to this, when the convective heat transfer processes between a solid and a surrounding medium is investigated, the heat transfer coefficient α is often unknown and must be determined. A new complex containing α is therefore introduced. This complex may be obtained by considering a heat transfer process associated with a liquid or gas film formed near a solid surface through which heat is transferred by conduction. In such a case, the equation may be written

$$\alpha(t_s - t_a) = -\lambda(\partial t/\partial x)_s, \tag{2.1.19}$$

which is formally identical to Eq. (2.1.1). In principle, however, both equations are different because Eq. (2.1.19) contains the heat conduction coefficient for the liquid or gas λ (but not for the solid) and the derivative $(\partial t/\partial x)_s$ is determined when approaching a surface from the liquid or gas side.

The treatment of this equation by the similarity theory methods leads to our already known expression $(\alpha/\lambda)l$. At first sight one may think that the result obtained differs from the Biot number only by the fact that it contains the heat conduction coefficient for a liquid. In reality the difference is more significant, as now α is an unknown quantity and, consequently, the complex corresponds to the category of relative variables or numbers. It is therefore advisable here to introduce a new notation and a new name for this group. Up to the present the name *Nusselt number* and the notation *Nu* are well established. The number Nu always serves as a function in heat transfer equations.

The most frequently used numbers and criteria of similarity are given in Table 2.1.

TABLE 2.1. CRITERIA AND SIMILARITY NUMBERS

Criteria	Numbers
1. Biot criterion $\text{Bi} = HR = R\alpha/\lambda$	Nusselt number $\text{Nu} = \alpha l/\lambda$
2. Kirpichev criterion $\text{Ki} = q_{\text{const}}R/\lambda(t_a - t_0)$	Fourier number $\text{Fo} = a\tau/R^2$
3. Kondratiev criterion $\text{Kn} = R^2(m/a)$	
4. Predvoditelev criterion $\text{Pd} = (d\theta_a/d\text{Fo})_{\text{max}}$	
5. Pomerantsev criterion $\text{Po} = WR^2/\lambda(t_a - t_0)$	
6. Fourier criterion $\text{Fo}' = ap/2\pi R^2$ where p = period of temperature oscillation	

Thus, the solution of a problem should be presented in the form of dimensionless quantities which define unknown dimensionless variables as functions of independent dimensionless variables with similarity criteria playing a role of constant parameters

$$Y_t = f(X_i, \pi_i, P_i)_{i=1,2,\ldots,n}, \tag{2.1.20}$$

where Y_t is the unknown variable; X_i are independent variables; π_i is the complex-type criterion; and P_i are parametric criteria.

If the form of the function of Eq. (2.1.20) is determined for any particular case with the help of a numerical solution of equations or by an experiment, then the result obtained is also valid for an infinite number of such phenomena, which are combined with the initial case into one group under the following requirements (which should be fulfilled in order that equal values of criteria really satisfy similar phenomena): (1) geometric similarity of systems; (2) similarity of their physical structure; (3) similarity of initial states; and (4) similarity of conditions on the surface of interaction between the system and the surrounding medium.

In conclusion it should be noted that any combination of the similarity criteria is also a similarity criterion. A product of a dimensionless variable by any combination of the similarity criteria is also a dimensionless variable.

The possibility of combining criteria and dimensionless variables is of great importance for solving heat and mass transfer problems.

2.2 Operational Calculus and Similarity Theory

The similarity criteria may be obtained from the governing differential equation and boundary conditions, using the methods of transforming from differential relations to algebraic ones. This method of a transformation is connected with operational calculus methods.

Equation (2.1.11) is a ratio of temperature differences transformed in time and along space coordinates, i.e., the Fourier number is a ratio of the transformed temperature differences.

Hence, there appears the idea that similarity criteria be defined, not by the ratio of the original functions themselves, but by their transforms. It is known that the Heaviside operator p, introduced with respect to a time variable τ, gives the following relations

$$dt/d\tau \doteqdot p\bar{t} - t_0; \qquad \frac{d^m t}{d\tau^m} \doteqdot p^m\bar{t} - p^{m-1}t(0) \cdots - t^{m-1}(0); \tag{2.2.1}$$

$$1/p \doteqdot \tau; \qquad 1/p^2 \doteqdot (1/2!)\tau^2; \qquad 1/p^m \doteqdot \tau^m/m!. \tag{2.2.2}$$

where $\doteqdot$ *is the sign of operational correspondence.*

Consequently, the replacement of derivatives by a ratio of one physical variable to the second variable raised to the mth power is clearly a transition from the original function to the transform following Carson-Heaviside. This corresponds to the essence of the analytical transformation itself. The operational methods are mathematical ones which transform symbols of one operation into those of another.

$$\frac{d^m t}{d\tau^m} \doteqdot p^m \bar{t} \doteqdot \frac{t}{\tau^m} m! \qquad \text{if} \quad t(0) = t'(0) = \cdots t^{m-1}(0) = 0. \tag{2.2.3}$$

The Carson–Heaviside method is an integral transformation method which makes it possible to dismember operations and to introduce fractional differentiators and integrators. For example,

$$p \doteqdot d/d\tau, \qquad p^{1/2} \doteqdot 1/(\pi\tau)^{1/2}.$$

Integration is replaced by an operator i and under certain conditions there exists an interrelation between i and the operator $p : p^{-1} = i \doteqdot \tau;\ p^{-2} = i \doteqdot \frac{1}{2}\tau^2$, etc. Therefore, to obtain the similarity criteria it is possible to use not only a system of differential equations but also systems of mixed integral-differential equations. In this sense the operational methods have certain advantages.

Usually the transformation methods by Carson–Heaviside or Laplace are used for nonstationary processes, i.e., transformation occurs along the time coordinate (integration takes place from 0 to ∞).

The Hankel and Fourier finite integral transforms, etc., are applied to bodies of finite dimensions. However, it is only in particular cases that we may write the relation analogous to

$$d^2 t/dx^2 \doteqdot p^2 \bar{t} \tag{2.2.4}$$

In addition, the cosine or sine Fourier transformation does not give the possibility of eliminating the derivative $\partial t/\partial x$ (or any odd derivative) as the integral

$$\int \frac{\partial^{2n+1} t}{\partial x^{2n+1}} \left\{ \frac{\sin px}{\cos px} \right\} dx$$

cannot be expressed in the form of the simple operational relation of type (2.2.4).

However, the finite Hankel integral transformation does make it possible to eliminate the set of terms of the form

$$F(t) = \frac{1}{r}\frac{\partial}{\partial r}\left(r\frac{\partial t}{\partial r}\right) - \frac{n^2 t}{r^2}, \tag{2.2.5}$$

and then the following relation is valid

$$F(t) \doteqdot - p^2\bar{t} + \varphi(p, t_s) \tag{2.2.6}$$

where $\varphi(p, t_s)$ is the function dependent on the boundary conditions.

Hence, the transition from differential equations to algebraic ones is not only a replacement of the partial derivative $\partial^m t/\partial x^m$ by the expression $p^m\bar{t}$ but also the introduction of additional relations taking into account the boundary conditions, i.e., the process of interaction between the body and the surrounding medium.

From the physical viewpoint this means a transition from actual values of the physical quantities (differential equations and single-valued conditions) to those averaged according to a specific statement of the physical problem by the operator transformation methods.

The Heaviside operational method has some advantage over integral transform methods from the viewpoint of its usage in the operator similarity method. In a process of development of operational calculus, the original Heaviside viewpoint was considerably enhanced by the works of Carson [11], Bromwich [5], Doetsch [25], and Van der Pol and Bremmer [122] who used the Laplace transform and the Mellin integral in their investigations. In 1946 the Polish mathematician Mikusinski [81] completely returned to the Heaviside viewpoint. The operator $p = d/d\tau$ is considered in the Heaviside–Mikusinski operational calculus, which makes it possible to reduce a differential equation to algebraic.

Recently Ditkin and Prudnikov [23] introduced the Bessel operator $B = (d/d\tau)\,\tau(d/d\tau)$ which is closely connected with the Bessel equation and makes it possible to solve some differential and integral equations.

Returning to the method of a transition from a real process to the simplest model scheme by the relation $d^m t/dx^m \sim t/x^m$, it may be noted that such a method allows one to obtain the similarity criteria of the whole class of phenomena because differential equations describe a class of similar phenomena. The replacement $d^m t/dx^m$ by $p^m\bar{t}$ taking into account the boundary conditions, permits us to obtain not only a complex of similarity criteria for the given law of interaction between a body and the surrounding medium but also to establish the interrelation between the similarity criteria and, consequently, to define the basic similarity criteria.

Hence, the transform solutions which may be obtained in the maiority of cases are basic initial relations to determine a relation between generalized variables.

It is known that the difficulty in analytical investigations lies in the inversion of transforms but not in obtaining a solution for the transform. Moreover, from the transform solution, it is possible to have some approximate solutions of relations from the tables of transforms and by means of simplifying approximation of the transform solution.

Thus, the operational methods when applied to differential equations together with the conditions of single-valuedness gives us the possibility of obtaining relations between averaged values of the basic similarity criteria of heat and mass transfer. The simultaneous application of the similarity-theory methods and the operational methods should make it possible in the future to develop the heat and mass transfer theory on the basis of the operational similarity methods.

CHAPTER

3

BASIC METHODS FOR SOLUTION OF BOUNDARY VALUE PROBLEMS

In Chapter 1 a differential equation for heat conduction was derived relating temperature, time, and body coordinates for an infinitesimal volume. This equation is a linear, homogeneous partial differential equation of the second order. In mathematical physics, a considerable number of books is devoted to the detailed analytical methods of solution of the classical heat conduction equation. These works will not be treated here. We shall deal only with the basic methods of solution, and special attention will be focused on summarizing the physical nature of the corresponding transformations. Mathematical rigor and analytical details will sometimes be sacrificed in our approach but our main aim is to demonstrate the methods of solution to engineering problems.

The readers who are interested in the mathematical analysis of the subject may refer to the appropriate works [41, 116].

3.1 Analysis of a Differential Equation for Heat Conduction

Without heat sources the differential heat conduction equation can be written as follows

$$\partial t/\partial \tau = a\, \nabla^2 t. \tag{3.1.1}$$

The solutions to this equation have the property of superposition identical with solutions to an ordinary homogeneous differential equation, i.e., if t_1 and t_2 are two particular solutions of the equation, the expression

$C_1t_1 + C_2t_2$ is also the solution to this equation at arbitrary values of constants C_1 and C_2. A partial differential equation of type (3.1.1) has an infinite number of particular solutions.

We illustrate this with an example. A homogeneous differential equation with constant coefficients (a, b, c, d, e, f) for some function t of two variables ξ and η has the form

$$a\frac{\partial^2 t}{\partial \xi^2} + b\frac{\partial^2 t}{\partial \xi\, \partial \eta} + c\frac{\partial^2 t}{\partial \eta^2} + d\frac{\partial t}{\partial \xi} + e\frac{\partial t}{\partial \eta} + ft = 0. \tag{3.1.2}$$

Then the substitution of

$$t = Ce^{k\xi + l\eta} \tag{3.1.3}$$

is a particular solution to this equation, viz:

$$\frac{\partial t}{\partial \xi} = kCe^{k\xi + l\eta}, \qquad \frac{\partial t}{\partial \eta} = lCe^{k\xi + l\eta}, \qquad \frac{\partial^2 t}{\partial \xi\, \partial \eta} = klCe^{k\xi + l\eta},$$

$$\frac{\partial^2 t}{\partial \xi^2} = k^2Ce^{k\xi + l\eta}, \qquad \frac{\partial^2 t}{\partial \eta^2} = l^2Ce^{k\xi + l\eta}.$$

Substituting these relationships into our equation gives, upon cancellation by $Ce^{k\xi + l\eta}$, the so-called equation of coefficients

$$ak^2 + bkl + cl^2 + dk + el + f = 0. \tag{3.1.4}$$

Hence, Eq. (3.1.3) is a particular solution for those values of k and l that satisfy equation of coefficients (3.1.4). Thus, we may take an arbitrary value of one of these two coefficients; however, the second one should be found from Eq. (3.1.4) i.e., we can obtain an infinite number of particular solutions.

The equation of coefficients is a quadratic equation, e.g., with respect to k (we consider k to be variable and l to be constant) and, depending on the value of the discriminant, we may obtain for k (1) two unequal real roots, (2) two equal real roots, and (3) two complex conjugate roots.

The result for the roots k depends on the physical nature of the process studied which is described by differential equation (3.1.2).

It should be noted that solution (3.1.3) may be written as a product of two functions as

$$t = Ce^{k\xi}e^{l\eta} = C\vartheta(\xi)\theta(\eta),$$

one of which, $\vartheta(\xi) = e^{k\xi}$, depends only on ξ and the other, $\theta(\eta) = e^{l\eta}$, depends only on η. However, there are certain solutions of Eq. (3.1.2) for which such a division is impossible.

3.2 Solution of the Equation by Classical Methods

a. Separation of Variables. The classical method of solution of the differential equation for heat conduction is that a group of particular solutions t_n that satisfy the equation and boundary condition are sought. Then by the superposition method, a set of these solutions is composed as

$$t = C_1 t_1 + C_2 t_2 + \cdots = \sum_{n=1}^{\infty} C_n t_n. \tag{3.2.1}$$

The coefficients C_n are found from the initial condition.

Strictly speaking, this property of superposition for an infinite series needs special substantiation since it is unreservedly valid only for a finite sum. Such substantiation would be that a uniform convergence of the series obtained upon differentiation of series (3.2.1), must be proved as well as the validity of the term-by-term integration of the series when determining the coefficients C_n. Monographs on mathematical physics are of use here.

A particular solution t is sought in the form of a product of two functions, one of which, $\theta(\tau)$, depends only on time, τ, and the ohter, $\vartheta(x, y, z)$, depends only on the coordinates, i.e.,

$$t = C\theta(\tau)\vartheta(x, y, z), \tag{3.2.2}$$

where C is the arbitrary constant.

If we substitute solution (3.2.2) into Eq. (3.1.1) we obtain

$$\theta'(\tau)\vartheta(x\ y, z) = a\theta(\tau)\,\nabla^2\vartheta(x, y, z).$$

This equality may also be written as

$$\frac{\theta'(\tau)}{\theta(\tau)} = a\,\frac{\nabla^2\vartheta(x, y, z)}{\vartheta(x, y, z)}. \tag{3.2.3}$$

The left-hand side of the equality may depend only on τ or may be a constant value, but it does not depend on the spatial coordinates. The right-hand side may depend only on the coordinates or be a constant value, but it does not depend on time. The equality should hold at any values of time and coordinates. It is possible only in the case, when the right- and left-hand sides of the equality are equal to some constant value D, i.e.,

$$\frac{\theta'(\tau)}{\theta(\tau)} = D = \text{const}, \tag{3.2.4}$$

$$\frac{a\nabla^2\vartheta(x, y, z)}{\vartheta(x, y, z)} = D = \text{const.} \tag{3.2.5}$$

Equation (3.2.4) may be integrated as

$$\theta(\tau) = e^{D\tau}. \tag{3.2.6}$$

We do not write the constant of integration since it may be attributed to the constant C.

The constant value D is chosen from physical considerations. For thermal processes tending to temperature equilibrium, and after some long period of time ($\tau \to \infty$), a certain temperature distribution should be established, when the value D cannot be positive but must be negative. If D is a positive value, then after a long period of time the temperature will be greater than any predicted value, i.e., it tends to infinity and this contradicts the physical nature of the process.

If the temperature of a body is a periodic function of time, e.g., in the case of the distribution of thermal waves in a body, the value D should be an imaginary value so that the periodic function of time may be obtained instead of the simple exponent (3.2.6).

Consider the first case, $D \leq 0$. Since the value D is still an arbitrary constant numerical value, we may assume

$$D = -ak^2, \tag{3.2.7}$$

where a is the thermal diffusivity and is inherently positive and k is some constant, determined from boundary conditions. Substituting these values for D, we obtain

$$\theta(\tau) = \exp[-ak^2\tau], \tag{3.2.8}$$

$$\nabla^2\vartheta(x, y, z) + k^2\vartheta(x, y, z) = 0. \tag{3.2.9}$$

Differential equation (3.2.9) is often referred to as the Pockel equation and is well known in mathematical physics.

Thus, applying the Fourier method, the heat conduction equation is reduced to the equation of the Pockel type, the solution of which is governed by the geometric shape of the body and the initial temperature distribution as well as by conditions of heat transfer between a body and the surrounding medium or surrounding bodies.

Let the solution of Eq. (3.2.9) be known under corresponding boundary conditions, i.e., the function $\vartheta(x, y, z)$ is found. Then the particular solution of the heat conduction equation may be written as

$$t = C \exp[- ak^2\tau]\,\vartheta(x, y, z). \qquad (3.2.10)$$

Solution (3.2.10) satisfies the differential equation for heat conduction at any values of C and k, i.e., it is a particular solution. Hence, giving various values to the constant C and k, we shall obtain an infinite number of particular solutions.

According to the principle of superposition, the general solution will be the sum of particular solutions as indicated by relationship (3.2.1). The constants k take a definite value, which is determined by the boundary conditions, and the constants C are determined from initial conditions.

In the simplest cases when ϑ depends only on one coordinate ξ (i.e., one-dimensional problems such as the symmetric temperature field in an infinite plate, cylinder, sphere), the solution of Eq. (3.2.9) may be presented as a sum of two particular solutions $\varphi(\xi)$ and $\psi(\xi)$, i.e.,

$$\vartheta(\xi) = \varphi(k\xi) + \psi(k\xi). \qquad (3.2.11)$$

This results from the fact that the general solution of any linear homogeneous differential equation of the second order

$$\vartheta'' + p(\xi)\vartheta' + q(\xi)\vartheta = 0 \qquad (3.2.12)$$

may be written in the form

$$\vartheta = C_1\vartheta_1 + C_2\vartheta_2 \qquad (3.2.13)$$

where C_1 and C_2 are constants and ϑ_1 and ϑ_2 are linear independent integrals of Eq. (3.2.12), i.e., integrals such that the ratio does not reduce to a constant

$$\vartheta_1/\vartheta_2 \neq \text{const}.$$

It is sufficient to know only one linear independent solution, e.g., ϑ_1. The second can then be found by the formula[1]

$$\vartheta_2 = \vartheta_1 \int \vartheta_1^{-2} \exp[- \int p\, d\xi]\, d\xi. \qquad (3.2.14)$$

We continue now the analysis of the particular solution of the differential equation for heat conduction. According to relationship (3.2.11) a particular solution (3.2.10) may be written as

$$t = C\exp[-ak^2\tau]\varphi(k\xi) + D\exp[-ak^2\tau]\psi(k\xi), \qquad (3.2.15)$$

i.e., it represents a sum or a linear combination of two eigenfunctions.

In a general case, the quantity k may assume fixed values defined from boundary conditions. The constants C and D are determined from the initial conditions.

The particular solution is not directly applicable to estimating the temperature field since the constants C and D cannot be determined from a particular solution. For example, at the initial instant ($\tau = 0$) the temperature may be constant, $t = t_0 =$ const, which does not follow from the particular solution (3.2.15).

If we assume $\tau = 0$ ($\exp[-ak^2\tau] = 1$), it follows that the constant t_0 should be equal to the variable $C\varphi(k\xi) + D\psi(k\xi)$, which cannot be the case. Therefore, to obtain the general heat conduction solution that satisfies the initial conditions as well, the sum of particular solutions is taken, in which the constants C and D have fixed values. The temperature at the initial instant may be a given function of the space variable ξ. Then, combining such particular solutions, we may approach the prescribed distribution as close as we desire. This is accomplished by the choice of appropriate values of C and D; such a way of choosing the constants C and D is usually called the satisfaction of the solution to the initial condition.

Thus, the first particular solution may be written as

$$t_1 = C_1\exp[-ak_1{}^2\tau]\,\varphi(k_1\xi) + D_1\exp[-ak_1{}^2\tau]\,\psi(k_1\xi),$$

---

[1] Formula (3.2.14) may be obtained by the following method: we assume $\vartheta_2 = \vartheta_1 z$ and substitution into Eq. (3.2.12) gives us

$$\vartheta_1 z'' + (2\vartheta_1' + p\vartheta_1)z' + (\vartheta_1'' + p\vartheta_1' + q\vartheta_1)z = \vartheta_1 z'' + (2\vartheta_1' + p\vartheta_1)z' = 0$$

(the second expression in brackets is equal to zero, as ϑ_1 is the solution of Eq. (3.2.12)).

We rewrite the above equation in the form

$$\frac{z''}{z'} = -\frac{2\vartheta_1' + p\vartheta_1}{\vartheta_1},$$

whence

$$\ln z' = -2\ln\vartheta_1 - \int p\,d\xi + \ln C_2$$

Integrating once more, we have

$$z = C_2\int \vartheta_1^{-2}\exp[-\int p\,d\xi]\,d\xi,$$

$$\vartheta = C_1\vartheta_1 + C_2\vartheta_1\int\vartheta_1^{-2}\exp[-\int p\,d\xi]\,d\xi = C_1\vartheta_1 + C_2\vartheta_2.$$

the second particular solution

$$t_2 = C_2 \exp[-ak_2{}^2\tau]\,\varphi(k_2\xi) + D_2 \exp[-ak_2{}^2\tau]\,\psi(k_2\xi),$$

and so on.

The general solution will have the form

$$t = \sum_{n=1}^{\infty} C_n\varphi(k_n\xi) \exp[-ak_n{}^2\tau] + \sum_{m=1}^{\infty} D_m\psi(k_m\xi) \exp[-ak_m{}^2\tau]. \quad (3.2.16)$$

It is necessary that the function $t_0(\xi)$ describing the initial temperature distribution be expanded into series with respect to eigenfunctions as

$$t_0(\xi) = \sum_{n=1}^{\infty} C_n\varphi(k_n\xi) + \sum_{m=1}^{\infty} D_m\psi(k_m\xi).$$

We illustrate the above steps of solution with the simplest example. The differential equation for heat conduction for an infinite plate has the form

$$\frac{\partial t(x,\tau)}{\partial \tau} = a\frac{\partial^2 t(x,\tau)}{\partial x^2}. \quad (3.2.17)$$

The particular solution of this equation will be found in the form of a product of two functions

$$t = C\theta(\tau)\vartheta(x).$$

Then, upon substitution into the differential equation, we obtain

$$\frac{\theta'(\tau)}{\theta(\tau)} = a\frac{\vartheta''(x)}{\vartheta(x)} = -ak^2. \quad (3.2.18)$$

The integration of the equation

$$\theta'(\tau)/\theta(\tau) = -ak^2 = \text{const.}$$

will give for the function $\theta(\tau)$

$$\theta(\tau) = \exp[-ak^2\tau].$$

The differential equation for the function $\vartheta(x)$ has the form

$$\vartheta''(x) = -k^2\vartheta(x). \quad (3.2.19)$$

Consequently, the function $\vartheta(x)$ should be such a function that its second derivative may equal the function itself multiplied by some value $(-k^2)$. It is easy to show that sin kx or cos kx may be such functions, viz:

$$\vartheta_1(x) = \sin kx,\ \vartheta_1'(x) = k \cos kx,\ \vartheta_1''(x) = -k^2 \sin kx = -k^2\vartheta_1(x);$$

$$\vartheta_2(x) = \cos kx,\ \vartheta_2'(x) = -k \sin kx,\ \vartheta_2''(x) = -k^2 \cos kx = -k^2\vartheta_2(x).$$

Thus, sin kx and cos kx are particular solutions of Eq. (3.2.19) whereby these solutions are linearly independent, since

$$\frac{\vartheta_1(x)}{\vartheta_2(x)} = \frac{\sin kx}{\cos kx} \neq \text{const}.$$

The general solution of Eq. (3.2.19) will be a sum of particular solutions

$$\vartheta(x) = C\vartheta_1(x) + D\vartheta_2(x) = C \sin kx + D \cos kx, \qquad (3.2.20)$$

where C and D are arbitrary constants.

The second particular solution, $\vartheta_2(x) = \cos kx$, may be obtained by formula (3.2.14) if we know the first solution $\vartheta_1(x) = \sin kx$, viz:

$$\vartheta_2(x) = \vartheta_1(x) \int \vartheta_1^{-2}(x) \exp[-\int p\, dx]\, dx = \vartheta_1(x) \int \vartheta_1^{-2}(x)\, dx$$

$$= \sin kx \int dx/\sin^2 kx = -(1/k) \sin kx \cot kx = -(1/k) \cos kx.$$

In this case $p(x) = 0$. The general solution will be the same:

$$\vartheta(x) = C\vartheta_1(x) + D'(\vartheta_2) = C \sin kx - \frac{D'}{k} \cos kx = C \sin kx + D \cos kx,$$

where $D = -(1/k)D'$ is an arbitrary constant.

The particular solution of the differential equation for heat conduction will have the form

$$t(x, \tau) = C \sin kx \exp[-ak^2\tau] + D \cos kx \exp[-ak^2\tau]. \qquad (3.2.21)$$

The constant k is determined from the boundary conditions, and the constants C and D are determined from the initial conditions; they acquire fixed values depending on the conditions of the problem. The detailed steps of calculation will be given when we are considering particular concrete problems. The general solution may be written as

$$t = \sum_{n=1}^{\infty} C_n \sin k_n x \exp[-ak_n^2\tau] + \sum_{m=1}^{\infty} D_m \cos k_m x \exp[-ak_m^2\tau]. \qquad (3.2.22)$$

b. The Method of Sources. The physical nature of the method of sources is that any process of heat propagation in a body by conduction may be described as combined processes of temperature leveling from numerous elementary heat sources distributed both in space and in time. The solution of heat conduction problems by this method reduces largely to an adequate choice of sources and their distribution.

The behavior of an elementary line source in an infinite body with one-dimensional heat flow is characterized by the expression

$$G(x, \xi, \tau) = \frac{b}{(4\pi a\tau)^{1/2}} \exp\left[-\frac{(x-\xi)^2}{4a\tau}\right], \tag{3.2.23}$$

which is called the source function on an infinite straight line. The function $G(x, \xi, \tau)$ satisfies the heat conduction equation, viz:

$$\frac{\partial G}{\partial \tau} = \frac{b}{(4\pi a\tau)^{1/2}} \left[\frac{(x-\xi)^2}{4a\tau^2} - \frac{1}{2\tau}\right] \exp\left[-\frac{(x-\xi)^2}{4a\tau}\right],$$

$$\frac{\partial^2 G}{\partial x^2} = \frac{b}{(4\pi a\tau)^{1/2}} \left[\frac{(x-\xi)^2}{4a^2\tau^2} - \frac{1}{2a\tau}\right] \exp\left[-\frac{(x-\xi)^2}{4a\tau}\right],$$

$$\text{i.e.,} \quad \frac{\partial G}{\partial \tau} = a\frac{\partial^2 G}{\partial x^2}.$$

Therefore, the function G is usually called the fundamental solution of the heat conduction equation. Direct checking proves that the function G gives the temperature at the point x if the initial temperature is zero and in the initial moment the quantity of heat $Q = bc\gamma$ is evolved at the point ξ.

The quantity of heat generated by our straight line source is

$$Q = c\gamma \frac{b}{\sqrt{\pi}} \int_{-\infty}^{\infty} \exp\left[-\frac{(x-\xi)^2}{4a\tau}\right] \frac{dx}{2(a\tau)^{1/2}} = \frac{bc\gamma}{\sqrt{\pi}} \int_{-\infty}^{\infty} \exp[-u^2]\, du = bc\gamma,$$

where

$$u = \frac{x-\xi}{2(a\tau)^{1/2}}, \qquad \int_{-\infty}^{\infty} \exp[-u^2]\, du = \sqrt{\pi}. \tag{3.2.24}$$

Consequently, the quantity of heat Q does not change with time and is numerically equal to a product of the area, bounded by the curve G, the abscissa axis x, and the volume heat capacity $c\gamma$. For small values of time, almost all heat is concentrated in the neighborhood of the point ξ.

The temperature distribution resulting from an instantaneous heat source for a body of finite dimensions and a one-dimensional heat flow may be represented as (see Chapter 9)

$$G_l(x, \xi, \tau) = \frac{2b}{l} \sum_{n=1}^{\infty} \sin \frac{n\pi x}{l} \sin \frac{n\pi\xi}{l} \exp\left[-\frac{n^2\pi^2}{l^2} a\tau\right]. \quad (3.2.25)$$

The function G_l represents the temperature distribution in an infinite plate $(0 < x < l)$ at the moment τ if the temperature at the initial moment is equal to zero and at this moment the quantity of heat $Q = bc\gamma$ is evolved instantaneously at the point ξ.

The application of the method of sources to the solution of several concrete problems will be shown later.

3.3 Integral Transform Methods

a. Operational Methods. Solutions obtained by classical methods are not always convenient for calculations. It is often necessary to obtain approximate solutions where a distinction is to be made between properties of the heat transfer process and physical constants of a body interacting with the surrounding medium. It is difficult to obtain these formulas from the classical solutions. As a result of engineering needs during the last decade, engineers and physicists began to widely apply the operational methods of solution. The basic rules and theorems of the operational calculus, obtained by Vashchenko-Zakharchenko [124] and independently by Heaviside [47], found an ever-growing use in electrical engineering thanks to Heaviside's works. This method turned out to be so efficient, that many hitherto unsolvable problems have now been solved. Moreover, the method permits solution of problems in a simpler form.

Subsequently, the operational methods found an application in thermal physics and chemical engineering for the solution of various problems of transient heat conduction and diffusion. In recent years these methods were extended to hydrodynamics, neutron transfer in absorbing media, etc.

Strict mathematical justification of the Heaviside operational method has been made in the works of Bromwich [5], Jeffreys [51], Efros and Danilevsky [28], Doetsch [25], Van der Pol [122], Ditkin [22], etc. At present they may be regarded as independent methods for solving equations of mathematical physics. By their consistency they are equivalent to the classical methods. The Heaviside operational method is equivalent to the method of the integral Laplace transform.

The Laplace transform method consists in the fact that not the function itself (inverse transform) but its modification (transform) is studied. This transformation is performed with the help of multiplication by an exponential function and its integration within certain limits. The Laplace transform is therefore an integral transform.

The integral transform $F(s)$ of the function $f(\tau)$ is determined by the formula

$$F(s) = \int_0^\infty f(\tau)e^{-s\tau}\, d\tau = L[f(\tau)], \tag{3.3.1}$$

where $f(\tau)$ is the inverse transform of the function, and $F(s)$ is its transform, that is also designated through $L[f(\tau)]$. Here s may also, be a compelx number, whereby it is assumed that the real part of it will be positive. To insure that the transform does exist, integral (3.3.1) should converge. This imposes certain limitations on the function $f(\tau)$ (for details see Chapter 14).

If the problem is solved in terms of the Laplace transforms, then, in the general case, the inversion of the transform (inverse transformation) is performed with the help of the inversion formula

$$f(\tau) = L^{-1}[F(s)] = \frac{1}{2\pi i}\int_{\sigma-i\infty}^{\sigma+i\infty} F(s)e^{s\tau}\, ds. \tag{3.3.2}$$

Integration is performed in the complex plane, $s = \xi + i\eta$ along the straight line $\sigma = \text{const}$, parallel to the imaginary axis. The real numbers ξ are so selected that all singularities of the integrand (3.3.2) lie in the left half-plane of the complex plane $\operatorname{Re} s \geq s_1 > \sigma_0$. The technique of such an integration is described in detail in manuals on the function theory of a complex variable. In many cases the inverse transformation may be performed without the contour integral by making use of tables [113].

Inversion of the transform is especially rapid if the transform coincides with one of the transforms in the table (see Appendix 5). The following formula may be used instead of (3.3.2) for inversion of the transform

$$f(\tau) = \lim_{n\to\infty}\left[\frac{(-1)^n}{n!}\left(\frac{n}{\tau}\right)^{n+1} F^{(n)}\left(\frac{n}{\tau}\right)\right]. \tag{3.3.2a}$$

In principle, this formula allows the function to be obtained using only differentials and a transition to the limit (see Chapter 14).

(a) If the transform represents the fractional function s

$$F(s) = \frac{\varphi(s)}{\psi(s)} = \frac{A_0 + A_1 s + A_2 s^2 + \cdots}{B_1 s + B_2 s^2 + \cdots}, \tag{3.3.3}$$

which is a quotient of two integer transcendental functions, whereby the denominator has a countable set of simple roots and does not contain a free term with the simultaneous condition that $A_0 \neq 0$ for which the existence of integral (3.3.2) of the function $F(s)$ is necessary. By the theorem of expansion we have

$$f(\tau) = L^{-1}\left[\frac{\varphi(s)}{\psi(s)}\right] = \sum_{n=1}^{\infty} \frac{\varphi(s_n)}{\psi'(s_n)} e^{s_n \tau}, \tag{3.3.4}$$

where s_n are simple roots of the function $\psi(s)$.

(b) If the transform $F(s)$ is a ratio of two polynomials (a fractional rational function) whereby the power of the polynomial $\varphi(s)$ is less than the power of the polynomial $\psi(s)$ which has multiple roots k at the point s_m, then

$$f(\tau) = L^{-1}\left[\frac{\varphi(s)}{\psi(s)}\right] = \sum_{m} \frac{1}{(k-1)!} \lim_{s\to s_m} \left\{\frac{d^{k-1}}{ds^{k-1}} \left[\frac{\varphi(s)(s-s_m)^k}{\psi(s)} e^{s\tau}\right]\right\}, \tag{3.3.4a}$$

where the sum is taken over all roots $F(s)$. If all roots $\psi(s)$ are simple, i.e., all k are equal to unity, then formula (3.3.4a) becomes (3.3.4).

Since in the present monograph the Laplace transform is used as a basic method for solution of heat conduction problems, this method is considered in detail in Chapter 14. Take the following example for illustration.

Example 1. The differential heat conduction equation for a one-dimensional heat flow in a plate has the form

$$\frac{\partial t(x, \tau)}{\partial \tau} = a \frac{\partial^2 t(x, \tau)}{\partial x^2}. \tag{3.3.5}$$

Let us use the Laplace transform to eliminate the variable τ:

$$L\left[\frac{\partial t(x, \tau)}{\partial \tau}\right] = aL\left[\frac{\partial^2 t(x, \tau)}{\partial x^2}\right] \tag{3.3.6}$$

whence we obtain an ordinary second-order differential equation with constant coefficients with respect to the transform

$$a\frac{d^2T(x, s)}{dx^2} - sT(x, s) + u(x) = 0. \tag{3.3.7}$$

Here $u(x)$ corresponds to the initial temperature $t(x, 0)$ which for the present example we shall take to be equal to zero:

$$t(x, 0) = u(x) = 0. \tag{3.3.8}$$

We rewrite Eq. (3.3.7) in the form

$$T''(x, s) - \frac{s}{a} T(x, s) = 0. \tag{3.3.9}$$

The solution to the differential equation (3.3.9) may be written directly, viz:

$$\begin{aligned} T(x, s) &= A \cosh \left(\frac{s}{a}\right)^{1/2} x + B \sinh \left(\frac{s}{a}\right)^{1/2} x \\ &= A_1 \exp\left[\left(\frac{s}{a}\right)^{1/2} x\right] + B_1 \exp\left[-\left(\frac{s}{a}\right)^{1/2} x\right], \end{aligned} \tag{3.3.10}$$

where A, B, $A_1 = \frac{1}{2}(A + B)$ and $B_1 = \frac{1}{2}(A - B)$ are constant with respect to x, but dependent on s.

If the boundary conditions are prescribed, then after the constants A and B or A_1 and B_1 have been determined, the inversion of the transform $t(x, \tau)$ is found with the help of the table of transforms or the expansion theorem.

We shall consider the same problem but with the initial temperature distribution as a function of x, i.e.,

$$t(x, 0) = u(x). \tag{3.3.11}$$

Application of the Laplace transform to eliminate the variable τ in the differential equation (3.3.6) yields the differential equation for the transform (3.3.7):

$$aT''(x, s) - sT(x, s) + u(x) = 0. \tag{3.3.12}$$

The solution of this nonuniform equation is easy by conventional methods, e.g., by variation of arbitrary constants described in textbooks on the theory of ordinary differential equations. The solution is of the form

$$\begin{aligned} T(x, s) = {} & A \cosh(s/a)^{1/2}x + B \sinh(s/a)^{1/2}x \\ & + (a/s)^{1/2} \cosh (s/a)^{1/2} x \int_0^x u(\xi) \sinh(s/a)^{1/2} \xi \, d\xi \\ & - (a/s)^{1/2} \sinh(s/a)^{1/2} x \int_0^x u(\xi) \cosh(s/a)^{1/2} \xi \, d\xi. \end{aligned} \tag{3.3.13}$$

Determining the arbitrary constants A and B from the boundary conditions reduces the solution to inversion of the transform $T(x, s)$.

If at the initial moment, the temperature at all the points is the same and equals t_0, i.e., $u(x) = t_0 = \text{const}$, then (3.3.13) yields

$$T(x, s) - (t_0/s) = A' \cosh (s/a)^{1/2}x + B' \sinh(s/a)^{1/2}x$$
$$= A_1' \exp[(s/a)^{1/2}x] + B_1' \exp[-(s/a)^{1/2}x] \qquad (3.3.14)$$

$$A' = T(0, s) - (t_0/s) = A - (t_0/s), \qquad (3.3.15)$$

$$B' = B \qquad (3.3.16)$$

$$A_1' = \tfrac{1}{2}(A' + B'), \qquad B_2' = \tfrac{1}{2}(A' - B'). \qquad (3.3.17)$$

This same result may be obtained if, in the differential equation (3.3.12) at constant initial temperature $u(x) = t_0 = \text{const}$, the variable $T(x, s) = U(x, s) - (t_0/s)$ is substituted. This would result in the transformation of Eq. (3.3.12) onto (3.3.9), the solution of which is known.

Since A', B', A_1', B_1' are constant with respect to x and determined from the boundary conditions, the subscripts may be omitted and the solution of the differential equation (3.3.12) at the initial constant temperature may be written in the form

$$T(x, s) - (t_0/s) = A \cosh(s/a)^{1/2}x + B \sinh (s/a)^{1/2}x$$
$$= A_1 \exp[(s/a)^{1/2}x] + B_1 \exp[-(s/a)^{1/2}x]. \qquad (3.3.18)$$

The constants A and B are determined from the corresponding boundary conditions.

We conclude the Chapter with the note that the biggest difficulty of the solution of the heat-conduction problem for different boundary conditions is the inversion of the transform T.

The application of the integral Laplace transform to the solution of differential heat conduction equations is frequently more advantageous than the classical methods of integration of differential equations and is usually superior to other methods of integral transform for the following reasons:

First, the technique of applying the integral Laplace transform is relatively independent of the nature of the problem and the shape of the body. The procedure of solution is more straightforward and does not require any special skill.

Second, the integral Laplace transforms are capable of handling problems with boundary conditions of the first, second, third, and fourth kinds without any new assumptions or transformations.

Third, the availability of a great number of simple theorems permits one to obtain adequate engineering results; in particular solutions in the form convenient for calculation at small and great values of time are available.

Fourth, the method yields particularly simple solutions of problems with simple initial conditions; the most efficient approach is the use of the Laplace transform with respect to the time coordinate as well as the space coordinate for infinite or semi-infinite bodies.

Fifth, the efficiency of the solution of various problems by the Laplace transform method is increased to a considerable extent by the availability of very detailed tables of transforms.

Although the integral Laplace transform possesses all these advantages, it should also be mentioned that it does have some defects. In particular, some difficulties arise when solving problems with initial conditions prescribed in the form of a function of space coordinates or some multidimensional problems. To overcome this difficulty a number of alternate integral transformations with respect to the space coordinates taking into consideration the geometry of a body were proposed. Such transformations were proposed by Deotsch, Sneddon, Tranter, and others. A number of works in this trend were performed in the Soviet Union (cf. [118, 41]).

If the transform is performed with respect to the space coordinate x, the integral transform of the function $f(x)$ may be presented as

$$[f(p)]_{F;H} = \int_0^\infty K(p, x) f(x)\, dx. \tag{3.3.19}$$

If the kernel of the transform $K(p, x)$ is taken in the form $(2/\pi)^{1/2} \sin px$ or of $(2/\pi)^{1/2} \cos px$, it is correspondingly referred to as the Fourier sine transform or the Fourier cosine transform. If the Bessel function $K(p, x) = xJ_\nu(px)$ is taken as the kernel of transform, it is referred to as the Hankel tranform. In a particular case, if the limits of integration change from $-\infty$ to $+\infty$ and the kernel has the form $K(p, x) = (1/2\pi)^{1/2} \exp(ipx)$ we obtain the complex integral Fourier transform. It is convenient to apply the complex integral Fourier transformation to infinite bodies. The Fourier sine transform is particularly useful when the value of the function is prescribed on the surface of a body, i.e., we have boundary conditions of the first kind, and the Fourier cosine transform is useful when we solve differential transfer problems with boundary conditions of the second kind. The Hankel transform is particularly convenient when the body has axial symmetry. Practical application of the above integral transforms will not pose any special difficulties after good tables of transforms have become available.

In those cases when the Fourier transform should be applied but the values of transforms are not available, the inverse transforms may be found by the following simple inversion formulas:

The complex Fourier transform:

$$f(x) = (1/2\pi)^{1/2} \int_{-\infty}^{\infty} [f(p)]_F\, e^{-ipx}\, dp. \tag{3.3.20}$$

The Fourier sine transform:

$$f(x) = (2/\pi)^{1/2} \int_0^\infty [f_s(p)]_{Fs} \sin px \, dp. \tag{3.3.21}$$

The Fourier cosine transform:

$$f(x) = (2/\pi)^{1/2} \int_0^\infty f_c(p)_{Fc} \cos px \, dp. \tag{3.3.22}$$

The Hankel transform:

$$f(r) = \int_0^\infty r[f_\nu(p)]_H \, J_\nu(pr) \, dp. \tag{3.3.23}$$

The peculiarity of the above transforms is that the upper limit of integration is equal to infinity. If in the Laplace transform (3.3.1), which in most cases is applied with respect to the time coordinae, the infinite limit of integration is caused by the very duration of an unsteady time process, then in the Fourier and Hankel transforms (3.3.20)–(3.3.23) with respect to the space coordinate, the presence of an infinite limit narrows the range of application of these methods. In other words, the integral transform (3.3.20)–(3.3.23) may be successfully applied only to problems of semi-infinite dimension. Moreover, it should be noted, that applying the Fourier transform, particularly the sine and cosine transforms, it is necessary to pay great attention to the convergence of integrals since here convergence conditions become more difficult than convergence conditions of corresponding integrals when applying the Laplace transform.

b. Finite Integral Transform. The scantiness of the Fourier, Hankel, and, to some extent, the Laplace transforms on the one hand, and a crying need for solution of the problems with finite region of the change of variables on the other hand led to creation of finite integral transform methods. Even for those problems which may be solved by the classical methods with the help of the Fourier or Fourier–Bessel series, the finite integral transform method may be preferable from the standpoint of simplicity of approach, although it is mathematically identical to the method of eigenfunctions.

First the idea of the finite integral transform method of the type

$$[f(p)]_{F,H} = \int_a^b K(p, x) f(x) \, dx \tag{3.3.24}$$

was proposed by Koshlyakov [57]. The theory of such integral transforms was more thoroughly developed by Grinberg [41] who generalized these methods for the case of step-by-step change of properties of the medium in the direction of that coordinate along which the transformation is performed. A detailed development of the integral transforms with the finite limits was made by Sneddon [109], Tranter [118], Deotsch [25], and others. The kernels $K(p, x)$ of the finite integral Fourier and Hankel transforms are to be found from the solution of the Sturm–Liouville boundary problem. If the eigenfunctions of this problem are designated by $X_n(x)$ and the weighting functions by $\varrho(x)$ within the range $[a, b]$, then we have (see Section 14.12)

$$K(p, x) = \varrho(x)X_n(x) \tag{3.3.25}$$

If the limit of integration lies between 0 and l, the kernels of finite sine and cosine Fourier transforms, as well as the Hankel transforms, have the form, respectively

$$K(p, x) = \sin(\mu x/l); \qquad K(p, x) = \cos(\mu x/l) \tag{3.3.26}$$

(with boundary conditions of the first and second kinds $\mu = n\pi$; and with boundary conditions of the third kind μ are the roots of the equation $\mu \tan \mu = \alpha l/\lambda$) and

$$K(p, x) = rJ_\nu(\mu x/l), \tag{3.3.27}$$

where μ is the root of the equation $J_\nu(\mu) = 0$ (boundary condition of the first kind). With boundary conditions of the third kind, μ is determined from equation

$$J_\nu(\mu) = -(\mu\lambda/\alpha l)J_\nu'(\mu). \tag{3.3.28}$$

The inversion formula is usually found with the help of expansion of the function into series with respect to the orthogonal functions of the corresponding Sturm–Liouville problem. Therefore, solutions obtained by these methods have the same principal disadvantages as the solutions obtained by the classical methods. Thus, the inversion formulas have the form: (see Section 14.12)

$$f(x) = \sum_{n=1}^{\infty} \frac{[f(\varrho)]_{F,H} X_n(x)}{\int_a^l \varrho(k) X_n^2(x)\, dx} \tag{3.3.29}$$

For particular cases:

(1) for the sine transform

$$f(x) = \frac{2}{l} \sum_{n=1}^{\infty} f_s(n) \sin \frac{n\pi x}{l}; \tag{3.3.30}$$

(2) for the cosine transform

$$f(x) = \frac{1}{l} f_c(0) + \frac{2}{l} \sum_{n=1}^{\infty} f_c(n) \cos \frac{n\pi x}{l}, \tag{3.3.31}$$

with boundary conditions of the second kind; and

$$f(x) = 2 \sum_{n=1}^{\infty} \frac{\mu_n^2 + (\alpha l/\lambda)^2}{l[\mu_n^2 + (\alpha l/\lambda)^2] + \alpha l/\lambda} f_c(\mu_n) \cos \frac{\mu_n x}{l} \tag{3.3.32}$$

with boundary conditions of the third kind, where the summation is conducted with respect to all the positive roots of the equation

$$\mu \tan \mu = \alpha l/\lambda.$$

For the Hankel transform,

$$f(x) = \frac{2}{l^2} \sum_{n=1}^{\infty} f_H(\mu_n) \frac{J_\nu(\mu_n x/l)}{[J_\nu'(\mu_n)]^2}, \tag{3.3.33}$$

where the summation is conducted with respect to all positive roots $J_\nu(\mu) = 0$ or

$$f(x) = \frac{1}{l^2} \sum_{n=1}^{\infty} f_H(\mu_n) \frac{\mu_n^2}{(\alpha l/\lambda)^2 + (\mu_n^2 - \nu^2)} \cdot \frac{J_\nu(\mu_n x/l)}{J_\nu^2(\mu_n)}, \tag{3.3.34}$$

where the summation is conducted with respect to the positive roots of (3.3.28).

The Hankel transform (3.3.27) is applied when solving heat conduction problem for a solid cylinder. For a hollow cylinder, the kernel $K(p, x)$ in the Hankel transform is taken in another form (for details see Chapters 4 and 5).

To overcome the above difficulties, various approximate integral transform methods were developed in which the direct transform and inverse transform were performed by approximate formulas. An idea has arisen to develop the method of the so-called Laplace finite integral transforms, or, which is more correct, the Green finite integral transforms. Let us dwell on this in more detail.

In a general case the Riemann–Mellin integral (see Chapter 14) is the inversion formula of the Laplace integral transform. This formula permits us to obtain the solutions in the form in which we are interested including those in closed form. The idea of the method is that the choice of the kernel of the integral transform $K(p, x)$ is performed in conformity with the differential equation and boundary conditions, i.e., taking into account the geometric form of the body and the law of its interaction with the surrounding medium. In other words Green's function is the kernel of the transform for the given problem. The transform of the function $f(x)$ is obtained with the help of the integral transform

$$[f(p)]_G = \int_0^l K(p, x) f(x)\, dx, \tag{3.3.35}$$

while the inverse transform is determined by formula (3.3.2) after first substituting $[f(p)]_G$ for $[f(s)]_L$.

Such a procedure of the integral transformation has its physical basis. The idea is that any integral transform taken with respect to the space coordinates is, from the physical point of view, an averaging of the considered physical value. It is quite natural that this averaging should take into account not only the character of the process and the shape of the body (with the form of the differential equation) but also the boundary conditions. In this case, the solution for the transform of the function will be of special interest since such a transformation in a physical sense will present transition from the analysis of actual values of considered functions (differential equation, univalence conditions) to the averaged values, made in conformity with the concrete situation of the physical problem. Thus, *the integral transform methods acquire a new and very essential advantage over the classical methods, since they give the possibility of obtaining a number of regularities of the proceeding of physical processes on the basis of the analysis of a solution for averaged values of the considered physical value* (*the analysis of the solution for the transform*). *This fact relates the given theoretical methods to the methods of the similarity theory.*

The integral transforms have specific advantages when solving the system of partial differential equations. The method of solution of the system of equations does not differ in principle from solution of individual equations and is carried out by a number of successive procedures. For example, for one-dimensional heat conduction problems dependent on the space coordinate and on time, it is necessary:

(i) to choose an appropriate integral transform or a group of integral transforms on the basis of the analysis of the differential equation and boundary conditions;

(ii) to multiply the differential equation and boundary conditions by the chosen kernel of the transform and integrate the expressions obtained over corresponding limits with respect to the variable which is to be excluded; as a result, we shall obtain a system of ordinary differential equations for the transform of the functions, which take into account the initial (when using the Laplace transform) or boundary (when using the Fourier transform) conditions, instead of the system of partial differential equations with respect to the inverse transform of the functions;

(iii) to solve an ordinary differential equation with respect to the transformed functions. (If the solution of the transformed equation still causes some difficulty, it should be attacked once more by an appropriate integral transform with respect to the second independent variable. As a result of the transformation, we obtain an algebraic equation, the solution of which is more elementary. Having found the expression for twice transformed functions, we apply to them the reverse transform. The solution obtained will be the desired solution of the differential equation.);

(iv) to define more exactly the expressions for the arbitrary constants which are contained in the solution of the equation, for which the end conditions of the considered problem are used;

(v) to invert the functions using the known relations between the transform of the function and the function itself or formulas of the reverse transform and, consequently, to find the final solution of the problem.

3.4 Methods of Numerical Solution of Heat Conduction Problems

The methods of mathematical physics, particularly those of integral transformations, allow the effective solution of a comparatively narrow range of problems of the transfer theory. When considering systems of differential equations with very general boundary conditions, we find the exact solution very difficult and with present methods it becomes impossible in the case of nonlinear problems. In these cases we resort to numerical methods of solution. At present the finite-difference method which is sometimes referred to as the net method, is one of the better procedures for the approximate solution of heat conduction equations in practice.

The finite-difference method is based on the replacement of the derivatives by their approximate values expressed through the values of a function at certain discrete points—nodal points. As a result of such transformations, the differential equation is replaced by an equivalent finite-difference relation whose solution is reduced to simple algebraic manipulations. The final result of the solution is given by an expression, according to which the value of a "future" potential (temperature) at a given nodal point is a function of time, which is its "present" potential and the "present" potential of neighboring node points. When calculating temperature fields, the repetition of the same operations facilitates the application of modern computing techniques, resulting in a considerable efficiency of work.

The approximate replacement of the first and second derivatives by the difference relations may be carried out simply as follows. Let the function $y = f(x)$ be given as plotted in Fig. 3.1. If the angle between the abscissa

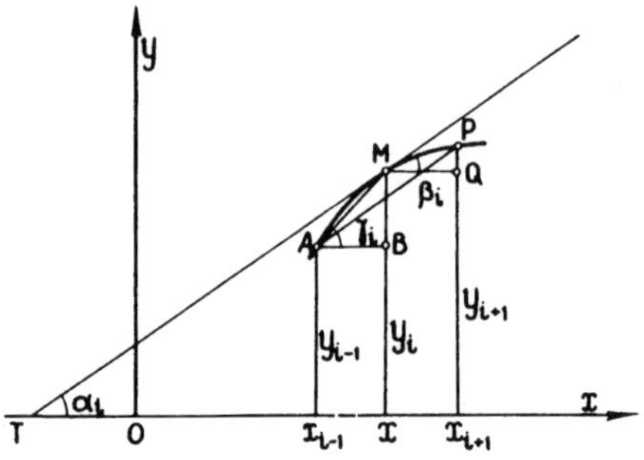

Fig. 3.1. Illustration of the determination of function $f(x)$.

and the tangent to the curve through the point $M(x_i, y_i)$ is designated by α_i, the derivative of the function at $x = x_i$ is given by the formula

$$y_i' = \tan \alpha_i. \tag{3.4.1}$$

Take two neighboring points $A(x_{i-1}, y_{i-1})$ and $P(x_{i+1}, y_{i+1})$ on the curve so that the differences $x_i - x_{i-1} = x_{i+1} - x_i = h$ (or, consider one of the secants MP or AM instead of the tangential MT) would be sufficiently small, and approximately replace α_i by β_i or γ_i. If the slope of MT is approximately replaced by that of AP then,

$$y_i' \approx \tan \beta_i = \frac{QP}{MQ} = \frac{y_{i+1} - y_i}{h} \tag{3.4.2}$$

or

$$y_i' \approx \tan \gamma_i = \frac{BM}{AB} = \frac{y_i - y_{i-1}}{h}. \tag{3.4.3}$$

An alternate expression is

$$y_i' \approx \frac{y_{i+1} - y_{i-1}}{2h}. \tag{3.4.4}$$

The right-hand sides of formulas (3.4.2) to (3.4.4) are respectively called forward difference method, backwards difference method, and symmetrical difference method.

The approximate value of the second derivative y_i'' of the function $y = f(x)$ at $x = x_i$ may be readily obtained if the curve on the section AP is replaced by a broken line AMP which has two slopes at M, where

$$y_i'' \approx \frac{1}{h}\left(\frac{y_{i+1} - y_i}{h} - \frac{y_i - y_{i-1}}{h}\right) = \frac{y_{i+1} - 2y_i + y_{i-1}}{h^2}. \tag{3.4.5}$$

Formulas (3.4.2) and (3.4.5), used to replace the derivatives by the difference quotients are surely not unique. Sometimes it is advisable to carry out another replacement. However, when integrating heat conduction equations, Eqs. (3.4.2) and (3.4.5) are used more frequently.

Let us consider for example a one-dimensional heat conduction equation for an insulated thin bar L in length:

$$\frac{\partial t(x, \tau)}{\partial \tau} = a\frac{\partial^2 t(x, \tau)}{\partial x^2} \qquad (0 \leq x \leq L). \tag{3.4.6}$$

Since the function $t(x, \tau)$ depends on two various x and τ, the rectangular network (Fig. 3.2) is used. We plot a segment with the length L on the ab-

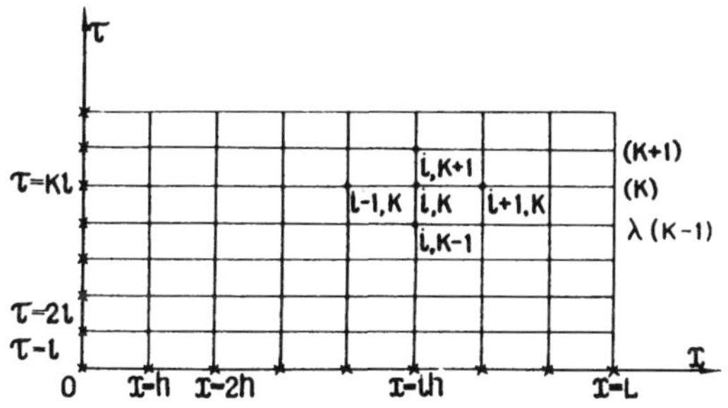

Fig. 3.2. Rectangular network.

scissa and divide it into n equal parts. The resulting step on the abscissa is designated through $h = L/n$. On the x axis, the resulting points of division have abscissas $x = 0$, $x = h$, ..., $x = L$.

The time values τ in equal intervals l are plotted on the ordinate. The straight lines parallel to the coordinate axis which pass through the obtained nodal points on the coordinates (in Fig. 3.2 they are marked by crosses) form a rectangular net. It is assumed that the values of t at the nodes lying on the coordinate axes, on the straight line parallel to the ordinate axis, and being at a distance L from it are specified by the initial and boundary conditions.

The problem of approximate numerical integration of equation (3.4.6) by the net method is to find an approximate value of the function t at each nodal point of the network.

We shall designate the actual value of the temperature at the point of the rod $x = ih$ at the moment $\tau = kl$ by $t_{i,k}$, i.e., at the point marked in Fig. 3.2 as i, k.

The partial derivatives $\partial t/\partial\tau$ and $\partial^2 t/\partial x^2$ at the point (ih, kl) are replaced by the difference quotients according to formulas (3.4.2)–(3.4.5), i.e., we assume

$$\frac{\partial t_{i,k}}{\partial \tau} = \frac{t_{i,k+1} - t_{i,k}}{l} + \varepsilon_1 , \tag{3.4.7}$$

$$\frac{\partial^2 t_{i,k}}{\partial x^2} = \frac{t_{i-1,k} - 2t_{i,k} + t_{i+1,k}}{h^2} + \varepsilon_2 , \tag{3.4.8}$$

where ε_1 and ε_2 are the residual terms tending to zero as l and h approach zero. At the nodal point (ih, kl) the differential equation (3.4.6) is replaced by the relation

$$\frac{t_{i,k+1} - t_{i,k}}{l} + \varepsilon_1 = a\left(\frac{t_{i-1,k} - 2t_{i,k} + t_{i+1,k}}{h^2} + \varepsilon_2\right), \tag{3.4.9}$$

or

$$t_{i,k+1} = \left(1 - \frac{2la}{h^2}\right) t_{i,k} + \frac{la}{h^2}(t_{i-1,k} + t_{i+1,k}) + lR \tag{3.4.10}$$

where $R = a\,\varepsilon_2 - \varepsilon_1$.

Omitting the residual term lR in Eq. (3.4.10) yields the difference equation

$$\vartheta_{i,k+1} = \left(1 - \frac{2la}{h^2}\right) \vartheta_{i,k} + \frac{la}{h^2}(\vartheta_{i-1,k} + \vartheta_{i+1,k}), \tag{3.4.11}$$

where the approximate value of the quantity $t_{i,k}$ at this nodal point (ih, kl) is designated as $\vartheta_{i,k}$.

Formula (3.4.11) allows the calculation of the value of ϑ at the nodal point of the horizontal row $(k + 1)$ using values of ϑ in the previous row (k). It is therefore possible, using (3.4.11), to find values of ϑ at nodal points of the first horizontal row (at $\tau = l$) with the help of the temperature values at the nodal points of the axis Ox (at $\tau = 0$). These values are known from the initial conditions. Thus, upon obtaining the values of ϑ in the first row, we find the values of ϑ at the nodal points of the second horizontal row (i.e., at $\tau = 2l$) by means of the same formula. This process of building up a table may be continued as far as one wishes since values of the temperature at the cross-points of straight lines $x = 0$ and $x = L$ will be known from the boundary conditions.

Formula (3.4.11) may be derived by applying the Fourier and Newton laws and performing heat balances on the elements of the body. At present, many investigators use this method of deriving the above calculation formulas.

Following Panov, we may rewrite formula (3.4.11) in the more convenient form (see the diagram in Fig. 3.3) of

$$\vartheta_A = \left(1 - \frac{2la}{h^2}\right)\vartheta_0 + \frac{la}{h^2}(\vartheta_1 + \vartheta_2). \tag{3.4.12}$$

By selecting the ratios between steps l and h in different ways, it is possible

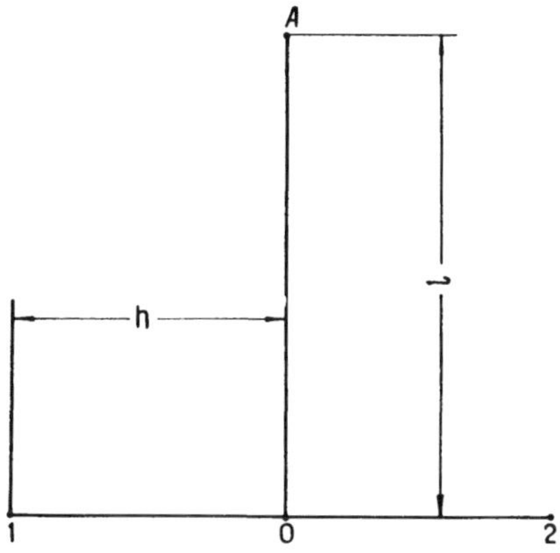

Fig. 3.3. Illustration of the derivation of formula (3.4.12).

to obtain some particular relations from (3.4.12). Thus, for example, for $l = h^2/3a$

$$\vartheta_A = \tfrac{1}{3}(\vartheta_1 + \vartheta_0 + \vartheta_2), \tag{3.4.13}$$

for $l = h^2/6a$

$$\vartheta_A = \tfrac{1}{6}(\vartheta_1 + 4\vartheta_0 + \vartheta_2), \tag{3.4.14}$$

for $l = h^2/12a$

$$\vartheta_A = \tfrac{1}{12}(\vartheta_1 + 10\vartheta_0 + \vartheta_2), \tag{3.4.15}$$

and generally for $l = h^2/pa$

$$\vartheta_A = \frac{\vartheta_1 + (p-2)\vartheta_0 + \vartheta_2}{p}. \tag{3.4.16}$$

For $p = 2$, formula (3.4.12) has the simplest form

$$\vartheta_A = \tfrac{1}{2}(\vartheta_1 + \vartheta_2). \tag{3.4.17}$$

This last formula,[2] called the Schmidt formula, has a great practical advantage over formula (3.4.14) and even more over formula (3.4.15) as, at the same h (i.e., same step size), the quantity l (i.e., time interval) is the largest for Eq. (3.4.17); consequently, the amount of computation work diminishes by a factor of three if we use Eq. (3.4.17) instead of (3.4.14) and by a factor of six if we use Eq. (3.4.17) instead of (3.4.15). It should be noted that, owing to its simplicity, formula (3.4.17) is widely used for the graphical solution of unsteady-state transfer problems [87]. Investigations show that at $p = 1$, we obtain a divergent calculation formula. In general, it should be noted that when solving unsteady-state partial differential parabolic equations we find that the problems of selecting the relations between h and l as well as the error of rounding-off are of the greatest significance since convergence and stability of the solution depend on them. The rigorous theoretical considerations show that

(1) formula (3.4.16) is valid only at $p \geq 2$;
(2) formula (3.4.17) gives the largest time step l when $p = 2$;
(3) the larger the value of p, the more closely formula (3.4.16) approaches the exact solution.

The network shown in Fig. 3.2 is convenient for numerical integration of Eq. (3.4.6) when the problem to be solved involves a boundary condition of the first kind, because, in this case, the boundary straight lines $x = 0$ and $x = L$ belong to the network itself. If the equation to be solved involves boundary conditions of the third kind, calculations and theoretical investigations show that for an increase in accuracy of determination of the potential on the boundaries, it is necessary to introduce additional nodes outside the actual region under consideration if we are to get an accurate determination of the potential on the boundaries. For example, when solving Eq. (3.4.6) with the boundary conditions

$$\left(\frac{\partial t}{\partial x}\right)_{x=0} = \frac{\alpha}{\lambda}\,[t(0, \tau) - t_a], \qquad \left(\frac{\partial t}{\partial x}\right)_{x=L} = -\frac{\alpha}{\lambda}\,[t(L, \tau) - t_a], \tag{3.4.18}$$

the network should be constructed so that the physical right-hand boundary lies in the middle between two straight lines $x = x_n$ and $x = x_{n+1}$, and the

[2] *Editor's note*: This formulation was apparently first suggested by E. Schmidt, *in* "Foppls Festschrift," Springer, Berlin, 1924 and was independently obtained by Panov in 1938.

left-hand boundary lies in the middle between the straight lines $x = x_0$ and $x = x_1$ (Fig. 3.4) ($x = h/2$). Taking this into account, we introduce values of $\vartheta_{n+1,k}$ and $\vartheta_{0,k}$, i.e., the values of the function for the points lying outside the region under consideration. The derivative $(\partial t/\partial x)_{x=L}$ entering

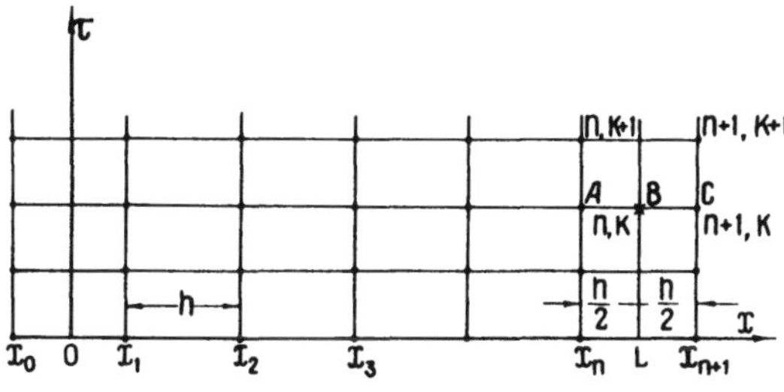

Fig. 3.4. Calculation scheme for unsteady heat conduction problem (boundary conditions of the third kind).

the second condition (3.4.18) at a point $B(L, kl)$, is replaced by the symmetrical difference quotient

$$\left(\frac{\partial t}{\partial x}\right)_{x=L} = \frac{t_{n+1,k} - t_{n,k}}{\frac{1}{2}h + \frac{1}{2}h} + \varepsilon', \tag{3.4.19}$$

and the temperature at the surface $t(L, kl)$, is taken as the arithmetic mean of the temperature at points A and C as

$$t(L, kl) = \tfrac{1}{2}(t_{n,k} + t_{n+1,k}). \tag{3.4.20}$$

Then condition (3.4.18) is written as

$$\frac{t_{n+1,k} - t_{n,k}}{h} + \varepsilon' = -\frac{\alpha}{\lambda}\left[\frac{t_{n,k} + t_{n+1,k}}{2} - t_a\right] \tag{3.4.21}$$

or, passing to the approximate potential ϑ, we obtain upon transformations,

$$\vartheta_{n+1,k} = \left[\vartheta_{n,k}\left(1 - \frac{\alpha h}{2\lambda}\right) + \frac{\alpha h}{\lambda} t_a\right] \cdot \left[1 + \frac{\alpha h}{2\lambda}\right]^{-1}. \tag{3.4.22}$$

The approximate values of the function at the nodes of the auxsiliary straight line $x = L + \frac{1}{2}h$ are obtained by this formula.

The temperature at the boundary straight line $x = L$ is defined by

$$\vartheta(L, kl) = \tfrac{1}{2}(\vartheta_{n,k} + \vartheta_{n+1,k}), \tag{3.4.23}$$

which upon transformation gives

$$\vartheta(L, kl) = \frac{\vartheta_{n,k} + \dfrac{h}{2}\dfrac{\alpha}{\lambda} t_a}{1 + \dfrac{h}{2}\dfrac{\alpha}{\lambda}} = \frac{2s\vartheta_{n,k} + ht_a}{2s + h}, \tag{3.4.24}$$

where

$$s = \lambda/\alpha. \tag{3.4.25}$$

The temperature at the nodes of the auxiliary straight line $x = -\frac{1}{2}h$ is

$$\vartheta_{0,k} = \frac{\left(1 - \dfrac{h}{2}\dfrac{\alpha}{\lambda}\right)\vartheta_{1,k} + h\dfrac{\alpha}{\lambda} t_a}{1 + \dfrac{h}{2}\dfrac{\alpha}{\lambda}}. \tag{3.4.26}$$

At the left-hand surface we obtain

$$\vartheta(0, kl) = \frac{\vartheta_{1,k} + \dfrac{h}{2}\dfrac{\alpha}{\lambda} t_a}{1 + \dfrac{h}{2}\dfrac{\alpha}{\lambda}} = \frac{2s\vartheta_{1,k} + ht_a}{2s + h}. \tag{3.4.27}$$

The methods for solving the differential heat conduction equation with sources do not differ in principle from the previous ones. The finite-difference method allows a successful solution of one-, two-, and three-dimensional problems. Mikeladze [77] completely investigated the case when the square network is selected as the coordinate system to describe the changes of variables x and y. Triangle and polar networks are also considered by Yushkov [128, 130] and others [101]. It should be noted that polar networks are especially convenient for solving axisymmetric problems.

Determination of the temperature field in the three-dimensional space with constant thermal properties is given by Yushkov [130] and with variable ones by Vashchenko–Zakharchenko [124] and Bromwich [5]. All these problems are considered by Sauliev [99] and Yushkov [130] in detail.

The finite-difference method as has been shown by Yushkov allows a successful solution of a system of differential heat conduction equations for either constant or variable coefficients. The method is called explicit as it gives the value of ϑ at the moment τ_{k+1} using the values at the moment τ_k.

In spite of the simplified calculation formulas, the method has an essential disadvantage due to the condition $p \geq 2n$, where n is the dimension of the space. Since here $l \sim h^2$, tracing the behavior of the solution over a rather long time interval, say up to $\mathrm{Fo} \sim 1$, demands very many time steps.

The detailed analysis of the solution shows that attempts at increasing the time step leads to an increase of the computational components which rapidly vary with respect to space variables. At the same time, for slowly varying (with respect to space coordinates and time) smooth inputs, the conditions for good approximation are preserved and no singularities are observed in this case. Since in a real process all singularities are smoothed very rapidly, and rapidly varying components of the solution damp quickly, the restriction for l/h^2 seems unnatural and may be attributed to peculiarities of the computational scheme.

Implicit schemes are free of such disadvantages. In these, a larger time step l may be used because at every time step, a set of equations is to be solved and the method is inherently more stable than the explicit scheme. In the case of a single space variable, the solution is obtained after accomplishing a small number of manipulations because of the three-diagonal-type matrix in this system. As an example, we shall consider a six-point scheme (see [99]) in which $\partial^2\vartheta/\partial x^2$ is approximated by the formula

$$\frac{\partial^2\vartheta}{\partial x^2} \approx \frac{(\vartheta_{k-1,i-1} - 2\vartheta_{k-1,i} + \vartheta_{k-1,i+1}) + (\vartheta_{k,i-1} - 2\vartheta_{k,i} + \vartheta_{k,i+1})}{2h^2}. \tag{3.4.28}$$

The time derivative is approximated according to formula (3.4.7).

It is easy to see that for a sufficiently smooth function the approximate formula for $(\partial/\partial\tau) - a(\partial^2/\partial x^2)$ composed from these expressions has an error $O(h^2 + l^2)$ for the point $x = ih$, $\tau = (k - \frac{1}{2})l$. Therefore, if the equation is inhomogeneous, the value for the right-hand side should be taken at this point. It is known for the equation

$$\frac{\vartheta_{k,i} - \vartheta_{k-1,i}}{l}$$
$$= a\,\frac{(\vartheta_{k-1,i-1} - 2\vartheta_{k-1,i} + \vartheta_{k-1,i+1}) + (\vartheta_{k,i-1} - 2\vartheta_{k,i} + \vartheta_{k,i+1})}{2h^2} + f_{k-1/2,i} \tag{3.4.29}$$

(this solution is stable and as $l \to 0$ reduces to the solution of the differential heat-conduction equation).

The reader is referred to Sauliev [99] and Dulnev [26], also for the method of appropriately matching the step sizes h and l, in addition to the discussion of numerous implicit methods and means of solving the set of equations obtained at each step.

Here we shall demonstrate how to solve the set of equations for each layer in the case of a single space variable only. It is easy to see that for the unknowns $\{\vartheta_{k,i}\}$ Eq. (3.4.29) may be written in the form

$$\vartheta_{k,i-1} - 2\left(1 + \frac{h^2}{al}\right)\vartheta_{k,i} + \vartheta_{k,i+1} = B_i, \tag{3.4.30}$$

where B_i are formed from $f_{k-1,i}$ and the known values of $\vartheta_{k-1,i}$.

Let the boundary conditions be of the first, second, or third kind having the general form

$$\vartheta_{k,0} = \alpha\vartheta_{k,1} + \beta, \qquad \vartheta_{k,N} = \gamma\vartheta_{k,N-1} + \delta. \tag{3.4.31}$$

In this case, we also assume $\alpha \leq 1$ and $\gamma \leq 1$. The relation of the form

$$\vartheta_{k,i} = p_i\vartheta_{k,i-1} + q_i \tag{3.4.32}$$

will be sought which satisfies the solution (3.4.30).

Substitution of (3.4.32) into (3.4.30) yields

$$\vartheta_{k,i-1} - \left[2\left(1 + \frac{h^2}{al}\right) - p_{i+1}\right]\vartheta_{k,i} = \beta_i - q_{i+1}. \tag{3.4.33}$$

We require that (3.4.33) follows from (3.4.31). Since the coefficients with the same ϑ and free terms must be proportional,

$$p_i = \left\{2\left(1 + \frac{h^2}{al}\right) - p_{i+1}\right\}^{-1}; \qquad q_i = p_i(B_1 - q_{i+1}). \tag{3.4.34}$$

These formulas allow determination of p_i $(i = N-1, N-2, \ldots, 1)$, beginning from $p_N = \gamma$. Here $|p_i| < \{1 + 2(h^2/al)\}^{-1}$.

Further, knowing p_i, we can compute q_i $(i = N-1, \ldots, 1; q_N = \delta)$.

Then taking (3.4.32) at $i = 1$ and the first relation (3.4.31), we shall solve the set of equations for $\vartheta_{k,0}$ and $\vartheta_{k,1}$; the inequalities $\alpha < 1$ and $|p_1| < \{1 + 2(h^2/al)\}^{-1}$ provide its solution. Then from relation (3.4.32), the remaining $\vartheta_{k,i}$ are found. Since estimates for p_i are obtained in these operations, essential accumulation and increase of errors are absent. This method is called a method of matching [37].

The rapid solution of the transfer equations depends today on the availability of high speed digital computers. On the other hand, transfer problems may also be solved on the analog computer or indeed by other analog devices.

The application of such analog models to transfer phenomena is based on the formal similarity in the analytical description of the new processes (as compared to the transfer process) which result from correspondence in behavior of the systems compared. This allows the study of transfer processes with the help of other processes occurring in a model. Naturally, the solution obtained on the analog model will not be of an analytical character; rather, an experimental determination of the desired solution can be obtained, such that it may be afterwards expressed in terms of the parameters of the initial problem. Models based on the hydraulic, electric, mechanical, and acoustic analogy of transfer processes are widely used at present.

The hydrodynamic analogy is based on the fact that the stream function and the velocity potential of an ideal liquid in an inviscid flow may be identified with the heat flux function and the temperature in a conducting body; this analogy was used by Moore and others to solve two-dimensional steady heat conduction problems [101]. Further, this model was widely used for a system with distributed sources [84].

In 1928, Emanuel, and later on, Budrin designed and constructed models based on the identity of the mathematical relations describing the temperature distribution in a solid and the pressure distribution in water moving through capillary pipes [6]. The resulting apparatus, called a hydraulic integrator, allowed the solution of unsteady heat conduction problems. Later on Lukiyanov developed some integrators for solving two- and three-dimensional heat conduction problems [74], and Budrin [6] developed hydrostatic integrators for solving nonlinear parabolic transfer equations.

Similarly to the hydrodynamic analogy method, Coyle [16] developed the method of air-aerodynamic analogy. The principle of operation of this apparatus is similar to that of Budrin's hydrostatic integrators. Here, the magnitude of the heat flow and the temperature in a heat conducting system corresponds to the amount of air and the pressure in the aerodynamic system. Other types of the hydrodynamic analogy were proposed, e.g., those based on the correspondence between heat transfer and liquid transfer in a porous body, i.e., between the law of Fourier and Darcy [48].

The value of hydrodynamic model is somewhat limited because of the large dimensions of the apparatus, its complexity of performance, and the difficulties in solving problems with variable thermal properties.

The electrical analogy is widely used to study phenomena of heat and mass transfer. The equipment necessary for this purpose may be portable, cheap, of rather simple components but made with great accuracy. Moreover, electrical engineering possesses high precision instruments for measuring electrical quantities.

The electrical analogy is based on the formal similarity of the differential heat conduction equations, on the one hand, and the electrical conductivity equations on the other. The extent of the analogy becomes clear from the comparison of electrical and thermal quantities and from the laws given in Table 3.1. The analysis of the corresponding equations and Table 3.1 shows that it is quite possible to reproduce the unsteady-state fields of a heat transfer potential by means of the electrical analogy for different boundary conditions as well as at different heat source distributions.

TABLE 3.1.

Process parameter	Type of transfer process	
	Electrical	Thermal
Potential	φ, W	t, °C
Motive force	$\nabla\varphi$, W/m	∇t, °C/m
Charge	Electrical charge	Entropy
Conductance	Electrical conductivity	Thermal conductivity
Resistance	Electrical resistance $R_e = 1/\lambda_e$, ohm	Thermal resistance $R_q = 1/\lambda$, deg hr/kcal
Current density	$j_e = \Delta\varphi/R_e$, A	$j_q = \Delta t/R_q$, W
Specific capacity	Electrical capacity, C_e, F	Heat capacity C_q, kcal/kg deg

The electrical analogy takes many different experimental forms. Those analog devices where the geometry of the original heat conducting body is reproduced and the model is made of a material of continuous conductivity are called geometric analogs or simulators of fields by means of a continuum method. If the analog model is constructed of an equivalent electric circuit with lumped constants, then the device is called an analog circuit.

An electrolytic bath serves as a convenient model in finding the potential field in a conductor. There has been wide application of electrolytic baths since they allow the establishment of uniformity of the properties of the electrolyte, the possibility of development of models of great sizes, and a comparatively easy access to inner points of the region in a liquid when three-dimensional fields are simulated.

Electrical analogs with liquid models are based on the application of ionic conductivity of electrolytes. An electrolyte with constant conductivity (weak solutions of salts, acids, and alkali, solutions of different vitriols, etc.)

is taken as a conductor. Models may be three- and two-dimensional. Their shape is identical with that of the original heat conducting body under consideration. The boundary of the analog should have a potential proportional to the temperature at the boundary of the original; this is carried out by the application of a metallic conductor, through which electric current is fed into the electrolyte. As an example, Langmuir conducted investigations on such a model for heat transfer through the walls of a shell in the form of a parallelepiped [73]. An electrolyte with variable concentration may be used, or a bath with a variable depth is created if it is necessary to transfer potentials in a nonuniform field [115].

Geometric analog models may be made of solid electric conducting materials or coatings. Models made of a thin sheet of an electric conducting material are also widely used for simulating plane-parallel fields. Metallic foils, metallized paper, or ordinary paper on which a layer of electrically conducting graphite with definite resistance (for example, teledeltos paper) is used as such a sheet. The sheet is cut in the same shape as the original. The electrodes are pasted on or painted on, using electrically conducting paint. The boundary potentials are taken corresponding to the values of the original problem. Sources may be inserted with the help of electrodes made of foil appropriately attached by conducting paste. Areas with different heat or mass conductivities are reproduced by perforating a sheet by square holes or by pasting separate portions from several layers of paper.

We generally resort to continuum methods when modeling three-dimensional problems. Dispersed media of different conductivity, e.g., a mixture of graphite power with quartz or colloidal masses (gelatine), are used [13].

In 1926 the Russian mathematician, S. A. Gershgorin, for the first time discussed the possibility of using electric grids for simulation. One of the advantages of grids, is that the coordinates of their points are electrical but not geometrical. This allows the solution of problems in any system of coordinates on rectangular network. The flexibility of the circuit creates great convenience when producing a model and insures its reliability in operation.

To obtain the analog model, the actual body under investigation is divided into a number of elementary volumes as in the finite-difference method. The value of the potential is obtained for a finite number of the points chosen, i.e., a continuous field of potentials in a body is replaced by their equivalent lumped values. The electric grid, which is sometimes referred to as an analog circuit, is composed of electric capacitors in parallel, which are fixed at the nodal points of the networks. Sources of current and substance are reproduced by switching power supplies into one or several nodal points of the network for the case of the lumped supplies or into all the nodal points for uniformly distributed sources.

Various types of universal electrical grid models (i.e., electrical integrators) have been designed and constructed to simulate steady- and unsteady-temperature fields. For example, to study plane steady-potential fields with boundary conditions of the first, second, and third kinds, electrical integrators were designed by Gutenmacher; he also designed integrators to study unsteady fields with the coefficients depending on the coordinates. These analogs allowed the solution of many important problems in different branches of science and eingneering [126].

Highly efficient analog devices have also been designed. We can mention the devices of Boiken (Holland), Pashkis and Liebman (USA), Lauson and Jackson (Great Britain), Fisher and Müller (Germany), and others.

Liebman's device [64] is of special interest for solving unsteady-state problems with the help of resistor matrices. Here it was possible to interrupt solutions, i.e., in Liebeman's device calculation was performed by means of explicit finite-difference space and time approximation. The latter, like Budrin's hydrostatic integrator, allows the determination of a temperature field taking into account the anisotropy and the change in thermal properties of a substance depending on temperature.

The shortcoming of Liebman's device is its "lack of universality". For each new problem it is necessary to change the value of all the resistors. This shortcoming is eliminated in new statistical (capacity and ohmic) integrators designed at the Thermal Problems Laboratory of the Kirov Kazakh State University [127]. It is possible to find a detailed review of the application of different electrical models for solving thermal problems [59, 16].

Other types of analogies are also used for investigation and calculation of heat and mass transfer processes. Particularly outside the USSR, the membrane analogy is widely used to solve problems on steady-state heat conduction with heat sources [101].

Methods of numerical solutions and analogies are generally used in those cases where it is impossible to obtain a closed solution; alternatively the solutions obtained are so complex that they cannot be used for practical calculations. The choice between the numerical solution methods and analogies depends on the particular problem, the demands for quantitative values, and the amount of time available for solving the stated problem and the degree of accuracy desired.

CHAPTER

4

NONSTATIONARY TEMPERATURE FIELD WITHOUT HEAT SOURCES: BOUNDARY CONDITION OF THE FIRST KIND

From the great number of problems which could be considered in this chapter we shall dwell only on the basic classical problems. *Under boundary conditions of the first kind the temperature of a body surface is given as a function of time. We shall study the simplest cases when the temperature of a body surface remains invariable during the entire heat transfer process.* This may be achieved by means of special apparatus which maintains constant temperature on the body surface, or may take place when heat transfer between a body and a surrounding medium with a constant temperature occurs according to the Newton law (boundary condition of the third kind), but with an infinitely great heat transfer coefficient α, when $\alpha/\lambda \to \infty$ precisely. In this case the boundary condition of the third kind reduces to that of the first kind in its simplest form.

Problems in which the body surface temperature is a specified function of time will be considered in Chapter 7 as a special case of more general problems.

All the solutions given in this chapter may be obtained from the corresponding solutions of the problems in Chapter 6, provided that in the latter the Biot number is set equal to infinity. From the methodical viewpoint we consider it advisable to present these simple problems into a special chapter to show the reader an orderly development of calculation techniques. The advantage of initially applying calculation methods to more simple problems is that it is easy for the reader to learn the calculation technique and to discover advantages and shortcomings of the various methods. In particular, we shall stress the advantages of the operational method of solution of problems as compared to the classical one.

Before studying the first basic problem (determination of a temperature field in a semi-infinite body) we shall consider a secondary problem.

4.1 Infinite Body

If there is a solid, the dimensions of which are very large in comparison with the region of interest, it may be considered infinite. It is only necessary that a noticeable change in a temperature field should occur in this portion of the body.

First, consider the problem when the temperature changes only along x; it does not change along y and z (i.e., $\partial t/\partial y = \partial t/\partial z \equiv 0$). Hence isothermal surfaces will represent planes parallel to a plane yz. At the initial time, the temperature distribution is given along x as some function $t(x, 0) = f(x)$. A temperature distribution at any time along x is to be found.

Our problem involves the solution of the differential heat conduction equation

$$\frac{\partial t(x, \tau)}{\partial \tau} = a\frac{\partial^2 t(x, \tau)}{\partial x^2} \qquad (\tau > 0; \quad -\infty < x < \infty), \tag{4.1.1}$$

with a given temperature distribution at the initial time instant

$$t(x, 0) = f(x) \tag{4.1.2}$$

Boundary conditions are absent but they may be replaced by the physical conditions

$$\frac{\partial t(+\infty, \tau)}{\partial x} = \frac{\partial t(-\infty, \tau)}{\partial x} = 0. \tag{4.1.3}$$

This problem may be solved by the Fourier method but the following limitations are imposed on the function $f(x)$: (1) the function $f(x)$ should be expressible in terms of the Fourier integral; (2) the function $f(x)$ should tend to zero at $x \to \infty$ as rapidly as possible so that the finite value of the integral $\int_{-\infty}^{\infty} |f(x)|^2\, dx$ may be preserved.

Hence we give here a solution according to the source method which does not impose these limitations on the function $f(x)$. The solution by the Fourier method will be given at the end of the present section.

In Chapter 3, Section 3.2 it is shown that the special solution of equation (4.1.1) has the form

$$t = \frac{C}{(4\pi a\tau)^{1/2}} \exp\left[-\frac{(x-\xi)^2}{4a\tau}\right]. \tag{4.1.4}$$

Inspection of expression (4.1.4) reveals that at a given time τ the temperature distribution curve along x has a maximum for $x = \xi$ (Fig. 4.1).

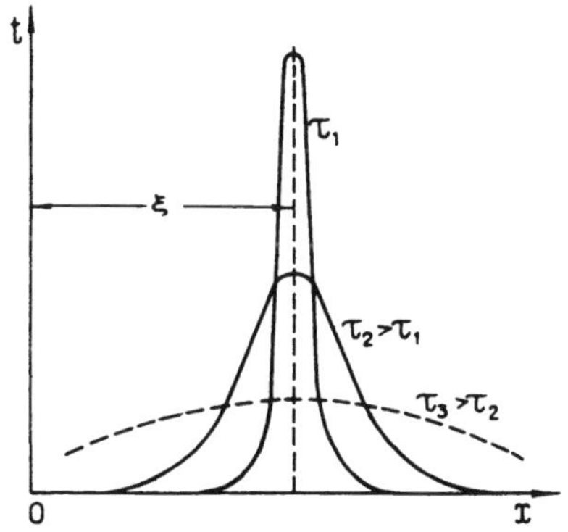

Fig. 4.1. Temperature distribution inside an infinite medium exposed to an instantaneous heat flux.

The area S under the curve, i.e., the area formed by the curve and the abscissa axis, is a finite value and equal to the integral of expression (4.1.4) within the limits $-\infty$ to ∞ (the origin of coordinates is taken at the point ξ).

$$S = C\int_{-\infty}^{\infty} \frac{1}{(4\pi a\tau)^{1/2}} \exp\left[-\frac{(x-\xi)^2}{4a\tau}\right] d(x-\xi)$$

$$= \frac{C}{\sqrt{\pi}} \int_{-\infty}^{\infty} \exp[-z^2]\, dz = C.$$

Here the variable z is introduced

$$z = \frac{x-\xi}{(4a\tau)^{1/2}}$$

and the value of the definite integral is known

$$\int_{-\infty}^{\infty} \exp[-z^2]\, dz = \sqrt{\pi}.$$

Thus the area under the curve equals a constant C. The ordinate of the curve at the maximum point is equal to $C/(4\pi a\tau)^{1/2}$. Hence, with an increase in time τ the ordinate decreases and the curve becomes more shallow (see Fig. 4.1) and vice versa, with a decrease in time τ the ordinate increases. At diminishing time ($\tau \to 0$) one obtains an infinitely narrow strip but its area remains equal to the constant C.

Using this property of expression (4.1.4), the given initial temperature distribution $t(x, 0) = f(x)$ inside an infinite body may be presented as a sum of separate particular solutions of form (4.1.4), i.e., the curve $f(x)$ may be replaced by a sum of the infiinte number of curves of the form

$$\lim_{\tau\to\infty} \frac{C}{(4\pi a\tau)^{1/2}} \exp\left[-\frac{(x-\xi)^2}{4a\tau}\right].$$

It is necessary to note that despite an infinitesimal width of a separate strip $d\xi$ (Fig. 4.2) the height of its quantity will be finite and equal to

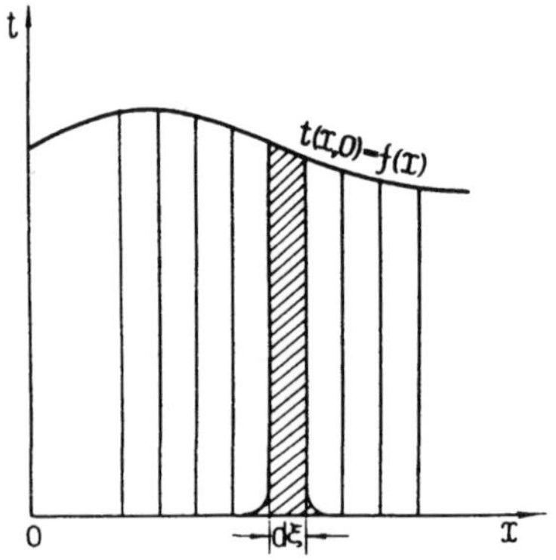

Fig. 4.2. Representation of the initial temperature distribution as a sum of particular solutions.

$f(\xi)$. The area of such a strip equal to C will be an infinitesimal quantity, i.e.,

$$t(\xi, 0)\, d\xi = f(\xi)\, d\xi = C.$$

The complete initial distribution inside such an infinite body will be equal to

$$\lim_{\tau\to 0} t(x, \tau) = \lim_{\tau\to 0} \frac{1}{(4\pi a\tau)^{1/2}} \int_{-\infty}^{\infty} f(\xi) \exp\left[-\frac{(x-\xi)^2}{4a\tau}\right] d\xi. \tag{4.1.5}$$

This relation will be valid not only for the initial time instant ($\tau \to 0$) but for any subsequent time interval, i.e., the general solution of our problem will be as follows

$$t(x, \tau) = \frac{1}{(4\pi a\tau)^{1/2}} \int_{-\infty}^{\infty} f(\xi) \exp\left[-\frac{(x-\xi)^2}{4a\tau}\right] d\xi. \tag{4.1.6}$$

The general solution of equation (4.1.6) may be rewritten by introducing a new variable u defined by the relation $\xi = x + 2(a\tau)^{1/2}u$. Then we shall have

$$t(x, \tau) = \frac{1}{\sqrt{\pi}} \int_{-\infty}^{\infty} f(x + 2(a\tau)^{1/2}u) \exp[-u^2]\, du. \qquad (4.1.7)$$

It may be shown that solution (4.1.7) satisfies the differential equation (4.1.1) and the initial condition, noting that at $\tau \to 0$ $f(x + 2(a\tau)^{1/2}u) \to f(x)$. Consequently, we shall have

$$t(x, 0) = \frac{1}{\sqrt{\pi}} \int_{-\infty}^{\infty} f(x) \exp[-u^2]\, du = f(x) \frac{1}{\sqrt{\pi}} \int_{-\infty}^{\infty} \exp[-u^2]\, du = f(x).$$

This approach may generalized to plane and space problems. By analogy, we have

$$t(x, y, \tau) = \frac{1}{4\pi a\tau} \int_{-\infty}^{\infty} \int_{-\infty}^{\infty} f(\xi, \eta) \exp\left[-\frac{(x-\xi)^2 + (y-\eta)^2}{4a\tau}\right] d\xi\, d\eta, \qquad (4.1.8)$$

$$t(x, y, z, \tau) = \frac{1}{(4\pi a\tau)^{3/2}} \int_{-\infty}^{\infty} \int_{-\infty}^{\infty} \int_{-\infty}^{\infty} f(\xi, \eta, \zeta) \times \exp\left[-\frac{(x-\xi)^2 + (y-\eta)^2 + (z-\zeta)^2}{4a\tau}\right] d\xi\, d\eta\, d\zeta. \qquad (4.1.9)$$

Our problem may be solved with the aid of the Fourier method also. In Section 3.2, it has been established that the general solution of a one-dimensional problem has the form

$$t(x, \tau) = \sum_{n=1}^{\infty} C_n \sin k_n x \exp[-a k_n^2 \tau] + \sum_{m=1}^{\infty} D_m \cos k_m x \exp[-a k_m^2 \tau].$$

Since a boundary condition is absent, values of k may be considered to form a continuous series of numbers, and each subsequent number differs from the previous one by an infinitesimal quantity dk. Then both sums will pass into a definite integral taken from 0 to ∞. In addition, the constants C_n and D_n will be some functions of k.

Thus we have

$$t(x, \tau) = \int_0^{\infty} \exp[-ak^2\tau][f_1(k) \sin kx + f_2(k) \cos kx]\, dk. \qquad (4.1.10)$$

To determine $f_1(k)$ and $f_2(k)$ we use the initial condition

$$t(x, 0) = f(x) = \int_0^{\infty} [f_1(k) \sin kx + f_2(k) \cos kx]\, dk. \qquad (4.1.11)$$

From the theory of the Fourier series it is known that if the predetermined function $f(x)$ satisfies definite conditions, then it may be expanded in the Fourier series which may be replaced by the Fourier integral

$$f(x) = \frac{1}{\pi} \int_0^\infty dk \int_{-\infty}^\infty f(\xi)\, [\sin k\xi \sin kx + \cos k\xi \cos kx]\, d\xi. \quad (4.1.12)$$

Comparing relations (4.1.11) and (4.1.12), we come to the conclusion that the derived functions $f_1(k)$ and $f_2(k)$ are equal to

$$f_1(k) = \frac{1}{\pi} \int_{-\infty}^\infty f(\xi) \sin k\xi\, d\xi, \quad f_2(k) = \frac{1}{\pi} \int_{-\infty}^\infty f(\xi) \cos k\xi\, d\xi.$$

The final solution of our one-dimensional problem for an infinite body in the Fourier form will be as follows

$$t(x, \tau) = \frac{1}{\pi} \int_0^\infty \exp[-ak^2\tau]\, dk \int_{-\infty}^\infty f(\xi) \cos[k(\xi - x)]\, d\xi. \quad (4.1.13)$$

Solution (4.1.6) or (4.1.7) for an infinite body has secondary importance for us. Both these solutions satisfy infinity conditions (4.1.3).

In our case at $x \to \pm\infty$, $t(x, \tau) \to 0$. A heat process in such an infinite body consists of a temperature leveling process starting from some time instant which has been taken as the initial one. This nonuniform temperature distribution may appear as a result of a momentary action of some heat source (instantaneous heat source), the power of which is proportional to $f(\xi)$. The method considered is therefore frequently called the method of point-by-point sources. Chapter 9 will deal in more detail with this problem. We shall now consider the basic problems.

4.2 Semi-Infinite Body

Consider a body bounded on one side by the plane yz and on the other stretching into infinity. Such a body is named a semi-infinite body. An infinitely long bar, the lateral surface of which has ideal thermal insulation, may serve as a semi-infinite body.

a. Statement of the Problem. *The temperature of a semi-infinite body at all points has a definite value given by some function $f(x)$, i.e., $t(x, 0) = f(x)$.* The problem of cooling such a body will be solved since the problem of heating may be always reduced to that of cooling by simply modifying the dimensionless temperature variable.

At the initial time the end of a bar has temperature t_a which is maintained constant during the entire heat transfer process. The temperature distribution over the bar length at any time and the heat loss through its end are sought.

From the mathematical viewpoint this problem may be formulated as follows. We have a differential equation

$$\frac{\partial t(x, \tau)}{\partial \tau} = a \frac{\partial^2 t(x, \tau)}{\partial x^2} \qquad (\tau > 0, \quad 0 < x < \infty), \qquad (4.2.1)$$

with boundary conditions

$$t(x, 0) = f(x), \qquad t(0, \tau) = t_a = \text{const}, \qquad (4.2.2)$$

$$\partial t(\infty, \tau)/\partial x = 0. \qquad (4.2.3)$$

A temperature gradient at the infinitely removed point is absent (see Fig. 4.3).

At first, to simplify the calculations, we assume $t_a = 0$.

It is necessary to determine $t(x, \tau)$.

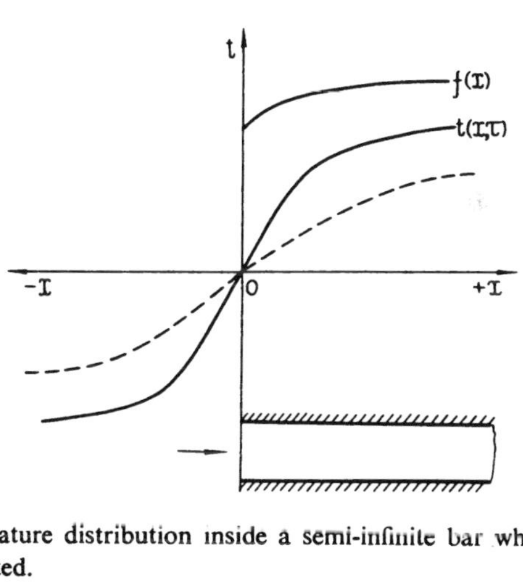

Fig. 4.3. Temperature distribution inside a semi-infinite bar when its lateral surface is thermally insulated.

b. Solution of the Problem by Classical Method. This problem may be solved on the basis of the previous one (for an infinite body). For this purpose let us prolong the bar in the negative x direction, i.e., it will be considered infinite (Fig. 4.3). The initial temperature at a point $x > 0$ is equal to $f(x)$ and that at a point $-x$ is chosen equal to $-f(x)$, i.e., the function of $f(x)$ is considered to be odd:

$$f(x) = -f(-x).$$

Proceeding from the symmetry considerations, the temperature distribution at subsequent time instants will be an odd function and for $x = 0$, its value will be always zero. Hence the surface condition is fulfilled.

If we replace x by ξ on the curve of the initial temperature distribution, the general solution will have the following form on the basis of the foregoing:

$$t(x, \tau) = \frac{1}{2(\pi a\tau)^{1/2}} \left\{ \int_0^\infty f(\xi) \exp\left[- \frac{(x-\xi)^2}{4a\tau} \right] d\xi \right.$$
$$\left. + \int_{-\infty}^0 -f(-\xi) \exp\left[- \frac{(x-\xi)^2}{4a\tau} \right] d\xi \right\}.$$

Write this solution in another form

$$t(x, \tau) = \frac{1}{2(\pi a\tau)^{1/2}} \int_0^\infty f(\xi) \left[\exp\left(- \frac{(x-\xi)^2}{4a\tau} \right) \right.$$
$$\left. - \exp\left(- \frac{(x+\xi)^2}{4a\tau} \right) \right] d\xi. \qquad (4.2.4)$$

Expression (4.2.4) is a general solution of our problem. If the initial temperature is constant and does not depend on x (temperature of a bar at the initial instant is uniform and equal to t_0), i.e.,

$$t(x, 0) = f(x) = t_0 = \text{const},$$

then the solution may simplified. By substituion of the first part of the integrand

$$\xi = x + 2u(a\tau)^{1/2},$$

and the second part

$$\xi = -x + 2u(a\tau)^{1/2},$$

we obtain

$$t(x, \tau) = t_0 \frac{1}{\sqrt{\pi}} \int_{-x/2(a\tau)^{1/2}}^{x/2(a\tau)^{1/2}} \exp[-u^2]\, du.$$

Since the function $\exp[-u^2]$ is a symmetrical function with respect to u, it may be written:

$$\frac{t(x,\tau)}{t_0} = \frac{2}{\sqrt{\pi}} \int_0^{x/2(a\tau)^{1/2}} \exp[-u^2]\, du = \operatorname{erf}\left(\frac{x}{2(a\tau)^{1/2}}\right), \tag{4.2.5}$$

where the integral

$$\frac{2}{\sqrt{\pi}} \int_0^u \exp[-u^2]\, du = \operatorname{erf}(u)$$

is called the Gauss error function.

The function erf(u) changes from 0 at $u = 0$, to 1 at $u \to \infty$ (in practice, this occurs when $u > 2.7$ since erf(2.7) = 0.9999).

If the end of the bar is not maintained at 0°C but at some temperature $t(0, \tau) = t_a =$ const, then by introducing a new variable $\vartheta = t - t_a$ our problem is reduced to the foregoing since

$$\vartheta(0,\tau) = t(0,\tau) - t_a = 0.$$

Hence the solution of this problem may be written thus:

$$\frac{t(x,\tau) - t_a}{t_0 - t_a} = \operatorname{erf}\left(\frac{x}{2(a\tau)^{1/2}}\right). \tag{4.2.6}$$

c. ***Solution of the Problem by the Operational Method.*** Applying the Laplace transformation to differential equation (4.2.1) results in

$$L\left[\frac{\partial t(x,\tau)}{\partial \tau}\right] = L\left[a\,\frac{\partial^2 t(x,\tau)}{\partial x^2}\right],$$

where

$$L[t(x,\tau)] = \int_0^\infty t(x,\tau)\, e^{-s\tau}\, d\tau = T(x,s).$$

In the left-hand side of the equation, the Laplace transformation should be taken from the first derivative. According to the basic theorem, it is equal to the product of the transform by an operator s minus the value of the function at the initial time instant, i.e.,

$$sT(x,s) - f(x) = a\,\frac{\partial^2}{\partial x^2}\{L[t(x,\tau)]\} = a\,\frac{d^2T(x,s)}{dx^2}. \tag{4.2.7}$$

Thus, differential partial equation (4.2.1) for the inverted transform of the function $t(x, \tau)$ turns into an ordinary differential equation for the transform $T(x, s)$, since $T(x, s)$ does not depend on τ. For this transition, the initial condition is used.

Rewriting Eq. (4.2.7) in the following form gives us

$$T''(x, s) - \frac{s}{a} T(x, s) + \frac{f(x)}{a} = 0. \tag{4.2.8}$$

We now introduce the special condition that the temperature of the bar before cooling is the same everywhere and equal to t_0, (i.e., $f(x) = t_0 =$ const). In this case Eq. (4.2.8) takes a more simple form

$$T''(x, s) - (s/a)[T(x, s) - (t_0/s)] = 0. \tag{4.2.9}$$

A general solution of this differential equation for the transform may be written thus (see Section 3.2)

$$T(x, s) - \frac{t_0}{s} = A_1 \exp[(s/a)^{1/2}x] + B_1 \exp[-(s/a)^{1/2}x] \tag{4.2.10}$$

where A_1 and B_1 are constants to be determined by the boundary conditions.

Using the Laplace transformation for the boundary conditions

$$L[t(0, \tau)] = 0, \qquad T(0, s) = 0, \tag{4.2.11}$$

$$L\left[\frac{\partial t(\infty, \tau)}{\partial x}\right] = 0 \qquad T'(\infty, s) = 0. \tag{4.2.12}$$

From condition (4.2.12) it follows that $A_1 = 0$ since otherwise the first term in the right-hand side of (4.2.10) increases infinitely with x, viz.

$$\begin{aligned} 0 &= T'(\infty, s) \\ &= (s/a)^{1/2}A_1 \exp[(s/a)^{1/2}(\infty)] - (s/a)^{1/2}B_1 \exp[-(s/a)^{1/2}(\infty)], \end{aligned}$$

whence it follows that $A_1 = 0$.

If we use condition (4.2.11)

$$0 - (t_0/s) = B_1 \exp[-(s/a)^{1/2} \cdot 0] = B_1; \qquad \text{i.e., } B_1 = -t_0/s,$$

then the solution for the transform will acquire the values

$$(t_0/s) - T(x, s) = (t_0/s) \exp[-(s/a)^{1/2}x].$$

To determine the inverse transform, the table of Laplace transformations is used, from which it is found that

$$L^{-1}[(1/s) \exp[-k\sqrt{s}\,]] = 1 - \operatorname{erf}(k/2\sqrt{\tau}\,).$$

In our problem $k = x/\sqrt{a}$. Consequently, the solution of the problem will be

$$t_0 - t(x, \tau) = t_0[1 - \mathrm{erf}(x/2(a\tau)^{1/2})],$$

whence

$$t(x, \tau)/t_0 = \mathrm{erf}(x/2\sqrt{a\tau}), \tag{4.2.13}$$

i.e., we obtain the same solution (4.2.5) for the case of a uniform initial temperature distribution over the bar length.

If the temperature of the bar end is not equal to zero but to $t_a = \text{const}$, then boundary condition (4.2.11) may be written as

$$L[t(0, \tau)] = L[t_a], \qquad T(0, s) = t_a/s. \tag{4.2.14}$$

Consequently, the constant $B_1 = -(t_0 - t_a)/s$ since $t_0 > t_a$. Then the solution for the transform has the form

$$(t_0/s) - T(x, s) = [(t_0 - t_a)/s]\exp[-(s/a)^{1/2}x]. \tag{4.2.15}$$

In a similar way the inverse Laplace transformation is obtained as

$$t_0 - t(x, \tau) = (t_0 - t_a)[1 - \mathrm{erf}\{x/2(a\tau)^{1/2}\}] = (t_0 - t_a)\,\mathrm{erfc}\{x/2(a\tau)^{1/2}\},$$

where

$$\mathrm{erfc}(u) = 1 - \mathrm{erf}(u).$$

This solution may be written thus

$$\frac{t(x, \tau) - t_a}{t_0 - t_a} = \mathrm{erf}\left(\frac{x}{2(a\tau)^{1/2}}\right), \tag{4.2.16}$$

i.e, an expression identical with (4.2.6) is obtained.

Solution (4.2.16) directly follows from (4.2.13) as it is only necessary to shift the temperature datum; however, detailed calculations have been given to show the application of the Laplace transformation to constant boundary conditions different from zero.

The original problem with a given initial temperature distribution $f(x)$ may be solved in a similar fashion and it is only necessary to proceed from differential equation (4.2.8) for the transform. As a final result the same solution (4.2.4) is obtained.

d. Solution by Fourier Transformation Method. In our problem at $x = 0$, $t = t_a$. The initial temperature is $t(x, 0) = t_0 = \text{const}$. The temperature gradient at an infinitely great distance does not change, i.e., $dt(\infty, \tau)/\partial x = 0$.

For convenience of calculations, the function $\vartheta(x, \tau) = t_0 - t(x, \tau)$ is introduced. Then the boundary conditions assume the form

$$\vartheta(x, 0) = 0, \qquad \vartheta(0, \tau) = t_0 - t_a = \text{const}, \qquad \frac{\partial\vartheta(\infty, \tau)}{\partial x} = 0. \qquad (4.2.17)$$

Since the surface temperature $(x = 0)$ is given, the Fourier sine transformation is used giving

$$F_s[\vartheta(x, \tau)] = \vartheta_F(\tau, p) = \int_0^\infty \vartheta(x, \tau) \sin px \, dx. \qquad (4.2.18)$$

Applying this transformation to the differential equation (4.2.1) we obtain

$$F_s\left[\frac{\partial t(x, \tau)}{\partial \tau}\right] = F_s\left[a \frac{\partial^2 t(x, \tau)}{\partial x^2}\right].$$

We first calculate the transformation of the right-hand side of the equation to obtain

$$\begin{aligned}\int_0^\infty \sin px \frac{\partial^2\vartheta(x, \tau)}{\partial x^2} dx &= - p\vartheta(x, \tau) \cos px \Big|_{x=0}^{x=\infty} \\ &\quad + \frac{\partial\vartheta(x, \tau)}{\partial x} \sin px \Big|_{x=0}^{x=\infty} - p^2 \int_0^\infty \vartheta(x, \tau) \sin px \, dx \\ &= p(t_0 - t_a) - p^2\vartheta_F(p, \tau), \end{aligned} \qquad (4.2.19)$$

as at $x = \infty$, $\vartheta(\infty, \tau) = 0$ and $\vartheta(0, \tau) = t_0 - t_a$; $\partial\vartheta(\infty, \tau)/\partial x = 0$.

Hence, we have

$$\frac{\partial\vartheta_F}{\partial\tau} = ap(t_0 - t_a) - ap^2\vartheta_F(p, \tau). \qquad (4.2.20)$$

Upon integration of Eq. (4.2.20) and taking into account the initial condition we obtain

$$\vartheta_F(p, \tau) = \frac{(t_0 - t_a)}{p} [1 - \exp(- p^2 a\tau)]. \qquad (4.2.21)$$

Using the transformation formula we obtain

$$\vartheta(x \ \tau)/(t_0 - t_a) = (2/\pi) \int_0^\infty [1 - \exp(- p^2 a\tau)] \sin px \, (dp/p).$$

It is known, that

$$\int_0^\infty (\sin px/p) \, dp = \pi/2,$$

$$(2/\pi) \int_0^\infty \exp(- p^2 a\tau) \sin px (dp/p) = \text{erf}\{x/2(a\tau)^{1/2}\}.$$

It follows that

$$t_0 - t(x, \tau)/(t_0 - t_a) = 1 - \text{erf}\{x/2(a\tau)^{1/2}\}, \qquad (4.2.22)$$

or

$$\{t(x, \tau) - t_a\}/(t_0 - t_a) = \text{erf}\{x/2(a\tau)^{1/2}\},$$

i.e., solution (4.2.16).

e. Determination of Heat Losses. The heat losses dQ_s from the bar end for time $d\tau$ through a unit area are given by

$$dQ_s = -\lambda\left(\frac{\partial t}{\partial x}\right)_{x=0} d\tau = -\lambda(t_0 - t_a)\left\{\frac{\partial}{\partial x}\left[\text{erf}\left(\frac{x}{2(a\tau)^{1/2}}\right)\right]\right\}_{x=0}$$

It is known that

$$\frac{\partial}{\partial x}\left[\text{erf}\left\{\frac{x}{2(a\tau)^{1/2}}\right\}\right] = \frac{1}{(\pi a\tau)^{1/2}} \exp\left[-\frac{x^2}{4a\tau}\right].$$

At $x = 0$ the exponential function is equal to unity. Then, we obtain

$$q = \frac{dQ_s}{\partial\tau} = -\frac{\lambda(t_0 - t_a)}{(\pi a\tau)^{1/2}} = -(\lambda c\gamma)^{1/2}(t_0 - t_a)\frac{1}{(\pi\tau)^{1/2}}\left(\frac{\text{kcal}}{\text{m}^2\ \text{hr}}\right). \qquad (4.2.23)$$

Thus, the rate of heat transfer from a surface unit or a heat flux is directly proportional to a temperature difference $(t_0 - t_a)$ to some thermal coefficient $(\lambda c\gamma)^{1/2}$ and is inversely proportional to $\sqrt{\tau}$. Consequently, at the first instant of time, the rate of heat transfer is infinitely large and then it gradually decreases. The thermal constant $(\lambda c\gamma)^{1/2}$ will be called the coefficient of heat activity of a body[1] or the coefficient of heat accumulation $\varepsilon = (\lambda c\gamma)^{1/2}$; it is usually measured in kcal/m^2 °C hr$^{1/2}$.

Relation (4.2.23) may be obtained by the operational method, viz.

$$L[q] = -\lambda\frac{\partial}{\partial x}\{L[t(x, \tau)]\}_{x=0} = -\lambda\frac{dT(0, s)}{dx}.$$

Differentiating solution (4.2.15) for the transform with respect to x, we obtain:

$$L[q] = -\lambda\frac{t_0 - t_a}{s}\left(\frac{s}{a}\right)^{1/2}\exp\left[-\left(\frac{s}{a}\right)^{1/2}\cdot 0\right] = -(\lambda c\gamma)^{1/2}(t_0 - t_a)\cdot\frac{1}{\sqrt{s}}.$$

[1] For the physical significance of the coefficient of heat activity of a body, see Chapter 10.

Using the table of the transforms we find:

$$q = -(\lambda c\gamma)^{1/2}(t_0 - t_a)\, 1/(\pi\tau)^{1/2}, \tag{4.2.24}$$

i.e., the same relation for the heat transfer rate is obtained.

The use of this method of determing the heat losses from the solution for the transform in many problems is even more simple.

The amount of heat given off during a finite interval, τ, *is determined by integration within the limits from* 0 *to* τ, i.e.,

$$Q_s = Q_{s,0} - \int_0^\tau \varepsilon(t_0 - t_a)(1/\{\pi\tau\}^{1/2})\, d\tau = Q_{s,0} - (2\varepsilon/\sqrt{\pi})(t_0 - t_a)\sqrt{\tau}\,. \tag{4.2.25}$$

Thus the amount of heat $(Q_{s,0} - Q_s)$ given off by a surface unit of the bar end is directly proportional to the square root of time, to the temperature difference $(t_0 - t_a)$, and to the coefficient of heat activity.

The amount of heat carried off by the whole bar, the end area of which is equal to S, will be

$$\Delta Q = (2\varepsilon/\sqrt{\pi})(t_0 - t_a)S\sqrt{\tau}\,. \tag{4.2.26}$$

f. Analysis of the Solution. We proceed to write solution (4.2.16) in dimensionless form. The relation $(t(x, \tau) - t_a)/(t_0 - t_a)$ is a relative excess temperature which we designate as θ, i.e.,

$$\theta = \frac{t(x, \tau) - t_a}{t_0 - t_a}. \tag{4.2.27}$$

The relation $a\tau/x^2$ is the homochronous number for processes of net heat conduction and is widely referred to as the Fourier number. For the x coordinate we designate it by Fo_x, i.e.,

$$\mathrm{Fo}_x = a\tau/x^2.$$

Then, solution (4.2.16) may be written

$$\theta = \operatorname{erf}\{1/2(\mathrm{Fo}_x)^{1/2}\}. \tag{4.2.28}$$

Figures 4.4 and 4.5 are charts of the solution (4.2.28). The ordinate represents the dimensionless temperature, θ, and the abscissa, the dimensionless time (the Fourier number). For the purpose of calculation the curves for the Fourier number from 0.02 to 1000 are constructed.

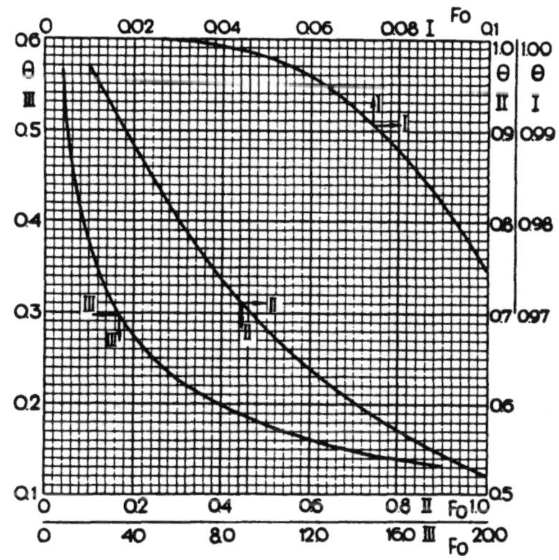

Fig. 4.4 Relation between the dimensionless excess temperature θ and the Fourier number Fo_x for a semi-infinite bar (at small values of the Fourier number).

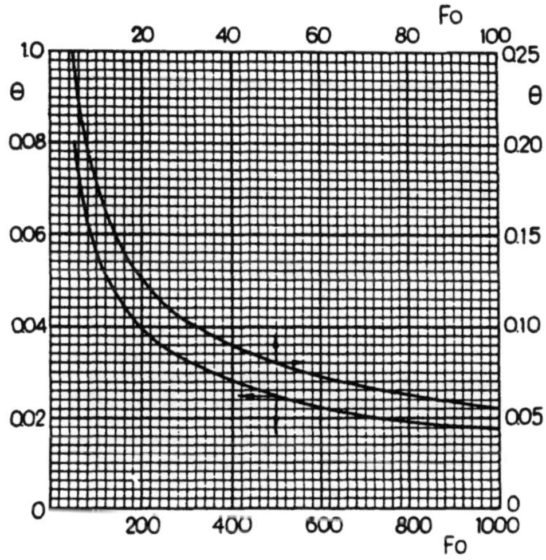

Fig. 4.5. Relation between the dimensionless excess temperature θ and the Fourier number Fo_x for a semi-infinite bar (at large values of the Fourier number).

Figure 4.6 provides diagrams of θ, $x\,(\partial\theta/\partial x)$, and $Fo_x(\partial\theta/\partial Fo_x)$ plotted versus $Fo_x^{-1/2}$. Figure 4.6 shows that at $1/2(Fo_x)^{1/2} = 1/\sqrt{2}$ the derivatives $x\,\partial\theta/\partial x$ and $\partial\theta/\partial Fo_x$ have a maximum.

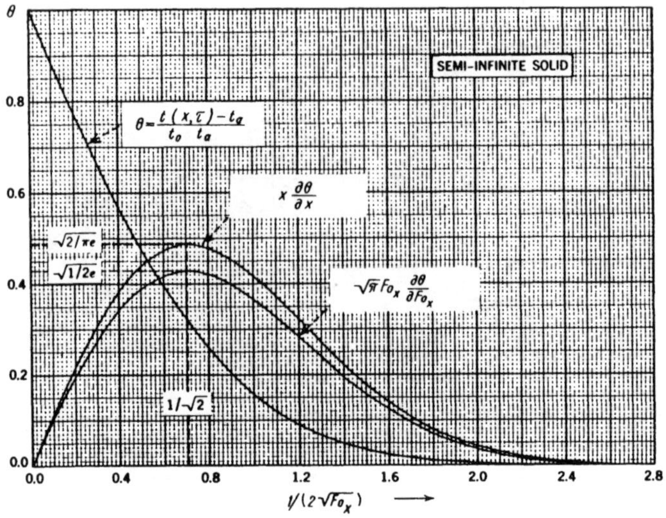

Fig. 4.6. Plot of θ, $x(\partial\theta/\partial x)$ and $\sqrt{\pi}\ \mathrm{Fo}_x(\partial\theta/\partial \mathrm{Fo}_x)$ versus local Fo_x for a semi-infinite bar [102].

We illustrate the use of the curves by means of a particular problem. It is desired to determine the temperature of the ground $t(x, \tau)$ at a depth $x = 0.5$ m if the ground surface is maintained for $\tau = 24$ hr at a temperature $t(0, \tau) = 1000°\mathrm{C}$. The temperature of the ground before heating $t(x, 0) = 20°\mathrm{C}$ and the thermal diffusivity of the ground $a = 2.5 \cdot 10^{-3}$ m²/hr.

This is a problem on heating of a semi-infinite body; it may be reduced to that on cooling by replacing a variable $\vartheta(x, \tau) = 1000°\mathrm{C} - t(x, \tau)$. Then $\vartheta(0, \tau) = 1000° - t(0, \tau) = 0°\mathrm{C}$; $\vartheta(x, 0) = 1000° - t(x, 0) = 980°\mathrm{C}$ const. Calculating the Fourier number for the coordinate $x = 0.5$ m:

$$\mathrm{Fo}_x = \frac{2.5 \cdot 10^{-3} \cdot 24}{25 \cdot 10^{-2}} = 0.24.$$

The relative temperature θ is found from Fig. 4.4 to be equal to 0.852; hence the ground temperature will be

$$\theta = \frac{\vartheta(x, \tau) - 0}{980° - 0} = 0.85; \qquad \vartheta(x, \tau) = 0.85 \cdot 980 = 833°,$$

$$t(x, \tau) = 1000 - \vartheta(x, \tau) = 1000 - 833 = 167°\mathrm{C}.$$

In the calculations any evaporation of moisture from ground is neglected. The temperature will be less in the presence of moisture evaporation, such problems with negative heat sources will be considered in Chapter 8.

4.3 Infinite Plate

An infinite plate is usually understood to be one such that the width and length are infinitely great compared to its thickness. Thus, an infinite plate represents a body restricted by two parallel planes. A change in temperature occurs only along x. Along the y and z axes the temperature remains invariable ($\partial t/\partial y = \partial t/\partial z = 0$). Hence this problem is one-dimensional.

a. Statement of Problem. *A temperature distribution over the plate thickness is given as some function $f(x)$. At the initial time, the bounding surfaces are instantaneously cooled to some temperature t_a which is maintained constant during the whole cooling process. Find the temperature distribution over the plate thickness and the heat rate at any instant.*

We place the origin of the coordinates at the center and designate the plate thickness by $2R$, i.e., R is half the plate thickness.

From the mathematical viewpoint the problem may be formulated in the following way. We have a differential equation

$$\frac{\partial t(x\ \tau)}{\partial \tau} = a\frac{\partial^2 t(x,\tau)}{\partial x^2} \qquad (\tau > 0; \quad -R < x < +R) \tag{4.3.1}$$

under conditions

$$t(x, 0) = f(x),$$

$$t(+R, \tau) = t_a = \text{const}, \tag{4.3.2}$$

$$t(-R, \tau) = t_a = \text{const}. \tag{4.3.3}$$

b. Solution of the Problem by the Method of Separation of Variables. Assume that the function $f(x)$ is even,[2] i.e., $f(x) = f(-x)$; therefore $(\partial f(x)/\partial x)_{x=0} = 0$. For this case, the boundary conditions (4.3.3) may be written

$$t(R, \tau) = t_a, \qquad \frac{\partial t(0,\tau)}{\partial x} = 0. \tag{4.3.3a}$$

The latter relation is a consequence of the symmetry condition of a temperature distribution curve at any instant; it must be satisfied during the whole cooling process, since heat transfer from boundary surfaces occurs uniformly.

A particular solution of differential equation (4.3.1) may be written as (see Eq. (3.2.21)

$$t(x, \tau) = C(\sin kx)\exp[-ak^2\tau] + D\cos kx\exp[-ak^2\tau]. \tag{4.3.4}$$

[2] The case when the function $f(x)$ is odd will be considered later in this section.

From the symmetry condition it follows that $C = 0$, viz:

$$\frac{\partial t(0, \tau)}{\partial x} = \lim_{x \to 0} (kC \cos kx - kD \sin kx) \exp[-ak^2\tau] = kC \exp[-ak^2\tau] = 0,$$

whence $C = 0$, since during the whole process of cooling $(0 < \tau < \infty)$ $\exp[-ak^2\tau]$ is not equal to zero.

This result may be obtained by analyzing the physical conditions imposed on the plate. The temperature distribution must be symmetrical relative to the axis of coordinates, consequently, it should be described by an even function. $\cos kx$ is such a function, whereas $\sin kx$ is an odd function and thus must be omitted from the solution.

It is now necessary to satisfy the second boundary condition. To simplify the calculations, we tentatively assume $t_a = 0$. We have

$$t(R, \tau) = D \cos kR \exp[-ak^2\tau] = 0,$$

from which it follows that

$$\cos kR = 0, \qquad kR = \tfrac{1}{2}\pi, \tfrac{3}{2}\pi, \tfrac{5}{2}\pi, \ldots,$$

$$k_n R = (2n - 1)\tfrac{1}{2}\pi, \tag{4.3.5}$$

i.e., k has not a single value but an infinite number.

Consequently, the general solution will be a sum of all the particular solutions, i.e.,

$$t(x, \tau) = \sum_{n=1}^{\infty} D_n \cos(2n - 1) \frac{\pi}{2} \frac{x}{R} \exp\left[-(2n - 1)^2 \frac{\pi^2}{4} \frac{a\tau}{R^2}\right]. \tag{4.3.6}$$

The constants D_n in each special solution will have their eigenvalues, since the sum of the particular temperature distributions for any given time should describe the real temperature distribution.

Thus, the superposition of the cosine curves must give the real temperature distribution, including the initial temperature distribution. Hence, assuming $\tau = 0$, we should obtain the given function $f(x)$, it being represented as the Fourier series

$$t(x, 0) = f(x) = \sum_{n=1}^{\infty} D_n \cos k_n x. \tag{4.3.7}$$

Now the trigonometric functions $\cos kx$ and $\sin kx$ form an orthogonal system of functions.

We note that the system of functions

$$f_1(x),\ f_2(x),\ f_3(x),\ \dots,\ f_n(x)$$

is referred to as an orthogonal over an interval (a, b) if the integral

$$\int_a^b f_i(x) f_j(x)\, dx = 0 \tag{4.3.8}$$

for values of i and j not equal to each other.

For example, it may be shown that the system of functions $\cos k_n x$ is orthogonal

$$I = \int_{-R}^{+R} \cos(k_m x)\cos(k_n x)\, dx \begin{cases} = 0 & \text{at } m \neq n \\ > 0 & \text{at } m = n. \end{cases} \tag{4.3.9}$$

Transforming the integrand by means of the following formula we have

$$\cos\alpha\cos\beta = \tfrac{1}{2}[\cos(\alpha - \beta) + \cos(\alpha + \beta)].$$

Then, the integral I will be equal to

$$I = \frac{\sin(k_m - k_n)R}{(k_m - k_n)} + \frac{\sin(k_m + k_n)R}{(k_m + k_n)}.$$

Using the trigonometric formula

$$\sin(\alpha \pm \beta) = \sin\alpha\cos\beta \pm \cos\alpha\sin\beta,$$

we obtain

$$I = \frac{2[k_m \sin k_m R \cos k_n R - k_n \cos k_m R \sin k_n R]}{(k_m^2 - k_n^2)}. \tag{4.3.10}$$

In our case $k_m R = (2m - 1)\frac{1}{2}\pi$, $k_n R = (2n - 1)\frac{1}{2}\pi$. Hence, $\cos k_m R = \cos k_n R = 0$. The numerator is zero and at $m \neq n$, the denominator differs from zero; consequently, the integral is equal to zero.

The special case when $m = n$ should be considered separately, since the numerator and denominator in the expression above are equal to zero. In this case we have

$$I = \int_{-R}^{+R} \cos^2 k_n x\, dx = 2\left[\frac{R}{2} + \frac{\sin 2k_n R}{4k_n}\right] = R, \tag{4.3.11}$$

i.e., the integral differs from zero.

These formulas are used to determine constant coefficients D_n. We multiply both sides of equality (4.3.7) by $\cos k_m x$ and integrate it from $-R$ to $+R$, i.e.,

$$\int_{-R}^{+R} t(x, 0) \cos k_m x \, dx = \int_{-R}^{+R} \sum_{n=1}^{\infty} D_n \cos k_n x \cos k_m x \, dx$$

$$= \sum_{n=1}^{\infty} \int_{-R}^{+R} D_n \cos k_n x \ \cos k_m x \, dx. \quad (4.3.12)$$

All the integrals in the right-hand side of this equality are equal to zero according to relation (4.3.9) with the exception of the case where $m = n$ which yields $I_{m=n} = R$.

Consequently

$$D_n = (1/R) \int_{-R}^{+R} t(x, 0) \cos k_n x \, dx = (2/R) \int_0^R t(x, 0) \cos k_n x \, dx. \quad (4.3.13)$$

In the derivation two necessary assumptions were not explicitly stated: (1) the integral (4.3.13) has a finite and definite value; (2) the integral of the infinite series is equal to the sum of integrals of separate terms of a series. These assumptions demand that the function $f(x)$ should satisfy the Dirichlet condition: the function $f(x)$ in a definite interval must be (1) single-valued, finite, and integrable; (2) have a finite number of maxima and minima, and (3) have a finite number of discontinuities.

Thus, the general solution of our problem (4.3.6) may be written as follows

$$t(x, \tau) = \sum_{n=1}^{\infty} \cos \mu_n \frac{x}{R} \exp\left[-\mu_n^2 \frac{a\tau}{R^2}\right] \cdot \frac{2}{R} \int_0^R f(x) \cos \mu_n \frac{x}{R} \, dx, \quad (4.3.14)$$

where

$$\mu_n = k_n R = (2n - 1)\tfrac{1}{2}\pi.$$

This solution is at the same time that of a problem on determination of a temperature field inside an infinite plate of thickness $l = R (0 < x < l)$ *when one of the surfaces is thermally insulated* (*at* $x = 0$ *a heat flow is absent as* $\partial t(0, \tau)/\partial x = 0$) *while the opposite surface* $x = l$ *is maintained at temperature* 0°C. At the initial time the temperature distribution is given as the function $f(x)$ which may be of any form but it must satisfy the Dirichlet conditions.

If the function $f(x)$ is odd, the solution to (4.3.1) subject to boundary conditions (4.3.2) and (4.3.3) (with $t_a = 0$) is obtained in a similar way, yielding

$$t(x, \tau) = \sum_{n=1}^{\infty} \sin \mu_n \frac{x}{R} \exp\left(-\mu_n^2 \frac{a\tau}{R^2}\right) \cdot \frac{2}{R} \int_0^R f(x) \sin \mu_n \frac{x}{R}\, dx, \tag{4.3.14'}$$

where $\mu_n = n\pi$.

Solution (4.3.14′) is at the same time applicable to the problem of cooling an infinite plate $l = R$ $(0 < x < l)$ thick when its surfaces are maintained at temperature 0°C (the origin of the coordinates is on one of the surfaces of the plate). At the initial time instant the temperature distribution is given as the arbitrary function $f(x)$ satisfying the Dirichlet conditions.

If the initial temperature distribution is uniform, i.e., $t(x, 0) = t_0 =$ const, then the integral (4.3.13) can be calculated as

$$\frac{2}{R} \int_0^R t_0 \cos \mu_n \frac{x}{R}\, dx = \frac{2t_0}{\mu_n} \sin \mu_n \frac{x}{R} \bigg|_0^R = \frac{2 \sin \mu_n}{\mu_n} t_0$$
$$= \frac{2t_0}{\mu_n} (-1)^{n+1},$$

since $\sin \mu_n = \sin(2n - 1)\frac{1}{2}\pi = \pm 1$, i.e., the sine acquires $+1$ or -1, respectively, depending on the value of the argument (for even values of n the sine is -1 and for odd ones, $+1$.)

Thus the solution of our problem may be written

$$\frac{t(x, \tau)}{t_0} = \sum_{n=1}^{\infty} \frac{2}{\mu_n} (-1)^{n+1} \cos \mu_n \frac{x}{R} \exp\left(-\mu_n^2 \frac{a\tau}{R^2}\right). \tag{4.3.15}$$

If the temperature of the limiting surfaces is not equal to zero but to t_a, then the solution may be rewritten thus

$$\theta = \frac{t(x, \tau) - t_a}{t_0 - t_a} = \sum_{n=1}^{\infty} \frac{2}{\mu_n} (-1)^{n+1} \cos \mu_n \frac{x}{R} \exp(-\mu_n^2\, \mathrm{Fo}). \tag{4.3.16}$$

c. Solution by the Operational Method. Upon use of the Laplace transformation, the differential heat conduction equation will be of the same form as in the previous problem, i.e.,

$$T''(x, s) \quad (s/a)T(x, s) + (t_0/a) = 0. \tag{4.3.17}$$

The initial condition $t(x, 0) = t_0 =$ constant was used when passing from the partial differential equation for the inverted transform $t(x, \tau)$ to equation (4.3.17) for the transform $T(x, s)$, viz, when applying the Laplace transformation to the arbitrary time temperature.

Boundary conditions (4.3.3a) after the transform will be

$$T(R, s) = t_a/s, \qquad T'(0, s) = 0. \tag{4.3.18}$$

The solution of differential equation (4.3.17) may be written as follows (see Section 3.3)

$$T(x, s) - (t_0/s) = A \cosh (s/a)^{1/2}x + B \sinh (s/a)^{1/2}x, \qquad (4.3.19)$$

where A and B are the constants determined from the boundary conditions (4.3.18). According to the symmetry condition, $B = 0$ since

$$\begin{aligned} T'(0, s) &= [(s/a)^{1/2}A \sinh(s/a)^{1/2}x + (s/a)^{1/2}B \cosh(s/a)^{1/2}x]_{x=0} \\ &= (s/a)^{1/2}B = 0, \end{aligned}$$

whence $B = 0$. The constant A is found from the first boundary condition (4.3.18)

$$T(R, s) = t_0/s + A \cosh(s/a)^{1/2}R = t_a/s$$

whence

$$A = -(t_0 - t_a)/s \cosh(s/a)^{1/2}R.$$

Thus, the solution for the transform will be of the following form

$$(t_0/s) - T(x, s) = \frac{(t_0 - t_a)\cosh(s/a)^{1/2}x}{s \cosh(s/a)^{1/2}R} = \frac{\varphi(s)}{\psi(s)}. \qquad (4.3.20)$$

It may be shown that the right-hand side of the equality is the ratio of two generalized polynomials[3] with respect to s, viz:

$$\begin{aligned} \varphi(s) &= (t_0 - t_a)\cosh(s/a)^{1/2}x \\ &= (t_0 - t_a)\left(1 + \frac{x^2}{2!a}s + \frac{x^4}{4!a^2}s^2 + \cdots\right), \end{aligned}$$

$$\psi(s) = s\left(1 + \frac{R^2}{2!a}s + \frac{R^4}{4!a^2}s^2 + \cdots\right).$$

Since the generalized polynomial $\psi(s)$ does not contain a constant (the first term is equal to s), all the conditions of the expansion theorem are fulfilled and it may be used when passing from the solution (4.3.20) for the transform to that for the inverse transform.

[3] An infinite convergent power series, the exponents of which are natural numbers, is referred to as a generalized polynomial or a polynomial of infinitely high power. Sometimes it will be simply called a polynomial.

The expansion theorem may be written

$$L^{-1}\left[\frac{\varphi(s)}{\psi(s)}\right] = \sum_{n=1}^{n} \frac{\varphi(s_n)}{\psi'(s_n)} \exp[s_n\tau], \tag{4.3.21}$$

where s_n are the roots of the polynomial $\psi(s)$.

We seek the roots of the function $\psi(s) = s\cosh(s/a)^{1/2}R$ (i.e., the values of s such that $\psi(s)$ is equal to zero). Then we have (1) the simple root $s = 0$, (2) an infinite number of simple roots determined by the following relation

$$i\sqrt{\frac{s_n}{a}}\, R = \mu_n = (2n-1)\frac{\pi}{2}, \tag{4.3.22}$$

whence

$$s_n = -\frac{a\mu_n^{\,2}}{R^2} = -\frac{(2n-1)^2\pi^2 a}{4R^2}. \tag{4.3.23}$$

Then determining $\psi'(s)$:

$$\psi'(s) = \frac{sR}{2(as)^{1/2}} \sinh\left(\frac{s}{a}\right)^{1/2} R + \cosh\left(\frac{s}{a}\right)^{1/2} R$$

$$= \frac{1}{2}\left(\frac{s}{a}\right)^{1/2} R \sinh\left(\frac{s}{a}\right)^{1/2} R + \cosh\left(\frac{s}{a}\right)^{1/2} R \tag{4.3.24}$$

We have:

$$\lim_{s\to 0} \psi'(s) = 1,$$

$$\lim_{s\to s_n} \psi'(s) = \tfrac{1}{2}\, i\mu_n \sinh i\mu_n = -\tfrac{1}{2}\,\mu_n \sin\mu_n = \psi'(s_n).$$

The quantity $\sin\mu_n$ is equal to $+1$ or -1, depending on the value of n, i.e.,

$$\sin\mu_n = (-1)^{n+1}$$

Next, we find the quantity $\varphi(s_n)$:

$$\varphi(0) = (t_0 - t_a),$$

$$\varphi(s_n) = (t_0 - t_a)\cosh(s_n/a)^{1/2}x = (t_0 - t_a)\cosh i\mu_n(x/R)$$

$$= (t_0 - t_a)\cos\mu_n(x/R).$$

Finally we obtain

$$L^{-1}[(t_0/s) - T(x, s)] = L^{-1}\left[\frac{(t_0 - t_a)\cosh(s/a)^{1/2}x}{s\cosh(s/a)^{1/2}R}\right]$$

$$t_0 - t(x, \tau) = (t_0 - t_a) - (t_0 - t_a)$$
$$\times \sum_{n=1}^{\infty} \frac{2}{\mu_n} (-1)^{n+1} \cos \mu_n \frac{x}{R} \exp\left[-\mu_n^2 \frac{a\tau}{R^2}\right],$$

i.e., the solution is identical with (4.3.16):

$$\theta = \frac{t(x, \tau) - t_a}{t_0 - t_a} = \sum_{n=1}^{\infty} A_n \cos \mu_n \frac{x}{R} \exp(-\mu_n^2 \,\mathrm{Fo}), \qquad (4.3.25)$$

where $A_n = (2/\mu_n)(-1)^{n+1}$ is the initial amplitude and $\mathrm{Fo} = a\tau/R^2$ is the Fourier number.

d. Analysis of the Solution. The relative excess temperature is a function of the relative coordinate and the Fourier number;

$$\theta = \varphi(x/R, \mathrm{Fo}).$$

Hence a cooling process consists of the progressive leveling of temperature over the plate thickness (the bounding surface temperature is the same all the time and equal to t_a), the rate of which is proportional to the thermal diffusivity coefficient. Such a heat transfer process is referred to as an internal process and the problem itself, an internal problem. A representation of the temperature distribution at different time intervals is given in Fig. 4.7.

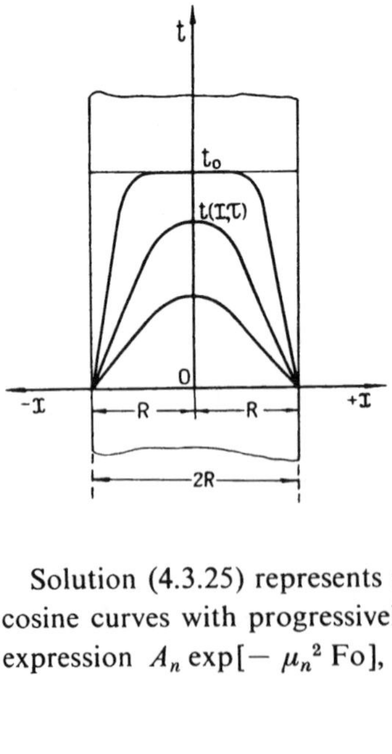

Fig. 4.7. Temperature distribution inside an infinite plate (internal symmetrical problem).

Solution (4.3.25) represents a convergent series, i.e., an algebraic sum of cosine curves with progressively attenuating amplitudes determined by the expression $A_n \exp[-\mu_n^2 \,\mathrm{Fo}]$, i.e., the amplitudes decrease both with an increase in μ_n and time. The frequency of such cosine curves increases with n as it is equal to $(2n - 1)/4$.

Figure 4.8 gives the temperature distribution over the plate thickness for Fo = 0.05 according to solution (4.3.25). The temperature distribution curve may be presented rather precisely as a sum of three cosine curves

$$\theta \approx \frac{4}{\pi} \exp[-\pi^2 \,\mathrm{Fo}/4] \cos\frac{\pi x}{2R} - \frac{4}{3\pi} \exp[-9\pi^2 \,\mathrm{Fo}/4] \cos\frac{3\pi x}{2R} + \frac{4}{5\pi} \exp[-25\pi^2 \,\mathrm{Fo}/4] \cos\frac{5\pi x}{2R}.$$

From Fig. 4.8 one can see that the summation of three cosine curves with various amplitudes and frequency gives us a temperature distribution

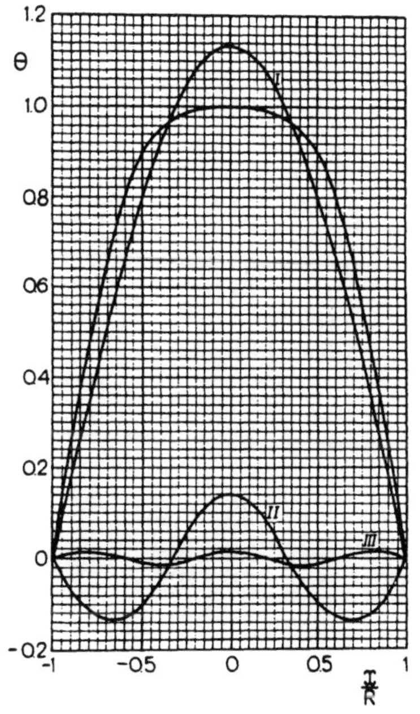

Fig. 4.8. Temperature distribution inside an infinite plate at a Fourier number Fo = 0.05 (symmetrical problem).

curve at the given dimensionless time instant corresponding to Fo = 0.05. For small values of Fo, it is necessary to take several terms of the series, since the amplitudes attenuate slowly; for large values of Fo, all the terms of a series are negligibly small, as compared to the first. This is clear since $\mu_1 < \mu_2 < \mu_3 < \cdots \mu_n = (2n - 1)\pi/2$, and consequently the exponential function $\exp(-\mu_n^2 \mathrm{Fo})$ quickly decreases with an increase in μ_n; for example, at Fo = 0.5 $\exp(-\mu_1^2 \mathrm{Fo}) = \exp(-\frac{1}{8}\pi^2) = 0.291$, and $\exp(-\mu_2^2 \mathrm{Fo})$

$= \exp(-\frac{9}{8}\pi^2) < 0.00004$). Therefore, starting from a definite value of Fo, one may restrict oneself to the first term of series (4.3.25). Such a solution will be convenient for practical calculations.

For small values of Fo, the solution (4.3.25) is less convenient. The operational method makes it possible to solve this problem in a form more applicable to small values of Fo.

Let us return to solution (4.3.20) for the transform. First we expand $1/\cosh(s/a)^{1/2}R$ into a series (see Appendix)

$$\frac{1}{\cosh(s/a)^{1/2}R} = 2(\exp[-(s/a)^{1/2}R] - \exp[-3(s/a)^{1/2}R] + \exp[-5(s/a)^{1/2}R] - \cdots)$$
$$= 2\sum_{n=1}^{\infty}(-1)^{n+1}\exp[-(2n-1)(s/a)^{1/2}R].$$

The solution (4.3.20) may be written

$$(t_0/s) - T(x, s) = [\{t_0 - t_a\}/s)\tfrac{1}{2}(\exp[(s/a)^{1/2}x] + \exp[-(s/a)^{1/2}x]) \times 2\sum_{n=1}^{\infty}(-1)^{n+1}\exp[-(2n-1)(s/a)^{1/2}R]$$
$$= [(t_0 - t_a)/s]\sum_{n=1}^{\infty}(-1)^{n+1}\{\exp[-((2n-1)R - x)(s/a)^{1/2}] + \exp[-((2n-1)R + x)(s/a)^{1/2}]\}. \quad (4.3.26)$$

For the inversion of the transforms the tabulated formula is used:

$$L^{-1}[(1/s)\exp[-k\sqrt{s}]] = \operatorname{erfc} k/2\sqrt{\tau}.$$

Then, the solution of our problem is obtained as follows

$$\frac{t_0 - t(x,\tau)}{t_0 - t_a} = \sum_{n=1}^{\infty}(-1)^{n+1}\left\{\operatorname{erfc}\left[\frac{(2n-1)R - x}{2(a\tau)^{1/2}}\right] + \operatorname{erfc}\left[\frac{(2n-1)R + x}{2(a\tau)^{1/2}}\right]\right\}. \quad (4.3.27)$$

This solution may be written as

$$\theta = \frac{t(x,\tau) - t_a}{t_0 - t_a} = 1 - \sum_{n=1}^{\infty}(-1)^{n+1} \times \left\{\operatorname{erfc}\frac{(2n-1) - (x/R)}{2\sqrt{\mathrm{Fo}}} + \operatorname{erfc}\frac{(2n-1) + (x/R)}{2\sqrt{\mathrm{Fo}}}\right\}. \quad (4.3.28)$$

Solution (4.3.27) satisfies the differential equation, initial and boundary conditions. At $\tau \to 0$, i.e., at small values of Fo, the arguments of the functions are large and the functions themselves are close to zero as $t(x, 0) = t_0$; at $x = R$ the sum is equal to 1, i.e., $t(R, \tau) = t_a$.

At small values of Fo, all the terms of the series are infinitesimal except for the first; thus the function erfc u quickly decreases with an increase in the argument; for example, at $u = 2.7$ erfc $u = 0.0001$, i.e., in practice it is equal to zero. The function erfc $u = 1 -$ erf u is tabulated and for engineering calculations the use of solution (4.3.28) does not involve difficulties.

Let us calculate the dimensionless temperature θ as a function of the Fourier number for the central plane of a plate ($x = 0$) according to the following approximate formulas taken from solutions (4.3.25) and (4.3.28)

$$\theta_c \approx \frac{4}{\pi}\left\{\exp\left[-\frac{\pi^2}{4}\,\mathrm{Fo}\right] - \frac{1}{3}\exp\left[-\frac{9\pi^2}{4}\,\mathrm{Fo}\right] + \frac{1}{5}\exp\left[-\frac{25\pi^2}{4}\,\mathrm{Fo}\right] - \frac{1}{7}\exp\left[-\frac{49\pi^2}{4}\,\mathrm{Fo}\right]\right\}, \tag{4.3.29}$$

$$\theta_c \approx 1 - 2\left\{\mathrm{erfc}\,\frac{1}{2\sqrt{\mathrm{Fo}}} - \mathrm{erfc}\,\frac{3}{2\sqrt{\mathrm{Fo}}} + \mathrm{erfc}\,\frac{5}{2\sqrt{\mathrm{Fo}}}\right\}, \tag{4.3.30}$$

and according to the exact formula

$$\theta_c = \frac{4}{\pi}\sum_{n=1}^{\infty}\frac{(-1)^{n+1}}{(2n-1)}\exp\left[-(2n-1)^2\frac{\pi^2}{4}\,\mathrm{Fo}\right]. \tag{4.3.25}$$

The calculated results, accurate to four decimal places, are given in Table 4.1.

From Table 4.1, it can be seen that the calculation by the approximate formula (4.3.30) will give complete coincidence with that obtained by the exact formula from small values of the Fourier number up to $\mathrm{Fo} < 0.4$. In the range of Fourier modulus from 0.4 to 1.8 ($0.4 < \mathrm{Fo} \leq 1.8$), a discrepancy is observed only in the fourth decimal. Errors become apparent for $\mathrm{Fo} > 2$. It should be noted that within the limits of Fo from 0.001 to 0.1, we need use only one term in the brackets of formula (4.3.30). Calculation by the approximate formula (4.3.29) evidently gives wrong results at small values of Fo from 0.001 to 0.08; beginning with a Fourier number of 0.08, the formula (4.3.29) yields correct results. Moreover, for $\mathrm{Fo} > 0.4$ one may restrict oneself to the first term of formula (4.3.29). The incorrect values obtained by formula (4.3.29) for small values of Fo are explained by the fact that relatively few terms of series (4.3.25) were used.

TABLE 4.1. TEMPERATURE OF INFINITE PLATE AT $x = 0$ (MIDDLE OF PLATE)[a]

Fo	$\theta_c(25)$	$\theta_c(29)$	$\theta_c(30)$	Fo	$\theta_c(25)$	$\theta_c(29)$	$\theta_c(30)$
0.001	1.0000	0.9332	1.0000	0.4	0.4745	0.4745	0.4744
0.004	1.0000	0.9591	1.0000	0.6	0.2897	0.2897	0.2896
0.010	1.0000	0.9850	1.0000	0.8	0.1769	0.1769	0.1768
0.020	1.0000	0.9978	1.0000	1.0	0.1080	0.1080	0.1079
0.040	0.9992	0.9991	0.9992	1.2	0.0659	0.0659	0.0660
0.050	0.9969	0.9971	0.9969	1.4	0.0402	0.0402	0.0402
0.060	0.9922	0.9923	0.9922	1.6	0.0246	0.0246	0.0244
0.080	0.9752	0.9752	0.9752	1.8	0.0150	0.0150	0.0146
0.100	0.9493	0.9493	0.9493	2.0	0.0092	0.0092	0.0082
0.2	0.7723	0.7723	0.7723	2.5	0.0026	0.0026	−0.0007

[a] Boundary surfaces ($x = R$, $x = -R$) held at temperature t_a, initial plate temperature t_0.

Thus, for small values of Fo it is necessary to use solution (4.3.27). By this example one can see a great advantage of the operational method which makes it possible to solve a problem in two forms: one is convenient for calculations at small values of Fo, another for large values of Fo.

Figure 4.9 gives curves of the dimensionless temperature distribution in a plate for various values of the Fourier number (0.005 to 1.5). From

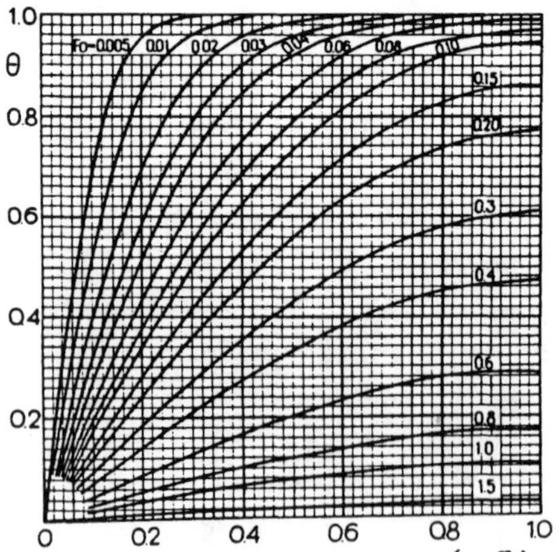

Fig. 4.9. Curves of dimensionless excess temperature distribution inside infinite plate. [(1 − x/R) is the abscissa and θ is the ordinate; the scale is the same in both cases] (symmetrical problem).

Fig. 4.9 one can see that the temperature in the middle of a plate begins to decrease rapidly, starting from Fo > 0.06.

A cooling process is essentially completed at Fo > 1.5.

Figure 4.10 furnishes plots of the dimensionless temperature $(t_0 - t)/(t_0 - t_a) = (1 - \theta)$ versus the Fourier number for various values of $(1 - x/R)$ from 0 to 1.

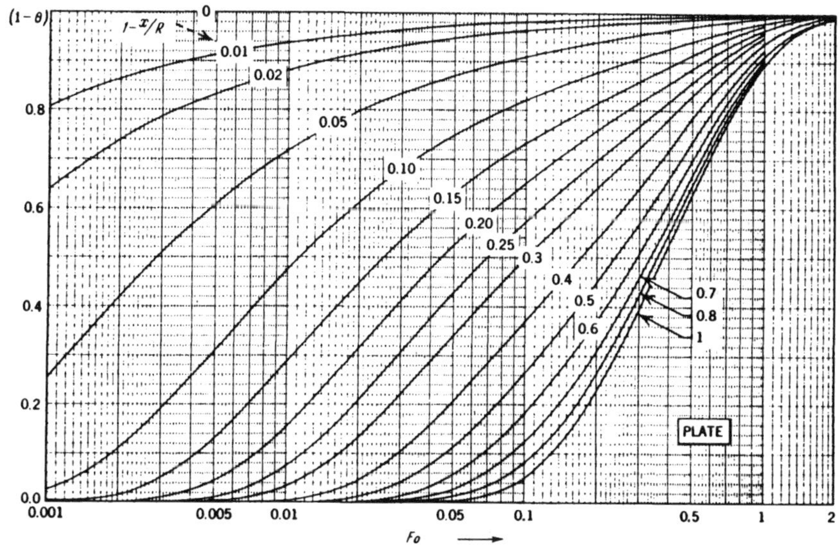

Fig. 4.10. Plot of dimensionless temperature $(t_0 - t)/(t_0 - t_a) = (1 - \theta)$ versus the Fourier number [102] (symmetrical problem).

These plots may serve as nomograms for engineering calculations. In addition, Table 4.2 illustrates values of the dimensionless excess temperature in the middle of a plate for various values of the Fourier number.

e. Determination of the Specific Heat Rate. The amount of heat in kilocalories lost by a plate is

$$\Delta Q = CM(t_0 - \bar{t}) = c\gamma V(t_0 - \bar{t}),$$

where $\bar{t}$ is the volumetric mean temperature of a plate. The specific heat rate (heat per unit volume; $kcal/m^3$) is

$$\Delta Q_v = c\gamma(t_0 - \bar{t}). \qquad (4.3.31)$$

TABLE 4.2. TEMPERATURE IN THE MIDDLE OF INFINITE PLATE[a]

$$\theta_c = \frac{4}{\pi} \sum_{n=1}^{\infty} \frac{1}{(2n-1)} (-1)^{n+1} \exp\left[-(2n-1)^2 \frac{\pi^2}{4} \text{Fo}\right]$$

$\frac{1}{4}$Fo	θ_c	$\frac{1}{4}$Fo	θ_c	$\frac{1}{4}$Fo	θ_c
0.001	1.0000	0.040	0.8458	0.079	0.5835
0.002	1.0000	0.041	0.8384	0.080	0.5778
0.003	1.0000	0.042	0.8310	0.081	0.5721
0.004	1.0000	0.043	0.8236	0.082	0.5665
0.005	1.0000	0.044	0.8162	0.083	0.5610
0.006	1.0000	0.045	0.8088	0.084	0.5555
0.007	1.0000	0.046	0.8015	0.085	0.5500
0.008	0.9998	0.047	0.7941	0.086	0.5447
0.009	0.9996	0.048	0.7868	0.087	0.5393
0.010	0.9992	0.049	0.7796	0.088	0.5340
0.011	0.9985	0.050	0.7723	0.089	0.5288
0.012	0.9975	0.051	0.7651	0.090	0.5236
0.013	0.9961	0.052	0.7579	0.091	0.5185
0.014	0.9944	0.053	0.7508	0.092	0.5134
0.015	0.9922	0.054	0.7437	0.093	0.5084
0.016	0.9896	0.055	0.7367	0.094	0.5034
0.017	0.9866	0.056	0.7297	0.095	0.4985
0.018	0.9832	0.057	0.7227	0.096	0.4936
0.019	0.9794	0.058	0.7158	0.097	0.4887
0.020	0.9752	0.059	0.7090	0.098	0.4839
0.021	0.9706	0.060	0.7022	0.099	0.4792
0.022	0.9657	0.061	0.6955	0.100	0.4745
0.023	0.9605	0.062	0.6888	0.102	0.4652
0.024	0.9550	0.063	0.6821	0.104	0.4561
0.025	0.9493	0.064	0.6756	0.106	0.4472
0.026	0.9433	0.065	0.6690	0.108	0.4385
0.027	0.9372	0.066	0.6626	0.110	0.4299
0.028	0.9308	0.067	0.6561	0.112	0.4215
0.029	0.9242	0.068	0.6498	0.114	0.4133
0.030	0.9175	0.069	0.6435	0.116	0.4052
0.031	0.9107	0.070	0.6372	0.118	0.3973
0.032	0.9038	0.071	0.6310	0.120	0.3895
0.033	0.8967	0.072	0.6249	0.122	0.3819
0.034	0.8986	0.073	0.6188	0.124	0.3745
0.035	0.8824	0.074	0.6128	0.126	0.3671
0.036	0.8752	0.075	0.6088	0.128	0.3600
0.037	0.8679	0.076	0.6009	0.130	0.3529
0.038	0.8605	0.077	0.5950	0.132	0.3460
0.039	0.8532	0.078	0.5892	0.134	0.3393

TABLE 4.2. (*continued*)

$\frac{1}{4}$Fo	θ_c	$\frac{1}{4}$Fo	θ_c	$\frac{1}{4}$Fo	θ_c
0.136	0.3326	0.198	0.1804	0.40	0.0246
0.138	0.3261	0.200	0.1769	0.42	0.0202
0.140	0.3198	0.205	0.1684	0.44	0.0166
0.142	0.3135	0.210	0.1602	0.46	0.016
0.144	0.3074	0.215	0.1525	0.48	0.0112
0.146	0.3014	0.220	0.1452	0.50	0.0092
0.148	0.2955	0.225	0.1382	0.52	0.0075
0.150	0.2897	0.230	0.1315	0.54	0.0062
0.152	0.2840	0.235	0.1252	0.56	0.0051
0.154	0.2785	0.240	0.1192	0.58	0.0042
0.156	0.2731	0.245	0.1134	0.60	0.0034
0.158	0.2677	0.250	0.1080	0.62	0.0028
0.160	0.2625	0.255	0.1028	0.64	0.0023
0.162	0.2574	0.260	0.0978	0.66	0.0019
0.164	0.2523	0.265	0.0931	0.68	0.0016
0.166	0.2474	0.270	0.0886	0.70	0.0013
0.168	0.2426	0.275	0.0844	0.72	0.0010
0.170	0.2378	0.280	0.0803	0.74	0.0009
0.172	0.2332	0.285	0.0764	0.76	0.0007
0.174	0.2286	0.290	0.0728	0.78	0.0006
0.176	0.2241	0.295	0.0693	0.80	0.0005
0.178	0.2198	0.300	0.0659	0.82	0.0004
0.180	0.2155	0.31	0.0597	0.84	0.0003
0.182	0.2113	0.32	0.0541	0.86	0.0003
0.184	0.2071	0.33	0.0490	0.88	0.0002
0.186	0.2031	0.34	0.0444	0.90	0.0002
0.188	0.1991	0.35	0.0402	0.92	0.0001
0.190	0.1952	0.36	0.0365	0.94	0.0001
0.192	0.1914	0.37	0.0330	0.96	0.0001
0.194	0.1877	0.38	0.0299	0.98	0.0001
0.196	0.1840	0.39	0.0271	1.00	0.0001

[a] Same conditions as in Table 4.1.

Thus, it is necessary to define the mean temperature of a plate. We have

$$\bar{t}(\tau) = \frac{1}{R} \int_0^R t(x, \tau)\, dx.$$

From solution (4.3.25) we may obtain

$$\bar{\theta} = \frac{\bar{t}(\tau) - t_a}{t_0 - t_a} = \sum_{n=1}^{\infty} B_n \exp[-\mu_n^2 \,\mathrm{Fo}], \qquad (4.3.32)$$

where

$$B_n = \frac{A_n \sin \mu_n}{\mu_n} = (-1)^{n+1} \frac{2 \sin \mu_n}{\mu_n^2} = \frac{2}{\mu_n^2} = \frac{8}{(2n-1)^2 \pi^2}.$$

Again a convergent series is obtained; the coefficients B_n rapidly decrease with an increase in μ_n.

We may also estimate the mean temperature from solution (4.3.20) for the transform. Using the following relation

$$\bar{T}(s) = \frac{1}{R} \int_0^R T(x, s)\, dx, \qquad t_0/s - \bar{T}(s) = \frac{(t_0 - t_a)}{s\left(\dfrac{s}{a}\right)^{1/2} R} \tanh\left(\frac{s}{a}\right)^{1/2} R. \tag{4.3.33}$$

Making use of the expansion theorem for determining the inverse transform, solution (4.3.33) may be rewritten as

$$\frac{t_0}{s} - \bar{T}(s) = \frac{(t_0 - t_a)\{\sinh (s/a)^{1/2} R/(s/a)^{1/2} R\}}{s \cosh(s/a)^{1/2} R} = \frac{\varphi(s)}{\psi(s)}.$$

Similarly we find the roots $\psi(s) = s \cosh (s/a)^{1/2} R$; as a result we have: $s_0 = 0$, $s_n = -\mu_n^2(a/R^2)$. The expression

$$\left[\frac{\sinh(s/a)^{1/2} R}{(s/a)^{1/2} R}\right] = \left[1 + \frac{R^2}{3!a} s + \frac{R^4}{5!a^2} s^2 + \cdots\right]$$

is a generalized polynomial, the first term of which is equal to unity.

In a similar fashion we evaluate

$$\psi'(s_n) = -\frac{1}{2} \mu_n \sin \mu_n,$$

$$\varphi(s_n) = (t_0 - t_a) \frac{\sinh i\mu_n}{i\mu_n} = (t_0 - t_a) \frac{\sin \mu_n}{\mu_n}.$$

Hence,

$$t_0 - \bar{t}(\tau) = t_0 - t_a - (t_0 - t_a) \sum_{n=1}^{\infty} \frac{2}{\mu_n^2} \exp[-\mu_n^2 \text{ Fo}],$$

or

$$\bar{\theta} = \frac{\bar{t}(\tau) - t_a}{t_0 - t_a} = \sum_{n=1}^{\infty} B_n \exp[-\mu_n^2 \text{ Fo}], \tag{4.3.34}$$

i.e., the expression identical with (4.3.32) is obtainable.

If we further analyze solution (4.3.33) for the transform, we note, for small values of time, or rather of the Fourier number, the quantity $(s/a)^{1/2} R$

is great. In this case $\tanh (s/a)^{1/2}R$ may be assumed to be unity ($\tanh (s/a)^{1/2}R \approx 1$ at $(s/a)^{1/2}R > 2$) and it may be written

$$\bar{T}(s) = \frac{t_0}{s} - \frac{(t_0 - t_a)}{s(s/a)^{1/2}R}. \tag{4.3.35}$$

According to the table of the transforms the inverse transform

$$\bar{t}(\tau) = t_0 - \frac{2(t_0 - t_a)}{R}\left(\frac{a\tau}{\pi}\right)^{1/2},$$

or

$$\bar{\theta} = 1 - 2\left(\frac{\text{Fo}}{\pi}\right)^{1/2} \tag{4.3.36}$$

The approximate solution (4.3.36) agrees very well with the exact solution (4.3.34) at small values of $\text{Fo} < 0.1$.

We now illustrate the use of these results by two numerical examples

(1) Let $\text{Fo} = 1/\pi^2 \approx 0.1$. Then, from solution (4.3.34) we have

$$\bar{\theta} = (8/\pi^2)(e^{-1/4} + (1/9)e^{-9/4} + (1/25)\,e^{-25/4} + \cdots) = (8/\pi^2)\,0.7906 = 0.6408.$$

From solution (4.3.36) we obtain:

$$\bar{\theta} = 1 - 2\left(\frac{1}{\pi}\right)^{3/2} = 0.6408.$$

There is a complete agreement.

(2) Assume $\text{Fo} = 2/\pi^2 \approx 0.2$. Then, from solution (4.3.34) we have $\bar{\theta} = 0.4926$ and from solution (4.3.36), $\bar{\theta} = 0.4921$. This agreement is quite satisfactory considering the fact Fo is as large as 0.2 in this case.

From this example one can see that at small values of Fo the approximate solution (4.3.36) may be used with sufficient degree of accuracy instead of a number of the terms of series (4.3.34); at large values of Fo, one may restrict oneself to the first term of the whole series (4.3.34), i.e.,

$$\bar{\theta} = (8/\pi^2)\exp[-\tfrac{1}{4}\pi^2\,\text{Fo}] \quad \text{at Fo} > 0.1. \tag{4.3.37}$$

Proceeding from the solution for the transform the method for obtaining approximate solutions has great advantages when solving complex problems. This is a major advantage of the operational method over the classical one.

A solution for the mean temperature $\bar{\theta}$, may be obtained in another way if we transform the solution (4.3.33) and use the following formula for the inversion of the transform:

$$L^{-1}[(1/s\sqrt{s})\exp[-k\sqrt{s}]] = 2\sqrt{\tau}\,\text{ierfc}\,(k/2\sqrt{\tau}).$$

Expanding $\tanh (s/a)^{1/2}R$ in series

$$\tanh(s/a)^{1/2}R = 1 - 2 \exp[-2(s/a)^{1/2}R] + 2 \exp[-4(s/a)^{1/2}R] - \cdots,$$

then we shall have from (4.3.33):

$$\frac{t_0}{s} - \bar{T}(s) = \frac{(t_0 - t_a)}{s(s/a)^{1/2}R} \{1 - 2 \sum_{n=1}^{\infty} (-1)^{n+1} \exp(-2n(s/a)^{1/2}R)\}.$$

Hence the solution for the inverse transform $\bar{t}(\tau)$ may be written thus:

$$t_0 - \bar{t}(\tau) = (t_0 - t_a)\left\{\frac{2}{R}\left(\frac{a\tau}{\pi}\right)^{1/2} - 2\sum_{n=1}^{\infty}(-1)^{n+1} \times\left[2\left(\frac{a\tau}{\pi R^2}\right)^{1/2}\exp\left(-\frac{4n^2R^2}{4a\tau}\right) - 2n \operatorname{erfc}\frac{nR}{(a\tau)^{1/2}}\right]\right\}.$$

Finally, the solution will be of the form

$$\bar{\theta} = \frac{\bar{t}(\tau) - t_a}{t_0 - t_a} = 1 - 2\left(\frac{\text{Fo}}{\pi}\right)^{1/2} + 4\sqrt{\text{Fo}} \sum_{n=1}^{\infty} (-1)^{n+1} \operatorname{ierfc}\frac{n}{\sqrt{\text{Fo}}}, \tag{4.3.38}$$

where $\operatorname{ierfc} u = (1/\sqrt{\pi}) \exp[-u^2] - u \operatorname{erfc} u$.

It also follows that at small values of time, or rather Fo, solution (4.3.38) transforms into approximate solution (4.3.36) as at $\text{Fo} \to 0$, $\exp[-n^2/\text{Fo})] \to 0$, and $\operatorname{erfc}(n/\sqrt{\text{Fo}}) \to 0$.

Solution (4.3.38) may be obtained directly from (4.3.28) using another calculation method. At first ΔQ_s may be determined by the formula

$$\Delta Q_s = -\int_0^\tau \lambda \frac{\partial t(R, \tau)}{\partial x} d\tau, \tag{4.3.39}$$

and then from relation (4.3.31) the mean temperature $\bar{t}(\tau)$ may also be determined. The result will be the same.

f. Analysis of the Solution. The relation between the dimensionless mean temperature $\bar{\theta}$ and the dimensionless time Fo in the form of (4.3.34) or (4.3.38) is widely used in problems on diffusion. Average concentration plays the same role as the mean temperature in such problems. Table 4.3 gives values of $\bar{\theta}$ for various values of the Fourier number and the corresponding calculation diagram of $\bar{\theta}$ versus the Fourier function is plotted in Fig. 4.11.

TABLE 4.3. Values of Functions[a]

$$\bar{\theta} = \frac{8}{\pi^2} \sum_{n=1}^{\infty} \frac{1}{(2n-1)^2} \exp\left[-(2n-1)^2\pi^2 \frac{\text{Fo}}{4}\right]$$

π^2 Fo/4	0.00	0.01	0.02	0.03	0.04	0.05	0.06	0.07	0.08	0.09
0.00	1.0000	0.9282	0.8984	0.8756	0.8563	0.8394	0.8240	0.8099	0.7968	0.7845
0.10	0.7728	0.7618	0.7512	0.7410	0.7312	0.7218	0.7127	0.7038	0.6952	0.6869
0.20	0.6787	0.6708	0.6631	0.6555	0.6481	0.6408	0.6337	0.6267	0.6199	0.6132
0.30	0.6066	0.6000	0.5936	0.5873	0.5812	0.5751	0.5691	0.5632	0.5573	0.5515
0.40	0.5458	0.5402	0.5346	0.5292	0.5237	0.5184	0.5131	0.5079	0.5028	0.4976
0.50	0.4926	0.4877	0.4827	0.4779	0.4730	0.4682	0.4636	0.4589	0.4543	0.4497
0.60	0.4452	0.4408	0.4363	0.4319	0.4277	0.4234	0.4192	0.4150	0.4108	0.4068
0.70	0.4027	0.3896	0.3847	0.3907	0.3868	0.3830	0.3792	0.3754	0.3716	0.3679
0.80	0.3643	0.3607	0.3570	0.3534	0.3499	0.3464	0.3430	0.3396	0.3362	0.3329
0.90	0.3296	0.3263	0.3230	0.3199	0.3166	0.3134	0.3104	0.3173	0.3042	0.3012
1.00	0.2982	0.2952	0.2923	0.2894	0.2865	0.2836	0.2809	0.2780	0.2751	0.2725
1.10	0.2698	0.2672	0.2645	0.2618	0.2592	0.2566	0.2541	0.2516	0.2491	0.2466
1.20	0.2441	0.2417	0.2393	0.2369	0.2346	0.2322	0.2300	0.2276	0.2253	0.2231
1.30	0.2209	0.2187	0.2165	0.2144	0.2122	0.2101	0.2081	0.2060	0.2039	0.2019
1.40	0.1999	0.1979	0.1959	0.1940	0.1920	0.1902	0.1882	0.1864	0.1845	0.1827
1.50	0.1808	0.1791	0.1773	0.1755	0.1738	0.1720	0.1703	0.1686	0.1670	0.1653
1.60	0.1836	0.1620	0.1604	0.1588	0.1573	0.1556	0.1541	0.1525	0.1511	0.1496
1.70	0.1481	0.1466	0.1452	0.1436	0.1423	0.1409	0.1394	0.1381	0.1367	0.1354
1.80	0.1340	0.1327	0.1313	0.1300	0.1287	0.1274	0.1262	0.1249	0.1237	0.1225
1.90	0.1213	0.1200	0.1188	0.1176	0.1165	0.1153	0.1142	0.1131	0.1119	0.1108
2.00	0.1097	0.1086	0.1076	0.1064	0.1054	0.1043	0.1033	0.1023	0.1012	0.1003
2.10	0.0993	0.0982	0.0973	0.0963	0.0954	0.0944	0.0935	0.0926	0.0916	0.0907
2.20	0.0898	0.0889	0.0880	0.0871	0.0863	0.0854	0.0846	0.0837	0.0829	0.0821
2.30	0.0843	0.0805	0.0797	0.0789	0.0781	0.0773	0.0765	0.0768	0.0751	0.0742
2.40	0.0735	0.0728	0.0721	0.0713	0.0707	0.0700	0.0692	0.0686	0.0678	0.0672
2.50	0.0665	0.0659	0.0653	0.0646	0.0640	0.0633	0.0627	0.0620	0.0614	0.0609
2.60	0.0602	0.0596	0.0590	0.0584	0.0579	0.0573	0.0567	0.0556	0.0556	0.0550
2.70	0.0545	0.0539	0.0534	0.0528	0.0524	0.0518	0.0513	0.0508	0.0503	0.0498
2.80	0.0493	0.0488	0.0483	0.0478	0.0473	0.0469	0.0460	0.0460	0.0455	0.0451
2.90	0.0446	0.0442	0.0437	0.0433	0.0429	0.0424	0.0420	0.0416	0.0412	0.0408
3.00	0.0404	0.0400	0.0396	0.0392	0.0387	0.0384	0.0380	0.0376	0.0373	0.0369
3.10	0.0365	0.0361	0.0357	0.0354	0.0350	0.0347	0.0344	0.0340	0.0336	0.0333
3.20	0.0331	0.0327	0.0324	0.0321	0.0318	0.0315	0.0311	0.0308	0.0306	0.0302
3.30	0.0292	0.0296	0.0293	0.0289	0.0287	0.0283	0.0281	0.0278	0.0276	0.0272
3.40	0.0271	0.0268	0.0265	0.0263	0.0260	0.0258	0.0255	0.0252	0.0250	0.0247
3.50	0.0245	0.0242	0.0240	0.0237	0.0235	0.0233	0.0230	0.0228	0.0226	0.0224

[a] Boundary conditions as in Table 4.1.

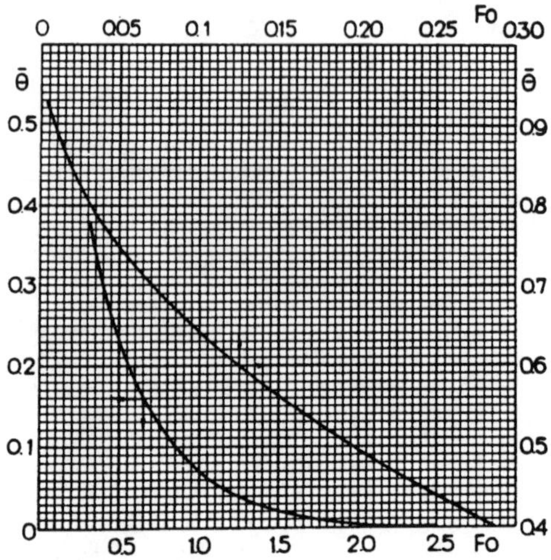

Fig. 4.11. Relation between the dimensionless mean excess temperature $\bar{\theta}$ and the Fourier number for an infinite plate (symmetrical problem).

We now find the dimensionless heating rate $d\bar{\theta}/d\text{Fo}$. According to equation (4.3.34) it is

$$-\,d\bar{\theta}/d\text{Fo} = 2\sum_{n=1}^{\infty} \exp[-\tfrac{1}{4}(2n-1)^2\pi^2\,\text{Fo}]. \qquad (4.3.40)$$

Using approximate formulas (4.3.36) and (4.3.37),

$$-\,d\bar{\theta}/d\text{Fo} = 1/(\pi\text{Fo})^{1/2} = (2/\pi)\,(1/\{1-\bar{\theta}\}) \qquad \text{for Fo} < 0.1, \qquad (4.3.41)$$

$$-\,d\bar{\theta}/d\text{Fo} = 2\exp(-\tfrac{1}{4}\pi^2\,\text{Fo}) = \tfrac{1}{4}\pi^2\bar{\theta} \qquad \text{for Fo} > 0.1, \qquad (4.3.42)$$

As Fo $= a\tau/R^2$, the heating rate $d\bar{\theta}/d\tau$ will be directly proportional to the thermal diffusivity and inversely proportional to the square of half the plate thickness. Table 4.4 illustrates values of $\bar{\theta}$ and $d\theta/d\text{Fo}$ for various values of the Fourier number.

All the above solutions are also valid for the case of cooling of an infinite plate when one bounding surface has ideal thermal insulation (absence of a heat flow which is characterized by the condition $\partial t(0,\tau)/\partial x = 0$) and at the initial time instant the opposite surface is instantaneously cooled to a constant temperature t_a i.e., $t(R,\tau) = t_a$ (the origin of coordinates is taken at the adiabatic surface). The initial temperature distribution over the plate thickness may be given as some function $f(x)$ or be uniform, i.e., $t(x, 0) = t_0 = \text{const}$. In this case R represents not half, but the whole plate thickness.

TABLE 4.4 RELATION BETWEEN $d\bar{\theta}/d\text{Fo}, \bar{\theta}$, AND Fo FOR AN INFINITE PLATE $(-R \le x \le R)$[a]

Fo	$\bar{\theta}$	$\frac{1}{2}\, d\bar{\theta}/d\text{Fo}$	Fo	$\bar{\theta}$	$\frac{1}{2}\, d\bar{\theta}/d\text{Fo}$
0.005	0.91978	3.9894	0.4	0.3021	0.3728
0.01	0.8849	2.9735	0.5	0.2353	0.2912
0.02	0.8401	1.9941	0.6	0.1844	0.2275
0.04	0.7729	1.3904	0.7	0.1475	0.1820
0.06	0.7236	1.1536	0.8	0.1127	0.1389
0.10	0.6430	0.8917	0.9	0.0900	0.1110
0.2	0.4960	0.6323	1.0	0.0687	0.0848
0.3	0.3859	0.4772	1.1	0.0537	0.0663

[a] Boundary surfaces held at temperature t_a and initial plate temperature t_0.

If at all time one surface of a plate is maintained at a constant temperature t_a = const, *and the opposite one, at a constant but different temperature, for example, at the initial temperature* t_0 = const, *then the boundary conditions may be written as* (Fig. 4.12)

$$t(0, \tau) = t_a = \text{const}, \qquad t(R, \tau) = t_0 = \text{const}. \tag{4.3.43}$$

The origin of coordinates is on the left surface of a plate $(0 < x < R)$.

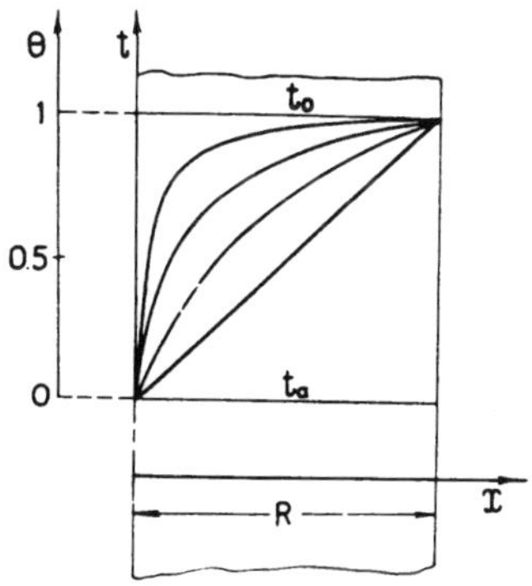

Fig. 4.12. Curves of the temperature distribution inside an infinite plate (asymmetrical problem).

A solution for the transform $T(x, s)$ has the form (see (4.3.19))

$$\frac{t_0}{s} - T(x, s) = \frac{t_0 - t_a}{s}\left(\frac{\sinh(s/a)^{1/2}(R - x)}{\sinh(s/a)^{1/2}R}\right). \tag{4.3.44}$$

The solution of the problem is obtained in two forms by using the same method. First, applying the expansion theorem, we obtain

$$\theta = \frac{t(x, \tau) - t_a}{t_0 - t_a} = \frac{x}{R} + \sum_{n=1}^{\infty} (-1)^{n+1} \frac{2}{\mu_n} \sin \mu_n \frac{R - x}{R} \times \exp(-\mu_n^2 \text{Fo}), \tag{4.3.45}$$

where $\mu_n = n\pi$ are the eigenvalues. In a stationary state ($\tau = \infty$), the temperature distribution will be a straight line passing through a point $(0, t_a)$ and (R, t_0). Second, developing $(\sinh(s/a)^{1/2}R)^{-1}$ in a series and using the table of the transforms it is found

$$\theta = \frac{t(x, \tau) - t_a}{t_0 - t_a} = \text{erf} \frac{x}{2(a\tau)^{1/2}} - \sum_{n=1}^{\infty} \left(\text{erfc} \frac{2nR + x}{2(a\tau)^{1/2}} - \text{erfc} \frac{2nR - x}{2(a\tau)^{1/2}} \right) \tag{4.3.46}$$

The first term of solution (4.3.46) represents a solution for a semi-infinite body since solution (4.3.16) may be obtained from this solution if we let $R = \infty$ [see (4.2.16)]. Hence, at small time values, or rather at small Fourier values, heat propagation in a plate occurs similarly as in a semi-infinite body. In this case (small values of Fo) the sum is negligibly small as compared to the first term. This follows directly from solution (4.3.46) if it is rewritten in the dimensionless form

$$\theta = \text{erf} \frac{(x/R)}{2\sqrt{\text{Fo}}} - \sum_{n=1}^{\infty} \left(\text{erfc} \frac{2n + (x/R)}{2\sqrt{\text{Fo}}} - \text{erfc} \frac{2n - (x/R)}{2\sqrt{\text{Fo}}} \right). \tag{4.3.47}$$

For small values of Fo the arguments of the function erfc are great and the functions themselves are close to zero. The sum may therefore be neglected.

The physical significance of heat propagation inside a plate at small values of Fo may be obtained from the analysis of solution (4.3.28), viz: if we move the origin of coordinates from the plate center to the left surface, i.e., we substitute a variable X so that $x + R = X$ and $2R \to \infty$, then solution (4.3.28) takes the form

$$\theta = \text{erf} \frac{X}{2(a\tau)^{1/2}} \tag{4.3.48}$$

This is clearly a solution for a semi-infinite body. In Fig. 4.13, the temperature θ is plotted versus the Fourier number for various values of the dimensionless coordinates of the plate from 0 to 1.

The solutions for an infinite plate have been analyzed in some detail to show the reader the great advantage of the operational method over the classical one. The solutions in the Laplace form, representing the known

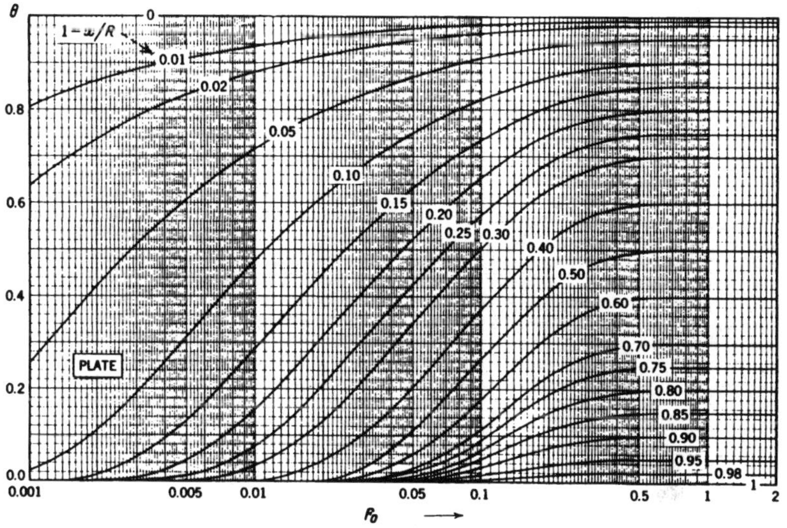

Fig. 4.13. Plot of the excess dimensionless temperature θ versus Fo for various values of $(1 - x/R)$ from 0 to 1 [102]. Boundary surfaces $x = 0$ at t_a, $x = R$ at t_0; initial plate temperature, t_0.

combination of the functions erf or erfc, are convenient for small values of the Fourier number. Vice versa, the solution in the Fourier form, consisting of the product of two functions (one is of the exponential form taking into account a temperature change in time; the other is of the trigonometric form characterizing a temperature change over the plate thickness) is convenient for large values of the Fourier number. In this case, one may limit oneself to the first term of a series, neglecting all the remainder.

The Laplace operational method makes it possible to obtain such approximate solutions with any degree of accuracy. These solutions are rather simple and may be used in engineering calculations with considerable success.

4.4 Sphere (Symmetrical Problem)

a. Statement of the Problem. *A problem for a spherical body may be formulated as follows. There is a spherical body of radius R with a known initial temperature distribution f(r). (An important particular case is where the initial temperature is constant and equal to t_0). At the initial time, the sphere surface is instantaneously cooled to some temperature equal to t_a which is* *maintained constant during the whole cooling process. The temperature distribution inside the sphere at any time instant and the specific heat rate are to be determined.* Since cooling occurs uniformly, isotherms inside the sphere may be represented by concentric spheres, i.e., the temperature depends only on a radius-vector r and time τ (see Fig. 4.14).

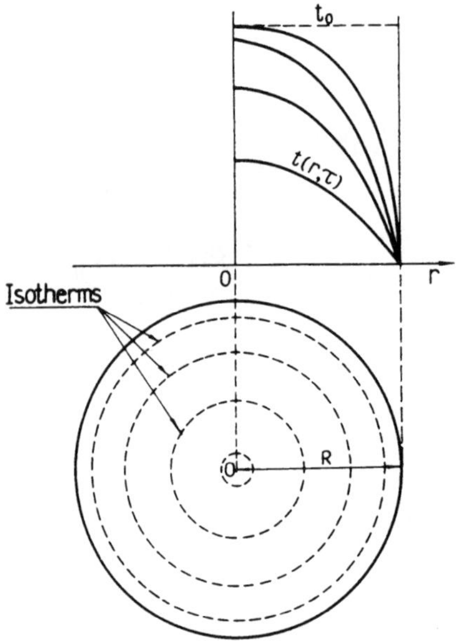

Fig. 4.14. Temperature distribution curves inside a sphere (symmetrical problem).

The differential heat conduction equation in a symmetrical problem has the form

$$\frac{\partial t(r, \tau)}{\partial \tau} = a\left(\frac{\partial^2 t(r, \tau)}{\partial r^2} + \frac{2}{r}\,\frac{\partial t(r, \tau)}{\partial r}\right)$$

$$(\tau > 0; \quad 0 < r < R) \tag{4.4.1}$$

at conditions

$$t(r, 0) = f(r), \tag{4.4.2}$$

$$t(R, \tau) = t_a = \text{const}, \tag{4.4.3}$$

$$t(0, \tau) \neq \infty, \tag{4.4.4}$$

$$\partial t(0, \tau)/\partial x = 0.$$

The last condition is the symmetry condition.

Differential equation (4.4.1) may be written

$$\frac{\partial[rt(r, \tau)]}{\partial \tau} = a\,\frac{\partial^2[rt(r, \tau)]}{\partial r^2}. \tag{4.4.5}$$

The equivalence of equations (4.4.1) and (4.4.5) may be shown to be

$$r\frac{\partial t}{\partial \tau} = a\left\{\frac{\partial}{\partial r}\left[\frac{\partial (rt)}{\partial r}\right]\right\}$$

$$= a\frac{\partial}{\partial r}\left(r\frac{\partial t}{\partial r} + t\right)$$

$$= a\left(r\frac{\partial^2 t}{\partial r^2} + \frac{\partial t}{\partial r} + \frac{\partial t}{\partial r}\right)$$

$$= a\left(r\frac{\partial^2 t}{\partial r^2} + 2\frac{\partial t}{\partial r}\right);$$

cancelling r we have

$$\frac{\partial t}{\partial \tau} = a\left(\frac{\partial^2 t}{\partial r^2} + \frac{2}{r}\frac{\partial t}{\partial r}\right).$$

b. Solution by Method of Separation of Variables. First of all, let us solve the problem by means of the method of separation of variables. If the product $rt(r, \tau)$ is replaced by ϑ, then Eq. (4.4.5) with respect to ϑ is identical with that for an infinite plate. The particular solution of this equation is known [see Eq. (4.3.4)]:

$$\vartheta = rt(r, \tau) = (C \sin kr + D \cos kr) \exp[-ak^2\tau], \qquad (4.4.6)$$

Consequently,

$$t(r, \tau) = (C\{\sin kr\}/r + D\{\cos kr\}/r) \exp[-ak^2\tau]. \qquad (4.4.7)$$

As the distribution of isotherms is symmetrical with respect to the center, and the temperature in the middle of the sphere ($r = 0$) cannot be infinitely large, then the constant D is equal to zero; the latter condition is necessary since at $r \to 0$, $\{\sin kr\}/r \to k$ and $\{\cos kr\}/r \to \infty$. This derivation also directly follows from solution (4.4.6), since at $r = 0$, $\vartheta(r, \tau)$ should be equal to zero, hence $D = 0$.

Thus the particular solution will be

$$t(r, \tau) = C\{(\sin kr)/r\} \exp[-ak^2\tau]. \qquad (4.4.8)$$

To simplify the calculations, we assume $t_a = 0$; it means that the temperature is taken relative to t_a as the datum. Then, imposing the boundary condition (4.4.3) on solution (4.4.8), we shall have

$$t(R, \tau) = C\{(\sin kR)/R\} \exp[-ak^2\tau] = t_a = 0.$$

Hence, $\sin kR = 0$, i.e., $kR = \pi, 2\pi, 3\pi, \ldots$ $(0 < \tau < \infty)$. Thus, there is an infinite number of values of k determined by the following equality:

$$k_n R = n\pi = \mu_n, \qquad n = 1, 2, 3, \cdots. \tag{4.4.9}$$

The values of μ_n are taken, starting from π since at $n = 0$ the appropriate term will be zero for all points $r < R$.

The general solution will be a sum of particular solutions

$$rt(r, \tau) = \sum_{n=v}^{\infty} C_n \sin k_n r \exp[-ak_n{}^2\tau]. \tag{4.4.10}$$

For the determination of constants C_n the initial condition is used

$$rt(r, 0) = rf(r) = \sum_{n=1}^{\infty} C_n \sin k_n r. \tag{4.4.11}$$

If the function $rf(r)$ satisfies the Dirichlet conditions, then it may be expanded in a Fourier series. We multiply both sides of equality (4.4.11) by $\sin k_m r$ and integrate it over the limits from 0 to R; then

$$\int_0^R rf(r)(\sin k_m r)\, dr = \sum_{n=1}^{\infty} \int_0^R C_n (\sin k_n r)(\sin k_m r)\, dr. \tag{4.4.12}$$

With the help of the method described in Section 4.3 it may be shown that

$$J_{(m\neq n)} = \int_0^R (\sin k_n r)(\sin k_m r)\, dr = \frac{k_m (\sin k_n R)(\cos k_m R) - k_n (\sin k_m R)(\cos k_n R)}{k_n{}^2 - k_m{}^2}; \tag{4.4.13}$$

$$J_{(m=n)} = \int_0^R (\sin^2 k_n r)\, dr = \frac{R}{2} - \frac{\sin 2k_n R}{4k_n}. \tag{4.4.14}$$

As $k_n = n\pi$, then $J_{(m\neq n)} = 0$ and $J_{(m=n)} = \frac{1}{2}R$, i.e., all the integrals are equal to zero except one when $m = n$ in which cases this integral is $\frac{1}{2}R$.

Thus, for the determination of the coefficients C_n the following relation is obtained:

$$C_n = (2/R) \int_0^R rf(r)(\sin k_n r)\, dr. \tag{4.4.15}$$

The general solution of our problem is as follows

$$t(r, \tau) = \sum_{n=1}^{\infty} \frac{\sin \mu_n (r/R)}{r} \frac{2}{R} \int_0^R rf(r) \left(\sin \mu_n \frac{r}{R}\right) dr \exp\left[-\mu_n{}^2 \frac{a\tau}{R^2}\right]. \tag{4.4.16}$$

For a uniform initial temperature distribution $t(r, 0) = f(r) = t_0$, we obtain

$$\frac{2}{R}\int_0^R rt_0\left(\sin\mu_n\frac{r}{R}\right)dr = -\frac{2Rt_0}{\mu_n}\cos\mu_n = (-1)^{n+1}\frac{2Rt_0}{\mu_n},$$

since $\cos\mu_n = \cos n\pi = (-1)^n$.

Finally we have

$$\frac{t(r, \tau)}{t_0} = \sum_{n=1}^{\infty}(-1)^{n+1}\frac{2}{\mu_n}\frac{R\sin\mu_n(r/R)}{r}\exp\left[-\mu_n^2\frac{a\tau}{R^2}\right]. \qquad (4.4.17)$$

If the temperature t_a is not zero, then the solution of this problem may be written

$$\theta = \frac{t(r, \tau) - t_a}{t_0 - t_a} = \sum_{n=1}^{\infty} A_n\frac{R\sin\mu_n(r/R)}{r\mu_n}\exp[-\mu_n^2\mathrm{Fo}], \qquad (4.4.18)$$

where

$$A_n = (-1)^{n+1}\cdot 2; \qquad \mu_n = n\pi. \qquad (4.4.19)$$

c. Solution by the Operational Method. Solution (4.4.18) is obtained by means of the method of the Laplace transformation. Applying the Laplace transformation to differential equation (4.4.5), we have:

$$[rT(r, s)]'' - (s/a)rT(r, s) + (rt_0/a) = 0. \qquad (4.4.20)$$

Solution (4.4.20) may be written by analogy with solution (4.3.17) as

$$rT(r, s) - (rt_0/s) = A\cosh(s/a)^{1/2}r + B\sinh(s/a)^{1/2}r. \qquad (4.4.21)$$

Since in the middle of the sphere ($r = 0$), the temperature $t(0, \tau)$ and its transform $T(0, s)$ cannot be infinitely large (i.e., it must be $\lim_{r\to 0} rT(r, s) \to 0$), then one can clearly see that $A = 0$.

Thus, we have:

$$T(r, s) - \frac{t_0}{s} = B\frac{\sinh(s/a)^{1/2}r}{r}. \qquad (4.4.22)$$

Boundary condition (4.4.3) for the transform will be of the form

$$T(R, s) = t_a/s \qquad (4.4.23)$$

Boundary condition (4.4.23) is now imposed on solution (4.4.22):

$$\frac{t_a}{s} = \frac{t_0}{s} + B\frac{\sinh(s/a)^{1/2}R}{R},$$

whence

$$B = -\frac{(t_0 - t_a)R}{s \sinh(s/a)^{1/2}R}. \tag{4.4.24}$$

Thus the solution (4.4.22) may be written

$$(t_0/s) - T(r, s) = \frac{(t_0 - t_a)R \sinh(s/a)^{1/2}r}{rs \sinh(s/a)^{1/2}R}. \tag{4.4.25}$$

The numerator $\varphi_1(s) = (t_0 - r_a)R \sinh(s/a)^{1/2}r$ and the denominator $\psi_1(s) = rs \sinh(s/a)^{1/2}R$ are not generalized polynomials with respect to s but they may be so reduced by multiplying or dividing by $(s/a)^{1/2}$, viz:

$$\frac{\sinh\left(\frac{s}{a}\right)^{1/2} r}{s \sinh\left(\frac{s}{a}\right)^{1/2} R} = \frac{r\left(\frac{s}{a}\right)^{1/2} + \frac{1}{3!}\left\{r\left(\frac{s}{a}\right)^{1/2}\right\}^3 + \cdots}{s\left[R\left(\frac{s}{a}\right)^{1/2} + \frac{1}{3!}\left(R\left(\frac{s}{a}\right)^{1/2}\right)^3 + \cdots\right]}$$

$$= \frac{r + \frac{1}{3!}\frac{r^3}{a}s + \cdots}{s\left[R + \frac{1}{3!}R^3\frac{s}{a} + \cdots\right]}. \tag{4.4.26}$$

From operational calculus it is known that if $\varphi(s)$ and $\psi(s)$ are generalized polynomials with respect to s where $\varphi(s) = s^k\varphi_1(s)$, $\psi(s) = s^k\psi_1(s)$, then

$$\lim_{s\to s_n} \frac{\varphi(s)}{\psi'(s)} = \lim_{s\to s_n} \frac{\varphi_1(s)}{\psi_1'(s)}, \qquad \text{if } s^k \neq 0, \tag{4.4.27}$$

where s_n are the roots of the equation $\psi_1(s) = 0$.

Therefore, except the first root $s = 0$ the inversion of the transform may be done according to the ordinary expansion theorem since in our case $\varphi_1(s)$ and $\psi_1(s)$ satisfy condition (4.4.27).

For determination of the roots $\psi_1(s)$ it is necessary to assume $\psi_1(s) = 0$:

$$\psi_1(s) = rs \sinh(s/a)^{1/2}R = 0.$$

Hence, we obtain the following roots: (1) $s = 0$ (zero root); (2) $\sinh(s/a)^{1/2}R = (1/i) \sin i(s/a)^{1/2}R = 0$, whence $i(s/a)^{1/2}R = \pi, 2\pi, 3\pi, \cdots n\pi$ or $s_n = -a\mu_n/R^2 = -an^2\pi^2/R^2$ as $\mu_n = n\pi$.

Using expansion theorem we obtain

$$L^{-1}\left[\frac{\varphi_1(s)}{\psi_1(s)}\right] = \sum_{n=1}^{\infty} \frac{\varphi_1(s_n)}{\psi_1'(s_n)} e^{s_n\tau},$$

$$\varphi_1(s_n) = (t_0 - t_a)\,\frac{R}{i}\sin\mu_n\frac{r}{R},$$

$$\psi_1'(s_n) = \lim_{s\to s_n}\left[r\sinh\left(\frac{s}{a}\right)^{1/2}R + \frac{1}{2}\,r\left(\frac{s}{a}\right)^{1/2}R\cosh\left(\frac{s}{a}\right)^{1/2}R\right]$$

$$= \frac{1}{2i}\,r\mu_n\cos\mu_n.$$

Consequently, we shall have

$$\sum_{n=1}^{\infty}\frac{\varphi_1(s_n)}{\psi_1'(s_n)}\,e^{s_n\tau}$$

$$= -\sum_{n=1}^{\infty}(-1)^{n+1}\frac{2R(t_0 - t_a)\sin\mu_n(r/R)}{r\mu_n}\exp\left[-\mu_n^2\frac{a\tau}{R^2}\right]. \tag{4.4.28}$$

Relation (4.4.26) is used for the zero root:

$$\varphi(s) = (t_0 - t_a)R\left[r + \frac{1}{3!}\,r^3\,\frac{s}{a} + \cdots\right],$$

$$\psi(s) = rs\left[R + \frac{1}{3!}\,R^3\,\frac{s}{a} + \cdots\right]. \tag{4.4.29}$$

Then

$$\lim_{s\to 0}\frac{\varphi(s)}{\psi'(s)} = (t_0 - t_a). \tag{4.4.30}$$

Finally, the solution of the problem may be written in such a form:

$$\frac{t_0 - t(r,\tau)}{t_0 - t_a} = 1 - \sum_{n=1}^{\infty}2(-1)^{n+1}\frac{R\sin\mu_n(r/R)}{r\mu_n}\exp\left[-\mu_n^2\frac{a\tau}{R^2}\right]. \tag{4.4.31}$$

which we rewrite in dimensionless quantities as

$$\theta = \frac{t(r,\tau) - t_a}{t_0 - t_a} = \sum_{n=1}^{\infty}A_n\frac{R\sin\mu_n(r/R)}{r\mu_n}\exp[-\mu_n^2\mathrm{Fo}], \tag{4.4.32}$$

where $A_n = 2(-1)^{n+1}$, $\mu_n = n\pi$, i.e., the solution identical with (4.4.18) is obtained. Solution (4.4.32) represents a quickly convergent series because the exponential function $\exp[-\mu_n^2\mathrm{Fo}]$ rapidly decreases with an increase in μ_n.

For small values of Fo, it is necessary to take a large number of terms of the series, resulting in considerable inconvenience in computations. For large values of Fo, a single term of a series may suffice and all the remainder may be neglected. Consequently, solution (4.4.32) is more convenient for large values of Fo.

For small values of Fo let us find a solution in a different form. Expanding $1/\sinh(s/a)^{1/2}R$ in series (see Appendix) we obtain

$$\frac{1}{\sinh(s/a)^{1/2}R} = 2\{\exp[-(s/a)^{1/2}R] + \exp[-3(s/a)^{1/2}R)] + \cdots\}$$
$$= 2\sum_{n=1}^{\infty} \exp[-(2n-1)(s/a)^{1/2}R].$$

In addition, using the formula

$$\sinh(s/a)^{1/2}r = \tfrac{1}{2}\{\exp[(s/a)^{1/2}r] - \exp[-(s/a)^{1/2}r]\}$$

solution (4.4.25) for the transform may be written as

$$\frac{t_0}{s} - T(r, s) = \frac{(t_0 - t_a)R}{sr} \sum_{n=1}^{\infty} (\exp\{-[(2n-1)R - r](s/a)^{1/2}\}$$
$$- \exp\{-[(2n-1)R + r](s/a)^{1/2}\}). \qquad (4.4.33)$$

Using the table of transforms, a solution for the inverse transform is obtained as

$$\theta = \frac{t(r, \tau) - t_a}{t_0 - t_a} = 1 - \frac{R}{r} \sum_{n=1}^{\infty} \left\{\operatorname{erfc} \frac{(2n-1) - (r/R)}{2\sqrt{\mathrm{Fo}}} - \operatorname{erfc} \frac{(2n-1) + (r/R)}{2\sqrt{\mathrm{Fo}}}\right\}. \qquad (4.4.34)$$

This solution is applicable to $r > 0$. For the temperature in the middle of the sphere ($r = 0$) let us derive a solution in the following way. From solution (4.4.25) we have

$$(t_0/s) - T(0, s) = [(t_0 - t_a)R/(as)^{1/2} \sinh(s/a)^{1/2}R]$$
$$= [2(t_0 - t_a)R/(as)^{1/2}] \sum_{n=1}^{\infty} \exp[-(2n-1)(s/a)^{1/2}R].$$

Then, using the relation

$$L^{-1}\{(1/\sqrt{s}) \exp[-k\sqrt{s}]\} = (1/\pi\tau)^{1/2} \exp[-k^2/4\tau]$$

we define

$$\theta_c = \frac{t(0, \tau) - t_a}{t_0 - t_a} = 1 - 2\sqrt{\frac{1}{\pi\,\mathrm{Fo}}} \sum_{n=1}^{\infty} \exp\left[-\frac{(2n-1)^2}{4\mathrm{Fo}}\right]. \qquad (4.4.35)$$

Solutions (4.4.34) and (4.4.35) show that solutions for an infinite plate and sphere may be presented in one and the some function.

Figure 4.15 gives curves of the temperature distribution as a function of the dimensionless coordinate r/R for various values of Fo from 0.005 to 0.4.

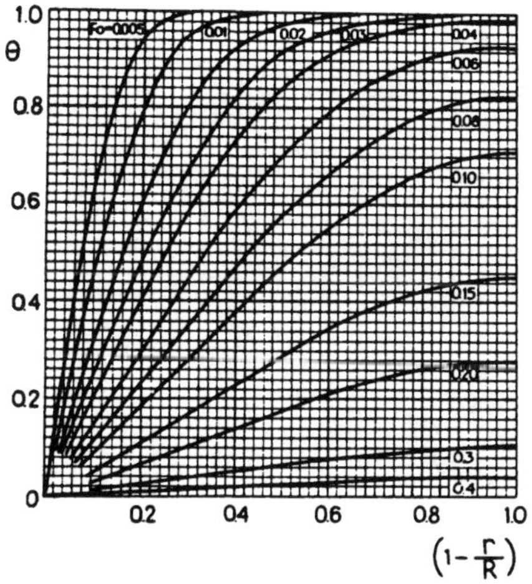

Fig. 4.15. Curves of the dimensionless excess temperature distribution inside sphere. [(1 − r/R) is the abscissa and θ is the ordinate.]

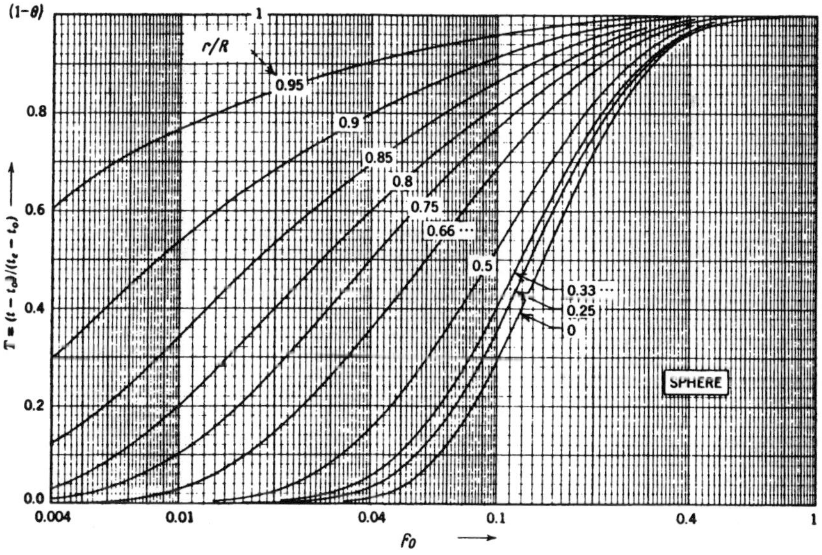

Fig. 4.16. Plot of dimensionless temperature $(t_0 - t)/(t_0 - t_a) = 1 - \theta$ versus the Fourier number Fo for various coordinates r/R from 0 to 1 for a sphere [102].

From Fig. 4.15, one can see that at Fo = 0.005 the temperature at a depth $r/R \leq 0.6$ differs only slightly from the initial one; starting from Fo > 0.04, a change in temperature in the middle of a sphere is noticeable. A cooling process almost ceases for Fo $\geq$ 0.4. In Fig. 4.16, the dimensionless temperature $(1 - \theta)$ is plotted versus Fo for various values of the dimensionless coordinates.

d. Determination of the Heat Losses. To determine the heat losses in a cooling process, we find the mean temperature over the volume of a sphere; this temperature is determined by the formula (see Section 1.7)

$$\bar{t}(\tau) = 3/R^3 \cdot \int_0^R r^2 t(r, \tau)\, dr. \tag{4.4.36}$$

If the appropriate solution (4.4.32) is substituted for $t(r, \tau)$ in this relation, we have upon integration

$$\bar{\theta} = \frac{\bar{t}(\tau) - t_a}{t_0 - t_a} = \sum_{n=1}^{\infty} B_n \exp[-\mu_n^2\, \mathrm{Fo}], \tag{4.4.37}$$

where $B_n = 6/\mu_n^2 = 6/n^2\pi^2$.

A solution for $\bar{\theta}$ may be obtained in another form. We define $\bar{T}(s)$ using solution (4.4.25) for the transform, namely:

$$\bar{T}(s) = \frac{3}{R^3} \int_0^R r^2 T(r, s)\, dr = \frac{t_0}{s} - \frac{3(t_0 - t_a)\sqrt{a}}{s\sqrt{s}\, R} \coth\left(\frac{s}{a}\right)^{1/2} R + \frac{3(t_0 - t_a)a}{s^2 R^2}. \tag{4.4.38}$$

We then develop $\coth(s/a)^{1/2}R$ in series to obtain

$$\begin{aligned} \coth(s/a)^{1/2} R &= 1 + 2\exp[-2(s/a)^{1/2} R] + 2\exp[-4(s/a)^{1/2} R] + \cdots \\ &= 1 + 2\sum_{n=1}^{\infty} \exp[-2n(s/a)^{1/2} R]. \end{aligned} \tag{4.4.39}$$

Upon the inversion, the solution is obtained as

$$\bar{\theta} = 1 + 3\mathrm{Fo} - 6\sqrt{\mathrm{Fo}}\, [(1/\sqrt{\pi}) + 2\sum_{n=1}^{\infty} \operatorname{ierfc} n/\sqrt{\mathrm{Fo}}\,]. \tag{4.4.40}$$

We now illustrate with a numerical example. Determine the sphere temperature in a cooling process at a point $r/R = 0.4$, as well as its average temperature at the dimensionless time 0.2 (Fo = 0.2) if the sphere was initially at 200°C and at the initial time instant it was placed into melting ice ($t_a = 0$).

We find the temperature at a point $r/R = 0.4$, according to solution (4.4.32):

$$\theta_1 = \frac{2}{\pi}\left[\frac{\sin 0.4\pi}{0.4}\exp[-0.2\pi^2] - \frac{1}{2}\,\frac{\sin 0.4\cdot 2\pi}{0.4}\exp[-0.2\cdot 4\pi^2)\right]$$

$$= 1\cdot 591\cdot 0.951\cdot 0.139 - 0.796\cdot 0.588\cdot 0.0004 \approx 0.21.$$

The same result is obtained from solution (4.4.34):

$$\theta_1 = 1 - \frac{1}{0.4}\left(\operatorname{erfc}\frac{1-0.4}{2\sqrt{0.2}} - \operatorname{erfc}\frac{1+0.4}{2\sqrt{0.2}}\right)$$

$$= 1 - 2.5(0.3427 - 0.0270) = 1 - 0.79 \approx 0.21.$$

This value might be obtained from Fig. 4.15, where we see that at $r/R = 0.4$ and $\text{Fo} = 0.2$, the relative temperature is equal to $\theta = 0.21$.

The sphere temperature at this point will be

$$t_1 = t_0\theta_1 = 200\cdot 0.21 = 42°.$$

The mean temperature $\bar{\theta}$ is determined by formula (4.4.37):

$$\bar{\theta}_1 = B_1\exp[-\mu_1^2\text{Fo}] + B_2\exp[-\mu_2^2\text{Fo}] = \frac{6}{\pi^2}\exp[-0.2\pi^2]$$

$$+ \frac{6}{4\pi^2}\exp[-0.8\pi^2] \approx \frac{6}{\pi^2}\exp[-0.2\pi^2]$$

$$= 0.6070\cdot 0.1389 \approx 0.084.$$

$\bar{\theta}$ may also be determined from Fig. 4.17; from this plot it follows that for $\text{Fo} = 0.2$, $\bar{\theta} = 0.085$.

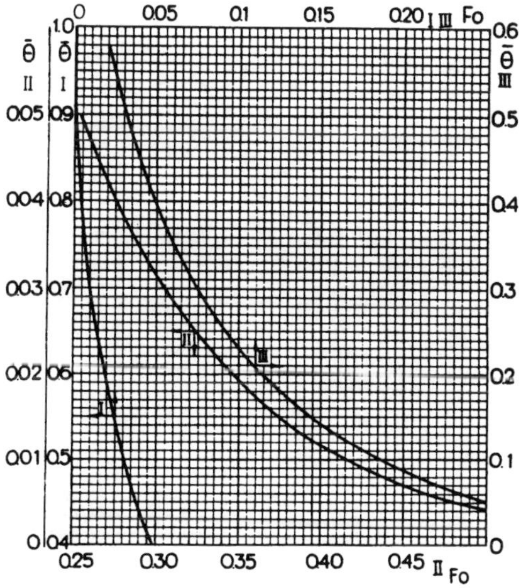

Fig. 4.17. Relation between mean dimensionless excess temperature and the Fourier number for a sphere (symmetrical problem).

TABLE 4.5. VALUES OF FUNCTIONS[a]

$$\theta_c = \sum_{n=1}^{\infty} (-1)^{n+1} \exp(-n^2\pi^2\text{Fo}) \quad \text{AND} \quad \bar{\theta} = \frac{6}{\pi^2} \sum_{n=1}^{\infty} \exp(-n^2\pi^2\text{Fo})$$

π^2Fo	θ_c	$\bar{\theta}$	π^2Fo	θ_c	$\bar{\theta}$
0.00	1.0000	1.0000	0.74	0.8531	0.2980
0.02	1.0000	0.8537	0.76	0.8418	0.2916
0.04	1.0000	0.7967	0.78	0.8303	0.2854
0.06	1.0000	0.7543	0.80	0.8186	0.2794
0.08	1.0000	0.7195	0.82	0.8068	0.2735
0.10	1.0000	0.6897	0.84	0.7950	0.2678
0.12	1.0000	0.6632	0.86	0.7831	0.2622
0.14	1.0000	0.6394	0.88	0.7711	0.2567
0.16	1.0000	0.6176	0.90	0.7591	0.2513
0.18	1.0000	0.5976	0.92	0.7471	0.2461
0.20	1.0000	0.5789	0.94	0.7351	0.2410
0.22	0.9999	0.5615	0.96	0.7232	0.2360
0.24	0.9998	0.5451	0.98	0.7112	0.2312
0.26	0.9995	0.5296	1.00	0.6994	0.2264
0.28	0.9990	0.5149	1.05	0.6700	0.2150
0.30	0.9983	0.5010	1.10	0.6413	0.2042
0.32	0.9972	0.4877	1.15	0.6132	0.1940
0.34	0.9957	0.4750	1.20	0.5860	0.1844
0.36	0.9938	0.4629	1.25	0.5596	0.1752
0.38	0.9913	0.4513	1.30	0.5340	0.1665
0.40	0.9883	0.4401	1.35	0.5095	0.1583
0.42	0.9846	0.4294	1.40	0.4858	0.1505
0.44	0.9804	0.4190	1.45	0.4631	0.1431
0.46	0.9755	0.4090	1.50	0.4413	0.1360
0.48	0.9700	0.3994	1.55	0.4204	0.1293
0.50	0.9639	0.3901	1.60	0.4005	0.1230
0.52	0.9573	0.3810	1.65	0.3814	0.1170
0.54	0.9500	0.3723	1.70	0.3631	0.1112
0.56	0.9422	0.3639	1.75	0.3457	0.1058
0.58	0.9339	0.3557	1.80	0.3291	0.1006
0.60	0.9251	0.3477	1.85	0.3133	0.0957
0.62	0.9158	0.3400	1.90	0.2981	0.0910
0.64	0.9062	0.3325	1.95	0.2837	0.0866
0.66	0.8962	0.3252	2.00	0.2700	0.0823
0.68	0.8858	0.3181	2.10	0.2445	0.0745
0.70	0.8752	0.3113	2.20	0.2213	0.0674
0.72	0.8643	0.3045	2.30	0.2003	0.0610

[a] Center dimensionless temperature and mean temperature, respectively for sphere, respectively (symmetrical problem).

TABLE 4.5. (*continued*)

π^2Fo	θ_c	$\bar{\theta}$	π^2Fo	θ_c	$\bar{\theta}$
2.40	0.1813	0.0552	4.00	0.0366	0.0111
2.50	0.1641	0.0499	4.50	0.0222	0.0068
2.60	0.1485	0.0452	5.00	0.0135	0.0041
2.70	0.1344	0.0409	5.50	0.0082	0.0025
2.80	0.1216	0.0370	6.00	0.0050	0.0015
2.90	0.1100	0.0335	6.50	0.0030	0.0009
3.00	0.0996	0.0303	7.00	0.0018	0.0006
3.20	0.0815	0.0248	7.50	0.0011	0.0003
3.40	0.0667	0.0203	8.00	0.0007	0.0002
3.60	0.0546	0.0166	8.50	0.0004	0.0001
3.80	0.0447	0.0136			

Consequently, the mean temperature of a sphere $\bar{t}(\tau) = \bar{\theta} = 17°C$.

If the sphere radius and its thermal coefficients are known, then it is possible to determine the cooling time and heat losses at this particular time by elementary formulas.

For engineering convenience, Table 4.5 gives values of θ_c [$\theta_c = \theta(0, \tau)$ is the temperature in the middle of a sphere] and of $\bar{\theta}$ for various values of the Fourier number. A plot $\bar{\theta} = f(\text{Fo})$ is constructed in Fig. 4.17.

4.5 Infinite Cylinder

If a cylinder of a length l is considerably greater than its diameter $2R$ (i.e. $l/2R \gg 1$), then it may be treated as an infinite cylinder whose length is infinitely great as compared to its diameter.

If the heat transfer between the cylinder surface and the surroundings occurs uniformly over the whole surface, its temperature will depend only on time and a radius (symmetrical problem).

a. Statement of the Problem. *An infinite cylinder is considered with some prescribed radial temperature distribution, i.e., in the form of the function $f(r)$. At the initial time instant the cylinder surface is instantaneously cooled to some temperature t_a which is maintained constant during the entire cooling process. The temperature distribution and specific heat rate are desired as a function of time.*

In our case, the differential heat conduction equation is written as (see Appendix III):

$$\frac{\partial t(r, \tau)}{\partial \tau} = a\left(\frac{\partial^2 t(r, \tau)}{\partial r^2} + \frac{1}{r}\frac{\partial t(r, \tau)}{\partial r}\right) \quad (\tau > 0;\ 0 < r < R) \qquad (4.5.1)$$

The boundary conditions are the following (Fig. 4.18):

$$t(r, 0) = f(r), \tag{4.5.2}$$

$$t(R, \tau) = t_a = \text{const}, \tag{4.5.3}$$

$$\partial t(0, \tau)/\partial r = 0, \qquad t(0, \tau) \neq \infty. \tag{4.5.4}$$

The last condition menas that temperature along the cylinder axis during the entire heat transfer process must be finite.

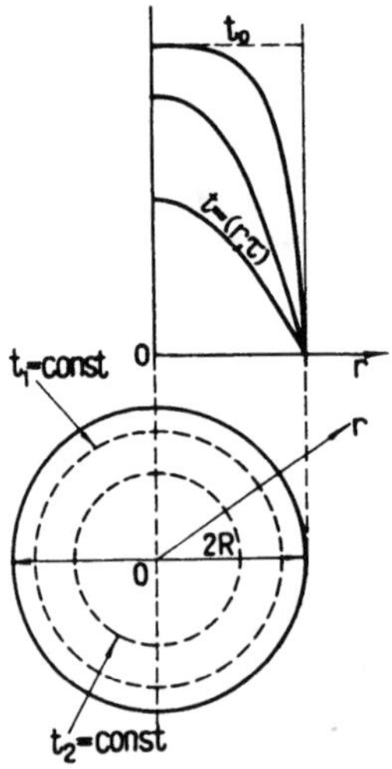

Fig. 4.18. Curves of temperature distribution inside an infinite cylinder (symmetrical problem).

b. Solution of the Problem by Separation of Variables. In Section 3.2, it is shown that a particular solution of the heat conduction equation by the separation method is as follows

$$t = \vartheta \exp[- ak^2\tau], \tag{4.5.5}$$

where ϑ is the solution of the differential equation

$$\nabla^2\vartheta + k^2\vartheta = 0.$$

In our case $\vartheta(r)$ should be a solution of the Bessel equation

$$\vartheta''(r) + (1/r)\vartheta'(r) + k^2\vartheta(r) = 0, \tag{4.5.6}$$

which may also be written

$$r\vartheta''(r) + \vartheta'(r) + k^2r\vartheta(r) = 0.$$

Let us find the solution of the Bessel equation

$$xu'' + u' + xu = 0 \tag{4.5.6a}$$

in the form of a power series

$$u = a_0 + a_1 x + a_2 x^2 + a_3 x^3 + \cdots . \tag{4.5.6b}$$

Differentiating each term of series (4.5.6b) and substituting u, u', u'' into the left-hand side of the expression (4.5.6a) we obtain

$$\begin{aligned} u &= a_0 + a_1 x + a_2 x^2 + a_3 x^3 + a_4 x^4 + \cdots , \\ u' &= a_1 + 2a_2 x + 3a_3 x^2 + 4a_4 x^3 + 5a_5 x^4 + \cdots , \\ u'' &= 2 \cdot 1 \cdot a_2 + 3 \cdot 2a_3 x + 4 \cdot 3a_4 x^2 + \cdots . \end{aligned}$$

Multiplying the first series by x, the second by 1, the third by x, and summing and gathering terms with the same powers of x, we have

$$a_1 + (a_0 + 2^2 a_2)x + (a_1 + 3^2 a_3)x^2 + (a_2 + 4^2 a_4)x^3 + \cdots . \tag{4.5.6c}$$

For expression (4.5.6c) to be equal to zero for all values of x, it is necessary that all of the coefficients of x be equal to zero, i.e.,

$$a_1 = 0; \qquad a_0 + 2^2 a_2 = 0; \qquad a_1 + 3^2 a_3 = 0; \qquad a_2 + 4^2 a_4 = 0, \ldots;$$
$$a_{n-2} + n^2 a_n = 0.$$

From these equalities it follows that all the coefficients with odd subscripts are zero (as $a_1 = 0$) and those with even ones are expressed through a_0 as

$$a_2 = -(1/2^2)a_0; \quad a_4 = -(1/4^2)a_2 = (1/2^2 4^2)a_0;$$
$$a_6 = -a_4/6^2 = -a_0/2^2 4^2 6^2, \cdots .$$

Substituting the above values of the coefficients into (4.5.6b) we obtain

$$u = a_0\left(1 - \frac{x^2}{2^2} + \frac{x^4}{2^2 4^2} - \frac{x^6}{2^2 4^2 6^2} + \cdots\right).$$

If $a_0 = 1$, then the particular integral of equation (4.5.6a) will be equal to the function:

$$J_0(x) = 1 - \frac{x^2}{2^2} + \frac{x^4}{2^2 4^2} - \frac{x^6}{2^2 4^2 6^2} \cdot \tag{4.5.6d}$$

This function is referred to as the Bessel function of the first kind and is of zero order.

The second particular solution of equation (4.5.6a) may be found by using the formula (see Eq. (3.2.14))

$$u_2 = u_1 \int u_1^{-2} \exp[- \int (1/x)\, dx]\, dx\,, \tag{4.5.6e}$$

where $u_1(x) = J_0(x)$ is the first particular solution, and $u_2(x)$ is the second linear particular solution independent of $u_1(x)$.

One finds that the second solution is

$$u_2(x) = J_0(x) \ln x + \frac{x^2}{2^2} - \frac{x^4}{2^2 4^2}\left(1 + \frac{1}{2}\right) + \frac{x^6}{2^2 4^2 6^2}\left(1 + \frac{1}{2} + \frac{1}{3}\right) - \cdots. \tag{4.5.6f}$$

Usually instead of the function $u_2(x)$, the $Y_0(x)$ is taken which is connected with $u_2(x)$ in such a way that

$$Y_0(x) = \frac{2}{\pi} u_2(x) + \frac{2}{\pi} J_0(x)\,(c - \ln 2),$$

where $c = 0.5772$ is the Euler constant. $Y_0(x)$ is the Bessel function of the second kind of zero order. Such a form of the function $Y_0(x)$ is taken to allow simple approximations at large values of x.

The particular solutions $u_1(x) = J_0(x)$ and $u_2(x)$ or $Y_0(x)$ are linearly independent since $Y_0(x)/J_0(x) \neq \text{const}$; the general integral of the Bessel equation (4.5.6a) is

$$u(x) = CJ_0(x) + DY_0(x)$$

where C and D are the arbitrary constants.

Equation (4.5.6) reduces to equation (4.5.6a) assuming $r = x/k$; the proof is left to the reader. Then the general integral of Eq. (4.5.6) will be

$$\vartheta(r) = CJ_0(kr) + DY_0(kr). \tag{4.5.7}$$

Since the temperature along the cylinder axis ($r = 0$) must be finite, then solution (4.5.7) cannot contain the Bessel function of the second kind which at $r \to 0$ tends to infinity (See Fig. 4.19). Hence, from the physical conditions of the problem, the constant D must be equal to zero ($D = 0$). Then, the particular solution of heat conduction equation (4.5.1) will be of the form

$$t = CJ_0(kr) \exp[- ak^2\tau]. \tag{4.5.8}$$

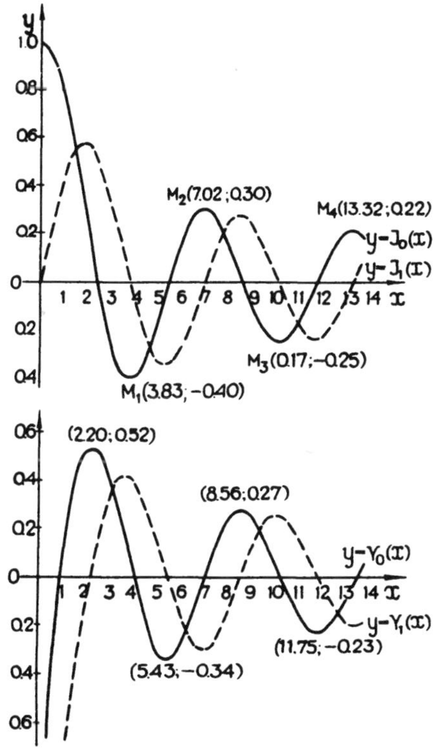

Fig. 4.19. Plot of the Bessel functions of the first and second kind.

The function $J_0(kr)$ is even, viz:

$$J_0(kr) = 1 - \frac{(kr)^2}{2^2} + \frac{(kr)^4}{2^2 4^2} - \frac{(kr)^6}{2^2 4^2 6^2} + \cdots.$$

It consequently satisfies condition (4.5.4) as

$$\begin{aligned} J_0'(kr) &= -k\left[\frac{kr}{2} - \frac{(kr)^3}{2^2 4} + \frac{(kr)^5}{2^2 4^2 6} - \cdots\right] \\ &= -kJ_1(kr), \end{aligned}$$

also at $r \to 0$, $J_0'(kr) \to 0$.

We find the constants k and C from the boundary and initial conditions. To simplify calculations we assume $t_a = 0$; it means that the datum of temperature is taken as t_a. Imposing boundary condition (4.5.3) on (4.5.8) we obtain

$$t_a = CJ_0(kR)\exp[-ak^2\tau] = 0.$$

Consequently, during the cooling process ($0 < \tau < \infty$) the following equality should be valid

$$J_0(kR) = 0. \tag{4.5.9}$$

This equality is referred to as a characteristic equation, from which eigenvalues k_n are defined.

The function $J_0(kR)$ is similar to a trigonometric function $\cos kR$ (Fig. 4.19); it has an infinite number of roots $k_nR = \mu_n$, namely, $\mu_1 = 2.4048$, $\mu_2 = 5.5201$, $\mu_3 = 8.6537$, etc. It should be noted that at large values of n the difference $\mu_{n+1} - \mu_n$ is close to π.

Thus from the characteristic equation it follows

$$k_nR = \mu_n, \qquad k_n = \mu_n/R. \tag{4.5.10}$$

Hence, there are an infinite number of particular solutions of the form

$$t = CJ_0(k_nr)\exp[-ak_n{}^2\tau]. \tag{4.5.11}$$

They will all be valid not only for differential equation (4.5.1) but also for boundary condition (4.5.3).

Such functions (4.5.11) are referred to as fundamental functions; a series consisting of them will be a general solution

$$t(r, \tau) = \sum_{n=1}^{\infty} C_nJ_0(k_nr)\exp[-ak_n{}^2\tau]. \tag{4.5.12}$$

For determination of the constants C_n the initial condition (4.5.2) is used, i.e.,

$$t(r, 0) = f(r) = \sum_{n=1}^{\infty} C_nJ_0(k_nr). \tag{4.5.13}$$

Equation (4.5.13) represents the Bessel transformation. For determination of the constants C_n the same method as described in the previous problem is used, but initially it should be proved that a system of functions $\sqrt{x}\,J_0(ax)$ $\sqrt{x}\,J_0(bx)$ is orthogonal.

Introducing the notations

$$y_1 = J_0(ax), \qquad y_2 = J_0(bx). \tag{4.5.14}$$

The functions $J_0(ax)$ and $J_0(bx)$ satisfy the appropriate differential equations and y_1 is therefore an integral of the equation

$$xy'' + y' + a^2xy = 0,$$

and y_2 is the integral of the equation

$$xy'' + y' + b^2xy = 0.$$

These equations may be written as

$$(xy')' = -a^2xy, \qquad (xy')' = -b^2xy.$$

Hence, we have

$$(xy_1')' = -a^2xy_1, \tag{4.5.15}$$

$$(xy_2')' = -b^2xy_2. \tag{4.5.16}$$

Multiplying equality (4.5.15) by y_2, equality (4.5.16) by y_1, and subtracting the second from the first, we obtain (also accounting for equality (4.5.14))

$$\begin{aligned} b^2xy_1y_2 - a^2xy_1xy_2 &= y_2(xy_1')' - y_1(xy_2')' \\ &= y_2'(xy_1')' - y_1(xy_2')' + y_2(xy_1')' - y_1'(xy_2')' \\ &= (y_2xy_1')' - (y_1xy_2')' = (y_2xy_1' - y_1xy_2')'. \end{aligned}$$

Rewriting this equality thus gives us

$$(b^2 - a^2)xy_1y_2 = (y_2xy_1' - y_1xy_2')'. \tag{4.5.17}$$

Upon integration of both sides of the equality from 0 to x we have:

$$(b^2 - a^2)\int_0^x xy_1y_2\,dx = xy_2y_1' - xy_1y_2'.$$

Passing to the former notation we obtain

$$\int_0^x xJ_0(ax)J_0(bx)\,dx = \frac{bxJ_0(ax)J_1(bx) - axJ_0(bx)J_1(ax)}{b^2 - a^2} \tag{4.5.18}$$

as

$$y_1' = aJ_0'(ax) = -aJ_1(ax),$$

$$y_2' = bJ_0'(bx) = -bJ_1(bx).$$

If $b = a$, then the right-hand side of (4.5.18) becomes indefinite of the type 0/0 which may be determined by means of the L'Hospital rule (differentiating the numerator and denominator of the fraction with respect to b and assuming a to be constant):

$$\int_0^x xJ_0^2(ax)\,dx = \lim_{b\to a} \frac{xJ_0(ax)J_1(bx) + bx^2J_0(ax)J_1'(bx) - ax^2J_0'(bx)J_1(ax)}{2b}$$
$$= \frac{1}{2a}\left\{xJ_0(ax)J_1(ax) + ax^2J_0(ax)\left[J_0(x) - \frac{J_1(ax)}{ax}\right]\right.$$
$$\left. + ax^2J_1^2(ax)\right\},$$

since

$$J_0'(ax) = -J_1(ax),$$
$$J_1'(ax) = J_0(ax) - (1/ax)J_1(ax).$$

Consequently, finally we have:

$$\int_0^x xJ_0^2(ax)\,dx = \tfrac{1}{2}x^2\,[J_0^2(ax) + J_1^2(ax)]. \qquad (4.5.19)$$

This formula is valid for all values of a and b and will be used later.

We multiply both sides of equality (4.5.13) by $rJ_0(k_mr)$ where k_mr are the roots of the function $J_0(k_mr)$ and integrate it within the limits from 0 to R:

$$\int_0^R rf(r)J_0(k_mr)\,dr = \int_0^R \sum_{n=1}^{\infty} C_nrJ_0(k_nr)J_0(k_mr)\,dr$$
$$= \sum_{n=1}^{\infty} C_n \int_0^R rJ_0(k_mr)J_0(k_nr)\,dr. \qquad (4.5.20)$$

According to equality (4.5.18), all the integrals of the right-hand side of the equality are equal to zero except when $m = n$. This follows from

$$\int_0^R rJ_0(k_n)J_0(k_mr)\,dr$$
$$= R\frac{k_mJ_0(k_nR)J_1(k_mR) - k_nJ_0(k_mR)J_1(k_nR)}{k_m^2 - k_n^2} = 0,$$

because $J_0(k_nR) = J_0(k_mR) = 0$. For $m = n$, according to formula (4.5.19), we have

$$\int_0^R rJ_0^2(k_nr)\,dr = \tfrac{1}{2}R^2\,J_1^2(k_nR).$$

Thus,

$$C_n = \frac{(2/R^2)\int_0^R rf(r)J_0(k_nr)\,dr}{J_1^2(k_nR)}. \qquad (4.5.21)$$

Finally, the solution of our problem will be of the form

$$t(r, \tau) = \sum_{n=1}^{\infty} \frac{J_0(\mu_n r/R)}{J_1^2(\mu_n)} \frac{2}{R^2} \times \int_0^R rf(r)J_0(\mu_n r/R)\, dr \exp[-\mu_n^2\{a\tau/R^2\}]. \qquad (4.5.22)$$

Consider the special case where $f(r) = t_0 = \text{const}$. Then it is necessary to calculate the integral

$$(2/R^2) \int_0^R t_0 r J_0(\mu_n r/R)\, dr.$$

Preliminarily, it will be shown that

$$\int xJ_0(ax)\, dx = (1/a)xJ_1(ax) + \text{const}. \qquad (4.5.23)$$

The function $y_1 = J_0(ax)$ is the integral of the equation

$$xy'' + y' + a^2xy = 0,$$

which may be written as

$$(xy')' + a^2xy = 0.$$

Consequently, we have:

$$(xy_1')' = -a^2xy_1,$$

$$y_1' = aJ_0'(ax) = -aJ_1(ax),$$

$$[-axJ_1(ax)]' = -a^2xJ_0(ax).$$

Upon integration of the last equality, there appears formula (4.5.23). Thus, we have

$$(2/R^2) \int_0^R t_0 r J_0(\mu_n r/R)\, dr = (2t_0/\mu_n)\, J_1(\mu_n). \qquad (4.5.24)$$

Then, the solution of our problem for $t_a \neq 0$ will be of the form:

$$\theta = \frac{t(r, \tau) - t_a}{t_0 - t_a} = \sum_{n=1}^{\infty} A_n J_0(\mu_n r/R) \exp[-\mu_n^2(a\tau/R^2)], \qquad (4.5.25)$$

where

$$A_n = 2/\mu_n J_1(\mu_n). \qquad (4.5.26)$$

Thus, the temperature distribution inside a cylinder depends on the Fourier number and the relative coordinate r/R.

$$\theta = \Psi\left(\mathrm{Fo}, \frac{r}{R}\right),$$

i.e., our problem is internal.

c. Solution by the Operational Method. Applying the Laplace transformation to a differential heat conduction equation, the ordinary differential Bessel equation for the transform $T(r, s)$ is obtained:

$$T''(r, s) + (1/r)T'(r, s) - (s/a)T(r, s) + (t_0/a) = 0.$$

Rewriting this equation as

$$rT''(r, s) + T'(r, s) - (s/a)r[T(r, s) - \{t_0/s\}] = 0. \qquad (4.5.27)$$

If Eq. (4.5.27) is compared with (4.5.6), one can see that the latter differs from equation (4.5.27) by a sign in front of the last term. In our case $k^2 = -s/a$, hence, $k = i(s/a)^{1/2}$.

The solution of the equation of form (4.5.27), referred to as the modified Bessel equation, consists of a sum of two particular solutions. The first solution is given by the modified Bessel function of the first kind, of zero order, or equivalently by the zero order Bessel function of the first kind with a purely imaginary argument

$$I_0(z) = J_0(iz) = 1 + \frac{z^2}{2^2} + \frac{z^4}{2^2 4^2} + \frac{z^6}{2^2 4^2 6^2} + \cdots. \qquad (4.5.28)$$

The function $I_0(z)$ in comparison with $J_0(z)$ plays the same role as a hyperbolic cosine $\cosh z$, as compared to the trigonometric $\cos z$.

The second particular solution is presented by the Bessel function of the second kind, of zero order, of a purely imaginary argument

$$K_0(z) = -[\log(\tfrac{1}{2}z) + C]I_0(z) + (\tfrac{1}{2}z)^2 + (1 + \tfrac{1}{2})\frac{(\tfrac{1}{2}z)^4}{(2!)^2} + (1 + \tfrac{1}{2} + \tfrac{1}{3})\frac{(\tfrac{1}{2}z)^6}{(3!)^2} + \cdots, \qquad (4.5.29)$$

where $C = 0.5772$ is the Euler constant.

Thus, the general solution of Eq. (4.5.27) may be written as

$$T(r, s) - \frac{t_0}{s} = AI_0\{(s/a)^{1/2}r\} + BK_0\{(s/a)^{1/2}r\}, \qquad (4.5.30)$$

where $I_0\{(s/a)^{1/2}r\} = J_0\{i(s/a)^{1/2}r\}$, A and B are the constants independent of r and are determined from the boundary conditions.

According to condition (4.5.4), the temperature along the cylinder axis ($r = 0$) cannot be infinity, and therefore the constant B is equal to zero because at $r \to 0$, $K_0(s/a)^{1/2}r) \to -\infty$. Hence, we have

$$T(r, s) - (t_0/s) = AI_0\{(s/a)^{1/2}r\}. \tag{4.5.31}$$

The function $I_0\{(s/a)^{1/2}r\}$ is even (see expansion (4.5.28)); it is valid for a symmetry condition.

Boundary condition (4.5.3) for the transform is written as

$$T(R, s) = t_a/s. \tag{4.5.32}$$

Imposing condition (4.5.32) on solution (4.5.31) we obtain

$$T(R, s) = (t_a/s) = t_0/s + AI_0\{(s/a)^{1/2}R\};$$

thence

$$A = -\frac{(t_0 - t_a)}{sI_0\{(s/a)^{1/2}R\}}. \tag{4.5.33}$$

Finally, the solution for the transform will be

$$(t_0/s) - T(r, s) = \frac{(t_0 - t_a)I_0\{(s/a)^{1/2}r\}}{sI_0\{(s/a)^{1/2}R\}} = \frac{\varphi(s)}{\psi(s)}. \tag{4.5.34}$$

Solution (4.5.34) represents a ratio of two power series with natural exponents with respect to s with the series in the denominator not containing a constant (the first term of the series is equal to s).

Thus, all the conditions of the expansion theorem are fulfilled and it may be applied to the inversion of transforms.

We determine the roots of $\psi(s)$ by equating the function to zero:

$$\psi(s) = sI_0\{(s/a)^{1/2})R\} = sJ_0\{i(s/a\}^{1/2}R) = 0. \tag{4.5.35}$$

Hence, (1) $s = 0$ (zero root), and (2) $i(s/a)^{1/2}R = \mu_1;\ \mu_2, \ldots, \mu_n$ are the roots of the Bessel function $J_0(\mu)$. Thus, there is an infinite number of roots for s; they are

$$s_n = -a\mu_n^2/R^2.$$

We now find $\psi'(s)$. We have

$$\begin{aligned}\psi'(s) &= I_0\left\{\left(\frac{s}{a}\right)^{1/2}R\right\} + \frac{sR}{2(as)^{1/2}}\, I_1\left\{\left(\frac{s}{a}\right)^{1/2}R\right\}\\ &= J_0\left(i\left(\frac{s}{a}\right)^{1/2}R\right) + \frac{1}{2i}\left(\frac{s}{a}\right)^{1/2}RJ_1\left(i\left(\frac{s}{a}\right)^{1/2}R\right),\end{aligned}$$

as

$$I_1(z) = dI_0(z)/dz; \qquad I_1(z) = (1/i)J_1(iz).$$

Consequently, we obtain:

$$L^{-1}\left[\frac{\varphi(s)}{\psi(s)}\right] = \frac{\varphi(0)}{\psi'(0)} + \sum_{n=1}^{\infty} \frac{\varphi(s_n)}{\psi'(s_n)} \exp[s_n\tau],$$

$$t_0 - t(x, \tau) = (t_0 - t_a) - \sum_{n=1}^{\infty} \frac{2(t_0 - t_a)}{\mu_n J_1(\mu_n)} J_0\left(\mu_n \frac{r}{R}\right) \exp\left[-\mu_n^2 \frac{2a\tau}{R^2}\right].$$

Finally we have

$$\theta = \frac{t(x, \tau) - t_a}{t_0 - t_a} = \sum_{n=1}^{\infty} A_n J_0(\mu_n r/R) \exp[-\mu_n^2 \,\mathrm{Fo}], \tag{4.5.36}$$

where

$$A_n = 2/\mu_n J_1(\mu_n).$$

Solution (4.5.36) is identical with (4.5.25).

Figure 4.20 gives the temperature distribution as a function of the dimensionless coordinate r/R for various values of Fo (0.005–0.8). From

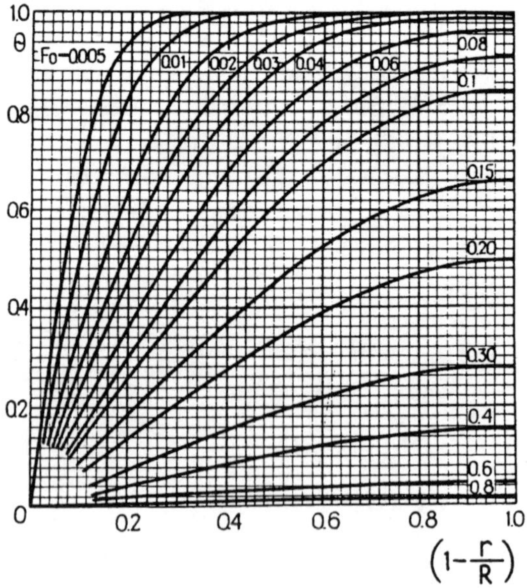

Fig. 4.20. Curves of the dimensionless excess temperature distribution inside an infinite cylinder. [$(1 - r)/R_0$) is the abscissa and θ is the ordinate.]

Fig. 4.20, one can see that a cooling process is essentially completed at $\mathrm{Fo} \geq 0.8$. These curves may serve as nomograms for determining θ for any specified values of Fo and r/R.

From the analysis of solution (4.5.36), it follows that the series rapidly converges since $\mu_1 < \mu_2 < \mu_3 < \cdots < \mu_n$. Also, with an increase in μ_n the initial amplitude A_n decreases, and the exponential function $\exp[-\mu_n^2\,\mathrm{Fo}]$ also sharply diminishes. Therefore, if small values of Fo are excluded from consideration, one term of series (4.5.36) is sufficient and the calculation formula (4.5.36) acquires a simple form.

For small time intervals the solution may be obtained in another form. Inspection of a solution (4.5.34) for the transform reveals that at small time values, the quantity $(s/a)^{1/2}R = qR$ $(q = (s/a)^{1/2})$ is large; then the following asymptotic approximation for the Bessel function may be used at large values of the argument.

$$I_0(z) \simeq \frac{1}{(2\pi z)^{1/2}}\, e^z \left(1 + \frac{1}{8z} + \frac{q}{128z^2} + \cdots\right). \tag{4.5.37}$$

Upon using this expansion for $I_0\{(s/a)^{1/2}R\}$ and $I_0\{(s/a)^{1/2}\,r\}$, the solution (4.5.34) for the transform may be written as

$$\begin{aligned}\frac{t_0}{s} - T(r, s) &\approx \frac{(t_0 - t_a)\sqrt{R}}{s\sqrt{r}} \exp\left[-\left(\frac{s}{a}\right)^{1/2}(R - r)\right] \\ &\quad \times \left\{\frac{1 + (\frac{1}{8}qr) + (9/128)(1/q^2r^2) + \cdots}{1 + (1/8qR) + (9/128q^2R)^2 + \cdots}\right\} \\ &= \frac{t_0 - t_a}{s}\left(\frac{R}{r}\right)^{1/2} \exp\left[-\left(\frac{s}{a}\right)^{1/2}(R - r)\right] \\ &\quad \times \left\{1 + \frac{R - r}{8qRr} + \frac{9R^2 - 7r^2 - 2Rr}{128q^2R^2r^2} + \cdots\right\}.\end{aligned}$$

According to the table of transforms, this solution will be of the form

$$\begin{aligned}\frac{t_0 - t(r, \tau)}{t_0 - t_a} &= \left(\frac{R}{r}\right)^{1/2} \operatorname{erfc} \frac{R - r}{2(a\tau)^{1/2}} + \frac{(R - r)(a\tau R)^{1/2}}{4Rr^{3/2}}\, \mathrm{i\,erfc}\, \frac{R - r}{2(a\tau)^{1/2}} \\ &\quad + \frac{(9R^2 - 7r^2 - 2Rr)a\tau}{32R^{3/2}\, r^{5/2}}\, \mathrm{i^2\,erfc}\, \frac{R - r}{2(a\tau)^{1/2}} + \cdots. \end{aligned} \tag{4.5.38}$$

We write this equation in dimensionless form as

$$\theta = \frac{t(r,\tau) - t_a}{t_0 - t_a} = 1 - \left(\frac{R}{r}\right)^{1/2} \operatorname{erfc} \frac{1 - r/R}{2\sqrt{\mathrm{Fo}}} + \frac{1}{4}\left(\frac{R}{r} - 1\right)$$

$$\times \left(\frac{R}{r}\right)^{1/2} \mathrm{Fo}\ \mathrm{i}\operatorname{erfc} \frac{1 - (r/R)}{2\sqrt{\mathrm{Fo}}} + \frac{1}{32}\left(\frac{R}{r}\right)^{1/2}\left[9\left(\frac{R}{r}\right)^2 - 7\right.$$

$$\left. - 2\left(\frac{R}{r}\right)\right] \mathrm{Fo}\ \mathrm{i}^2 \operatorname{erfc} \frac{1 - (r/R)}{2\sqrt{\mathrm{Fo}}} + \cdots. \qquad (4.5.39)$$

This solution is valid for small values of Fo and for $r > 0$. At the small values of Fo when the argument of the function i erfc and i^2 erfc is large and the functions themselves are close to zero (see Appendix), solution (4.5.39) becomes similar to that for a semi-infinite body. Consequently, cooling of a cylinder at the initial values of time occurs similarly to that of a semi-infinite body.

The solution along the axis ($r = 0$) may be obtained from that for the transform assuming $I_0\{(s/a)^{1/2} r\} = 1$. Then, we shall have

$$(t_0/s) - T(r, s) = (t_0 - t_a)(2\pi R)^{1/2} \frac{1}{a^{1/4} s^{3/4}} \exp\left[-\left(\frac{s}{a}\right)^{1/2} R\right]$$

$$\times \left\{1 + \frac{1}{8qR} + \frac{9}{128q^2R^2} + \cdots\right\}^{-1}$$

$$\approx \frac{t_0 - t_a}{a^{1/4}s^{3/4}} (2\pi R)^{1/2} \exp\left[-\left(\frac{s}{a}\right)^{1/2} R\right].$$

Using the table of transforms we obtain

$$\theta \approx 1 - \frac{1}{(\pi \mathrm{Fo})^{1/2}} \exp\left[-\frac{1}{8\mathrm{Fo}}\right] K_{1/4}\left(\frac{1}{8\mathrm{Fo}}\right). \qquad (4.5.40)$$

Table 4.6 gives numerical values of the relative excess temperature along the cylinder axis for various values of the Fourier number. The necessary calculations were made according to formulas (4.5.39) and (4.5.36). The relation $(t_0 - t)/(t_0 - t_a) = f(\mathrm{Fo})$ is plotted in Fig. 4.21 for different values of the dimensionless coordinates r/R. These may be used for design purposes.

d. Determination of Heat Losses. First, we find the mean cylinder temperature by means of the formula

$$\bar{t}(\tau) = (2/R^2) \int_0^R rt(r, \tau)\, dr.$$

TABLE 4.6. Values of $\theta_c = \sum_{n=1}^{\infty} \frac{2}{\mu^n} \left[\frac{1}{J_1(\mu_n)} \right] \exp(-\mu_n^2 \text{Fo})$[a]

Fo	θ_c	Fo	θ_c	Fo	θ_c
0.005	1.0000	0.205	0.4875	0.41	0.1496
0.010	1.0000	0.210	0.4738	0.42	0.1412
0.015	1.0000	0.215	0.4605	0.43	0.1332
0.020	1.0000	0.220	0.4475	0.44	0.1258
0.025	0.9999	0.225	0.4349	0.45	0.1220
0.030	0.9995	0.230	0.4227	0.46	0.1187
0.035	0.9985	0.235	0.4107	0.47	0.1057
0.040	0.9963	0.240	0.3991	0.48	0.0998
0.045	0.9926	0.245	0.3878	0.49	0.0942
0.050	0.9871	0.250	0.3768	0.50	0.0887
0.055	0.9798	0.255	0.3662	0.52	0.0792
0.060	0.9705	0.260	0.3558	0.54	0.0704
0.065	0.9596	0.265	0.3457	0.56	0.0628
0.070	0.9470	0.270	0.3359	0.58	0.0560
0.075	0.9330	0.275	0.3263	0.60	0.0499
0.080	0.9177	0.280	0.3170	0.62	0.0444
0.085	0.9015	0.285	0.3080	0.64	0.0396
0.090	0.8844	0.290	0.2993	0.66	0.0352
0.095	0.8666	0.295	0.2908	0.68	0.0314
0.100	0.8484	0.300	0.2825	0.70	0.0280
0.105	0.8297	0.305	0.2744	0.72	0.0249
0.110	0.8109	0.310	0.2666	0.74	0.0222
0.115	0.7919	0.315	0.2590	0.76	0.0198
0.120	0.7729	0.320	0.2517	0.78	0.0176
0.125	0.7540	0.325	0.2445	0.80	0.0157
0.130	0.7351	0.330	0.2375	0.85	0.0117
0.135	0.7164	0.335	0.2308	0.90	0.0088
0.140	0.6980	0.340	0.2242	0.95	0.0066
0.145	0.6798	0.345	0.2178	1.00	0.0049
0.150	0.6618	0.350	0.2116	1.05	0.0037
0.155	0.6442	0.355	0.2056	1.10	0.0028
0.160	0.6269	0.360	0.1997	1.15	0.0021
0.165	0.6100	0.365	0.1940	1.20	0.0016
0.170	0.5934	0.370	0.1885	1.25	0.0012
0.175	0.5771	0.375	0.1831	1.30	0.0009
0.180	0.5613	0.380	0.1779	1.35	0.0007
0.185	0.5458	0.385	0.1728	1.40	0.0005
0.190	0.5306	0.390	0.1679	1.50	0.0003
0.195	0.5159	0.395	0.1631	1.60	0.0002
0.200	0.5015	0.400	0.1585	1.70	0.0001

[a] Center line temperature of infinite cylinder (symmetrical problem).

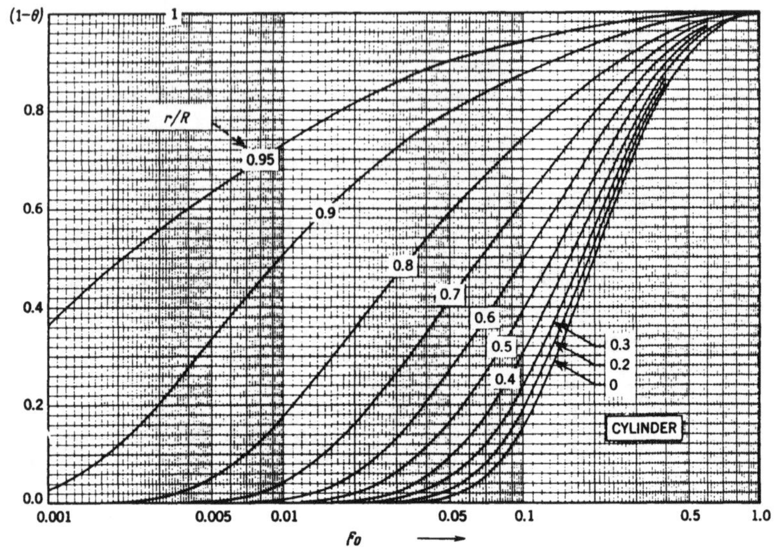

Fig. 4.21. Plot of dimensionless excess temperature $(1 - \theta)$ versus the Fourier number for various values of the coordinate for a cylinder [102].

Substituting the appropriate expression from Eq. (4.5.36) for $t(r, \tau)$ and taking into account equality (4.5.24), we obtain upon transformation

$$\bar{\theta} = \frac{\bar{t}(\tau) - t_a}{t_0 - t_a} = \sum_{n=1}^{\infty} B_n \exp[-\mu_n^2 \mathrm{Fo}], \tag{4.5.41}$$

where $B_n = 4/\mu_n^2$, i.e., a solution similar to that for the plate and sphere is obtained. The difference is that for a plate $\mu_n = (2n - 1)(\pi/2)$, $B_n = 2/\mu_n^2$ and for a sphere, $\mu_n = n\pi$, $B_n = 6/\mu_n^2$, respectively. Thus, in the cases considered a decrease in the mean temperature is described by a simple exponential law.

The relation between $\bar{\theta}$ and the Fourier number is depicted in Fig. 4.22 which may serve as a calculation plot.

Let us now solve a particular problem. A glass long bar heated to 120°C is placed into melting ice, 0°C. Determine the temperature inside the bar after a minute of cooling as well as the amount of heat lost if the bar diameter is 2 cm.

Consider the bar to be an infinite cylinder. The thermal coefficients of glass are the following $\lambda = 0.64$ kcal/m°Chr, $c = 0.16$ kcal/kg°C and $\gamma = 2500$ kg/m³; then $a = 1.6$ 10^{-3} m²/hr.

We find the Fourier number to be

$$\mathrm{Fo} = \frac{1 \cdot 6 \cdot 10^{-3} \cdot 1}{1 \cdot 10^{-4} \cdot 60} = 0.266 \approx 0.27,$$

as $\tau = 1/60$ hr, $R = 1$ cm $= 0.01$ m.

We now determine the dimensionless temperature by formula (4.5.36) for $r = 0$. Since Fo = 0.27, one may restrict oneself to the first term of the series (4.5.36), i.e.,

$$\theta_c = A_1 \exp[-\mu_1^2 \text{Fo}] \frac{2}{\mu_1 J_1(\mu_1)} = \frac{2}{(2.4) J_1(2.4)} \exp[-(2.4)^2 \cdot 0.27]$$
$$= 1.6 \cdot 0.211 = 0.34.$$

Here, it is taken into account that $J_1(2.4) = 0.52$; $\exp[-(2.4)^2 \cdot 0.27] = 0.211$.

The quantity A_1 might be determined directly from Table 6.10 for $\text{Bi} = \infty$ (see Chapter 6).

The calculated value $\theta_c = 0.34$ may be checked from Table 4.6, from which one can see that for Fo = 0.27, $\theta \approx 0.34$. The temperature along the cylinder axis at $t_a = 0$ will be

$$t(0, \tau) = t_0 \theta_c = 0.34 \cdot 120 = 41°.$$

For calculation of the heat rate let us determine the mean temperature by formula (4.5.41) as

$$\bar{\theta} = B_1 \exp[-\mu_1^2 \text{Fo}] = \frac{4}{(2 \cdot 4)^2} \exp[-(2.4)^2 \cdot 0.27] = 0.147 \approx 0.15.$$

Let us check this from the plot in Fig. 4.22. Hence at Fo = 0.27 $\bar{\theta} = 0.15$, i.e., the calculation is correct.

The mean temperature becomes

$$\bar{t}(\tau) = \bar{\theta}(t_0) = 0.15 \cdot 120 = 18°.$$

The specific heat rate is

$$\Delta Q_v = c\gamma(t_0 - \bar{t}) = 0.16 \cdot 2500(120 - 18) = 40{,}800 \text{ (kcal/m}^3\text{)}.$$

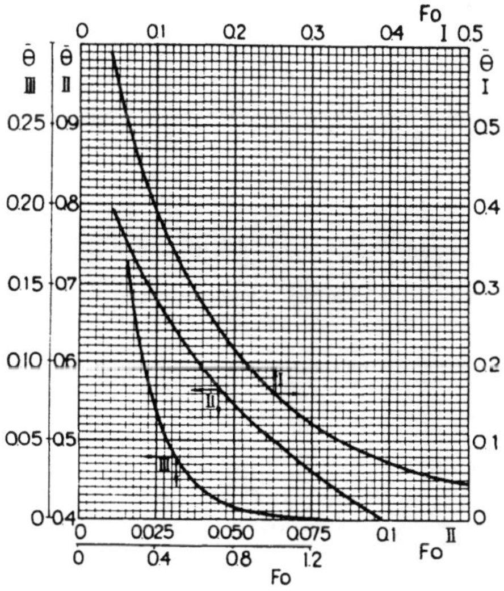

Fig. 4.22. Relation between the dimensionless excess mean temperature and the Fourier number for an infinite cylinder.

4.6 Infinite Hollow Cylinder

a. Statement of the Problem. *Consider an infinite hollow cylinder (cylindrical tube). At the initial time instant, the temperature of the tube is a given function of the radius f(r); the temperature of the external and internal surfaces is maintained constant during the whole cooling process. The temperature distribution at any time is desired.*

The boundary conditions are (see Fig. 4.23)

$$t(r, 0) = f(r), \tag{4.6.1}$$

$$t(R, \tau) = t_2 = \text{const}, \tag{4.6.2}$$

$$t(R_0, \tau) = t_1 = \text{const}, \tag{4.6.3}$$

where R is the radius of the external surface and R_0 is the radius of the internal surface.

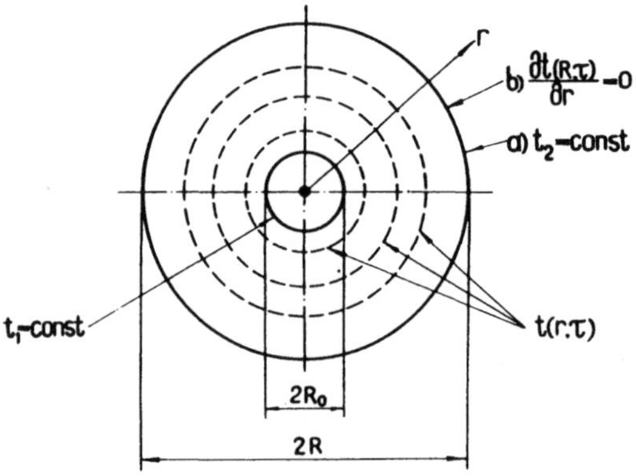

Fig. 4.23. Infinite hollow cylinder.

b. Solution of the Problem by Method of Separation of Variables. First of all, to simplify the problem assume $t_1 = t_2 = 0$. In Section 4.5 it has been shown that the solution of the Bessel equation assumes the form

$$\vartheta(r) = CJ_0(kr) + DY_0(kr).$$

However, here $D = 0$ cannot be assumed as the function $Y_0(kr)$ varies over a range of $R_0 < r < R$ and there is no special difficulty with either function in that they are both finite over the range of interest.

Consequently, the solution of the heat conduction equation will take the form

$$t(r, \tau) = [CJ_0(kr) + DY_0(kr)] \exp[-ak^2\tau]. \tag{4.6.4}$$

Boundary conditions (4.6.2) and (4.6.3) are used to determine constants C and D. Assuming $t_1 = t_2 = 0$,

$$CJ_0(kR_0) + DY_0(kR_0) = 0, \tag{4.6.5}$$

$$CJ_0(kR) + DY_0(kR) = 0. \tag{4.6.6}$$

The coefficients C and D obviously cannot be simultaneously equal to zero as in this case the temperature is zero (a trivial solution). A system of two linear homogeneous equations (4.6.5) and (4.6.6) with respect to the unknowns C and D will have a solution different from zero in the case when the determinant of the system equals zero

$$\begin{vmatrix} J_0(kR_0), & Y_0(kR_0) \\ J_0(kR), & Y_0(kR) \end{vmatrix} = 0 \tag{4.6.7}$$

or

$$J_0(kR)Y_0(kR_0) - J_0(kR_0)Y_0(kR) = 0.$$

Hence, the proper values of the problem or eigenvalues k_n are obtained from equation (4.6.7).

Characteristic equation (4.6.7) has no complex roots but an infinite number of positive real roots k_n. Consequently, a general solution will acquire the values

$$t(r, \tau) = \sum_{n=1}^{\infty} [C_n J_0(k_n r) + D_n Y_0(k_n r)] \exp[-ak_n{}^2\tau]. \tag{4.6.8}$$

From Eq. (4.6.5) it follows

$$\frac{D}{C} = -\frac{J_0(kR_0)}{Y_0(kR_0)}.$$

Then, solution (4.6.8) may be written

$$t(r, \tau) = \sum_{n=1}^{\infty} \frac{C_n}{Y_0(k_n R_0)} \exp[-ak_n{}^2\tau] [J_0(k_n r)Y_0(k_n R_0) - J_0(k_n R_0)Y_0(k_n r)]. \tag{4.6.9}$$

We introduce the symbols

$$A_n = C_n/Y_0(k_n R_0),$$

$$V_0(k_n r) = J_0(k_n r)Y_0(k_n R_0) - J_0(k_n R_0)Y_0(k_n r). \tag{4.6.10}$$

Solution (4.6.9) then takes the form

$$t(r, \tau) = \sum_{n=1}^{\infty} A_n \exp[-ak_n^2\tau]V_0(k_nr). \tag{4.6.11}$$

The initial condition ($\tau = 0$) is used for determining the constant A_n:

$$f(r) = \sum_{n=1}^{\infty} A_n V_0(k_nr). \tag{4.6.12}$$

If the function $f(r)$ can be developed in series whose terms are combinations of Bessel's functions (4.6.10), then it is assumed that Eq. (4.6.12) may be integrated term-by-term.

Multiplying the left- and right-hand sides of Eq. (4.6.12) by $rV_0(k_mr)$ and integrating from R_0 to R, we obtain

$$\int_{R_0}^{R} rf(r)V_0(k_mr)\,dr = \sum_{n=1}^{\infty} A_n \int_{R_0}^{R} rV_0(k_nr)V_0(k_mr)\,dr. \tag{4.6.13}$$

We will now prove that all terms of the series (4.6.13) for all $n \neq m$ are equal to zero.

Using formula (4.5.17a) we have

$$(b^2 - a^2)\int_0^x xy_1y_2\,dx = xy_2y_1' - xy_1y_2'.$$

We assume

$$y_1 = AJ_0(ax) + BY_0(ax) = V_0(ax),$$

$$y_2 = AJ_0(bx) + BY_0(bx) = V_0(bx).$$

A formula similar to (4.5.18) is obtained

$$\int xV_0(ax)V_0(bx)\,dx$$
$$= -\frac{bxV_0(ax)V_1(bx) - axV_0(bx)V_1(ax)}{b^2 - a^2} + \text{const.} \tag{4.6.14}$$

If we assume $a = k_n$, $b = k_m$, $x \equiv r$, then

$$\int_{R_0}^{R} rV_0(k_nr)V_0(k_mr)\,dr$$
$$= \frac{[k_mrV_0(k_nr)V_1(k_mr) - k_nrV_0(k_mr)V_1(k_nr)]_{R^0}^{R}}{k_m^2 - k_n^2}$$
$$= \frac{1}{k_m^2 - k_n^2}[k_mRV_0(k_nR)V_1(k_mR)$$
$$- k_nRV_0(k_mR)V_1(k_nR)$$
$$- k_mR_0V_0(k_nR_0)V_1(k_mR_0)$$
$$+ k_nR_0V_0(k_mR_0)V_1(k_nR_0)], \tag{4.6.15}$$

where $V_0(k_n r)$ is defined by formula (4.6.10) and $V_1(k_n r)$ by

$$V_1(k_n r) = J_1(k_n r) Y_0(k_n R_0) - J_0(k_n R_0) Y_1(k_n r). \tag{4.6.16}$$

According to characteristic equation (4.6.7) the quantities $V_0(k_n R)$ and $V_0(k_m R)$ are zero. The first two terms of the right-hand side of Eq. (4.6.15) are therefore equal to zero.

The third and the fourth terms of Eq. (4.6.15) are also zero as $V_0(k_n R_0)$ and $V_0(k_m R_0)$ are zero according to formula (4.6.7)

$$V_0(kR_0) = J_0(kR_0) Y_0(kR_0) - J_0(kR_0) Y_0(kR_0) = 0. \tag{4.6.16a}$$

Hence, when $n \neq m$ the integral of Eq. (4.6.15) [and of (4.6.13)] equals zero; it has a finite value at $n = m$, where

$$\int_{R_0}^{R} rf(r) V_0(k_n r)\, dr = A_n \int_{R_0}^{R} r V_0^2(k_n r)\, dr. \tag{4.6.17}$$

Similarly, as in Section 4.5, the following formula may be obtained from formula (4.6.14):

$$\begin{aligned} \int_{R_0}^{R} r V_0^2(k_n r)\, dr &= \{\tfrac{1}{2} r^2 [V_0^2(k_n r) + V_1^2(k_n r)]\}_{R_0}^{R} \\ &= \tfrac{1}{2} R^2 [V_0^2(k_n R) + V_1^2(k_n R)] \\ &\quad - \tfrac{1}{2} R_0^2 [V_0^2(k_n R_0) + V_1^2(k_n R_0)]. \end{aligned} \tag{4.6.18}$$

As $V_0(k_n R) = V_0(k_n R_0) = 0$, formula (4.6.18) yields:

$$\int_{R_0}^{R} r V_0^2(k_n r)\, dr = \tfrac{1}{2} [R^2 V_1^2(k_n R) - R_0^2 V_1^2(k_n R_0)]. \tag{4.6.19}$$

Formula (4.6.19) may be transformed to a somewhat simpler form. From the theory of Bessel functions, it is known

$$J_1(x) Y_0(x) - J_0(x) Y_1(x) = 2/\pi x. \tag{4.6.20}$$

Then

$$V_1(k_n R_0) = J_1(k_n R_0) Y_0(k_n R_0) - J_0(k_n R_0) Y_1(k_n R_0) = 2/\pi k_n R_0. \tag{4.6.21}$$

Characteristic equation (4.6.7) is used for determining $V_1(k_n R)$.

$$Y_0(kR_0) = \frac{J_0(kR_0) Y_0(kR)}{J_0(kR)},$$

$$V_1(k_nR) = J_1(k_nR)Y_0(k_nR_0) - J_0(k_nR_0)Y_1(k_nR)$$
$$= \frac{J_1(k_nR)J_0(k_nR_0)Y_0(k_nR)}{J_0(k_nR)} - J_0(k_nR_0)Y_1(k_nR)$$
$$= \frac{J_0(k_nR_0)}{J_0(k_nR)}[J_1(k_nR)Y_0(k_nR) - J_0(k_nR)Y_1(k_nR)].$$

Using Eq. (4.6.20) we have

$$V_1(k_nR) = \frac{J_0(k_nR_0)}{J_0(k_nR)} \frac{2}{\pi(k_nR)}. \tag{4.6.22}$$

Substituting Eqs. (4.6.21) and (4.6.22) into (4.6.19) we obtain

$$\int_{R_0}^{R} rV_0^2(k_nr)\,dr = \frac{2[J_0^2(k_nR_0) - J_0^2(k_nR)]}{\pi^2k_n^2J_0^2(k_nR)}. \tag{4.6.23}$$

Then, the constants A_n will be:

$$A_n = \frac{\pi^2k_n^2J_0^2(k_nR)}{2[J_0^2(k_nR_0) - J_0^2(k_nR)]} \int_{R_0}^{R} rf(r)V_0(k_nr)\,dr. \tag{4.6.24}$$

Thus, the solution of the simplified problem acquires the values

$$t(r, \tau) = \frac{\pi^2}{2} \sum_{n=1}^{\infty} \left\{\left[\frac{k_n^2J_0^2(k_nR)V_0(k_nr)}{J_0^2(k_nR_0) - J_0^2(k_nR)} \times \int_{R_0}^{R} rf(r)V_0(k_nr)\,dr\right] \exp[-ak_n^2\tau]\right\}. \tag{4.6.25}$$

Returning to our original problem where t_1 and t_2 are assumed not to be equal to zero, we seek a solution of our problem in such a form as

$$t(r, \tau) = \theta(r) + \vartheta(r, \tau). \tag{4.6.26}$$

The function $\theta(r)$ must satisfy the differential equation

$$\frac{d^2\theta}{dr^2} + \frac{1}{r}\frac{d\theta}{dr} = 0 \tag{4.6.27}$$

and the boundary conditions

$$\theta(R_0) = t_1, \qquad \theta(R) = t_2. \tag{4.6.28}$$

The function $\vartheta(r, \tau)$ should satisfy differential equation (4.5.1) and the boundary conditions

$$\vartheta(R_0, \tau) = 0, \qquad \vartheta(R, \tau) = 0. \tag{4.6.29}$$

as well as the initial condition

$$\vartheta(0, r) = f(r) - \theta(r). \tag{4.6.30}$$

It is obvious that the function $t(r, \tau)$ will satisfy the differential equation and boundary conditions (4.6.2) and (4.6.3):

$$\begin{aligned} t(R_0, \tau) &= \theta(R_0) + \vartheta(R_0, \tau) = t_1 + 0 = t_1, \\ t(R, \tau) &= \theta(R) + \vartheta(R, \tau) = t_2 + 0 = t_2, \\ t(r, 0) &= \theta(r) + \vartheta(r, 0) \\ &= \theta(r) + f(r) - \theta(r) = f(r). \end{aligned}$$

A new variable $d\theta/dr = z$ is introduced to solve Eq. (4.6.27). Then Eq. (4.6.27) assumes the form

$$\frac{dz}{dr} + \frac{1}{r} z = 0. \tag{4.6.31}$$

The solution of Eq. (4.6.31) has the form

$$z = B/r = d\theta/dr, \tag{4.6.32}$$

and upon integration we have

$$\theta(r) = B \ln r + C. \tag{4.6.33}$$

The constants B and C are determined from boundary conditions (4.6.28) as

$$B = \frac{t_2 - t_1}{\ln(R/R_0)}, \qquad C = \frac{t_1 \ln R - t_2 \ln R_0}{\ln(R/R_0)}. \tag{4.6.34}$$

Upon substitution of Eq. (4.6.34) into (4.6.33) we obtain

$$\theta(r) = \frac{t_1 \ln(R/r) + t_2 \ln(r/R_0)}{\ln(R/R_0)}. \tag{4.6.35}$$

The function $\theta(r)$ represents a temperature distribution in a hollow cylinder in a stationary state. The solution $\vartheta(r, \tau)$ may be obtained from Eq.

(4.6.25) substituting $[f(r) - \theta(r)]$ for $f(r)$:

$$\vartheta(r, \tau) = \frac{\pi^2}{2} \sum_{n=1}^{\infty} \frac{k_n^2 J_0(k_n R) V_0(k_n r)}{J_0^2(k_n R_0) - J_0^2(k_n R)} \exp[-ak_n^2\tau] \times \left\{ \int_{R_0}^{R} rf(r) V_0(k_n r)\, dr - \int_{R_0}^{R} r\theta(r) V_0(k_n r)\, dr \right\}. \quad (4.6.36)$$

The second integral in Eq. (4.6.36) may be calculated if $\theta(r)$ is replaced by equation (4.6.35).

We integrate it term-by-term assuming

$$u = t_1 \ln(R/r) + t_2 \ln(r/R_0),$$

$$\vartheta = (R/k_n) V_0(k_n r); \qquad du = (t_2 - t_1)(dr/r); \qquad d(\vartheta) = rV_0(k_n r)\, dr:$$

$$\int_{R_0}^{R} r\theta(r) V_0(k_n r)\, dr = \{1/\ln(R/R_0)\} \{t_1 \ln(R/r) + t_2 \ln(r/R_0)] \times (r/k_n) V_1(k_n r)\Big|_{R_0}^{R} - \frac{t_2 - t_1}{k_n} \int_{R_0}^{R} V_1(k_n r)\, dr\}$$

$$= \frac{1}{\ln(R/R_0)} \left\{ t_2 \frac{R}{k_n} V_1(k_n R) \ln \frac{R}{R_0} - t_1 \frac{R_0}{k_n} V_1(k_n R_0) \times \ln \frac{R}{R_0} + \frac{t_2 - t_1}{k_n^2} \Big[V_0(k_n r) \Big]_{R_0}^{R} \right\}.$$

Here the formula $dV_0(kr)/dr = -kV_1(kr)$ was used. Using formulas for $V_0(k_n R)$ and $V_1(k_n R)$ we have:

$$\int_{R_0}^{R} r\vartheta(r) V_0(k_n r)\, dr = t_2 \frac{2}{\pi} \frac{R}{k_n^2 R} \frac{J_0(k_n R_0)}{J_0(k_n R)} - \frac{2t_1 R_0}{\pi k_n^2 R_0}$$

$$= \frac{2[t_2 J_0(k_n R_0) - t_1 J_0(k_n R)]}{\pi k_n^2 J_0(k_n R)}. \quad (4.6.37)$$

Finally, the following solution is obtained

$$t(r, \tau) = \frac{1}{\ln(R/R_0)} \left[t_1 \ln \frac{R}{r} + t_2 \ln \frac{r}{R_0} \right] + \sum_{n=1}^{\infty} \frac{V_0(k_n r) \exp[-ak_n^2\tau]}{J_0^2(k_n R_0) - J_0^2(k_n R)} \times \left\{ \frac{\pi^2}{2} k_n^2 J_0^2(k_n R) \int_{R_0}^{R} rf(r) V_0(k_n r)\, dr - \pi J_0(k_n R) \times [t_2 J_0(k_n R_0) - t_1 J_0(k_n R)] \right\}. \quad (4.6.38)$$

Introducing the symbols $\mu_n = k_n R_0$; $R/R_0 = m$,

$$\text{Fo} = a\tau/R_0^2.$$

Then, solution (4.6.38) may be written as

$$t(r, \tau) = \frac{1}{\ln m}\left[t_1 \ln \frac{R}{r} + t_2 \ln \frac{r}{R_0}\right] + \sum_{n=1}^{\infty} \frac{V_0(\mu_n r/R_0) \exp[-\mu_n^2 \mathrm{Fo}]}{J_0^2(\mu_n) - J_0^2(\mu_n m)}$$
$$\times \{(\pi/2R_0^2)\mu_n^2 J_0(\mu_n m) \int_{R_0}^{R} rf(r) V_0\left(\mu_n \frac{r}{R_0}\right) dr$$
$$- \pi J_0(\mu_n m)[t_2 J_0(\mu_n) - t_1 J_0(m\mu_n)]\}. \quad (4.6.39)$$

The roots μ_n are determined from the characteristic equation

$$J_0(\mu)\, Y_0(m\mu) - J_0(m\mu)\, Y_0(\mu) = 0. \quad (4.6.40)$$

The first five values of μ are presented in Table 4.7 for value of m from 1.2 to 4.0.

TABLE 4.7. Values of the Roots μ_n of the Characteristic Equation

$$J_0(\mu)Y_0(m\mu) - Y_0(\mu)J_0(m\mu) = 0$$

m	μ_1	μ_2	μ_3	μ_4	μ_5
1.2	15.7014	31.4126	47.1217	62.8304	78.5385
1.5	6.2702	12.5598	18.8451	25.1294	31.4133
2.0	3.1230	6.2734	9.4182	12.5614	15.7040
2.5	2.0732	4.1773	6.2754	8.3717	10.4672
3.0	1.5485	3.1291	4.7038	6.2767	7.8487
3.5	1.2339	2.5002	3.7608	5.0196	6.2776
4.0	1.0244	2.0809	3.1322	4.1816	5.2301

Let us now consider a particular case. The initial temperature of all the points of a tube is assumed to be constant, i.e.,

$$t(r, 0) = f(r) = t_0 = \text{const}. \quad (4.6.41)$$

Then, using formulas (4.6.21) and (4.6.22) the integral is

$$\int_{R_0}^{R} rf(r) V_0(k_n r)\, dr = \frac{t_0 r}{k_n} V_1(k_n r) \Big|_{R_0}^{R}$$
$$= \frac{2t_0}{\pi k_n^2 J_0(k_n R)} [J_0(k_n R_0) - J_0(k_n R)]. \quad (4.6.42)$$

Solution (4.6.38) acquires the values

$$t(r, \tau) = \frac{1}{\ln m}\left(t_1 \ln \frac{R}{r} + t_2 \ln \frac{r}{R_0}\right) + \pi \sum_{n=1}^{\infty} \left\{ \frac{J_0(\mu_n m) V_0(\mu_n r/R)}{J_0(\mu_n) + J_0(m\mu_n)} \right.$$

$$\left. \times \exp[-\mu_n^2 \mathrm{Fo}] \left[t_0 - \frac{t_2 J_0(\mu_n) - t_1 J_0(m\mu_n)}{J_0(\mu_n) - J_0(m\mu_n)} \right] \right\}. \qquad (4.6.43)$$

Figure 4.24 furnishes calculation diagrams for the case when the temperature of the internal cylinder surface is constant and equal to the initial temperature $t_1 = t_0 = \text{const}$, and the temperature at the external cylinder surface at the initial instant assumes the value of t_a ($t_2 = t_a$) and is maintained constant during the whole process of heating ($t_a > t_0$). The diagrams are plotted for various values of the dimensionless coordinate r/R [or equivalently, $\frac{1}{2}(1 + \{R_0/R\})$].

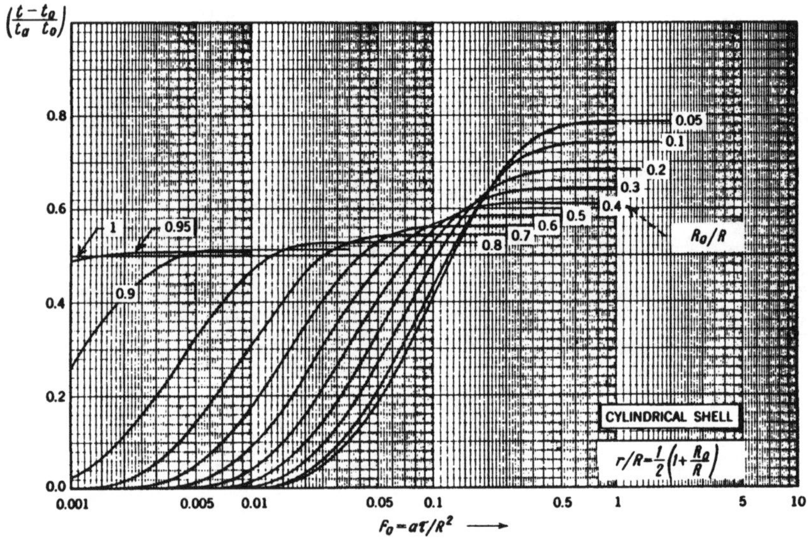

Fig. 4.24. Plot of dimensionless temperature $(t - t_0)/(t_a - t_0)$ versus Fourier number Fo for different ratios R_0/R at $r/R = \frac{1}{2}(1 + \{R_0/R\})$ [102].

In Fig. 4.25, similar calculation diagrams are presented for the case of a uniform initial temperature distribution: $t(\tau, 0) = t_0 = \text{const}$ and for the boundary conditions $t_1 = t_a = \text{const}$, $t_2 = t_0 = \text{const}$. In the diagrams shown in Figs. 4.24 and 4.25, the Fourier number is defined by the relation $\mathrm{Fo} = a\tau/R^2$.

If $t_1 = t_2 = t_a$, then the expression in square brackets under the sum-

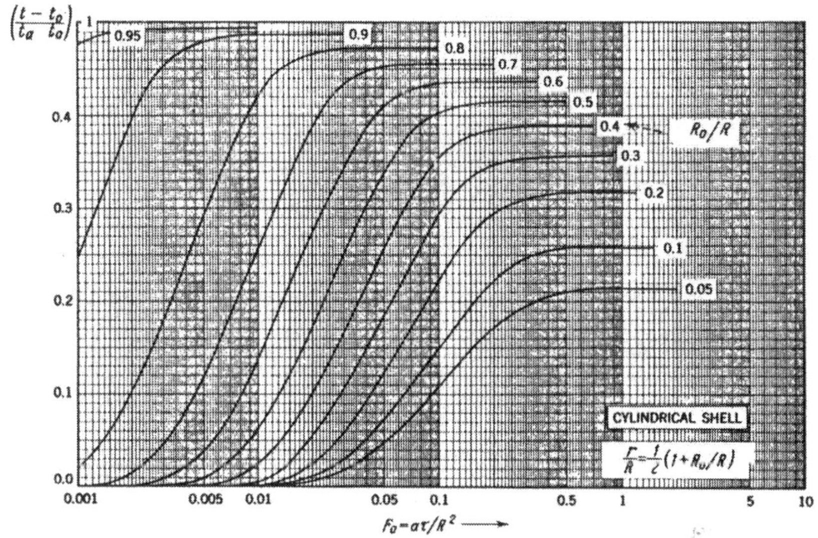

Fig. 4.25. Plot of the dimensionless temperature $(t - t_0)/(t_a - t_0)$ versus Fo for different ratios R_0/R at $r/R = \frac{1}{2}(1 - \{R_0/R\})$ [102].

mation sign will be equal to $t_0 - t_a$, and the first addend of formula (4.6.43) will be t_a.

Then, we have

$$\theta = \frac{t(r, \tau) - t_a}{t_0 - t_a} = \pi \sum_{n=1}^{\infty} \frac{J_0(m\mu_n) V_0(\mu_n r/R_0)}{J_0(\mu_n) + J_0(m\mu_n)} \exp[-\mu_n^2 \,\mathrm{Fo}]. \tag{4.6.44}$$

In some solutions the function $V_0(kr)$ is replaced by $U_0(kr)$. This is due to the fact that the relation between the constants D/C is defined from equation (4.6.6) rather than from (4.6.5). In this case we obtain

$$t(r, \tau) = \sum_{n}^{\infty} \frac{C_n}{Y_0(k_n R)} \exp[-ak_n^2 r][J_0(k_n r) Y_0(k_n R) - J_0(k_n R) Y_0(k_n r)$$
$$= \sum_{n}^{\infty} B_n \exp[-ak_n^2 r]\, U_0(k_n r).$$

The following simple relationship exists between these functions:

$$V_0(kr) = \frac{J_0(kR_0)}{J_0(kR)} U_0(kr). \tag{4.6.45}$$

Obviously

$$V_0(\mu_n r/R_0) = \frac{J_0(\mu_n)}{J_0(m\mu_n)} U_0(\mu_n r/R_0). \tag{4.6.46}$$

c. A Second Problem of Cooling a Cylindrical Tube. In this case, the external surface is thermally insulated and the internal one is maintained at a constant temperature t_a. The initial temperature distribution over the tube-wall thickness is assumed to be uniform.

Mathematically the problem may be written as

$$\frac{\partial t(r, \tau)}{\partial \tau} = \frac{a}{r} \frac{\partial}{\partial r} \left(r \frac{\partial t}{\partial r} \right), \quad R_0 < r < R, \quad \tau > 0, \tag{4.6.47}$$

$$t(r, 0) = t_0 = \text{const}, \tag{4.6.48}$$

$$\frac{\partial t(R, \tau)}{\partial r} = 0, \qquad t(R_0, \tau) = t_a = \text{const}. \tag{4.6.49}$$

Since the initial temperature is constant and does not depend on the coordinate, for solving this problem it is advisable to use the Hankel transformation method

$$t_H(p) = \int_{R_0}^{R} r t(t, \tau) V_0(p_k r)\, dr, \tag{4.6.50}$$

where p_k are the roots of a transcendental equation

$$J_1(pR) Y_0(pR_0) - J_0(pR_0) Y_1(pR) = 0, \tag{4.6.51}$$

$$V_0(pr) = J_0(pr) Y_0(pR_0) - J_0(pR_0) y_0(pr). \tag{4.6.52}$$

We apply the Hankel transformation to Eq. (4.6.47). Changing the order of sequence of differentiation and integration, the left-hand side of the equation will have the form:

$$\int_{R_0}^{R} r V_0(pr) \frac{\partial t(r, \tau)}{\partial \tau}\, dr = \frac{\partial t_H(p)}{\partial \tau}. \tag{4.6.53}$$

The right-hand side of Eq. (4.6.47) is integrated termwise

$$a \int_{R_0}^{R} V_0(pr) \frac{\partial}{\partial r} \left(r \frac{\partial t(r, \tau)}{\partial r} \right) dr = a \left[V_0(pr) r \frac{\partial t(r, \tau)}{\partial r} \right]_{r=R_0}^{r=R} - a \int_{R_0}^{R} r \frac{\partial t(r, \tau)}{\partial r} V_0(pr)\, dr. \tag{4.6.54}$$

The expression in square brackets is equal to zero as at $r = R$, $[\partial t(R, \tau)/\partial r] = 0$ according to the condition and at $r = R_0$ the function $V_0(pR_0) = 0$ according to formula (16a).

The integral in Eq. (4.6.54) is integrated term-by-term as

$$- ap[tV_0'(pr)t(r, \tau)]_{r=R_0}^{r=R} + ap \int_{R_0}^{R} t(r, \tau)[V_0'(pr) + prV_0''(pr)]\, dr. \quad (4.6.55)$$

Furthermore,

$$- V_0'(pr) = V_1(pr) = J_1(pr)Y_0(pR_0) - J_0(pR_0)Y_1(pr). \quad (4.6.56)$$

From formula (4.6.16) it follows that $V_1(pR) = 0$.

In addition

$$V_0'(pr) + prV_0''(pr) = - rpV_0(pr), \quad (4.6.57)$$

since $y = V_0(pr)$ is the solution of the Bessel equation at $pr = x$.

Thus,

$$a \int_{R_0}^{R} V_0(pr) \frac{\partial}{\partial r} \left(r \frac{\partial t(r, \tau)}{\partial r} \right) dr = - apR_0V_1(pR_0)t(R_0\tau)$$

$$- ap^2 \int_{R_0}^{R} rV_0(pr)t(r, \tau)\, dr. \quad (4.6.58)$$

According to Eq. (4.6.21)

$$V_1(pR_0) = 2/\pi pR_0 \quad (4.6.59)$$

and according to the condition $t(R_0, \tau) = t_a$, the differential Hankel transformed equation is obtained as

$$\frac{dt_H(p, \tau)}{\partial \tau} + ap^2t_H(p, \tau) - \frac{2at_a}{\pi} = 0. \quad (4.6.60)$$

The general integral of this linear equation has the form

$$t_H(p, \tau) = \exp\left[- \int ap^2\, d\tau\right] \cdot \left[A - \int (2at_a/\pi) \exp\left[\int ap^2\, d\tau\right]\right]$$

$$= A \exp[- ap^2\tau] - \frac{2t_a}{\pi p^2}. \quad (4.6.61)$$

The constant A is defined from the initial condition, to which the Hankel transformation is applied

$$\int_{R_0}^{R} rV_0(pr)t(r, 0)\, dr = t_0 \int_{R_0}^{R} rV_0(pr)\, dr = (t_0/p)[rV_1(pr)]_{r=R_0}^{r=R}$$

$$= (t_0/p)[RV_1(pR) - R_0V_1(pR_0)]. \quad (4.6.62)$$

Taking into account Eq. (4.6.59) and that $V_1(pR) = 0$ we can obtain

$$t_H(p, 0) = - 2t_0/\pi p^2. \quad (4.6.63)$$

Now, evaluating Eq. (4.6.61) at $\tau = 0$ we have

$$A = \frac{2(t_a - t_0)}{\pi p^2}, \tag{4.6.64}$$

$$t_H(p, \tau) = \frac{2(t_a - t_0)}{\pi p^2} \exp[-ap^2\tau] - \frac{2t_a}{\pi p^2}. \tag{4.6.65}$$

Making use of the transformation formula

$$H^{-1}[f(p_n)] = \frac{\pi^2}{2} \sum_{n=1}^{\infty} \frac{p_n^2 J_1(p_n R) V_0(p_n r)}{J_0^2(p_n R_0) - J_1^2(p_n R)} f(p_n) \tag{4.6.66}$$

we obtain

$$\begin{aligned} H^{-1}&[t_H(p, \tau)] \\ &= t(r, \tau) \\ &= \pi \sum_{n=1}^{\infty} \frac{p_n^2 J_1^2(p_n R) V_0(p_n r)}{p_n^2[J_0^2(p_n R_0) - J_1^2(p_n R)]} \\ &\quad \times \{(t_a - t_0) \exp[-ap_n^2\tau] - t_a\}. \end{aligned} \tag{4.6.67}$$

The formula $f(r)$ was developed above in series form:

$$\begin{aligned} t_0 = f(r) &= \sum_{n=1}^{\infty} C_n V_0(k_n r) \\ &= -\sum_{n=1}^{\infty} \frac{\pi t_0 J_1^2(k_n R) V_0(k_n r)}{J_0^2(k_n R_0) - J_1^2(k_n R)}. \end{aligned} \tag{4.6.68}$$

Hence, the second sum is equal to t_0.
Thus, finally we have

$$\begin{aligned} \theta &= \frac{t(r, \tau) - t_a}{t_a - t_0} \\ &= \pi \sum_{n=1}^{\infty} \frac{J_1^2(\mu_n m) V_0(\mu_n r/R_0)}{J_0^2(\mu_n) - J_1^2(\mu_n m)} \exp[-\mu_n^2 \,\mathrm{Fo}]. \end{aligned} \tag{4.6.69}$$

4.7 Parallelepiped

Statement of the Problem. *There is a plate of finite size $2R_1 \times 2R_2 \times 2R_3$, the temperature of which is the same everywhere and equal to t_0. At time instant $\tau = 0$ all the surfaces of a plate are instantaneously cooled to some temperature $t_a < t_0$ which is maintained constant during the whole cooling process. We wish to determine the temperature distribution at any time as well as the mean temperature required for determination of heat losses.*

We place the origin of the coordinates in the center of the parallelepiped (Fig. 4.26); then our problem may be formulated mathematically as follows.

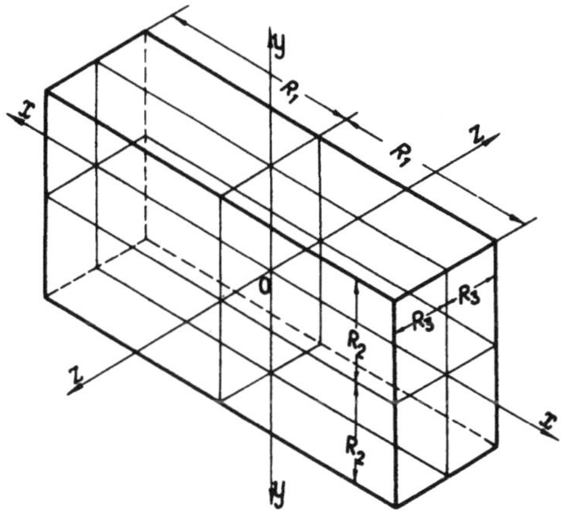

Fig. 4.26. Parallelepiped.

We have

$$\frac{\partial t(x, y, z, \tau)}{\partial \tau} = a \left[\frac{\partial^2 t(x, y, z, \tau)}{\partial x^2} + \frac{\partial^2 t(x, y, z, \tau)}{\partial y^2} + \frac{\partial^2 t(x, y, z, \tau)}{\partial z^2} \right]$$

$$(\tau > 0; \quad -R_1 < x < +R_1; \quad -R_2 < y < +R_2; \quad -R_3 < z < +R_3) \tag{4.7.1}$$

$$t(x, y, z, 0) = t_0 = \text{const}, \tag{4.7.2}$$

$$t(\pm R_1, y, z, \tau) = t_a, \qquad t(x, \pm R_2, z, \tau) = t_a, \qquad t(x, y, \pm R_3, \tau) = t_a. \tag{4.7.3}$$

We now prove that the solution of this problem may be presented as the product of the solution for three infinite plates, the thickness of each plate being equal to $2R_1$, $2R_2$ and $2R_3$, respectively, i.e.,

$$\frac{t(x, y, z, \tau) - t_a}{t_0 - t_a} = \frac{t(x, \tau) - t_a}{t_0 - t_a} \times \frac{t(y, \tau) - t_a}{t_0 - t_a} \times \frac{t(z, \tau) - t_a}{t_0 - t_a}. \tag{4.7.4}$$

In addition, temperature $t(x, \tau)$, $t(y, \tau)$ and $t(z, \tau)$ are determined by the solution of differential equations

$$\frac{\partial (x, \tau)}{\partial \tau} = a \frac{\partial^2 t(x, \tau)}{\partial x^2}, \qquad \frac{\partial t(y, \tau)}{\partial \tau} = a \frac{\partial^2 t(y, \tau)}{\partial y^2}, \qquad \frac{\partial t(z, \tau)}{\partial \tau} = a \frac{\partial^2 t(z, \tau)}{\partial z^2} \tag{4.7.5}$$

with boundary conditions

$$t(x, 0) = t(y, 0) = t(z, 0) = t_0 = \text{const}, \tag{4.7.6}$$

$$t(\pm R_1, \tau) = t_a; \qquad t(\pm R_2, \tau) = t_a, \qquad t(\pm R_3, \tau) = t_a. \tag{4.7.7}$$

Equation (4.7.4) may be written thus:

$$t(x, y, z, \tau) = t_a + \frac{1}{(\Delta t)^2}[t(x, \tau) - t_a][t(y, \tau) - t_a][t(z, \tau) - t_a], \tag{4.7.8}$$

where $\Delta t = t_0 - t_a$. For the proof, we will substitute solution (4.7.8) into differential equation (4.7.1). Upon transformation we obtain

$$\begin{aligned} &[t(y, \tau) - t_a][t(z, \tau) - t_a]\left\{\frac{\partial t(x, \tau)}{\partial \tau} - a\frac{\partial^2 t(x, \tau)}{\partial x^2}\right\} \\ &\quad + [t(x, \tau) - t_a][t(z, \tau) - t_a]\left\{\frac{\partial t(y, \tau)}{\partial \tau} - a\frac{\partial^2 t(y, \tau)}{\partial y^2}\right\} \\ &\quad + [t(x, \tau) - t_a][t(y, \tau) - t_a]\left\{\frac{\partial t(z, \tau)}{\partial \tau} - a\frac{\partial^2 t(z, \tau)}{\partial z^2}\right\} = 0. \end{aligned} \tag{4.7.9}$$

Since $t(x, \tau)$, $t(y, \tau)$ and $t(z, \tau)$ are valid for differential equation (4.7.5), then all the values in brackets in Eq. (4.7.9) will be equal to zero; hence, Eq. (4.7.9) becomes indentity. Thus, solution (4.7.8) satisfies differential equation (4.7.1).

Substituting solution (4.7.8) into initial condition (4.7.2) gives us

$$t_0 = t_a + \frac{1}{(\Delta t)^2}[t(x, 0) - t_a][t(y, 0) - t_a][t(z, 0) - t_a];$$

then, according to condition (4.7.6), we obtain the identity

$$t_0 - t_a = (\Delta t)^{-2}(\Delta t)^3 = \Delta t = t_0 - t_a.$$

Hence, our solution (4.7.8) is valid for the initial condition.

Substitution of solution (4.7.8) into boundary conditions (4.7.3) gives us

$$\begin{aligned} t(R_1, y, z, \tau) &= t_a = t_a + (\Delta t)^{-2}[t(R_1, \tau) - t_a] \\ &\qquad \times [t(y, \tau) - t_a][t(z, \tau) - t_a]; \\ t(x, R_2, z, \tau) &= t = t_a + (\Delta t)^{-2}[t(x, \tau) - t_a] \\ &\qquad \times [t(R_2, \tau) - t_a][t(z, \tau) - t_a] \\ t(x, y, R_3, \tau) &= t_a = t_a + (\Delta t)^{-2}[t(x, \tau) - t_a][t(y, \tau) - t_a] \\ &\qquad \times [t(R_3, \tau) - t_a]. \end{aligned} \tag{4.7.10}$$

As $t(R_1, \tau) = t(R_2, \tau) = t(R_3, \tau) = t_a$, then in each equality of relations (4.7.10), one of the quantities in square brackets is equal to zero and all three solutions yield the identity

$$t_a \equiv t_a.$$

Hence, solution (4.7.8) also satisfies the boundary conditions; thus, according to the uniqueness theorem this solution fits our problem.

Thus, the solution of our problem may be written in the form of (4.7.8) or (4.7.4) as

$$\theta = \frac{t(x, y, z, \tau) - t_a}{t_0 - t_a} = \sum_{n=1}^{\infty} \sum_{m=1}^{\infty} \sum_{k=1}^{\infty} A_n A_m A_k \cos \mu_n(x/R_1) \cos \mu_m(y/R_2)$$
$$\times \cos \mu_k(z/R_3) \exp[-(\mu_n{}^2 K_1{}^2 + \mu_m{}^2 K_2{}^2 + \mu_k{}^2 K_3{}^2) \text{ Fo}], \qquad (4.7.11)$$

where

$$A_n = (-1)^{n+1}(2/\mu_n); \qquad A_m = (-1)^{m+1}(2/\mu_m); \qquad A_k = (-1)^{k+1}(2/\mu_k)$$
$$\mu_n = (2n-1)(\pi/2); \qquad \mu_m = (2m-1)(\pi/2); \qquad \mu_k = (2k-1)(\pi/2).$$

Fo is the Fourier number ($\text{Fo} = a\tau/R^2$); R is the generalized size where

$$\frac{1}{R^2} = \frac{1}{R_1{}^2} + \frac{1}{R_2{}^2} + \frac{1}{R_3{}^2}; \qquad K_i = \frac{R}{R_i} \quad (i = 1, 2, 3).$$

It is possible to use the corresponding solutions for an infinite plate otherwise and to obtain a solution for a parallelepiped in the following form:

$$\theta(x, y, z, \tau) = \theta(x, \tau)\theta(y, \tau)\theta(z, \tau). \qquad (4.7.12)$$

The mean temperature of the parallelepiped is determined by means of expression (4.7.12) as

$$\bar{\theta} = \sum_{n=1}^{\infty} \sum_{m=1}^{\infty} \sum_{k=1}^{\infty} B_n B_m B_k \exp[-(\mu_n{}^2 K_1{}^2 + \mu_m{}^2 K_2{}^2 + \mu_k{}^2 K_3{}^2) \text{ Fo}], \quad (4.7.13)$$

where

$$B_n = 2/\mu_n{}^2, \quad B_m = 2/\mu_m{}^2, \quad B_k = 2/\mu_k{}^2.$$

Thus, the problem for a parallelepiped is reduced to that for an infinite plate. The analysis and calculations are not therefore given here.

4.8 Finite Cylinder

a. Statement of the Problem. *Consider a cylinder, of diameter 2R and length 2l. The cylinder temperature is everywhere the same and equal to t_0. At the initial time instant, the cylinder surface (lateral and end surfaces) is instantaneously cooled to some temperature t_a which is maintained constant during the whole cooling process. The temperature distribution inside a cylinder at any time as well as the mean temperature as the time function is to be determined.*

The determination of a temperature field of a finite cylinder when its temperature is a function of only three variables (time, radius, and z coordinate) involves a solution of a differential heat conduction equation

$$\frac{\partial t(r, z, \tau)}{\partial \tau} = a\left\{\frac{\partial^2 t(r, z, \tau)}{\partial r^2} + \frac{1}{r}\frac{\partial t(r, z, \tau)}{\partial r} + \frac{\partial^2 (r, z, \tau)}{\partial z^2}\right\}$$

$$(\tau > 0; \quad 0 < r < R; \quad -l < z < +l), \qquad (4.8.1)$$

at the initial

$$t(r, z, 0) = t_0 = \text{const}, \qquad (4.8.2)$$

and boundary conditions

$$t(r, \pm l, \tau) = t_a, \qquad t(R, z, \tau) = t_a. \qquad (4.8.3)$$

The origin of coordinates is at the center of a cylinder (Fig. 4.27).

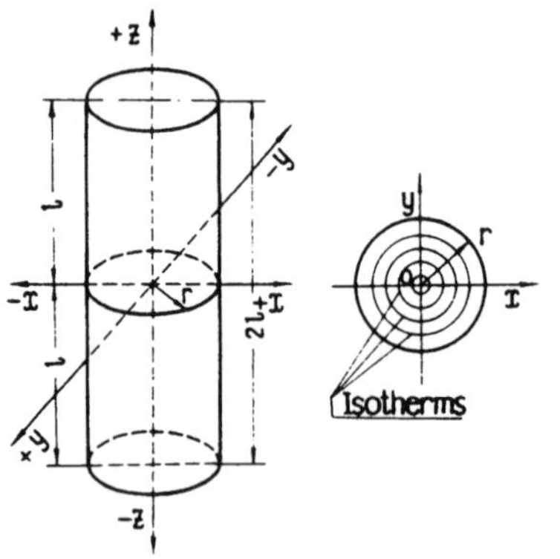

Fig. 4.27. Finite cylinder.

b. Solution of the Problem. In a manner similar to that in the previous paragraph it may be proved that the solution of our problem $\theta(r, z, \tau)$ has the form

$$\theta(r, z, \tau) = \theta(r, \tau)\theta(z, \tau), \qquad (4.8.4)$$

where $\theta(r, \tau)$ is the solution for an infinite cylinder and $\theta(z, \tau)$ is the solution for an infinite plate. Their intersection forms a finite cylinder. In addition, the initial and boundary conditions remain the same:

$$t(r, 0) = t(z, 0) = t_0 = \text{const}; \quad t(R, \tau) = t(\pm l, \tau) = t_a = \text{const}. \tag{4.8.5}$$

Thus, the solution of our problem will be

$$\theta = \frac{t(r, z, \tau) - t_a}{t_0 - t_a} = \sum_{n=1}^{\infty} \sum_{m=1}^{\infty} A_n A_m J_0\left(\mu_n \frac{r}{R}\right) \cos \mu_m \frac{z}{l} \exp[-(\mu_n^2 + \mu_m^2 K_l^2)\mathrm{Fo}], \tag{4.8.6}$$

where $A_n = 2/\mu_n J_1(\mu_n)$, μ_n are the roots of the Bessel function of the first kind of zero order

$$A_m = (-1)^{m+1} \frac{2}{\mu_m}; \quad \mu_m = (2m - 1)\frac{\pi}{2}; \quad K_l = \frac{R}{l}; \quad \mathrm{Fo} = \frac{a\tau}{R^2}.$$

The mean cylinder temperature is

$$\bar{\theta} = \sum_{n=1}^{\infty} \sum_{m=1}^{\infty} B_n B_m \exp[-(\mu_n^2 + \mu_m^2 K_l^2)\,\mathrm{Fo})], \tag{4.8.7}$$

where

$$B_n = \frac{4}{\mu_n^2}, \quad B_m = \frac{2}{\mu_m^2}.$$

We will now carry out a representative calculation. A steel disk 20 cm in diameter and 12 cm in thickness is heated to 300°C. At the initial time instant, it is placed into melting ice. We are to determine the temperature at the disk center 2 min after cooling.

The disk-surface temperature during the cooling process is considered to be constant and equal to $t_a = 0$°C. We assume the thermal diffusivity to be equal to $a = 45 \cdot 10^{-3}$ m²/hr.

The solution of this problem may be written in the form

$$\theta(r, z, \tau) = \theta(r, \tau)\theta(z, \tau)$$

where $\theta(r, \tau)$ and $\theta(z, \tau)$ are the solutions of a problem for an infinite cylinder and an infinite plate, respectively.

We use the tables for determination of θ_{cyl}. For this purpose we first calculate the Fourier number for a cylinder and a plate:

$$\mathrm{Fo}_l = \frac{a\tau}{l^2} = \frac{45 \cdot 10^{-3} \cdot 1}{36 \cdot 10^{-4} \cdot 30} = 0.416 \approx 0.42$$

because $l = 6$ cm $= 0.06$ m, $\tau = 2$ min $= 1/30$ hr;

$$\mathrm{Fo}_R = \frac{a\tau}{R^2} = \frac{45 \cdot 10^{-3}}{10^{-2} \cdot 30} = 0.15,$$

as $R = 10$ cm $= 0.1$ m.

From Table 4.2, one can see that for Fo = 0.42, the relative temperature in the middle of an infinite plate is equal to $\theta_{cp} = 0.45$, and from Table 4.6, for Fo = 0.12, the temperature of the whole axis line of an infinite cylinder is equal to $\theta_{cc} = 0.67$. Hence, the dimensionless temperature of a finite cylinder is

$$\theta_c = \theta_{cp} \cdot \theta_{cc} = 0.45 \cdot 0.67 = 0.3$$

whence

$$t = 300 \cdot 0.3 \approx 90° \ C.$$

4.9 Heating Problems

In heating problems the cooling of a body with a given initial temperature is considered subject to the condition that at the initial time the body surface assumes some constant temperature which is maintained constant during the whole cooling process ($t_s = t_a =$ const). *The problem for the heating of a body with some prescribed initial temperature may be reduced to that for cooling by means of a simple replacement of a variable when at the initial time instant the surface temperature instantaneously becomes constant and equals t_a ($t_a > t_0$).*

We have the problem for cooling

$$\partial t/\partial \tau = a\,\nabla^2 t; \qquad t(0) = t_0; \qquad t_s = t_a \qquad (t_s < t_0).$$

Introducing a variable $\vartheta = t_0 - t$, we have

$$\frac{\partial \vartheta}{\partial \tau} = a\,\nabla^2 \vartheta; \qquad \vartheta(0) = 0; \qquad \vartheta_s = t_0 - t_a = \vartheta_a.$$

The problem for body cooling is obtained when the initial temperature $\vartheta(0)$ is zero and the body surface temperature is equal to $\vartheta_a(0) = t_0 - t_a =$ const. Hence, all the formulae derived will also be valid for problems on body heating. However, by θ it should be understood that with cooling

$$\theta = \frac{t - t_a}{t_0 - t_a} \qquad (t_0 > t_a),$$

and with heating

$$\theta = \frac{t_a - t}{t_a - t_0} = 1 - \frac{t - t_0}{t_a - t_0} \qquad (t_a > t_0).$$

Thus, in transition to a heating problem in the solution for body cooling the dimensionless quantity θ should be replaced by $(t_a - t)/(t_a - t_0)$ or $(1 - \{t - t_0\}/\{t_a - t_0\})$. This method is used to calculate the problem in Section 4.2.

CHAPTER

5

BOUNDARY CONDITION OF THE SECOND KIND

A process of heat transfer to a body being heated in high-temperature furnaces takes place mainly by radiation; in the majority of such cases convective transfer may be neglected.

The heat flow absorbed by a body surface from heated walls and a roof is directly proportional to the difference of the fourth powers of absolute temperatures of the exchanging surfaces

$$q_s = \sigma C(t_{\text{rad}}^4 - t_s^4) \tag{5.0.1}$$

where σ is the Stefan–Boltzmann constant and C is the coefficient which depends upon the radiative heat-absorption capacity of the body surface and the arrangement or irradiated and radiative bodies relative to one another. Suffix s stands for a body surface and rad a radiating surface.

Heat exchange with heated gases is governed by a modified law, in which t_{rad}^4 and t_s^4 have multipliers ε_{gas} and $\varepsilon_{\text{body}}$ respectively, referred to as emissivities of the radiative gases at the gas temperature and at the surface temperature. It is often assumed that the values of ε_{gas} and $\varepsilon_{\text{body}}$ are almost the same, so an average value of ε may therefore be taken and factored out. Then, the ordinary relation for a law of heat transfer by radiation is obtained.

Thus the heat flux transferred to the body surface is a certain function of time $f(\tau)$ which is given as

$$q_s = f(\tau) \tag{5.0.2}$$

In certain particular cases, the boundary condition (5.1.2) may be simplified.

In the heat transfer theory for furnaces, it is customary that all the radiation sources be replaced by one having some average temperature referred to as a furnace temperature (t_a).

If temperature of a body surface (t_s) is considerably less than that of a furnace (t_a), the second term in brackets may be neglected and a constant heat flow absorbed by a body surface is obtained

$$q_c \approx \sigma C(t_a^4) \times \text{const.} \tag{5.0.3}$$

This boundary condition is a specific case (simplest) of the second kind (i.e., the heat flow to a body surface is prescribed) when a heat flow is constant. Solution of problems with a variable heat flow $q_s = f(\tau)$ *may be obtained from the corresponding solution for a constant heat flow with the help of the Duhamel theorem or by the Fourier and Hankel integral transformation.*

5.1 Semi-Infinite Body

By way of an example of a semi-infinite body, we consider a long bar with its lateral surface insulated and of a size such that the thickness and width of it are negligible compared to its length. In the previous chapter, heat transmission at small values of the Fourier number was shown to occur similarly to heat propagation in a semi-infinite body.

a. Statement of the Problem. *Consider a semi-infinite body having an initial temperature* t_0. *A bounding surface is heated by a constant heat flow* $q_c =$ *const. The temperature varies in one direction. We are to find the temperature distribution in this direction at any time instant.*

We have

$$\frac{\partial t(x, \tau)}{\partial \tau} = a \frac{\partial^2 t(x, \tau)}{\partial x^2} \quad (\tau > 0; \quad 0 < x < \infty), \tag{5.1.1}$$

$$t(x, 0) = t_0 = \text{const}, \tag{5.1.2}$$

$$\lambda \frac{\partial t(0, \tau)}{\partial x} + q_c = 0, \tag{5.1.3}$$

$$t(\infty, \tau) = t_0, \qquad \partial t(\infty, \tau)/\partial x = 0. \tag{5.1.4}$$

b. Solution of the Problem by the Classical Method. This problem may be reduced to that of heat conduction with the first kind of boundary condition considered in Chapter 4, Section 2.

Instead of a variable t, we introduce a new variable q (heat flux) determined by the relation

$$q(x, \tau) = -\lambda \frac{\partial t(x, \tau)}{\partial x}. \tag{5.1.5}$$

Differentiating Eq. (5.1.1) with respect to x gives us

$$\frac{\partial}{\partial x}\left[\frac{\partial t(x, \tau)}{\partial \tau}\right] = \frac{\partial}{\partial x}\left[a\,\frac{\partial^2 t(x, \tau)}{\partial x^2}\right]. \tag{5.1.6}$$

Then differential equation (5.1.6) may be written

$$\frac{\partial}{\partial \tau}\left[\frac{\partial t(x, \tau)}{\partial x}\right] = a\,\frac{\partial^2}{\partial x^2}\left[\frac{\partial t(x, \tau)}{\partial x}\right]$$

or

$$\frac{\partial q(x, \tau)}{\partial \tau} = a\,\frac{\partial^2 q(x, \tau)}{\partial x^2}, \tag{5.1.7}$$

i.e., an ordinary differential equation for a one-dimensional problem, only now the variable q replaces t.

For the new variable q, the initial and boundary conditions have the form

$$q(x, 0) = 0 \tag{5.1.8}$$

(moreover, to simplify calculations we assume that $t_0 = 0$)

$$q(0, \tau) = q_c = \text{const}, \tag{5.1.9}$$

$$q(\infty, \tau) = 0. \tag{5.1.10}$$

We know the solution of Eq. (5.1.7) for conditions (5.1.8)–(5.1.10)[1] (see Chapter 4, Section 2) viz:

$$q(x, \tau)/q_c = \operatorname{erfc} x/2(a\tau)^{1/2} \tag{5.1.11}$$

To determine $t(x, \tau)$, we substitute into expression (5.1.5) the expression for $q(x, \tau)$ from (5.1.11); and integrate between x and ∞ to obtain

$$t(x, \tau) = \frac{q_c}{\lambda}\int_x^\infty \operatorname{erfc}\frac{x}{2(a\tau)^{1/2}}\,dx = \frac{2q_c}{\lambda}(a\tau)^{1/2}\,\mathrm{i}\operatorname{erfc}\frac{x}{2(a\tau)^{1/2}}, \tag{5.1.12}$$

[1] In Chapter 4, Section 2, is the solution of a problem for cooling a semi-infinite bar. The problem for heating is obtained from that for cooling by replacing θ by $(t_a - t)/(t_a - t_0)$.

where

$$\mathrm{i\,erfc}\, u = \int_u^\infty \mathrm{erfc}\, w\, dw = \frac{1}{\sqrt{\pi}} \exp[-u^2] - u\, \mathrm{erfc}\, u. \qquad (5.1.13)$$

c. Solution by the Operational Method. The solution of differential equation (5.1.1) for the transform $T(x, s)$ has the form

$$T(x, s) - (t_0/s) = A_1 \exp[(s/a)^{1/2}x] + B_1 \exp[-(s/a)^{1/2}x]. \qquad (5.1.14)$$

Boundary conditions (5.1.3) and (5.1.4) for the transform may be written as

$$\lambda T'(0, s) + (q_c/s) = 0, \qquad (5.1.15)$$

$$T'(\infty, s) = 0. \qquad (5.1.16)$$

It follows from condition (5.1.16) that $A_1 = 0$, for as $x \to \infty$ the temperature gradient tends to zero and a body temperature cannot be infinitely great (at $x \to \infty$, $t(\infty, \tau) \to t_0$).

The constant B_1 is determined from boundary condition (5.1.15). We have

$$-\left(\frac{s}{a}\right)^{1/2} B_1 + \frac{q_c}{\lambda s} = 0.$$

Whence

$$B_1 = \frac{q_c}{\lambda s \left(\frac{s}{a}\right)^{1/2}}.$$

Hence, solution (5.1.14) acquires the form:

$$T(x, s) - \frac{t_0}{s} = \frac{q_c \sqrt{a}}{\lambda s^{3/2}} \exp\left[-\left(\frac{s}{a}\right)^{1/2} x\right]. \qquad (5.1.17)$$

To find the inverse transform, we use the table of transforms according to which

$$L^{-1}\left[\frac{1}{s^{1+1/2n}} e^{-k\sqrt{s}}\right] = (4\tau)^{n/2}\, \mathrm{i}^n\, \mathrm{erfc} \frac{k}{2\sqrt{\tau}}. \qquad (5.1.18)$$

Finally we have

$$t(x, \tau) - t_0 = \frac{2q_c}{\lambda} (a\tau)^{1/2}\, \mathrm{i\,erfc} \frac{x}{2(a\tau)^{1/2}} \qquad (5.1.19)$$

i.e., the solution is identical with (5.1.12).

It is seen from the above example that the solution by the operational method rapidly leads to results.

d. Analysis of the Solution and Heat Rate. We now introduce a new criterion $q_c x/\lambda(t_a - t_0)$, where t_a is the average temperature of a heating furnace, and call it the Kirpichev criterion:

$$\mathrm{Ki}_x = \frac{q_c x}{\lambda(t_a - t_0)}. \tag{5.1.20}$$

The Kirpichev criterion is the ratio of the heat flux (q_c) through the bar end to the maximum possible heat flux at the point x provided that the temperature gradient at this point is maximum and equal to $(t_a - t_0)/x$.

The solution (5.1.19) may be written

$$\theta = \frac{t(x, \tau) - t_0}{t_a - t_0} = \mathrm{Ki}_x \,\mathrm{Fo}_x \,\mathrm{i\,erfc}\frac{1}{2(\mathrm{Fo}_x)^{1/2}}, \tag{5.1.21}$$

where $\mathrm{Fo}_x = a\tau/x^2$ is the Fourier number.

Figure 5.1 furnishes calculation diagrams. The generalized variable $\theta/\mathrm{Ki}_x \,\mathrm{Fo}_x$, the dimensionless temperature gradient $2x(\partial\theta/\partial x)/\mathrm{Ki}_x$ and the dimensionless heating rate $2\sqrt{\pi}\,\mathrm{Fo}_x(\partial\theta/\partial \mathrm{Fo}_x)$ are all plotted versus the number $1/(2(\sqrt{\mathrm{Fo}_x}))$. These diagrams allow approximate calculations.

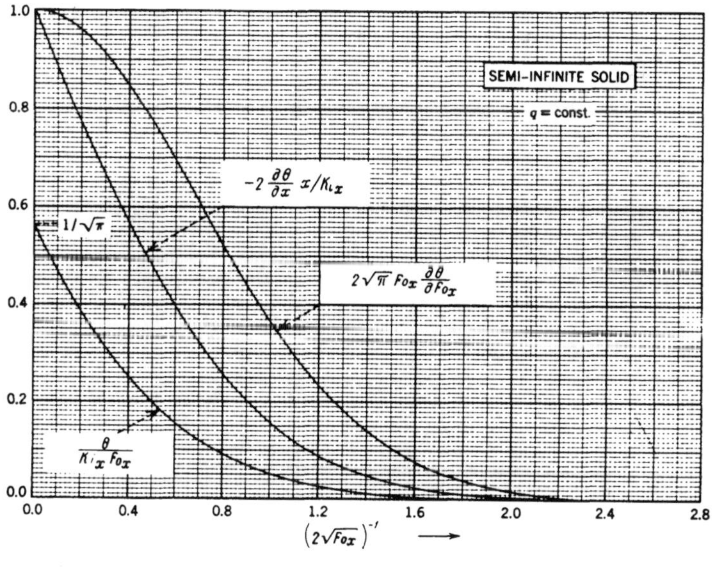

Fig. 5.1. Temperature, temperature gradient, and heating rate versus $1/2(\mathrm{Fo}_x)^{1/2}$ [102].

Additional insight into the meaning of the Kirpichev criterion may be obtained. We note that

$$q_s = \sigma C(t_a^2 + t_s^2)(t_a + t_s)(t_a - t_s) = \alpha_{\text{rad}}(t_a - t_s). \qquad (5.1.22)$$

For a given t_a, a maximum heat flow will occur when $t_s = t_0$. Thus, in this case

$$\text{Ki}_x = \alpha_{\text{rad}} x/\lambda. \qquad (5.1.23)$$

i.e., the Kirpichev criterion is similar to the Biot criterion.

Thus, the criterion Ki_x is numerically equal to the ratio of the internal (x/λ) and the external $1/\alpha_{\text{rad}}$ thermal resistances if the radiant heat flow is replaced by an approximate relation of the Newton law for cooling.

Hence solution (5.1.19) may be written

$$\theta = \frac{t(x, \tau) - t_0}{t_a - t_0} = \text{Ki}_x\left[2\left(\frac{\text{Fo}_x}{\pi}\right)^{1/2} \cdot \exp\left[-\frac{1}{4\text{Fo}_x}\right] - \text{erfc}\,\frac{1}{2(\text{Fo}_x)^{1/2}}\right]. \qquad (5.1.24)$$

The amount of heat required for heating the bar is prescribed in terms of the heat flux q_c, i.e.,

$$dQ_s/d\tau = q_c = \text{const.} \qquad (5.1.25)$$

Hence, the heat consumption per unit area of the bar-end surface (ΔQ_s) will be directly proportional to time:

$$\Delta Q_s = q_c\tau = c\gamma(t_a - t_s). \qquad (5.1.26)$$

5.2 Infinite Plate

a. Statement of the Problem. *Consider an infinite plate $2R$ in thickness at initial temperature t_0. The plate is uniformly heated from both sides by a constant heat flow (for example, it is heated in a furnace at a rather high temperature t_a). The temperature distribution across the plate thickness at any time instant is to be found.*

The differential equation will be the same as in Chapter 4. For a symmetrical problem, the initial and boundary conditions have the form

$$t(x, 0) = t_0 = \text{const}, \qquad (5.2.1)$$

$$-\frac{\partial t(R, \tau)}{\partial x} + \frac{q_c}{\lambda} = 0, \qquad (5.2.2)$$

$$\frac{\partial t(0, \tau)}{\partial x} = 0. \qquad (5.2.3)$$

b. Solution of the Problem by Classical Method. To solve the problem, we introduce a new variable $q(x, \tau)$ by relation (5.1.5). Then, for the new variable, we shall obtain differential equation (5.1.7) similar to the differential heat conduction equation.

The initial and boundary conditions for the new variable will be

$$q(x, 0) = 0, \tag{5.2.4}$$

$$q(R, \tau) = q_c, \tag{5.2.5}$$

$$q(0, \tau) = 0. \tag{5.2.6}$$

The last condition follows from the symmetry condition (5.2.3).

Thus, a kind of a problem on "heating" of an infinite plate with zero initial "temperature" is obtained when one boundary surface is maintained at "zero temperature" and the opposite at temperature equal to q_c.

The solution of essentially the same problem is given in Chapter 4, Section 3.

To solve our problem, $(R - x)$ in solution (4.3.45) should be replaced by x and θ by $(1 - \theta)$, i.e., the problem for cooling should be changed to that for heating.

Upon such a replacement, we obtain

$$\frac{q(x, \tau)}{q_c} = \frac{x}{R} - \sum_{n=1}^{\infty} (-1)^{n+1} \frac{2}{\mu_n} \sin \mu_n \frac{x}{R} \exp[-\mu_n^2 \, \mathrm{Fo}], \tag{5.2.7}$$

where $\mu_n = n\pi$ are the eigenvalues.

The temperature field is found according to the formula

$$t(x, \tau) = \frac{1}{\lambda} \int q(x, \tau)\, dx + \varphi(\tau) + C, \tag{5.2.8}$$

where $\varphi(\tau)$ is some function of time and C is the constant of integration. Substituting solution (5.2.7) into (5.2.8) and integrating we obtain

$$t(x, \tau) = \frac{q_c x^2}{2\lambda R} + \frac{q_c R}{\lambda} \sum_{n=1}^{\infty} \frac{2}{\mu_n^2} (-1)^{n+1} \cos \mu_n \frac{x}{R} \exp[-\mu_n^2 \mathrm{Fo}] + \varphi(\tau) + C. \tag{5.2.9}$$

To determine $\varphi(\tau)$ and C, we use

$$q_c = c\gamma R \frac{d\bar{t}(\tau)}{d\tau} \tag{5.2.10}$$

where $\bar{t}(\tau)$ is the average temperature of an infinite plate defined by the formula

$$\bar{t}(\tau) = \frac{1}{R}\int_0^R t(x, \tau)\, dx. \tag{5.2.11}$$

Relation (5.2.11) may be written in an integral form as

$$\bar{t}(\tau) = \frac{q_c}{c\gamma R}\tau + t_0 \tag{5.2.12}$$

We obtain $\bar{t}(\tau)$ from Eq. (5.2.9) as

$$\bar{t}(\tau) = \frac{q_c R}{6\lambda} + \varphi(\tau) + C, \tag{5.2.13}$$

since the integral of the sum with respect to x between 0 and R is equal to zero ($\sin \mu_n = 0$).

Comparing (5.2.12) and (5.2.13) we find

$$\varphi(\tau) = \frac{q_c}{c\gamma R}\tau, \qquad C = t_0 - \frac{q_c R}{6\lambda}.$$

The final solution of our problem will be of the form:

$$t(x, \tau) - t_0$$

$$= \frac{q_c}{\lambda}\left[\frac{a\tau}{R} - \frac{R^2 - 3x^2}{6R} + R\sum_{n=1}^{\infty}(-1)^{n+1}\frac{2}{\mu_n^2}\cos\mu_n\frac{x}{R}\exp[-\mu_n^2\,\mathrm{Fo}]\right]. \tag{5.2.14}$$

Thus, to solve the main problem it was first necessary to find a solution of the auxiliary problem for $q(x, \tau)$. Moreover, additional relation (5.2.12) had to be used to determine the constants of integration. All of this greatly complicates the solution; therefore, the problems considered below will be solved by integral transformations.

c. Solution of the Problem by the Operational Method. In the case of an infinite plate, the general solution of a one-dimensional problem for the transform $T(x, s)$ is the following

$$T(x, s) - \frac{t_0}{s} = A\cosh\left(\frac{s}{a}\right)^{1/2} x + B\sinh\left(\frac{s}{a}\right)^{1/2} x. \tag{5.2.15}$$

The initial temperature of the plate is uniform everywhere and equal to t_0. Boundary conditions (5.2.2) and (5.2.3) for the transform will have the form

$$- T'(R, s) + \frac{q_c}{\lambda s} = 0, \tag{5.2.16}$$

$$T'(0, s) = 0. \tag{5.2.17}$$

It follows from the symmetry condition (5.2.17) that $B = 0$ (the temperature distribution is symmetrical with respect to the central line).

The constant A is determined from condition (5.2.16), i.e.,

$$- \left(\frac{s}{a}\right)^{1/2} A \sinh\left(\frac{s}{a}\right)^{1/2} R + \frac{q_c}{\lambda s} = 0, \quad A = \frac{q_c}{\lambda s (s/a)^{1/2} \sinh(s/a)^{1/2} R}.$$

Consequently, solution (5.2.15) assumes the form

$$T(x, s) - \frac{t_0}{s} = \frac{q_c \cosh(s/a)^{1/2} x}{\lambda s (s/a)^{1/2} \sinh(s/a)^{1/2} R} = \frac{\Phi(s)}{\psi(s)}. \tag{5.2.18}$$

Solution (5.2.18) is the ratio of two generalized polynomials as

$$\Phi(s) = q_c \left(1 + \frac{1}{2!} \frac{x^2}{a} s + \frac{1}{4!} \frac{x^4}{a^2} s^2 + \cdots\right),$$

$$\psi(s) = \lambda s^2 \left(\frac{R}{a} + \frac{1}{3!} \frac{R^3}{a^2} s + \frac{1}{5!} \frac{R^5}{a^3} s^2 + \cdots\right) = \lambda s^2 \varphi(s),$$

where $\varphi(s)$ is the expression in parentheses. The power series $\psi(s)$ does not contain the constant, i.e., all the conditions of the expansion theorem are fulfilled.

To determine the roots s_n of the generalized polynomial $\psi(s)$, the latter is equated to zero:

$$\psi(s) = \lambda s (s/a)^{1/2} \sinh(s/a)^{1/2} R = \lambda s^2 \varphi(s) = 0. \tag{5.2.19}$$

Hence we find $s_0 = 0$ (double root) and $s_n = - a\mu_n^2/R^2$, which is an infinite number of roots since

$$\sinh(s/a)^{1/2} R = (1/i) \sin i(s/a)^{1/2} R = 0, \qquad i(s_n/a)^{1/2} R = n\pi = \mu_n,$$

where $n = 1, 2, 3, \ldots$.

Using the expansion theorem, we obtain

$$\frac{\Phi(0)}{\psi'(0)} = \lim_{s \to 0} \left\{\frac{d}{ds} \left[\frac{\Phi(s)}{\lambda \varphi(s)} e^{s\tau}\right]\right\}$$

$$= \lim_{s \to 0} \left[\tau e^{s\tau} \frac{\Phi(s)}{\lambda \varphi(s)} + e^{s\tau} \frac{\Phi'(s)}{\lambda \varphi(s)} - e^{s\tau} \frac{\Phi(s) \varphi'(s)}{\lambda [\varphi(s)]^2}\right]$$

$$= \frac{q_c a \tau}{\lambda R} + \frac{q_c x^2}{2 \lambda R} - \frac{q_c R}{6 \lambda},$$

since

$$\Phi(0) = q_c\,, \qquad \varphi(0) = R/a\,, \qquad \Phi'(0) = x^2/2a\,, \qquad \varphi'(0) = R^2/6a^2.$$

Further,

$$\sum_{n=1}^{\infty} \frac{\Phi(s_n)}{\psi'(s_n)} e^{s_n\tau} = -\frac{q_c R}{\lambda} \sum_{n=1}^{\infty} \frac{2}{\mu_n^2 \cos\mu_n} \cos\mu_n \frac{x}{R} \exp\left[-\mu_n^2 \frac{a\tau}{R^2}\right],$$

since

$$\psi'(s) = \lambda\left[\frac{3}{2}\left(\frac{s}{a}\right)^{1/2} \sinh\left(\frac{s}{a}\right)^{1/2} R + \frac{sR}{2a}\cosh\left(\frac{s}{a}\right)^{1/2} R\right],$$

$$\psi'(s_n) = \lambda \frac{s_n R}{2a} \cosh\left(\frac{s}{a}\right)^{1/2} R.$$

Hence the solution of our problem will be

$$t(x, \tau) - t_0 = \frac{q_c}{\lambda}\left[\frac{a\tau}{R} - \frac{R^2 - 3x^2}{6R} + R\sum_{n=1}^{\infty}(-1)^{n+1}\frac{2}{\mu_n^2}\cos\mu_n\frac{x}{R}\exp[-\mu_n^2\,\mathrm{Fo}]\right]. \tag{5.2.20}$$

Solution (5.2.20) is identical with (5.2.14).

d. Analysis of Solution and Heat Rate. We introduce the Kirpichev criterion

$$\mathrm{Ki} = \frac{q_c R}{\lambda(t_a - t_0)},$$

where t_a is the average temperature of the furnace. Then our solution acquires the form

$$\theta = \frac{t(x, \tau) - t_0}{(t_a - t_0)} = \mathrm{Ki}\left[\mathrm{Fo} - \frac{1}{6}\left(1 - 3\frac{x^2}{R^2}\right) + \sum_{n=1}^{\infty}(-1)^{n+1}\frac{2}{\mu_n^2} \times \cos\mu_n\frac{x}{R}\exp[-\mu_n^2\,\mathrm{Fo}]\right]. \tag{5.2.21}$$

Thus, the relative temperature will be a function of Ki, Fo, and x/R, i.e.,

$$\theta = f\{\mathrm{Ki}, \mathrm{Fo}, (x/R)\}. \tag{5.2.22}$$

With the increase of time, or rather of the number Fo, the series decreases rapidly and at some value of $\mathrm{Fo} > \mathrm{Fo}_1$ the series becomes negligible, leaving only the first two terms of solution (5.2.21). From this time on, the temperature at any point of the plate will be a linear function of the time and the temperature distribution across the plate thickness is described by a parabolic law, i.e., quasi-stationary conditions for a temperature-gradient field are observed. For the coordinate $x = 0$ (the middle of the plate) quasi-stationary conditions appear to within $\frac{1}{2}\%$ at Fo > 0.5 (see Chapter 6, Section 10).

Our problem also may be solved in a form convenient for small values of Fo.

Using the expansion $[\sinh(s/a)^{1/2}R]^{-1}$ in series (see Appendix I) the solution for the transform is written

$$T(x, s) - \frac{t_0}{s} = \frac{q_c\sqrt{a}}{\lambda s^{3/2}} \sum_{n=1}^{\infty} \left\{\exp\left[-\left(\frac{s}{a}\right)^{1/2}((2n-1)R - x)\right] + \exp\left[-\left(\frac{s}{a}\right)^{1/2}((2n-1)R + x)\right]\right\}. \tag{5.2.23}$$

Then, when applying the table of transforms and, in particular, relation (5.2.18) of the previous section, the general solution of our problem will have the form

$$t(x, \tau) - t_0 = \frac{2q_c}{\lambda}(a\tau)^{1/2} \sum_{n=1}^{\infty} \left[\text{i erfc}\,\frac{(2n-1)R - x}{2(a\tau)^{1/2}} + \text{i erfc}\,\frac{(2n-1)R + x}{2(a\tau)^{1/2}}\right]. \tag{5.2.24}$$

Solution (5.1.19) may be obtained from solution (5.2.24) if the variable $x + R = X$ is substituted and $2R = \infty$ is assumed. In criterial form, solution (5.2.24) may be written as

$$\theta = \frac{t(x, \tau) - t_0}{t_a - t_0} = 2\text{Ki}(\text{Fo})^{1/2} \sum_{n=1}^{\infty} \left[\text{i erfc}\,\frac{(2n-1) - (x/R)}{2(\text{Fo})^{1/2}} + \text{i erfc}\,\frac{(2n-1) + (x/R)}{2(\text{Fo})^{1/2}}\right]. \tag{5.2.25}$$

At small values of Fo, all the terms of the series are negligibly small compared to the first one, so that we may confine ourselves to it. Then, the calculation formula acquires a simple form.

Figure 5.2 gives the curves of the temperature distribution inside the plate for various values of the criterion Fo (from 0.05 to 1.1) and Fig. 5.3 shows the temperature at the surface and the middle of the plate during the heating process.

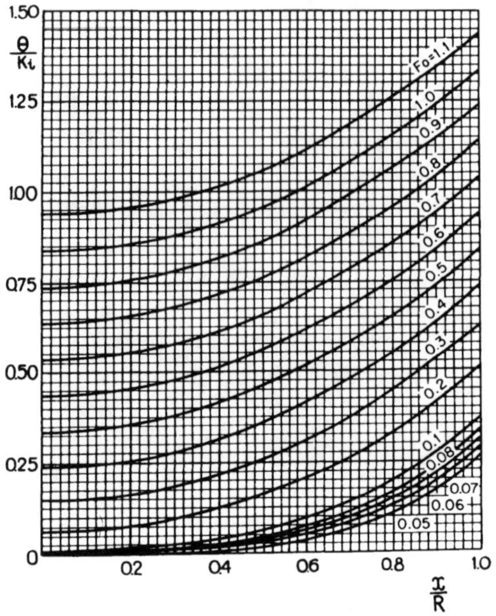

Fig. 5.2. Temperature field of an infinite plate with constant heat flux at the surface.

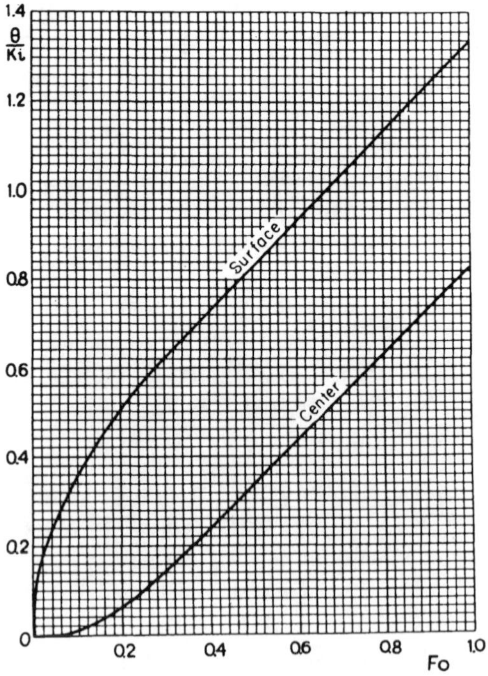

Fig. 5.3. Relation between θ/Ki and Fo for the surface and center of the plate.

We now give an illustrative example. In a drying chamber of infrared radiation, a ceramic plate ($2R = 4$ cm) is heated for 15 min. The average chamber temperature is 400°C and the radiant flux is $q_c = 2500$ kcal/m^2 hr (the thermal coefficients of the plate are assumed to be equal to $\lambda = 0.25$ kcal/m hr °C, $a = 4.8 \cdot 10^{-4}$ m^2/hr). The initial temperature of the plate is 20°C. The temperature at the surface and in the center of the plate is to be found.

First calculate the Kirpichev criterion and Fourier number,

$$\mathrm{Ki} = \frac{2500 \cdot 0.02}{0.25(400 - 20)} \approx 0.53, \qquad \mathrm{Fo} = \frac{4.8 \cdot 10^{-4} \cdot 15}{4 \cdot 10^{-4} \cdot 60} = 0.3$$

using $R = 2$ cm $= 0.02$ m.

Calculate the value of θ_c/Ki for the plate center ($x = 0$) according to the formula (5.2.21)

$$\theta_c/\mathrm{Ki} = \mathrm{Fo} - \tfrac{1}{6} + (2/\pi^2)\exp[-\pi^2\mathrm{Fo}] = 0.3 - 0.1667 + 0.2026 \quad \exp[-2.961] = 0.144.$$

We check the value obtained by referring to Fig. 5.3. From Fig. 5.3 it follows that for $\mathrm{Fo} = 0.3$, $\theta/\mathrm{Ki} = 0.14$, i.e., the calculation is correct. It is of interest to calculate this value by formula (5.2.25)

$$\begin{aligned}\theta_c/\mathrm{Ki} &= 2(\mathrm{Fo})^{1/2}\left(\mathrm{i\,erfc}\,\frac{1}{2(\mathrm{Fo})^{1/2}} + \mathrm{i\,erfc}\,\frac{1}{2(\mathrm{Fo})^{1/2}}\right) \\ &= 2(0.3)^{1/2}\, 2\mathrm{i\,erfc}(0.913) \\ &= 1.095 \cdot 0.132 = 0.144.\end{aligned}$$

The value $\mathrm{i\,erfc}(0.913)$ is determined from the table ($\mathrm{i\,erfc}(0.913) \approx 0.066$). Thus, one and the same value is obtained. So,

$$t(0, \tau) = t_0 + \theta_c(t_a - t_0) = 20° + 380 \cdot 0.53 \cdot 0.144 = 49\ °\mathrm{C}.$$

We next determine the temperature at the plate surface. According to formula (5.2.21) we have

$$\begin{aligned}\theta_s/\mathrm{Ki} &= \mathrm{Fo} + \frac{1}{3} - \frac{2}{\pi^2}\exp[-\pi^2\mathrm{Fo}] \\ &= 0.3 + 0.3333 - 0.2026 \cdot \exp[-2.961] \approx 0.623,\end{aligned}$$

and, according to formula (5.2.25) we obtain

$$\theta_s/\mathrm{Ki} = 2(\mathrm{Fo})^{1/2}\left(\mathrm{i\,erfc}\,0 + 2\,\mathrm{i\,erfc}\,\frac{2}{2(\mathrm{Fo})^{1/2}}\right) = 1.0954(0.564 + 0.005) \approx 0.623,$$

since $\mathrm{i\,erfc}\,0 = 0.5642$.

Thus, both formulas lead to one and the same result. We check this by Fig. 5.3, where we find that for

$$\mathrm{Fo} = 0.3, \qquad \frac{\theta}{\mathrm{Ki}} = 0.62.$$

The temperature at the plate surface is finally determined to be

$$t(R, \tau) = t_0 + (t_a - t_0)\theta_s = 20° + 380 \cdot 0.53 \cdot 0.623 \approx 145\ °\mathrm{C}.$$

The amount of heat transferred to the plate is determined from Eq. (5.1.26).

e. Solution by Fourier Integral Transform Method. *Consider a more general case when the heat flow q_c is the function of time $q_c = f(\tau)$.* For a generality of the problem, the nonuniform initial distribution of temperature is taken

$$t(x, 0) = f(x). \tag{5.2.26}$$

The boundary conditions may be written as

$$-\lambda \frac{\partial t(R, \tau)}{\partial x} + q(\tau) = 0, \tag{5.2.27}$$

$$\frac{\partial t(0, \tau)}{\partial x} = 0. \tag{5.2.28}$$

Let us make the integral Fourier cosine-transformation

$$T_c(n, \tau) = \int_0^R t(x, \tau) \cos(n\pi x/R)\, dx, \tag{5.2.29}$$

where $n = 0, 1, 2, 3, \ldots$ and $T_c(n, \tau)$ is the transform of the function $t(x, \tau)$ which satisfies the Dirichlet conditions.

Inversion of the transform of the function is accomplished by the formula

$$t(x, \tau) = (1/R)T_c(0, \tau) + (2/R) \sum_{n=1}^{\infty} T_c(n, \tau) \cos(n\pi x/R). \tag{5.2.30}$$

Multiplying both parts of the differential heat conduction equation by $\cos(n\pi x/R)$ and integrating between 0 and R, we obtain

$$\int_0^R \frac{\partial t(x, \tau)}{\partial \tau} \cos \frac{n\pi x}{R}\, dx = \int_0^R a \frac{\partial^2 t(x, \tau)}{\partial x^2} \cos \frac{n\pi x}{R}\, dx. \tag{5.2.31}$$

The expression for the partial derivative of the second order is

$$\int_0^R \frac{\partial^2 t(x, \tau)}{\partial x^2} \cos \frac{n\pi x}{R}\, dx = (-1)^n \frac{\partial t(R, \tau)}{\partial x} - \frac{\partial t(0, \tau)}{\partial x} - \frac{n^2\pi^2}{R^2} T_c(n, \tau). \tag{5.2.32}$$

The use of boundary conditions (5.2.27)–(5.2.28) yields

$$\frac{dT_c(n, \tau)}{\partial \tau} + \frac{an^2\pi^2}{R^2} T_c(n, \tau) = (-1)^n \frac{a}{\lambda} q(\tau). \tag{5.2.33}$$

The solution of this simple equation gives

$$T_c(n, \tau) = \exp\left[-\frac{an^2\pi^2\tau}{R^2}\right] \left[C(n) + (-1)^n \frac{a}{\lambda} \int_0^\tau q(\vartheta) \exp\left[\frac{an^2\pi^2}{R^2}\right] d\vartheta\right]. \tag{5.2.34}$$

To determine the constant $C(n)$, the initial condition (5.2.26) is used:

$$T_c(n, 0) = C(n) = \int_0^R t(x, 0) \cos \frac{n\pi x}{R}\, dx = \int_0^R f(x) \cos \frac{n\pi x}{R}\, dx. \quad (5.2.35)$$

Consequently, the solution for the transform will have the form

$$T_c(n, \tau) = (-1)^n \frac{a}{\lambda} \int_0^\tau q(\vartheta) \exp\left[-\frac{an^2\pi^2}{R^2}(\tau - \vartheta)\right] d\vartheta + \exp\left[-\frac{an^2\pi^2\tau}{R^2}\right] \int_0^R f(x) \cos \frac{n\pi x}{R}\, dx. \quad (5.2.36)$$

To accomplish the inversion of the transform by formula (5.2.30), it is more convenient if solution (5.2.36) is rewritten

$$T_c(n, \tau) = T(0, \tau) + T_n(n, \tau), \quad (5.2.37)$$

where in the second term, $n = 1, 2, 3, \ldots$. Then

$$T_c(n, \tau) = \int_0^R f(x)\, dx + \frac{a}{\lambda} \int_0^\tau q(\vartheta)\, d\vartheta + \exp\left[-\frac{an^2\pi^2\tau}{R^2}\right] \int_0^R f(x) \cos \frac{n\pi x}{R}\, dx + (-1)^n \frac{a}{\lambda} \int_0^\tau q(\vartheta) \exp\left[-\frac{an^2\pi^2}{R^2}(\tau - \vartheta)\right] d\vartheta. \quad (5.2.38)$$

The original function is now obtained by formula (5.2.30) as

$$t(x, \tau) = \frac{1}{R}\left\{\int_0^R f(x)\, dx + \frac{a}{\lambda} \int_0^\tau q(\vartheta)\, d\vartheta\right\} + \frac{2}{R} \sum_{n=1}^{\infty} \cos \frac{n\pi x}{R} \exp\left[-\frac{an^2\pi^2\tau}{R^2}\right] \times \int_0^R f(x) \cos \frac{n\pi x}{R}\, dx + \frac{2a}{R\lambda} \sum_{n=1}^{\infty} (-1)^n \cos \frac{n\pi x}{R} \times \int_0^R q(\vartheta) \exp\left[-\frac{an^2\pi^2}{R^2}(\tau - \vartheta)\right] d\vartheta. \quad (5.2.39)$$

Solution (5.2.39) is more general than (5.2.14) and (5.2.20), and both of these solutions may be obtained from (5.2.39).

If the temperature distribution at the initial moment is uniform,

$$f(x) = t_0 = \text{const},$$

solution (5.2.39) yields

$$t(x, \tau) - t_0 = \frac{a}{\lambda R} \int_0^\tau q(\vartheta)\, d\vartheta + \frac{2a}{\lambda R} \sum_{n=1}^{\infty} (-1)^n \cos \mu_n \frac{x}{R}$$
$$\times \int_0^\tau q(\vartheta) \exp\left[- \frac{a\mu_n^2}{R^2} (\tau - \vartheta) \right] d\vartheta \qquad (5.2.40)$$

where $\mu_n = n\pi$.

In deriving this expression, we took into account that for $n \neq 0$,

$$\int_0^R t_0 \cos \mu_n \frac{x}{R}\, dx = 0,$$

i.e., the first series in solution (5.2.39) becomes zero. If, in addition, the heat flow at the plate surface is uniform, i.e.,

$$q(\tau) = q_c = \text{const},$$

then solution (5.2.40) gives

$$t(x, \tau) - t_0 = \frac{q_c}{\lambda} \left[\frac{a\tau}{R} + 2\, R \sum_{n=1}^{\infty} (-1)^n \frac{1}{\mu_n^2} \cos \mu_n \frac{x}{R} \right.$$
$$\left. + 2R \sum_{n=1}^{\infty} (-1)^{n+1} \frac{1}{\mu_n^2} \cos \mu_n \frac{x}{R} \exp[-\mu_n^2\, \mathrm{Fo}] \right]. \qquad (5.2.41)$$

From the theory of Fourier series, it is known that

$$2R \sum_{n=1}^{\infty} (-1)^n \frac{1}{\mu_n^2} \cos \mu_n \frac{x}{R} = - \frac{R^2 - 3x^2}{6R}. \qquad (5.2.42)$$

Then we have finally,

$$t(x, \tau) - t_0 = \frac{q_c}{\lambda} \left[\frac{a\tau}{R} - \frac{R^2 - 3x^2}{6R} \right.$$
$$\left. + R \sum_{n=1}^{\infty} (-1)^{n+1} \frac{2}{\mu_n^2} \cos \mu_n \frac{x}{R} \exp[-\mu_n^2\, \mathrm{Fo}] \right], \qquad (5.2.43)$$

which is identical with (5.2.14) and (5.2.20).

5.3 Sphere (Symmetrical Problem)

a. Statement of Problem. *This problem is similar to the previous one, but instead of an infinite plate, we have a sphere. The surface of the sphere is heated uniformly (symmetrical problem) at a constant heat flux* $q_s = q_c$

$= const.$ *The radial temperature distribution at any time instant and the specific heat rate are to be found.*

The differential equation is given in Chapter 4, Section 4.

The initial condition is the following:

$$t(r, 0) = t_0 \tag{5.3.1}$$

Since the solution is obtained by the operational method, the boundary conditions are written for both the transform and the inverse transform as

$$-\frac{\partial t(R, \tau)}{\partial r} + \frac{q_c}{\lambda} = 0, \qquad -T'(R, s) + \frac{q_c}{\lambda s} = 0. \tag{5.3.2}$$

$$\frac{\partial t(0, \tau)}{\partial r} = 0, \qquad T'(0, s) = 0, \tag{5.3.3}$$

$$t(0, \tau) \neq \infty, \qquad T(0, s) \neq \infty. \tag{5.3.4}$$

b. Solution of the Problem. The solution of the differential heat conduction equation for the transform, when isotherms are located symmetrically with respect to the sphere center (i.e., when conditions (5.3.3) and (5.3.4) are taken into account), has the form (see solution (4.4.22))

$$T(r, s) - \frac{t_0}{s} = B\frac{\sinh(s/a)^{1/2}r}{r}. \tag{5.3.5}$$

The constant B is determined from boundary condition (5.3.2) i.e.,

$$-B\left(\frac{s}{a}\right)^{1/2}\frac{1}{R}\cosh\left(\frac{s}{a}\right)^{1/2}R + \frac{B}{R^2}\sinh\left(\frac{s}{a}\right)^{1/2}R + \frac{q_c}{\lambda} = 0,$$

hence

$$B = \frac{q_c R^2}{\lambda[(s/a)^{1/2}R\cosh(s/a)^{1/2}R - \sinh(s/a)^{1/2}R)]}. \tag{5.3.6}$$

Hence, the solution for the transform will be of the form

$$T(r, s) - \frac{t_0}{s} = \frac{q_c R(\sinh(s/a)^{1/2}r)/(s/a)^{1/2}r}{\lambda s[\cosh(s/a)^{1/2}R - (1/(s/a)^{1/2}R)\sinh(s/a)^{1/2}R]} = \frac{\Phi(s)}{\psi(s)}. \tag{5.3.7}$$

It represents the relation of two generalized polynomials

$$\Phi(s) = \frac{q_c R\sinh(s/a)^{1/2}r}{\lambda(s/a)^{1/2}r} = \frac{q_c R}{\lambda}\left(1 + \frac{1}{3!}\frac{r^2}{a}s + \frac{1}{5!}\frac{r^4}{a^2}s^2 + \cdots\right),$$

$$\begin{aligned}\psi(s) &= s\left[\cosh\left(\frac{s}{a}\right)^{1/2} R - \frac{1}{(s/a)^{1/2}R}\sinh\left(\frac{s}{a}\right)^{1/2} R\right] \\ &= s^2\left[\left(\frac{1}{2!}\,\frac{R^2}{a} + \frac{1}{4!}\,\frac{R^4}{a^2}\,s + \cdots\right)\right. \\ &\quad \left. - \left(\frac{1}{3!}\,\frac{R^2}{a} + \frac{1}{5!}\,\frac{R^4}{a^2}\,s + \cdots\right)\right] = s^2\varphi(s),\end{aligned}$$

where $\varphi(s)$ is the expression in square brackets which is a generalized polynomial with respect to s. Thus, the conditions of the expansion theorem are fulfilled.

To find the roots of the expression $\psi(s)$, we equate it to zero. Hence we obtain $s = 0$ (double root) and $s_n = -a\mu_n{}^2/R^2$, which is the infinite number of simple roots determined from the characteristic equation

$$\tan\mu = \mu\,, \tag{5.3.8}$$

which is obtained as follows:

$$\begin{aligned}&\cosh(s/a)^{1/2}R - (1/(s/a)^{1/2}R)\sinh(s/a)^{1/2}R \\ &\quad = \cos i(s/a)^{1/2}R - (1/i(s/a)^{1/2}R)\sin i(s/a)^{1/2}R = 0.\end{aligned}$$

Designating $i(s/a)^{1/2}R = \mu$, we obtain $\cos\mu - (1/\mu)\sin\mu = 0$, i.e., characteristic equation (5.3.8). Applying the expansion theorem (the case of multiple roots) to the root $s = 0$, we find

$$\begin{aligned}\frac{\Phi(0)}{\psi'(0)} &= \lim_{s\to 0}\left[\frac{d}{ds}\left(e^{s\tau}\,\frac{\Phi(s)}{\varphi(s)}\right)\right] \\ &= \lim_{s\to 0}\left[\tau e^{s\tau}\,\frac{\Phi(s)}{\varphi(s)} + e^{s\tau}\,\frac{\Phi'(s)}{\varphi(s)} - e^{s\tau}\,\frac{\Phi(s)\varphi'(s)}{[\varphi(s)]^2}\right] \\ &= \frac{q_cR}{\lambda}\left(\frac{3a\tau}{R^2} + \frac{r^2}{2R^2} - \frac{3}{10}\right),\end{aligned}$$

since

$$\Phi(0) = \frac{q_cR}{\lambda}, \qquad \varphi(0) = \frac{1}{3a}R^2, \qquad \Phi'(0) = \frac{1}{6}\,\frac{r^2}{a}, \qquad \varphi'(0) = \frac{1}{30}\,\frac{R^4}{a^2}.$$

Then, we apply initially, the expansion theorem to the simple roots s_n.
Initially, we find

$$\begin{aligned}\psi'(s) &= \left[\cosh\left(\frac{s}{a}\right)^{1/2} R - \frac{1}{(s/a)^{1/2}R}\sinh\left(\frac{s}{a}\right)^{1/2} R\right] \\ &\quad + \frac{1}{2}\left(\frac{s}{a}\right)^{1/2} R\sinh\left(\frac{s}{a}\right)^{1/2} R - \frac{1}{2}\cosh\left(\frac{s}{a}\right)^{1/2} R + \frac{1}{2}\,\frac{\sinh\left(\frac{s}{a}\right)^{1/2} R}{\left(\frac{s}{a}\right)^{1/2} R},\end{aligned}$$

$$= -\tfrac{1}{2}\mu_n \sin \mu_n$$
$$= -\tfrac{1}{2}\mu_n^2 \cos \mu_n .$$

$$\psi'(s_n) = \frac{1}{2}\left[\frac{1}{i}\left(\frac{s_n}{a}\right)^{1/2} R \sin i\left(\frac{s_n}{a}\right)^{1/2} R\right].$$

For the latter equation, we took into account the characteristic equation (5.3.8). For the quantity $\Phi(s_n)$,

$$\Phi(s_n) = \frac{q_c R}{\lambda}\,\frac{\sin i(s_n/a)^{1/2} r}{i(s_n/a)^{1/2} r} = \frac{q_c R^2 \sin \mu_n \dfrac{r}{R}}{\lambda r \mu_n}.$$

Consequently, the solution of our problem will be of the form

$$t(r, \tau) - t_0 = \frac{q_c R}{\lambda}\left[\frac{3a\tau}{R^2} - \frac{3R^2 - 5r^2}{10R^2} - \sum_{n=1}^{\infty} \frac{2}{\mu_n^2 \cos \mu_n}\,\frac{R \sin \mu_n \dfrac{r}{R}}{r\mu_n} \exp\left(-\mu_n^2 \frac{a\tau}{R^2}\right)\right]. \tag{5.3.9}$$

The roots of the characteristic equation (5.3.8) are a series of values independent of the criterion Ki, viz: $\mu_1 = 4.4934$, $\mu_2 = 7.7253$, $\mu_3 = 10.9041$, $\mu_4 = 17.2208$, etc. (see Table 6.5 for Bi = 0).

c. Analysis of the Solution and the Specific Heat Rate. We write the solution in a criterial form as

$$\theta = \frac{t(r, \tau) - t_0}{t_a - t_0}$$
$$= \mathrm{Ki}\left[3\,\mathrm{Fo} - \frac{1}{10}\left(3 - 5\frac{r^2}{R^2}\right) - \sum_{n=1}^{\infty} \frac{2}{\mu_n^2 \cos \mu_n}\right.$$
$$\left. \times \frac{R \sin \mu_n \dfrac{r}{R}}{r\mu_n} \exp[-\mu_n^2\,\mathrm{Fo}]\right], \tag{5.3.10}$$

where Ki is the Kirpichev criterion.

The series in solution (5.3.10) converges rapidly and, from some value of $\mathrm{Fo} > \mathrm{Fo}_1$, it may be neglected compared to the first two terms in the brackets. From this value, the temperature at any point of the sphere will be a linear function of time and the temperature distribution will be parabolic.

We next find an approximate solution for small values of Fo. We return to solution (5.3.7), which may be rewritten as

$$T(r, s) - \frac{t_0}{s} = \frac{q_c R^2}{\lambda r} \frac{\sinh(s/a)^{1/2} r}{s[(s/a)^{1/2} R \cosh(s/a)^{1/2} R - \sinh(s/a)^{1/2} R]}. \quad (5.3.11)$$

At small values of Fo the value of $(s/a)^{1/2}R$ is great. At large values of $(s/a)^{1/2}R$ (> 6.0), it is known that $\sinh u = \cosh u = \frac{1}{2}e^u$, $\tanh u = \coth u = 1$ to within the third decimal place. Then, for values of r close to R, $\sinh (s/a)^{1/2} r$ may also be replaced by $\frac{1}{2} \exp[(s/a)^{1/2} r]$, i.e.,

$$T(r, s) - \frac{t_0}{s} \approx \frac{q_c R^2}{\lambda [rs(s/a)^{1/2} R - 1]} \exp[-(s/a)^{1/2}(R - r)]. \quad (5.3.12)$$

Using the table of transforms (See Appendix 5, formula (56)), we find

$$\theta = \frac{t(r, \tau) - t_0}{t_a - t_0}$$
$$= \mathrm{Ki} \frac{R}{r} \left[\exp\left[\mathrm{Fo} - \frac{R - r}{R}\right] \mathrm{erfc}\left(\frac{1 - \dfrac{r}{R}}{2(\mathrm{Fo})^{1/2}} - (\mathrm{Fo})^{1/2} \right) - \mathrm{erfc} \frac{1 - \dfrac{r}{R}}{2(\mathrm{Fo})^{1/2}} \right]. \quad (5.3.13)$$

For the sphere center ($r = 0$), the solution (5.3.11) may be written

$$T(0, s) - \frac{t_0}{s} = \frac{q_c R}{\lambda s [\cosh(s/a)^{1/2} R - (1/(s/a)^{1/2} R) \sin(s/a)^{1/2} R]}$$
$$\approx \frac{2 q_c R}{\lambda \sqrt{s} (\sqrt{s} - \{\sqrt{a}/R\})} \exp[-(s/a)^{1/2} R]. \quad (5.3.14)$$

Then using the table of transforms (see Appendix 5, formula (57)), we have

$$\theta \approx 2\,\mathrm{Ki} \left\{ [\exp(\mathrm{Fo} - 1)]\, \mathrm{erfc}\left(\frac{1}{2(\mathrm{Fo})^{1/2}} - (\mathrm{Fo})^{1/2} \right) \right\}. \quad (5.3.15)$$

From the exact solution (5.3.10) and the approximate solutions (5.3.13) and (5.3.15), it follows that the relative excess temperature is directly proportional to the Kirpichev criterion and is also dependent on the number Fo and the relative coordinate r/R. Thus the relation θ/Ki is a function of Fo and r/R alone

$$\frac{\theta}{\mathrm{Ki}} = f\left(\mathrm{Fo}, \frac{r}{R}\right). \quad (5.3.16)$$

In Fig. 5.4, the curves of the distribution of the quantity θ/Ki are shown as a functional relative coordinate r/R for various values of the Fourier

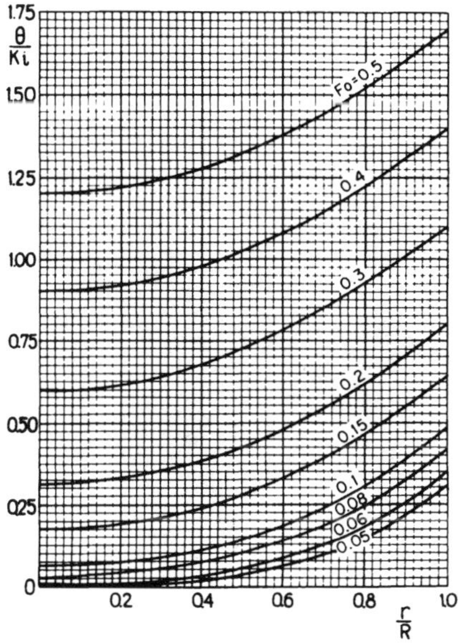

Fig. 5.4. Temperature field of a sphere for q_c = const (symmetrical problem).

number (from 0.05 to 0.5). From Fo = 0.5, a heating process becomes quasi stationary: the temperature of any point increases in a linear fashion and the temperature distribution follows the parabolic law. In Fig. 5.5, the quantity θ/Ki is plotted versus the Fourier number for the surface and the center of the sphere. These graphs considerably simplify the calculation.

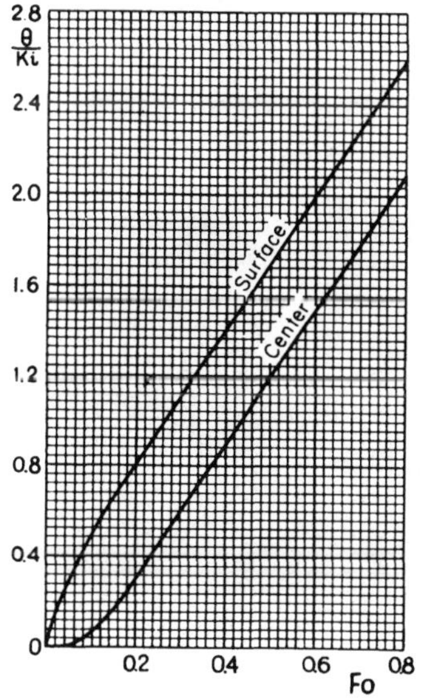

Fig. 5.5. Relation between θ/Ki and Fo for the surface and center of a sphere.

The specific heat rate per unit volume is determined by the formula

$$\Delta Q_v = (3/R)q_c\tau. \tag{5.3.17}$$

d. The Solution for the Case $q(\tau)$, $t_0 = f(\tau)$. Consider a more general problem with the following boundary conditions

$$t(r, 0) = f(r), \tag{5.3.18}$$

$$-\lambda \frac{\partial t(R, \tau)}{\partial r} + q(\tau) = 0, \tag{5.3.19}$$

$$\partial t(0, \tau)/\partial r = 0. \tag{5.3.20}$$

Use the finite integral transform, we have

$$T_F(p, \tau) = \int_0^R rt(r, \tau) \frac{\sin pr}{p} dr, \tag{5.3.21}$$

where p is the root of the characteristic equation

$$\sin pR - pR \cos pR = 0. \tag{5.3.22}$$

The inversion of the transform $T_F(p, \tau)$ to obtain $t(r, \tau)$ is carried out by the formula

$$t(r, \tau) = \frac{3}{R^3} T_F(0, \tau) + \frac{2}{R} \sum_{n=1}^{\infty} \frac{p_n}{\sin^2 p_n R} \frac{\sin p_n r}{r} T_F(p_n, \tau). \tag{5.3.23}$$

Application of transform (5.3.21) to the differential heat conduction equation and taking into account condition (5.3.20), gives us

$$\int_0^R \left(\frac{\partial^2 t(r, \tau)}{\partial r^2} + \frac{2}{r} \frac{\partial t(r, \tau)}{\partial r}\right) r \frac{\sin pr}{p} dr$$
$$= \left[\frac{\partial t(r, \tau)}{\partial r} r \frac{\sin pr}{p}\right]_{r=R} - p^2 T_F(p, \tau). \tag{5.3.24}$$

From (5.3.19) it follows that

$$\int_0^R \left(\frac{\partial^2 t(r, \tau)}{\partial r^2} + \frac{2}{r} \frac{\partial t(r, \tau)}{\partial r}\right) r \frac{\sin pr}{p} dr$$
$$= \frac{q(\tau)}{\lambda} R \frac{\sin pR}{p} - p^2 T_F(p, \tau). \tag{5.3.25}$$

Multiplying now all the terms of the differential heat conduction equation by $r(\sin pr)/p$ and integrating with respect to r between 0 and R, we obtain on the basis of (5.3.25)

$$\frac{dT_F(p, \tau)}{d\tau} + ap^2 T_F(p, \tau) = \frac{a}{\lambda} R \frac{\sin pR}{p} q(\tau). \tag{5.3.26}$$

Solution of Eq. (5.3.26) gives

$$T_F(p, \tau) = \exp[-ap^2\tau] \left\{ C(p) + \frac{a}{\lambda} R \frac{\sin pR}{p} \times \int_0^\tau q(\vartheta) \exp[ap^2\vartheta] \, d\vartheta \right\}. \tag{5.3.27}$$

To determine $C(p)$, we make use of initial condition (5.3.18) and obtain

$$C(p) = \int_0^R rf(r) \frac{\sin pr}{p} \, dr. \tag{5.3.28}$$

Then solution (5.3.27) may be written as

$$T_F(p, \tau) = \exp[-ap^2\tau] \left\{ \int_0^R rf(r) \frac{\sin pr}{p} \, dr + \frac{a}{\lambda} R \frac{\sin pR}{p} \int_0^\tau q(\vartheta) \exp[ap^2\vartheta] \, d\vartheta \right\}. \tag{5.3.29}$$

For convenience of the transition to the inverse transform, we find

$$T_F(0, \tau) = \int_0^R r^2 f(r) \, dr + \frac{a}{\lambda} R^2 \int_0^\tau q(\vartheta) \, d\vartheta. \tag{5.3.30}$$

Substitution of $T_F(0, \tau)$ and $T_F(p, \tau)$ into Eq. (5.3.23) gives the solution

$$\begin{aligned} t(r, \tau) = {} & \frac{3}{R^3} \int_0^R r^2 f(r) \, dr + \frac{3a}{\lambda R} \int_0^\tau q(\vartheta) \, d\vartheta \\ & + \sum_{n=1}^{\infty} \frac{p_n}{\sin^2 p_n R} \frac{\sin p_n r}{r} \exp[-a{p_n}^2\tau] \\ & \times \frac{2}{R} \int_0^R rf(r) \frac{\sin p_n r}{p_n} \, dr + \frac{a}{\lambda} \sum_{n=1}^{\infty} \frac{Rp_n \sin p_n R}{p_n \sin^2 p_n R} \\ & \times \frac{\sin p_n r}{r} \exp[-a{p_n}^2\tau] \\ & \times \frac{2}{R} \int_0^\tau q(\vartheta) \exp[a{p_n}^2\vartheta] \, d\vartheta. \end{aligned} \tag{5.3.31}$$

Designating $\mu_n = p_n R$; $\text{Fo} = a\tau/R^2$ and substituting from the characteristic equation $\sin\mu = \mu\cos\mu$, we arrive at the final form of the solution:

$$
\begin{aligned}
t(r,\tau) = {} & \frac{3}{R^3}\int_0^R r^2 f(r)\,dr + \frac{3a}{\lambda R}\int_0^\tau q(\vartheta)\,d\vartheta \\
& + \sum_{n=1}^{\infty} \frac{2}{\mu_n^2\cos^2\mu_n} \frac{\mu_n \sin\mu_n r/R}{rR} \\
& \times \exp[-\mu_n^2\,\text{Fo}] \int_0^R r f(r) \frac{\sin\mu_n r/R}{\mu_n}\,dr \\
& + \frac{a}{\lambda}\sum_{n=1}^{\infty} \frac{2}{\mu_n\cos\mu_n} \frac{\sin\mu_n r/R}{r} \exp[-\mu_n^2\,\text{Fo}] \\
& \times \int_0^\tau q(\vartheta)\exp\left(\mu_n^2\frac{\vartheta a}{R^2}\right)d\vartheta.
\end{aligned}
\tag{5.3.32}
$$

In the specific case

$$t(r,0) = t_0 = \text{const}, \qquad q(\tau) = q_c = \text{const}.$$

Solution (5.3.9) is found from solution (5.3.32).

5.4 Infinite Cylinder

a. Statement of the Problem. *For an infinite cylinder with radius R, the statement of our problem is the same. The whole surface of the cylinder is heated uniformly by constant heat flow (symmetrical problem).*

The differential heat conduction equation for an infinite cylinder when the temperature depends only on r and τ, is given in Chapter 4, Section 5.

Initial and boundary conditions are identical with conditions (5.3.1)–(5.3.4) of the previous problem.

b. Solution of the Problem. The solution of the differential equation for the initial condition (5.3.1) and conditions (5.3.3) and (5.3.4) may be written (see solution (4.5.31)

$$T(r,s) - (t_0/s) = AI_0\{(s/a)^{1/2}r\} \tag{5.4.1}$$

where $I_0\{(s/a)^{1/2}r\}$ is the modified Bessel function of the first kind and zeroth order and A is the constant with respect to r. The constant A is determined from the boundary condition

$$-\left(\frac{s}{a}\right)^{1/2} AI_0'\left\{\left(\frac{s}{a}\right)^{1/2}R\right\} + \frac{q_c}{\lambda s} = 0.$$

We obtain

$$A = \frac{q_c}{\lambda s(s/a)^{1/2} I_1\{(s/a)^{1/2} R\}},$$

as $I_0'(z) = I_1(z)$ is the modified Bessel function of the first kind and the first order.

Hence, the solution of the transform may be written

$$T(r, s) - \frac{t_0}{s} = \frac{q_c R I_0\{(s/a)^{1/2} r\}}{\lambda s(s/a)^{1/2} R I_1\{(s/a)^{1/2} R\}} = \frac{\Phi(s)}{\psi(s)}. \tag{5.4.2}$$

This solution is a ratio of two generalized polynomials with respect to s; the polynomial $\psi(s)$ does not contain a constant and the polynomial $\Phi(s)$ has a constant equal to $q_c R$, i.e.,

$$\Phi(s) = q_c R I_0 \left\{ \left(\frac{s}{a} \right)^{1/2} r \right\}$$

$$= q_c R \left(1 + \frac{r^2}{2^2 a} s + \frac{r^4}{2^2 4^2 a^2} s^2 + \cdots \right), \tag{5.4.3}$$

$$\psi(s) = \lambda s \left(\frac{s}{a} \right)^{1/2} R I_1 \left\{ \left(\frac{s}{a} \right)^{1/2} R \right\}$$

$$= \lambda s^2 \left(\frac{1}{2} \frac{R^2}{a} + \frac{1}{2^2 4} \frac{R^4}{a^2} s + \cdots \right)$$

$$= \lambda s^2 \varphi(s), \tag{5.4.4}$$

where $\varphi(s)$ is the expression in brackets being a polynomial with respect to s.

We now use the expansion theorem to find the fundamental roots of the expression. For this purpose, the latter is equated with zero:

$$\psi(s) = \lambda s(s/a)^{1/2} R I_1\{(s/a)^{1/2} R\} = \lambda s^2 \varphi(s) = 0.$$

Hence, we obtain $s = 0$ (double root), $I_1\{(s/a)^{1/2} R\} = (1/i) J_1\{i(s/a)^{1/2} R\} = 0$, i.e., the infinite number of roots $s_n = -a\mu_n^2/R^2$, where $i(s/a)^{1/2} R = \mu$ are the roots of the function $J_1(\mu)$ (see Table 5.1).

We now find the value of $\Phi(0)/\psi'(0)$ as

$$\frac{\Phi(0)}{\psi'(0)} = \lim_{s \to 0} \left[\frac{d}{ds} \left(e^{s\tau} \frac{\Phi(s)}{\lambda \varphi(s)} \right) \right] = \frac{q_c R}{\lambda} \left(\frac{2a\tau}{R^2} + \frac{1}{2} \frac{r^2}{R^2} - \frac{1}{4} \right),$$

TABLE 5.1. ROOTS OF CHARACTERISTIC EQUATIONS $J_0(\mu) = 0$ AND $J_1(\mu) = 0$

n	Roots μ_n of equation $J_0(\mu) = 0$	Roots μ_n of equation $J_1(\mu) = 0$	n	Roots μ_n of equation $J_0(\mu) = 0$	Roots μ_n of equation $J_1(\mu) = 0$
1	2.4048	3.8317	6	18.0711	19.6159
2	5.5201	7.0156	7	21.2116	22.7601
3	8.6537	10.1735	8	24.3525	25.9037
4	11.7915	13.3237	9	27.4935	29.0468
5	14.9309	16.4706	10	30.6346	32.1897

since

$$\Phi(0) = q_c R, \qquad \varphi(0) = \frac{R^2}{2a}, \qquad \Phi'(0) = q_c R\,\frac{r^2}{4a}, \qquad \varphi'(0) = \frac{R^4}{16a^2}.$$

Further, we determine $\psi'(s_n)$ as

$$\psi'(s) = \frac{3}{2}\lambda\left(\frac{s}{a}\right)^{1/2} I_1\left\{\left(\frac{s}{a}\right)^{1/2} R\right\} + \frac{\lambda s R^2}{2a} I_1'\left\{\left(\frac{s}{a}\right)^{1/2} R\right\},$$

$$\lim_{s\to s_n} \psi'(s_n) = \frac{\lambda s_n R^2}{2a} I_1'\left\{\left(\frac{s_n}{a}\right)^{1/2} R\right\} = -\frac{\mu_n^2}{2}\lambda J_1'(\mu_n).$$

since $I_1'(z) = J_1'(iz)$. Thus, we obtain

$$\sum_{n=1}^{\infty} \frac{\Phi(s_n)}{\psi'(s_n)} e^{s_n \tau} = -\frac{q_c R}{\lambda} \sum_{n=1}^{\infty} \frac{2}{\mu_n^2 J_1'(\mu_n)} J_0\left(\mu_n \frac{r}{R}\right) \exp\left[-\mu_n^2 \frac{a\tau}{R^2}\right].$$

Using the recurrence formula, we obtain

$$\mu_n J_1'(\mu_n) = \mu_n J_0(\mu_n) - J_1(\mu_n),$$

since, according to the characteristic equation $J_1(\mu_n) = 0$,

$$J_1'(\mu_n) = J_0(\mu_n).$$

Hence, the solution of our problem may be written

$$\begin{aligned}\theta &= \frac{t(r,\tau) - t_0}{t_a - t_0} \\ &= \mathrm{Ki}\Big[2\mathrm{Fo} - \tfrac{1}{4}\left(1 - 2\frac{r^2}{R^2}\right) \\ &\quad - \sum_{n=1}^{\infty} \frac{2}{\mu_n^2 J_0(\mu_n)} J_0\left(\mu_n \frac{r}{R}\right) \exp[-\mu_n^2\,\mathrm{Fo}]\Big].\end{aligned} \tag{5.4.5}$$

c. Analysis of the Solution and Specific Heat Rate. The series in solution (5.4.5) converges rapidly because μ_n are large quantities (see Table 5.1). Therefore, from a certain value of Fo $>$ Fo_1, the series may be neglected and the temperature at any point of the cylinder will be a linear function of time and the temperature distribution will be parabolic (quasi-stationary condition).

Figure 5.6 gives distribution curves of the dimensionless quantity θ/Ki for various values of Fo. From Fig. 5.6 it follows that from Fo $= 0.6$,

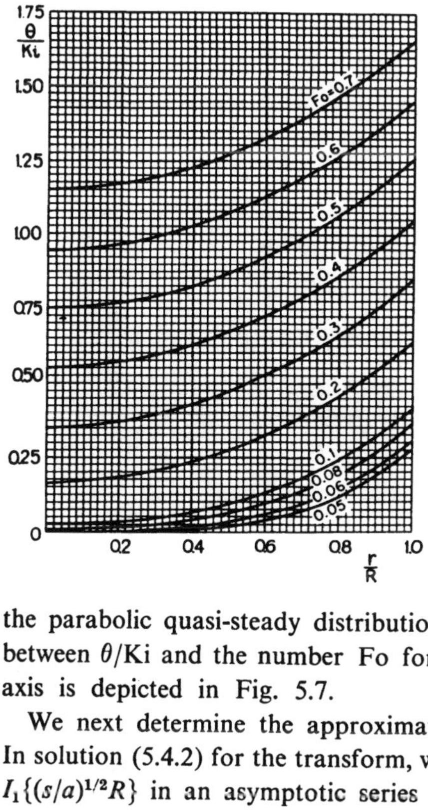

Fig. 5.6. Temperature field of an infinite cylinder for $q_c = \mathrm{const}$ (symmetrical problem).

the parabolic quasi-steady distribution of θ/Ki is obtained. The relation between θ/Ki and the number Fo for the surface and along the cylinder axis is depicted in Fig. 5.7.

We next determine the approximate solution for small values of Fo. In solution (5.4.2) for the transform, we expand the functions $I_0\{(s/a)^{1/2}R\}$ $I_1\{(s/a)^{1/2}R\}$ in an asymptotic series

$$I_0(u) \simeq \frac{1}{(2\pi u)^{1/2}}\, e^u\left(1 + \frac{1}{8u} + \frac{9}{128u^2} + \cdots\right),$$

$$I_1(u) \simeq \frac{1}{(2\pi u)^{1/2}}\, e^u\left(1 - \frac{3}{8u} - \frac{15}{128u^2} - \cdots\right).$$

Restricting our attention to the first two terms, we may write solution (5.4.2) as

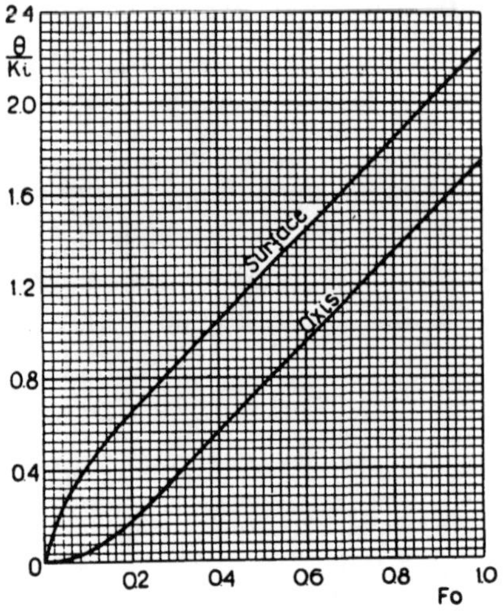

Fig. 5.7. Relation between θ/Ki and Fo for the surface and the cylinder axis.

$$T(r, s) - \frac{t_0}{s} \approx \frac{q_c}{\lambda}\left[\frac{(aR)^{1/2}}{s(rs)^{1/2}} + \frac{(R + 3r)a}{8rs^2(rR)^{1/2}}\right] \exp\left[-\left(\frac{s}{a}\right)^{1/2}(R - r)\right]. \tag{5.4.6}$$

Making use of the table of transforms, we obtain the approximate solution

$$\theta \approx \mathrm{Ki}\left[2\left(\frac{R}{r}\,\mathrm{Fo}\right)^{1/2} \mathrm{i\,erfc}\,\frac{1 - (r/R)}{2(\mathrm{Fo})^{1/2}} + \frac{\{1 + 3(r/R)\}\,\mathrm{Fo}\,R^2}{2r(Rr)^{1/2}}\,\mathrm{i}^2\,\mathrm{erfc}\,\frac{1 - (r/R)}{2(\mathrm{Fo})^{1/2}} + \cdots\right], \tag{5.4.7}$$

where

$$\mathrm{i\,erfc}\,u = \frac{1}{\sqrt{\pi}}\exp[-u^2] - u\,\mathrm{erfc}\,u, \qquad \mathrm{i}^2\,\mathrm{erfc}\,u = \tfrac{1}{4}\,[\mathrm{erfc}\,u - 2u\,\mathrm{i\,erfc}\,u].$$

The temperature at the cylinder surface ($r = R$) will change with time, as

$$\theta_s \approx 2\mathrm{Ki}\{\mathrm{Fo}/\pi\}^{1/2} + \tfrac{1}{2}\mathrm{Ki}\,\mathrm{Fo}.$$

Hence, at the beginning of the heating process, the temperature of the cylinder surface increases according to a parabolic law and then to a linear law (a quasi-stationary condition); thus, there is some analogy with a change in the average temperature $\bar{t}(\tau)$ when an infinite plate is heated (see Section 5.3). In the last case, the average temperature first increases according to a parabolic law and then to an exponential law.

The specific heat rate ΔQ_v may be determined by the formula

$$\Delta Q_v = (2/R)q_c\tau. \tag{5.4.8}$$

d. The Solution for the Case $q(\tau)$, $t_0 = f(r)$. Next, let us consider a more general problem with the conditions

$$t(r, 0) = f(r), \tag{5.4.9}$$

$$-\lambda \frac{\partial t(R, \tau)}{\partial r} + q(\tau) = 0, \tag{5.4.10}$$

$$\frac{\partial t(0, \tau)}{\partial r} = 0. \tag{5.4.11}$$

To solve this problem, we make use of the finite integral Hankel transform

$$T_H(p, \tau) = \int_0^R rt(r, \tau) J_0(rp)\, dr, \tag{5.4.12}$$

where $J_0(z)$ is the Bessel function of the first kind and the zeroth order and p is the root of the characteristic equation

$$J_0'(p, R) = 0.$$

The transition from the transform $T_H(p, \tau)$ to $t(r, \tau)$ is carried out by the formula

$$t(r, \tau) = \frac{2}{R^2} T_H(0, \tau) + \frac{2}{R^2} \sum_{n=1}^{\infty} T_H(p_n, \tau) \frac{J_0(p_n r)}{J_0^2(p_n R)}. \tag{5.4.13}$$

Applying the transform formula (5.4.12) to the differential heat condition equation and taking into account conditions (5.4.10) and (5.4.11) we obtain

$$\begin{aligned} &\int_0^R \left(\frac{\partial^2 t(r, \tau)}{\partial r^2} + \frac{1}{r} \frac{\partial t(r, \tau)}{\partial r} \right) rJ_0(rp)\, dr \\ &= RJ_0(Rp) \frac{\partial t(R, \tau)}{\partial r} - p^2 T_H(p, \tau) \\ &= RJ_0(Rp) \frac{q(\tau)}{\lambda} - p^2 T_H(p, \tau). \end{aligned} \tag{5.4.14}$$

Multiplication of all the terms of the differential heat conduction equation by $rJ_0(pr)$ and integration with respect to r between 0 and R with regard for (5.4.14) yield

$$\frac{dT_H(p, \tau)}{d\tau} + ap^2 T_H(p, \tau) = aR \frac{q(\tau)}{\lambda} J_0(pR). \tag{5.4.15}$$

The solution of this ordinary differential equation will be

$$T_H(p, \tau) = \exp[-ap^2\tau]\Big[C(p) + R\frac{a}{\lambda}J_0(pR)\int_0^\tau q(\vartheta)\exp[ap^2\vartheta]\,d\vartheta\Big]. \qquad (5.4.16)$$

To determine $C(p)$, we use condition (5.4.9). From Eq. (5.4.16) at $\tau \to 0$ we obtain

$$T_H(p, 0) = C(p).$$

On the other hand, according to formula (5.4.12) the initial condition (5.4.9) may be written as

$$T_H(p, 0) = \int_0^R t(r, 0)rJ_0(pr)\,dr = \int_0^R f(r)rJ_0(pr)\,dr. \qquad (5.4.17)$$

Hence

$$C(p) = \int_0^R f(r)rJ_0(pr)\,dr. \qquad (5.4.18)$$

Substitution of (5.4.18) into (5.4.16) gives

$$T_H(p, \tau) = \exp[-ap^2\tau]\Big[\int_0^R f(r)rJ_0(pr)\,dr + R\frac{a}{\lambda}J_0(pR)\int_0^\tau q(\vartheta)\exp[ap^2\vartheta]\,d\vartheta\Big]. \qquad (5.4.19)$$

Inversion of the transform $T_H(p, \tau)$ is carried out by formula (5.4.13). Initially we find $T_H(0, \tau)$ as

$$T_H(0, \tau) = \int_0^R f(r)r\,dr + \frac{a}{\lambda}R\int_0^\tau q(\vartheta)\,d\vartheta. \qquad (5.4.20)$$

Substitution of $T_H(0, \tau)$ and $T_H(p, \tau)$ into (5.4.13) gives

$$t(r, \tau) = \frac{2}{R^2}\int_0^R f(r)r\,dr + \frac{2a}{\lambda R}\int_0^\tau q(\vartheta)\,d\vartheta + \sum_{n=1}^\infty \frac{J_0(p_n r)}{J_0^2(p_n R)}\exp[-ap_n^2\tau]\frac{2}{R^2}\int_0^R f(r)rJ_0(p_n r)\,dr + \frac{a}{\lambda}\sum_{n=1}^\infty \frac{J_0(p_n R)J_0(p_n r)}{J_0^2(p_n R)}\exp[-ap_n^2\tau]\frac{2}{R}\int_0^\tau q(\vartheta)\exp[ap_n^2\vartheta]\,d\vartheta. \qquad (5.4.21)$$

Using the symbols $\mu_n = p_n R$, $\mathrm{Fo} = a\tau/R^2$ we obtain the solution of the problem in the form

$$t(r,\tau) = \frac{2}{R^2}\int_0^R r f(r)\,dr + \frac{2a}{\lambda R}\int_0^\tau q(\vartheta)\,d\vartheta$$
$$+ \sum_{n=1}^{\infty} \frac{J_0\{\mu_n(r/R)\}}{J_0^2(\mu_n)} \exp[-\mu_n^2\,\mathrm{Fo}]\,\frac{2}{R^2}\int_0^R r f(r) J_0\left(\mu_n \frac{r}{R}\right) dr$$
$$+ \frac{a}{\lambda}\sum_{n=1}^{\infty} \frac{J_0\left(\mu_n \frac{r}{R}\right)}{J_0(\mu_n)} \exp[-\mu_n^2\,\mathrm{Fo}]\,\frac{2}{R}\int_0^\tau q(\vartheta)\exp\left(\mu_n \frac{a\vartheta}{R^2}\right) d\vartheta, \tag{5.4.22}$$

where μ_n are the roots of the characteristic equation

$$J_0'(\mu) = J_1(\mu) = 0. \tag{5.4.23}$$

From the general solution (5.4.22) we obtain solution (5.4.5) as a specific case. For this purpose it is sufficient to assume

$$t(r,0) = f(r) = t_0 = \mathrm{const}, \qquad q(\tau) = q_c = \mathrm{const},$$

$$t(r,\tau) - t_0 = \frac{q_c R}{\lambda}\left[2\mathrm{Fo} - \frac{1}{4}\left(1 - 2\frac{r^2}{R^2}\right) - \sum_{n=1}^{\infty} \frac{2}{\mu_n^2 J_0(\mu_n)} J_0\left(\mu_n \frac{r}{R}\right)\exp[-\mu_n^2\,\mathrm{Fo}]\right]. \tag{5.4.24}$$

If we substitute $\mathrm{Ki} = q_c R/\lambda(t_a - t_0)$, solution (5.4.24) will be identical with solution (5.4.5).

5.5 Hollow Infinite Cylinder

a. Statement of the Problem. *Consider a hollow infinite cylinder. The initial temperature distribution $f(r)$ is prescribed. At the external and internal surfaces the heat flows are given as functions of time. The temperature distribution at any time is to be found.*

The differential heat condition equation is given in Chapter 4, Section 5. The initial and boundary conditions are

$$t(r,0) = f(r), \tag{5.5.1}$$

$$-\lambda\frac{\partial t(R_1,\tau)}{\partial r} + q_1(\tau) = 0, \tag{5.5.2}$$

$$-\lambda\frac{\partial t(R_2,\tau)}{\partial r} + q_2(\tau) = 0. \tag{5.5.3}$$

b. Solution of the Problem. To solve this problem, we use the finite integral Hankel transform

$$H[t(r, \tau)] = T_H(p, \tau) = \int_{R_1}^{R_2} t(r, \tau) M(p, r) r \, dr, \tag{5.5.4}$$

where the kernel of the transform $M(p, r)$ has the form

$$M(p, r) = \frac{\pi}{2} \left[J_0(pr) \frac{Y_1(pR_1)}{J_1(pR_1)} - Y_0(pr) \right]. \tag{5.5.5}$$

$Y_1(z)$ and $Y_0(z)$ are the Bessel functions of the second kind and of the first and the zeroth order.

The parameter p is determined from the characteristic equation

$$J_1(pR_1) Y_1(pR_2) = Y_1(pR_1) J_1(pR_2). \tag{5.5.6}$$

The inversion formula for this transformation is the expansion of the function $t(r, \tau)$ into series with respect to the orthogonal function of the kernel of the transform

$$t(r, \tau) = \sum_p A_p M(p, r), \tag{5.5.7}$$

where

$$A_p = T_H(p, \tau) \Big/ \int_{R_1}^{R_2} M^2(p, r) r \, dr.$$

Applying transform formula (5.5.4) to the right-hand side of the differential heat conduction equation with regard for conditions (5.5.2) and (5.5.3) we obtain

$$(R_2/\lambda) q_2(\tau) M(p, R_2) - (R_1/\lambda) q_1(\tau) M(p, R_1)$$

$$+ p^2 \int_{R_1}^{R_2} r t(r, \tau) M(p, r) \, dr = F(p, \tau) + p^2 T_H(p, \tau) \tag{5.5.8}$$

where $F(p, \tau)$ designates the first two terms. Using Eq. (5.5.8) we transform the differential heat conduction equation into the linear equation of the first power as

$$\frac{dT_H(p, \tau)}{d\tau} = a[F(p, \tau) + p^2 T(p, \tau)]. \tag{5.5.9}$$

The initial condition for this equation will acquire the form

$$T_H(p, 0) = H[f(r)] = \int_{R_1}^{R_2} r f(r) M(p, r) \, dr. \tag{5.5.10}$$

Equation (5.5.9) has the solution

$$T_H(p, \tau) = \exp[ap^2\tau]\left\{a\int_0^\tau F(p, \vartheta)\exp[-ap^2\vartheta]\,d\vartheta + H[f(r)]\right\}. \quad (5.5.11)$$

Inversion of the transform $T_H(p, \tau)$ is carried out using formula (5.5.7):

$$t(r, \tau) = \sum_{n=0}^{\infty} \frac{M(p_n, r)}{\int_{R_1}^{R_2} M^2(p_n, r) r\,dr} \exp[ap_n^2\tau]\left\{H[f(r)] + a\int_0^\tau F(p_n, \vartheta)\exp[-ap_n^2\vartheta]\,d\vartheta\right\}. \quad (5.5.12)$$

The first term of the series sum corresponding to the zero root $p = 0$ has the form

$$\lim_{p\to 0} \frac{M(p, r)}{\int_{R_1}^{R_2} rM^2(p, r)\,dr} \exp[ap^2\tau]\left\{H[f(r)] + a\int_0^\tau F(p, \vartheta)\exp[-ap^2\vartheta]\,d\vartheta\right\}$$

$$= \frac{\left\{\int_{R_2}^{R_1} f(r)\,dr + \frac{a}{\lambda}\int_0^\tau [R_2 q_2(\vartheta) - R_1 q_1(\vartheta)]\,d\vartheta\right\}}{\int_{R_1}^{R_2} r\,dr}. \quad (5.5.13)$$

For all the remainder $(p \neq 0)$

$$\int_{R_1}^{R_2} M^2(p, r) r\,dr = -\frac{J_1^2(pR_1) - J_1^2(pR_2)}{2p^2 J_1^2(pR_1) J_1^2(pR_2)}, \quad (5.5.14)$$

$$F(p, \tau) = \frac{1}{p\lambda}\left[\frac{q_2(\tau)}{J_1(pR_2)} - \frac{q_1(\tau)}{J_1(pR_1)}\right]. \quad (5.5.15)$$

Thus, the solution will have the form:

$$\begin{aligned} t(r, \tau) = {} & \frac{2}{R_2^2 - R_1^2}\left\{\int_{R_1}^{R_2} rf(r)\,dr + \frac{a}{\lambda}\int_0^\tau [R_2 q_2(\vartheta) - R_1 q_1(\vartheta)]\,d\vartheta\right\} \\ & + \pi \sum_{n=1}^{\infty} p_n^2 \frac{J_1^2(p_n R_1) J_1^2(p_n R_2)}{J_1^2(p_n R_1) - J_1^2(p_n R_2)}\left[J_0(p_n r)\frac{Y_1(p_n R_1)}{J_1(p_n R_1)} - Y_0(p_n r)\right] \\ & \times \left\{\frac{\pi}{2}\int_{R_1}^{R_2} rf(r)\left[J_0(p_n r)\frac{Y_1(p_n R_1)}{J_1(p_n R_1)} - Y_0(p_n r)\right] dr \right. \\ & \left. - \frac{a}{p_n\lambda}\int_0^\tau \left[\frac{q_2(\vartheta)}{J_1(p_n R_2)} - \frac{q_1(\vartheta)}{J_1(p_n R_1)}\right]\exp[ap_n^2\vartheta]\,d\vartheta\right\}\exp[-ap_n^2\tau]. \end{aligned} \quad (5.5.16)$$

We now find the specific solution for the boundary counditions

$$t(r, 0) = f(r) = t_0 = \text{const}, \tag{5.5.17}$$

$$q_1(\tau) = 0; \qquad q_2(\tau) = q_c = \text{const}. \tag{5.5.18}$$

The condition $q_1(\tau) = 0$ means that the surface is thermally insulated $[\partial t(R_1, \tau)/\partial r = 0]$. For simplification we use the relation

$$\pi \sum_{n=1}^{\infty} \frac{J_1^2(p_n R_1) J_1^2(p_n R_2)}{p_n J_1(p_n R_2)[J_1^2(p_n R_1) - J_1^2(p_n R_2)]} \times \left[J_0(p_n r) \frac{Y_1(p_n R_1)}{J_1(p_n R_1)} - Y_0(p_n r)\right]$$
$$= \frac{R_2^2}{R_2^2 - R_1^2}\left[\frac{1}{4}(R_2^2 - 2r^2) + R_1^2\left(\ln\frac{r}{R_1} + \frac{R_2^2}{R_2^2 - R_1^2}\ln\frac{R_1}{R_2} + \frac{3}{4}\right)\right]. \tag{5.5.19}$$

Finally we have

$$t(r, \tau) - t_0 = \frac{q}{\lambda} R_2 \left\{ \frac{R_2^2}{R_2^2 - R_1^2}\left[2\text{Fo} - \frac{1}{4}\left(1 - 2\frac{r^2}{R_2^2}\right) - \frac{R_1^2}{R_2^2}\left(\ln\frac{r}{R_1} + \frac{R_2^2}{R_2^2 - R_1^2}\ln\frac{R_1}{R_2} + \frac{3}{4}\right)\right]\right.$$
$$+ \sum_{n=1}^{\infty} \frac{\pi}{\mu_n} \frac{J_1\left(\mu_n \frac{R_1}{R_2}\right) J_1(\mu_n)}{J_1^2\left(\mu_n \frac{R_1}{R_2}\right) - J_1^2(\mu_n)} \left[J_0\left(\mu_n \frac{r}{R_2}\right) Y_1\left(\mu_n \frac{R_1}{R_2}\right)\right.$$
$$\left.\left. - Y_0\left(\mu_n \frac{r}{R_2}\right) J_1\left(\mu_n \frac{R_1}{R_2}\right)\right] \exp[-\mu_n^2 \text{Fo}]\right\}. \tag{5.5.20}$$

where $\text{Fo} = a\tau/R_2^2$, $\mu_n = p_n R_2$ are the roots of the characteristic equation (5.5.6), which may be written as

$$J_1\{\mu(R_1/R_2)\} Y_1(\mu) = Y_1\{\mu(R_1/R_2)\} J_1(\mu). \tag{5.5.21}$$

Solution (5.5.20) is symmetrical relative to the boundary conditions. From solution (5.5.20), at $R_1 \to 0$, the solution is obtained for an infinite solid cylinder.

CHAPTER

6

BOUNDARY CONDITION OF THE THIRD KIND

This chapter is a generalization and further development of Chapter 4. The problems considered here are more general and the problems of Chapter 4 result from these as special cases, assuming Bi $= \infty$.

Some problems of Chapter 4 made it possible to establish that the transform method is superior to the classical method of separation of variables provided we have a uniform initial temperature distribution. If at the initial time instant, the temperature of a body depends upon its coordinates (non-uniform initial distribution), the results are attained more rapidly by the classical method or by the integral Fourier and Hankel transforms than by the operational one. The classical method will, therefore, be used to solve such problems.

The operational method is advantageous in that it allows the obtaining of an effective solution for small values of Fo in which special functions are often absent.

Main attention is paid to the so-called classical bodies (semi-infinite rod, plate, cylinder, and sphere) because, in a special section, it will be shown that the average volume temperature of a body of complex configuration and the average time temperature history of its surface may be given in terms of the corresponding temperatures of classical bodies if the dimensionless times are chosen properly. This principle is of great importance for practical calculations.

To systematize the calculations, we introduce a single system of notation for all the problems, and give eigenvalues as well as tabulate initial heat amplitudes A_n and coefficients B_n entering the relation for a mean temperature. Using such tables, diagrams, and nomograms we may accomplish actual calculations rapidly and with sufficient accuracy.

This chapter deals with problems on heating. In the next chapter these problems will be further developed (i.e., the temperature of the medium changes with time). It will be shown that a heating problem may always be changed to a cooling one.

We now consider a body with the prescribed initial temperature distribution as some function $f(x, y, z, 0)$. At the initial time the body is placed in a medium with a constant temperature $t_a > t(x, y, z, 0)$. Heat transfer between the body surface and surrounding medium occurs according to the Newton law. Newton's law not only describes the boundary condition in the presence of convection, but also simultaneous radiant and convective heat transfer to the first approximation when the temperature difference Δt ($\Delta t = t_a - t_s$) is sufficiently small. Thus, this is a problem of body heating under boundary conditions of the third kind.

$$\partial t/\partial\tau = a\,\nabla^2 t\,, \tag{6.1}$$

$$t(x, y, z, 0) = f(x, y, z)\,, \tag{6.2}$$

$$-(\nabla t)_s + H(t_a - t_s) = 0 \qquad (t_a > t_s)\,, \tag{6.3}$$

where subscript s denotes the body surface.

Replacing the variable

$$t_a - t = \vartheta\,,$$

we have

$$\partial\vartheta/\partial\tau = a\,\nabla^2\vartheta\,, \tag{6.1$'$}$$

$$\vartheta(x, y, z, 0) = t_a - t(x, y, z, 0) = t_a - f(x, y, z) = \varphi(x, y, z)\,, \tag{6.2$'$}$$

$$(\nabla\vartheta)_s + H\vartheta_s = 0\,, \tag{6.3$'$}$$

i.e., the problem of cooling a body in a medium with zero temperature ($\vartheta_a = 0$) is obtained when the initial temperature of the body is given as some function $\varphi(x, y, z)$. If the initial temperature of the body is the same at all its points, i.e., $f(x, y, z) = t_0 = \text{const}$, then all the solutions obtained for heating the body in the form of the relative temperature θ, depending upon the Biot criterion and Fourier number and relative co-ordinates, will also be valid for cooling the body. However, the quantity θ is defined as

$$\theta = \frac{t - t_0}{t_a - t_0} = \Phi\left(\text{Fo}, \text{Bi}, \frac{x}{R_1}, \frac{y}{R_2}, \frac{z}{R_3}\right) \tag{6.4}$$

$$\theta = \frac{t_0 - t}{t_0 - t_a} = \Phi\left(\text{Fo, Bi}, \frac{x}{R_1}, \frac{y}{R_2}, \frac{z}{R_3}\right) \tag{6.5}$$

with heating and cooling, respectively. The last value may also be written as

$$\theta = \frac{t_0 - t}{t_0 - t_a} = 1 - \frac{t - t_a}{t_0 - t_a}. \tag{6.6}$$

6.1 Semi-Infinite Body

a. Statement of the Problem. *Consider a semi-infinite rod, the lateral surface of which is thermally insulated. The temperature throughout the rod is the same and equals t_0 (initial temperature). At the initial time instant, its end is placed in medium with a constant temperature $t_a > t_0$. Heat transfer between the surrounding medium and the noninsulated end of the rod proceeds according to Newton's law (boundary condition of the third kind). The temperature distribution along the rod length at any time and the specific heat flow through its end are to be found* (Fig. 6.1).

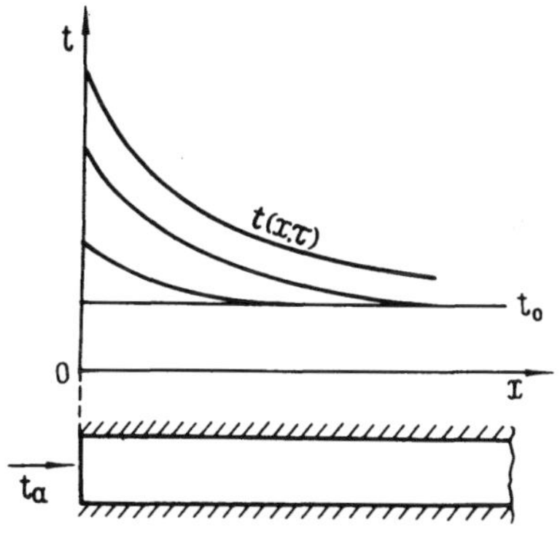

Fig. 6.1. Temperature-distribution curves in a semi-infinite rod with a thermally insulated side surface.

As there are no heat losses from the lateral surfaces of the rod then it may be considered a semi-infinite body where heat propagation takes place in only one direction. Thus, the determination of a temperature fields involves a solution of the differential equation

$$\frac{\partial t(x, \tau)}{\partial \tau} = a\frac{\partial^2 t(x, \tau)}{\partial x^2} \qquad (\tau > 0; \quad 0 < x < \infty), \tag{6.1.1}$$

with the initial condition

$$t(x, 0) = t_0 = \text{const}\,, \qquad (6.1.2)$$

and boundary conditions

$$\lambda \frac{\partial t(0, \tau)}{\partial x} + \alpha[t_a - t(0, \tau)] = 0\,, \qquad (6.1.3)$$

and

$$t(\infty, \tau) = t_0\,, \quad \frac{\partial t(\infty, \tau)}{\partial x} = 0. \qquad (6.1.4)$$

The heat transfer coefficient α is assumed to be constant.

b. Solution of the Problem. The problem will be solved by the operational method because with a uniform initial temperature distribution the operational method yields results more quickly.

If the Laplace transformation is applied to differential equation (6.1.1), as has been done before, then the equation for the transform accounting for the initial condition will be of the form

$$T''(x, s) - (s/a)\, T(x, s) + (t_0/a) = 0\,. \qquad (6.1.5)$$

Boundary conditions for the transform will be

$$T'(0, s) + H[(t_a/s) - T(0, s)] = 0\,, \qquad (6.1.6)$$

$$T'(\infty, s) = 0\,, \qquad (6.1.7)$$

since

$$L[t(0, \tau)] \doteq T(0, s)\,, \qquad L[t_a] = t_a/s\,.$$

In Eq. (6.1.6) $H = \alpha/\lambda$ is a relative heat transfer coefficient.

The solution of Eq. (6.1.5) in a general form may be written

$$T(x, s) - (t_0/s) = A_1 \exp[(s/a)^{1/2} x] + B_1 \exp[-(s/a)^{1/2} x]. \qquad (6.1.8)$$

From condition (6.1.7) it follows that $A_1 = 0$; i.e.,

$$0 = (s/a)^{1/2} A_1 \exp(\infty) - (s/a)^{1/2} B_1 \exp[(-\infty)]\,,$$

from which it is clear that $A_1 = 0$. From the physical viewpoint it means that the temperature at an infinitely great distance from the rod end does not change during the entire heat-transfer process, i.e., at $x \to \infty$, $t \to t_0$.

The constant B_1 is determined from the boundary condition (6.1.6)

$$-(s/a)^{1/2}B_1 + H[(t_a/s) - (t_0/s) - B_1] = 0,$$

whence

$$B_1 = \frac{(t_a - t_0)}{s(1 + (1/H)(s/a)^{1/2})}. \qquad (6.1.9)$$

Then the solution for the transform assumes the form

$$T(x, s) - \frac{t_0}{s} = \frac{(t_a - t_0)}{s(1 + (1/H)(s/a)^{1/2})} \exp\left[-\left(\frac{s}{a}\right)^{1/2} x\right]. \qquad (6.1.10)$$

The table of transforms is used to determine the inverse transform

$$L^{-1}\left[\frac{1}{s(1 + (1/C)\sqrt{s})} \exp[-k\sqrt{s}]\right]$$

$$= \operatorname{erfc}\left(\frac{k}{2\sqrt{\tau}}\right) - \exp[Ck + C^2\tau] \operatorname{erfc}\left(\frac{k}{2\sqrt{\tau}} + C\sqrt{\tau}\right).$$

Consequently, the solution of our problem may be written

$$\theta = \frac{t(x, \tau) - t_0}{t_a - t_0}$$

$$= \operatorname{erfc} \frac{x}{2(a\tau)^{1/2}} - \exp[Hx + H^2 a\tau] \cdot \operatorname{erfc}\left\{\frac{x}{2(a\tau)^{1/2}} + H(a\tau)^{1/2}\right\}. \qquad (6.1.11)$$

c. ***Analysis of the Solution.*** *If the heat transfer coefficient α is very large, then $H \to \infty$. From boundary condition* (6.1.3) *it follows that $t(0, \tau) = t_a =$ $= const$, i.e., the temperature of the rod end becomes immediately equal to that of the surrounding medium.* In this case our problem is similar to that in Chapter 4, Section 2. The function erfc z decreases rapidly with an increase in z and at $z \geq 2.8$ is practically equal to zero. Therefore, at $H \to \infty$, the second term of solution (6.1.11) tends to zero (it may be shown by expanding the indeterminate term according to L'Hopital's rule), and the solution assumes the form

$$\theta = \operatorname{erfc}[x/2(a\tau)^{1/2}]. \qquad (6.1.12)$$

Solution (6.1.12) is identical with that on cooling a semi-infinite rod given in Chapter 4, Section 2; however, in the case of cooling, $\theta = (t_0 - t)/(t_0 - t_a)$, where $t_0 > t_a$.

Returning to the solutions (6.1.11), we find that the dimensionless temperature of the rod end $t(0, \tau)$ is defined as

$$\theta_e = \frac{t(0, \tau) - t_0}{t_a - t_0} = 1 - \exp[H^2 a\tau] \operatorname{erfc} H(a\tau)^{1/2}. \tag{6.1.13}$$

At large values of time, it is very difficult to use solution (6.1.13) owing to a sharp increase in the exponential function. Therefore, it is necessary to find an effective solution for large values of time.

In the Appendix it is shown that

$$\exp[u^2] \operatorname{erfc} u \simeq \frac{1}{\sqrt{\pi}} \left(\frac{1}{u} - \frac{1}{2u^3} + \frac{3}{4u^5} - \cdots \right).$$

Consequently, an approximate solution, convenient for large values of time, may be written as

$$\theta_e = 1 - \frac{1}{H(\pi a\tau)^{1/2}} \left(1 - \frac{1}{2H^2 a\tau} + \frac{3}{4H^4 a^2 \tau^2} - \cdots \right). \tag{6.1.13a}$$

To write the solution in a criterial form, we designate the Fourier number for this point with the coordinate x as $\text{Fo}_x = a\tau/x^2$ and the Biot criterion as Bi_x ($\text{Bi}_x = Hx$).

Thus, the solution (6.1.11) of our problem may be written as

$$\theta = \operatorname{erfc} \frac{1}{2(\text{Fo}_x)^{1/2}} - \exp[\text{Bi}_x + (\text{Bi})_x{}^2 \text{Fo}_x] \times \operatorname{erfc}\left\{ \frac{1}{2(\text{Fo})_x{}^{1/2}} + \text{Bi}_x(\text{Fo}_x)^{1/2} \right\}. \tag{6.1.14}$$

The product $\text{Bi}_x(\text{Fo}_x)^{1/2}$ gives a new generalized argument and a new dimensionless number for the temperature field in semi-infinite bodies

$$N_{\alpha,\tau} = \text{Bi}_x(\text{Fo}_x)^{1/2} = \frac{\alpha\sqrt{\tau}}{(\lambda c\gamma)^{1/2}} = \frac{\alpha}{\varepsilon}\sqrt{\tau},$$

where ε is a coefficient of thermal activity of the body. The number $N_{\alpha,\tau}$ is numerically equal to the ratio of the heat transferred to a unit surface of a body per unit time in the presence of a unit temperature difference between a surface and surrounding medium, to the coefficient of thermal activity of a body.

For engineering calculations, Fig. 6.2 gives the plot of the dimensionless temperature versus the number $\text{Bi}_x(\text{Fo}_x)^{1/2}$ for various values of the number $1/2(\text{Fo}_x)^{1/2}$.

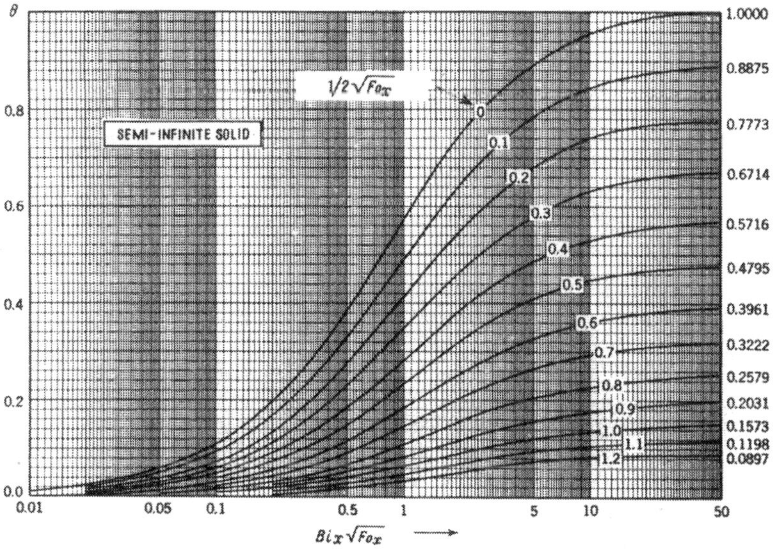

Fig. 6.2. The dimensionless excess temperature versus the number $\mathrm{Bi}_x(\mathrm{Fo}_x)^{1/2}$ for various Fourier numbers (semi-infinite rod) [102].

d. Calculation of the Heat Flow. The heat flux through the rod end is

$$q = -\lambda \frac{\partial t(0, \tau)}{\partial x}. \tag{6.1.15}$$

If Eq. (6.1.11) is differentiated with respect to x and it is assumed that $x = 0$, then we have.

$$\frac{\partial t(0, \tau)}{\partial x} = -H(t_a - t_0)\exp[H^2 a\tau]\operatorname{erfc} H(a\tau)^{1/2}. \tag{6.1.16}$$

This relation could be obtained in an alternate way. Differentiating solution (6.1.10) for the transform with respect to x and assuming $x = 0$, we have

$$T'(0, s) = -\frac{t_a - t_0}{(as)^{1/2}\left\{1 + \dfrac{1}{H}\left(\dfrac{s}{a}\right)^{1/2}\right\}}. \tag{6.1.17}$$

With the help of the tables of transforms, the same relation (6.1.16) is obtained from Eq. (6.1.17). Consequently, the heat flux is

$$q = \alpha(t_a - t_0)\exp[H^2 a\tau]\operatorname{erfc} H(a\tau)^{1/2}. \tag{6.1.18}$$

From relation (6.1.18), it follows that at the initial time the magnitude of the heat flow is at a maximum and then it progressively decreases, tending to zero at $\tau \to \infty$. At the initial instant the heat flux is at a maximum and equal to

$$q_{\max} = \alpha(t_a - t_0),$$

which directly follows from the boundary condition.

The amount of heat transferred in heating the rod for a given time period τ_1 may be determined by integration of expression (6.1.18) with respect to τ between 0 and τ_1 and multiplying this result by the surface area of the rod end.

6.2 Semi-Infinite Rod without Thermal Insulation of Its Surface

a. Statement of the Problem. *Consider the same problem but with the lateral surface not thermally insulated, i.e., heat transfer between the lateral surface of the rod and the surrounding medium occurs according to the Newton law. The temperature of the medium surrounding the lateral surface of the rod is assumed to be constant and equal to its initial temperature.*

If the height and width of the rod are small compared to its length and the heat conductivity is large, then the temperature drop over the height and width of the rod may be considered to be zero, i.e., $\partial t/\partial y = \partial t/\partial z = 0$. Thus, the problem stated is reduced to a one-dimensional problem, with a temperature drop occurring only in the x direction (Fig. 6.3). Heat transfer from the lateral surface of the rod into the surrounding medium should be taken into account in the differential equation itself as a negative heat source.

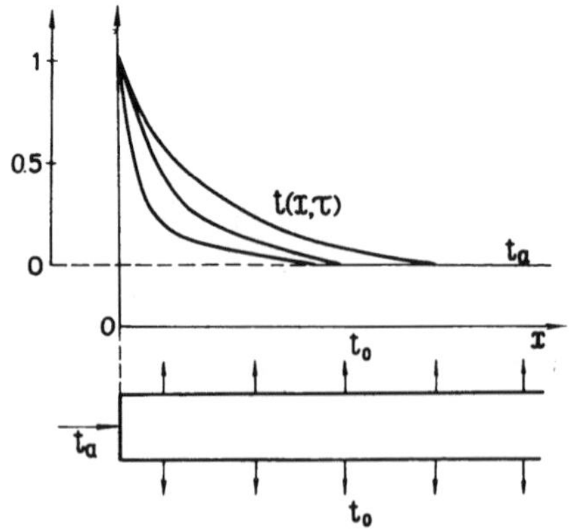

Fig. 6.3. Temperature distribution in a semi-infinite rod without thermal insulation of the side surface.

Thus, the differential heat conduction equation may be written as

$$c\gamma \frac{\partial t(x, \tau)}{\partial \tau} - \lambda \frac{\partial^2 t(x, \tau)}{\partial x^2} - w \qquad (\tau > 0; \quad 0 < x < \infty), \qquad (6.2.1)$$

where w is the amount of heat given off into the surrounding medium per unit volume of the rod per unit time.

If the cross-sectional area of the rod is designated by S, the periphery of the cross section by P, and the length of a rather small portion of the rod by l, then

$$w = (\alpha/Sl)[t(x, \tau) - t_0]\cdot Pl = \alpha[t(x, \tau) - t_0](1/h),$$

where α is the heat transfer coefficient (kcal/m^2 hr deg). $h = S/P$ is the ratio between the cross-sectional area of a rod and the cross-sectional perimeter (for a rod with a cylinder cross section $h = \frac{1}{2}R$, for a rod with a square cross-section $h = \frac{1}{4}a$ where a is the side of a square, etc.), normally measured in meters.

Thus, we have the differential equation of the form

$$\frac{\partial t(x, \tau)}{\partial \tau} = a \frac{\partial^2 t(x, \tau)}{\partial x^2} - \frac{\alpha}{c\gamma h} [t(x, \tau) - t_0]. \qquad (6.2.2)$$

To simplify the problem, we assume $H = \alpha/\lambda = \infty$ for the rod end. This means that the rod-end temperature immediately becomes constant and equal to t_a. Then the initial and boundary conditions may be written

$$t(x, 0) = t_0, \qquad (6.2.3)$$

$$t(0, \tau) = t_a, \qquad (6.2.4)$$

$$t(\infty, \tau) = t_0, \qquad \partial t(\infty, \tau)/\partial x = 0. \qquad (6.2.5)$$

The solution of a problem for a finite value of H for a finite rod will be given in Section 6.4.

b. Solution of the Problem. The differential equation for the transform $T(x, s)$ has the form

$$T''(x, s) - \frac{s}{a} T(x, s) + \frac{t_0}{a} - \frac{\alpha}{\lambda h} T(x, s) + \frac{\alpha t_0}{\lambda h s} = 0. \qquad (6.2.6)$$

Boundary conditions for the transform will be written as

$$T(0, s) = t_a/s, \qquad (6.2.7)$$

$$T'(\infty, s) = 0. \qquad (6.2.8)$$

Equation (6.2.6) may be rewritten as

$$T''(x, s) - \left(\frac{s}{a} + \frac{\alpha}{\lambda h}\right)\left[T(x, s) - \frac{t_0}{s}\right] = 0. \tag{6.2.9}$$

The solution of Eq. (6.2.9) imposing condition (6.2.8) will be of the form

$$T(x, s) - \frac{t_0}{s} = B_1 \exp\left[-\left(\frac{s}{a} + \frac{\alpha}{\lambda h}\right)^{1/2} x\right]. \tag{6.2.10}$$

The constant B_1 is determined from boundary condition (6.2.7)

$$B_1 = (t_a - t_0)/s.$$

Hence, the solution for the transform has the form

$$T(x, s) - \frac{t_0}{s} = \frac{t_a - t_0}{s} \exp\left[-\left(\frac{s}{a} + \frac{\alpha}{\lambda h}\right)^{1/2} x\right]. \tag{6.2.11}$$

The table of transforms is used for the inversion of the transform $T(x, s)$ to $t(x, \tau)$. It is known that

$$L^{-1}\left[\frac{1}{s}\exp[-k(s+b)^{1/2}]\right] = \frac{1}{2}\left[\exp[-k\sqrt{b}\,]\operatorname{erfc}\left(\frac{k}{2\sqrt{\tau}} - (b\tau)^{1/2}\right) + \exp[k\sqrt{b}\,]\operatorname{erfc}\left\{\frac{k}{2\sqrt{\tau}} + (b\tau)^{1/2}\right\}\right].$$

Consequently, the solution of our problem will be

$$\begin{aligned}\theta &= \frac{t(x, \tau) - t_0}{t_a - t_0} \\ &= \frac{1}{2}\left[\exp\left[-\left(\frac{\alpha}{\lambda h}\right)^{1/2} x\right] \operatorname{erfc}\left\{\frac{x}{2(a\tau)^{1/2}} - \left(\frac{\alpha a\tau}{\lambda h}\right)^{1/2}\right\}\right. \\ &\quad \left. + \exp\left[\frac{\alpha}{\lambda h}\right]^{1/2} x \operatorname{erfc}\left\{\frac{x}{2(a\tau)^{1/2}} + \left(\frac{\alpha a\tau}{\lambda h}\right)^{1/2}\right\}\right].\end{aligned} \tag{6.2.12}$$

c. Analysis of the Solution. If there is no heat transfer from the side surface of the rod; i.e., $\alpha/\lambda h = 0$, then solution (6.1.12) is obtained from the solution (6.2.12), viz:

$$\theta = \operatorname{erfc}\,[x/2(a\tau)^{1/2}]. \tag{6.2.13}$$

This is the solution for a semi-infinite body when the temperature of the bounding surface is constant and equal to t_a.

Introducing the Fourier number and Biot criterion

$$\mathrm{Fo} = a\tau/h^2, \qquad \mathrm{Bi} = \alpha h/\lambda\,,$$

our solution may be written

$$\begin{aligned} \theta &= \frac{t(x,\tau) - t_0}{t_a - t_0} \\ &= \frac{1}{2}\left\{\exp\left[-(\mathrm{Bi})^{1/2}\frac{x}{h}\right]\operatorname{erfc}\left\{\frac{x/h}{2(\mathrm{Fo})^{1/2}} - (\mathrm{Fo}^*)^{1/2}\right\}\right. \\ &\quad \left. + \exp\left[(\mathrm{Bi})^{1/2}\frac{x}{h}\right]\operatorname{erfc}\left\{\frac{x/h}{2(\mathrm{Fo})^{1/2}} + (\mathrm{Fo}^*)^{1/2}\right\}\right\}. \end{aligned} \tag{6.2.14}$$

Here $\mathrm{Fo}^* = \mathrm{Bi\ Fo}$ is a new dimensionless group which characterizes heat transfer between a body and the surrounding medium for a negligible transverse temperature drop inside a body [the so-called external problem (see Section 6.3)].

In our case, the temperature drop over the height and width of the rod is absent, despite heat transfer between it and the surrounding medium, therefore

$$\mathrm{Fo}^* = \mathrm{Bi\ Fo} = \alpha\tau/c\gamma h. \tag{6.2.15}$$

Thus, the first term in the argument of the erfc in (6.2.14) characterizes heat transfer due to net heat conduction in the x direction and the second term the heat transfer from the lateral surface of the rod to the surrounding medium when a temperature gradient is absent in the transverse directions. In the stationary state ($\tau = \infty$), there will be a particular temperature distribution over the rod length. At $\tau \to \infty$, the first function $\operatorname{erfc} z_1 = 1 - \operatorname{erf} z_1$, will tend to 2, since $\operatorname{erf} z_1 \to -1$ and the second function $\operatorname{erfc} z_2 \to 0$. Hence, in a stationary state the temperature distribution will be described by a simple exponential function

$$\theta_\infty = \exp\left[-\left(\frac{\alpha}{\lambda h}\right)^{1/2} x\right] = \exp\left[-(\mathrm{Bi})^{1/2}\frac{x}{h}\right]. \tag{6.2.16}$$

At small values of time, or rather $\mathrm{Fo} \to 0$, the quantity Fo^* may be neglected, as compared to the first term of the argument of the erfc. In addition, at small values of time, the influence of the heating will have spread over a small distance, hence x/h is also small. In this case, both exponential functions are close to unity. Then, solution (6.2.14) will be close to solution (6.2.13), i.e., at the first time instants the heating of the rod occurs as though heat transfer from the lateral surface is absent.

d. Heat Flux. A heat flux (kcal/m² hr) is determined by the value of the temperature gradient at the end of the rod. It will be calculated by the operational method.

From solution (6.2.11) we obtain the transform

$$T'(0, s) = -\frac{t_a - t_0}{s}\left(\frac{s}{a} + \frac{\alpha}{\lambda h}\right)^{1/2} \tag{6.2.17}$$

Inversion of the transform to $T'(0, \tau)$ is made according to the tables of transforms, viz

$$L^{-1}\left[\frac{(s+b)^{1/2}}{s}\right] = \frac{1}{(\pi\tau)^{1/2}}\, e^{-b\tau} + \sqrt{b}\, \mathrm{erf}(b\tau)^{1/2}.$$

Then, we shall have

$$-t'(0, \tau) = (t_a - t_0)\left[\frac{1}{(\pi a\tau)^{1/2}} \exp\left[-\frac{\alpha a\tau}{\lambda h}\right] + \left(\frac{\alpha}{\lambda h}\right)^{1/2} \mathrm{erf}\left(\frac{\alpha a\tau}{\lambda h}\right)^{1/2}\right].$$

Hence, the heat flux is

$$q = -\lambda\frac{\partial t(0, \tau)}{\partial x} = \frac{\lambda(t_a - t_0)}{h}\left[\frac{1}{(\pi\,\mathrm{Fo})^{1/2}} \exp[-\mathrm{Fo}^*] + (\mathrm{Bi})^{1/2}\,\mathrm{erf}(\mathrm{Fo}^*)^{1/2}\right]. \tag{6.2.18}$$

In a state close to the stationary condition ($\mathrm{Fo} \to \infty$), the first term in square brackets of relation (6.2.18) tends to zero, and the second to $(\mathrm{Bi})^{1/2}$, so the heat flux is

$$q = \frac{\lambda(t_a - t_0)}{h}(\mathrm{Bi})^{1/2}. \tag{6.2.19}$$

Thus, we find that the specific heat flow is directly proportional to the heat flow through a rod with length h when the temperature of its boundary surfaces is equal to t_a and t_0, respectively. The greater the Biot criterion, the greater the proportionality factor and, consequently, the heat rate. To illustrate this, some computations will be carried out.

One end of the long steel rod, 140 mm in diameter, is heated in a furnace at 800°C. The temperature of the rod before heating is equal to that of the surrounding medium ($t_0 = 20$°C). We are to determine the temperature at a distance of 17.25 cm from the end of the rod after exposure to heating for 15 min. The heat transfer coefficient is assumed to be equal to $\alpha = 140$ kcal/m² hr deg. The other coefficients are equal to $\lambda = 40$ kcal/m hr deg, $a = 45\cdot 10^{-3}$ m²/hr.

First of all, we calculate the criteria required for the computations. The characteristic dimension is

$$h = \tfrac{1}{2}R = 3.5 \quad \text{cm} = 0.035 \quad \text{m};$$

the Fourier number

$$\text{Fo} = \frac{a\tau}{h^2} = \frac{45 \cdot 10^{-8} \cdot 1}{12.25 \cdot 10^{-4} \cdot 4} = 9.2;$$

the Biot criterion

$$\text{Bi} = \frac{\alpha h}{\lambda} = \frac{140 \cdot 0.035}{40} = 0.125;$$

the modified Fourier number

$$\text{Fo}^* = \text{Bi Fo} = 0.125 \cdot 9.2 = 1.15;$$

and the relative coordinate

$$\frac{x}{h} = \frac{17.5}{3.5} = 5.$$

We determine the relative excess temperature by solution (6.2.14) as

$$\begin{aligned}\theta &= \tfrac{1}{2}\Big[\exp[-(0.125)^{1/2} \cdot 5] \operatorname{erfc}\left[\frac{5}{2(9.2)^{1/2}} - (1.15)^{1/2}\right] \\ &\quad + \exp[(0.125)^{1/2} \cdot 5] \\ &\quad \times \operatorname{erfc}\left\{\frac{5}{2(9.2)^{1/2}} + (1.15)^{1/2}\right\}\Big] \\ &= \tfrac{1}{2}(0.170 \cdot 1.2744 + 0.043) = 0.130.\end{aligned}$$

Then we obtain

$$t - t_0 = \theta(t_a - t_0) = 0.130 \cdot 780 = 101°\text{C},$$

$$t = 101 + 20 = 121°\text{C}.$$

Thus, the rod temperature is about 121°.

If heat transfer from the lateral surface is neglected, the temperature of the rod for the coordinate $x/h = 5$ determined by the formula is

$$\theta = \operatorname{erfc}\frac{x/h}{2(\text{Fo})^{1/2}} = \operatorname{erfc}\frac{5}{2(9.2)^{1/2}} = 1 - 0.756 = 0.244.$$

Hence the rod temperature is

$$t = 20° + 0.244 \cdot 780 = 210°\text{C}.$$

Thus, the temperature is overestimated approximately by 90° if we neglect the convective heat transfer from the lateral surfaces.

6.3 Infinite Plate

a. Statement of the Problem. *Consider an infinite plate $2R$ in thickness. The initial temperature distribution is given as some function $t(x, 0) = f(x)$. At the initial time instant, the plate is placed into a medium with a constant temperature $t_a > t(x, 0)$. Heat transfer occurs between bounding surfaces of the plate and the surrounding medium according to the Newton law.*

The temperature distribution across the plate thickness as well as the specific heat flow is to be found. We have

$$\frac{\partial t(x, \tau)}{\partial \tau} = a \frac{\partial^2 t(x, \tau)}{\partial x^2} \qquad (\tau > 0; \ -R < x < +R), \tag{6.3.1}$$

$$t(x, 0) = f(x), \tag{6.3.2}$$

$$-\lambda \frac{\partial t(R, \tau)}{\partial x} + \alpha[t_a - t(R, \tau)] = 0, \tag{6.3.3}$$

$$+\lambda \frac{\partial t(-R, \tau)}{\partial x} + \alpha[t_a - t(-R, \tau)] = 0. \tag{6.3.4}$$

The origin of coordinates is chosen in the middle of the plate (Fig. 6.4). We shall attack the problem by two methods. With a nonuniform temperature distribution, it is better to use the classical Fourier method.

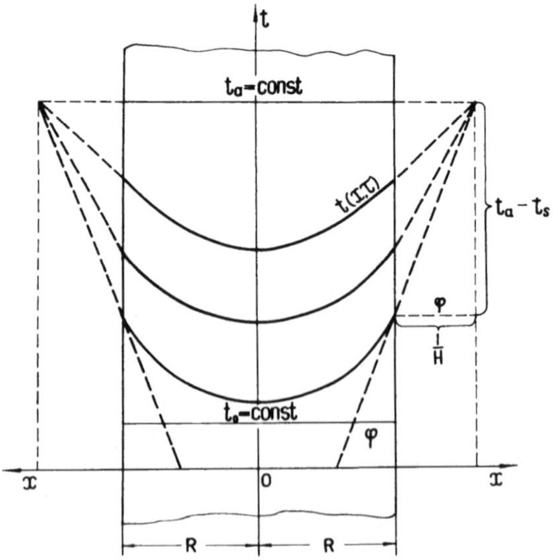

Fig. 6.4. Temperature distribution in an infinite plate (symmetrical problem).

To simplify calculations, the function $f(x)$ is assumed to be even with respect to x, then our problem becomes symmetrical as heat transfer between both surfaces and the surrounding medium occurs equally. In this case, instead of condition (6.3.4) we may write

$$\partial t(0, \tau)/\partial x = 0. \qquad (6.3.4')$$

b. Solution by Separation of Variables. We reduce the problem of heating to that of cooling by replacing the variable $t_a - t(x, \tau) = \vartheta(x, \tau)$. Then, the differential equation for $\vartheta(x, \tau)$ will be identical with differential equation (6.3.1). The particular solution of equation (6.3.1) with condition (6.3.4′) has the form (see Chapter 3, Section 2)

$$\vartheta(x, \tau) = D \cos kx \exp[-k^2 a\tau]. \qquad (6.3.5)$$

Solution (6.3.5) must also satisfy boundary condition (6.3.3), which for the new variable ϑ takes the form

$$\frac{\partial\vartheta(R, \tau)}{\partial x} + H\vartheta(R, \tau) = 0. \qquad (6.3.6)$$

We obtain

$$-kD \sin kR \exp[-k^2 a\tau] + HD \cos kR \exp[-k^2 a\tau] = 0. \qquad (6.3.7)$$

It is possible to divide through by $D \exp[-k^2 a\tau]$ because $0 < \tau < \infty$, and a trigonometric equation for determining the constant k is obtained

$$\cot kR = \frac{k}{H} = \frac{kR}{HR} = \frac{kR}{\mathrm{Bi}}. \qquad (6.3.8)$$

The quantity $HR = (\alpha/\lambda)R = \mathrm{Bi}$ is the Biot criterion.

We designate kR by μ (i.e., $\mu = kR$). From an analysis of Eq. (6.3.8) it can be seen that μ has an infinite number of values. Roots of Eq. (6.3.8) may be most easily determined graphically. If the left-hand side of the equation $\cot\mu$ is designated by y_1 $(y_1 = \cot\mu)$ and the right-hand side by y_2 $(y_2 = (1/\mathrm{Bi})\mu)$, and the values of the roots μ of the characteristic equations, are given by intersection of the cotangent curve y_1 with the straight line y_2 (see Fig. 6.5). From Fig. 6.5 it can be seen that there is an infinite number of μ, each subsequent solution being greater than the previous one

$$\mu_1 < \mu_2 < \mu_3 < \cdots < \mu_n.$$

The greater n, the closer μ_n approaches the number $(n - 1)\pi$.

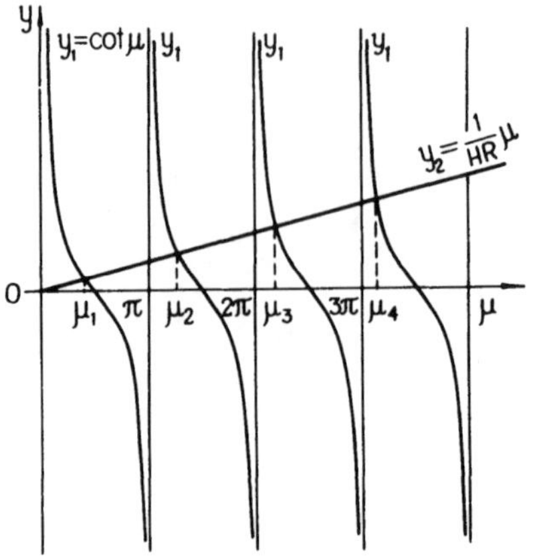

Fig. 6.5. Graphical method for determining the roots of the characteristic equation.

The slope of the straight line y_2 is equal to $1/\text{Bi}$. At $\text{Bi} \to \infty$ the tangent of an inclination angle will, therefore, be zero and the straight line y_2 coincides with the abscissa axis. Then $\mu_n = (2n - 1)\frac{1}{2}\pi$, i.e., eigenvalues for a cooling problem at a constant temperature on the bounding surfaces (see Chapter 4, Section 3) are obtained.

If the Biot criterion tends to zero, the tangent of an inclination angle of the straight line y_2 tends to infinity, which implies that the straight line y_2 coincides with the ordinate axis. Then, the roots of the equation $\mu_n = (n - 1)\pi$ are equal where $n = 1, 2, \cdots$, i.e., $\mu_1 = 0$, $\mu_2 = \pi$, etc.

Thus, characteristic equation (6.3.8) may be written as

$$\cot \mu = (1/\text{Bi})\mu . \tag{6.3.9}$$

The first six roots μ_n are given in Table 6.1 for various values of the Biot criterion (from 0 to ∞). The tabulated values of μ_n are accurate to four decimal places. In the majority of cases, one and occasionally two roots μ_n have to be used for large and intermediate values of Fo. For small values of Fo the solution is given in another form.

Thus, a general solution will be equal to the sum of all particular solutions

$$\vartheta(x, \tau) = \sum_{n=1}^{\infty} D_n \cos \mu_n \frac{x}{R} \exp\left[-\mu_n^2 \frac{a\tau}{R^2}\right]. \tag{6.3.10}$$

The constants D_n are defined from initial condition (6.3.2) which for the variable ϑ may be written as

$$\vartheta(x, 0) = t_a - t(x, 0) = t_a - f(x) = f_1(x). \tag{6.3.11}$$

TABLE 6.1. Root of Characteristic Equation $\cot \mu = (1/\mathrm{Bi})\,\mu$

Bi	μ_1	μ_2	μ_3	μ_4	μ_5	μ_6
0	0.0000	3.1416	6.2832	9.4248	12.5664	15.7080
0.001	0.0316	3.1419	6.2833	9.4249	12.5665	15.7080
0.002	0.0447	3.1422	6.2835	9.4250	12.5665	15.7081
0.004	0.0632	3.1429	6.2838	9.4252	12.5667	15.7082
0.006	0.0774	3.1435	6.2841	9.4254	12.5668	15.7083
0.008	0.0893	3.1441	6.2845	9.4256	12.5670	15.7085
0.01	0.0998	3.1448	6.2848	9.4258	12.5672	15.7086
0.02	0.1410	3.1479	6.2864	9.4269	12.5680	15.7092
0.04	0.1987	3.1543	6.2895	9.4290	12.5696	15.7105
0.06	0.2425	3.1606	6.2927	9.4311	12.5711	15.7118
0.08	0.2791	3.1668	6.2959	9.4333	12.5727	15.7131
0.1	0.3111	3.1731	6.2991	9.4354	12.5743	15.7143
0.2	0.4328	3.2039	6.3148	9.4459	12.5823	15.7207
0.3	0.5218	3.2341	6.3305	9.4565	12.5902	15.7270
0.4	0.5932	3.2636	6.3461	9.4670	12.5981	15.7334
0.5	0.6533	3.2923	6.3616	9.4775	12.6060	15.7397
0.6	0.7051	3.3204	6.3770	9.4879	12.6139	15.7460
0.7	0.7506	3.3477	6.3923	9.4983	12.6218	15.7524
0.8	0.7910	3.3744	6.4074	9.5087	12.6296	15.7587
0.9	0.8274	3.4003	6.4224	9.5190	12.6375	15.7650
1.0	0.8603	3.4256	6.4373	9.5293	12.6453	15.7713
1.5	0.9882	3.5422	6.5097	9.5801	12.6841	15.8026
2.0	1.0769	3.6436	6.5783	9.6296	12.7223	15.8336
3.0	1.1925	3.8088	6.7040	9.7240	12.7966	15.8945
4.0	1.2646	3.9352	6.8140	9.8119	12.8678	15.9536
5.0	1.3138	4.0336	6.9096	9.8928	12.9352	16.0107
6.0	1.3496	4.1116	6.9924	9.9667	12.9988	16.0654
7.0	1.3766	4.1746	7.0640	10.0339	13.0584	16.1177
8.0	1.3978	4.2264	7.1263	10.0949	13.1141	16.1675
9.0	1.4149	4.2694	7.1806	10.1502	13.1660	16.2147
10.0	1.4289	4.3058	7.2281	10.2003	13.2142	16.2594
15.0	1.4729	4.4255	7.3959	10.3898	13.4078	16.4474
20.0	1.4961	4.4915	7.4954	10.5117	13.5420	16.5864
30.0	1.5202	4.5615	7.6057	10.6543	13.7085	16.7691
40.0	1.5325	4.5979	7.6647	10.7334	13.8048	16.8794
50.0	1.5400	4.6202	7.7012	10.7832	13.8666	16.9519
60.0	1.5451	4.6353	7.7259	10.8172	13.9094	17.0026
80.0	1.5514	4.6543	7.7573	10.8606	13.9644	17.0686
100.0	1.5552	4.6658	7.7764	10.8871	13.9981	17.1093
∞	1.5708	4.7124	7.8540	10.9956	14.1372	17.2788

We have

$$f_1(x) = \sum_{n=1}^{\infty} D_n \cos \mu_n \frac{x}{R}.$$

If the function $f_1(x)$ satisfies the Dirichlet conditions, it may be expanded into a Fourier series. We multiply both parts of the equation by $\cos \mu_m(x/R)\, dx$ and integrate between $-R$ and $+R$:

$$\int_{-R}^{R} f_1(x) \cos \mu_m \frac{x}{R}\, dx = \sum_{n=1}^{\infty} D_n \int_{-R}^{R} \cos \mu_n \frac{x}{R} \cos \mu_m \frac{x}{R}\, dx. \quad (6.3.12)$$

In Chapter 4, Section 3, it was shown that integrals on the right-hand side of relation (6.3.12) are equal to

$$\int_{-R}^{R} \cos \mu_n \frac{x}{R} \cos \mu_m \frac{x}{R}\, dx$$

$$= \frac{2R[\mu_m \sin \mu_m \cos \mu_n - \mu_n \cos \mu_m \sin \mu_n]}{\mu_m{}^2 - \mu_n{}^2}. \quad (6.3.13)$$

(See formula (4.3.10), assuming $kR = \mu$.)

Since μ_m and μ_n are the roots of the characteristic equation (6.3.9), it may be written

$$\mathrm{Bi} = \frac{\mu_n \sin \mu_n}{\cos \mu_n}, \qquad \mathrm{Bi} = \frac{\mu_m \sin \mu_m}{\cos \mu_m}.$$

We then find

$$\mu_m \sin \mu_m \cos \mu_n = \mathrm{Bi} \cos \mu_n \cos \mu_m$$

$$\mu_n \sin \mu_n \cos \mu_m = \mathrm{Bi} \cos \mu_n \cos \mu_m.$$

Hence, the expression in square brackets in equality (6.3.13) is equal to zero at $m \neq n$. Integral (6.3.13) is, therefore, zero if $m \neq n$.

For $m = n$, integral (6.3.13) is given by

$$\int_{-R}^{R} \cos^2 \mu_n \frac{x}{R}\, dx = R\left(1 + \frac{\sin 2\mu_n}{2\mu_n}\right) = R\left(1 + \frac{\sin \mu_n \cos \mu_n}{\mu_n}\right)$$

$$= \frac{R}{\mu_n}(\mu_n + \sin \mu_n \cos \mu_n). \quad (6.3.14)$$

(See Chapter 4, formula (4.3.11).)

Hence, all the integrals in the right-hand side of equality (6.3.12) are equal to zero, except when $m = n$. The last integral is determined by formula (6.3.14). Thus, the constant coefficients D_n are

$$D_n = \frac{\int_{-R}^{R} f_1(x)\cos\mu_n(x/R)\,dx}{\int_{-R}^{R}\cos^2\mu_n(x/R)\,dx}$$

$$= \frac{\mu_n}{\mu_n + \sin\mu_n\cos\mu_n}\,\frac{1}{R}\int_{-R}^{R} f_1(x)\cos\mu_n\frac{x}{R}\,dx.$$

A general solution of our problem may be written

$$\vartheta(x,\tau) = t_a - t(x,\tau)$$

$$= \sum_{n=1}^{\infty}\left\{\frac{\mu_n}{\mu_n + \sin\mu_n\cos\mu_n}\cos\mu_n\frac{x}{R}\right.$$

$$\left.\times\frac{2}{R}\int_0^R f_1(x)\cos\mu_n\frac{x}{R}\,dx\right\}\exp\left[-\mu_n^2\frac{a\tau}{R^2}\right]. \qquad (6.3.15)$$

Solution (6.3.15) is at the same time a solution to the problem on heating of an infinite plate $l = R$ in thickness, when its one surface ($x = 0$) is thermally insulated and the opposite one ($x = l$) is heated due to heat transfer with the surrounding medium.

If the function $f_1(x)$ is an odd function with respect to x, then instead of particular solution (6.3.5) the following particular solution should be taken:

$$\vartheta(x,\tau) = C\sin kx\exp[-k^2a\tau].$$

Then, similar computations would lead to

$$t_a - t(x,\tau) = \sum_{m=1}^{\infty}\left\{\frac{\mu_m\sin\mu_m(x/R)}{\mu_m - \sin\mu_m\cos\mu_m}\right.$$

$$\left.\times\frac{2}{R}\int_0^R f_1(x)\sin\mu_m\frac{x}{R}\,dx\right\}\exp[-\mu_m^2\,\mathrm{Fo}], \qquad (6.3.15')$$

where μ_m are the roots of the characteristic equation

$$\tan\mu = -(1/\mathrm{Bi})\mu.$$

Solution (6.3.15′) is at the same time a solution of the problem of heating an infinite plate with the thickness $l = R$ when its one surface ($x = 0$) is maintained at a constant temperature $t_a = \mathrm{const}$ and the opposite surface ($x = l$) is heated according to the Newton law (the function $f_1(x)$ is any function satisfying the Dirichlet condition).

In the general case when both surfaces are heated and the initial temperature is any function $f_1(x)$, the solution of the problem may be written as

$$\vartheta(x, \tau) = \sum_{n=1}^{\infty} \frac{\mu_n \cos \mu_n(x/l) + \text{Bi} \sin \mu_n(x/l)}{2\text{Bi} + \mu_n^2 + \text{Bi}^2}$$
$$\times \left\{\frac{2}{l} \int_0^l f_1(x)\left[\mu_n \cos \mu_n \frac{x}{l} + \text{Bi} \sin \mu_n \frac{x}{l}\right] dx\right\}$$
$$\times \exp\left[-\mu_n^2 \frac{a\tau}{l^2}\right], \tag{6.3.16}$$

where $l = 2R$ is the thickness of the plate; μ_n are the roots of the characteristic equation

$$\cot \mu = \frac{\mu^2 - \text{Bi}^2}{2\mu \text{ Bi}}. \tag{6.3.17}$$

The origin of coordinates is on one of the surfaces of the plate $(0 < x < l)$.

Returning to solution (6.3.15) and considering a uniform initial temperature distribution we have

$$t(x, 0) = f(x) = t_0 = \text{const},$$
$$f_1(x) = t_a - f(x) = t_a - t_0 = \text{const}.$$

Then, the integral in solution (6.3.15) will be

$$(t_a - t_0)\frac{2}{R} \int_0^R \cos \mu_n \frac{x}{R}\, dx = \frac{2}{\mu_n}(t_a - t_0) \sin \mu_n,$$

and the solution takes the form

$$\frac{t_a - t(x, \tau)}{t_a - t_0} = \sum_{n=1}^{\infty} \left\{\frac{2 \sin \mu_n}{\mu_n + \sin \mu_n \cos \mu_n} \cos \mu_n \frac{x}{R}\right\} \exp\left[-\mu_n^2 \frac{a\tau}{R^2}\right]. \tag{6.3.18}$$

Writing this solution in *dimensionless quantities* we have

$$\theta = \frac{t(x, \tau) - t_0}{t_a - t_0} = 1 - \sum_{n=1}^{\infty} \left\{A_n \cos \mu_n \frac{x}{R}\right\} \exp[-\mu_n^2 \text{Fo}]. \tag{6.3.19}$$

where

$$A_n = \frac{2 \sin \mu_n}{\mu_n + \sin \mu_n \cos \mu_n}. \tag{6.3.20}$$

c. Solution by Operational Method. A solution of differential equation (6.3.1) for the transform $T(x, s)$ under the condition that the initial temperature is constant and equal t_0 may be written as (see Chapter 3, Section 3)

$$T(x, s) - \frac{t_0}{s} = A \cosh \left(\frac{s}{a}\right)^{1/2} x + B \sinh \left(\frac{s}{a}\right)^{1/2} x. \qquad (6.3.21)$$

Boundary conditions for the transform will be of the form

$$-T'(R, s) + H[(t_a/s) - T(R, s)] = 0 \qquad (6.3.22)$$

$$T'(0, s) = 0. \qquad (6.3.23)$$

From the symmetry condition (6.3.23) it follows that $B = 0$ (the temperature distribution is symmetrical with respect to the ordinate axis).

We subsitute solution (6.3.21) into boundary condition (6.3.22) to obtain

$$-A\left(\frac{s}{a}\right)^{1/2} \sinh\left(\frac{s}{a}\right)^{1/2} R + \frac{Ht_a}{s} - \frac{Ht_0}{s} - HA \cosh\left(\frac{s}{a}\right)^{1/2} R = 0$$

thence, the constant A is defined as

$$A = \frac{t_a - t_0}{s\left[\cosh\left(\frac{s}{a}\right)^{1/2} R + \frac{1}{H}\left(\frac{s}{a}\right)^{1/2} \sinh\left(\frac{s}{a}\right)^{1/2} R\right]}. \qquad (6.3.24)$$

Thus, the solution for the transform will be

$$T(x, s) - \frac{t_0}{s} = \frac{(t_a - t_0) \cosh \left(\frac{s}{a}\right)^{1/2} x}{s\left[\cosh\left(\frac{s}{a}\right)^{1/2} R + \frac{1}{H}\left(\frac{s}{a}\right)^{1/2} \sinh\left(\frac{s}{a}\right)^{1/2} R\right]} = \frac{\Phi(s)}{\psi(s)}. \qquad (6.3.25)$$

Solution (6.3.25) is a ratio of two generalized polynomials. The quantity $\cosh (s/a)^{1/2}R$ represents a polynomial with respect to s, and the quantity $(s/a)^{1/2} \sinh (s/a)^{1/2}R$ is also a polynomial, viz.

$$\left(\frac{s}{a}\right)^{1/2} \sinh\left(\frac{s}{a}\right)^{1/2} R = \left(\frac{s}{a}\right)^{1/2}\left\{\left(\frac{s}{a}\right)^{1/2} R + \frac{1}{3!}\left[\left(\frac{s}{a}\right)^{1/2} R\right]^3 \cdots\right\}$$

$$= \frac{s}{a} R + \frac{1}{3!}\left(\frac{s}{a}\right)^2 R^3 + \cdots.$$

Thus, solution (6.3.25) satisfies conditions of the expansion theorem because the polynomial $\psi(s)$ does not contain a constant.

We are to determine the root of the function $\psi(s)$, for which we equate it with zero as

$$\psi(s) = s\left[\cosh\left(\frac{s}{a}\right)^{1/2} R + \frac{1}{H}\left(\frac{s}{a}\right)^{1/2} \sinh\left(\frac{s}{a}\right)^{1/2} R\right]$$

$$= s\left[\cos i\left(\frac{s}{a}\right)^{1/2} R + \frac{1}{H}\left(\frac{s}{a}\right)^{1/2} \frac{1}{i} \sin i\left(\frac{s}{a}\right)^{1/2} R\right] = 0; \qquad (6.3.26)$$

here the relation $\cosh z = \cos iz$, $\sinh z = (1/i) \sin iz$ is used.

So we obtain (1) $s = 0$ (zero root) and (2) an infinite number of roots s_n determined by the equation

$$\cos \mu + \frac{\mu}{i^2 HR} \sin \mu = 0,$$

where $i(s/a)^{1/2} R = \mu$; since $i^2 = -1$, we can obtain

$$\cot \mu = (1/\text{Bi})\mu. \qquad (6.3.27)$$

This characteristic equation is identical with equation (6.3.9); it has been analyzed above.

We now determine $\psi'(s)$ and substitute the corresponding value of the root $s_n = -a\mu_n{}^2/R^2$ into it:

$$\psi'(s) = \left[\cosh\left(\frac{s}{a}\right)^{1/2} R + \frac{1}{H}\left(\frac{s}{a}\right)^{1/2} \sinh\left(\frac{s}{a}\right)^{1/2} R\right]$$
$$+ \frac{1}{2}\left(\frac{s}{a}\right)^{1/2} R \sinh\left(\frac{s}{a}\right)^{1/2} R + \frac{1}{2H}\left(\frac{s}{a}\right)^{1/2} \sinh\left(\frac{s}{a}\right)^{1/2} R$$
$$+ \frac{1}{2}\frac{sR}{aH} \cosh\left(\frac{s}{a}\right)^{1/2} R.$$

At $s = s_n$, the expression in square brackets is zero according to equality (6.3.26). Hence,

$$\lim_{s \to s_n} \psi'(s) = -\tfrac{1}{2}\mu_n \sin \mu_n - \frac{1}{2HR}\mu_n \sin \mu_n - \tfrac{1}{2}\mu_n{}^2 \frac{1}{HR} \cos \mu_n$$
$$= -\tfrac{1}{2}(\mu_n \sin \mu_n + \sin \mu_n \cot \mu_n + \mu_n \cos \mu_n \cot \mu_n)$$
$$= -\frac{1}{2 \sin \mu_n}(\mu_n \sin^2 \mu_n + \sin \mu_n \cos \mu_n + \mu_n \cos^2 \mu_n)$$
$$= -\frac{\sin \mu_n \cos \mu_n + \mu_n}{2 \sin \mu}.$$

In addition, we have

$$\psi'(0) = 1\,, \qquad \Phi(0) = (t_a - t_0)\,, \qquad \Phi(s_n) = (t_a - t_0)\cos \mu_n(x/R)\,.$$

Finally, we obtain

$$\frac{t(x,\tau) - t_0}{t_a - t_0} = 1 - \sum_{n=1}^{\infty} \left\{\frac{2\sin \mu_n}{\mu_n + \sin \mu_n \cos \mu_n} \cos \mu_n \frac{x}{R}\right\} \exp\left[-\mu_n^2 \frac{a\tau}{R^2}\right]. \tag{6.3.28}$$

Solution (6.3.28) is identical with (6.3.18).

d. Analysis of the Solution. We can rewrite solution (6.3.28) in dimensionless quantities as

$$\theta = \frac{t(x,\tau) - t_0}{t_a - t_0} = 1 - \sum_{n=1}^{\infty} \left\{A_n \cos \mu_n \frac{x}{R}\right\} \exp[-\mu_n^2\, \mathrm{Fo}], \tag{6.3.29}$$

where

$$A_n = \frac{2\sin \mu_n}{\mu_n + \sin \mu_n \cos \mu_n} = (-1)^{n+1} \frac{2\mathrm{Bi}(\mathrm{Bi}^2 + \mu_n^2)^{1/2}}{\mu_n(\mathrm{Bi}^2 + \mathrm{Bi} + \mu_n^2)}, \tag{6.3.30}$$

since sin μ_n and cos μ_n may be replaced by μ_n and Bi from the characteristic equation.

From solution (6.3.29), it can be seen that the relative excess temperature θ is a function of the Fourier number, the relative coordinate x/R and the Biot criterion, since the initial thermal amplitudes A_n are single-valued functions of the Biot criterion (see formula (6.3.30))

$$\theta = \Psi\{x/R, \mathrm{Bi}, \mathrm{Fo}\}.$$

Numerical values of the first six thermal amplitudes A_n are given in Table 6.2.

For practical engineering calculations, Figs. 6.6a–6.7b present nomograms taken from [102] for determining the temperature of a plate surface θ_s and at the plate center θ_c for prescribed values of Fo and Bi.

If the Biot criterion tends to infinity, the temperature of the plate surface immediately becomes equal to that of the surrounding medium, because from boundary condition (6.3.3), it follows that

$$t_a - t(R,\tau) = \lim_{\mathrm{Bi}\to\infty} \left[\frac{R}{\mathrm{Bi}} \frac{\partial t(R,\tau)}{\partial x}\right] = 0. \tag{6.3.31}$$

And the eigenvalues μ_n will be equal to

$$\mu_n = (2n - 1)\tfrac{1}{2}\pi\,,$$

TABLE 6.2. VALUES OF THE CONSTANTS

$$A_n = (-1)^{n+1} \frac{2\,\mathrm{Bi}(\mathrm{Bi}^2 + \mu_n^2)^{1/2}}{\mu_n(\mathrm{Bi}^2 + \mathrm{Bi} + \mu_n^2)}$$

Bi	A_1	A_2	A_3	A_4	A_5	A_6
0	1.0000	−0.0000	0.0000	−0.0000	0.0000	−0.0000
0.001	1.0002	−0.0002	0.0000	−0.0000	0.0000	−0.0000
0.002	1.0004	−0.0004	0.0001	−0.0000	0.0000	−0.0000
0.004	1.0008	−0.0008	0.0002	−0.0001	0.0001	−0.0000
0.006	1.0012	−0.0012	0.0003	−0.0001	0.0001	−0.0000
0.008	1.0015	−0.0016	0.0004	−0.0002	0.0001	−0.0001
0.01	1.0020	−0.0020	0.0005	−0.0002	0.0001	−0.0001
0.02	1.0030	−0.0040	0.0010	−0.0004	0.0003	−0.0002
0.04	1.0065	−0.0080	0.0020	−0.0009	0.0005	−0.0003
0.06	1.0099	−0.0119	0.0030	−0.0013	0.0007	−0.0004
0.08	1.0130	−0.0158	0.0040	−0.0018	0.0010	−0.0006
0.10	1.0159	−0.0197	0.0050	−0.0022	0.0013	−0.0008
0.20	1.0312	−0.0381	0.0100	−0.0045	0.0025	−0.0016
0.30	1.0450	−0.0555	0.0148	−0.0067	0.0038	−0.0024
0.40	1.0581	−0.0719	0.0196	−0.0089	0.0050	−0.0032
0.50	1.0701	−0.0873	0.0243	−0.0110	0.0063	−0.0040
0.60	1.0813	−0.1025	0.0289	−0.0132	0.0075	−0.0048
0.70	1.0918	−0.1154	0.0335	−0.0153	0.0087	−0.0056
0.80	1.1016	−0.1282	0.0379	−0.0175	0.0100	−0.0064
0.90	1.1107	−0.1403	0.0423	−0.0196	0.0112	−0.0072
1.00	1.1192	−0.1517	0.0466	−0.0217	0.0124	−0.0080
1.5	1.1537	−0.2013	0.0667	−0.0318	0.0184	−0.0119
2.0	1.1784	−0.2367	0.0848	−0.0414	0.0241	−0.0157
3.0	1.2102	−0.2881	0.1154	−0.0589	0.0351	−0.0231
4.0	1.2287	−0.3215	0.1396	−0.0750	0.0451	−0.0300
5.0	1.2403	−0.3442	0.1588	−0.0876	0.0543	−0.0366
6.0	1.2478	−0.3604	0.1740	−0.0991	0.0626	−0.0427
7.0	1.2532	−0.3722	0.1861	−0.1089	0.0701	−0.0483
8.0	1.2569	−0.3812	0.1959	−0.1174	0.0768	−0.0535
9.0	1.2598	−0.3880	0.2039	−0.1246	0.0828	−0.0583
10.0	1.2612	−0.3934	0.2104	−0.1309	0.0881	−0.0676
15.0	1.2677	−0.4084	0.2320	−0.1514	0.1072	−0.0795
20.0	1.2699	−0.4147	0.2394	−0.1621	0.1182	−0.0901
30.0	1.2717	−0.4198	0.2472	−0.1718	0.1291	−0.1015
40.0	1.2723	−0.4217	0.2502	−0.1759	0.1340	−0.1069
50.0	1.2727	−0.4227	0.2517	−0.1779	0.1365	−0.1098
60.0	1.2728	−0.4232	0.2526	−0.1791	0.1379	−0.1115
80.0	1.2730	−0.4237	0.2535	−0.1803	0.1394	−0.1132
100.0	1.2731	−0.4239	0.2539	−0.1808	0.1405	−0.1141
∞	1.2732	−0.4244	0.2546	−0.1819	0.1415	−0.1157

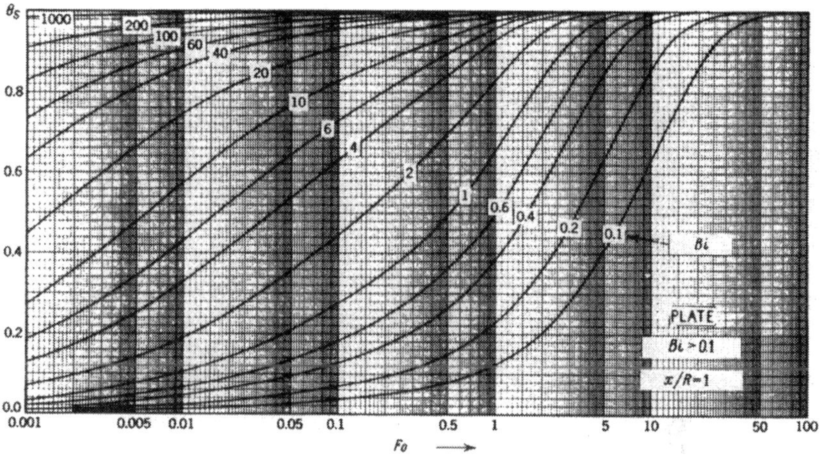

Fig. 6.6a. Graphs for determination of the dimensionless excess temperature for a plate θ_s for high values of Bi ($0.1 < \text{Bi} < 1000$) [102].

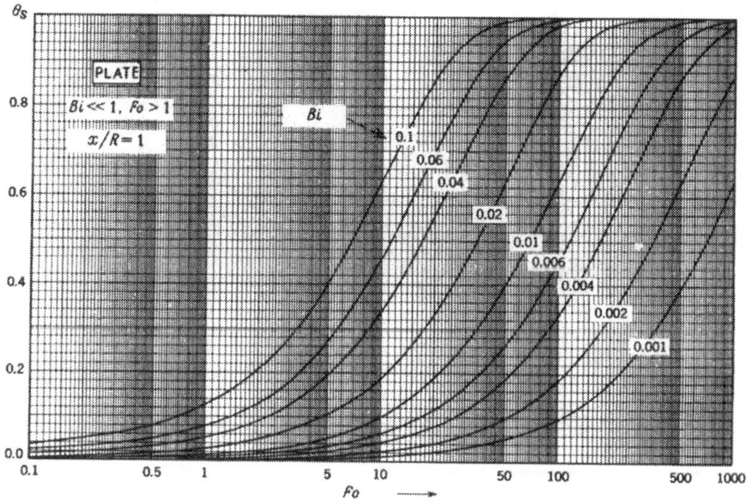

Fig. 6.6b. Graphs for determination of the dimensionless excess temperature for a plate with small values of Bi ($0.001 < \text{Bi} < 0.1$) [102].

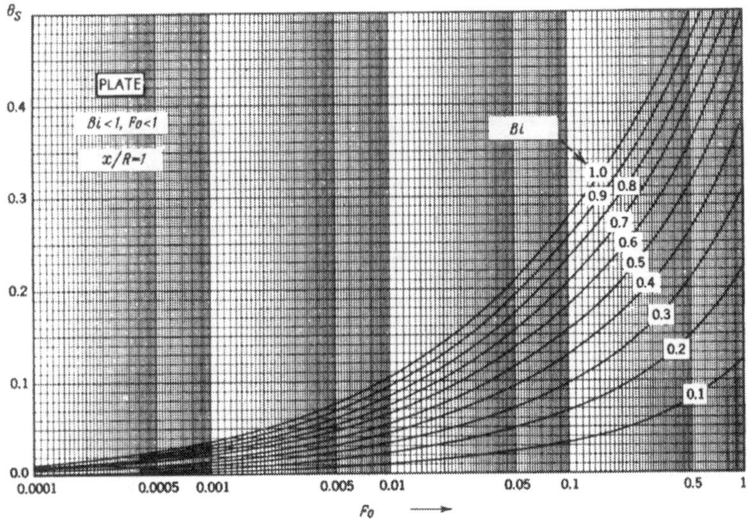

Fig. 6.6c. Graphs for determination of dimensionless excess temperature θ_s at the place surface for small values of Fo (0.0001 < Fo < 1) and moderate values of Bi (0.1 < Bi < 1.0) [102].

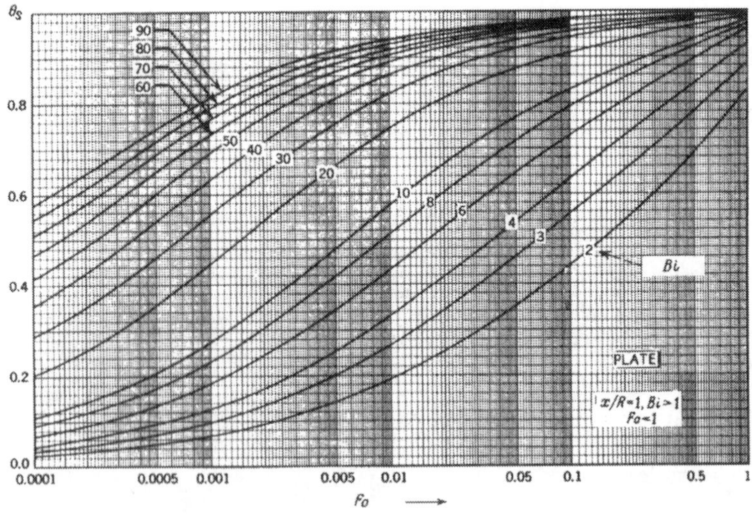

Fig. 6.6d. Graphs for determination of dimensionless excess temperature θ_s at the place surface for small values of Fo (0.0001 < Bi < 1) and high values of Bi (2 < Bi < < 100) [102].

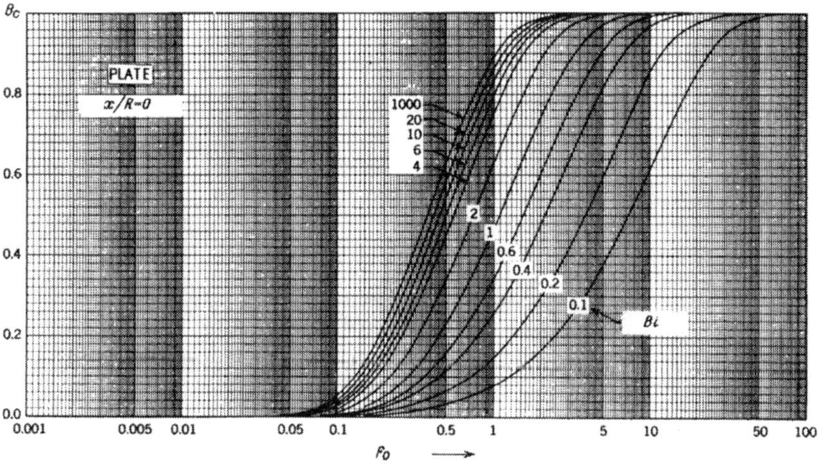

Fig. 6.7a. Graphs for determination of dimensionless excess temperature in the plate center θ_c for high values of Bi (0.1 < Bi < 1000) [102].

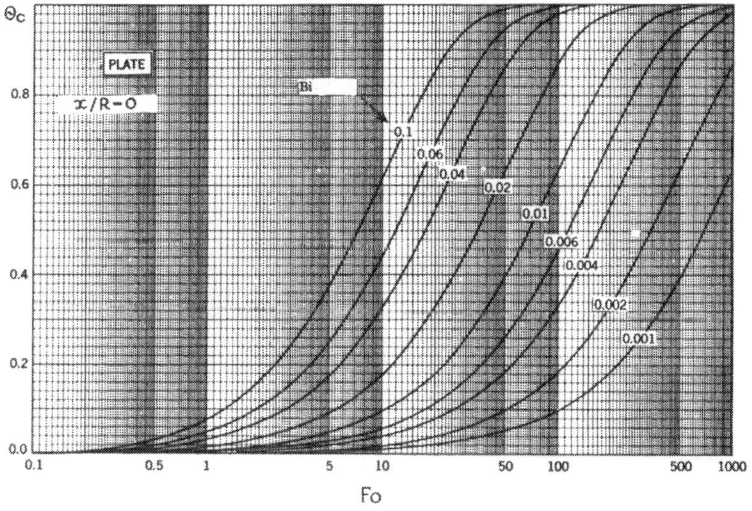

Fig. 6.7b. Graphs for determination of dimensionless excess temperature in the plate center θ_c for small values of Bi (0.001 < Bi < 0.1) [102].

i.e., μ_n will not depend on the thickness of a plate. From formula (6.3.30) it follows that at $\mathrm{Bi} \to \infty$

$$A_n = \frac{2 \sin \mu_n}{\mu_n} = (-1)^{n+1} \frac{2}{\mu_n} = (-1)^{n+1} \frac{4}{(2n-1)\pi}, \tag{6.3.32}$$

since $\cos (2n-1)\frac{1}{2}\pi = 0$. Then solution (6.3.29) assumes the form

$$\theta = 1 - \sum_{n=1}^{\infty} (-1)^{n+1} \frac{4}{(2n-1)\pi} \cos \frac{(2n-1)\pi x}{2R} \exp\left[-\frac{(2n-1)^2\pi^2}{4} \mathrm{Fo}\right]. \tag{6.3.33}$$

This solution is identical with (4.3.16); it is only necessary to assume $\theta_{\mathrm{cool}} = 1 - \theta_{\mathrm{heat}}$.

The intensity in heating $d\theta/d\tau$ may be found by differentiating equation (6.3.33); it will be inversely proportional to the second power of the characteristic plate dimension and directly proportional to the thermal diffusivity, as

$$\frac{d\theta}{d\tau} = \frac{a}{R^2} \sum_{n=1}^{\infty} \mu_n^2 A_n \cos \mu_n \frac{x}{R} \exp[-\mu_n^2 \mathrm{Fo}].$$

Hence, the heating rate of a plate at a given point will be determined by the thermal inertia body properties and depend only on the velocity of heat displacement inside the plate (internal problem).

If the Biot criterion is small ($\mathrm{Bi} < 0.1$), then all the terms of the series are negligibly small in comparison to the first, because at $\mu_n \to (n-1)\pi$, $A_n \to 0$ with the exception of A_1 which is equal to

$$A_1 = \lim_{\mu_1 \to 0} \left[\frac{2 \dfrac{\sin \mu_1}{\mu_1}}{1 + \cos \mu_1 \dfrac{\sin \mu_1}{\mu_1}} \right] = 1.$$

For small values of μ_1, $\tan \mu_1$ may be replaced by μ_1, then from the characteristic equation we obtain

$$\mu_1^2 = \mathrm{Bi}.$$

Hence, solution (6.3.29) acquires the values

$$\theta = 1 - \cos(\mathrm{Bi})^{1/2} x/R \exp[-\mathrm{Bi}\,\mathrm{Fo}]. \tag{6.3.34}$$

In this case, the heating rate

$$\frac{d\theta}{d\tau} = \frac{\alpha}{c\gamma R} \cos(\mathrm{Bi})^{1/2} \frac{x}{R} \exp[- \mathrm{Fo}^*] \qquad (6.3.35)$$

is directly proportional to the heat transter coefficient and inversely proportional to the first power of the characteristic dimension of the plate. Thus, the rate of heating is determined by that of heat transfer from the surrounding medium to a plate surface (external problem).

If the criterion Bi is larger than 0.1 or less than 100 ($0.1 < \mathrm{Bi} < 100$), then μ_n is a function of Bi i.e., it depends on the plate thickness. In this case, heating rate is inversely proportional to the nth power of the plate thickness ($1 < n < 2$) and determined by both the heat transfer rate inside the material and that through a boundary layer (boundary-value problem).

In Fig. 6.8 all three cases are plotted as temperature-distribution curves at various time instants and one can see that in the first case the plate surface temperature is equal to that of the medium, starting from the very commencement of heating. The temperature difference between the central plane and the plate surface for Fig. 6.8a is greater than for the other two

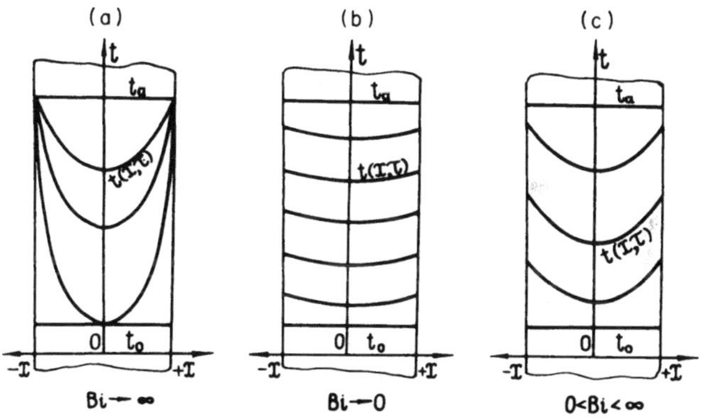

Fig. 6.8. Temperature distribution in an infinite plate for various values of Bi: (a) internal problem (Bi → ∞); (b) external problem (Bi → 0); (c) boundary-value problem ($0 < \mathrm{Bi} < \infty$).

cases. In Fig. 6.8b (Bi → 0) the temperature of the surface is close to that of the central layer, and the temperature drop between them is the smallest. Figure 6.8c is the intermediate case falling between the first two.

If the temperature-distribution curves are extended beyond the plate surface in the direction of tangents, then they all intersect at one point located on a line $t_a = \text{const}$ (see Chapter 1, Section 6).

Thus, if the plate thickness is increased by $2/H$ (by $1/H$ from both sides) and it is assumed that in this fictitious layer a temperature distribution occurs according to a straight-line law, then all the distribution curves will intersect at one point and a boundary-value problem reduces to a generalized internal problem.

Returning to solution (6.3.29) of our problem, we note that

$$\mu_1 < \mu_2 < \cdots < \mu_n$$

and consequently (6.3.29) quickly converges and, starting from a certain value of Fo_1, all the terms of these series becomes infinitesimal as compared to the first term, so they may be neglected (it will be shown that to within 0.25% at $x/R = 0$ and $\mathrm{Bi} = 1$, all the terms of the series may be neglected, if $\mathrm{Fo} \geq 0.55$).

Hence, at $\mathrm{Fo} \geq \mathrm{Fo}_1$, solution (6.3.29) is considerably simplified and has the form

$$\theta = 1 - A_1 \cos \mu_1(x/R) \exp[-\mu_1^2 \mathrm{Fo}]. \tag{6.3.36}$$

For convenience in calculations Figs. 6.9 and 6.10 present diagrams of $\mu_1 = f(\mathrm{Bi})$ and $A_1 = \varphi(\mathrm{Bi})$ for values of Bi from 0 to 20 (at $\mathrm{Bi} > 0.1$, μ_1 may be calculated according to formula (6.10.9) developed later in this chapter). For small values of Bi, a number of terms of the series must be taken, which leads to certain difficulties in computations.

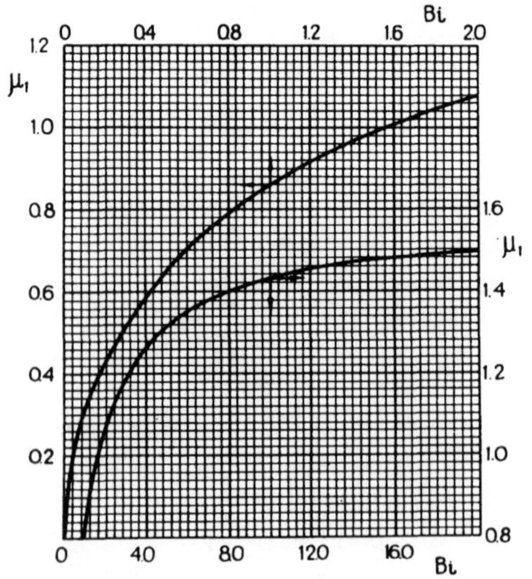

Fig. 6.9. The first root μ_1 of the characteristic equation versus Bi for a plate.

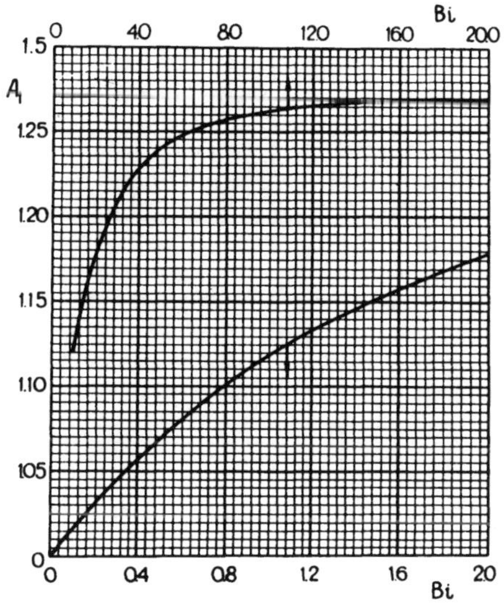

Fig. 6.10. Coefficient A_1 versus Bi for an infinite plate.

Let us now determine an approximate solution of our problem applicable to small values of Fo. Solution (6.3.25) for the transform may be written

$$T(x, s) - \frac{t_0}{s} = \frac{t_a - t_0}{s}$$

$$\times \left[\frac{\exp[(s/a)^{1/2}x] + \exp[-(s/a)^{1/2}x]}{\exp[(s/a)^{1/2}R]+\exp[-(s/a)^{1/2}R]+(1/H)(s/a)^{1/2}\{\exp[(s/a)^{1/2}R]-\exp[-(s/a)^{1/2}R]\}}\right]$$

$$= \frac{t_a - t_0}{s(1 + (1/H)(s/a)^{1/2})}\{\exp[-(s/a)^{1/2}(R - x)] + \exp[-(s/a)^{1/2}(R + x)]\}$$

$$\times [1 + \{[1 - (1/H)(s/a)^{1/2}]/[1 + (1/H)(s/a)^{1/2}]\} \exp[-2(s/a)^{1/2}R]]^{-1} \qquad (6.3.37)$$

If the expression in the square brackets is expanded into a geometric series and only the first term of the series is taken, then we obtain

$$T(x, s) - \frac{t_0}{s}$$

$$\approx \frac{(t_a - t_0)}{s(1 + (1/H)(s/a)^{1/2})}$$

$$\times \{\exp[-(s/a)^{1/2}(R - x)] + \exp[-(s/a)^{1/2}(R + x)]\}. \qquad (6.3.38)$$

The inversion of the transform is accomplished according to the relation used in Section 6.1

$$\theta = \frac{t(x, \tau) - t_0}{t_a - t_0}$$
$$\approx \operatorname{erfc} \frac{R - x}{2(a\tau)^{1/2}} - \exp[H(R - x) + aH^2\tau] \operatorname{erfc}\left(\frac{R - x}{2(a\tau)^{1/2}} + H(a\tau)^{1/2}\right)$$
$$+ \operatorname{erfc} \frac{R + x}{2(a\tau)^{1/2}} - \exp[H(R + x) + aH^2\tau] \operatorname{erfc}\left\{\frac{R + x}{2(a\tau)^{1/2}} + H(a\tau)^{1/2}\right\}. \tag{6.3.39}$$

If the origin of coordinates is displaced from the middle of the plate to the left surface, (i.e., a variable $x + R = X$ is substituted) and provided that $2R \to \infty$, then solution (6.3.39) converts into solution (6.1.11).

Rewriting solution (6.3.39) in criterial form we have

$$\theta \approx \operatorname{erfc}\frac{1 - (x/R)}{2(\mathrm{Fo})^{1/2}} - \exp\left[\mathrm{Bi}\left(1 - \frac{x}{R}\right) + \mathrm{Bi}^2\,\mathrm{Fo}\right]$$
$$\times \operatorname{erfc}\left\{\frac{1 - (x/R)}{2(\mathrm{Fo})^{1/2}} + \mathrm{Bi}\,(\mathrm{Fo})^{1/2}\right\}$$
$$+ \operatorname{erfc}\frac{1 + (x/R)}{2(\mathrm{Fo})^{1/2}} - \exp\left[\mathrm{Bi}\left(1 + \frac{x}{R}\right) + \mathrm{Bi}^2\,\mathrm{Fo}\right]$$
$$\times \operatorname{erfc}\left(\frac{1 + (x/R)}{2(\mathrm{Fo})^{1/2}} + \mathrm{Bi}(\mathrm{Fo})^{1/2}\right). \tag{6.3.40}$$

Solution (6.3.40) is approximate and applicable to small values of Fo; it in effect replaces the sum of a large number of terms of a series of the ordinary solution (6.3.29).

We now give a numerical example to illustrate the effectiveness of solution (6.3.40) at small values of Fo. At small values of the Fourier number, the temperature of the body center does not change in practice, and a change in the temperature of the plate surface ($x = R$) is, therefore, of the greatest interest. In 1932, Pöschl [95] calculated the temperature of such a plate surface with cooling for small values of Fo (from 0.0003 to 0.01) for various values of Bi (from 0.1 to 2000). These calculation results are given in Table 6.3.

Computations were made according to solution (6.3.29). Pöschl noted that he had to take 36 terms of a series at Fo = 0.0003. Indeed, such bulky computations cause astonishment.

We next calculate $(1 - \theta_s)$ according to our approximate formula (6.3.40). For the plate surface only the first two terms of solution (6.3.40) need be taken, i.e.,

$$1 - \theta_s \approx \exp\left[\{\mathrm{Bi}\,(\mathrm{Fo})^{1/2}\}^2\right] \operatorname{erfc} \mathrm{Bi}\,(\mathrm{Fo})^{1/2}. \tag{6.3.41}$$

This approximate solution (6.3.41) is convenient for calculations at small values of the number $\mathrm{Bi}(\mathrm{Fo})^{1/2}$; at large values of $\mathrm{Bi}(\mathrm{Fo})^{1/2}$ the following relation is used

$$e^{u^2} \operatorname{erfc} u \simeq \frac{1}{\sqrt{\pi}}\left(\frac{1}{u} - \frac{1}{2u^3} + \frac{3}{4u^5} - \cdots\right),$$

TABLE 6.3. RELATIVE SURFACE TEMPERATURE OF A PLATE $(1-\theta_s) = \psi(\text{Fo}, \text{Bi})$

Bi	Fourier number, Fo				
	0.0003	0.0010	0.0025	0.0050	0.0100
0.1	0.999	0.997	0.995	0.993	0.989
0.5	0.996	0.983	0.975	0.963	0.948
1	0.980	0.965	0.947	0.926	0.897
4	0.927	0.872	0.809	0.747	0.670
10	0.883	0.726	0.615	0.522	0.428
20	0.705	0.555	0.441	0.336	0.256
50	0.468	0.309	0.211	0.154	0.111
100	0.287	0.171	0.111	0.079	0.056
200	0.157	0.088	0.057	0.040	0.028
500	0.066	0.036	0.023	0.016	0.011
1000	0.033	0.018	0.011	0.008	0.006
2000	0.017	0.009	0.006	0.004	0.003

i.e.,

$$1-\theta_s \approx \frac{1}{\sqrt{\pi}}\left(\frac{1}{\text{Bi(Fo)}^{1/2}} - \frac{1}{2\{\text{Bi(Fo)}^{1/2}\}^3} + \cdots\right). \tag{6.3.42}$$

To calculate $(1-\theta_s)$ for Fo = 0.0003 and Bi = 1000 we first determine that the number $\text{Bi(Fo)}^{1/2}$ is equal to

$$\text{Bi(Fo)}^{1/2} = (0.0003)^{1/2}\cdot 1000 = 17.32.$$

We use formula (6.3.42) to calculate

$$1-\theta_s \approx \frac{1}{\sqrt{\pi}}\left(\frac{1}{17.32} - \frac{1}{2\cdot(17.32)^3}\right) \approx 0.033.$$

Thus, a value is obtained, which corresponds to the tabular one which was calculated by using a great number of series terms.

For Fo = 0.0100 and Bi = 1000 the number $\text{Bi(Fo)}^{1/2}$ is equal to 100, then

$$1-\theta_s = \frac{1}{100\sqrt{\pi}} \approx 0.006.$$

From Table 6.3 one can find that at Fo = 0.0100 and Bi = 1000, $(1-\theta_s) = 0.006$, i.e., is correct to within 1%.

Taking next a small value of the Biot criterion (Bi = 0.5), at Fo = 0.0010 we shall have

$$\text{Bi(Fo)}^{1/2} = 0.5\,(0.0010)^{1/2} = 0.0158.$$

According to formula (6.3.41) we find

$$1 - \theta_s = \exp(0.00025)\,\mathrm{erfc}(0.0158) \approx 0.983.$$

which also agrees with the tabular data.

From the above computations one can see that approximate solution (6.3.41) gives quite satisfactory results and tremendously decreases the labor of computations required by the classical solution (6.3.29). This is the greatest advantage of the operational method over the ordinary one: it permits one to obtain approximate solutions for various values of the criteria Fo and Bi.

e. Specific Heat Rate. To determine the heating rate, we define the mean plate temperature by the formula

$$\bar{t}(\tau) = \frac{1}{R}\int_0^R t(x, \tau)\,dx. \tag{6.3.43}$$

TABLE 6.4. THE VALUES OF CONSTANTS

$$B_n = \frac{2\,\mathrm{Bi}^2}{\mu_n^2(\mathrm{Bi}^2 + \mathrm{Bi} + \mu_n^2)}$$

Bi	B_1	B_2	B_3	B_4	B_5	B_6
∞	0.8106	0.0901	0.0324	0.0165	0.0100	0.0067
50.0	0.8250	0.0899	0.0323	0.0161	0.0095	0.0061
30.0	0.8354	0.0893	0.0315	0.0152	0.0086	0.0053
15.0	0.8565	0.0885	0.0279	0.0120	0.0060	0.0033
10.0	0.8743	0.0839	0.0236	0.0090	0.0040	0.0020
9.0	0.8796	0.0821	0.0222	0.0081	0.0035	0.0017
8.0	0.8859	0.0797	0.0205	0.0072	0.0030	0.0015
7.0	0.8932	0.0766	0.0185	0.0062	0.0025	0.0012
6.0	0.9021	0.0723	0.0162	0.0051	0.0020	0.0009
5.0	0.9130	0.0664	0.0135	0.0040	0.0015	0.0007
4.0	0.9264	0.0582	0.0104	0.0029	0.0010	0.0005
3.0	0.9430	0.0468	0.0070	0.0019	0.0006	0.0003
2.0	0.9635	0.0313	0.0037	0.0009	0.0003	0.0001
1.5	0.9749	0.0220	0.0023	0.0005	0.0002	0.0001
1.0	0.9862	0.0124	0.0011	0.0002	0.0001	
0.9	0.9882	0.0105	0.0009	0.0002	0.0001	
0.8	0.9903	0.0088	0.0007	0.0001		
0.7	0.9920	0.0070	0.0006	0.0001		
0.6	0.9939	0.0054	0.0004	0.0001		
0.5	0.9955	0.0040	0.0003	0.0001		
0.4	0.9973	0.0027	0.0002	0.0001		
0.3	0.9982	0.0016	0.0001			
0.2	0.9995	0.0007				
0.1	1.0000	0.0002				

If the corresponding expression from solution (6.3.29) is substituted for $t(x, \tau)$, we have on integrating

$$\bar{\theta}(\tau) = \frac{\bar{t}(\tau) - t_0}{t_a - t_0} = 1 - \sum_{n=1}^{\infty} B_n \exp[-\mu_n^2 \, \mathrm{Fo}] \tag{6.3.44}$$

where

$$B_n = \frac{A_n \sin \mu_n}{\mu_n} = \frac{2\,\mathrm{Bi}^2}{\mu_n^2(\mathrm{Bi}^2 + \mathrm{Bi} + \mu_n^2)}. \tag{6.3.45}$$

All the coefficients B_n are positive, and quickly decrease with an increase in μ_n. In Table 6.4 the first six values of B_n are given as a function of the Biot criterion. These values of B_n are accurate to four decimal places.

At Bi $\rightarrow \infty$, the coefficient B_n will be equal to

$$B_n = \frac{2}{\mu_n^2} = \frac{8}{\pi^2(2n-1)^2}.$$

For convenience of computations, Fig. 6.11 gives diagrams of $B_1 = f(\mathrm{Bi})$

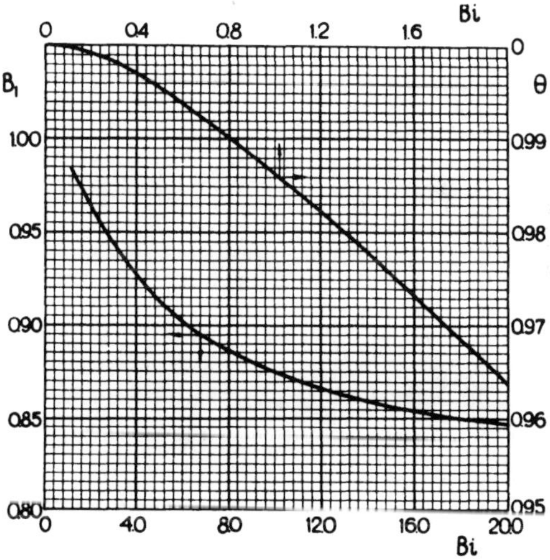

Fig. 6.11. Coefficient B_1 versus Bi for an infinite plate.

for values of the Biot criterion from 0 to 20. The specific heat rate (kcal/m³) for the given time τ is determined by the formula

$$\Delta Q_v = c\gamma[\bar{t}(\tau) - \bar{t}_0]. \tag{6.3.46}$$

f. Parabolic Initial Temperature Distribution. Next consider the problem of cooling a plate in a medium with temperature $\vartheta_a = 0$ when the initial temperature distribution is given according to a parabolic law:

$$\vartheta(x, 0) = \vartheta_c - (\vartheta_c - \vartheta_s)(x^2/R^2)$$

where ϑ_c and ϑ_s are temperatures at the center and surface of the plate, respectively. The solution of such a problem is widely used in the diffusion theory as well as in drying technique.

The solution of this problem from (6.3.15) will be of the form

$$\vartheta(x, \tau) = \sum_{n=1}^{\infty} \frac{\mu_n \cos \mu_n(x/R)}{\mu_n + \sin \mu_n \cos \mu_n} \exp[-\mu_n^2 \,\mathrm{Fo}] \times \frac{2}{R} \int_0^R f_1(x) \cos \mu_n \frac{x}{R}\, dx, \tag{6.3.47}$$

where $f_1(x) = \vartheta(x, 0)$ is the temperature distribution at the initial time instant.

The integral in solution (6.3.47) may be determined explicitly. Using the formula

$$\int u^2 \cos u\, du = 2u \cos u + (u^2 - 2) \sin u\,,$$

we have

$$\frac{2}{R} \int_0^R \vartheta_c \cos \mu_n \frac{x}{R}\, dx = \frac{2 \sin \mu_n}{\mu_n} \vartheta_c$$

and

$$\frac{2}{R} \int_0^R (\vartheta_c - \vartheta_s) \frac{x^2}{R^2} \cos \mu_n \frac{x}{R}\, dx = \frac{2(\vartheta_c - \vartheta_s) \sin \mu_n}{\mu_n} \left(\frac{2}{\mu_n} \cot \mu_n + 1 - \frac{2}{\mu_n^2} \right).$$

$(1/\mu_n) \cot \mu_n$ may be replaced by Bi according to the characteristic equation; then upon the necessary transformations, the solution takes the form

$$\vartheta(x, \tau) = \sum_{n=1}^{\infty} \left[\vartheta_s - 2(\vartheta_c - \vartheta_s) \left(\frac{1}{\mathrm{Bi}} - \frac{1}{\mu_n^2} \right) \right] A_n \times \cos \mu_n \frac{x}{R} \exp[-\mu_n^2 \,\mathrm{Fo}], \tag{6.3.48}$$

where A_n are the constant coefficients defined by relation (6.3.30).

To conclude this section, a particular calculation is given. The temperature in the middle and at the surface of a rubber plate 6 mm in thickness is to be determined after two minutes of heating in an air stream at $t = 150°C$. The initial plate temperature is 20°C.

The thermal coefficients of rubber are assumed to be $\lambda = 0.21$ kcal/m hr deg, $c = 0.36$ kcal/kg deg, $\gamma = 1100$ kg/cm³, $a = 5.3 \times 10^{-4}$ m²/hr, and $\alpha = 28$ kcal/m² hr deg. First we determine the dimensionless independent quantities.

The Fourier number is

$$\mathrm{Fo} = \frac{a\tau}{R^2} = \frac{5.3 \cdot 10^{-4} \cdot 2}{60 \cdot 9 \cdot 10^{-6}} = 1.96,$$

as $R = 3$ mm $= 0.003$ m. For the Biot criterion,

$$\mathrm{Bi} = \frac{\alpha R}{\lambda} = \frac{28}{0.21} 0.003 = 0.4.$$

Since the criterion $\mathrm{Fo} = 1.96$ is greater than unity, we may restrict ourselves to the first term of the series in solution (6.3.29), i.e.,

$$\theta = \frac{t(x, \tau) - t_0}{t_a - t_0} = 1 - A_1 \cos \mu_1 \frac{x}{R} \exp[-\mu_1^2 \mathrm{Fo}].$$

Then from Tables 6.1 and 6.2 it follows $\mu_1 = 0.5932 \approx 0.593$, $A_1 = 1.0581$.

Thus, we have

$$\theta_s = \frac{t(R, \tau) - 20}{150 - 20} = 1 - 1.058 \cos 0.593 \exp[-0.3516 \cdot 1.96]$$

$$= 1 - 0.440 = 0.560.$$

The value obtained is verified by that shown in Fig. 6.6a. For $\mathrm{Fo} \approx 1.96$ and $\mathrm{Bi} = 0.4$ this nomogram gives $\theta_s = 0.56$.

The relative excess temperature in the middle of the plate is

$$\theta_c = \frac{t(0, \tau) - 20}{150 - 20} = 1 - 1.058 \exp[-0.3516 \cdot 1.96] = 1 - 0.529 = 0.471.$$

According to the nomogram in Fig. 6.7a we find for $\mathrm{Fo} = 1.96$, $\mathrm{Bi} = 0.4$ that $\theta_c \approx 0.47$. Thus we have

$$t_s = 20° + 130\theta_s = 20° + 130 \cdot 0.56 = 92.8°\mathrm{C}$$

$$t_c = 20° + 130\theta_c = 20° + 61.2 = 81.2°\mathrm{C}.$$

To determine the specific heat rate we calculate the mean temperature of the plate $\bar{t}(\tau)$. We have

$$\bar{\theta} = \frac{\bar{t}(\tau) - t_0}{t_a - t_0} = 1 - B_1 \exp[-\mu_1^2 \mathrm{Fo}]$$

$$= 1 - 0.997 \exp[-0.689] = 1 - 0.498 = 0.502 \approx 0.5.$$

The constant B_1 is determined from Table 6.4. Hence,

$$\bar{t} = 20 + 130 \cdot 0.5 = 85°.$$

The specific heat rate will be

$$\Delta Q_v = c\gamma[\bar{t}(\tau) - t_0] = 0.36 \cdot 1100(85 - 20) = 25740 \text{ kcal/m}^3$$
$$= 25.74 \text{ cal/cm}^3.$$

This amount of heat was transferred to 1 cm^3 of the plate during the two minute period.

g. Nonsymmetrical Heating. To conclude this section we consider nonsymmetrical heating of an infinite plate with boundary conditions

$$t(x, 0) = t_0 = \text{const}, \tag{6.3.49}$$

$$+\frac{\partial t(0, \tau)}{\partial x} + H_1[t_0 - t(0, \tau)] = 0, \tag{6.3.50}$$

$$-\frac{\partial t(R, \tau)}{\partial x} + H_2[t_a - t(R, \tau)] = 0 \tag{6.3.51}$$

where $H_1 = \alpha_1/\lambda$, $H_2 = \alpha_2/\lambda$.

Using the Laplace transform with respect to τ, the solution of differential equation (6.3.1) for the transform may be written as follows

$$T(x, s) - \frac{t_0}{s} = A \cosh\left(\frac{s}{a}\right)^{1/2} x + B \sinh\left(\frac{s}{a}\right)^{1/2} x. \tag{6.3.52}$$

Boundary conditions (6.3.50) and (6.3.51) for the transform will have the form

$$-T'(0, s) + H_1[T(0, s) - (t_0/s)] = 0, \tag{6.3.53}$$

$$T'(R, s) + H_2[T(R, s) - (t_a/s)] = 0. \tag{6.3.54}$$

The constants A and B in solution (6.3.52) will be determined from boundary conditions (6.3.53) and (6.3.54). The resulting values of A and B will be substituted into (6.3.52) to give

$$T(x, s) - \frac{t_0}{s}$$
$$= \frac{(t_a - t_0)\left[\cosh\left(\frac{s}{a}\right)^{1/2} x + H_1\left(\frac{a}{s}\right)^{1/2} \sinh\left(\frac{s}{a}\right)^{1/2} x\right]}{s\left[\left(1+\frac{H_1}{H_2}\right)\cosh\left(\frac{s}{a}\right)^{1/2} R + \left\{\frac{1}{H_2}\left(\frac{s}{a}\right)^{1/2} + H_1\left(\frac{a}{s}\right)^{1/2}\right\}\sinh\left(\frac{s}{a}\right)^{1/2} R\right]}$$
$$= \frac{\Phi(s)}{\psi(s)}. \tag{6.3.55}$$

Equation (6.3.55) is the ratio of two generalized polynomials where $\psi(s)$ does not contain the constant, i.e., the conditions of the expansion theorem are fulfilled.

Equate the function $\psi(s)$ to zero to find its roots.

$$s\left[\left(1+\frac{H_1}{H_2}\right)\cos i\left(\frac{s}{a}\right)^{1/2}R+\left(\frac{1}{H_2}\left(\frac{s}{a}\right)^{1/2}+H_1\left(\frac{s}{a}\right)^{-1/2}\right)\frac{1}{i}\sin i\left(\frac{s}{a}\right)^{1/2}R\right]=0. \tag{6.3.56}$$

Thence we have (1) $s = 0$, and (2) an infinite number of roots s_n evaluated from the characteristic equation

$$\cot\mu=\left(\frac{\mu}{\mathrm{Bi}_2}-\frac{\mathrm{Bi}_1}{\mu}\right)\left(1+\frac{\mathrm{Bi}_1}{\mathrm{Bi}_2}\right)^{-1} \tag{6.3.57}$$

where $i(s/a)^{1/2}R = \mu$, $\mathrm{Bi}_1 = H_1R$, $\mathrm{Bi}_2 = H_2R$. The points of intersection of the cotangent curve $y_1 = \cot\mu$ with the hyperbola of the form

$$y_2=\left(\frac{\mu}{\mathrm{Bi}_2}-\frac{\mathrm{Bi}_1}{\mu}\right)\Big/\left(1+\frac{\mathrm{Bi}_1}{\mathrm{Bi}_2}\right)$$

give the roots μ_n of Eq. (6.3.57), their number is infinitely large. In pairs they have the same absolute values and opposite signs. Since $s_n = -a\mu_n^2/R^2$ only positive values of the roots μ_n need be considered to determine s_n. Applying the expansion theorem to the case of simple roots we find

$$\theta=\frac{t(x,\tau)-t_0}{t_a-t_0}=\frac{1+\mathrm{Bi}_1x/R}{(1+\mathrm{Bi}_1+\mathrm{Bi}_1/\mathrm{Bi}_2)}$$
$$-\sum_{n=1}^{\infty}A_n\left(\cos\mu_n\frac{x}{R}+\frac{\mathrm{Bi}_1}{\mu_n}\sin\mu_n\frac{x}{R}\right)\exp[-\mu_n^2\,\mathrm{Fo}], \tag{6.3.58}$$

where A_n is the initial thermal amplitude equal to

$$A_n=\left\{\left(1+\frac{\mathrm{Bi}_1}{\mathrm{Bi}_2}\right)\frac{\sin\mu_n\cos\mu_n+\mu_n}{2\sin\mu_n}+\frac{\mathrm{Bi}_1}{\mu_n}\sin\mu_n\right\}^{-1}. \tag{6.3.59}$$

The specific heat rate is easily determined by the ordinary method.

If $\mathrm{Bi}_1 = \mathrm{Bi}_2$ is assumed, the characteristic equation (6.3.57) becomes identical to (6.3.17). Then from (6.3.15), solution (6.3.58) is obtained, if we assume $f_1(x)$ equal to $t_0 = \text{const}$. In Fig. 6.12, $\theta = f(\mathrm{Fo})$ is plotted for the case when the temperature at one surface of the plate ($x = R$) is maintained constant and equal to the initial temperature

$$t(R,\tau)=t_0=\text{const}. \tag{6.3.60}$$

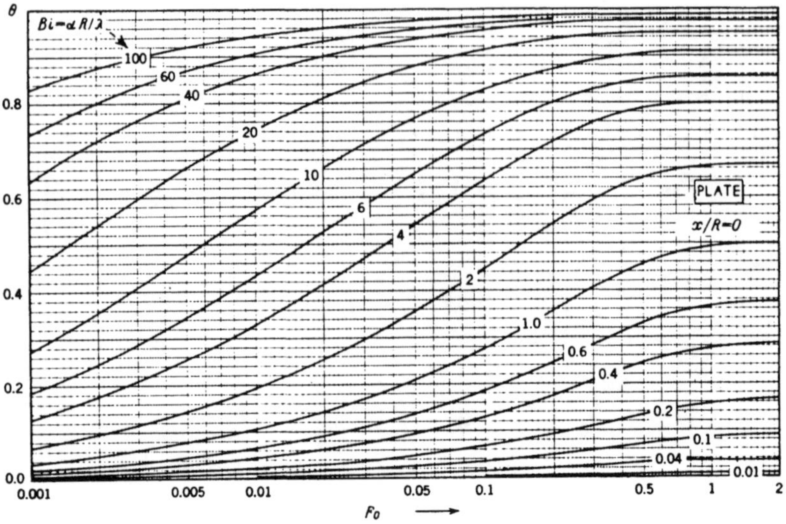

Fig. 6.12. Dimensionless excess temperature θ versus Fo for an infinite plate with one surface maintained at a constant temperature equal to the initial one [102].

On the opposite surface of the plate, the heat transfer follows the Newton law

$$-\frac{\partial t(0, \tau)}{\partial x} + H[t_a - t(0, \tau)] = 0.$$

6.4 Finite Rod without Thermal Insulation of Its Lateral Surface

a. Statement of the Problem. *Consider a finite rod 2R in length in thermal equilibrium with the surrounding medium with the temperature t_0, i.e., the rod temperature is the same everywhere and equal to that of the surrounding medium. At the initial time instant the rod ends are brought in contact with a new medium, the temperature of which is $t_a > t_0$ (for example, they are immersed in heated oil). The lateral surface of the rod gives off heat to the surrounding medium with temperature t_0 (absence of thermal insulation of a lateral surface). The temperature distribution along the rod length at any time instant as well as the heat rate are to be determined under the assumption that temperature drop takes place only in the direction of the rod length. Heat transfer follows the Newton law.* Our problem is similar to that in Section 6.2, but here the rod is of finite length.

We place the origin of coordinates at the center of the rod (Fig. 6.13). Then, on the basis of computations similar to those in Section 6.2, the differential heat conduction equation is written as

$$\frac{\partial t(x, \tau)}{\partial \tau} = a\frac{\partial^2 t(x, \tau)}{\partial x^2} - \frac{\alpha_1}{c\gamma h}[t(x, \tau) - t_0]$$

$$(\tau > 0; \; -R < x < +R) \qquad (6.4.1)$$

where h is the ratio of the area of the rod cross-section to its cross-section periphery. To generalize this problem, the heat transfer coefficient (α_1) for the lateral surface is assumed not to be equal to that (α_2) for the rod ends ($\alpha_1 \neq \alpha_2$), since this condition is more realistic in describing the present heating process.

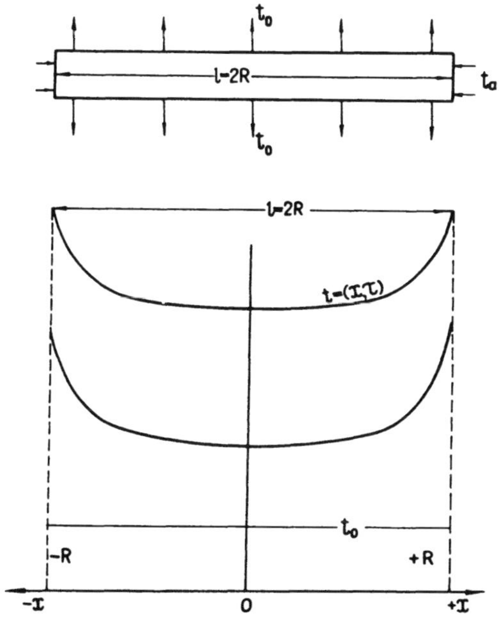

Fig. 6.13. Temperature distribution in a finite rod without thermal insulation of its side surface.

The boundary conditions may be written thus

$$t(x, 0) = t_0 = \text{const}, \qquad (6.4.2)$$

$$-\lambda\frac{\partial t(R, \tau)}{\partial x} + \alpha_2[t_a - t(R, \tau)] = 0 \qquad (6.4.3)$$

$$\frac{\partial t(0, \tau)}{\partial x} = 0. \qquad (6.4.4)$$

b. Solution of the Problem. Let us solve this problem by the operational method. If the Laplace transformation is applied to (6.4.1), we shall have (see Section 6.1)

$$T''(x, s) - \left(\frac{s}{a} + \frac{\alpha_1}{\lambda h}\right)\left[T(x, s) - \frac{t_0}{s}\right] = 0. \tag{6.4.5}$$

The general solution of Eq. (6.4.5) is

$$T(x, s) - \frac{t_0}{s} = A\cosh\left(\frac{s}{a} + \frac{\alpha_1}{\lambda h}\right)^{1/2} x + B\sinh\left(\frac{s}{a} + \frac{\alpha_1}{\lambda h}\right)^{1/2} x. \tag{6.4.6}$$

Boundary conditions for the transform are

$$-T'(R, s) + H_2[(t_a/s) - T(R, s)] = 0, \tag{6.4.7}$$

$$T'(0, s) = 0, \tag{6.4.8}$$

where $H_2 = \alpha_2/\lambda$.

From the symmetry condition (6.4.8) it follows that $B = 0$. The constant A is found from condition (6.4.7)

$$-\left(\frac{s}{a} + \frac{\alpha_1}{\lambda h}\right)^{1/2} A\sinh\left(\frac{s}{a} + \frac{\alpha_1}{\lambda h}\right)^{1/2} R + \frac{H_2 t_a}{s} - AH_2\cosh\left(\frac{s}{a} + \frac{\alpha_1}{\lambda h}\right)^{1/2} R - \frac{H_2 t_0}{s} = 0$$

from which

$$A = \frac{t_a - t_0}{s} \times \frac{1}{\cosh\left(\frac{s}{a} + \frac{\alpha_1}{\lambda h}\right)^{1/2} R + \frac{1}{H_2}\left(\frac{s}{a} + \frac{\alpha_1}{\lambda h}\right)^{1/2} \sinh\left(\frac{s}{a} + \frac{\alpha_1}{\lambda h}\right)^{1/2} R}. \tag{6.4.9}$$

Hence, the solution for the transform will be of the form

$$T(x, s) - \frac{t_0}{s} = \frac{(t_a - t_0)\cosh\left(\frac{s}{a} + \frac{\alpha_1}{\lambda h}\right)^{1/2} x}{s\left[\cosh\left(\frac{s}{a} + \frac{\alpha_1}{\lambda h}\right)^{1/2} R + \frac{1}{H_2}\left(\frac{s}{a} + \frac{\alpha_1}{\lambda h}\right)^{1/2} \sinh\left(\frac{s}{a} + \frac{\alpha_1}{\lambda h}\right)^{1/2} R\right]}. \tag{6.4.10}$$

Before turning to the inverse transform, consider a more simple problem. Assume $H_2 = \infty$; then from boundary condition (6.4.3) it follows that

$$t(\pm R, \tau) = t_a, \tag{6.4.11}$$

i.e., from the very beginning of the heating process the ends of the rod take the medium temperature t_a.

The solution for the transform with $H_2 = \infty$ is

$$T(x, s) - \frac{t_0}{s} = \frac{(t_a - t_0)\cosh\left(\dfrac{s}{a} + \dfrac{\alpha_1}{\lambda h}\right)^{1/2} x}{s\cosh\left(\dfrac{s}{a} + \dfrac{\alpha_1}{\lambda h}\right)^{1/2} R} = \frac{\Phi(s)}{\psi(s)}. \qquad (6.4.12)$$

Here $\Phi(s)$ and $\psi(s)$ are generalized polynomials with respect to s, which has already been proved many times; using the expansion theorem, we find

$$\psi(s) = s\cosh\left(\frac{s}{a} + \frac{\alpha_1}{\lambda h}\right)^{1/2} R = 0$$

(1) $s = 0$ (zero root),

(2) $i\left(\dfrac{s_n}{a} + \dfrac{\alpha_1}{\lambda h}\right)^{1/2} R = \mu_n = (2n-1)\dfrac{\pi}{2}, \qquad s_n = -\left(\dfrac{a\mu_n^2}{R^2} + \dfrac{\alpha_1 a}{\lambda h}\right).$

The value of $\psi'(s)$ at $s = s_n$ is

$$\psi'(s) = \cosh\left(\frac{s}{a} + \frac{\alpha_1}{\lambda h}\right)^{1/2} R + \frac{sR}{2a}\left(\frac{s}{a} + \frac{\alpha_1}{\lambda h}\right)^{-1/2} \sinh\left(\frac{s}{a} + \frac{\alpha_1}{\lambda h}\right)^{1/2} R,$$

$$\psi'(s_n) = \frac{1}{2\mu_n}\sin\mu_n\left[\mu_n^2 + \frac{\alpha_1}{\lambda h}R^2\right].$$

Consequently, the solution of the simplified problem has the form

$$\theta = \frac{t(x, \tau) - t_0}{t_a \quad t_0} = \frac{\cosh(\mathrm{Bi})^{1/2}\dfrac{x}{h}}{\cosh(\mathrm{Bi})^{1/2}\dfrac{R}{h}} - \sum_{n=1}^{\infty}(-1)^{n+1}\frac{2\mu_n}{\mu_n^2 + \mathrm{Bi}\left(\dfrac{R}{h}\right)^2}\cos\mu_n\frac{x}{R}$$
$$\times \exp\left[-\left(\mu_n^2 + \mathrm{Bi}\frac{R^2}{h^2}\right)\mathrm{Fo}\right], \qquad (6.4.13)$$

where $\mathrm{Bi} = \alpha_1 h/\lambda$ is the Biot criterion, $\mathrm{Fo} = a\tau/R^2$ is the Fourier number.

If the rod is thermally insulated (i.e., heat transfer from the lateral surface is absent), then $\mathrm{Bi} = 0$; thus

$$\theta = 1 - \sum_{n=1}^{\infty}(-1)^{n+1}\frac{2}{\mu_n}\cos\mu_n\frac{x}{R}\exp[-\mu_n^2\,\mathrm{Fo}]. \qquad (6.4.14)$$

Equation (6.4.14) is also a solution for an infinite plate with constant temperature at its boundary surfaces.

This is caused by the fact that heat propagation in a finite rod of which the lateral surface is thermally insulated is similar to that of an infinite plate.

Solution (6.4.13) may be obtained in another way if solution (6.4.12) is transformed by the method described in Section 4.3. Then, we shall have

$$T(x, s) - \frac{t_0}{s} = \frac{t_a - t_0}{s} \sum_{n=1}^{\infty} (-1)^{n+1}\left\{\exp\left[-\left(\frac{s}{a} + \frac{\alpha_1}{\lambda h}\right)^{1/2} [(2n-1)R - x]\right]\right.$$
$$\left. + \exp\left[-\left(\frac{s}{a} + \frac{\alpha_1}{\lambda h}\right)^{1/2} [(2n-1)R + x]\right]\right\}. \qquad (6.4.15)$$

Passing to the inverse transform, we obtain

$$\theta = \frac{1}{2} \sum_{n=1}^{\infty} (-1)^{n+1} \left[\exp\left[-\left(2n - 1 - \frac{x}{R}\right)(\mathrm{Bi})^{1/2}\frac{R}{h}\right]\right.$$
$$\times \operatorname{erfc}\left(\frac{2n-1-\dfrac{x}{R}}{2(\mathrm{Fo})^{1/2}} - (\mathrm{Bi\ Fo})^{1/2}\frac{R}{h}\right)$$
$$\left. + \exp\left[\left(2n - 1 - \frac{x}{R}\right)(\mathrm{Bi})^{1/2}\frac{R}{h}\right] \operatorname{erfc}\left(\frac{2n-1-\dfrac{x}{R}}{2(\mathrm{Fo})^{1/2}} + (\mathrm{Bi\ Fo})^{1/2}\frac{R}{h}\right)\right]$$
$$+ \frac{1}{2} \sum_{n=1}^{\infty} (-1)^{n+1} \left\{\exp\left[-\left(2n - 1 + \frac{x}{R}\right)(\mathrm{Bi})^{1/2}\frac{R}{h}\right]\right.$$
$$\times \operatorname{erfc}\left(\frac{2n-1+\dfrac{x}{R}}{2(\mathrm{Fo})^{1/2}} - (\mathrm{Bi\ Fo})^{1/2}\frac{R}{h}\right)$$
$$\left. + \exp\left[\left(2n - 1 + \frac{x}{R}\right)(\mathrm{Bi})^{1/2}\frac{R}{h}\right] \operatorname{erfc}\left(\frac{2n-1+\dfrac{x}{R}}{2(\mathrm{Fo})^{1/2}} + (\mathrm{Bi\ Fo})^{1/2}\frac{R}{h}\right)\right\}.$$
$$(6.4.16)$$

If in solution (6.4.16), it is assumed $\mathrm{Bi} = 0$, then solution (4.3.28) is obtained. If the origin of coordinates is removed from the middle to the left end, i.e., the substitution $x + R = X$ and $R - x = 2R - X$ is made and $2R = \infty$ is assumed, then solution (6.4.14) will be identical with (6.2.14).

Let us return to the problem stated (i.e., (6.4.1)–(6.4.4)). It may be shown that solution (6.4.10) for the transform satisfies the conditions of the expansion theorem, i.e., the numerator and denominator are generalized polynomials with respect to s, where the polynomial of the denominator does contain a constant due to the presence of the multiplier s.

Using the expansion theorem and making similar transformations as made in the previous section, the solution of our problem will be of the form

$$\theta = \frac{t(x, \tau) - t_0}{t_a - t_0} = \frac{\cosh(\mathrm{Bi}_1)^{1/2} \dfrac{x}{h}}{\cosh(\mathrm{Bi}_1)^{1/2} \dfrac{R}{h} + \dfrac{R\sqrt{\mathrm{Bi}_1}}{h\,\mathrm{Bi}_2} \sinh(\mathrm{Bi}_1)^{1/2} \dfrac{R}{h}}$$

$$- \sum_{n=1}^{\infty} \frac{\mu_n^2}{\mu_n^2 + \mathrm{Bi}_1 \dfrac{R^2}{h^2}} A_n \cos \mu_n \frac{x}{R} \exp\left[- \left(\mu_n^2 + \mathrm{Bi}_1 \frac{R^2}{h^2} \right) \mathrm{Fo} \right], \tag{6.4.17}$$

where $\mathrm{Bi}_1 = \alpha_1 h/\lambda$ and $\mathrm{Bi}_2 = \alpha_2 h/\lambda$. The values of $A_n = (2 \sin \mu_n)/(\mu_n + \sin \mu_n \cos \mu_n)$ are the initial heat amplitudes and are given in Table 6.2; they are the functions of the criterion $\mathrm{Bi}_2 = (\alpha_2/\lambda)R$. The eigenvalues μ_n are determined by

$$\cot \mu = \frac{1}{\mathrm{Bi}_2} \mu .$$

*c. **Analysis of the Solution.*** If $\mathrm{Bi}_2 = \infty$ (it means that the temperature of the rod ends is constant), then solution (6.4.17) becomes identical with solution (6.4.13) as $A_n = (-1)^{n+1}(2/\mu_n)$ where $\mu_n = (2n - 1)(\pi/2)$.

If $\mathrm{Bi}_1 = 0$, then we have

$$\theta = 1 - \sum_{n=1}^{\infty} A_n \cos \mu_n \frac{x}{R} \exp[- \mu_n^2 \mathrm{Fo}], \tag{6.4.18}$$

i.e., the solution for an infinite plate when heat transfer between boundary surfaces takes place according to the Newton law. This solution is identical with solution (6.3.29).

In the stationary state ($\mathrm{Fo} = \infty$) series (6.4.17) is equal to zero, and temperature distribution along the rod length is described by the hyperbolic cosine law (law of catenary line).

If there are no heat losses from the lateral surfaces, then in the stationary state there is a uniform temperature distribution equal to $\theta = 1\,[t(x, \infty) = t_a = \mathrm{const}]$.

*d. **Specific Heat Rate.*** Using relation (6.4.17) we determine the mean temperature of the rod in order to evaluate the specific heat rate. Then, we shall have

$$\bar{\theta} = \frac{\bar{t}(\tau) - t_0}{t_a - t_0} = \frac{\tanh(\mathrm{Bi}_1)^{1/2} \dfrac{R}{h}}{(\mathrm{Bi}_1)^{1/2} \dfrac{R}{h} + \dfrac{\mathrm{Bi}_1}{\mathrm{Bi}_2} \dfrac{R^2}{h^2} \tanh(\mathrm{Bi}_1)^{1/2} \dfrac{R}{h}}$$

$$- \sum_{n=1}^{\infty} \frac{\mu_n^2}{\mu_n^2 + \mathrm{Bi}_1 (R^2/h^2)} B_n \exp\left[- \left(\mu_n^2 + \mathrm{Bi}_1 \frac{R^2}{h^2} \right) \mathrm{Fo} \right], \qquad (6.4.19)$$

where B_n are the constant coefficients determined by the relation (6.3.45).

As an example we make the following computation. The relative excess temperature in the rod center is to be determined when its ends are maintained at a constant temperature ($\theta_s = 1$) for various values of Bi_1 and Fo. Heat transfer occurs from the lateral surface to the surrounding medium, the temperature of which is equal to the initial rod temperature. $R/h = 10$.

For computations, we use solution (6.4.13). We have

$$\theta_c = \frac{t(R, \tau) - t_0}{t_a - t_0} = \frac{1}{\cosh((\mathrm{Bi}_1)^{1/2} \cdot 10)} - \frac{\pi}{\frac{1}{4}\pi^2 + 100\,\mathrm{Bi}_1}$$

$$\times \exp\left[- \left(\frac{\pi^2}{4} + 100\,\mathrm{Bi}_1 \right) \mathrm{Fo} \right]$$

$$+ \frac{3\pi}{\dfrac{9}{4}\pi^2 + 100\,\mathrm{Bi}_1} \exp\left[- \left(\frac{9}{4}\pi^2 + 100\,\mathrm{Bi}_1 \right) \mathrm{Fo} \right] - \cdots. \qquad (6.4.20)$$

From formula (6.4.20) we can see that at large values of Bi (at $R/h = 10$)

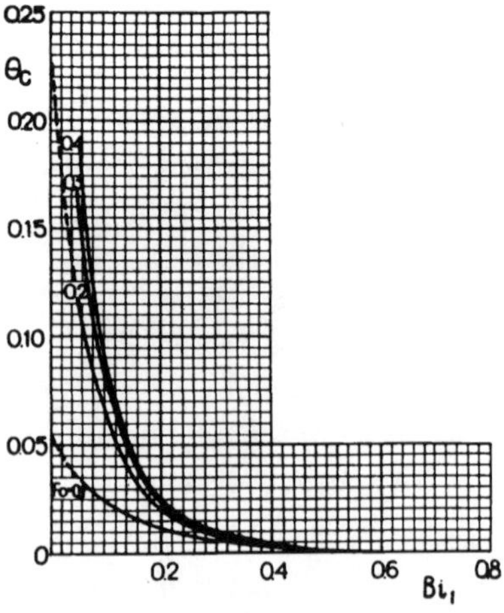

Fig. 6.14. Dimensionless excess temperature in the rod center versus Bi for various values of Fo with $R/h = 10$.

$\theta_c = 0$. This means that the convective heat transfer predominates over transfer by conduction through the rod ends because the rod is thin and long. The criterion Bi is therefore taken from 0.05 to 1.0. We calculate θ_c for the criterion Fo $= 0.1$, 0.2, 0.3, and 0.4, and the results are given in Fig. 6.14 where θ_c represents the ordinate axis and the criterion Bi_1, the abscissa axis. From Fig. 6.14 it can be seen that starting from $\mathrm{Bi}_1 \approx 0.4$ all the curves become one line. For small values of the Biot criterion ($\mathrm{Bi}_1 < 0.1$) the curves rise sharply with increase in Fo.

If $\mathrm{Bi}_1 = 0$ then the previously tabulated solution (6.4.14) results directly from solution (6.4.13) (see Chapter 4, Table 2).

6.5 Sphere (Symmetrical Problem)

a. Statement of the Problem. *Consider a spherical body of radius R with the given initial temperature distribution as the function f(r). At the initial instant the sphere is placed into a medium with a constant temperature $t_a > t(r, 0)$. The temperature distribution inside the sphere at any time and the specific heat rate are to be determined provided that the temperature at any point of the sphere is a function only of time and radius r.* This latter condition corresponds to the uniform heating of the sphere surface and isothermal surfaces represent concentric spheres (symmetrical problem).

In this case the differential heat conduction equation may be written thus

$$\frac{\partial[rt(r, \tau)]}{\partial \tau} = a\,\frac{\partial^2[rt(r, \tau)]}{\partial r^2} \quad (\tau > 0;\ 0 < r < R). \qquad (6.5.1)$$

Initial and boundary conditions are the following:

$$t(r, 0) = f(r)\,, \qquad (6.5.2)$$

$$-\frac{\partial t(R, \tau)}{\partial r} + H[t_a - t(R, \tau)] = 0\,, \qquad (6.5.3)$$

$$\frac{\partial t(0, \tau)}{\partial r} = 0\,, \quad t(0, \tau) \neq \infty. \qquad (6.5.4)$$

In the presence of a nonuniform initial temperature distribution, the solution by the Fourier method yields results more rapidly.

b. Solution by the Method of Separation of Variables. We introduce the variable $t_a - t(r, \tau) = \vartheta(r, \tau)$ in order to reduce the problem of heating to that of cooling and thus we can use the particular solution obtained in Chapter 4, Section 4.

The particular solution of Eq. (6.5.1) under the symmetry condition (6.5.4) has the form (see solution (4.4.8))

$$\vartheta(r, \tau) = t_a - t(r, \tau) = C\frac{\sin kr}{r}\exp[-k^2 a\tau]. \tag{6.5.5}$$

The quantity k is determined from boundary condition (6.5.3) which for the variable $\vartheta(r, \tau)$ assumes the form

$$\frac{\partial\vartheta(R, \tau)}{\partial r} + H\vartheta(R, \tau) = 0. \tag{6.5.6}$$

We now introduce boundary condition (6.5.6) into solution (6.5.5) to obtain

$$-C\frac{\sin kR}{R^2}\exp[-ak^2\tau] + C\frac{k\cos kR}{R}\exp[-ak^2\tau]$$
$$+C\frac{H}{R}\sin kR\exp[-ak^2\tau] = 0. \tag{6.5.7}$$

Cancelling through by $(C/R)\exp[-ak^2\tau]$ we obtain

$$\left(H - \frac{1}{R}\right)\sin kR + k\cos kR = 0,$$

and from this

$$\tan kR = -kR/(HR - 1). \tag{6.5.8}$$

Equation (6.5.8) is a trigonometric equation which has an infinite number of roots (kR). We designate these characteristic roots by $\mu = kR$ and the Biot criterion by HR. Then, the left-hand side of the characteristic equation $y_1 = \tan\mu$ represents a tangent curve and the right-hand side and y_2 a straight line, with a slope equal to $(\mathrm{Bi} - 1)^{-1}$, i.e., at values of $\mathrm{Bi} > 1$ the straight line is located in the fourth octant and at values of $\mathrm{Bi} < 1$, in the first one. Points of intersection of the tangent curve y_1 with the straight line y_2 give values of roots μ (Fig. 6.15).

From Fig. 6.15 it can be seen that there is an infinite number of roots μ, each subsequent root being greater than the previous one

$$\mu_1 < \mu_2 < \mu_3 < \cdots < \mu_n. \tag{6.5.9}$$

At $\mathrm{Bi} = \infty$, the slope of the straight line y_2 is zero and the straight line coincides with the abscissa axis; then the eigenvalues μ will be proportional to π, i.e.,

$$\mu_n = n\pi \quad \text{at } \mathrm{Bi} = \infty.$$

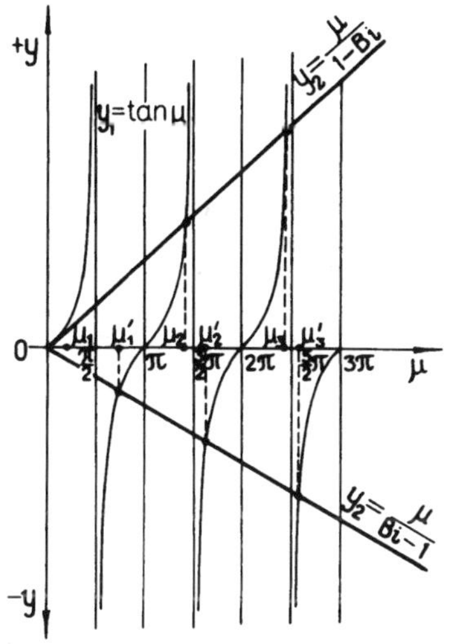

Fig. 6.15. Graphical method for determining the roots of characteristic equation for a sphere.

If the Biot criterion is unity (Bi = 1) then the straight line y_2 coincides with the ordinate axis. In this case $\mu_1 = \frac{1}{2}\pi$ and $\mu_2 = \frac{3}{2}\pi$, $\mu_3 = \frac{5}{2}\pi$, etc., i.e., $\mu_n = (2n - 1)\frac{1}{2}\pi$.

If Bi → 0, then the slope of the straight line is equal to unity, $\mu_1 \to (3\ \mathrm{Bi})^{1/2}$ because from the characteristic equation it follows

$$\cot \mu_1 = -\frac{\mathrm{Bi} - 1}{\mu_1} = \frac{1}{\mu_1} - \frac{\mu_1}{3} - \frac{\mu_1^3}{3^2 \cdot 5} - \cdots.$$

Restricting ourselves to the first two terms of this series for $\cot \mu_1$ we obtain

$$\mu_1^2 = 3\ \mathrm{Bi}. \tag{6.5.10}$$

The remaining roots ($\mu_2, \mu_3, \cdots$) are determined from the equation

$$\tan \mu = \mu \quad \text{at } \mathrm{Bi} \to 0. \tag{6.5.11}$$

Thus, the characteristic equation may be written as follows

$$\tan \mu = -\frac{1}{\mathrm{Bi} - 1}\mu. \tag{6.5.12}$$

The first six roots μ_n are given in Table 6.5 for various values of the Biot criterion. These values of μ_n are accurate to four decimal places. In the

TABLE 6.5. THE ROOTS OF THE CHARACTERISTIC EQUATION

$$\tan \mu = -\frac{1}{\mathrm{Bi} - 1}\mu$$

Bi	μ_1	μ_2	μ_3	μ_4	μ_5	μ_6
0.0	0.0000	4.4934	7.7253	10.9041	14.0662	17.2208
0.005	0.1224	4.4945	7.7259	10.9046	14.0666	17.2210
0.01	0.1730	4.4956	7.7265	10.9050	14.0669	17.2213
0.02	0.2445	4.4979	7.7278	10.9060	14.0676	17.2219
0.03	0.2991	4.5001	7.7291	10.9069	14.0683	17.2225
0.04	0.3450	4.5023	7.7304	10.9078	14.0690	17.2231
0.05	0.3854	4.5045	7.7317	10.9087	14.0697	17.2237
0.06	0.4217	4.5068	7.7330	10.9096	14.0705	17.2242
0.07	0.4551	4.5090	7.7343	10.9105	14.0712	17.2248
0.08	0.4860	4.5112	7.7356	10.9115	14.0719	17.2254
0.09	0.5150	4.5134	7.7369	10.9124	14.0726	17.2260
0.10	0.5423	4.5157	7.7382	10.9133	14.0733	17.2266
0.15	0.6609	4.5268	7.7447	10.9179	14.0769	17.2295
0.20	0.7593	4.5379	7.7511	10.9225	14.0804	17.2324
0.30	0.9208	4.5601	7.7641	10.9316	14.0875	17.2382
0.40	1.0528	4.5822	7.7770	10.9408	14.0946	17.2440
0.50	1.1656	5.6042	7.7899	10.9499	14.1017	17.2498
0.60	1.2644	4.6261	7.8028	10.9591	14.1088	17.2556
0.70	1.3525	4.6479	7.8156	10.9682	14.1159	17.2614
0.80	1.4320	4.6696	7.8384	10.9774	14.1230	17.2672
0.90	1.5044	4.6911	7.8412	10.9865	14.1301	17.2730
1.0	1.5708	4.7124	7.8540	10.9956	14.1372	17.2788
1.1	1.6320	4.7335	7.8667	11.0047	14.1443	17.2845
1.2	1.6687	4.7544	7.8794	11.0137	14.1513	17.2903
1.3	1.7414	4.7751	7.8920	11.0228	14.1584	17.2961
1.4	1.7906	4.7956	7.9046	11.0318	14.1654	17.3019
1.5	1.8366	4.8158	7.9171	11.0409	14.1724	17.3076
1.6	1.8798	4.8358	7.9295	11.0498	14.1795	17.3134
1.7	1.9203	4.8556	7.9419	11.0588	14.1865	17.3192
1.8	1.9586	4.8751	7.9542	11.0677	14.1935	17.3249
1.9	1.9947	4.8943	7.9665	11.0767	14.2005	17.3306
2.0	2.0288	4.9132	7.9787	11.0856	14.2075	17.3364
2.5	2.1746	5.0037	8.0385	11.1296	14.2421	17.3649
3.0	2.2889	5.0870	8.0962	11.1727	14.2764	17.3932
4.0	2.4557	5.2329	8.2045	11.2560	14.3434	17.4490
5.0	2.5704	5.3540	8.3029	11.3349	14.4080	17.5034
6.0	2.6537	5.4544	8.3914	11.4086	14.4699	17.5562
7.0	2.7165	5.5378	8.4703	11.4773	14.5288	17.6072
8.0	2.7654	5.6078	8.5406	11.5408	14.5847	17.6567
9.0	2.8044	5.6669	8.6031	11.5994	14.6374	17.7032

TABLE 6.5. (*continued*)

Bi	μ_1	μ_2	μ_3	μ_4	μ_5	μ_6
10.0	2.8363	5.7172	8.6587	11.6532	14.6870	17.7481
11.0	2.8628	5.7606	8.7083	11.7027	14.7335	17.7908
16.0	2.9476	5.9080	8.8898	11.8959	14.9251	17.9742
21.0	2.9930	5.9921	9.0019	12.0250	15.0625	18.1136
31.0	3.0406	6.0831	9.1294	12.1807	15.2380	18.3018
41.0	3.0651	6.1311	9.1987	12.2688	15.3417	18.4180
51.0	3.0801	6.1606	9.2420	12.3247	15.4090	18.4953
61.0	3.0901	6.1805	9.2715	12.3632	15.4559	18.5497
81.0	3.1028	6.2058	9.3089	12.4124	15.5164	18.6209
101.0	3.1105	6.2211	9.3317	12.4426	15.5537	18.6650
∞	3.1416	6.2832	9.4248	12.5664	15.7080	18.8496

majority of cases we can get an adequate answer if we restrict ourselves to one, seldom two, value, of μ_n.

Turning to solution (6.5.5), since there is an infinite number of constants $k_n = \mu_n/R$ determined by characteristic equation (6.5.12), the general solution of our problem may be written as the sum of all the particular solutions

$$\vartheta(r, \tau) = \sum_{n=1}^{\infty} C_n \frac{\sin \mu_n(r/R)}{r} \exp\left[-\mu_n^2 \frac{a\tau}{R^2}\right]. \tag{6.5.13}$$

The constants C_n are determined from the initial condition

$$\vartheta(r, 0) = t_a - t(r, 0) = t_a - f(r) = f_1(r). \tag{6.5.14}$$

We have

$$f_1(r) = \sum_{n=1}^{\infty} C_n \frac{\sin \mu_n(r/R)}{r}. \tag{6.5.15}$$

With certain restrictions on the function $f_1(r)$ which have been stated before, C_n may be determined by the Fourier method.

Multiplying both parts of the equality by $r \sin \mu_m(r/R)\, dr$ where μ_m is the mth root of characteristic equation (6.5.12) and integrating from 0 to R, we obtain

$$\begin{aligned}\int_0^R r f_1(r) \sin \mu_m(r/R)\, dr &= \int_0^R \sum_{n=1}^{\infty} C_n \sin \mu_m(r/R) \sin \mu_n(r/R)\, dr \\ &= \sum_{n=1}^{\infty} C_n \int_0^R \sin \mu_m(r/R) \sin \mu_n(r/R)\, dr. \end{aligned} \tag{6.5.16}$$

In Chapter 4 it was shown that any integral of the right-hand side of equality (6.5.16) may be presented as

$$I = \int_0^R \sin \mu_m \frac{r}{R} \sin \mu_n \frac{r}{R} \, dr$$

$$= \frac{R(\mu_m \sin \mu_n \cos \mu_m - \mu_n \sin \mu_m \cos \mu_n)}{\mu_n^2 - \mu_m^2}$$

$$= \frac{R\mu_n \sin \mu_m \cos \mu_n}{\mu_n^2 - \mu_m^2} \left(\frac{\mu_m}{\mu_n} \tan \mu_n \cot \mu_m - 1 \right). \quad (6.5.17)$$

From characteristic equation (6.5.12) it may be shown

$$\frac{\mu_m}{\mu_n} \tan \mu_n \cot \mu_m = \frac{\mu_m}{\mu_n} \frac{\mu_n}{\mu_m} = 1 .$$

Hence, integral (6.5.17) is zero at $m \neq n$. If $m = n$, integral (6.5.17) will be

$$\int_0^R \sin^2 \mu_n \frac{r}{R} \, dr = R\left(\frac{1}{2} - \frac{\sin 2\mu_n}{4\mu_n} \right) = \frac{R}{2\mu_n} (\mu_n - \sin \mu_n \cos \mu_n). \quad (6.5.18)$$

Thus, all the integrals in relation (6.5.16) are equal to zero, except when $m = n$. Hence, the constant C_n is

$$C_n = \frac{\int_0^R r f_1(r) \sin \mu_n \frac{r}{R} \, dr}{\int_0^R \sin^2 \mu_n \frac{r}{R} \, dr} = \frac{2\mu_n}{\mu_n - \sin \mu_n \cos \mu_n} \frac{1}{R} \int_0^R r f_1(r) \sin \mu_n \frac{r}{R} \, dr.$$

The general solution of our problem has the form

$$\vartheta(x, \tau) = t_a - t(x, \tau) = \sum_{n=1}^{\infty} \frac{2\mu_n}{\mu_n - \sin \mu_n \cos \mu_n} \frac{\sin \mu_n (r/R)}{rR}$$

$$\times \int_0^R r f_1(r) \sin \mu_n (r/R) \, dr \exp[-\mu_n^2 \, \mathrm{Fo}]. \quad (6.5.19)$$

If at the initial instant the temperature of a sphere does not depend on r

$$t(r, 0) = f_1(r) = t_0 = \text{const}$$

the integral in solution (6.5.19) may be calculated explicitly:

$$\frac{1}{R} \int_0^R (t_a - t_0) r \sin \mu_n \frac{r}{R} \, dr = \frac{(t_a - t_0)R}{\mu_n^2} (\sin \mu_n - \mu_n \cos \mu_n)$$

since

$$\int x \sin x \, dx = \sin x - x \cos x.$$

Finally, the solution of the problem may be written thus:

$$\theta = \frac{t(r, \tau) - t_0}{t_a - t_0} = 1 - \sum_{n=1}^{\infty} \frac{2(\sin \mu_n - \mu_n \cos \mu_n)}{\mu_n - \sin \mu_n \cos \mu_n} \frac{R \sin \mu_n(r/R)}{r\mu_n} \exp[-\mu_n^2 \mathrm{Fo}]. \tag{6.5.20}$$

c. ***Solution by the Operational Method.*** The solution of differential equation (6.5.1) for the transform $T(r, s)$ under condition (6.5.4) has the form (see solution (4.4.22))

$$T(r, s) - \frac{t_0}{s} = B \frac{\sinh\left(\frac{s}{a}\right)^{1/2} r}{r}. \tag{6.5.21}$$

Boundary condition (6.5.3) for the transform will be of the form

$$-T'(R, s) + H[(t_a/s) - T(R, s)] = 0. \tag{6.5.22}$$

We determine the constant B to be

$$-\frac{1}{R} B\left(\frac{s}{a}\right)^{1/2} \cosh\left(\frac{s}{a}\right)^{1/2} R + \frac{B}{R^2} \sinh\left(\frac{s}{a}\right)^{1/2} R$$
$$+ \frac{Ht_a}{s} - \frac{Ht_0}{s} - \frac{HB}{R} \sinh\left(\frac{s}{a}\right)^{1/2} R = 0$$

from which we have

$$B = \frac{(t_a - t_0)HR^2}{s\left[(HR - 1)\sinh\left(\frac{s}{a}\right)^{1/2} R + \left(\frac{s}{a}\right)^{1/2} R \cosh\left(\frac{s}{a}\right)^{1/2} R\right]}. \tag{6.5.23}$$

Solution (6.5.21) takes the form

$$T(r, s) - \frac{t_0}{s} = \frac{(t_a - t_0)HR^2 \sinh\left(\frac{s}{a}\right)^{1/2} r}{rs\left[(HR - 1)\sinh\left(\frac{s}{a}\right)^{1/2} R + \left(\frac{s}{a}\right)^{1/2} R \cosh\left(\frac{s}{a}\right)^{1/2} R\right]}$$
$$= \frac{\Phi_1(s)}{\psi_1(s)}. \tag{6.5.24}$$

The numerator $\Phi_1(s)$ and denominator $\psi_1(s)$ of solution (6.5.24) are not generalized polynomials with respect to s but they may be reduced to such polynomials $\Phi(s)$ and $\psi(s)$ by multiplying both by $(s/a)^{1/2}$. In this case the relation

$$\frac{\Phi(s_n)}{\psi'(s_n)} = \frac{\Phi_1(s_n)}{\psi_1(s_n)} \tag{6.5.25}$$

may be used, where $\Phi(s) = s^k\Phi_1(s)$, $\psi(s) = s^k\psi_1(s)$ at $s^k \neq 0$, $\Phi(s)$ and $\psi(s)$ are generalized polynomials with respect to s.

We determine the roots s_n for which $\psi_1(s)$ is equated to zero as

$$\psi_1(s) = rs\left[(HR-1)\sinh\left(\frac{s}{a}\right)^{1/2}R + \left(\frac{s}{a}\right)^{1/2}R\cosh\left(\frac{s}{a}\right)^{1/2}R\right] = 0.$$

Thence, we shall have $s = 0$, (zero root), and

$$(HR-1)\,\text{sihn}\left(\frac{s}{a}\right)^{1/2}R + \left(\frac{s}{a}\right)^{1/2}R\cosh\left(\frac{s}{a}\right)^{1/2}R$$
$$= (HR-1)\frac{1}{i}\sin i\left(\frac{s}{a}\right)^{1/2}R + \left(\frac{s}{a}\right)^{1/2}R\cos i\left(\frac{s}{a}\right)^{1/2}R = 0.$$

Putting $i(s/a)^{1/2}R = \mu$, we then have $s_n = -a\mu_n{}^2/R^2$, and μ_n is determined by the characteristic equation

$$(HR-1)\sin\mu + \mu\cos\mu = 0$$

or

$$\tan\mu = -\frac{\mu}{\text{Bi}-1}, \tag{6.5.26}$$

where Bi $= HR$ is the Biot criterion.

To determine the value of $\Phi(0)/\psi'(0)$ for the first (zero) root, solution (6.5.24) is transformed so that it represents the ratio of two generalized polynomials

$$\frac{\Phi_1(s)}{\psi_1(s)}$$
$$= \frac{(t_a - t_0)\,\text{Bi}\,R\left(r + \frac{1}{3!}\,\frac{s}{a}r^3 + \frac{1}{5!}\,\frac{s^2}{a^2}r^5 + \cdots\right)}{rs\left[(\text{Bi}-1)\left(R+\frac{1}{3!}\,\frac{s}{a}R^3+\cdots\right)+\left(R+\frac{1}{2!}\,\frac{s}{a}R^3+\frac{1}{4!}\,\frac{s^2}{a^2}R^5+\cdots\right)\right]}$$
$$= \frac{\Phi(s)}{\psi(s)};$$

As a result we obtain

$$\lim_{s\to 0}\frac{\Phi(s)}{\psi'(s)} = (t_a - t_0).$$

For the remaining roots $\psi_1'(s)$ we find

$$\psi_1'(s) = r[\sim] + r\Big[(\text{Bi} - 1)\frac{1}{2}\left(\frac{s}{a}\right)^{1/2} R\cosh\left(\frac{s}{a}\right)^{1/2} R$$
$$+ \frac{1}{2}\left(\frac{s}{a}\right)^{1/2} R\cosh\left(\frac{s}{a}\right)^{1/2} R + \frac{1}{2}\,\frac{s}{a}R^2\sinh\left(\frac{s}{a}\right)^{1/2} R\Big]$$

where $[\sim]$ denotes the expression in the square brackets in the denominator of solution (6.5.24); it is equal to zero at $s = s_n$. Thus,

$$\psi_1'(s_n) = \frac{1}{2}\,\frac{r\mu_n}{i}\,[(\text{Bi} - 1)\cos\mu_n + \cos\mu_n - \mu_n\sin\mu_n]$$
$$= \frac{r\mu_n}{2i}\,[\text{Bi}\cos\mu_n - \mu_n\sin\mu_n]$$
$$= \frac{r\mu_n}{2i\sin\mu_n}\,(\sin\mu_n\cos\mu_n - \mu_n),$$

$$\Phi_1(s_n) = (t_a - t_0)\,\frac{\text{Bi}\,R}{i}\,\sin\mu_n\,\frac{r}{R}$$
$$= \frac{(t_a - t_0)\,R}{i\sin\mu_n}\,(\sin\mu_n - \mu_n\cos\mu_n)\sin\mu_n\,\frac{r}{R}.$$

So the solution of our problem has the form

$$\theta = \frac{t(r,\tau) - t_0}{t_a - t_0} = 1 - \sum_{n=1}^{\infty} A_n\,\frac{R\sin\mu_n\dfrac{r}{R}}{r\mu_n}\exp[-\mu_n^2\,\text{Fo}], \quad (6.5.27)$$

$$A_n = \frac{2(\sin\mu_n' - \mu_n\cos\mu_n)}{\mu_n - \sin\mu_n\cos\mu_n}. \quad (6.5.28)$$

Thus, a solution which is identical with solution (6.5.20) is obtained. Initial thermal amplitudes are single-valued functions of the Biot criterion. Therefore, it is more convenient in computations to use an expression for A_n in which trigonometric functions are replaced by μ_n and by the Biot criterion according to the characteristic equation. Thus, for A_n it may be written

$$A_n = (-1)^{n+1}\,\frac{2\text{Bi}\,[\mu_n^2 + (\text{Bi} - 1)^2]^{1/2}}{(\mu_n^2 + \text{Bi}^2 - \text{Bi})}. \quad (6.5.29)$$

Six values of A_n calculated according to this formula, accurate to four decimal places, are presented in Table 6.6.

TABLE 6.6. THE VALUES OF THE CONSTANTS

$$A_n = (-1)^{n+1} \frac{2\mathrm{Bi}\{\mathrm{Bi} - 1)^2 + \mu_n{}^2\}^{1/2}}{\mu_n{}^2 + \mathrm{Bi}^2 - \mathrm{Bi}}$$

Bi	A_1	A_2	A_3	A_4	A_5	A_6
0.000	1.0000	−0.0000	0.0000	−0.0000	0000	−0.0000
0.005	1.0025	−0.0023	0.0013	−0.0009	0.0007	−0.0006
0.01	1.0035	−0.0046	0.0026	−0.0018	0.0014	−0.0012
0.02	1.0055	−0.0091	0.0052	−0.0037	0.0029	−0.0023
0.03	1.0089	−0.0137	0.0078	−0.0055	0.0043	−0.0035
0.04	1.0121	−0.0182	0.0104	−0.0074	0.0057	−0.0047
0.05	1.0147	−0.0227	1.0130	−0.0092	0.0071	−0.0058
0.06	1.0181	−0.0273	0.0156	−0.0110	0.0085	−0.0070
0.07	1.0206	−0.0318	0.0183	−0.0129	0.0100	−0.0081
0.08	1.0239	−0.0363	0.0209	−0.0147	0.0114	−0.0093
0.09	1.0266	−0.0409	0.0235	−0.0166	0.0128	−0.0105
0.10	1.0297	−0.0454	0.0260	−0.0184	0.0142	−0.0116
0.15	1.0443	−0.0679	0.0390	−0.0276	0.0214	−0.0174
0.20	1.0592	−0.0894	0.0520	−0.0368	0.0285	−0.0232
0.30	1.0880	−0.1345	0.0779	−0.0551	0.0427	−0.0349
0.40	1.1164	−0.1781	0.1036	−0.0734	0.0569	−0.0465
0.50	1.1440	−0.2216	0.1292	−0.0916	0.0710	−0.0580
0.60	1.1713	−0.2633	0.1546	−0.1098	0.0852	−0.0696
0.70	1.1978	−0.3048	0.1799	−0.1270	0.0998	−0.0812
0.80	1.2237	−0.3455	0.2050	−0.1460	0.1134	−0.0927
0.90	1.2488	−0.3854	0.2299	−0.1640	0.1275	−0.1042
1.0	1.2732	−0.4244	0.2546	−0.1819	0.1415	−0.1157
1.1	1.2970	−0.4626	0.2792	−0.1997	0.1555	−0.1272
1.2	1.3200	−0.4999	0.3035	−0.2175	0.1694	−0.1387
1.3	1.3424	−0.5364	0.3276	−0.2352	0.1833	−0.1501
1.4	1.3640	−0.5720	0.3515	−0.2528	0.1972	−0.1616
1.5	1.3848	−0.6067	0.3752	−0.2703	0.2110	−0.1730
1.6	1.4051	−0.6405	0.3986	−0.2878	0.2248	−0.1843
1.7	1.4247	−0.6735	0.4218	−0.3051	0.2385	−0.1957
1.8	1.4436	−0.7063	0.4447	−0.3228	0.2522	−0.2078
1.9	1.4618	−0.7368	0.4674	−0.3395	0.2659	−0.2183
2.0	1.4793	−0.7673	0.4899	−0.3565	0.2795	−0.2296
2.5	1.5579	−0.9073	0.5980	−0.4365	0.3449	−0.2855
3.0	1.6223	−1.0288	0.6993	−0.5205	0.4122	−0.3405
4.0	1.7201	−1.2253	0.8811	−0.6719	0.5384	−0.4476
5.0	1.7870	−1.3733	1.0363	−0.8095	0.6570	−0.5501
6.0	1.8338	−1.4860	1.1673	−0.9333	0.7702	−0.6428
7.0	1.8673	−1.5731	1.2776	−1.0437	0.8695	−0.7398
8.0	1.8920	−1.6409	1.3703	−1.1415	0.9633	−0.8264
9.0	1.9106	−1.6949	1.4482	−1.2280	1.0489	−0.9073

TABLE 6.6. (*continued*)

Bi	A_1	A_2	A_3	A_4	A_5	A_6
10	1.9249	−1.7381	1.5141	−1.3042	1.1269	−0.9827
11	1.9364	−1.7732	1.5698	−1.3713	1.1677	−1.0527
16	1.9663	−1.8766	1.7489	−1.6058	1.4633	−1.3305
21	1.9801	−1.9235	1.8385	−1.7360	1.6256	−1.5149
31	1.9905	−1.9626	1.9186	−1.8616	1.7950	−1.7225
41	1.9948	−1.9780	1.9515	−1.9161	1.8732	−1.8263
51	1.9964	−1.9856	1.9680	−1.9441	1.9145	−1.8802
61	1.9974	−1.9901	1.9773	−1.9601	1.9387	−1.9135
81	1.9985	−1.9942	1.9869	−1.9769	1.9644	−1.9492
101	1.9993	−1.9962	1.9915	−1.9850	1.9767	−1.9667
∞	2.0000	−2.0000	2.0000	−2.0000	2.0000	−2.0000

For practical calculations, Figures 6.16 and 6.17 depict nomograms for the determination of θ_s (relative excess temperature of a sphere surface) and θ_c (relative excess temperature at a sphere center) as a function of the Fourier numbers and Biot criteria.

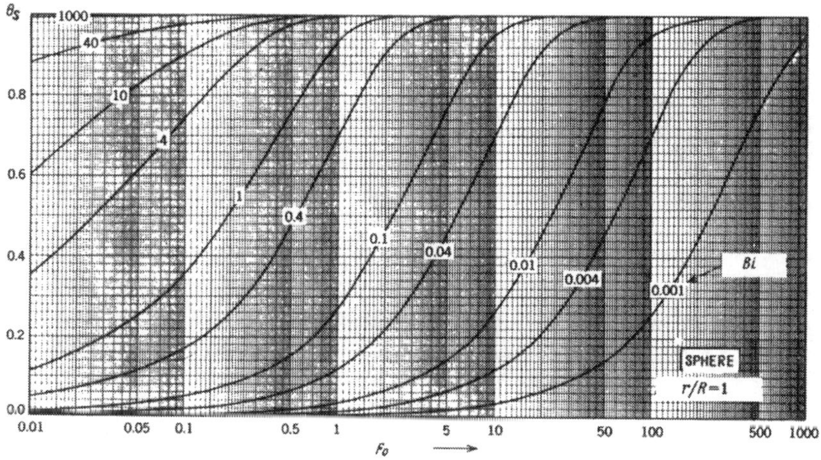

Fig. 6.16. Plots for determination of dimensionless excess temperature at the sphere surface [102].

d. Analysis of the Solution. If Bi → ∞, then according to characteristic equation (6.5.26)

$$\mu_n = \pi n \,, \tag{6.5.30}$$

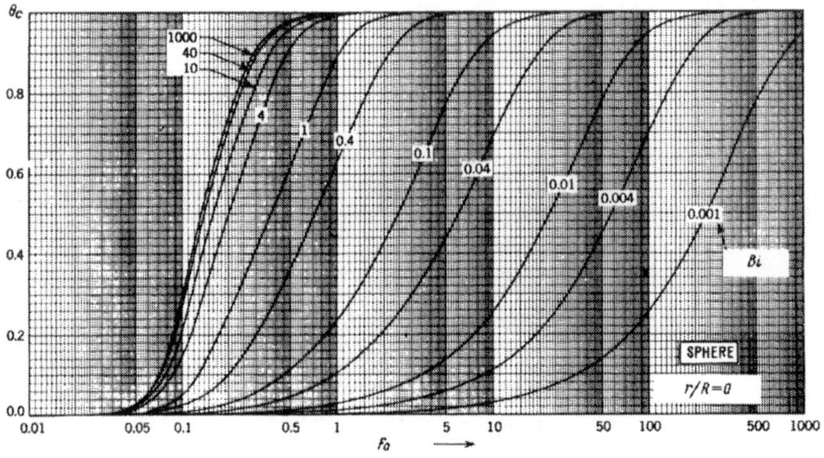

Fig. 6.17. Plots for determination of dimensionless excess temperature in the sphere center [102].

and the initial thermal amplitudes are

$$A_n = (-1)^{n+1} \cdot 2 = 2(-1)^{n+1}. \tag{6.5.31}$$

Then solution (6.5.27) becomes identical with solution (4.4.32) if in the latter it is assumed that $\theta_{\text{cool}} = 1 - \theta_{\text{heat}}$ where $\theta_{\text{cool}} = (t - t_a)/(t_0 - t_a)$ ($t_0 > t_a$, the sphere is cooled) and $\theta_{\text{heat}} = (t - t_0)/(t_a - t_0)$ ($t_a > t_0$, the sphere is heated). If Bi = 1, then

$$\mu_n = (2n - 1)\tfrac{1}{2}\pi, \quad A_n = (-1)^{n+1}(2/\mu_n). \tag{6.5.32}$$

The heating rate will thus depend only on thermal inertia properties of a material.

At small values of Bi (Bi → 0) all the coefficients $A_n \to 0$, except the first ($A_1 = 1$) and $\mu_1^2 = 3$ Bi (see relation (6.5.10)). Then solution (6.5.27) for small values of Bi may be written as

$$\theta = 1 - \frac{R \sin(3\,\text{Bi})^{1/2} \dfrac{r}{R}}{r(3\,\text{Bi})^{1/2}} \exp[-3\,\text{Bi Fo}]. \tag{6.5.33}$$

From solution (6.5.27) it follows that owing to inequality (6.5.9) the series converges rapidly except for very small values of Fo, since the exponential functions $\exp(-\mu_n^2\,\text{Fo})$ rapidly decrease with an increase in μ_n. Therefore, if small values of Fo are not considered, we may restrict ourselves to one term of the series. Then the solution assumes the form

$$\theta = 1 - A_1 \frac{R \sin \mu_1 (r/R)}{r\mu_1} \exp[-\mu_1^2 \mathrm{Fo}], \quad \text{at } \mathrm{Fo} > \mathrm{Fo}_1. \tag{6.5.34}$$

For convenience in practical calculations, Figs. 6.18 and 6.19 present graphs $\mu_1 = f(\mathrm{Bi})$ and $A_1 = f(\mathrm{Bi})$ for various values of the Biot criterion (from 0 to 20). At $\mathrm{Bi} > 0.1$ the root μ_1 may be calculated by formula (6.10.15).

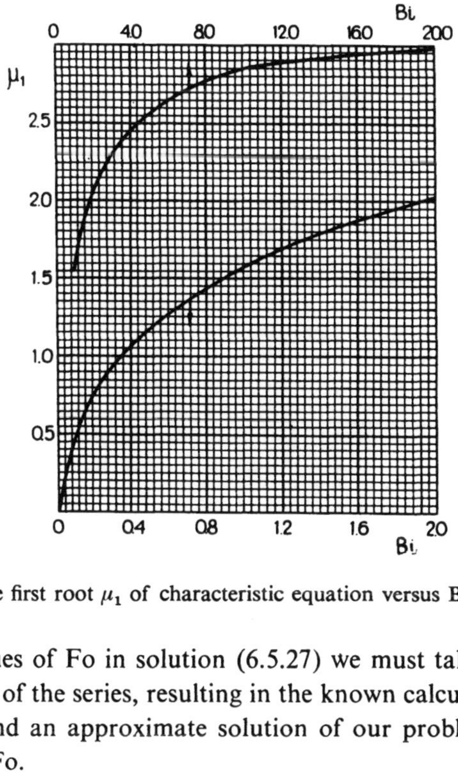

Fig. 6.18. The first root μ_1 of characteristic equation versus Bi for a sphere.

For small values of Fo in solution (6.5.27) we must take a considerable number of terms of the series, resulting in the known calculation difficulties. Therefore, we find an approximate solution of our problem applicable to small values of Fo.

The solution for the transform may be written thus

$$T(r, s) - \frac{t_0}{s} = \frac{\mathrm{Bi}(t_a - t_0)R}{rs[(\mathrm{Bi} - 1) + qR]} (\exp[-q(R - r)] - \exp[-q(R + r)])$$

$$\times \left[1 + \frac{qR - (\mathrm{Bi} - 1)}{qR + (\mathrm{Bi} - 1)} \exp(-2qR)\right]^{-1}$$

$$\approx \frac{\mathrm{Bi}\, R(t_a - t_0)}{rs[(\mathrm{Bi} - 1) + qR]} (\exp[(-q(R-r)] - \exp[-q(R+r)]), \tag{6.5.35}$$

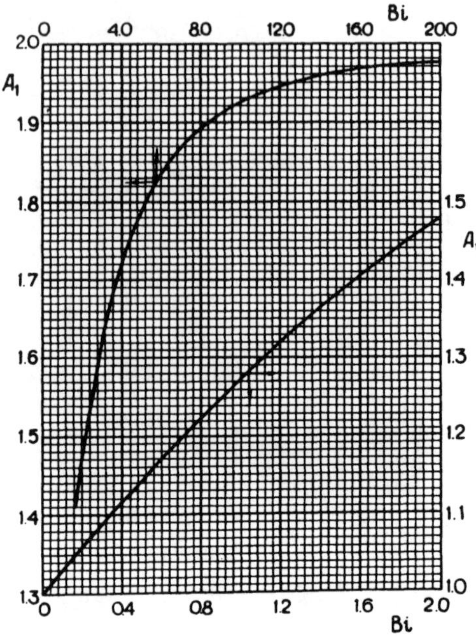

Fig. 6.19. The coefficient A_1 versus Bi for a sphere.

where $q = (s/a)^{1/2}$. To arrive at this result we expanded the expression in square brackets into series and restricted ourselves to the first term, as at the small values of Fo the quantity $qR = (s/a)^{1/2}R$ is large.

The inversion of the transform is made according to the table of transforms (see Appendix)

$$\theta = \frac{t(r, \tau) - t_0}{t_a - t_0} \approx \pm \frac{\text{Bi}\, R}{r(\text{Bi} - 1)} \left\{ \text{erfc} \frac{1 \mp (r/R)}{2(\text{Fo})^{1/2}} - \exp\left[(\text{Bi} - 1)^2 \,\text{Fo} + (\text{Bi} - 1)\left(1 \mp \frac{r}{R}\right) \right] \times \text{erfc} \left(\frac{1 \mp (r/R)}{2(\text{Fo})^{1/2}} + (\text{Bi} - 1)(\text{Fo})^{1/2} \right) \right\}. \quad (6.5.36)$$

Here the following notation is introduced

$$[(\pm)A(\mp z)] = + A(- z) - A(+ z).$$

An exact solution may be found for Bi = 1. Relation (6.5.35) may be written in such a form

$$T(r, s) - \frac{t_0}{s} = \frac{(t_a - t_0)R}{rs(s/a)^{1/2}R} \sum_{n=1}^{\infty} (-1)^{n+1}$$
$$\times \{\exp[-q[(2n-1)R-r]] - \exp[-q[(2n-1)R+r]]\}. \tag{6.5.37}$$

From the table of transforms we have

$$L^{-1}\left[\frac{1}{s\sqrt{s}} \exp[-k\sqrt{s}\,]\right] = 2\sqrt{\tau}\; i\,\mathrm{erfc}\,\frac{k}{2\sqrt{\tau}}. \tag{6.5.38}$$

Then our solution takes the form

$$\theta = 2\frac{R}{r}(\mathrm{Fo})^{1/2} \sum_{n=1}^{\infty} (-1)^{n+1}\left\{i\,\mathrm{erfc}\,\frac{2n-1-(r/R)}{2(\mathrm{Fo})^{1/2}} - i\,\mathrm{erfc}\,\frac{2n-1+(r/R)}{2(\mathrm{Fo})^{1/2}}\right\}. \tag{6.5.39}$$

If in solution (6.5.36), $r = 0$ is assumed, an indeterminant form is obtained. Accordingly, we now determine an approximate solution for the sphere center ($r = 0$) from the solution of the transform $T(r, s)$ which in this case may be written as

$$T(r, s) - \frac{t_0}{s}$$
$$= \frac{\mathrm{Bi}\; qR(t_a - t_0)}{s[(\mathrm{Bi} - 1)\sinh qR + qR\cosh qR]}$$
$$\approx 2\,\mathrm{Bi}(t_a - t_0)\left[1 - \frac{\mathrm{Bi} - 1}{(\mathrm{Bi} - 1) + qR}\right]\frac{1}{s}\exp[-qR], \tag{6.5.40}$$

since at large values of qR one may assume $\sinh qR = \cosh qR = \frac{1}{2}e^{qR}$. Using the table of transforms we find

$$\theta_c \approx 2\,\mathrm{Bi}\exp[(\mathrm{Bi} - 1)^2\,\mathrm{Fo} + (\mathrm{Bi} - 1)]\,\mathrm{erfc}\left\{\frac{1}{2(\mathrm{Fo})^{1/2}} + (\mathrm{Bi} - 1)(\mathrm{Fo})^{1/2}\right\}. \tag{6.5.41}$$

Approximate solutions (6.5.36) and (6.5.41) are valid for the small values of Fo; at great values of Fo it is necessary to use solution (6.5.27).

At the small values of Fo, the sphere-center temperature is almost constant. A change in the surface temperature is therefore of the greatest interest. This temperature changes over wide ranges at small values of Fo and at large values of Bi.

Table 6.7 presents the values of $(1 - \theta_s)$ for Fo from 0.0003 to 0.0050 when Bi changes from 0.1 to 1000. Pöschl obtained the values $(1 - \theta_s)$

TABLE 6.7. RELATIVE TEMPERATURE AT THE SPHERE SURFACE $1 - \theta_s = \psi(\mathrm{Fo}, \mathrm{Bi})$

Bi	Fo				
	0.0003	0.0005	0.0010	0.0025	0.0050
0.1	0.999	0.998	0.996	0.994	0.991
0.5	0.991	0.988	0.982	0.971	0.959
1	0.981	0.975	0.964	0.944	0.920
4	0.925	0.905	0.868	0.805	0.734
10	0.830	0.787	0.717	0.603	0.502
20	0.701	0.638	0.545	0.413	0.315
50	0.463	0.392	0.300	0.197	0.137
100	0.284	0.227	0.164	0.103	0.071
200	0.154	0.131	0.084	0.052	0.035
1000	0.031	0.026	0.015	0.010	0.007

from solution (6.5.27) by means of lengthy calculation (see Section 6.3). Our approximate solutions yield the same result with very simple calculations. We shall illustrate the above by examples.

Let Bi = 1 and Fo = 0.0003. Then, using formula (6.5.39) we obtain

$$\theta_s = 2(\mathrm{Fo})^{1/2}\,\mathrm{i\,erfc}\,0 = 2(\mathrm{Fo})^{1/2}(1/\sqrt{\pi}) \tag{6.5.42}$$

as

$$\mathrm{i\,erfc}\,u = (1/\sqrt{\pi})\,e^{-u^2} - u\,\mathrm{erfc}\,u\,, \tag{6.5.43}$$

and the next terms of formula (6.5.39) are negligibly small as compared to the first. Further,

$$1 - \theta_s = 1 - 2(0.0003)^{1/2}\,1/\sqrt{\pi} = 0.981\,, \tag{6.5.44}$$

which fully coincides with the data in Table 6.7.

Let Fo = 0.0025 and Bi = 10.0. Making use of solution (6.5.36) which may be written thus

$$1 - \theta_s = 1 - \frac{\mathrm{Bi}}{\mathrm{Bi} - 1}\,[1 - \exp[(\mathrm{Bi} - 1)^2\mathrm{Fo}]\,\mathrm{erfc}(\mathrm{Bi} - 1)(\mathrm{Fo})^{1/2}]. \tag{6.5.45}$$

The remaining terms in solution (6.5.45) are neglected as they are negligibly small as compared to the ones taken. We have

$$(\mathrm{Bi} - 1)(\mathrm{Fo})^{1/2} = 0.45.$$

Hence, we have

$$1 - \theta_s = 1 - (10.0/9.0)(1 - \exp[0.2025]\,\mathrm{erfc}\,0.45) = 0.6028. \tag{6.5.46}$$

In this case, according to Table 6.7, $(1 - \theta_s)$ is equal to 0.603, i.e., there is a very good agreement.

Thus, the approximate solutions yield quite satisfactory results and replace the bulky calculations obtained from formula (6.5.27).

e. Specific Heat Rate. We determine the mean sphere temperature by

$$\bar{t}(\tau) = (3/R^3) \int_0^R r^2 t(r, \tau)\, dr. \tag{6.5.47}$$

If the corresponding expression from solution (6.5.27) is substituted for $t(r, \tau)$ and the following formula is used

$$\int u \sin u \, du = \sin u - u \cos u,$$

then upon integration we obtain

$$\bar{\theta} = \frac{\bar{t}(\tau) - t_0}{t_a - t_0} = 1 - \sum_{n=1}^{\infty} B_n \exp[- \mu_n^2 \, \mathrm{Fo}], \tag{6.5.48}$$

where B_n are the constant coefficients determined by the relation

$$B_n = \frac{6\, \mathrm{Bi}^2}{\mu_n^2 (\mu_n^2 + \mathrm{Bi}^2 - \mathrm{Bi})}. \tag{6.5.49}$$

The first six constant coefficients B_n are presented in Table 6.8 for various values of the Fourier number.

Since we usually need to use only the first coefficient graphs of $B_1 = f(\mathrm{Bi})$ for values of Bi (from 0 to 20) are plotted in Fig. 6.20. Two curves of different scale along the abscissa axis are plotted for convenience. The specific heat rate ($\mathrm{kcal/m^3}$) is determined by the formula

$$\Delta Q_v = c\gamma[\bar{t}(\tau) - t_0]. \tag{6.5.50}$$

For small time values, the specific heat rate may be determined by the relation

$$\Delta Q_v = - \frac{3}{R^3} \int_0^\tau \lambda \frac{\partial t(R, \tau)}{\partial r} \, d\tau, \tag{6.5.51}$$

where $\partial t(R, \tau)/\partial r$ is calculated similarly as in the previous paragraphs.

Having nomograms and Tables 6.5, 6.6, and 6.8, as well as the corresponding graphs for μ_1, A_1, and B_1, it is possible to make particular calculations quickly and easily. Consequently, no additional calculations will be made.

TABLE 6.8. The Values of the Constants

$$B_n = \frac{6\mathrm{Bi}^2}{\mu_n^2(\mu_n^2 + \mathrm{Bi}^2 - \mathrm{Bi})}$$

Bi	B_1	B_2	B_3	B_4	B_5	B_6
∞	0.6079	0.1520	0.0675	0.0380	0.0243	0.0169
51.0	0.6427	0.1518	0.0671	0.0380	0.0236	0.0158
21.0	0.6886	0.1510	0.0652	0.0363	0.0180	0.0108
10.1	0.7667	1.1496	0.0485	0.0196	0.0091	0.0047
9.0	0.7737	0.1453	0.0455	0.0175	0.0079	0.0040
8.0	0.7889	0.1396	0.0408	0.0152	0.0067	0.0033
7.0	0.8068	0.1319	0.0360	0.0128	0.0055	0.0027
6.0	0.8280	0.1215	0.0305	0.0104	0.0044	0.0021
5.0	0.8533	0.1075	0.0245	0.0079	0.0032	0.0015
4.0	0.8829	0.0890	0.0180	0.0055	0.0021	0.0010
3.0	0.9171	0.0655	0.0115	0.0034	0.0013	0.0006
2.5	0.9353	0.0520	0.0085	0.0024	0.0009	0.0004
2.0	0.9534	0.0380	0.0057	0.0016	0.0006	0.0003
1.9	0.9569	0.0352	0.0052	0.0014	0.0005	0.0002
1.8	0.9605	0.0325	0.0047	0.0013	0.0005	0.0002
1.7	0.9641	0.0297	0.0043	0.0011	0.0004	0.0002
1.6	0.9678	0.0270	0.0038	0.0010	0.0004	0.0002
1.5	0.9707	0.0243	0.0034	0.0009	0.0003	0.0001
1.4	0.9739	0.0217	0.0030	0.0008	0.0003	0.0001
1.3	0.9770	0.0192	0.0026	0.0007	0.0003	0.0001
1.2	0.9800	0.0167	0.0022	0.0006	0.0002	0.0001
1.1	0.9828	0.0144	0.0019	0.0005	0.0002	0.0001
1.0	0.9855	0.0122	0.0016	0.0004	0.0001	0.0001
0.90	0.9881	0.0101	0.0013	0.0003	0.0001	
0.80	0.9905	0.0081	0.0010	0.0003	0.0001	
0.70	0.9926	0.0064	0.0008	0.0002	0.0001	
0.60	0.9944	0.0048	0.0005	0.0001		
0.50	0.9959	0.0034	0.0004	0.0001		
0.40	0.9974	0.0022	0.0003	0.0001		
0.30	0.9985	0.0013	0.0001			
0.20	0.9994	0.0006	0.0001			
0.15	0.9996	0.0003				
0.10	0.9997	0.0001				
0.09	1.0000	0.0001				
0.08	1.0000	0.0001				

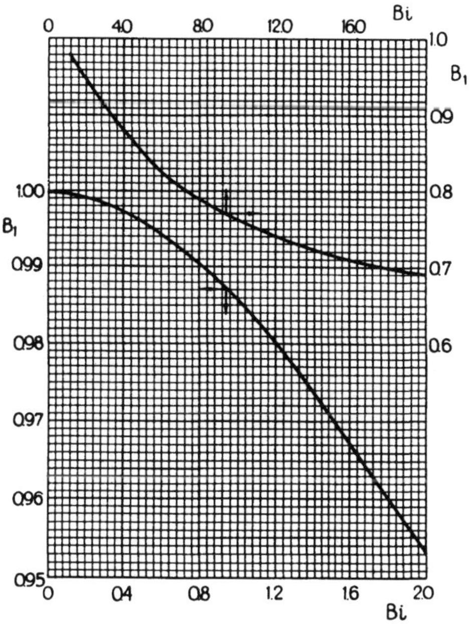

Fig. 6.20. The coefficient B_1 versus Bi for a sphere.

6.6 Infinite Cylinder

a. Statement of the Problem. *Consider an infinite cylinder with radius R and with a prescribed radial temperature distribution in the form of some function f(r). Isotherms are assumed to be represented by coaxial cylindrical surfaces, i.e., the temperature in the cylinder depends only on its radius and time. At the initial moment the cylinder is placed in a medium with a constant temperature* $t_a > t(r, 0)$. *The temperature distribution in the cylinder at any moment of time, as well as the specific heat flow rate, is to be found.*

The differential heat conduction equation for an infinite cylinder was given in Chapter 4, Section 5. The initial and boundary conditions may be written as

$$t(r, 0) = f(r), \tag{6.6.1}$$

$$-\frac{\partial t(R, \tau)}{\partial r} + H[t_a - t(R, \tau)] = 0, \tag{6.6.2}$$

$$\frac{\partial t(0, \tau)}{\partial r} = 0, \quad t(0, \tau) \neq \infty. \tag{6.6.3}$$

b. Solution by the Method of Separation of Variables. Let us reduce our problem of heating a cylinder to that of cooling by replacing the variable, i.e., we assume $\vartheta(r, \tau) = t_a - t(r, \tau)$. Then the initial and boundary conditions will take the form

$$\vartheta(r, 0) = t_a - f(r) = f_1(r), \tag{6.6.4}$$

$$\frac{\partial \vartheta(R, \tau)}{\partial r} + H\vartheta(R, \tau) = 0, \tag{6.6.5}$$

$$\frac{\partial \vartheta(0, \tau)}{\partial r} = 0, \quad \vartheta(0, \tau) \neq \infty. \tag{6.6.6}$$

A particular solution of the differential heat conduction equation for an infinite cylinder when the isotherms are distributed coaxially with respect to the cylinder axis has the form

$$\vartheta(r, \tau) = [CJ_0(kr) + DY_0(kr)] \exp[-k^2 a\tau], \tag{6.6.7}$$

where $J_0(kr)$ and $Y_0(kr)$ are zeroth-order Bessel functions of the first and second kind, respectively, and C and D are constants (see Chapter 4, Section 5). It follows from Eq. (6.6.6) that $D = 0$ (see Chapter 4, Section 5); the constant C will be determined later from the initial condition.

Substituting boundary condition (6.6.5) into particular solution (6.6.7) gives as

$$-kCJ_1(kR) \exp[-k^2 a\tau] + HCJ_0(kR) \exp[-k^2 a\tau] = 0$$

since

$$\frac{dJ_0(kr)}{dr} = -kJ_1(kr)$$

dividing by $C \exp[-k^2 a\tau]$ $(0 < \tau < \infty)$, we obtain

$$\frac{J_0(kR)}{J_1(kR)} = \frac{kR}{HR} = \frac{kR}{\mathrm{Bi}}. \tag{6.6.8}$$

Equation (6.6.8) is transcendental; we may obtain its solution graphically. Denote kR by μ $(\mu = kR)$. The function $J_0(\mu)/J_1(\mu)$ becomes zero at those points for which $J_0(\mu) = 0$, i.e., at the points $\nu_1, \nu_2, \cdots, \nu_n$ are the roots of the function $J_0(\mu)$. At those points at which the function $J_1(\mu)$ becomes zero, the function $J_0(\mu)/J_1(\mu)$ undergoes a discontinuity and becomes equal $\pm\infty$. We denote the roots of the function $J_1(\mu)$ by $\varkappa_n$.

Let us construct the curves $y_1 = J_0(\mu)/J_1(\mu)$ that intersect the x axis at points ν_n and have asymptotes parallel to the y axis at points $\varkappa_n$ (Fig. 6.21). The curves $y_1 = J_0(\mu)/J_1(\mu)$ resemble a cotangent curve but differ in the phase.

Let us then plot straight line $y_2 = (1/\mathrm{Bi})\,\mu$. The points of intersection of the line y_2 with the curves y_1 give the values of the characteristic roots. It is seen from Fig. 6.21 that there is an infinite number of roots μ_n, all of which lie between limits ν_n and $\varkappa_n$ ($\nu_n < \mu_n < \varkappa_n$). If $\mathrm{Bi} \to \infty$, the straight line coincides with the x axis and the roots μ_n become equal to the roots ν_n ($\mu_n = \nu_n$), i.e., they do not depend on the Bi criterion.

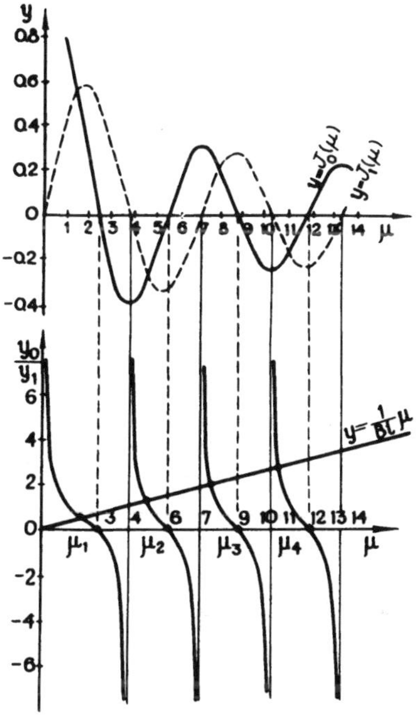

Fig. 6.21. Graphical method for determination of the roots of characteristic equation for a cylinder.

At $\mathrm{Bi} \to 0$ the value $\mu_1 \to (2\,\mathrm{Bi})^{1/2}$ which follows from the characteristic equation if the functions $J_0(\mu)$ and $J_1(\mu)$ are developed into series and if we restrict the development to the first terms of the series. Indeed

$$\frac{J_0(\mu)}{J_1(\mu)} = \frac{1}{\mathrm{Bi}}\,\mu = \frac{1 - \frac{1}{2^2}\,\mu^2 + \cdots}{\frac{1}{2}\,\mu - \frac{1}{2^2 4}\,\mu^3 + \cdots},$$

from which

$$\mu_1^{\,2} = 2\,\mathrm{Bi}.$$

The remaining roots are determined from the equation

$$J_1(\mu) = 0,$$

and, hence, do not depend on the radius of the cylinder. In the case where $0 < \text{Bi} < \infty$, the roots μ_n depend on Bi and, consequently, on the radius of the cylinder. The first six roots $\mu_1, \mu_2, \cdots, \mu_6$ are given in Table 6.9 for different values of the Bi criterion.

TABLE 6.9. THE ROOTS OF THE CHARACTERISTIC EQUATION

$$\frac{J_0(\mu)}{J_1(\mu)} - \frac{1}{\text{Bi}}\mu$$

Bi	μ_1	μ_2	μ_3	μ_4	μ_5	μ_6
0.0	0.0000	3.8317	7.0156	10.1735	13.3237	16.4706
0.01	0.1412	3.8343	7.0170	10.1745	13.3244	16.4712
0.02	0.1995	3.8369	7.0184	10.1754	13.3252	16.4718
0.04	0.2814	3.8421	7.0213	10.1774	13.3267	16.4731
0.06	0.3438	3.8473	7.0241	10.1794	13.3282	16.4743
0.08	0.3960	3.8525	7.0270	10.1813	13.3297	16.4755
0.10	0.4417	3.8577	7.0298	10.1833	13.3312	16.4767
0.15	0.5376	3.8706	7.0369	10.1882	13.3349	16.4797
0.20	0.6170	3.8835	7.0440	10.1931	13.3387	16.4828
0.30	0.7465	3.9091	7.0582	10.2029	13.3462	16.4888
0.40	0.8516	3.9344	7.0723	10.2127	13.3537	16.4949
0.50	0.9408	3.9594	7.0864	10.2225	13.3611	16.5010
0.60	1.0184	3.9841	7.1004	10.2322	13.3686	16.5070
0.70	1.0873	4.0085	7.1143	10.2419	13.3761	16.5131
0.80	1.1490	4.0325	7.1282	10.2519	13.3835	16.5191
0.90	1.2048	4.0562	7.1421	10.2613	13.3910	16.5251
1.0	1.2558	4.0795	7.1558	10.2710	13.3984	16.5312
1.5	1.4569	4.1902	7.2233	10.3188	13.4353	16.5612
2.0	1.5994	4.2910	7.2884	10.3658	13.4719	16.5910
3.0	1.7887	4.4634	7.4103	10.4566	13.5434	16.6499
4.0	1.9081	4.6018	7.5201	10.5423	13.6125	16.7073
5.0	1.9898	4.7131	7.6177	10.6223	13.6786	16.7630

Thus, the characteristic equation (6.6.8) may be written as

$$\frac{J_0(\mu)}{J_1(\mu)} = \frac{1}{\text{Bi}}\mu. \tag{6.6.9}$$

We return to Eq. (6.6.7). A general solution will be the sum of all particular solutions

$$\vartheta(r, \tau) = \sum_{n=1}^{\infty} C_n J_0\left(\mu_n \frac{r}{R}\right) \exp\left[-\mu_n^2 \frac{a\tau}{R^2}\right]. \tag{6.6.10}$$

The constants C_n are determined from the initial condition which will be written as

$$\vartheta(r, 0) = f_1(r) = \sum_{n=1}^{\infty} C_n J_0\left(\mu_n \frac{r}{R}\right). \tag{6.6.11}$$

With the restrictions stated in Chapter 4, Section 5, we may develop the function $f_1(r)$ into series. For this purpose we shall multiply both sides of Eq. (6.6.11) by $rJ_0\{\mu_m(r/R)\}\,dr$ and integrate with respect to r between 0 and R (we recall that the Bessel functions themselves are not orthogonal but rather the product $\sqrt{r}J_0\{\mu_n(r/R)\}$. Then, assuming the term-by-term integration of the series we shall have

$$\int_0^R rf_1(r)J_0\{\mu_m(r/R)\}\,dr = \sum_{n=1}^{\infty} C_n \int_0^R rJ_0\{\mu_n(r/R)\}J_0\{\mu_m(r/R)\}\,dr. \tag{6.6.12}$$

Let us show that all the terms of Eq. (6.6.12) become zero for $m \neq n$. According to Eq. (4.5.20) we may write

$$\int_0^R rJ_0\{\mu_n(r/R)\}J_0\{\mu_m(r/R)\}\,dr = \frac{R[\mu_m J_0(\mu_n)J_1(\mu_m) - \mu_n J_0(\mu_m)J_1(\mu_n)]}{\mu_m^2 - \mu_n^2}. \tag{6.6.13}$$

From the characteristic equation, we have $\mu_n J_1(\mu_n) = J_0(\mu_n)\,\mathrm{Bi}$. Then we may write

$$\mu_m J_0(\mu_n)J_1(\mu_m) = \mathrm{Bi}\, J_0(\mu_n)J_0(\mu_m),$$

$$\mu_n J_0(\mu_m)J_1(\mu_n) = \mathrm{Bi}\, J_0(\mu_n)J_0(\mu_m).$$

Integral (6.6.13) is clearly equal to zero for $m \neq n$. If $m = n$, let us use Eq. (4.5.19) which we shall write as

$$\int_0^R rJ_0^2\{\mu_n(r/R)\}\,dr = (R^2/2)[J_0^2(\mu_n) + J_1^2(\mu_n)]. \tag{6.6.14}$$

Thus from Eq. (6.6.12) under conditions (6.6.13) and (6.6.14) we obtain

$$C_n = \frac{\int_0^R rf_1(r)J_0\{\mu_n(r/R)\}\,dr}{\int_0^R rJ_0^2\{\mu_n(r/R)\}\,dr} = \frac{2}{R^2}\frac{1}{[J_0^2(\mu_n) + J_1^2(\mu_n)]}\int_0^R rf_1(r)J_0\{\mu_n(r/R)\}\,dr.$$

The final form of the general solution of our problem is

$$\vartheta(r, \tau) = t_a - t(r, \tau)$$
$$= \sum_{n=1}^{\infty} \frac{J_0\{\mu_n (r/R)\}}{[J_0^2(\mu_n) + J_1^2(\mu_n)]}$$
$$\times \frac{2}{R^2} \int_0^R r f_1(r) J_0\left(\mu_n \frac{r}{R}\right) dr \exp\left[-\mu_n^2 \frac{a\tau}{R^2}\right]. \tag{6.6.15}$$

If $f_1(r) = t_a - t_0 = \text{const}$, the integral in Eq. (6.6.15) is equal to (see formula (4.5.24))

$$(2/R^2) \int_0^R (t_a - t_0) r J_0\{\mu_n(r/R)\}\, dr = (2/\mu_n) J_1(\mu_n). \tag{6.6.16}$$

Then solution (6.6.15) may be written as

$$\theta = \frac{t(r, \tau) - t_0}{t_a - t_0} = 1 - \sum_{n=1}^{\infty} A_n J_0\{\mu_n(r/R)\} \exp[-\mu_n^2 \,\mathrm{Fo}], \tag{6.6.17}$$

where

$$A_n = \frac{2\, J_1(\mu_n)}{\mu_n[J_0^2(\mu_n) + J_1^2(\mu_n)]}$$

are constant coefficients (the so called initial thermal amplitudes) determined by the Bi criterion.

*c. **Operational Solution.*** It was shown in Chapter 4, Section 5, that for a symmetric problem, a solution for the transform $T(r, s)$ had the form

$$T(r, s) - (t_0/s) = A I_0\left\{\left(\frac{s}{a}\right)^{1/2} r\right\}, \tag{6.6.18}$$

where A is a constant independent of r and $I_0\{(s/a)^{1/2} r\} = J_0\{i(s/a)^{1/2} r\}$ is the modified zeroth-order Bessel function of the first kind. The transition from the modified functions to the ordinary Bessel functions may be made by means of the relationship

$$I_\nu(z) = i^{-\nu} J_\nu(iz). \tag{6.6.19}$$

The constant A is found from boundary condition (6.6.2) which for the transform $T(r, s)$ will take the form

$$-T'(R, s) + H\left[\frac{t_a}{s} - T(R, s)\right] = 0. \tag{6.6.20}$$

Let us substitute boundary condition (6.6.20) into solution (6.6.18)

$$-A\left(\frac{s}{a}\right)^{1/2} I_0'\left\{\left(\frac{s}{a}\right)^{1/2} R\right\} + (Ht_a/s) - (Ht_0/s) - AHI_0\left\{\left(\frac{s}{a}\right)^{1/2} R\right\} = 0.$$

Then

$$A = \frac{(t_a - t_0)}{s\left[I_0\left\{\left(\frac{s}{a}\right)^{1/2} R\right\} + \frac{1}{H}\left(\frac{s}{a}\right)^{1/2} I_1\left\{\left(\frac{s}{a}\right)^{1/2} R\right\}\right]},$$

where

$$I_1(z) = I_0'(z) = \frac{1}{2}z + \frac{1}{2^2 4}z^3 + \frac{1}{2^2 4^2 6}z^5 + \cdots$$

is an odd function with respect to $z[I_1(-z) = -I_1(z)]$; it plays the same role as $\sinh z$. Hence, solution (6.6.18) will have the form

$$T(r, s) - \frac{t_0}{s} = \frac{(t_a - t_0) I_0\left\{\left(\frac{s}{a}\right)^{1/2} r\right\}}{s\left[I_0\left\{\left(\frac{s}{a}\right)^{1/2} R\right\} + \frac{1}{H}\left(\frac{s}{a}\right)^{1/2} I_1\left\{\left(\frac{s}{a}\right)^{1/2} R\right\}\right]} = \frac{\Phi(s)}{\psi(s)}. \tag{6.6.21}$$

Solution (6.6.21) is a single-valued function of s and represents the ratio of generalized polynomials, the polynomial of the denominator not containing a constant, i.e., conditions of the expansion theorem are observed. Let us obtain the roots of $\psi(s)$. For this purpose we shall equate it with zero:

$$\psi(s) = s\left[I_0\left\{\left(\frac{s}{a}\right)^{1/2} R\right\} + \frac{1}{H}\left(\frac{s}{a}\right)^{1/2} I_1\left\{\left(\frac{s}{a}\right)^{1/2} R\right\}\right] = s\varphi(s) = 0, \tag{6.6.22}$$

where the expression in brackets is denoted by $\varphi(s)$. From Eq. (6.6.22) we shall find the roots $s_0 = 0$ and $\varphi(s) = 0$. Then $s_n = -a\mu_n^2/R^2$ where $\mu = i(s/a)^{1/2}R$. The constants μ are determined from

$$\begin{aligned} &I_0\left\{\left(\frac{s}{a}\right)^{1/2} R\right\} + \frac{1}{H}\left(\frac{s}{a}\right)^{1/2} I_1\left\{\left(\frac{s}{a}\right)^{1/2} R\right\} \\ &= J_0\left\{i\left(\frac{s}{a}\right)^{1/2} R\right\} + \frac{1}{iH}\left(\frac{s}{a}\right)^{1/2} J_1\left\{i\left(\frac{s}{a}\right)^{1/2} R\right\} \\ &= J_0(\mu) - \frac{1}{HR}\mu J_1(\mu) = 0. \end{aligned} \tag{6.6.23}$$

Such a transcendental equation has an infinite number of roots μ_n which may be obtained by a graphical solution (see above). Hence the characteristic equation has the form

$$\{J_0(\mu)/J_1(\mu)\} = (1/\text{Bi})\,\mu. \tag{6.6.24}$$

Let us find the subsidiary values

$$\psi'(s_n) = \varphi(s) + s\varphi'(s)$$

$$= \varphi(s) + \frac{sR}{2(as)^{1/2}}\left[1 + \frac{1}{HR}\, I_0'\left\{\left(\frac{s}{a}\right)^{1/2} R\right\} + \frac{1}{H}\left(\frac{s}{a}\right)^{1/2} I_1'\left\{\left(\frac{s}{a}\right)^{1/2} R\right\}\right]$$

$$\frac{\Phi(0)}{\psi'(0)} = (t_a - t_0),$$

$$\frac{\Phi(s_n)}{\psi'(s_n)} \exp[s_n\tau] = -\frac{2(t_a - t_0)HRJ_0\left(\mu_n \dfrac{r}{R}\right)}{[(HR + 1)J_1(\mu_n) + \mu_n J_1'(\mu_n)]\mu_n} \times \exp\left[-\mu_n^2 \frac{a\tau}{R^2}\right],$$

since $I_1'(z) = J_1'(iz)$.

Hence the solution of our problem has the form

$$\theta = \frac{t(r,\tau) - t_0}{t_a - t_0} = 1 - \sum_{n=1}^{\infty} A_n J_0\left(\mu_n \frac{r}{R}\right) \exp[-\mu_n^2\,\text{Fo}] \tag{6.6.25}$$

where

$$A_n = \frac{2\,\text{Bi}}{[(\text{Bi} + 1)J_1(\mu_n) + \mu_n J_1'(\mu_n)]\mu_n} = \frac{2J_1(\mu_n)}{\mu_n[J_0^2(\mu_n) + J_1^2(\mu_n)]}. \tag{6.6.26}$$

The latter transformation is made on the basis of the recurrence formula

$$\mu_n J_1'(\mu_n) = \mu_n J_0(\mu_n) - J_1(\mu_n)$$

and the characteristic equation (6.6.24).

The constant coefficients A_n can be calculated from relation (6.6.26) which may be transformed to

$$A_n = \frac{2\,\text{Bi}}{J_0(\mu_n)[\mu_n^2 + \text{Bi}^2]}. \tag{6.6.27}$$

With the help of this formula, the first six coefficients A_n can be calculated, and are given in Table 6.10.

TABLE 6.10. THE VALUES OF THE CONSTANTS

$$A_n = \frac{2\,\mathrm{Bi}}{(\mu_n^2 + \mathrm{Bi}^2)J_0(\mu_n)}$$

Bi	$+A_1$	$-A_2$	$+A_3$	$-A_4$	$+A_5$	$-A_6$
0.0	1.0000	0.0000	0.0000	0.0000	0.0000	0.0000
0.01	1.0031	0.0034	0.0013	0.0008	0.0005	0.0004
0.02	1.0049	0.0067	0.0027	0.0015	0.0010	0.0007
0.04	1.0102	0.0135	0.0052	0.0031	0.0021	0.0015
0.06	1.0150	0.0201	0.0081	0.0046	0.0031	0.0023
0.08	1.0199	0.0268	0.0110	0.0062	0.0041	0.0030
0.10	1.0245	0.0333	0.0135	0.0077	0.0051	0.0037
0.15	1.0366	0.0497	0.0202	0.0116	0.0077	0.0056
0.20	1.0482	0.0653	0.0269	0.0154	0.0103	0.0075
0.30	1.0711	0.0972	0.0401	0.0231	0.0155	0.0112
0.40	1.0931	0.1277	0.0582	0.0307	0.0205	0.0150
0.50	1.1142	0.1571	0.0662	0.0383	0.0256	0.0187
0.60	1.1345	0.1857	0.0790	0.0458	0.0307	0.0224
0.70	1.1539	0.2132	0.0917	0.0533	0.0358	0.0261
0.80	1.1724	0.2398	0.1043	0.0608	0.0408	0.0298
0.90	1.1902	0.2654	0.1167	0.0682	0.0459	0.0335
1.0	1.2071	0.2901	0.1289	0.0756	0.0509	0.0372
1.5	1.2807	0.4008	0.1877	0.1117	0.0756	0.0554
2.0	1.3377	0.4923	0.2422	0.1404	0.0998	0.0732
3.0	1.4192	0.6309	0.3384	0.2114	0.1463	0.1084
4.0	1.4698	0.7278	0.4184	0.2699	0.1898	0.1420
5.0	1.5029	0.7973	0.4842	0.3220	0.2301	0.1735
6.0	1.5253	0.8484	0.5382	0.3679	0.2672	0.2038
7.0	1.5409	0.8869	0.5825	0.4080	0.3010	0.2317
8.0	1.5523	0.9225	0.6189	0.4430	0.3316	0.2579
9.0	1.5611	0.9393	0.6491	0.4735	0.3593	0.2826
10.0	1.5677	0.9575	0.6784	0.5000	0.3843	0.3042
15.0	1.5853	1.0091	0.7519	0.5901	0.4760	0.3913
20.0	1.5918	1.0309	0.7889	0.6382	0.5303	0.4461
30.0	1.5964	1.0488	1.8195	0.6827	0.5853	0.5062
40.0	1.5988	1.0550	0.8335	0.7018	0.6133	0.5390
50.0	1.5995	1.0587	0.8396	0.7112	0.6227	0.5544
60.0	1.6009	1.0589	0.8428	0.7165	0.6301	0.5642
80.0	1.6012	1.0599	0.8463	0.7212	0.6398	0.5770
100.0	1.6014	1.0631	0.8505	0.7245	0.6415	0.5850
∞	1.6021	1.0648	0.8558	0.7296	0.6485	0.5896

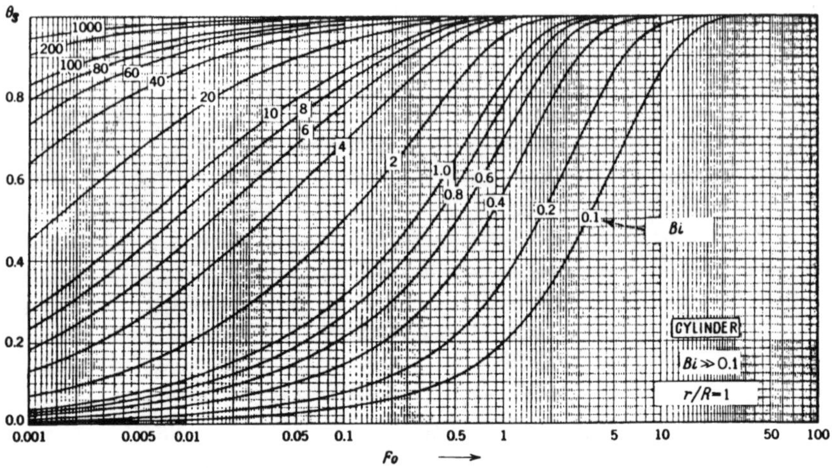

Fig. 6.22a. Graphs for determination of dimensionless excess temperature at the cylinder surface for Bi from 0.1 to 1000 [102].

For engineering calculations, Figs. 6.22a–6.23b give nomograms for determination of θ_s (the surface temperature of the cylinder) and of θ_c (the temperature in the center of the cylinder) as a function of Fo and Bi.

d. Analysis of the Solution. It has been shown that at Bi $\to \infty$, the roots μ_n were determined by the equation $J_0(\mu_n) = 0$, i.e., they are the roots of the function $J_0(\mu_n)$. In this case, the coefficients A_n, following formula

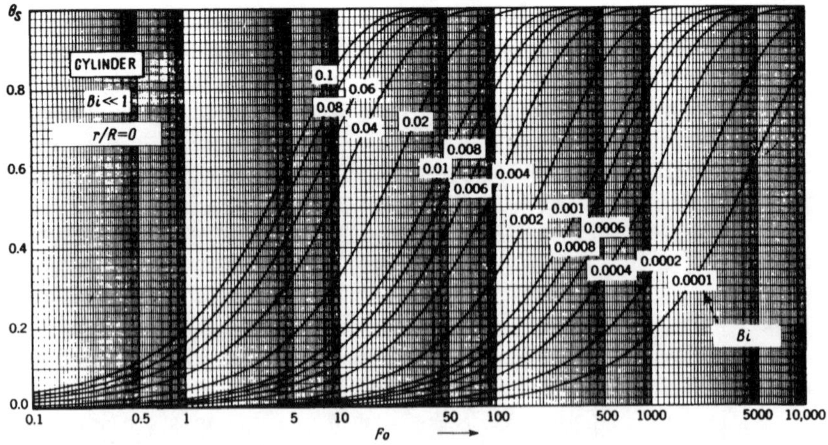

Fig. 6.22b. Graphs for determination of dimensionless excess temperature at the cylinder surface for small Bi (0.0001 < Bi < 0.1) [102].

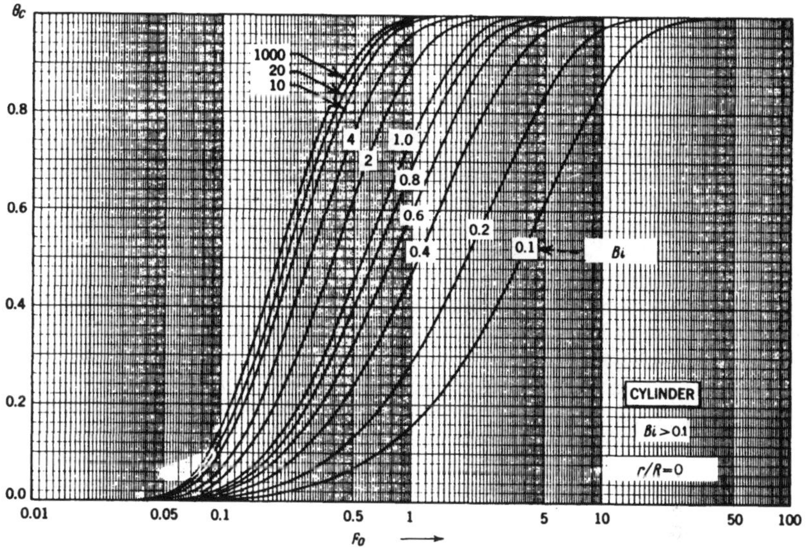

Fig. 6.23a. Graphs for determination of dimensionless excess temperature in the cylinder center for Bi from 0.1 to 1000 [102].

(6.6.26), will be equal to

$$A_n = 2/\mu_n J_1(\mu_n).$$

Solution (6.6.25) becomes identical with solution (4.5.36) if in the latter θ is replaced by $(1 - \theta)$, since in this case, the problem of cooling a cylinder

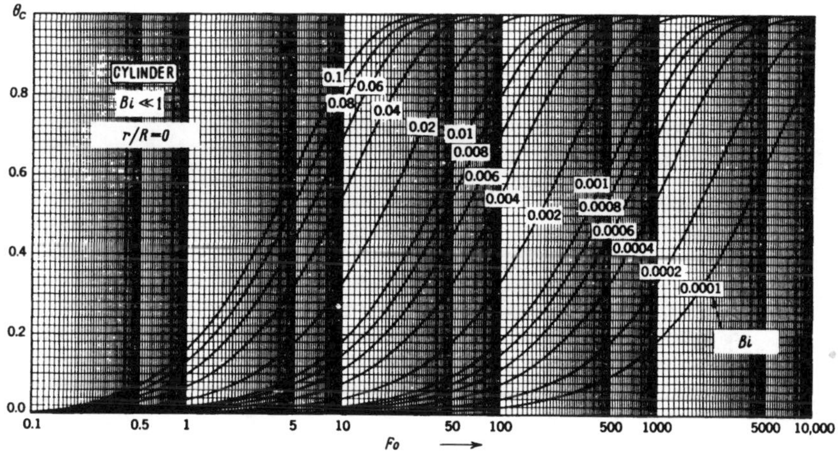

Fig. 6.23b. Graphs for determination of dimensionless excess temperature in the cylinder center for small values of Bi $(0.0001 < \mathrm{Bi} < 0.1)$ [102].

is replaced by that of heating. It follows from boundary condition (6.6.3), that the surface temperature of the cylinder $t(R, \tau)$ rapidly becomes equal to the ambient temperature t_a and the whole process of heating reduces to a temperature leveling inside the cylinder (internal problem). In the stationary state (Fo $= \infty$) the temperature at any point of the cylinder is equal to the ambient temperature.

At low values of Bi (Bi $\to 0$), with the exception of $A_1 \to 1$ all the coefficients A_n approach zero (see formula (6.6.26)), since $J_1(\mu_n) \to 0$; and $J_1(z)/z \to 1$ when $z \to 0$; in addition, $\mu_1^2 = 2$ Bi (see relation (6.6.8)). Then, for low values of Bi Eq. (6.6.25) may be written as

$$\theta = 1 - J_0\left((2\mathrm{Bi})^{1/2} \frac{r}{R}\right) \exp[-2 \mathrm{\,Bi\,Fo}]. \tag{6.6.28}$$

In this case, the temperature drop inside the cylinder will be small and the process of heating depends only on heat transfer between the surrounding medium and the surface of the cylinder (external problem). In all intermediate cases, the process of heating depends both on the heat transfer rate inside the cylinder and the rate of heat exchange with the surrounding medium (boundary value problem). The characteristic numbers μ_n depend on Bi and, consequently, on the radius of the cylinder. The rate of heating will be, therefore, inversely proportional to the nth power of the cylinder radius, where n lies between the limits 1 and 2 ($1 < n < 2$). The series (6.6.25)

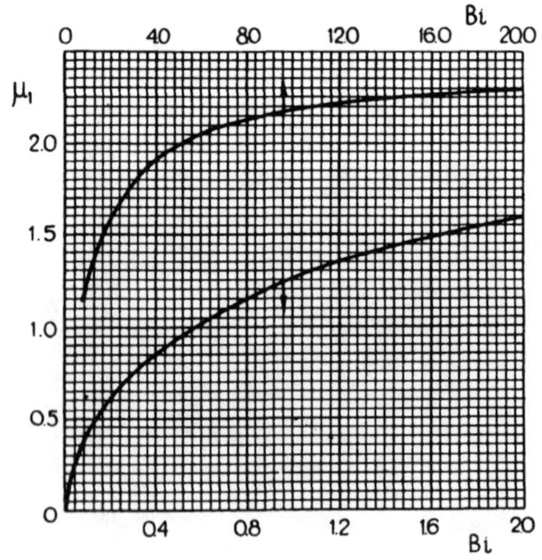

Fig. 6.24. The first root μ_1 of the characteristic equation versus Bi for an infinite cylinder.

converges quite rapidly as on the basis of inequality

$$\mu_1 < \mu_2 < \mu_3 < \cdots < \mu_n ,$$

the exponential function exp $[-\mu_n^2 \text{ Fo}]$ decreases rapidly with an increase in μ_n. If, therefore, we exclude from consideration small values of Fo, we may restrict ourselves to one term of the series and neglect the remainder. Now solution (6.6.25) takes the form

$$\theta = 1 - A_1 J_0\{\mu_1(r/R)\} \exp[-\mu_1^2 \text{ Fo}], \qquad \text{at Fo} > \text{Fo}_1.$$

See Section 6.11 for the limiting value of Fo_1.

For practical computations, graphs of $\mu_1 = f(\text{Bi})$ and $A_1 = f(\text{Bi})$ for values of Bi from 0 to 20 are given in Figs. 6.24 and 6.25 (for Bi > 0.1 the root μ_1 may be calculated by formula (6.10.15)).

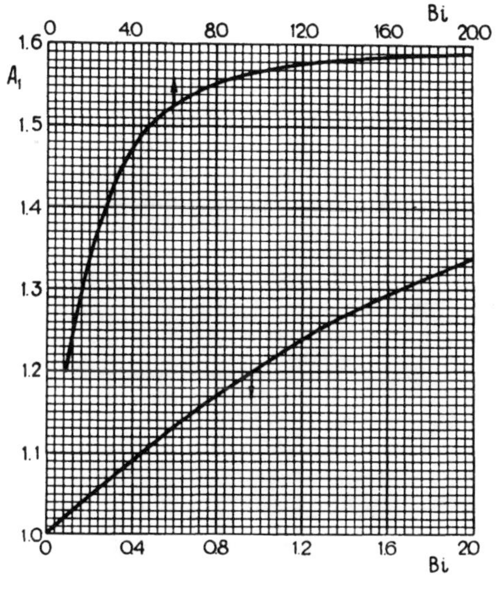

Fig. 6.25. The coefficient A_1 versus Bi for an infinite cylinder.

If the Fourier number is small, many terms of the series must be taken, thus making the application of solution (6.6.25) to engineering calculations difficult.

Let us now find an approximate solution applicable to small values of Fo. For the transform in solution (6.6.21), let us develop the functions $I_0[(s/a)^{1/2}R]$ and $I_1[(s/a)^{1/2}R]$ in an asymptotic series (see Appendix 1) since for small values of Fo the value of $(s/a)^{1/2}R$ is great

$$T(r, s) - \frac{t_0}{s} = \frac{(t_a - t_0)a}{s\sqrt{s}} \exp[-q(R-r)] \times \frac{\left(1 + \frac{1}{8qr} + \frac{9}{128q^2r^2} + \cdots\right)\left(\frac{R}{r}\right)^{1/2} H}{1 + \frac{1}{qR}\left(HR - \frac{3}{8}\right) + \frac{1}{8q^2R^2}\left(HR - \frac{15}{16}\right) + \cdots}$$

$$= \frac{(t_a - t_0)a}{s\sqrt{s}} \exp[-q(R-r)] \times \left[1 + \frac{1}{qR}\left(\frac{R}{8r} + \frac{3}{8} - HR\right) + \cdots\right]\left(\frac{R}{r}\right)^{1/2} H,$$

where $q = (s/a)^{1/2}$. Using the table of transforms we find

$$\theta = \frac{t(r, \tau) - t_0}{t_a - t_0} = 2\,\mathrm{Bi}\left(\frac{R}{r}\,\mathrm{Fo}\right) \mathrm{i\,erfc}\frac{1 - \frac{r}{R}}{2(\mathrm{Fo})^{1/2}} + 4\left(\frac{R}{r}\right)^{1/2} \mathrm{Fo\,Bi}\left(\frac{R}{8r} + \frac{3}{8} - \mathrm{Bi}\right) \mathrm{i^2\,erfc}\frac{1 - \frac{r}{R}}{2(\mathrm{Fo})^{1/2}} + \cdots, \quad (6.6.29)$$

where

$$\mathrm{i\,erfc}\,u = \frac{1}{\sqrt{\pi}} \exp[-u^2] - u\,\mathrm{erfc}\,u,$$

$$\mathrm{i^2\,erfc}\,u = \frac{1}{4}\left[(1 + 2u^2)\,\mathrm{erfc}\,u - \frac{2}{\sqrt{\pi}}\,u \exp[-u^2]\right].$$

In the development into the asymptotic series, we assume that not only is the value of $(s/a)^{1/2}R$ great, but also that of $(s/a)^{1/2}r$. For small values of r, such an expansion will therefore not be valid. In this case $(r \to 0)$, the function $I_0(qR)$ may be developed into an asymptotic series and the function $I_0(qr)$ into a power series, i.e.,

$$I_0(qr) = 1 + \frac{1}{2^2} q^2r^2 + \cdots.$$

After some manipulations we obtain

$$\theta_c = 1 - 4\,\mathrm{Bi\,Fo} \exp\left[-\frac{1}{4\,\mathrm{Fo}}\right]. \quad (6.6.30)$$

Solutions (6.6.29) and (6.6.30) are valid for small values of Bi. For large values of Bi, we shall find another approximate solution for the surface temperature of the cylinder.

For the surface of the cylinder ($r = R$), solution (6.6.21) may be written

$$T(r, s) - \frac{t_0}{s} = \frac{(t_a - t_0)}{s\left[1 + \dfrac{qR}{HR}\,\dfrac{I_1(qR)}{I_0(qR)}\right]} \approx \frac{(t_a - t_0)}{s\left[1 + \dfrac{qR}{HR} - \dfrac{1}{2HR}\right]},$$

since

$$\frac{I_1(z)}{I_0(z)} = \frac{1 - \dfrac{3}{8z} + \cdots}{1 + \dfrac{1}{8z} + \dfrac{9}{128z^2} + \cdots}.$$

Using the table of transforms we find

$$\theta_s \approx \frac{\text{Bi}}{\text{Bi} - \frac{1}{2}}\,[1 - \exp[\text{Fo}(\text{Bi} - \tfrac{1}{2})^2]\,\text{erfc}(\text{Bi} - \tfrac{1}{2})(\text{Fo})^{1/2}]. \qquad (6.6.31)$$

For a sphere, solution (6.6.31) is similar to solution (6.5.41), with the exception that we use the multiplier ($\text{Bi} - \frac{1}{2}$) instead of ($\text{Bi} - 1$).

Now we may show in the same manner as in Sections 6.3 and 6.5, that approximate solutions (6.6.29)–(6.6.31) give very satisfactory results, replacing cumbersome calculations by formula (6.6.25). In Table 6.11, the

TABLE 6.11. Relative Temperature at the Cylinder Surface $1 - \theta_s = \psi(\text{Fo}, \text{Bi})$

Bi	Fourier number, Fo				
	0.0003	0.0005	0.0010	0.0025	0.0050
0.1	0.999	0.998	0.996	0.994	0.992
0.5	0.991	0.988	0.983	0.972	0.960
1	0.981	0.975	0.965	0.945	0.923
4	0.926	0.905	0.868	0.815	0.740
10	0.831	0.789	0.720	0.611	0.514
20	0.703	0.643	0.551	0.421	0.325
50	0.465	0.394	0.305	0.205	0.147
100	0.286	0.230	0.168	0.107	0.075
200	0.156	0.122	0.087	0.054	0.038
500	0.064	0.050	0.034	0.022	0.016
1000	0.032	0.025	0.017	0.011	0.008
2000	0.016	0.012	0.009	0.006	0.004

values of $1 - \theta_s$ are given (θ_s is the relative excess surface temperature of the cylinder) for small values of the Fourier numbers at various values of Bi. Table 6.11 is reproduced here from the work of Pöschel already mentioned. Approximate calculations by formulas (6.6.29) and (6.6.31) give good agreement of the results with the tabulated data.

e. Specific Heat Rate. Let us find the mean temperature of the cylinder by formula

$$\bar{t}(\tau) = \frac{2}{R^2} \int_0^R rt(r, \tau)\, dr. \tag{6.6.32}$$

in which the expression from solution (6.6.25) should be substituted for $t(r, \tau)$. Then, taking into account relation (6.6.16) we shall have

$$\bar{\theta} = \frac{\bar{t}(\tau) - t_0}{t_a - t_0} = 1 - \sum_{n=1}^{\infty} B_n \exp[-\mu_n^2 \,\mathrm{Fo}], \tag{6.6.33}$$

where

$$B_n = \frac{2 J_1(\mu_n)}{\mu_n} A_n \quad \text{or} \quad B_n = \frac{4\,\mathrm{Bi}^2}{\mu_n^2(\mu_n^2 + \mathrm{Bi}^2)}. \tag{6.6.34}$$

i.e., coefficients B_n depend on the Bi criterion. The first six coefficients B_n are contained in Table 6.12.

The series in Eq. (6.6.33) converges quite rapidly, therefore, except for very small values of the Fourier number, we shall restrict ourselves to the first term of the series. For convenience of calculation of the first approxi-

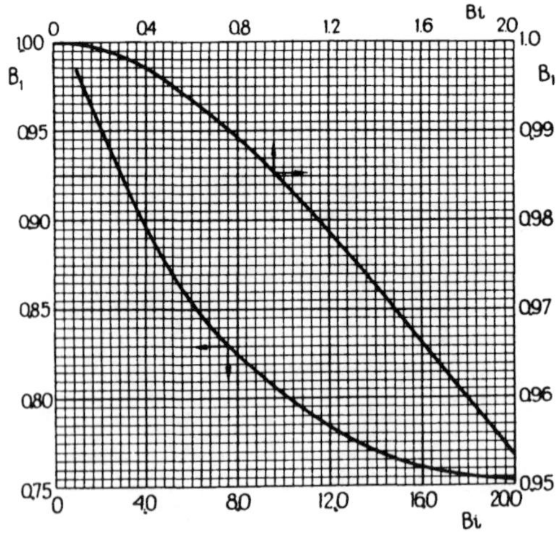

Fig. 6.26. The coefficient B_1 versus Bi for an infinite cylinder.

TABLE 6.12. THE VALUES OF THE CONSTANTS

$$B_n = \frac{4\,\mathrm{Bi}^2}{\mu_n^2(\mu_n^2 + \mathrm{Bi}^2)}$$

Bi	B_1	B_2	B_3	B_4	B_5	B_6
∞	0.6917	0.1313	0.0534	0.0288	0.0179	0.0122
50.0	0.7170	0.1309	0.0530	0.0284	0.0172	0.0113
30.0	0.7359	0.1289	0.0529	0.0268	0.0155	0.0097
10.0	0.8041	0.1260	0.0387	0.0152	0.0070	0.0035
9.0	0.8133	0.1229	0.0361	0.0137	0.0061	0.0030
8.0	0.8242	0.1187	0.0331	0.0120	0.0052	0.0025
7.0	0.8375	0.1132	0.0296	0.0103	0.0043	0.0021
6.0	0.8532	0.1057	0.0254	0.0084	0.0034	0.0016
5.0	0.8721	0.0953	0.0207	0.0064	0.0025	0.0012
4.0	0.8984	0.0813	0.0156	0.0045	0.0017	0.0008
3.0	0.9225	0.0625	0.0103	0.0028	0.0010	0.0005
2.0	0.9535	0.0388	0.0053	0.0013	0.0005	0.0002
1.5	0.9694	0.0240	0.0032	0.0008	0.0003	0.0001
1.0	0.9843	0.0136	0.0015	0.0003	0.0001	0.0001
0.90	0.9868	0.0114	0.0012	0.0003	0.0001	
0.80	0.9893	0.0093	0.0010	0.0002	0.0001	
0.70	0.9916	0.0074	0.0007	0.0002	0.0001	
0.60	0.9936	0.0056	0.0006	0.0001		
0.50	0.9955	0.0040	0.0004			
0.40	0.9970	0.0026	0.0003			
0.30	0.9983	0.0015	0.0001			
0.20	0.9992	0.0007				
0.15	0.9995	0.0004				
0.10	0.9998	0.0002				
0.08	0.9999	0.0001				

mation, curves of $B_1 = f(\mathrm{Bi})$ are given in Fig. 6.26 for the values of Bi from 0 to 20.

The specific heat rate is found by the conventional formula

$$\Delta Q_v = c\gamma[\bar{t}(\tau) - t_0]. \qquad (6.6.35)$$

6.7 Infinite Hollow Cylinder

a. Statement of the Problem. *Consider an infinite hollow cylinder with radii R_1 and R_2 with an assigned initial temperature distribution in the form of some function $f(r)$. The temperature of the cylinder depends on the radius and time.*

At the initial time moment the cylinder is put into the medium with constant temperature t_a. The heat transfer coefficients for the external and internal surfaces are different ($\alpha_1 \neq \alpha_2$). The temperature distribution in the cylinder at any moment is to be found.

The differential heat conduction equation is given in Chapter 4, Section 5. The initial and boundary conditions may be written as

$$t(r, 0) = f(r), \tag{6.7.1}$$

$$\frac{\partial t(R_1, \tau)}{\partial r} + \frac{\alpha_1}{\lambda} [t_a - t(R_1, \tau)] = 0, \tag{6.7.2}$$

$$-\frac{\partial t(R_2, \tau)}{\partial r} + \frac{\alpha_2}{\lambda} [t_a - t(R_2, \tau)] = 0. \tag{6.7.3}$$

b. Solution by Separation of Variables. First our problem of heating the cylinder is reduced to that of cooling by replacing the variable

$$\vartheta(r, \tau) = t_a - t(r, \tau). \tag{6.7.4}$$

We then find that the particular solution of the differential heat conduction equation is similar to that for a hollow infinite cylinder with boundary conditions of the first kind.

As a result of all the manipulations, the solution of our problem will be of the form

$$\vartheta(r, \tau) = t_a - t(r, \tau) = \sum_{n=1}^{\infty} E_n \exp[- a p_n{}^2 \tau] W_0(p_n, r), \tag{6.7.5}$$

where p_n are the roots of the characteristic equation

$$\begin{aligned}[(\alpha_1/\lambda) J_0(pR_1) + pJ_1(pR_1)] [(\alpha_2/\lambda) Y_0(pR_2) - pY_1(pR_2)] \\ -[(\alpha_2/\lambda) J_0(pR_2) - pJ_1(pR_2)] [(\alpha_1/\lambda) Y_0(pR_1) - pY_1(pR_1)] = 0\end{aligned} \tag{6.7.6}$$

$W_0(p_n, r)$ is the linear combination of the Bessel functions of the form

$$\begin{aligned}W_0(p_n, r) = &- [(\alpha_1/\lambda) Y_0(p_n R_1) + p_n Y_1(p_n R_1)] J_0(p_n r) \\ &+ [p_n J_1(p_n R_1) + (\alpha_1/\lambda) J_0(p_n R_1)] Y_0(p_n r).\end{aligned} \tag{6.7.7}$$

E_n are the coefficients of the expansion

$$f_1(r) = t_a - f(r) \tag{6.7.8}$$

into the series in functions $W_0(p_1, r)$, $W_0(p_2, r)$, $\cdots$,

$$f_1(r) = \sum_{n=1}^{\infty} E_n W_0(p_n, r). \tag{6.7.9}$$

The calculation shows that

$$E_n = (\pi^2 p_n^2/2)\,[(\alpha_2/\lambda)J_0(p_n R_2) - p_n J_1(p_n R_2)]^2 \int_{R_1}^{R_2} r f_1(r) W_0(p_n r)\,dr$$
$$\times \left\{\left(p_n^2 + \frac{\alpha_2^2}{\lambda^2}\right)\left[\frac{\alpha_1}{\lambda} J_0(p_n R_1) + p_n J_1(p_n R_1)\right]^2\right.$$
$$\left. - \left(p_n^2 + \frac{\alpha_1^2}{\lambda^2}\right)\left[\frac{\alpha^2}{\lambda} J_0(p_n R_2) - p_n J_1(p_n R_2)\right]^2\right\}^{-1}. \tag{6.7.10}$$

The solution may be written in another form if we designate $p_n R_1 = \mu_n$ and $\mathrm{Fo} = a\tau/R_1^2$. We suggest that the reader write the solution of the problem and the characteristic equation (6.7.6) in this form. If in solution (6.7.5) we assume $\alpha_1 = \infty$ and $\alpha_2 = 0$, we obtain the solution of the problem considered in Chapter 4.

If the initial temperature distribution is assumed uniform,

$$t(r, 0) = f(r) = t_0 = \text{const}, \tag{6.7.11}$$

then from (6.7.5), Carslaw and Jaeger's solution of the problem is obtained [8].

The given problem may be solved by the integral transform method. It will be considered below for the solution of a more general problem when the medium temperature is a function of time.

6.8 Finite Cylinder

a. Statement of the Problem. *Consider a cylinder with radius R and length 2l the temperature of which is t_0. At the initial moment, the cylinder is put into a medium with a constant temperature $t_a > t_0$. The temperature distribution at any moment under the conditions of a symmetric problem is to be found* (see Fig. 4.27a). We have

$$\frac{\partial t(r, z, \tau)}{\partial \tau} = a\left(\frac{\partial^2 t(r, z, \tau)}{\partial r^2} + \frac{1}{r}\frac{\partial t(r, z, \tau)}{\partial r} + \frac{\partial^2 t(r, z, \tau)}{\partial z^2}\right)$$
$$(\tau > 0;\quad 0 < r < R;\quad -l < z < +l), \tag{6.8.1}$$

$$t(r, z, 0) = t_0 = \text{const}, \tag{6.8.2}$$

$$-\frac{\partial t(R, z, \tau)}{\partial r} + H[t_a - t(R, z, \tau)] = 0, \tag{6.8.3}$$

$$\frac{\partial t(0, z, \tau)}{\partial r} = 0; \quad t(0, z, \tau) \neq \infty, \tag{6.8.4}$$

$$-\frac{\partial t(r, l, \tau)}{\partial z} + H[t_a - t(r, l, \tau)] = 0, \tag{6.8.5}$$

$$\frac{\partial t(r, 0, \tau)}{\partial z} = 0. \tag{6.8.6}$$

The origin of the coordinates is the center of the cylinder.

b. Solution of the Problem. Let us prove that the relative temperature $\{t_a - t(r, z, \tau)\}/(t_a - t_0)$ at any point of the cylinder is equal to

$$\frac{t_a - t(r, z, \tau)}{t_a - t_0} = \frac{t_a - t(r, \tau)}{t_a - t_0} \, \frac{t_a - t(z, \tau)}{t_a - t_0}, \tag{6.8.7}$$

where $t(r, \tau)$ and $t(z, \tau)$ are temperatures at the same point in an infinite cylinder and a plate, the intersection of which forms a cylinder of finite dimensions.

The initial and boundary conditions for an infinite cylinder and a plate remain the same as for a cylinder of finite dimensions:

$$t(r, 0) = t(z, 0) = t_0, \tag{6.8.8}$$

$$-\frac{\partial t(R, \tau)}{\partial r} + H[t_a - t(R, \tau)] = 0, \tag{6.8.9}$$

$$-\frac{\partial t(l, \tau)}{\partial z} + H[t_a - t(l, \tau)] = 0, \tag{6.8.10}$$

$$\frac{\partial t(0, \tau)}{\partial r} = \frac{\partial t(0, \tau)}{\partial z} = 0. \tag{6.8.11}$$

Let us write Equation (6.8.7) in the form

$$t(r, z, \tau) = t_a - \frac{1}{\Delta t} [t_a - t(r, \tau)] \, [t_a - t(z, \tau)], \tag{6.8.12}$$

where $\Delta t = t_a - t_0$, and substitute it into the differential equation (6.8.1). After some manipulation, we obtain

$$[t_a - t(r, \tau)] \left\{ \frac{\partial t(z, \tau)}{\partial \tau} - a \frac{\partial^2 t(z, \tau)}{\partial z^2} \right\}$$

$$+ [t_a - t(z, \tau)] \left\{ \frac{\partial t(r, \tau)}{\partial \tau} - a \left(\frac{\partial^2 t(r, \tau)}{\partial r^2} + \frac{1}{r} \frac{\partial t(r, \tau)}{\partial r} \right) \right\} = 0. \tag{6.8.13}$$

Expressions in brackets are equal to zero since $t(r, \tau)$ and $t(z, \tau)$ are the solutions of the corresponding equations. Hence, relation (6.8.12) satisfies Eq. (6.8.1).

Substituting Eq. (6.8.12) into initial condition (6.8.2) gives us

$$t(r, z, 0) = t_a - \frac{1}{\Delta t}\,[t_a - t(r, 0)]\,[t_a - t(z, 0)] = t_0. \qquad (6.8.14)$$

We shall then obtain the identity

$$t(r, z, 0) = t_a - \frac{1}{\Delta t}\,(t_a - t_0)(t_a - t_0) = t_0,$$

since

$$\Delta t = t_a - t_0.$$

Thus solution (6.8.12) conforms to the initial condition. Substituting Eq. (6.8.12) into boundary conditions (6.8.3) and (6.8.5) gives us

$$\left\{-\frac{\partial t(R, \tau)}{\partial r} + H[t_a - t(R, \tau)]\right\} \frac{t_a - t(z, \tau)}{\Delta t} = 0, \qquad (6.8.15)$$

$$\left\{-\frac{\partial t(l, \tau)}{\partial z} + H[t_a - t(l, \tau)]\right\} \frac{t_a - t(r, \tau)}{\Delta t} = 0. \qquad (6.8.16)$$

The expressions in brackets are equal to zero owing to conditions (6.8.9) and (6.8.10), hence, solution (6.8.12) satisfies boundary conditions.

Thus, solution (6.8.7) satisfies the differential equation, initial and boundary conditions and, according to the uniqueness theorem is the solution of our problem. Thus

$$\begin{aligned} 1 - \theta &= \frac{t_a - t(r, z, \tau)}{t_a - t_0} \\ &= \sum_{n=1}^{\infty} \sum_{m=1}^{\infty} A_{n,1} A_{m,2} J_0\!\left(\mu_{n,1} \frac{r}{R}\right) \cos \mu_{m,2} \frac{z}{l} \\ &\quad \times \exp\left[-\left(\frac{\mu_{n,1}^2}{R^2} + \frac{\mu_{m,1}^2}{l^2}\right) a\tau\right], \end{aligned} \qquad (6.8.17)$$

where $A_{n,1}$, $A_{m,2}$ are constant coefficients determined by the formulas

$$A_{n,1} = \frac{2\,\mathrm{Bi}_1}{J_0(\mu_{n,1})\,[\mu_{n,1}^2 + \mathrm{Bi}_1{}^2]},$$

$$A_{m,2} = (-1)^{m+1} \frac{2\,\mathrm{Bi}_2\,(\mathrm{Bi}_2{}^2 + \mu_{m,2}^2)^{1/2}}{\mu_{m,2}(\mathrm{Bi}_2{}^2 + \mathrm{Bi}_2 + \mu_{m,2}^2)},$$

where $\mu_{n,1}$, $\mu_{m,2}$ are the roots of the corresponding characteristic equations.

It should be noted that the heat transfer coefficient for the lateral surface may differ from the heat transfer coefficient for the face surface; the same may be said about the other thermal coefficients. Thus, solution (6.8.17) will be valid for an anisotropic body.

For low values of Fo, we may take the corresponding approximate relationships from the solutions for an infinite plate and an infinite cylinder, in as much as the solution of our problem consists of the product of solutions of these more simple problems.

The mean temperature of a finite cylinder is found by the formula

$$\bar{t}(\tau) = \frac{2}{R^2 l} \int_0^R \int_0^l rt(r, z, \tau)\, dr\, dz .$$

If instead of $t(r, z, \tau)$, the corresponding expression from solution (6.8.17) is substituted, upon integration we obtain

$$\bar{\theta} = 1 - \sum_{n=1}^{\infty} \sum_{m=1}^{\infty} B_{n,1} B_{m,2} \exp\left[-\left(\frac{\mu_{n,1}^2}{R^2} + \frac{\mu_{m,2}^2}{l^2} \right) a\tau \right], \tag{6.8.18}$$

where

$$B_{m,2} = \frac{2\,\mathrm{Bi}_2{}^2}{\mu_{m,2}^2(\mathrm{Bi}_2{}^2 + \mathrm{Bi}_2 + \mu_{m,2}^2)}, \qquad B_{n,1} = \frac{4\,\mathrm{Bi}_1{}^2}{\mu_{n,1}^2(\mu_{n,1}^2 + \mathrm{Bi}_1{}^2)}.$$

The specific heat rate is found by the previously developed relations.

6.9 Finite Plate

a. Statement of the Problem. *Consider a plate with dimensions $2R_1 \times 2R_2 \times 2R_3$ (a parallelepiped), the temperature of which is t_0. At the initial moment, the plate is put into medium with constant temperature $t_a > t_0$, the temperature distribution and specific heat rate at any moment are to be found.*

Let us take the origin of the coordinates at the center of the parallelepiped (see Chapter 4, Fig. 4.26), thus R_1, R_2, R_3 represent half the dimension of the plate in three directions (along the axes x, y, and z).

For such a three-dimensional problem, we have

$$\frac{\partial t(x, y, z, \tau)}{\partial \tau} = a\, \nabla^2 t(x, y, z, \tau)$$

$$(\tau > 0;\quad -R_1 < x < +R_1;\quad -R_2 < y < +R_2;\quad -R_3 < z < +R_3), \tag{6.9.1}$$

$$t(x, y, z, 0) = t_0 = \text{const}, \tag{6.9.2}$$

$$\mp \frac{\partial t(\pm R_1, y, z, \tau)}{\partial x} + H[t_a - t(\pm R_1, y, z, \tau)] = 0, \tag{6.9.3}$$

$$\mp \frac{\partial t(x, \pm R_2, z, \tau)}{\partial y} + H[t_a - t(x, \pm R_2, z, \tau)] = 0, \tag{6.9.4}$$

$$\mp \frac{\partial t(x, y, \pm R_3, \tau)}{\partial z} + H[t_a - t(x, y, \pm R_3, \tau)] = 0. \tag{6.9.5}$$

The temperature distribution is symmetrical with respect to the center of the plate.

b. Solution of the Problem. In a manner similar to that of the previous problem, we may prove that the solution of the given problem may be presented in the form of the product of solutions for three infinite plates, the intersection of which gives the parallelepiped, i.e.,

$$\frac{t_a - t(x, y, z, \tau)}{t_a - t_0} = \frac{t_a - t(x, \tau)}{t_a - t_0} \, \frac{t_a - t(y, \tau)}{t_a - t_0} \, \frac{t_a - t(z, \tau)}{t_a - t_0}, \tag{6.9.6}$$

where $t(x, \tau)$, $t(y, \tau)$, $t(z, \tau)$ are the temperatures of three infinite plates, respectively.

Solutions for $t(x, \tau)$, $t(y, \tau)$, $t(z, \tau)$ satisfy corresponding differential equations, boundary conditions (6.9.2), and boundary conditions similar to (6.9.3)–(6.9.5). The proof is also the same and is left for the reader.

As a result, the solution of our problem may be written

$$\theta = \frac{t(x, y, z, \tau) - t_0}{t_a - t_0} = 1 - \sum_{n=1}^{\infty} \sum_{m=1}^{\infty} \sum_{k=1}^{\infty} A_{n,1} A_{m,2} A_{k,3}$$
$$\times \cos \mu_{n,1} \frac{x}{R_1} \cos \mu_{m,2} \frac{y}{R_2} \cos \mu_{k,3} \frac{z}{R_3}$$
$$\times \exp\left[-\left(\frac{\mu_{n,1}^2}{R_1^2} + \frac{\mu_{m,2}^2}{R_2^2} + \frac{\mu_{k,3}^2}{R_3^2}\right) a\tau\right], \tag{6.9.7}$$

where

$$A = \frac{2 \sin \mu}{\mu + \sin \mu \cos \mu} = (-1)^{(n,m,k)+1} \frac{2\,\mathrm{Bi}(\mathrm{Bi}^2 + \mu^2)^{1/2}}{\mu\,(\mathrm{Bi}^2 + \mathrm{Bi} + \mu^2)},$$

$$\cot \mu = (1/\mathrm{Bi})\, \mu, \tag{6.9.8}$$

$$\mathrm{Bi}_i = \frac{\alpha}{\lambda} R_i \quad (i = 1, 2, 3). \tag{6.9.9}$$

If the length $2R_2$ and width $2R_3$ are great as compared to the thickness

$2R_1$ ($2R_2 = 2R_3 \to \infty$), then solution (6.9.7) will become solution (6.3.29) for an infinite plate.

For calculation of the specific heat rate, it is necessary to determine the mean temperature of the plate by the formula

$$\bar{t}(\tau) = \frac{1}{R_1 R_2 R_3} \int_0^{R_1} \int_0^{R_2} \int_0^{R_3} t(x, y, z, \tau)\, dx\, dy\, dz. \tag{6.9.10}$$

Substitution of the corresponding expression from solution (6.9.7) instead of $t(x, y, z, \tau)$ gives

$$\bar{\theta} = 1 - \sum_{n=1}^{\infty} \sum_{m=1}^{\infty} \sum_{k=1}^{\infty} B_{n,1} B_{m,2} B_{k,3} \exp\left[-\left(\frac{\mu_{n,1}^2}{R_1^2} + \frac{\mu_{m,2}^2}{R_2^2} + \frac{\mu_{k,3}^2}{R_3^2}\right) a\tau\right], \tag{6.9.11}$$

where

$$B = \frac{2\,\mathrm{Bi}^2}{\mu^2(\mathrm{Bi}^2 + \mathrm{Bi} + \mu^2)}. \tag{6.9.12}$$

For practical computations we may use the appropriate graphs for $\mu_{1,i} = f(\mathrm{Bi}_i)$, $A_{1,i} = f(\mathrm{Bi}_i)$ and $\mathrm{Bi}_{1,i} = f(\mathrm{Bi}_i)$ where $i = 1, 2, 3$. For more accurate calculations, the values of μ, A, and B should be taken from the corresponding tables.

For low values of Fo, the corresponding approximate solutions for an infinite plate may be taken (see Section 6.3).

Solution (6.9.7) is also valid in the case when in all three directions, x, y, z, the thermal coefficients will be different, i.e., in the case of an anisotropic body.

6.10 Analysis of the Generalized Solution

Let us analyze the solutions of the problems considered. For bodies of the simplest geometry (a plate, sphere, cylinder, parallelepiped) the solution of the problem of heating in a medium with constant temperature (boundary condition of the third kind) may be written as

$$\frac{t_a - t}{t_a - t_0} = \sum_{n=1}^{\infty} \prod_{i=1}^{3} A_{n,i} \Phi\left(\mu_{n,i} \frac{x_i}{R_i}\right) \exp\left[-\left(\mu_{n,i}^2 \frac{R_v^2}{R_i^2}\right) \mathrm{Fo}_v\right], \tag{6.10.1}$$

where $A_{n,i} = (A_{n,1} A_{n,2} A_{n,3})$ are initial thermal amplitudes dependent on the initial temperature distribution and the geometry of the body, $\Phi\{\mu_{n,i}(x_i/R_i)\}$, is the function accounting for the change of temperature along the coor, dinates ($x = x_1$, $y = x_2$, $z = x_3$), R_1, R_2, R_3 are the body dimensions-

R_v is the generalized dimension of a body equal to the ratio of body volume V to its surface S ($R_v = V/S$). (For an infinite plate $R_v = R$, for an infinite cylinder $R_v = \frac{1}{2}R$, for a sphere $R_v = \frac{1}{3}R$, and $\mu_{n,i}$ are the roots of the characteristic equations.) It should be noted that

$$\mu_{1,i} < \mu_{2,i} < \mu_{3,i} < \cdots < \mu_{n,i}. \tag{6.10.2}$$

$\text{Fo}_v = a\tau/R_v^2$ is the Fourier number, in which a generalized dimension is taken as the characteristic dimension of the body.

Owing to inequality (6.10.2), each succeeding root, with the exception of small values of Fo_v, will be vanishingly small as compared to the previous one, and the sum of all the roots will differ only negligibly from the value of the first term. Therefore, beginning from a certain value of the Fourier number Fo_1 we may restrict ourselves to the first term, i.e.,

$$\frac{t_a - t}{t_a - t_0} = \prod_{i=1}^{3} A_{1,i}\Phi\left(\mu_{1,i}\frac{x_i}{R_i}\right)\exp\left[-\left(\mu_{1,i}^2\frac{R_v^2}{R_i^2}\right)\text{Fo}_v\right]^3 \quad \text{at} \quad \text{Fo}_v > \text{Fo}_1. \tag{6.10.3}$$

Beginning from this value of Fo_1, the relationship between $(t_a - t)$ and time τ will be described by a simple exponent.

Taking a logarithm of (6.10.3) we obtain

$$\ln\frac{(t_a - t)}{t_a - t_0} = \sum_{i=1}^{3}\left\{\ln A_{1,i}\Phi\left(\mu_{1,i}\frac{x_i}{R_i}\right) - \left(\mu_{1,i}\frac{R_v}{R_i}\right)^2\frac{a\tau}{R_v^2}\right\}. \tag{6.10.4}$$

Thus, a graph of $\ln(t_a - t)$ versus time will have the form of a straight line. With long time heating ($\text{Fo}_v \to \infty$) the temperature at all points of the body is the same and equal to t_a (steady state).

Consequently, the whole process of heating may be divided into three stages. In the first stage, the main role is played by the initial temperature distribution. Any irregularity in the initial distribution affects the temperature distribution in the following moments. The relation between $(t_a - t)$ *and* τ *is described by series* (6.10.1). *The second stage is referred to as the regular regime. The relation between* $(t_a - t)$ *and* τ *is described by a simple exponent* (Fig. 6.27). *The temperature distribution inside the body is described by the function* Φ *and does not depend on the initial distribution, since the values of* $A_{1,i}$ *enter as multipliers, i.e., they determine the scale but not the essence of the phenomenon. The third stage corresponds to the steady state* ($\text{Fo}_v = \infty$) *at which temperature at all the points is equal to the ambient temperature.*

In Fig. 6.27, graphs are given of $\ln(t_a - t)$ versus τ for the surface and the center of the body. It is seen from this figure that in the stage of the regular regime these graphs have the form of a straight line. If at the initial moment the temperature at all the points is the same and equal to t_0, then the curves should initiate from one point.

As the surface layers are heated more rapidly than the central ones, in the first stage for the central layer the curve $\ln(t_a - t) = f(\tau)$ approaches a tangent to the ordinate axis, and for the surface layers to the abscissa axis (see Fig. 6.27).

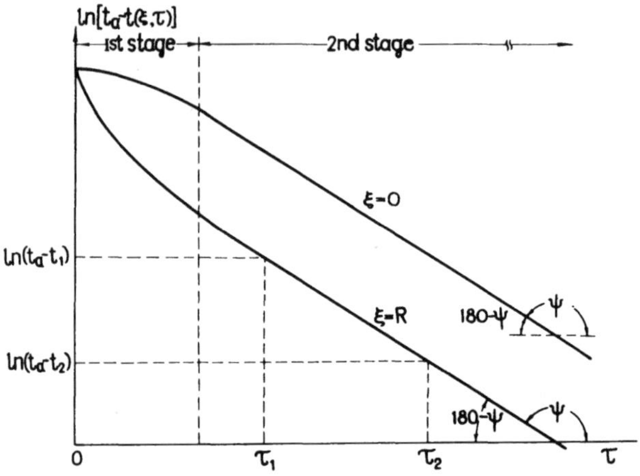

Fig. 6.27. The temperature difference logarithm as a function of time in heating process.

The foregoing analysis is valid for bodies of any form. Temkin [114] has shown that the problem of heating a body of a complex shape may be reduced to the problem of heating a simple shape (a plate, cylinder, sphere) by introduction of an approximate similarity number.

The slope of the straight line in the regular regime stage will be

$$\tan(180 - \psi) = -\tan\psi = \frac{\ln(t_a - t_1) - \ln(t_a - t_2)}{\tau_2 - \tau_1} = m = \text{const.} \tag{6.10.5}$$

The constant m is the rate of change of the logarithm of the excess temperature with time, i.e.,

$$-\frac{\partial[\ln(t_a - t)]}{\partial\tau} = m. \tag{6.10.6}$$

It is the same for all the points of the body as well as for the average over the volume temperature ($\bar{t}$) and is called the rate of heating or cooling of a body. From Eq. (6.10.4) we have

$$m = \sum_{i=1}^{3} \left(\mu_{1,i} \frac{R_v}{R_i}\right)^2 \frac{a}{R_v^2}. \tag{6.10.7}$$

Consequently, the numerical value of m is determined by the thermal coefficients, dimensions, and the form of the body.

On the basis of equality (6.10.6) and the boundary condition of the third kind, for a body of any form, the following equality holds which is valid in the regular regime stage of heating

$$c\gamma V \frac{d\bar{t}}{d\tau} = \alpha S(t_a - t_s) = c\gamma V(t_a - \bar{t})m. \tag{6.10.8}$$

Thence we obtain

$$m = \frac{\alpha S}{c\gamma V} \frac{t_a - t_s}{t_a - \bar{t}} = \frac{\alpha a}{\lambda R_v} \Psi = \frac{a}{R_v^2} \mathrm{Kn} \tag{6.10.9}$$

where $\mathrm{Kn} = \mathrm{Bi}_v \Psi = (\alpha/\lambda)R_v \Psi$ is the Kondratiev criterion[1] and

$$\Psi = (t_a - t_s)/(t_a - \bar{t}) \tag{6.10.10}$$

is the parametric number, characterizing the nonuniformity of the temperature field as it is equal to the ratio of the actual excess temperature of the surface of the body and the volume average excess temperature. If the temperature distribution in the body is uniform ($\mathrm{Bi}_v \to 0$) then $\Psi = 1$. The less uniform the temperature, the lower the value of Ψ. At $\Psi = 0$ the nonuniformity of the temperature distribution is highest ($\mathrm{Bi}_v \to \infty$ and $t_s \to t_a$).

Thus the Kondratiev number not only characterizes the uniformity of the temperature field but also the intensity of interaction of the body surface with the surrounding medium.

If $\mathrm{Bi}_v \to 0$ (this condition is practically attained if $\mathrm{Bi}_v < 0.1$), then $t_s \to \bar{t}$ ($\Psi = 1$). Consequently, the Kondratiev criterion will be equal to

$$(\mathrm{Kn})_0 = \mathrm{Bi}_v = \sum_{i=1}^{3} (\mu_{1,i}^2)_0 \left(\frac{R_v}{R_i}\right)^2, \tag{6.10.11}$$

[1] The criterion $\mathrm{Bi}_v \Psi$ is the basic value defining the character of heat exchange of a body in the regular regime and is named after the outstanding thermal physicist G. M. Kondratiev who was the first to study the laws of cooling a body at this stage. In contrast to the Kossovich criterion (Ko) the criterion $\mathrm{Bi}_v \Psi$ is designated through Kn.

i.e., the Kondratiev criterion is equal to the Biot criterion, where the generalized dimension is taken as the characteristic dimension.

In this case the heating parameter m is equal to

$$(m)_0 = \frac{a}{R_v^2} (\text{Kn})_0 = \frac{\alpha}{c\gamma R_v}. \tag{6.10.12}$$

The value of μ_1 as functions of Bi for the characteristic equations of the main bodies is given in the corresponding tables (for example, Tables 6.1, 6.5, 6.9). We may also find an approximate formula for calculation of μ_1 depending on the Bi criterion (see below).

From (6.10.9) an important relation of the regular regime theory is obtained:

$$\text{Kn} = \Psi \, \text{Bi}_v = \frac{mR_v^2}{a} = \sum_{i=1}^{3} \left(\mu_{1,i} \frac{R_v}{R_i}\right)^2. \tag{6.10.13}$$

Thus, the number Kn is characterized by the shape of the body and its eigenvalues $\mu_{1,1}$, $\mu_{1,2}$, $\mu_{1,3}$, and consequently, by the Biot criterion, as the eigenvalues are the functions of the Biot criterion.

The curves $\text{Kn} = f(\text{Bi}_v)$ for quite different geometries (sphere, parallelepiped, cylinder, etc.) turned out to be so close to each other that prac-

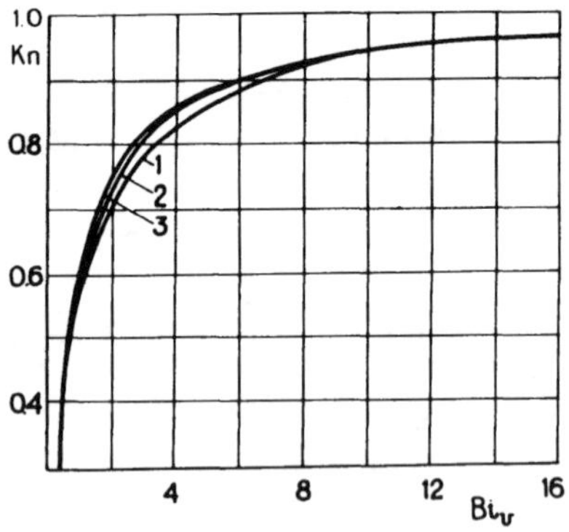

Fig. 6.28. Universal approximate relation $\text{Kn} = f(\text{Bi}_v)$. 1: plate; 2: sphere; 3: cylinder.

tically the entire family of curves may be replaced by one averaged curve (Fig. 6.28). Analytically this is expressed by Yaryshev's relation

$$\Psi = \frac{\text{Kn}}{\text{Bi}_v} = \frac{1}{(\text{Bi}_v^2 + 1.437\,\text{Bi}_v + 1)^{1/2}}.$$

If $\mathrm{Bi}_v \to 0$ (practically, $\mathrm{Bi}_v < 0.1$), $t_s \to t$ ($\Psi = 1$). Hence the Kondratiev criterion will be equal to the Biot criterion $(\mathrm{Kn})_0 = \mathrm{Bi}_v$ and the rate of heating $m = \alpha/c\gamma R_v$, i.e., the Newtonian law of heating is obtained. If $\mathrm{Bi}_v \to \infty$ (practically, $\mathrm{Bi}_v > 100$), the Kondratiev criterion will be a constant quantity $(\mathrm{Kn})_\infty = \sum_{i=1}(\mu_{1,i}^2)_\infty (R_v^2/R_i^2) = \mathrm{const}$. In this case, the rate of heating will be directly proportional to thermal diffusivity (the first Kondratiev theorem) $(m)_\infty = a/R_v^2 \, (\mathrm{Kn})_\infty$. Thus the criterion Kn lies in the limits between zero and some constant value $(\mathrm{Kn})_\infty$, which is characterized by the shape of the body. Kondratiev suggested that the character of the heating kinetics just considered be named the regular regime of the first kind.[2] Its first development is given in Figs. 6.29 and 6.30 for the center of a plate and a cylinder as a function of Fo, Bi_v, the dimensionless temperature ratio of the center $t(0, \mathrm{Fo}) = (t_a - t_c)/(t_a - t_0)$ and the permissible calculation error ε. It is seen from these figures that at small values of the

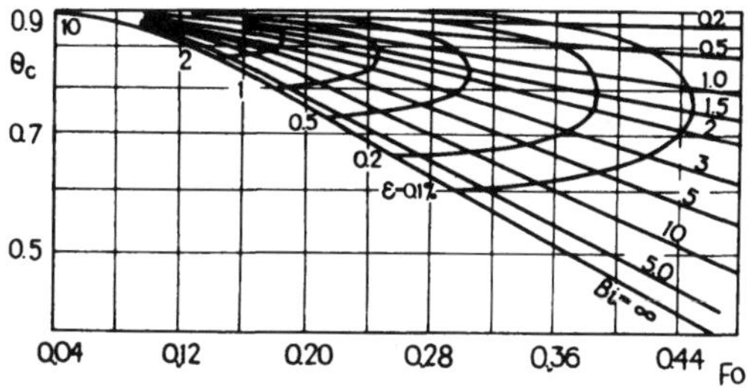

Fig. 6.29. The dependence of regular thermal regime onset on Fo, Bi and allowable calculation error ε (%) for the plate center.

[2] There is no relation between the boundary conditions of the first, second, and third kind, and the regular regime of the first, second, and third kind. Boundary conditions are determined independently, and regular regime introduced by G. M. Kondratiev describes the change of the temperature field in time.

When the temperature at any point of the body changes with time according to the exponential law, such a regime of cooling or heating is referred to as the regime of the first kind. This is usually the situation when a body is cooled with conditions of the first or the third kind. It is important that the medium temperature (boundary conditions of the third kind) or the surface temperature (boundary conditions of the first kind) would remain constant. In case of the regular regime of the second kind, the temperature at any point of a body changes similarly with the medium temperature. In case of the regular regime of the second kind, if the medium temperature increases according to the linear law, the temperature at any point of the body also increases following the linear law; or when the surface temperature increases according to the linear law (boundary conditions of the first kind), the temperature at all points of the body will also increase following the linear law.

Biot criterion the central points of the plate and cylinder enter the regular regime stage rapidly enough. With an increase of the Biot number the relative time to reach the regular regime rises, reaches maximum at Bi ≈ 2 and with further increase of Bi somewhat decreases again. With equal ε and Bi, the onset of the regular regime comes earlier on the cylinder axis than in the center of the plate.

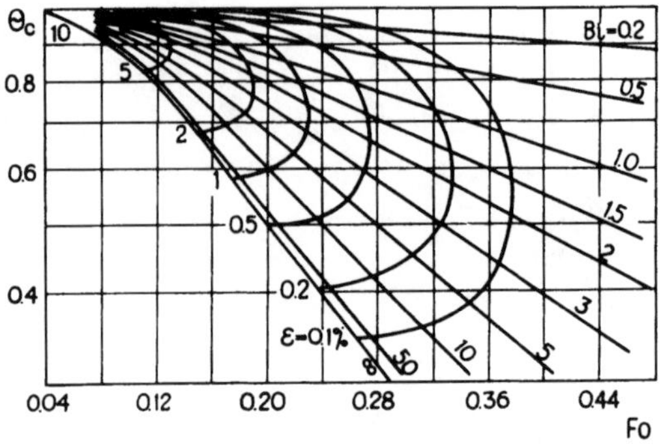

Fig. 6.30. The dependence of regular thermal regime onset for the axis of an infinite cylinder θ_c on Fo, Bi and allowable calculation error ε (%).

In Fig. 6.31 are given the graphs of $\ln[\{(\mu_1^2)_\infty/\mu_1^2\} - 1]$ versus ln Bi for the main geometries. It is seen from Fig. 6.31 that the points are nicely distributed on a straight line. Hence we may write

$$\frac{(\mu_1^2)_\infty}{\mu_1^2} - 1 = \frac{A}{\mathrm{Bi}^k} \tag{6.10.14}$$

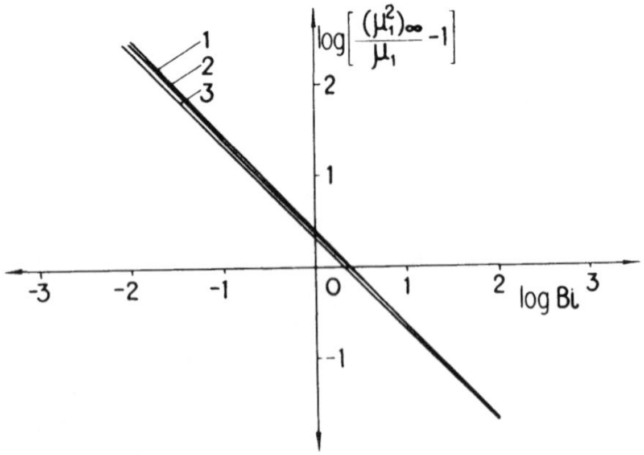

Fig. 6.31. The dimensionless quantity $[\{(\mu_1^2)_\infty/\mu_1^2\} - 1]$ versus Bi. 1: sphere, 2: cylinder, 3: plate.

or

$$\mu_1^2 = (\mu_1^2)_\infty \left[\frac{1}{1 + A/\mathrm{Bi}^k}\right], \tag{6.10.15}$$

where $(\mu_1)_\infty$ is the value of μ_1 at Bi $= \infty$ (eigenvalues $(\mu_1)_\infty$ are constant; for example, for a sphere $(\mu_1)_\infty = \pi$, for a cylinder $(\mu_1)_\infty = 2.405$, and so on).

From Fig. 6.31 the constants A and k may be found. For an infinite plate $A = 2.24$, $k = 1.02$; for an infinite cylinder $A = 2.45$, $k = 1.04$; for a sphere $A = 2.70$, $k = 1.07$.

Thus, making use of relation (6.10.15) we may calculate the Kondratiev criterion by formulas (6.10.11) and (6.10.13) for any body.

The basic relation of the regular regime

$$t_a - t = \prod_{i=1}^{3} A_{1,i}\Phi\{\mu_{1,i}(x_i/R_i)\} \exp[- \mathrm{Kn}\ \mathrm{Fo}_v] \tag{6.10.16}$$

was used by G. N. Kondratiev and his co-workers as the basis for developing new methods of determining thermal coefficients of various materials. However, application of (6.10.16) for boundary conditions of the third kind (Bi $\neq \infty$) to determine thermal conductivity and thermal diffusivity is very difficult.

6.11 Estimation of Approximation

Application of the solutions obtained in Sections 6.3–6.8 to practical calculations requires estimation of the accuracy of the approximate relations when some terms of the series are discarded from the solution. It is especially necessary for calculations in the case of small values of the Fo number, when a considerable number of terms of the series should be taken. Here we shall give a concrete example of the method of estimation.

Let us take solution (6.3.29) for an infinite plate and assume that $x = R$ (surface temperature of the plate), i.e.,

$$1 - \theta_s = \sum_{n=1}^{\infty} A_n \cos \mu_n \exp[- \mu_n^2\ \mathrm{Fo}] \tag{6.11.1}$$

where

$$A_n = (-1)^{n+1} \frac{2\ \mathrm{Bi}(\mathrm{Bi}^2 + \mu_n^2)^{1/2}}{\mu_n(\mathrm{Bi}^2 + \mathrm{Bi} + \mu_n^2)} \tag{6.11.2}$$

and μ_n are the roots of the characteristic equation

$$\tan \mu = \mathrm{Bi}/\mu. \tag{6.11.3}$$

Let us substitute into solution (6.12.1) a corresponding expression for $\cos \mu_n$ using the characteristic equation:

$$|\cos \mu_n| = (1 + \tan^2 \mu_n)^{-1/2} = \{1 + (\mathrm{Bi}^2/\mu_n^2)\}^{-1/2} = \mu_n(\mu_n^2 + \mathrm{Bi}^2)^{-1/2}.$$

Consequently, we obtain, (after taking into account that $\cos \mu_n = (-1)^{n+1} |\cos \mu_n|$),

$$1 - \theta_s = 2\mathrm{Bi} \sum_{n=1}^{\infty} \frac{1}{\mu_n^2 + \mathrm{Bi}(1 + \mathrm{Bi})} \exp[-\mu_n^2 \mathrm{Fo}]. \qquad (6.11.4)$$

We designate the remaining terms of the series (6.12.4) by φ_k as

$$\varphi_k = 2\mathrm{Bi} \sum_{n=k+1}^{\infty} \frac{1}{\mu_n^2 + \mathrm{Bi}(1 + \mathrm{Bi})} \exp[-\mu_n^2 \mathrm{Fo}]. \qquad (6.11.5)$$

From characteristic equation (6.12.3), it is seen that each value of μ_n differs from the previous one by somewhat less than π (see Fig. 6.32). At $\mathrm{Bi} \to 0$ each value of μ_n is larger than the previous one by exactly π, and at $\mathrm{Bi} \to \infty$,

$$\mu_n = (2n - 1)\tfrac{1}{2}\pi.$$

In Fig. 6.32 the form of the curve φ_k is shown in dependence on μ_n. From the curve of Fig. 6.32 it follows that

$$\varphi_k < \frac{2\mathrm{Bi}}{\pi} \int_{k+1/2}^{\infty} \frac{1}{\mu^2 + (1 + \mathrm{Bi})\,\mathrm{Bi}} \exp[-\mu^2 \mathrm{Fo}]\, d\mu. \qquad (6.11.6)$$

Since the area a is greater than the area b we may write

$$\frac{1}{\pi} \int_{k+1/2}^{k+3/2} (\sim) > \varphi_{k+1},$$

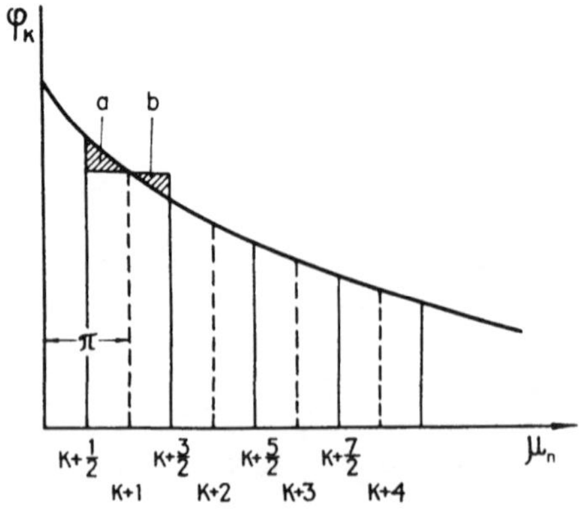

Fig. 6.32. Estimations of the solution approximation.

where φ_k is the term of series (6.11.5)

$$\int_{k+1/2}^{\infty}(\sim) = \int_{k+1/2}^{k+3/2}(\sim) + \int_{k+3/2}^{k+5/2}(\sim) + \cdots,$$

Designating $\beta = (1 + \mathrm{Bi})\,\mathrm{Bi}$, we obtain:

$$\varphi_k < \frac{2\mathrm{Bi}}{\pi}\int_{\mu_m}^{\infty}\frac{d\mu}{\mu^2+\beta}\exp[-\mu^2\,\mathrm{Fo}]$$

$$= \frac{2\mathrm{Bi}}{\pi}\int_{\mu_m^2+\beta}^{\infty}\frac{1}{2\nu(\nu-\beta)^{1/2}}\exp[-\nu\,\mathrm{Fo} + \beta\,\mathrm{Fo}]\,d\nu, \quad (6.11.7)$$

where

$$\mu^2 + \beta = \nu.$$

The inequality (6.11.7) may be increased as

$$\varphi_k < \frac{2\,\mathrm{Bi}}{2\pi(\nu-\beta)^{1/2}}\exp[\beta\,\mathrm{Fo}]\int_{\mu_m^2+\beta}^{\infty}\frac{1}{\nu}\exp[-\nu\,\mathrm{Fo}]\,d\nu. \quad (6.11.8)$$

Finally, we obtain

$$\varphi_k < \frac{\mathrm{Bi}}{\pi\mu_{k+1/2}}e^{\beta\mathrm{Fo}}\int_{\mathrm{Fo}(\mu_{k+1/2}^2+\beta)}^{\infty}\frac{e^{-u}}{u}\,du. \quad (6.11.9)$$

The value of the integral in expression (6.12.9) may be taken from corresponding tables.

Example. It is necessary to find the limiting value of the Fo number such that the whole series of (6.11.4) with the exception of the first term can be neglected when the required accuracy is 0.25%.

Let Bi = 1.00; then $\mu_1 = 0.86$, $\mu_2 = 3.426$ (see Table 6.1). In the present case, $k = 1$; consequently, $\mu_{3/2} = 2.14$ ($\mu_{3/2}$ is found as the arithmetic mean of μ_1 and μ_2), and $\beta = 2$. We have

$$0.0025 \le \frac{1}{2.14\,\pi}\exp[2\mathrm{Fo}]\int_{6.6\mathrm{Fo}}^{\infty}(e^{-u}/u)\,du, \quad (6.11.10)$$

$$0.01685\exp(-2\mathrm{Fo}) \le \int_{6.6\mathrm{Fo}}^{\infty}(e^{-u}/u)\,du = -\mathrm{Ei}(-6.6\,\mathrm{Fo}), \quad (6.11.11)$$

where

$$\mathrm{Ei}(-z) = -\int_{z}^{\infty}\frac{e^{-z}}{z}\,dz. \quad (6.11.12)$$

Assigning various values of Fo, we may calculate separately the right- and left-hand sides of inequality (6.11.11). As a result, we find a value of Fo for which both the left- and right-hand sides of this inequality give the same numerical values.

For the present case, we find: $\mathrm{Fo} \geq 0.55$. A more accurate solution may be obtained graphically. Thus, starting from $\mathrm{Fo} \geq 0.55$, we may restrict ourselves (with 0.25% accuracy) to one term of the whole series (6.11.1).

Pöschl gives the following example: in order to calculate the value of θ_s for $\mathrm{Fo} = 0.0003$ accurate to three decimal places it is necessary to take 36 terms of series (6.11.1).

In a similar way, we may show that the estimation of the series

$$\varphi_n = \sum_{n=1}^{\infty} (1/\mu_n^2) \exp[-\mu_n^2 \mathrm{Fo}] \tag{6.11.13}$$

may be expressed with the help of the following relation:

$$\varphi_k \leq 1/2\pi\mu_{k+1/2} \int_{\mathrm{Fo}\mu_{k+1/2}^2}^{\infty} (e^{-u}/u)\, du. \tag{6.11.14}$$

Inequality (6.11.14) makes it possible for us to obtain the following estimation of the series: within 0.25% accuracy, we may neglect all the terms of series (6.11.13) if $\mathrm{Fo} \geq 1.0$ (in the calculation we assume $k = 0$, $\mu_1 = \frac{1}{2}\pi$).

Consider the method of successive approximations to calculate the roots of the characteristic equation. In the first rough approximation, we find graphically the roots of the characteristic equation, e.g., for a cylinder as

$$\frac{J_0(\mu)}{J_1(\mu)} = \frac{1}{\mathrm{Bi}} \mu. \tag{6.11.15}$$

Taking a logarithm of Eq. (6.11.15)

$$\ln |J_1(\mu)| + \ln \mu - \ln \mathrm{Bi} = \ln |J_0(\mu)|. \tag{6.11.16}$$

We let the value of the first root be equal to μ_1^0 and substitute this into the left-hand side of Eq. (6.11.16). Using the left-hand side thus calculated, we find μ_1' from the right-hand side by formula

$$\ln |J_1(\mu_1^0)| + \ln \mu_1^0 - \ln \mathrm{Bi} = \ln |J_0(\mu_1')|.$$

We then substitute μ_1' into the left-hand side of Eq. (6.11.16) and from the right-hand side we find μ_1'' by formula

$$\ln |J_1(\mu_1')| + \ln \mu_1' - \ln \mathrm{Bi} = \ln |J_0(\mu_1'')|.$$

This process is followed from equation

$$\ln |J_1(\mu_1^{(n)})| + \ln \mu_1^{(n)} - \ln \mathrm{Bi} = \ln |J_0(\mu_1^{(n+1)})|$$

until we find

$$\mu_1^{(n+1)} = \mu_1^{(n)}$$

to the limits of the prescribed accuracy. The same calculations should also be fulfilled for the remaining roots.

CHAPTER

7

TEMPERATURE FIELDS WITHOUT HEAT SOURCES WITH VARIABLE TEMPERATURE OF THE SURROUNDING MEDIUM

In this chapter, those problems are considered in which the temperature of the body surface is a prescribed function of time. To generalize, we shall first take the case when the medium temperature changes with time according to the following law $t_a = f(\tau)$. Then, assuming that the Biot criterion is infinitely large ($\text{Bi} \to \infty$), we obtain the solution of the problem for the case where the temperature of the body surface varies with time ($t_s = f(\tau)$). Consequently, the solutions of this chapter may be considered a generalization of those of Chapter 4, since the problems considered therein are particular cases of the present problems.

First, problems with the most simple law of ambient temperature change (linear law) are considered; then those with more complicated laws. Here the problems of temperature waves are considered as well. The chapter ends with some generalization and the derivation of the Duhamel theorem by the operational method. In contrast to the order adopted in previous chapters, we consider here first the problem of an infinite plate, sphere, and cylinder. The problem of a semi-infinite body is considered in Section 7.7.

7.1 Infinite Plate. Ambient Temperature as a Linear Function of Time

a. Statement of the Problem. *Consider a plate* $2R$ *thick which is in thermal equilibrium with the surrounding medium, i.e., it has a temperature equal to the temperature of the surrounding medium* t_0. *At the initial time the medium* *is heated with the constant rate b* (deg/hr), *i.e., the ambient temperature is a linear function of time*: $t_a(\tau) = t_0 + b\tau$. *Heat is transferred between the surface of the plate and the surrounding medium by the Newton law. The temperature distribution over the thickness of the plate at any time as well as the specific heat rate is to be found.*

The boundary and initial conditions may be written as

$$t(x, 0) = t_0 = \text{const}\,, \tag{7.1.1}$$

$$\frac{\partial t(0, \tau)}{\partial x} = 0\,, \tag{7.1.2}$$

$$-\frac{\partial t(R, \tau)}{\partial x} + H[(t_0 + b\tau) - t(R, \tau)] = 0\,. \tag{7.1.3}$$

b. Solution of the Problem. We shall obtain the solution of the problem by the operational method. The solution for the transform $T(x, s)$ of the differential heat conduction equation for the case of an infinite plate under conditions (7.1.1) and (7.1.2) has the form (Chapter 6, Section 3).

$$T(x, s) - \frac{t_0}{s} = A\cosh\left(\frac{s}{a}\right)^{1/2} x\,. \tag{7.1.4}$$

The constant A is found from boundary condition (7.1.3), which for the transform will have the form

$$-T'(R, s) + \frac{Ht_0}{s} + \frac{Hb}{s^2} - HT(R, s) = 0\,, \tag{7.1.5}$$

since $L[Hb\tau] = Hb/s^2$. Substituting boundary condition (7.1.5) into solution (7.1.4) gives us

$$-A\left(\frac{s}{a}\right)^{1/2}\sinh\left(\frac{s}{a}\right)^{1/2} R + \frac{Ht_0}{s} + \frac{Hb}{s^2} - \frac{Ht_0}{s} - AH\cosh\left(\frac{s}{a}\right)^{1/2} R = 0\,. \tag{7.1.6}$$

Let us determine the constant A from equality (7.1.6) and substitute the expression obtained into solution (7.1.4)

$$T(x, s) - \frac{t_0}{s} = \frac{b\cosh\left(\frac{s}{a}\right)^{1/2} x}{s^2\left(\cosh\left(\frac{s}{a}\right)^{1/2} R + \frac{1}{H}\left(\frac{s}{a}\right)^{1/2}\sinh\left(\frac{s}{a}\right)^{1/2} R\right)}$$

$$= \frac{\Phi(s)}{\psi(s)} = \frac{\Phi(s)}{s^2\varphi(s)}\,. \tag{7.1.7}$$

where $\varphi(s)$ is the expression in brackets in the denominator.

As was shown in Chapter 6, Section 3, $\Phi(s)$ and $\psi(s)$ are generalized polynomials with respect to s. Since the polynomial $\psi(s)$ does not contain the constant, the condition of the expansion theorem is fulfilled. Equating $\psi(s) = s^2\varphi(s)$ to zero, we find the roots (see Chapter 6, Section 3) $s_0 = 0$ (double root) and $s_n = -a\mu_n^2/R^2$, where $i(s/a)^{1/2}R = \mu$ are simple roots, determined from the characteristic equation

$$\cot \mu = \frac{1}{\mathrm{Bi}}\,\mu .$$

Thus, we find the original function in the same way as in Chapter 6, Section 3 with the exception of the zero root.

Since the zero root is double, we apply the expansion theorem for multiple roots (in the present case $k = 2$),

$$\begin{aligned} L^{-1}\left[\frac{\Phi(0)}{\psi(0)}\right] &= \lim_{s\to 0}\left[\frac{d}{ds}\left(\frac{\Phi(s)s^2}{\psi(s)}\,e^{s\tau}\right)\right] \\ &= \lim_{s\to 0}\left[\frac{d}{ds}\left(\frac{\Phi(s)}{\varphi(s)}\,e^{s\tau}\right)\right] \\ &= \lim_{s\to 0}\left[\frac{\tau e^{s\tau}\Phi(s)}{\varphi(s)} + \frac{e^{s\tau}\Phi'(s)}{\varphi(s)} - \frac{e^{s\tau}\Phi(s)\varphi'(s)}{[\varphi(s)]^2}\right] \\ &= b\tau + \frac{bx^2}{2a} - \frac{bR^2}{2a} - \frac{bR^2}{2aHR} - \frac{bR^2}{2aHR} \\ &= b\tau + \frac{b}{2a}\left[x^2 - R^2\left(1 + \frac{2}{HR}\right)\right], \end{aligned}$$

since

$$\Phi(0) = b, \qquad \varphi(0) = 1, \qquad \Phi'(0) = \lim_{s\to 0}\frac{bx\sinh(s/a)^{1/2}x}{2(as)^{1/2}} = \frac{bx^2}{2a};$$

$$\varphi'(s) = \frac{R\sinh(s/a)^{1/2}R}{2R(as)^{1/2}} + \frac{\sinh(s/a)^{1/2}R}{2H(as)^{1/2}} + \frac{R^2}{2aHR}\cosh(s/a)^{1/2}R.$$

Substituting the remaining roots we find the result in the usual way as

$$\begin{aligned} &\sum_{n=1}^{\infty}\frac{\Phi(s)}{\psi'(s_n)}\,e^{s_n\tau} \\ &= -\frac{bR^2}{a}\sum_{n=1}^{\infty}\frac{2\sin\mu_n}{(\mu_n + \sin\mu_n\cos\mu_n)\mu_n^2}\cos\mu_n\frac{x}{R}\exp\left[-\mu_n^2\frac{a\tau}{R^2}\right]. \end{aligned}$$

Thus the solution of our problem will be

$$t(x, \tau) = t_0 + b\tau - \frac{b}{2a}\left[R^2\left(1 + \frac{2}{\mathrm{Bi}}\right) - x^2\right]$$
$$+ \frac{bR^2}{a}\sum_{n=1}^{\infty}\frac{A_n}{\mu_n^2}\cos\mu_n\frac{x}{R}\exp[-\mu_n^2\,\mathrm{Fo}], \quad (7.1.8)$$

where A_n are the initial thermal amplitudes determined by relation (6.3.30).

c. Analysis of the Solution. Let us introduce the criterion characterizing the rate of ambient temperature rise as

$$\mathrm{Pd} = \left(\frac{d\theta_a}{d\mathrm{Fo}}\right)_{\max} \quad (7.1.9)$$

and name it the Predvoditelev criterion (Pd).

In the present case, the rate of heating $dt_a(\tau)/d\tau$ is constant during the whole process and is equal to b since $t_a(\tau) = t_0 + b\tau$. Hence, the Predvoditelev criterion is equal to

$$\mathrm{Pd} = \frac{R^2}{a}\frac{d}{d\tau}\left(\frac{t_a(\tau)}{t_0}\right) = \frac{bR^2}{at_0}. \quad (7.1.10)$$

The dimensionless ambient temperature θ_a was taken with respect to the initial temperature; it may also be taken with respect to some average temperature or to the final one if the latter is known. Thus the solution of our problem will be of the form

$$\theta = \frac{t(x, \tau) - t_0}{t_0} = \mathrm{Pd}\left\{\mathrm{Fo} - \frac{1}{2}\left[\left(1 + \frac{2}{\mathrm{Bi}}\right) - \frac{x^2}{R^2}\right]\right.$$
$$\left. + \sum_{n=1}^{\infty}\frac{A_n}{\mu_n^2}\cos\mu_n\frac{x}{R}\exp[-\mu_n^2\,\mathrm{Fo}]\right\}. \quad (7.1.11)$$

It is seen from (7.1.11) that the dimensionless temperature at any point of the plate is directly proportional to the Predvoditelev criterion, and, hence, to the rate of medium heating. The dimensionless value θ/Pd is a function of Bi, Fo, and x/R

$$\theta/\mathrm{Pd} = \Psi\{\mathrm{Fo}, \mathrm{Bi}, x/R\}. \quad (7.1.12)$$

It is seen from Eq. (7.1.11) that the series converges rapidly; with time its contribution becomes increasingly smaller and beginning from a definite value $\mathrm{Fo} > \mathrm{Fo}_1$, it may be neglected. Then the temperature at any point of the plate will be a linear function of time, and the temperature distribution over the thickness will be parabolic. Such a condition of heating is called quasi stationary since the temperature gradient field will be steady (i.e., the temperature gradient at any given point does not change with time).

If we let $\mathrm{Bi} \to \infty$, then according to boundary condition (7.1.3), the surface temperature of the plate instantaneously becomes equal to the ambient temperature and then changes by the linear law

$$t(\pm R, \tau) = t_0 + b\tau . \tag{7.1.13}$$

Hence a problem is obtained with a boundary condition of the first kind (see Chapter 4). To solve such a problem it is only necessary in (7.1.11) to put $\mathrm{Bi} = \infty$, i.e.,

$$\frac{\theta}{\mathrm{Pd}} = \mathrm{Fo} - \frac{1}{2}\left(1 - \frac{x^2}{R^2}\right) + \sum_{n=1}^{\infty} \frac{A_n}{\mu_n^{\,2}} \cos \mu_n \frac{x}{R} \exp[-\mu_n^{\,2}\,\mathrm{Fo}], \tag{7.1.14}$$

where

$$\mu_n = (2n - 1)\tfrac{1}{2}\pi, \qquad A_n = (-1)^{n+1} \frac{4}{(2n-1)\pi} .$$

Equation (7.1.14) may be obtained in another form. Let us assume in (7.1.7) for the transform that $H = \infty$ which corresponds to the case $\mathrm{Bi} = \infty$. Then we expand $(1/\cosh (s/a)^{1/2} R)$ into a series (see Eq. 4.3.26). Then we shall have

$$T(x, s) - \frac{t_0}{s} = \frac{b}{s^2} \sum_{n=1}^{\infty} (-1)^{n+1} \times \{\exp[-[(2n-1)R-x](s/a)^{1/2}] + \exp[-[(2n-1)R+x](s/a)^{1/2}]\}. \tag{7.1.15}$$

Using Eq. (53) from the table of transforms (Appendix 5), we obtain

$$\frac{\theta}{\mathrm{Pd}} = 4\,\mathrm{Fo} \sum_{n=1}^{\infty} (-1)^{n+1} \times \left[\mathrm{i}^2 \operatorname{erfc} \frac{(2n-1) - (x/R)}{2(\mathrm{Fo})^{1/2}} + \mathrm{i}^2 \operatorname{erfc} \frac{(2n-1) + (x/R)}{2(\mathrm{Fo})^{1/2}}\right]. \tag{7.1.16}$$

This solution is most convenient for small values of Fo, since in this case we may confine ourselves to one term of series (7.1.16).

In Fig. 7.1 the generalized function $\theta/\mathrm{Pd\,Fo}$ is plotted versus the Fourier number for various values of the dimensionless coordinate.

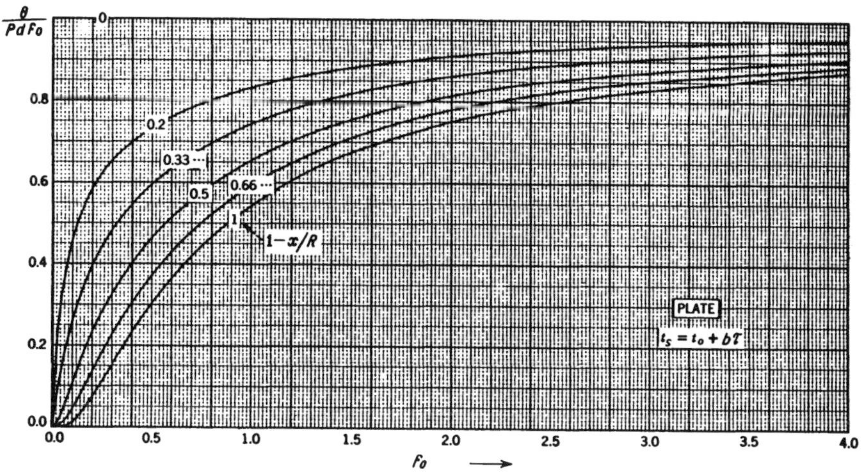

Fig. 7.1. θ/Pd Fo versus Fo for various values of x/R from 0 to 1 for a plate [102].

d. Specific Heat Rate. The specific heat rate is found by the conventional relationship

$$\Delta Q_v = c\gamma[\bar{t}(\tau) - t_0]. \tag{7.1.17}$$

Integrating Eq. (7.1.11) with respect to x/R between 0 and 1, we obtain

$$\bar{\theta} = \frac{\bar{t}(\tau) - t_0}{t_0} = \text{Pd}\left[\text{Fo} - \left(\frac{1}{3} + \frac{1}{\text{Bi}}\right) + \sum_{n=1}^{\infty} \frac{B_n}{\mu_n^2} \exp[-\mu_n^2 \text{Fo}]\right], \tag{7.1.18}$$

where B_n are the constant coefficients determined from Eq. (6.3.45).

The specific heat rate may be found by another method. The amount of heat transferred per unit time to a unit surface of the plate from the surrounding medium according to the Newton law will be equal to

$$q = \frac{dQ_s}{d\tau} = \alpha[t_a(\tau) - t(R, \tau)].$$

Then the specific heat rate will be

$$\Delta Q_v = \frac{\alpha S}{V} \int_0^\tau [t_a(\tau) - t(R, \tau)]\, d\tau, \tag{7.1.19}$$

where S/V is the ratio of the surface to the volume for an infinite plate (it is equal to $1/R$). The ambient temperature t_a is a linear function of time, i.e.,

$$t_a(\tau) = t_0 + b\tau = t_0(1 + \text{Pd Fo}). \tag{7.1.20}$$

Then the difference $[t_a(\tau) - t(R, \tau)]$ will be equal to

$$(t_0/\text{Bi})\,\text{Pd} - t_0\,\text{Pd} \sum_{n=1}^{\infty} (1/\mu_n^2) A_n \cos\mu_n \exp[-\mu_n^2\,\text{Fo}]. \quad (7.1.21)$$

Substituting this expression into (7.1.19), upon integration and necessary rearrangement, we obtain

$$\Delta Q_v = c\gamma t_0\,\text{Pd}\left\{\text{Fo} - \sum_{n=1}^{\infty} \frac{1}{\mu_n^2} B_n[1 - \exp[-\mu_n^2\,\text{Fo}]]\right\}, \quad (7.1.22)$$

where B_n are constant coefficients determined from equation (6.3.45). Since $\sum_{n=1}^{\infty} (1/\mu_n^2) B_n = \frac{1}{3} + (1/\text{Bi})$, relation (7.1.22) becomes identical with (7.1.17).

7.2 Sphere. Ambient Temperature as a Linear Function of Time

a. Statement of the Problem. *The statement of the problem is the same as the previous one.* With concentric distribution of isotherms relative to the center the boundary conditions are the following

$$t(r, 0) = t_0 = \text{const}, \quad (7.2.1)$$

$$\frac{\partial t(0, \tau)}{\partial r} = 0, \qquad t(0, \tau) \neq \infty, \quad (7.2.2)$$

$$-\frac{\partial t(R, \tau)}{\partial r} + H[t_0 + b\tau - t(R, \tau)] = 0. \quad (7.2.3)$$

b. Solution of the Problem. The solution of differential equation (6.5.1) for the transform $T(r, s)$ under conditions (7.2.1) and (7.2.2) has the form

$$T(r, s) - \frac{t_0}{s} = \frac{B}{r} \sinh\left(\frac{s}{a}\right)^{1/2} r. \quad (7.2.4)$$

Boundary condition (7.2.3) for the transform is the following

$$-T'(R, s) + \frac{Ht_0}{s} + \frac{Hb}{s^2} - HT(R, s) = 0. \quad (7.2.5)$$

Substituting boundary condition (7.2.5) into solution (7.2.4)

$$-\frac{B}{R}\left(\frac{s}{a}\right)^{1/2} \cosh\left(\frac{s}{a}\right)^{1/2} R + \frac{B}{R^2} \sinh\left(\frac{s}{a}\right)^{1/2} R + \frac{Ht_0}{s}$$
$$+ \frac{Hb}{s^2} - \frac{Ht_0}{s} - H\frac{B}{R} \sinh\left(\frac{s}{a}\right)^{1/2} R = 0. \quad (7.2.6)$$

From equation (7.2.6) we find the constant B and substitute into solution (7.2.4) to obtain

$$T(r, s) - \frac{t_0}{s} = \frac{bHR^2 \sinh(s/a)^{1/2}r}{rs^2[(HR-1)\sinh(s/a)^{1/2}R + (s/a)^{1/2}R\cosh(s/a)^{1/2}R]}$$
$$= \frac{\Phi_1(s)}{\psi_1(s)}. \tag{7.2.7}$$

The numerator $\Phi_1(s)$ and denominator $\psi_1(s)$ are not generalized polynomials with respect to s; but they may be transformed into generalized polynomials if they are multiplied by $s^{1/2}$.

Then we may use the relationship

$$\frac{\Phi_1(s_n)}{\psi_1'(s_n)} = \frac{\Phi(s_n)}{\psi'(s_n)} \tag{7.2.8}$$

and apply the expansion theorem for the ratio $\Phi_1(s)/\psi_1(s)$ if the roots s_n differ from zero. To find the roots s_n we shall equate $\psi_1(s)$ to zero:

$$\psi_1(s) = rs^2\left[(HR-1)\sinh\left(\frac{s}{a}\right)^{1/2}R + \left(\frac{s}{a}\right)^{1/2}R\cosh\left(\frac{s}{a}\right)^{1/2}R\right] = 0.$$

The bracketed expression will give us the roots $s_n = -a\mu_n{}^2/R^2$ $(i(s/a)^{1/2}R = \mu)$ which are determined from the characteristic equation

$$\tan\mu = -\frac{1}{\mathrm{Bi}-1}\mu$$

(see Chapter 6, Section 5, for details). Hence, we have

$$\sum_{n=1}^{\infty}\frac{\Phi_1(s_n)}{\psi_1'(s_n)}e^{s_n\tau}$$
$$= \frac{2bR^2}{a}\sum_{n=1}^{\infty}\frac{R(\sin\mu_n - \mu_n\cos\mu_n)}{r\mu_n{}^3(\mu_n - \sin\mu_n\cos\mu_n)}\sin\mu_n\frac{r}{R}\exp\left[-\mu_n{}^2\frac{a\tau}{R^2}\right]. \tag{7.2.9}$$

For the zero root ($s = 0$) we transform solution (7.2.7) for which purpose $\sinh z$ and $\cosh z$ are expanded in series. Then we shall have

$$\frac{\Phi_1(s)}{\psi_1(s)} = \frac{bHR\left(1 + \frac{1}{3!}r^2\frac{s}{a} + \frac{1}{5!}r^4\frac{s^2}{a^2} + \cdots\right)}{s^2\left\{(HR-1)\left[1 + \frac{1}{3!}R^2\frac{s}{a} + \cdots\right] + \left[1 + \frac{1}{2!}R^2\frac{s}{a} + \cdots\right]\right\}}$$
$$= \frac{f(s)}{s^2\varphi(s)}, \tag{7.2.10}$$

where $f(s)$ is the numerator of relation (7.2.10) and $\varphi(s)$ is the brackete expression in the denominator.

We apply the expansion theorem for multiple roots, i.e.,

$$\lim_{s\to 0}\left[\frac{d}{ds}\left(\frac{f(s)}{\varphi(s)}e^{s\tau}\right)\right] = \lim_{s\to 0}\left\{\tau e^{s\tau}\frac{f(s)}{\varphi(s)} + e^{s\tau}\frac{f'(s)}{\varphi(s)} - e^{s\tau}\frac{f(s)\varphi'(s)}{[\varphi(s)]^2}\right\}$$

$$= b\tau + \frac{br^2}{6a} - \frac{2bR^2}{6aHR} - \frac{bR^2}{6a}, \tag{7.2.11}$$

since

$$f(0) = bHR, \qquad \varphi(0) = HR, \qquad \varphi'(0) = \frac{2R^2}{6a} + \frac{HRR^2}{6a}.$$

The final solution of our problem will have the form

$$\theta = \frac{t(r,\tau) - t_0}{t_0}$$

$$= \mathrm{Pd}\left\{\mathrm{Fo} - \frac{1}{6}\left[\left(1+\frac{2}{\mathrm{Bi}}\right) - \frac{r^2}{R^2}\right] + \sum_{n=1}^{\infty}\frac{A_n}{\mu_n^2}\,\frac{R\sin\mu_n\frac{r}{R}}{r\mu_n}\exp[-\mu_n^2\,\mathrm{Fo}]\right\}, \tag{7.2.12}$$

where $\mathrm{Bi} = HR$, $\mathrm{Pd} = bR^2/at_0$ is the Predvoditelev criterion, and A_n are the initial thermal amplitudes determined from relation (6.5.28) or (6.5.29).

c. ***Analysis of the Solution.*** From solution (7.2.12) it is seen that the relative excess temperature θ is directly proportional to the criterion Pd, i.e., the rate of heating of the surrounding medium effects the rise of the body temperature at every point.

The series in solution (7.2.12) converges rapidly and therefore for quasi-stationary conditions determined by $\mathrm{Fo} > \mathrm{Fo}_1$ it may be neglected. The value of Fo_1 is determined by the method explained in Chapter 6, Section 11.

At $\mathrm{Bi} \to \infty$, the temperature of the sphere surface will be a linear function of time $t(R,\tau) = t_a + b\tau$. Solution (7.2.12) then takes the form

$$\frac{\theta}{\mathrm{Pd}} = \mathrm{Fo} - \frac{1}{6}\left(1 - \frac{r^2}{R^2}\right) + \sum_{n=1}^{\infty}\frac{A_n}{\mu_n^2}\,\frac{R\sin\mu_n(r/R)}{r\mu_n}\exp[-\mu_n^2\,\mathrm{Fo}], \tag{7.2.13}$$

where

$$A_n = (-1)^{n+1}2, \quad \mu_n = n\pi.$$

Let us evaluate the series in (7.2.13) for the center of the sphere ($r = 0$). Solution (7.2.13) is written

$$\theta/\text{Pd} = \text{Fo} - \tfrac{1}{6}[1 - \varphi(\tau)] \tag{7.2.14}$$

where

$$\varphi(\tau) = (12/\pi^2) \sum_{n=1}^{\infty} (-1)^{n+1}(1/n^2) \exp[-n^2\pi^2 \text{ Fo}]. \tag{7.2.15}$$

Let us require that $\varphi(\tau) < 0.005$, which is $\tfrac{1}{2}\%$ of 1. Then using the method stated in Chapter 6, Section 11, we find that $\varphi(\tau) \leq 0.005$ if $\text{Fo} \geq 0.54$.

For the point with the coordinate $r/R = 0.5$ the value of $\varphi(\tau, 0.5)$ corresponding to the series in solution (7.2.13) at $\text{Fo} = 0.5$ is equal to 0.0008, which is 0.11% of 0.75.

For small values of Fo, we may obtain the solution in another form. Let us assume in solution (7.2.7) that $\text{Bi} = HR \rightarrow \infty$ then after some manipulations we shall have (see Chapter 4, Section 4)

$$\begin{aligned} T(r, s) - \frac{t_0}{s} &= \frac{bR \sinh(s/a)^{1/2} r}{rs^2 \sinh(s/a)^{1/2} R} \\ &= \frac{bR}{rs^2} \sum_{n=1}^{\infty} \{\exp[-(s/a)^{1/2}[(2n-1)R - r]] \\ &\quad - \exp[-(s/a)^{1/2}[(2n-1)R + r]]\}. \end{aligned}$$

Using relation (56) of the table of transforms we find

$$\frac{\theta}{\text{Pd}} = 4\text{Fo} \sum_{n=1}^{\infty} \frac{R}{r} \left[\text{i}^2 \operatorname{erfc} \frac{(2n-1) - \dfrac{r}{R}}{2(\text{Fo})^{1/2}} - \text{i}^2 \operatorname{erfc} \frac{(2n-1) + \dfrac{r}{R}}{2(\text{Fo})^{1/2}} \right]. \tag{7.2.16}$$

Figure 7.2 furnishes plots of the generalized variable $\theta/\text{Pd Fo}$ versus the Fourier number for different dimensionless coordinates r/R. For small values of Fo we may confine ourselves to the first term of the series (7.2.16). The value of the function $\text{i}^2 \operatorname{erfc}$ is taken from a graph or from tables.

For $\text{Bi} = \infty$, for the center of the sphere we may write (see Chapter 4, Section 4)

$$T(r, s) - \frac{t_0}{s} = \frac{2Rb}{s(as)^{1/2}} \sum_{n=1}^{\infty} \exp\left[-\left(\frac{s}{a}\right)^{1/2} R(2n-1)\right].$$

Then, using relation (52) of the table of transforms (see Appendix 5) we obtain

$$\theta = 4 \text{ Pd}(\text{Fo})^{1/2} \sum_{n=1}^{\infty} \text{i} \operatorname{erfc}\left(\frac{2n-1}{2(\text{Fo})^{1/2}}\right). \tag{7.2.17}$$

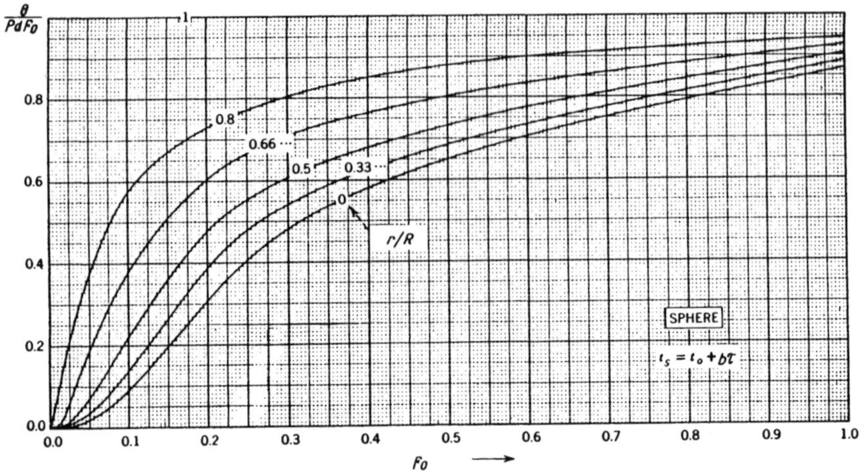

Fig. 7.2. θ/Pd Fo versus Fo for various values of r/R from 0 to 1 for a sphere [102].

d. Specific Heat Rate. To determine the specific heat rate let us find the mean temperature of the sphere

$$\bar{\theta} = \frac{\bar{t}(\tau) - t_0}{t_0} = \text{Pd}\left[\text{Fo} - \frac{1}{15}\left(1 + \frac{5}{\text{Bi}}\right) + \sum_{n=1}^{\infty} \frac{1}{\mu_n^2} B_n \exp[-\mu_n^2 \text{Fo}]\right], \tag{7.2.18}$$

where B_n are constant coefficients determined from relation (6.5.49). The remainder of the calculation is carried out in the usual manner.

7.3 Infinite Cylinder. Ambient Temperature as a Linear Function of Time

a. Statement of the Problem. *This problem is similar to the previous one, but the heated body is taken in the form of an infinite cylinder.* For a symmetric problem, the differential heat conduction equation of an infinite cylinder has been presented more than once in previous chapters. The boundary conditions are similar to conditions (7.2.1)–(7.2.3) of the previous problem, therefore they are not given here.

b. Solution of the Problem. The solution for the transform $T(r, s)$ with the initial constant temperature of the cylinder t_0 in the case of a symmetric temperature field relative to the cylinder axis has the form (see Chapter 6, Section 6)

$$T(r, s) - (t_0/s) = AI_0[(s/a)^{1/2}r]. \tag{7.3.1}$$

The constant A is determined from the boundary condition similar to condition (7.2.5) of the previous problem:

$$-(s/a)^{1/2} AI_0'[(s/a)^{1/2}R] + (Hb/s^2) - HAI_0[(s/a)^{1/2}R] = 0.$$

Having determined the constant A from this equality, solution (7.3.1) may be written

$$T(r, s) - \frac{t_0}{s} = \frac{bI_0((s/a)^{1/2}r)}{s^2[I_0((s/a)^{1/2}R) + (1/H)(s/a)^{1/2}I_1((s/a)^{1/2}R)]}$$

$$= \frac{\Phi(s)}{\psi(s)} = \frac{\Phi(s)}{s^2\varphi(s)}. \qquad (7.3.2)$$

Solution (7.3.2) is the ratio of two generalized polynomials $\Phi(s)$ and $\psi(s)$ inasmuch as the latter expression does not contain a constant (see Chapter 6, Section 6). Equating $\psi(s)$ with zero, we find the roots of the polynomial $\psi(s)$ as $s = 0$ (double root) and $s_n = -a\mu_n^2/R^2$, which is infinite number of simple roots, determined from the characteristic equation

$$\frac{J_0(\mu)}{J_1(\mu)} = \frac{1}{\mathrm{Bi}}\mu. \qquad (7.3.3)$$

This equation is obtained, if $\varphi(s)$ is equated with zero. Let us use the expansion theorem

$$\lim_{s\to 0}\left\{\frac{d}{ds}\left[e^{s\tau}\frac{\Phi(s)}{\varphi(s)}\right]\right\}$$

$$= \lim_{s\to 0}\left\{\tau e^{s\tau}\frac{\Phi(s)}{\varphi(s)} + e^{s\tau}\frac{\Phi'(s)}{\varphi(s)} - e^{s\tau}\frac{\Phi(s)\varphi'(s)}{[\varphi(s)]^2}\right\}$$

$$= b\tau + \frac{br^2}{4a} - \frac{bR^2}{4a} - \frac{bR^2}{4aHR} - \frac{bR^2}{4aHR}$$

$$= b\tau - \frac{b}{4u}\left[R^2\left(1 + \frac{2}{HR}\right) - r^2\right],$$

since

$$\Phi(0) = b, \qquad \varphi(0) = 1, \qquad \Phi'(0) = \lim_{s\to 0}\left(\frac{rbI_1((s/a)^{1/2}r)}{2\,(as)^{1/2}}\right) = \frac{br^2}{4a},$$

$$I_1(z) = \frac{1}{2}z + \frac{1}{2^4\cdot 4}z^3 + \cdots,$$

$$\varphi'(s) = \frac{R}{2(as)^{1/2}}\,[(1 + (1/HR))I_1((s/a)^{1/2}R) + (1/H)(s/a)^{1/2}I_1'((s/a)^{1/2}R)].$$

Further

$$\frac{\Phi(s_n)}{\psi'(s_n)} e^{s_n \tau} = \frac{\Phi(s_n)}{s_n^2 \varphi'(s_n)} e^{s_n \tau}$$

$$= \frac{bR^2}{a} \frac{1}{\mu_n^2} A_n J_0\left(\mu_n \frac{r}{R}\right) \exp\left[-\mu_n^2 \frac{a\tau}{R^2}\right],$$

since $\psi'(s) = 2s\varphi(s) + s^2\varphi'(s)$. *Hence, the solution of our problem will have the form*

$$\theta = \frac{t(r, \tau) - t_0}{t_0}$$

$$= \mathrm{Pd}\left\{\mathrm{Fo} - \frac{1}{4}\left[\left(1 + \frac{2}{\mathrm{Bi}}\right) - \frac{r^2}{R^2}\right]\right.$$

$$\left. + \sum_{n=1}^{\infty} \frac{A_n}{\mu_n^2} J_0\left(\mu_n \frac{r}{R}\right) \exp[-\mu_n^2 \,\mathrm{Fo}]\right\}, \tag{7.3.4}$$

where A_n are the initial thermal amplitudes determined by relation (6.6.27).

c. Analysis of the Solution. *If* Bi $\to \infty$, *the surface temperature of the cylinder will be a linear function of time, i.e.,* $t(R, \tau) = t_0 + b\tau$. In this case the solution of the problem will have the form

$$\frac{\theta}{\mathrm{Pd}} = \mathrm{Fo} - \frac{1}{4}\left(1 - \frac{r^2}{R^2}\right) + \sum_{n=1}^{\infty} \frac{1}{\mu_n^2} A_n J_0\left(\mu_n \frac{r}{R}\right) \exp[-\mu_n^2 \,\mathrm{Fo}], \tag{7.3.5}$$

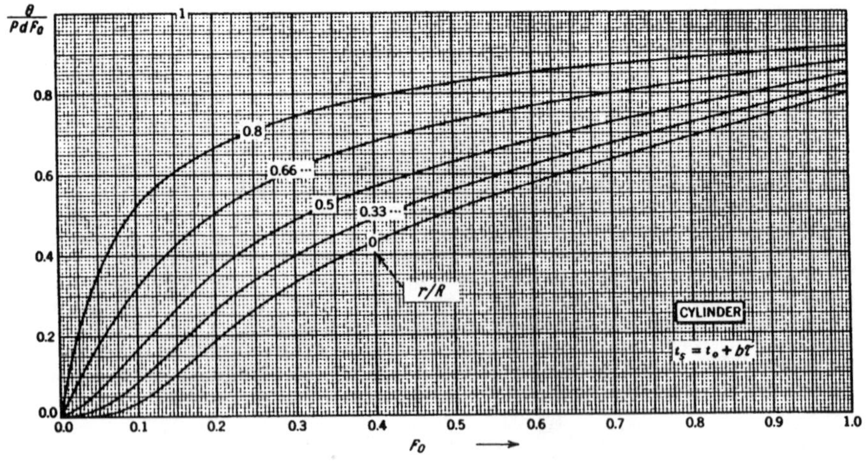

Fig. 7.3. $\theta/\mathrm{Pd}\,\mathrm{Fo}$ versus the number Fo for various r/R from 0 to 1 for an infinite cylinder [102].

where μ_n are the roots of the function $J_0(\mu)$ since $J_0(\mu) = 0$ at $\text{Bi} = \infty$. Figure 7.3 presents plots of the generalized variable $\theta/\text{Pd Fo}$ versus the Fourier number for different values of the relative coordinate r/R.

For the quasi-stationary state in solution (7.3.5), we may neglect the series. For small values of Fo we may find an approximate solution. Applying the method of expansion of the function $I_0(z)$ into asymptotic series, after some manipulations we obtain the solution for the transform in the case $\text{Bi} \to \infty$ in the following form (for details see Chapter 4, Section 5)

$$T(r, s) - \frac{t_0}{s} \approx \frac{b}{s^2}\left(\frac{R}{r}\right)^{1/2}\left(1 + \frac{R - r}{8rR(s/a)^{1/2}}\right)\exp\left[-\left(\frac{s}{a}\right)(R - r)\right]. \tag{7.3.6}$$

Using relation (54) of the table of transforms (see Appendix 5), we find

$$\theta \approx \text{Pd}\left(\frac{R}{r}\right)^{1/2}\left[4\text{Fo i}^2 \operatorname{erfc}\frac{1 - (r/R)}{2(\text{Fo})^{1/2}} + \frac{R - r}{8r}(4\,\text{Fo})^{3/2}\,\text{i}^3 \operatorname{erfc}\frac{1 - (r/R)}{2(\text{Fo})^{1/2}}\right]. \tag{7.3.7}$$

The values of the functions $\text{i}^2 \operatorname{erfc} x$ and $\text{i}^3 \operatorname{erfc} x$ may be determined from Appendix 6.

d. Specific Heat Rate. Let us find the mean temperature $\bar{t}(\tau)$ of the cylinder to calculate the specific heat rate. Applying the integral relation (6.6.32) we obtain

$$\bar{\theta} = \frac{\bar{t}'(\tau) - t_0}{t_0} = \text{Pd}\left[\text{Fo} - \frac{1}{8}\left(1 + \frac{4}{\text{Bi}}\right) + \sum_{n=1}^{\infty}\frac{1}{\mu_n^2}B_n \exp[-\mu_n^2\,\text{Fo}]\right], \tag{7.3.8}$$

where B_n are the constants determined from relation (6.6.34). The first six values of B_n are given in Chapter 6, Table 6.12.

Comparing the problems given above with the problems of Chapter 5, we see that they are similar. In both cases after a definite interval determined by the inequality $\text{Fo} > \text{Fo}_1$ a quasi-stationary state sets in and $dt/d\tau$ becomes a constant value (the temperature at any point of the body is a linear function of time). Therefore under quasi-stationary conditions in one-dimensional problems, the temperature distribution obeys the law of a parabola. The difference between the problems lies in the fact that in the present problems the rate of heating $d\bar{t}/d\tau$ (derivative of the mean temperature with respect to time) becomes a constant value only after a definite interval (i.e., when $\text{Fo} > \text{Fo}_1$) whereas in the problems of Chapter 5, the rate of heating is a constant value from the very beginning of the process of heating, which directly follows from the boundary condition.

Thus, for the specific heat rate we have

$$q_c = \frac{V}{S} c\gamma \frac{d\bar{t}(\tau)}{d\tau},$$

where V/S is the ratio of the volume to the surface of the body.

7.4 Infinite Plate, Sphere, and Cylinder. Ambient Temperature as an Exponential Function of Time

a. Statement of the Problem. *Consider an isotropic body (infinite plate, sphere, and cylinder) which is in thermal equilibrium with the surrounding medium. At the initial time, the ambient temperature rises by the law* $t_a(\tau) = t_m - (t_m - t_0)e^{-k\tau}$, *where* t_m *is the maximum ambient temperature* $t_a(\infty)=t_m$ *and* k *is a constant. The temperature distribution inside the body and the specific heat rate are to be found.*

We shall consider this problem in detail for a plate and for the other geometries we shall give the final results only.

We have an ordinary differential heat conduction equation for an infinite plate with the thickness $2R$. The origin of coordinates is in the middle of the plate, relative to which the curves of the temperature distribution are symmetrical. We have the initial condition

$$t(x, 0) = t_0 = \text{const}, \tag{7.4.1}$$

the condition of symmetry

$$\frac{\partial t(0, \tau)}{\partial x} = 0, \quad t(0, \tau) \neq \infty \tag{7.4.2}$$

and the boundary condition

$$-\frac{\partial t(R, \tau)}{\partial x} + H[t_m - (t_m - t_0)\, e^{-k\tau} - t(R, \tau)] = 0. \tag{7.4.3}$$

b. Solution of the Problem. Let us solve the problem by the operational method. The solution of the equation for the transform $T(x, s)$ under conditions (7.4.1) and (7.4.2) has the form

$$T(x, s) - \frac{t_0}{s} = A \cosh\left(\frac{s}{a}\right)^{1/2} x. \tag{7.4.4}$$

Boundary condition (7.4.3) for the transform is written as

$$-T'(R,s)+\frac{Ht_m}{s}-\frac{H(t_m-t_0)}{s+k}-HT(R,s)=0\,, \qquad (7.4.5)$$

since

$$L[e^{-k\tau}]=1/(s+k).$$

Solution (7.4.4) must satisfy the boundary condition (7.4.5), namely

$$-A\left(\frac{s}{a}\right)^{1/2}\sinh\left(\frac{s}{a}\right)^{1/2}R+\frac{H}{s}(t_m-t_0)-\frac{H(t_m-t_0)}{s+k}-HA\cosh\left(\frac{s}{a}\right)^{1/2}R=0.$$

Having determined the constant A from this equality and then substituting the expression obtained into solution (7.4.4) we have

$$T(x,s)-(t_0/s)=\frac{(t_m-t_0)\cosh(s/a)^{1/2}x}{s[\cosh(s/a)^{1/2}R+(1/H)(s/a)^{1/2}\sinh(s/a)^{1/2}R]\{(s/k)+1\}}$$
$$=\frac{\Phi(s)}{\psi(s)}. \qquad (7.4.6)$$

The numerator and denominator are generalized polynomials with respect to s; the polynomial $\psi(s)$ does not contain a constant (its first term is equal to s). The polynomial $\psi(s)$ has the following simple roots $s=0$; $s=-k$; $s_n=-a\mu_n{}^2/R^2$ where $i(s/a)^{1/2}R=\mu$ resulting in an infinite number of roots determined from the characteristic equation (for details see Chapter 6, Section 3). Applying the expansion theorem for the case of simple roots, we find

$$\theta=\frac{t(x,\tau)-t_0}{t_m-t_0}=1-\frac{\cos(k/a)^{1/2}x}{\cos(k/a)^{1/2}R-(1/H)(k/a)^{1/2}\sin(k/a)^{1/2}R}\,e^{-k\tau}$$
$$-\sum_{n=1}^{\infty}\frac{A_n}{\{1-(a\mu_n{}^2/kR^2)\}}\cos\mu_n\frac{x}{R}\exp\left[-\mu_n{}^2\frac{a\tau}{R^2}\right], \qquad (7.4.7)$$

where A_n are the constant thermal amplitudes determined from relation (6.3.30).

c. ***Analysis of the Solution.*** The Predvoditelev criterion Pd, introduced in Section 7.1, in the present case will be equal to

$$\mathrm{Pd}=kR^2/a\,, \qquad (7.4.8)$$

since

$$\left(\frac{dt_a(\tau)}{d\tau}\right)_{\max} = k(t_m - t_0); \quad \left\{\frac{d}{d\tau}\left[\frac{t_a(\tau)R^2}{a(t_m - t_0)}\right]\right\}_{\max} = \frac{kR^2}{a}.$$

The maximum rate of heating of the medium will correspond to the initial moment of time; with time it decreases and at $\tau \to \infty$, $dt_a(\tau)/d\tau \to 0$, the ambient temperature becomes constant and equal to t_m.

Solution (7.4.7) *may be written in dimensionless values as*

$$\theta = 1 - \frac{\cos(\mathrm{Pd})^{1/2}\, x/R}{\cos(\mathrm{Pd})^{1/2} - 1/\mathrm{Bi}\, (\mathrm{Pd})^{1/2} \sin(\mathrm{Pd})^{1/2}} \exp[-\mathrm{Pd}\ \mathrm{Fo}]$$
$$- \sum_{n=1}^{\infty} \frac{A_n}{1 - (\mu_n^2/\mathrm{Pd})} \cos \mu_n \frac{x}{R} \exp[-\mu_n^2\ \mathrm{Fo}]. \qquad (7.4.9)$$

Thus the relative temperature is a function of Fo, Bi, Pd and x/R, namely

$$\theta = \Psi(\mathrm{Fo}, \mathrm{Bi}, \mathrm{Pd}, x/R). \qquad (7.4.10)$$

If we set Pd $= \infty$ ($k \to \infty$), the ambient temperature is constant and equal to t_m ($t_a = t_m$). Then solution (7.4.9) becomes solution (6.3.29).

Setting Bi $= \infty$ *means that the surface temperature of a plate is an exponential function of time (a boundary condition of the first kind),*

$$t(\pm R, \tau) = t_m - (t_m - t_0)e^{-k\tau},$$

then solution (7.4.9) transforms into the equation

$$\theta = 1 - \frac{\cos(\mathrm{Pd})^{1/2} x/R}{\cos(\mathrm{Pd})^{1/2}} \exp[-\mathrm{Pd}\ \mathrm{Fo}]$$
$$- \sum_{n=1}^{\infty} \frac{A_n}{\left(1 - \dfrac{\mu_n^2}{\mathrm{Pd}}\right)} \cos \mu_n \frac{x}{R} \exp[-\mu_n^2\ \mathrm{Fo}] \qquad (7.4.11)$$

where $\mu_n = (2n - 1)\frac{1}{2}\pi$.

d. Specific Heat Rate. To determine the specific heat rate we find the mean temperature as

$$\bar{\theta} = \frac{\bar{t}(\tau) - t_0}{t_m - t_0} = 1 - \frac{1}{(\mathrm{Pd})^{1/2}[\cot(\mathrm{Pd})^{1/2} - (1/\mathrm{Bi})(\mathrm{Pd})^{1/2}]} e^{-\mathrm{Pd}\mathrm{Fo}}$$
$$- \sum_{n=1}^{\infty} \frac{B_n}{\{1 - (\mu_n^2/\mathrm{Pd})\}} \exp[-\mu_n^2\ \mathrm{Fo}], \qquad (7.4.12)$$

where B_n are the constant coefficients determined by relation (6.3.45). Further calculations are carried out in the usual way.

e. Solution for the Sphere. The corresponding solution for a sphere is

$$\theta = 1 - \frac{R \text{ Bi} \sin\{(\text{Pd})^{1/2}(r/R)\}}{r[(\text{Bi} - 1)\sin(\text{Pd})^{1/2} + (\text{Pd})^{1/2}\cos(\text{Pd})^{1/2}]} \exp[-\text{Pd Fo}]$$
$$- \sum_{n=1}^{\infty} \frac{A_n}{1 - (\mu_n^2/\text{Pd})} \frac{R \sin \mu_n(r/R)}{r\mu_n} \exp[-\mu_n^2 \text{ Fo}], \qquad (7.4.13)$$

where A_n are the initial thermal amplitudes determined by relation (6.5.29). The mean temperature is given by

$$\bar{\theta} = 1 - \frac{3 \text{ Bi}(\tan(\text{Pd})^{1/2} - (\text{Pd})^{1/2})}{\text{Pd}[(\text{Bi}-1)\tan(\text{Pd})^{1/2} + (\text{Pd})^{1/2}]} e^{-\text{PdFo}}$$
$$- \sum_{n=1}^{\infty} \frac{B_n}{1 - (\mu_n^2/\text{Pd})} \exp[-\mu_n^2 \text{ Fo}], \qquad (7.4.14)$$

where B_n are constant coefficients determined by relation (6.5.49).

f. Solution for the Infinite Cylinder

$$\theta = 1 - \frac{J_0\{(\text{Pd})^{1/2}(r/R)\}}{[J_0((\text{Pd})^{1/2}) - (1/\text{Bi})(\text{Pd})^{1/2}J_1((\text{Pd})^{1/2})]} \exp[-\text{Pd Fo}]$$
$$- \sum_{n=1}^{\infty} \frac{A_n}{1 - (\mu_n^2/\text{Pd})} J_0\left(\mu_n \frac{r}{R}\right) \exp[-\mu_n^2 \text{ Fo}], \qquad (7.4.15)$$

where A_n are the coefficients determined by relation (6.6.27). The mean temperature is

$$\bar{\theta} = 1 - \frac{2 J_1(\text{Pd})^{1/2}}{[(\text{Pd})^{1/2}J_0(\text{Pd})^{1/2} - (1/\text{Bi}) \text{ Pd } J_1(\text{Pd})^{1/2}]} \exp[-\text{Pd Fo}]$$
$$- \sum_{n=1}^{\infty} \frac{1}{1 - (\mu_n^2/\text{Pd})} B_n \exp[-\mu_n^2 \text{ Fo}], \qquad (7.4.16)$$

where B_n are the constant coefficients determined by relation (6.6.34).

7.5 Heating of Moist Bodies (Infinite Plate, Sphere, and Infinite Cylinder)

a. Statement of the Problem. *In this section are considered the heating of moist bodies in a medium with constant temperature when moisture evaporates at the surface.*

It is known from the theory of drying that the rate of evaporation (the amount of moisture evaporated per unit time from unit surface of a body) is in the first period constant and then in the second period changes with time according to an exponential law. Thus, in the first approximation for the rate of evaporating m we may write

$$m = m_0 e^{-k\tau}, \tag{7.5.1}$$

where k is the drying coefficient (1/hr) and m_0 is the maximum rate of heating (kg/m^2 hr). If we set $k = 0$, then $m = m_0 = \text{const}$, i.e., we shall obtain the constant rate of evaporation corresponding to the first period of drying.

Consider the problem of heating a moist infinite plate in a medium with constant temperature $t_a = \text{const}$. The initial and boundary conditions may be written as

$$t(x, 0) = t_0, \tag{7.5.2}$$

$$-\lambda \frac{\partial t(R, \tau)}{\partial x} + \alpha[t_a - t(R, \tau)] - \varrho m_0 e^{-k\tau} = 0, \tag{7.5.3}$$

$$\frac{\partial t(0, \tau)}{\partial x} = 0, \quad t(0, \tau) \neq \infty, \tag{7.5.4}$$

where ϱ is the specific heat of evaporation in cal/kg.

Boundary condition (7.5.3) may be rewritten as

$$-\frac{\partial t(R, \tau)}{\partial x} + H\left[\left(t_a - \frac{\varrho m_0}{\alpha} e^{-k\tau}\right) - t(R, \tau)\right] = 0, \tag{7.5.5}$$

i.e., this boundary condition is similar to boundary condition (7.4.3), but in the present case the value $\varrho m_0/\alpha$ is used instead of $(t_m - t_0)$.

b. Solution of the Problem. The solution for the transform has the form

$$T(x, s) - \frac{t_0}{s} = \frac{\left[(t_a - t_0)(s + k) - \frac{s}{a}\varrho m_0\right] \cosh\left(\frac{s}{a}\right)^{1/2} x}{s(s + k)\left[\cosh\left(\frac{s}{a}\right)^{1/2} R + \frac{1}{H}\left(\frac{s}{a}\right)^{1/2} \sinh\left(\frac{s}{a}\right)^{1/2} R\right]}. \tag{7.5.6}$$

With the help of the expansion theorem, we find the solution for the inverse transform

$$\theta = \frac{t(x,\tau) - t_0}{t_a - t_0}$$

$$= 1 - \frac{\theta_{wb} \cos\{(\mathrm{Pd})^{1/2}(x/R)\}}{\cos(\mathrm{Pd})^{1/2} - (1/\mathrm{Bi})(\mathrm{Pd})^{1/2} \sin(\mathrm{Pd})^{1/2}} e^{-\mathrm{Pd}\,\mathrm{Fo}}$$

$$- \sum_{n=1}^{\infty} \left[1 - \frac{\theta_{wb}}{(1 - (1/\mu_n^2)\,\mathrm{Pd})}\right] A_n \cos \mu_n \frac{x}{R} \exp[-\mu_n^2\,\mathrm{Fo}], \qquad (7.5.7)$$

where $\mathrm{Pd} = kR^2/a$ is the Predvoditelev criterion,[1] $\theta_{wb} = \varrho m_0/\alpha(t_a - t_0)$ is a characteristic temperature which, as it will be shown below, is equal to the ratio of $(t_a - t_{wb})$ to the excess ambient temperature, i.e.,

$$\theta_{wb} = \frac{t_a - t_{wb}}{t_a - t_0}, \qquad (7.5.8)$$

where t_{wb} is the wet bulb temperature.

Thus the temperature of the plate is a function of a number of criteria, i.e.,

$$\theta = \Phi(\mathrm{Bi}, x/R, \mathrm{Fo}, \mathrm{Pd}, \theta_{wb}). \qquad (7.5.9)$$

If the rate of evaporation is a constant value $m = m_0 = \text{const}$ (the first period of drying), then the Predvoditelev number is equal to zero (the drying coefficient $k = 0$). Then solution (7.5.7) will acquire the form

$$\theta = 1 - \theta_{wb} - \sum_{n=1}^{\infty} (1 - \theta_{wb}) A_n \cos \mu_n(x/R) \exp[-\mu_n^2\,\mathrm{Fo}]. \qquad (7.5.10)$$

As a first approximation, we may argue that in steady state ($\mathrm{Fo} = \infty$) the wet body temperature (i.e., extending the first period of drying to a steady state) is equal to the wet bulb tèmperature, i.e., $\theta_\infty = 1 - \theta_{wb}$ or

$$t_{wb} - t_0 = t_a - t_0 - (\varrho m/\alpha),$$

thence

$$\varrho m/\alpha = t_a - t_{wb}. \qquad (7.5.11)$$

[1] We may obtain this by starting from (7.1.9), where $\mathrm{Pd} = (d\theta_a/d\,\mathrm{Fo})_{\max}$. Here $\theta_a = t_a^*/(t_a - t_{wb})$ and $t_a^* = t_a - (\varrho m_0/\alpha)e^{-k\tau}$. Hence

$$\mathrm{Pd} = \frac{R^2}{a}\left[\frac{d}{d\tau}\left(\frac{t_a^*}{t_a - t_{wb}}\right)\right]_{\max} = \frac{R^2}{a}\left[\frac{d}{d\tau}\left(t_a - \frac{\varrho m_0}{\alpha}\right)e^{-k\tau} \Big/ (t_a - t_{wb})\right]_{\max}$$

$$= \frac{kR^2}{a}\left[\frac{\varrho m_0\, e^{-k\tau}}{\alpha(t_a - t_{wb})}\right]_{\tau\to 0} = \frac{kR^2}{a}.$$

Thus we obtain a new expression for the number θ_{wb} which indicates the meaning of the new parametric criterion.

Solution (7.5.10), applicable to heating a moist body with a constant rate of evaporation from the surface, may be written as follows

$$\theta' = \frac{t(x, \tau) - t_0}{t_{wb} - t_0} = 1 - \sum_{n=1}^{\infty} A_n \cos \mu_n \frac{x}{R} \exp[-\mu_n^2 \,\mathrm{Fo}]. \quad (7.5.12)$$

Expression (7.5.12) coincides in form with the solution for heating a dry plate, but here the ambient temperature t_a is replaced by the wet bulb temperature t_{wb}. Hence the problem of heating a moist body with a constant rate of evaporation at its surface may be reduced to that of heating a dry body with the replacement of t_a by t_{wb}. It also follows directly from the boundary condition, which may be written as

$$-\frac{\partial t(R, \tau)}{\partial x} + H\left[\left(t_a - \frac{\varrho m}{\alpha}\right) - t(R, \tau)\right] = 0 \quad (7.5.13)$$

or

$$-\frac{\partial t(R, \tau)}{\partial x} + H[t_{wb} - t(R, \tau)] = 0. \quad (7.5.13a)$$

Similar solutions for a sphere and infinite cylinder will have the form

$$\theta = 1 - \frac{R\theta_{wb} \,\mathrm{Bi} \sin\{(\mathrm{Pd})^{1/2}(r/R)\}}{r[(\mathrm{Bi}-1) \sin(\mathrm{Pd})^{1/2} + (\mathrm{Pd})^{1/2} \cos(\mathrm{Pd})^{1/2}]} \exp[-\mathrm{Pd}\,\mathrm{Fo}]$$
$$- \sum_{n=1}^{\infty} \left[1 - \frac{\theta_{wb}}{(1-(1/\mu_n^2)\,\mathrm{Pd})}\right] A_n \frac{R \sin \mu_n(r/R)}{r\mu_n} \exp[-\mu_n^2 \,\mathrm{Fo}], \quad (7.5.14)$$

and

$$\theta = 1 - \frac{\theta_{wb} J_0\left((\mathrm{Pd})^{1/2} \frac{r}{R}\right)}{[J_0(\mathrm{Pd})^{1/2} - (1/\mathrm{Bi})(\mathrm{Pd})^{1/2} J_1(\mathrm{Pd})^{1/2}} \exp[-\mathrm{Pd}\,\mathrm{Fo}]$$
$$- \sum_{n=1}^{\infty} \left[1 - \frac{\theta_{wb}}{(1-(1/\mu_n^2)\,\mathrm{Pd})}\right] A_n \cos \mu_n \frac{r}{R} \exp[-\mu_n^2 \,\mathrm{Fo}]. \quad (7.5.15)$$

If we compare solutions (7.5.7), (7.5.14), and (7.5.15) with solutions (7.4.9), (7.4.13), and (7.4.15) we find a great resemblance. The second term of the present solutions differs from the second term of the solutions of the previous section only by the factor θ_{wb}. The third term of the solution has another factor associated with the coefficient A_n: instead of a multiplier

$$\frac{1}{[1 - (1/\mathrm{Pd})\mu_n^2]},$$

we have here the multiplier

$$\left[1 - \frac{\theta_{wb}}{\{1-(1/\mu_n^2)\,\mathrm{Pd}\}}\right].$$

Therefore the mean temperatures for our problem may be obtained directly from solutions (7.4.12), (7.4.14), and (7.4.16) if the replacements mentioned are made.

c. Asymmetrical Problem. *Let the evaporation rate be constant, but different on both surfaces of the plate (an asymmetric problem).*

We have:

$$t(x, 0) = t_0, \tag{7.5.16}$$

$$-\frac{\partial t(R,\tau)}{\partial x} + H\left[t_a - \frac{\varrho m_1}{\alpha} - t(+R,\tau)\right] = 0,$$

$$+\frac{\partial t(-R,\tau)}{\partial x} + H\left[t_a - \frac{\varrho m_2}{\alpha} - t(-R,\tau)\right] = 0, \tag{7.5.17}$$

where m_1 and m_2 are evaporation rates on the opposite surfaces of the plate (kg/m^2 hr).

The solution of the equation has the form

$$t(x,\tau) = t_a - \frac{\varrho(m_1+m_2)}{2\alpha} - \sum_{n=1}^{\infty}\left[(t_a - t_0) - \frac{\varrho(m_1+m_2)}{2\alpha}\right]$$
$$\times A_n \cos\mu_n\frac{x}{R}\exp[-\mu_n^2\,\mathrm{Fo}] - \frac{\varrho(m_1-m_2)\,\mathrm{Bi}\,x}{2\alpha(1+\mathrm{Bi})R}$$
$$-\sum_{k=1}^{\infty}\frac{\varrho(m_2-m_1)}{2\alpha}A_{k,1}\sin\mu_{k,1}\frac{x}{R}\exp[-\mu_{k,1}^2\,\mathrm{Fo}], \tag{7.5.18}$$

where

$$A_n = \frac{2\sin\mu_n}{\mu_n + \sin\mu_n\cos\mu_n} = (-1)^{n+1}\frac{2\,\mathrm{Bi}(\mathrm{Bi}^2+\mu_n^2)^{1/2}}{\mu_n(\mathrm{Bi}^2+\mathrm{Bi}+\mu_n^2)},$$

$$A_{k,1} = \frac{2\cos\mu_{k,1}}{\mu_{k,1} - \sin\mu_{k,1}\cos\mu_{k,1}} = (-1)^{k+1}\frac{2\,\mathrm{Bi}(\mathrm{Bi}^2+\mu_{k,1}^2)^{1/2}}{\mu_{k,1}(\mathrm{Bi}^2+\mathrm{Bi}+\mu_{k,1}^2)}, \tag{7.5.19}$$

where μ_n and $\mu_{k,1}$ are the roots of the characteristic equations

$$\cot\mu = (1/\mathrm{Bi})\,\mu, \qquad \tan\mu_1 = -(1/\mathrm{Bi})\,\mu_1. \tag{7.5.20}$$

The first six values of the roots μ_n and coefficients A_n are given in Chapter 6, Tables 1 and 2. In Tables 7.1 and 7.2, the first six values of $\mu_{k,1}$ and $A_{k,1}$ are given for various values of the Biot criterion (from 0 to ∞).

TABLE 7.1. THE ROOTS OF THE CHARACTERISTIC EQUATION $\tan \mu = -\mu/\mathrm{Bi}$

Bi	$\mu_{1,1}$	$\mu_{2,1}$	$\mu_{3,1}$	$\mu_{4,1}$	$\mu_{5,1}$	$\mu_{6,1}$
0	1.5708	4.7124	7.8540	10.9956	14.1372	17.2788
0.1	1.6320	4.7335	7.8667	11.0047	14.1443	17.2845
0.2	1.6887	4.7544	7.8794	11.0137	14.1513	17.2903
0.3	1.7414	4.7751	7.8920	11.0228	14.1584	17.2961
0.4	1.7906	4.7956	7.9046	11.0318	14.1654	17.3019
0.5	1.8366	4.8158	7.9171	11.0409	14.1724	17.3076
0.6	1.8798	4.8358	7.9295	11.0498	14.1795	17.3134
0.7	1.9203	4.8556	7.9419	11.0588	14.1865	17.3192
0.8	1.9586	4.8751	7.9542	11.0677	14.1935	17.3249
0.9	1.9947	4.8943	7.9665	11.0767	14.2005	17.3306
1.0	2.0288	4.9132	7.9787	11.0856	14.2075	17.3364
1.5	2.1746	5.0037	8.0385	11.1296	14.2421	17.3649
2.0	2.2889	5.0870	8.0962	11.1727	14.2764	17.3932
3.0	2.4557	5.2329	8.2045	11.2560	14.3434	17.4490
4.0	2.5704	5.3540	8.3029	11.3349	14.4080	17.5034
5.0	2.6537	5.4544	8.3914	11.4086	14.4699	17.5562
6.0	2.7165	5.5378	8.4703	11.4773	14.5288	17.6072
7.0	2.7654	5.6078	8.5406	11.5408	14.5847	17.6562
8.0	2.8044	5.6669	8.6031	11.5994	14.6374	17.7032
9.0	2.8363	5.7172	8.6587	11.6532	14.6870	17.7481
10.0	2.8628	5.7606	8.7083	11.7027	14.7335	17.7908
15.0	2.9476	5.9080	8.8898	11.8959	14.9251	17.9742
20.0	2.9930	5.9921	9.0019	12.0250	15.0625	18.1136
30.0	3.0406	6.0831	9.1294	12.1807	15.2380	18.3018
40.0	3.0651	6.1311	9.1986	12.2688	15.3417	18.4180
50.0	3.0801	6.1606	9.2420	12.3247	15.4090	18.4953
60.0	3.0901	6.1805	9.2715	12.3632	15.4559	18.5497
80.0	3.1028	6.2058	9.3089	12.4124	15.5164	18.6209
100.0	3.1105	6.2211	9.3317	12.4426	15.5537	18.6650
∞	3.1416	6.2832	9.4248	12.5664	15.7080	18.8496

With the steady state established (Fo $= \infty$) *we have*

$$t(x) = t_a - \frac{\varrho}{2\alpha}\left[(m_1 + m_2) + \frac{x}{R}(m_1 - m_2)\frac{\mathrm{Bi}}{1 + \mathrm{Bi}}\right], \qquad (7.5.21)$$

i.e., there is a linear variation of the temperature over the thickness of the plate.

If $m_1 = m_2 = m$, then from (7.5.21) we obtain

$$t(x) = t_a - (\varrho m/\alpha) = t_{wb} = \text{const.}$$

TABLE 7.2. THE VALUES OF THE CONSTANTS

$$A_{k,1} = (-1)^{k+1} \frac{2\mathrm{Bi}(\mathrm{Bi}^2 + \mu_{k,1}^2)^{1/2}}{\mu_{k,1}(\mathrm{Bi}^2 + \mathrm{Bi} + \mu_{k,1}^2)}$$

Bi	$A_{1,1}$	$A_{2,1}$	$A_{3,1}$	$A_{4,1}$	$A_{5,1}$	$A_{6,1}$
0	0.0000	−0.0000	0.0000	−0.0000	0.0000	−0.0000
0.1	0.0721	−0.0089	0.0032	−0.0017	0.0010	−0.0007
0.2	0.1303	−0.0175	0.0064	−0.0033	0.0020	−0.0013
0.3	0.1779	−0.0259	0.0096	−0.0049	0.0030	−0.0020
0.4	0.2172	−0.0341	0.0127	−0.0065	0.0040	−0.0027
0.5	0.2514	−0.0420	0.0158	−0.0082	0.0050	−0.0033
0.6	0.2803	−0.0497	0.0189	−0.0098	0.0059	−0.0040
0.7	0.3065	−0.0571	0.0219	−0.0114	0.0069	−0.0047
0.8	0.3156	−0.0643	0.0249	−0.0129	0.0079	−0.0053
0.9	0.3471	−0.0713	0.0278	−0.0145	0.0089	−0.0060
1.0	0.3646	−0.0781	0.0305	−0.0160	0.0098	−0.0066
1.5	0.4298	−0.1088	0.0446	−0.0237	0.0146	−0.0099
2.0	0.4726	−0.1348	0.0576	−0.0311	0.0193	−0.0129
3.0	0.5254	−0.1756	0.0805	−0.0448	0.0281	−0.0192
4.0	0.5562	−0.2052	0.0957	−0.0571	0.0365	−0.0251
5.0	0.5759	−0.2270	0.1159	−0.0682	0.0442	−0.0310
6.0	0.5892	−0.2435	0.1293	−0.0779	0.0513	−0.0360
7.0	0.5987	−0.2561	0.1404	−0.0865	0.0578	−0.0409
8.0	0.6056	−0.2659	0.1496	−0.0941	0.0638	−0.0455
9.0	0.6108	−0.2736	0.1574	−0.1007	0.0691	−0.0498
10.0	0.6148	−0.2789	0.1639	−0.1065	0.0739	−0.0538
15.0	0.6256	−0.2978	0.1844	−0.1265	0.0920	−0.0694
20.0	0.6300	−0.3056	0.1945	−0.1375	0.1028	−0.0797
30.0	0.6336	−0.3122	0.2034	−0.1479	0.1140	−0.0911
40.0	0.6348	−0.3147	0.2068	−0.1524	0.1191	−0.0966
50.0	0.6354	−0.3160	0.2088	−0.1546	0.1218	−0.0995
60.0	0.6358	−0.3167	0.2098	−0.1559	0.1234	−0.1015
80.0	0.6361	−0.3175	0.2105	−0.1573	0.1250	−0.1034
100.0	0.6363	−0.3177	0.2113	−0.1579	0.1258	−0.1043
∞	0.6366	−0.3183	0.2122	−0.1591	0.1274	−0.1060

The mean temperature of the plate is

$$\bar{\theta} = \frac{\bar{t}(\tau) - t_0}{t_a - t_0}$$

$$= 1 - \frac{\varrho(m_1 + m_2)}{2\alpha(t_a - t_0)} - \sum_{n=1}^{\infty} \left[1 - \frac{\varrho(m_1 + m_2)}{2\alpha(t_a - t_0)}\right] B_n \exp[-\mu_n^2 \,\mathrm{Fo}]. \quad (7.5.22)$$

In Figs. 7.4 and 7.5 the graphs $\mu_{1,1} = f(\mathrm{Bi})$ and $A_{1,1} = f(\mathrm{Bi})$ are given using the data of Tables 7.1 and 7.2.

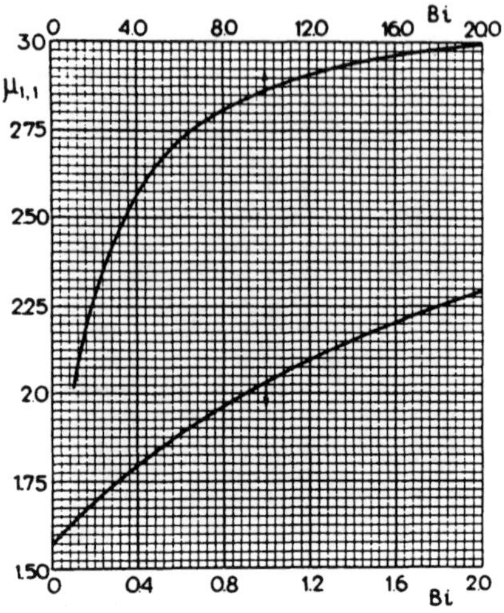

Fig. 7.4. The first root $\mu_{1,1}$ of the characteristic equation versus the Biot criterion for a plate (asymmetric problem).

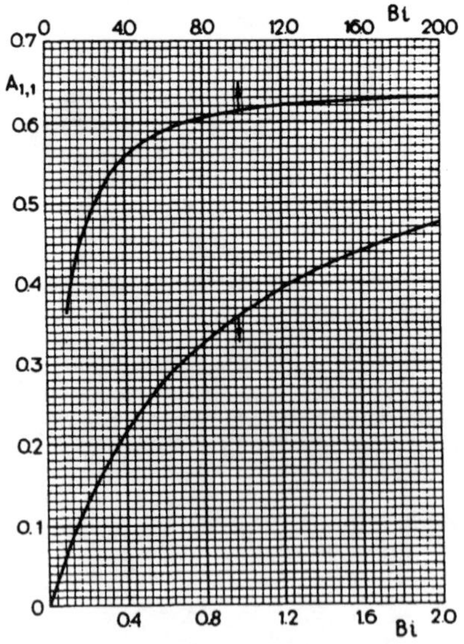

Fig. 7.5. Coefficient $A_{1,1}$ versus the Biot criterion for a plate (asymmetric problem).

7.6 Thermal Waves. Infinite Plate, Semi-Infinite Body, Sphere, and Infinite Cylinder. Ambient Temperature as a Simple Harmonic Function of Time

In many thermal processes the ambient temperature is a periodic function of time. Then the temperature distribution in a solid body heated in this medium will be similar to the distribution of displacement of oscillating points exposed to a wave process in an elastic medium. Usually these problems are therefore referred to as the problems of distribution of thermal waves, the latter being understood in the macroscopic sense of the word.

We shall consider problems of heating a body (infinite plate, sphere, and infinite cylinder) when the ambient temperature changes by the law of simple harmonic oscillation, i.e., it undergoes a cosine or sine variation.

Usually, the solution of problems on thermal waves is given for the quasi-stationary state, i.e., it is assumed that the process continues so long that the initial temperature distribution has lost its influence on the process history (problems without initial conditions). The solutions are given here for a general case accounting for the initial conditions from which, as a special case, the known solutions are obtained for the quasi-stationary state.

Let us consider in detail the solution of the problem for an infinite plate.

a. Infinite Plate. Statement of the Problem. *Consider an infinite plate with the thickness $2R$ at temperature $t_0 = 0°$. At the initial time, it is placed into a medium, the temperature of which changes according to the law of simple harmonic oscillation*

$$t_a(\tau) = t_m \cos 2\pi\nu\tau\,, \tag{7.6.1}$$

where ν is the frequency of oscillation and t_m is the maximum ambient temperature, i.e., the amplitude of the ambient temperature oscillation. Heat is transferred between the surfaces of the plate and the surrounding medium by Newton's law of cooling. The temperature distribution along the thickness of the plate at any moment and the specific heat rate are to be found.

We have

$$t(x, 0) = t_0 = 0\,, \tag{7.6.2}$$

$$\partial t(0, \tau)/\partial x = 0\,, \quad t(0, \tau) \neq \infty \tag{7.6.3}$$

$$-\frac{\partial t(R, \tau)}{\partial x} + H[t_m \cos 2\pi\nu\tau - t(R, \tau)] = 0\,. \tag{7.6.4}$$

The problem is symmetric, the origin of coordinates being in the center of the plate.

b. Solution of the Problem. The solution for the transform under condition (7.6.2) and (7.6.3) will be

$$T(x, s) = A \cosh(s/a)^{1/2}x. \tag{7.6.5}$$

The constant A is determined from the boundary condition (7.6.4) which for the transform will have the form

$$-T'(R, s) + H\left[\frac{st_m}{s^2 + 4\pi^2\nu^2} - T(R, s)\right] = 0, \tag{7.6.6}$$

since

$$L[\cos 2\pi\nu\tau] = \frac{s}{s^2 + (2\pi\nu)^2}.$$

Substituting the solution into condition (7.6.6) we have

$$-A\left(\frac{s}{a}\right)^{1/2} \sinh\left(\frac{s}{a}\right)^{1/2} R + \frac{sHt_m}{s^2 + 4\pi^2\nu^2} - HA \cosh\left(\frac{s}{a}\right)^{1/2} R = 0. \tag{7.6.7}$$

Having determined the constant A from equality (7.6.7) and substituting the expression obtained into solution (7.6.5), we shall have

$$\begin{aligned} T(x, s) &= \frac{st_m \cosh(s/a)^{1/2}x}{(s^2+4\pi^2\nu^2)[\cosh(s/a)^{1/2}R+(1/H)(s/a)^{1/2}\sinh(s/a)^{1/2}R]} \\ &= \frac{\Phi(s)}{\psi(s)}. \end{aligned} \tag{7.6.8}$$

Solution (7.6.8) represents the ratio of two generalized polynomials which satisfy the conditions of the expansion theorem. Equating the polynomial of the denominator with zero, i.e., $\psi(s) = 0$, we find the roots: $s_1 = i2\pi\nu$, $s_2 = -i2\pi\nu$ and $s_n = -a\mu_n^2/R^2$, which is an infinite number of roots determined from an ordinary characteristic equation.

If we use the expansion theorem (the case of simple roots), then the solution of our problem will be obtained in the form

$$\begin{aligned} \theta &= \frac{t(x, \tau)}{t_m} \\ &= \frac{1}{2}\left[\frac{\cosh(i\{\omega/a\})^{1/2}x}{\cosh(i\{\omega/a\})^{1/2}R+(1/H)(i\{\omega/a\})^{1/2}\sinh(i\{\omega/a\})^{1/2}R} e^{i\omega\tau}\right. \\ &\quad \left. + \frac{\cosh(-i\{\omega/a\})^{1/2}x}{\cosh(-i\{\omega/a\})^{1/2}R+(1/H)(-i\{\omega/a\})^{1/2}\sinh(-i\{\omega/a\})^{1/2}R} e^{-i\omega\tau}\right] \\ &\quad - \sum_{n=1}^{\infty} A_n\left(\frac{\mu_n^4}{\mu_n^4 + (\omega^2/a^2)R^4}\right) \cos \mu_n \frac{x}{R} \exp\left[-\mu_n^2 \frac{a\tau}{R^2}\right], \end{aligned} \tag{7.6.9}$$

where $\omega = 2\pi\nu$ is the circular frequency ($\nu = 1/P$, P is the period of oscillation) and A_n are the coefficients determined from the known relation for an infinite plate.

c. Analysis of the Solution. The maximum rate of change the of ambient temperature $[dt_a(\tau)/d\tau]_m$ is equal to $2\pi\nu t_m$. Then the Predvoditelev criterion will be equal to

$$\mathrm{Pd} = \left(\frac{d\theta_a}{d\mathrm{Fo}}\right)_{\max} = \frac{2\pi\nu}{a} R^2 = \frac{\omega}{a} R^2. \tag{7.6.10}$$

Hence, solution (7.6.9) may be rewritten as

$$\theta = \tfrac{1}{2}\{N_i \exp[i\,\mathrm{Pd}\,\mathrm{Fo}] + N_{-i}\exp[-i\,\mathrm{Pd}\,\mathrm{Fo}]\}$$
$$- \sum_{n=1}^{\infty}\left[\frac{\mu_n{}^4}{\mu_n{}^4 + \mathrm{Pd}^2}\right] A_n \cos\mu_n \frac{x}{R}\exp[-\mu_n{}^2\,\mathrm{Fo}], \tag{7.6.11}$$

where

$$N_i = \frac{\cosh(i\,\mathrm{Pd})^{1/2}(x/R)}{[\cosh(i\,\mathrm{Pd})^{1/2} + (1/\mathrm{Bi})(i\,\mathrm{Pd})^{1/2}\sinh(i\,\mathrm{Pd})^{1/2}]}, \tag{7.6.12}$$

$$N_{-i} = \frac{\cosh(-i\,\mathrm{Pd})^{1/2}\, x/R}{[\cosh(-i\,\mathrm{Pd})^{1/2} + (1/\mathrm{Bi})(-i\,\mathrm{Pd})^{1/2}\,\sinh(-i\,\mathrm{Pd})^{1/2}]}. \tag{7.6.13}$$

The sum in solution (7.6.11) decreases with time and beginning from some value $\mathrm{Fo} > \mathrm{Fo}_1$ (quasi-stationary state) it becomes negligible as compared with the two first terms. *Then for the quasi-stationary state we may write*

$$\theta = \tfrac{1}{2}\{N_i \exp[i\,\mathrm{Pd}\,\mathrm{Fo}] + N_{-i}\exp[-i\,\mathrm{Pd}\,\mathrm{Fo}]\}$$
$$= \tfrac{1}{2}[(N_i + N_{-i})\cos\mathrm{Pd}\,\mathrm{Fo} + i(N_i - N_{-i})\sin\mathrm{Pd}\,\mathrm{Fo}]$$
$$= (N_i \cdot N_{-i})^{1/2}\cos\left[\mathrm{Pd}\,\mathrm{Fo} - \arctan\left(i\frac{N_i - N_{-i}}{N_i + N_{-i}}\right)\right]. \tag{7.6.14}$$

From solution (7.6.14) it follows that the temperature at any point of the plate undergoes simple harmonic oscillation with the same frequency, but the period of oscillation lags behind the period of the ambient temperature oscillation by the value $\arctan\{i(N_i - N_{-i})/(N_i + N_{-i})\}$ since $\theta_a = t_a/t_m = \cos\mathrm{Pd}\,\mathrm{Fo}$. (The Predvoditelev criterion is directly proportional to the frequency.)

The amplitude of temperature oscillations at any point of the plate decreases with depth. The maximum amplitude corresponds, therefore, to the surface of the plate ($x = R$), but it is less than the amplitude of the ambient temperature oscillations by the value $(N_i N_{-i})^{1/2}$.

Let Bi $= \infty$; *then the temperature of the surface at all times is equal to the temperature of the medium, which undergoes cosine variations, i.e.,*

$$t(\pm R, \tau) = t_m \cos 2\pi\nu\tau = t_m \cos \text{Pd Fo}.$$

In this case

$$N_i = \frac{\cosh\{(i\,\text{Pd})^{1/2}(x/R)\}}{\cosh(i\,\text{Pd})^{1/2}}, \qquad N_{-i} = \frac{\cosh\{(-i\,\text{Pd})^{1/2}(x/R)\}}{\cosh(-i\,\text{Pd})^{1/2}}. \tag{7.6.15}$$

The value of $\sqrt{\pm i}$ may be found in the following way. We have $(1 + i)^2 = 1 + 2i - 1 = 2i$, from which

$$\sqrt{i} = \pm\sqrt{\tfrac{1}{2}}\,(1 + i).$$

In a similar way we obtain

$$\sqrt{-i} = \pm\sqrt{\tfrac{1}{2}}\,(1 - i).$$

The sign $\pm$ for solutions (7.6.12) and (7.6.13) may be omitted since cosh z is an even function, and z stands as a multiplier before the odd function sinh z. Therefore, when we take the $+$ sign or the $-$ sign, we obtain the same result irrespective of the sign taken.

Using the formulas

$$\begin{aligned}\cosh(i\,\text{Pd})^{1/2}\,x/R &= \cosh\{(\tfrac{1}{2}\,\text{Pd})^{1/2}\,(1 + i)\,(x/R)\}\\ &= \cosh\{(\tfrac{1}{2}\,\text{Pd})^{1/2}\,(x/R)\}\cos\{(\tfrac{1}{2}\,\text{Pd})^{1/2}\,(x/R)\}\\ &\quad + i\sinh\{(\tfrac{1}{2}\,\text{Pd})^{1/2}\,(x/R)\}\sin\{(\tfrac{1}{2}\,\text{Pd})^{1/2}\,(x/R)\},\end{aligned}$$

we may rewrite solution (7.6.14) in precisely the same form but the value of supplementary quantities in it will be the following for the case Bi $= \infty$.

$$(N_i N_{-i})^{1/2} = \left[\frac{f_c^2(\{\tfrac{1}{2}\,\text{Pd})^{1/2}\,(x/R)\} + f_s^2\{(\tfrac{1}{2}\,\text{Pd})^{1/2}\,(x/R)\}}{f_c^2(\tfrac{1}{2}\,\text{Pd})^{1/2} + f_s^2(\tfrac{1}{2}\,\text{Pd})^{1/2}}\right]^{1/2}, \tag{7.6.16}$$

$$-\arctan\left(i\frac{N_i - N_{-i}}{N_i + N_{-i}}\right)$$

$$= \arctan\frac{f_c(\tfrac{1}{2}\,\text{Pd})^{1/2} f_s((\tfrac{1}{2}\,\text{Pd})^{1/2}\,\{x/R\}) - f_s(\tfrac{1}{2}\,\text{Pd})^{1/2} f_c((\tfrac{1}{2}\,\text{Pd})^{1/2}\,\{x/R\})}{f_c(\tfrac{1}{2}\,\text{Pd})^{1/2} f_c((\tfrac{1}{2}\,\text{Pd})^{1/2}\,\{x/R\}) + f_s(\tfrac{1}{2}\,\text{Pd})^{1/2} f_s((\tfrac{1}{2}\,\text{Pd})^{1/2}\,\{x/R\})}, \tag{7.6.17}$$

where $f_c(z) = \cosh z$, $f_s(z) = \sinh z \sin z$. The values of the hyperbola-trigonometric functions may be found in Krylov's monograph.[1]

[1] A. Krylov, "On the Calculation of Bars on Elastic Foundation." Leningrad, 1931.

TABLE 7.3. THE VALUES OF THE CONSTANTS A_c, B_c, A_s, AND B_s

z	A_c	B_c	A_s	B_s
0.01	1.000	0.0001	0.0071	0.0071
0.025	1.000	0.0003	0.0177	0.0177
0.05	1.000	0.001	0.035	0.035
0.075	1.000	0.003	0.050	0.050
0.10	1.000	0.005	0.070	0.071
0.15	1.000	0.011	0.106	0.106
0.20	1.000	0.020	0.140	0.142
0.25	1.000	0.031	0.175	0.179
0.30	1.000	0.045	0.209	0.215
0.35	0.999	0.062	0.242	0.253
0.40	0.999	0.080	0.275	0.290
0.45	0.998	0.095	0.307	0.329
0.50	0.997	0.125	0.339	0.369
0.55	0.996	0.151	0.369	0.408
0.60	0.994	0.180	0.398	0.449
0.65	0.993	0.212	0.427	0.492
0.70	0.990	0.245	0.454	0.534
0.75	0.987	0.281	0.479	0.578
0.80	0.983	9.320	0.505	0.625
0.85	0.978	0.362	0.526	0.670
0.90	0.973	0.404	0.547	0.718
0.95	0.966	0.451	0.566	0.771
1.00	0.958	0.499	0.583	0.819
1.05	0.950	0.549	0.599	0.871
1.10	0.938	0.601	0.612	0.924
1.15	0.927	0.658	0.622	0.980
1.20	0.916	0.705	0.629	1.026
1.25	0.898	0.776	0.636	1.096
1.30	0.881	0.839	0.639	1.155
1.35	0.862	0.906	0.639	1.216
1.40	0.834	0.970	0.636	1.271
1.50	0.790	1.108	0.621	1.410
1.60	0.727	1.258	0.590	1.550
1.70	0.654	1.411	0.546	1.692
1.80	0.566	1.571	0.484	1.839
1.90	0.460	1.741	0.402	1.995
2.0	0.340	1.911	0.302	2.151
2.1	0.199	2.087	0.180	2.312
2.2	0.038	2.261	0.036	2.477
2.3	−0.146	2.440	−0.135	2.636
2.4	−0.355	2.616	−0.291	2.798
2.5	−0.587	2.787	−0.554	2.954

TABLE 7.3. (*continued*)

z	A_c	B_c	A_s	B_s
2.6	− 0.851	2.953	− 0.808	3.107
2.7	− 1.105	3.093	− 1.057	3.234
2.8	− 1.468	3.259	− 1.413	3.386
2.9	− 1.821	3.389	− 1.761	3.504
3.0	− 2.206	3.502	− 2.143	3.604
3.2	− 3.086	3.656	− 3.021	3.639
3.4	− 4.129	3.695	− 4.057	3.752
3.6	− 5.304	3.553	− 5.239	3.602
3.8	− 6.629	3.256	− 6.557	3.239
4.0	− 8.093	2.589	− 8.038	2.607
4.2	− 9.628	1.660	− 9.578	1.669
4.4	−11.227	0.353	−11.183	0.355
4.6	−12.839	− 1.393	−12.800	− 1.393
4.8	−14.437	− 3.715	−14.404	− 3.724
5.0	−15.851	− 6.568	−15.824	− 6.759
5.5	−17.939	−16.982	−17.984	−16.996
6.0	−15.794	−30.897	−15.787	−30.911
6.5	− 5.796	−49.149	− 5.795	−49.159
7.0	16.616	−68.601	16.614	−68.601
7.5	55.532	−83.365	55.529	−83.369
8.0	115.788	−83.919	115.786	−83.921
8.5	196.188	−54.969	196.186	−54.970
9.0	289.033	23.791	289.019	23.791
9.5	375.199	175.924	275.197	175.925
10.0	414.391	417.288	414.525	417.289

We believe that the calculation of coefficients N_i and N_{-i} by formulas (7.6.16) and (7.6.17) is less convenient than the calculation directly by formula (7.6.15).

In order to simplify calculations, Table 7.3 gives the values of coefficients A_s, B_s, A_c, and B_c with the help of which we can calculate the hyperbolic functions in formula (7.6.15) as

$$\cosh z\sqrt{\pm i} = A_c \pm iB_c, \qquad \sinh z\sqrt{\pm i} = A_s \pm iB_s.$$

d. A Semi-Infinite Body. From solution (7.6.11) we may obtain the solutions for a semi-infinite body. Let us transpose the origin of coordinates from the middle of the plate to the left surface, i.e., we shall replace the variable x by $X - R$ and assume $2R \to \infty$ ($x = X - R$).

We rewrite the expression (7.6.12) as

$$N_i = \frac{\exp\left[-\left(i\frac{\omega}{a}\right)^{1/2}(R-x)\right] + \exp\left[-\left(i\frac{\omega}{a}\right)^{1/2}(R+x)\right]}{\left(1+\frac{1}{H}\left(i\frac{\omega}{a}\right)^{1/2}\right)\left[1+\left\{\frac{1-\frac{1}{H}\left(i\frac{\omega}{a}\right)^{1/2}}{1+\frac{1}{H}\left(i\frac{\omega}{a}\right)^{1/2}}\right\}\exp\left[-2R\left(i\frac{\omega}{a}\right)^{1/2}\right]\right]}.$$

At $R-x=2R-X$, $R+x=X$, the value N_i will be equal to

$$\lim_{2R\to\infty} N_i = \frac{1}{\left(1+\frac{1}{H}\left(i\frac{\omega}{a}\right)^{1/2}\right)}\exp\left[-X\left(i\frac{\omega}{a}\right)^{1/2}\right].$$

In a similar way we shall find

$$\lim_{2R\to\infty} N_{-i} = \frac{1}{\left(1+\frac{1}{H}\left(-i\frac{\omega}{a}\right)^{1/2}\right)}\exp\left[-X\left(-i\frac{\omega}{a}\right)^{1/2}\right].$$

Then we obtain

$$\theta = \frac{t(X,\tau)}{t_m} = \frac{1}{2}\left\{\frac{1}{1+\frac{1}{H}(1+i)\left(\frac{1}{2}\frac{\omega}{a}\right)^{1/2}}\exp\left[i\omega\tau - X(1+i)\left(\frac{1}{2}\frac{\omega}{a}\right)^{1/2}\right]\right.$$

$$\left. + \frac{1}{1+\frac{1}{H}(1-i)\left(\frac{1}{2}\frac{\omega}{a}\right)^{1/2}}\exp\left[-i\omega\tau - X(1-i)\left(\frac{1}{2}\frac{\omega}{a}\right)^{1/2}\right]\right\}$$

$$= \exp\left[-X\left(\frac{1}{2}\frac{\omega}{a}\right)^{1/2}\right]\frac{1}{(1+\delta)^2+\delta^2}\left[(1+\delta)\cos\left(X\left(\frac{1}{2}\frac{\omega}{a}\right)^{1/2} - \omega\tau\right)\right.$$

$$\left. - \delta\sin\left(X\left(\frac{1}{2}\frac{\omega}{a}\right)^{1/2} - \omega\tau\right)\right]$$

$$= \frac{1}{[(1+\delta)^2+\delta^2]^{1/2}}\exp\left[-X\left(\frac{1}{2}\frac{\omega}{a}\right)^{1/2}\right]\cos\left(X\left(\frac{1}{2}\frac{\omega}{a}\right)^{1/2} - \omega\tau + \arctan\frac{\delta}{1+\delta}\right) \tag{7.6.18}$$

where

$$\delta = \frac{1}{H}\left(\frac{1}{2}\frac{\omega}{a}\right)^{1/2} = \frac{1}{H}\left(\frac{\pi}{aP}\right)^{1/2}P,$$

and P is the period of oscillation ($P = 1/\nu$). Solution (7.6.18) may be rewritten

$$\theta = A_0 \exp\left[-X\left(\frac{\pi}{aP}\right)^{1/2}\right] \cos\left[\frac{2\pi}{P}\tau - \left(X\left(\frac{\pi}{aP}\right)^{1/2} + M\right)\right], \quad (7.6.19)$$

where the value

$$A_0 = \left(1 + \frac{2}{H}\left(\frac{\pi}{aP}\right)^{1/2} + \frac{2\pi}{H^2 aP}\right)^{-1/2} \quad (7.6.20)$$

is the maximum amplitude of temperature oscillation equal to the amplitude of the temperature oscillation of the bounding surface, and the value

$$M = \arctan\left(1 + H\left(\frac{aP}{\pi}\right)^{1/2}\right)^{-1} \quad (7.6.21)$$

represents the shift of temperature oscillation of the bounding surface as compared to the ambient temperature oscillation (Fig. 7.6).

This solution may be expressed in terms of generalized variables

$$\theta = A_0 \exp\left[-\left(\tfrac{1}{2}\mathrm{Pd}_x\right)^{1/2}\right] \cos\left[\mathrm{Fo}_x/\mathrm{Fo}_x' - \left(\left(\tfrac{1}{2}\mathrm{Pd}_x\right)^{1/2} + M\right)\right], \quad (7.6.22)$$

where

$$A_0 = [1 + (\sqrt{2}/\mathrm{Bi}^*) + (1/\mathrm{Bi}^*)^2]^{-1/2}, \quad (7.6.23)$$

$$M = \arctan(1 + \mathrm{Bi}^*\sqrt{2})^{-1}, \quad (7.6.24)$$

where $\mathrm{Pd}_x = \omega X^2/a$ is the local Pd criterion for the coordinate X. $\mathrm{Fo}_x = a\tau/X^2$ is a generalized local argument (dimensionless time) and $\mathrm{Fo}_x' = aP/2\pi X^2$ is a dimensionless number characterizing the periodic state. For the ratio $\mathrm{Fo}_x/\mathrm{Fo}_x' = \omega\tau$, Bi^* is a generalized argument for the steady-periodic state

$$\mathrm{Bi}^* = \alpha/(\lambda c\gamma\omega)^{1/2}. \quad (7.6.25)$$

We shall now clarify the physical significance of the number Bi^*. The instantaneous heat flux at the surface is

$$q(0, \tau) = -\lambda(\partial t/\partial X)_{X=0} = -\lambda t_m(\omega/2a)^{1/2}(\sin\omega\tau - \cos\omega\tau). \quad (7.6.26)$$

We now make use of the relation

$$\cos(\omega\tau + \tfrac{1}{4}\pi) = (\cos\omega\tau - \sin\omega\tau)(\sqrt{2})^{-1}.$$

Then the value of instantaneous heat flux at the surface

$$q(0, \tau) = (\lambda c\gamma\omega)^{1/2} t_m \cos(\omega\tau + \tfrac{1}{4}\pi)$$
$$= q_m \cos(\omega\tau + \tfrac{1}{4}\pi), \quad (7.6.27)$$

where q_m is the maximum specific heat flux or the amplitude of specific heat flux, which is equal to

$$q_m = t_m(\lambda c\gamma\omega)^{1/2}. \tag{7.6.28}$$

The ratio of amplitudes of heat fluxes and temperature

$$\frac{q_m}{t_m} = (\lambda c\gamma\omega)^{1/2}. \tag{7.6.29}$$

Thus the value of $(\lambda c\gamma\omega)^{1/2}$ *equals the maximum instantaneous heat flux at the body surface at an amplitude of the wall temperature of unity* ($t_m = 1$°C).

Thus, the dimensionless number Bi* *is the ratio of the steady-state heat flux* $\alpha\,\Delta t$ *at a unit temperature difference* ($\Delta t = 1$) *to* $(\lambda c\gamma\omega)^{1/2}$ *the maximum steady-periodic heat flux at a unit temperature amplitude* ($t_m = 1$). *In other words,* Bi* *is the modified Biot criterion for the steady-periodic state.*

Returning to formula (7.6.23), we note that for Bi* $\to \infty$ the dimensionless amplitude A_0 is unity, and $M = 0$.

In Fig. 7.6, the variation of relative ambient temperature θ_a and the relative temperature of the boundary surface of the body ($X = 0$) are shown. It is seen from Fig. 7.6 that the amplitude of the relative ambient temperature

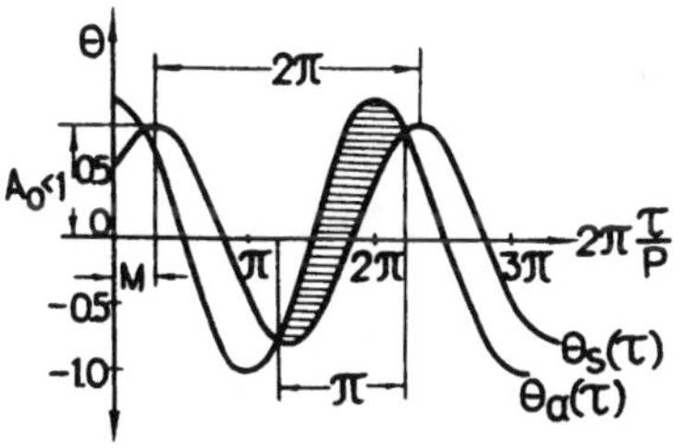

Fig. 7.6. Time changes of dimensionless temperatures of the surrounding medium and the body surface.

oscillations is equal to 1 and moreover on the surface of the body $A_0 < 1$. If the number Bi $= \infty$ ($H = \infty$), then $A_0 = 1$ (see relationship (7.6.20)). In Fig. 7.7, the plot of A_0 versus Bi* is presented which shows that with Bi* $= 0$, $A_0 = 0$. The phase shift between cosine curves θ_a and θ_s is equal to M; at Bi $= \infty$, this shift will be equal to π (see Eq. 7.6.21, Fig. 7.6). Thus, the temperature of any point of the body undergoes a harmonic oscillation.

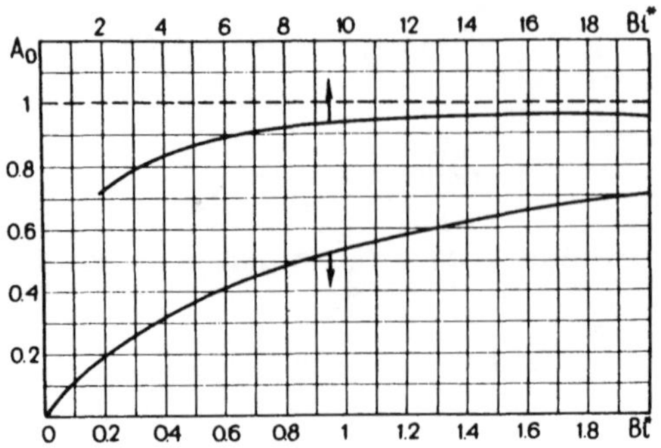

Fig. 7.7. The surface temperature oscillation amplitude versus Bi*.

For any given time ($\tau = \text{const}$) the temperature distribution along the depth of the body takes place according to the cosine law with gradually attenuating amplitude $A_0 \exp[-X(\pi/aP)^{1/2}]$ (Fig. 7.8).

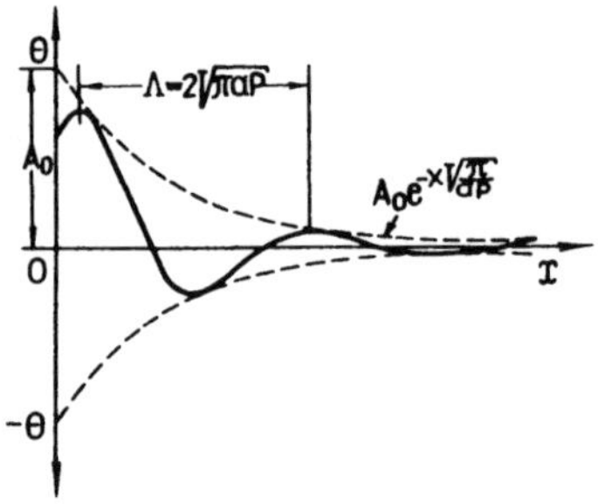

Fig. 7.8. Dimensionless temperature distribution along the plate depth at a certain moment.

The length of the wave Λ may be found as the distance between two points which differ in phase by 2π. It follows from solution (7.6.19) that the length of the wave is equal to

$$\Lambda = 2(\pi a P)^{1/2} = 2(\pi a/\nu)^{1/2}, \tag{7.6.30}$$

since $X(\pi/aP)^{1/2} = 2\pi$. *Hence, the wave length characterizing the depth of penetration of thermal waves is directly proportional to the square root of the product of heat conduction coefficient and the period of oscillation.*

It is known from the theory of oscillations that the speed of wave propagation is equal to the wave length divided by the period of oscillation. In the present case, the speed of thermal wave (speed with which any point of the wave moves) is equal to

$$u = \Lambda/P = 2(\pi a/P)^{1/2} = 2(\pi a\nu)^{1/2}, \tag{7.6.31}$$

i.e., the speed of thermal wave propagation increases with frequency and with the increase of the thermal diffusivity coefficient.

In Fig. 7.9, θ is plotted versus $\mathrm{Fo}_x/\mathrm{Fo}_x'$ for various values of the generalized argument $1/2(\mathrm{Fo}_x')^{1/2}$ from 0 to 180° at $\mathrm{Bi}^* = \infty$.

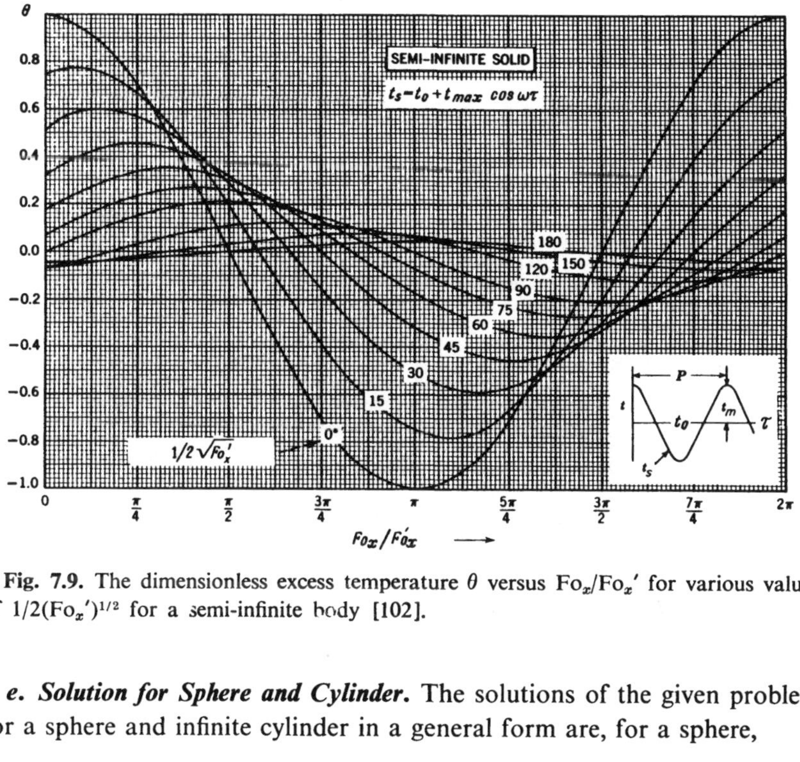

Fig. 7.9. The dimensionless excess temperature θ versus $\mathrm{Fo}_x/\mathrm{Fo}_x'$ for various values of $1/2(\mathrm{Fo}_x')^{1/2}$ for a semi-infinite body [102].

e. Solution for Sphere and Cylinder. The solutions of the given problem for a sphere and infinite cylinder in a general form are, for a sphere,

$$\theta = \tfrac{1}{2}\,(N_i \exp[i\,\mathrm{Pd\,Fo}] + N_{-i} \exp[-i\,\mathrm{Pd\,Fo}]) - \sum_{n=1}^{\infty} \frac{\mu_n^4}{\mu_n^4 + \mathrm{Pd}^2} A_n \frac{R \sin \mu_n(r/R)}{r\mu_n} \exp[-\mu_n^2\,\mathrm{Fo}], \tag{7.6.32}$$

where

$$N_i = \frac{R\,\mathrm{Bi}\sinh\{(-i\,\mathrm{Pd})^{1/2}\,(r/R)\}}{r[(\mathrm{Bi}-1)\sinh(i\,\mathrm{Pd})^{1/2} + (i\,\mathrm{Pd})^{1/2}\cosh(i\,\mathrm{Pd})^{1/2}}, \tag{7.6.33}$$

$$N_{-i} = \frac{R\,\mathrm{Bi}\sinh\{(-i\,\mathrm{Pd})^{1/2}(r/R)\}}{r[(\mathrm{Bi}-1)\sinh(-i\,\mathrm{Pd})^{1/2} + (-i\,\mathrm{Pd})^{1/2}\cosh(-i\,\mathrm{Pd})^{1/2}]}, \tag{7.6.34}$$

and where A_n are constants determined by relation (6.5.29), and for an infinite cylinder the solution is

$$\theta = \tfrac{1}{2}(N_i \exp[i \,\mathrm{Pd}\, \mathrm{Fo}] + N_{-i} \exp[-i \,\mathrm{Pd}\, \mathrm{Fo}]) - \sum_{n=1}^{\infty} \frac{\mu_n^4}{\mu_n^4 + \mathrm{Pd}^2} A_n J_0(\mu_n\{r/R\}) \exp[-\mu_n^2 \,\mathrm{Fo}], \quad (7.6.35)$$

where

$$N_i = \frac{I_0((i\,\mathrm{Pd})^{1/2}\{r/R\})}{[I_0(i\,\mathrm{Pd})^{1/2} + (1/\mathrm{Bi})(i\,\mathrm{Pd})^{1/2}\, I_1(i\,\mathrm{Pd})^{1/2}]}, \quad (7.6.36)$$

$$N_{-i} = \frac{I_0((-i\,\mathrm{Pd})^{1/2}\,\{r/R\})}{[I_0(-i\,\mathrm{Pd})^{1/2} + (1/\mathrm{Bi})\,(-i\,\mathrm{Pd})^{1/2}\, I_1(-i\,\mathrm{Pd})^{1/2}]}, \quad (7.6.37)$$

and where A_n are the constants determined by relation (6.6.27).

f. Specific Heat Rate. The specific heat rate for any interval of time $\Delta\tau = \tau_2 - \tau_1$ is found by the conventional formula

$$\Delta Q_v = c\gamma[\bar{\theta}(\tau_2) - \bar{\theta}(\tau_1)]t_m.$$

The mean temperature over the volume is equal to

$$\bar{\theta} = \tfrac{1}{2}\,(\bar{N}_i \exp[i\,\mathrm{Pd}\,\mathrm{Fo}] + \bar{N}_{-i} \exp[-i\,\mathrm{Pd}\,\mathrm{Fo}]) - \sum_{n=1}^{\infty} \frac{\mu_n^4}{\mu_n^4 + \mathrm{Pd}^2} B_n \exp[-\mu_n^2\,\mathrm{Fo}], \quad (7.6.38)$$

where B_n are the constant coefficients determined by relations from Chapter 6, Sections 2, 5, and 6 and the coefficients $\bar{N}_i$ and $\bar{N}_{-i}$ are equal, respectively.

For an infinite plate

$$\bar{N}_i = \frac{1}{(i\,\mathrm{Pd})^{1/2} \coth(i\,\mathrm{Pd})^{1/2} + (1/\mathrm{Bi})\, i\,\mathrm{Pd}}, \quad (7.6.39)$$

$$\bar{N}_{-i} = \frac{1}{(-i\,\mathrm{Pd})^{1/2} \coth(-i\,\mathrm{Pd})^{1/2} - (i/\mathrm{Bi})\,\mathrm{Pd}}; \quad (7.6.40)$$

for a sphere

$$\bar{N}_i = \frac{3\,\mathrm{Bi}((i\,\mathrm{Pd})^{1/2} - \tanh(i\,\mathrm{Pd})^{1/2})}{i\,\mathrm{Pd}[(\mathrm{Bi} - 1)\tanh(i\,\mathrm{Pd})^{1/2} + (i\,\mathrm{Pd})^{1/2}]}, \quad (7.6.41)$$

$$\bar{N}_{-i} = \frac{3\,\mathrm{Bi}((-i\,\mathrm{Pd})^{1/2} - \tanh(-i\,\mathrm{Pd})^{1/2})}{-i\,\mathrm{Pd}[(\mathrm{Bi} - 1)\tanh(-i\,\mathrm{Pd})^{1/2} + (-i\,\mathrm{Pd})^{1/2}]}; \quad (7.6.42)$$

and for an infinite cylinder

$$\bar{N}_i = \frac{2\, I_1(i\,\mathrm{Pd})^{1/2}}{[I_0(i\,\mathrm{Pd})^{1/2} + (1/\mathrm{Bi})(i\,\mathrm{Pd})^{1/2}\, I_1(i\,\mathrm{Pd})^{1/2}](i\,\mathrm{Pd})^{1/2}}, \tag{7.6.43}$$

$$\bar{N}_{-i} = \frac{2\, I_1(-i\,\mathrm{Pd})^{1/2}}{(-i\,\mathrm{Pd})^{1/2}[I_0(-i\,\mathrm{Pd})^{1/2} + (1/\mathrm{Bi})(-i\,\mathrm{Pd})^{1/2} I_1(-i\,\mathrm{Pd})^{1/2}]}. \tag{7.6.44}$$

For the quasi-stationary state, the mean temperature will be a periodic function of time. Using similar transformations, we have

$$\bar{\theta} = (\bar{N}_i \bar{N}_{-i})^{1/2} \cos(\{2\pi/P\}\tau - \bar{M}), \tag{7.6.45}$$

where

$$\bar{M} = \arctan i \frac{\bar{N}_i - \bar{N}_{-i}}{\bar{N}_i + \bar{N}_{-i}}. \tag{7.6.46}$$

Thence

$$\begin{aligned} \Delta Q_v &= c\gamma t_m (\bar{N}_i \bar{N}_{-i})^{1/2} \left[\cos\left(\frac{2\pi\tau_2}{P} - \bar{M}\right) - \cos\left(\frac{2\pi\tau_1}{P} - \bar{M}\right)\right] \\ &= 2c\gamma t_m (\bar{N}_i \bar{N}_{-i})^{1/2} \sin \pi\nu(\tau_2 - \tau_1) \sin[\bar{M} - \pi\nu(\tau_2 + \tau_1)]. \end{aligned} \tag{7.6.47}$$

Thus the specific heat rate changes in time by the law of simple harmonic oscillation with the same period as the period of ambient temperature oscillation but with the phase shift.

It follows from relation (7.6.47) that the amount of heat transferred during the period of oscillation ($\tau_2 - \tau_1 = P = 1/\nu$) is equal to zero. We shall now determine the heat transferred for the time interval $\Delta\tau$ equal to a half period (i.e., $\Delta\tau = 1/2\nu$). For this time increment, $\Delta\tau = 1/2\nu$, the value of of $\sin \pi\nu(\tau_2 - \tau_1)$ is unity. The reference point of τ_1 is chosen so that $\sin[\bar{M} - \pi\nu(\tau_2 + \tau_1)$ be unity, i.e.,

$$\tau_1 = (1/2\pi\nu)(\bar{M} - \pi). \tag{7.6.48}$$

Then the heat transferred during the half period, $\Delta\tau = 1/2\nu$, is

$$\Delta Q_v = 2c\gamma t_m (\bar{N}_i \bar{N}_{-i})^{1/2}. \tag{7.6.49}$$

The quantity $(\bar{N}_i \bar{N}_{-i})^{1/2}$ is a function of Pd and Bi, i.e., it depends on the heat transfer coefficient α, frequency ω, and thermal properties (a, λ, γ).

We shall analyze formula (7.6.49) for the case of an infinite plate for $\mathrm{Bi} \to \infty$. In this case, the temperature of the wall surface varies according to the law of harmonic oscillations

$$t_s(\tau) = t_m \cos \omega\tau. \tag{7.6.50}$$

If the wall thickness is large or the temperature changes very quickly, then the temperature oscillations across the wall should damp completely before the midplane of the wall (see Fig. 7.10a). Then either half of the wall be haves as a body of an infinite thickness (half-space). Thus, a plate of large thickness represents these extreme cases.

In the opposite case (a very thin wall or extremely slow temperature changes), the whole thickness of the wall responds to the temperature oscillations of the surface, with no amplitude decrease and time lag. Under these conditions the temperature across the whole thickness is uniform (independent of x), i.e., $t_x = t_s$. Then the heat accumulated during the half-period $\Delta\tau = \tau_2 - \tau_1$ by a unit volume of the wall is

$$(\Delta Q_v)_{2R\to 0} = c\gamma\left[\int_{\tau_1}^{\tau_2} (t_m \cos \omega\tau)\, d\tau\right] = 2\, t_m c\gamma. \qquad (7.6.51)$$

Relation (7.6.51) is similar to the simple calorimetric formula $\Delta Q_v = c\gamma\,\Delta t$ for heating a body by Δt. In the case of interest $\Delta t = 2t_m$. Consequently, formula (7.6.51) defines the heat quantity received by a unit volume of the wall when it is uniformly heated from $-t_m$ to $+t_m$.

All actual walls fall between these two extreme cases of very thick and very thin walls. Decreasing the wall thickness starting from a "thick" wall finally leads to the case when oscillations propagating from both sides come in contact in the midplane of the wall, and mutual penetration of oscillations begins. Instantaneous temperature distribution for this case is shown in Fig. 7.10b, and the solution of the problem has been discussed.

According to (7.6.49), *the heat transfer during the half period is*

$$\Delta Q_v = 2c\gamma t_m K_\varphi, \qquad (7.6.52)$$

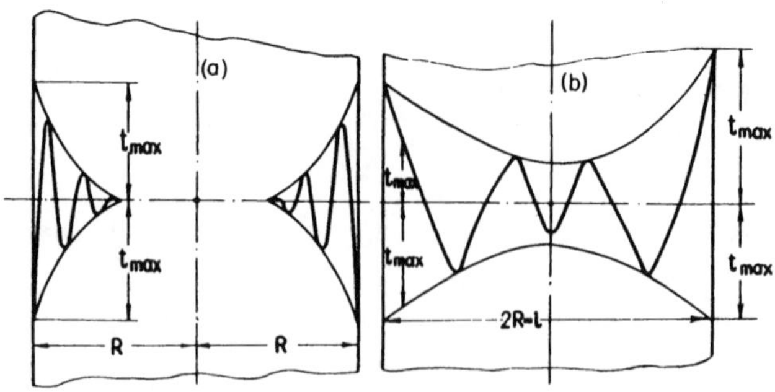

Fig. 7.10. Penetration of temperature waves into a wall of (a) considerable thickness, (b) medium thickness.

where K_φ is a factor equal to

$$K_\varphi = [\{(i\text{Pd})^{1/2}\coth(i\text{Pd})^{1/2}\}(-i\text{Pd})^{1/2}\coth(-i\text{Pd})^{1/2}]^{-1/2}. \qquad (7.6.53)$$

Formula (7.6.52) differs from (7.6.51) by the factor K_φ. It shows the fraction of heat accumulated by the wall, $2R$ thick, compared with an infinitely thin wall ($2R \to 0$)

$$K_\varphi = \Delta Q_v/(\Delta Q_v)_{2R\to 0}. \qquad (7.6.54)$$

TABLE 7.4. The Dependence of K_φ on the Number Pd

Pd	K_φ	Pd	K_φ
0.0	1.00	15	0.260
0.5	0.96	20	0.225
1	0.92	30	0.180
2	0.79	40	0.150
3	0.66	50	0.135
4	0.55	60	0.120
5	0.48	80	0.105
7	0.40	100	0.096
10	0.32	∞	0.00

The quantity K_φ is therefore called a heat-utilization factor. The factor K_φ depends on the number Pd (see Table 7.4). Figure 7.11 shows that at Pd $= 0$, $K_\varphi = 1$ (the case of an infinitely thin wall). At Pd $\to \infty$, $K_\varphi = 0$.

It is of interest to analyze the formula describing the heat losses of an infinitely thick wall (half-space). *At high Pd* ($2R \to \infty$), K_φ *approaches*

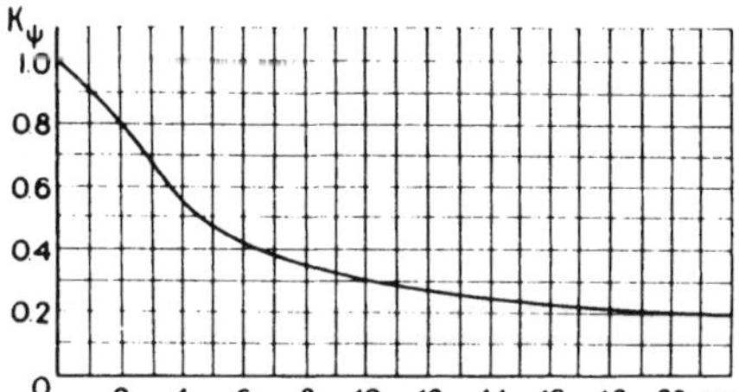

Fig. 7.11. The heat accumulation coefficient versus the Predvoditelev criterion in symmetric heating of walls of finite thickness.

$\frac{1}{2}(\text{Pd})^{1/2}$. *Hence the form of formula* (7.6.52) *becomes*

$$(\Delta Q_v)_{2R\to\infty} = 2c\gamma t_m/2(\text{Pd})^{1/2} = 2c\gamma t_m(a/\omega)^{1/2}(1/l), \qquad (7.6.55)$$

where l is the wall thickness ($l = 2R$).

For a half-space, the heat quantity absorbed by a unit surface area of the wall for a half period is calculated as a value describing heat accumulation, i.e.,

$$\Delta Q_s = \Delta Q_v l = 2c\gamma t_m(a/\omega)^{1/2}. \qquad (7.6.56)$$

Relation (7.6.56) has the following physical significance: ΔQ_s equals the heat amount received per unit surface by a wall layer with a thickness of $(a/\omega)^{1/2}$ ($(a/\omega)^{1/2}$ is measured in units of length) when it is uniformly heated throughout the whole thickness from $-t_m$ to t_m. Thus $(a/\omega)^{1/2}$ characterizes the conventional thickness of a semi-infinite body uniformly heated at the steady-periodic state. $(a/\omega)^{1/2}$ will be denoted by ξ. It is a value characterising assimilation of heat by a homogeneous wall. Then the factor ξ is approximately 9 times less than the temperature wave length Λ ($\xi = 0.11\ \Lambda$), since

$$\Lambda = (8\pi)^{1/2}(a/\omega)^{1/2} = (8\pi)^{1/2}\xi. \qquad (7.6.57)$$

This value is inversely proportional to $\sqrt{\omega}$ *and directly proportional to* $\sqrt{a}$. *Therefore the value of* ξ *diminishes as the frequency of temperature oscillations increases. At high frequency, assimilation of heat is insignificant. At constant frequency* ($\omega = \text{const}$), ξ *depends on thermal diffusivity alone.* For example, with an oscillation period of 24 hr. ($\omega = \pi/12$), the heat assimilation coefficient for cork slabs ($\xi = 0.039m$) is approximately 3.5 times less than that for marble slabs ($\xi = 0.137m$).

If heat is transferred through an air layer by heat conduction, a heat-assimilation coefficient is very large ($\xi = 0.543$, $\omega = \pi/12$, $t = 20°\text{C}$). The heat-assimilation coefficient determines the rate of damping temperature oscillations in the wall thickness.

At a depth of X_n the temperature oscillations decrease by a factor of n compared with oscillations at the surface. This depth is equal to

$$X_n = \sqrt{2}\,\xi \ln n = \Lambda f(n). \qquad (7.6.58)$$

The values of the function $f(n)$ are presented in Table 7.5 which shows that at $n = 2$, $f(n) = 0.110$. Thus, at $n = 2$, $X_n = 0.110$, $\Lambda = \xi$, i.e., the value of ξ is equal to the depth of the layer X_n at which temperature oscillations become half the value of those at the surface.

TABLE 7.5. THE FUNCTION $f(n)$ OF TEMPERATURE OSCILLATION ATTENUATION

n	2	4	10	20	50	100	1000
$f(n)$	0.110	0.221	0.367	0.477	0.623	0.733	1.100

The heat accumulated by a unit surface area of a semi-infinite wall for a half period at Bi* $\neq 0$ *is*

$$(Q_s)_{l\to\infty} = 2c\gamma t_m A_0(a/\omega)^{1/2}, \tag{7.6.59}$$

i.e., a relation is obtained which is similar to (7.6.56), since the surface temperature oscillation amplitude is $t_m A_0$.

The heat rate may be calculated in another way. The rate of heat flow is equal to

$$dQ_v/d\tau = \alpha(S/V)[t_a(\tau) - t_s(\tau)], \tag{7.6.60}$$

where $t_s(\tau)$ is the temperature of the body surface (at $x = R$ or at $r = R$, S/V is the ratio of the body surface to its volume).

Thus if we draw a curve of ambient temperature history

$$t_a(\tau) = t_m \cos 2\pi\nu\tau, \tag{7.6.61}$$

and the curve of surface temperature history

$$t_s(\tau) = t_m\theta_s = t_m(N_{i,s}N_{-i,s})^{1/2}\cos(2\pi\nu\tau - M_s) \tag{7.6.62}$$

(the subscript s designates the value of the corresponding quantities at $x = R$ or at $r = R$), then the area between these two curves gives the value proportional to the specific heat rate for the given interval of time. For example, in Fig. 7.6 the shaded area will give a value proportional to the specific heat rate during the half period.

If we substitute into Eq. (7.6.60) a corresponding expression for $t_a(\tau)$ and $t_s(\tau)$ from relations (7.6.61) and (7.6.62) and integrate with respect to τ between τ_1 to τ_2, we obtain a formula for the specific heat rate similar to (7.6.47).

7.7 Semi-Infinite Body. Ambient Temperature as a Function of Time

a. Statement of the Problem. *As an example of a semi-infinite body, consider a thin long rod with the thermally insulated side surface. At the initial moment the end of the rod is placed into a medium the temperature of which is some given function of time* $t_a(\tau) = f(\tau)$. *Heat transfer occurs between the non-insulated end of the rod and the surrounding medium by convection. The temperature distribution along the length of the rod is to be found.* The differential heat conduction equation for a one-dimensional problem is known. The boundary conditions are the following

$$t(x, 0) = 0\,, \tag{7.7.1}$$

$$[\partial t(0, \tau)/\partial x] + H[f(\tau) - t(0, \tau)] = 0\,, \tag{7.7.2}$$

$$t(\infty, \tau) = 0\,. \tag{7.7.3}$$

b. Solution of the Problem. The solution of the differential equation for the transform $T(x, s)$ under the conditions (7.7.1) and (7.7.3) will have the form

$$T(x, s) = B \exp[-(s/a)^{1/2}x]\,. \tag{7.7.4}$$

The constant B is determined from the boundary condition (7.7.2), which for the transform will have the form

$$T'(0, s) + HF(s) - HT(0, s) = 0\,, \tag{7.7.5}$$

where $F(s)$ is the transform of the function $f(\tau)$, i.e.,

$$F(s) = L[f(\tau)]\,. \tag{7.7.6}$$

Solution (7.7.4), after determination of the constant, will be

$$T(x, s) = F(s)\frac{1}{1+(1/H)(s/a)^{1/2}}\exp[-(s/a)^{1/2}x] = F(s)\Phi(s)\,. \tag{7.7.7}$$

For the transition from the transform, we shall use the theorem of multiplication of transforms.

The inversion of the transform $\Phi(s)$ is known (see formula (56) in Appendix 5):

$$L^{-1}[\Phi(s)] = L^{-1}\left[\frac{1}{1+(1/H)(s/a)^{1/2}}\exp\left[-\left(\frac{s}{a}\right)^{1/2}x\right]\right] = \varphi(\tau)\,,$$

$$\varphi(\tau) = H\left(\frac{a}{\pi\tau}\right)^{1/2} \exp\left[-\frac{x^2}{4a\tau}\right]$$
$$- aH^2 \exp[Hx + aH^2\tau] \operatorname{erfc}\left(\frac{x}{2\,(a\tau)^{1/2}} + H(a\tau)^{1/2}\right).$$

Then, according to the theorem of multiplication of transforms we may write

$$L^{-1}[\Phi(s)F(s)] = \int_0^\tau \varphi(\vartheta)f(\tau - \vartheta)\,d\vartheta = \int_0^\tau \varphi(\tau - \vartheta)f(\vartheta)\,d\vartheta\,,$$

where $\varphi(\tau)$ is the inversed transform $\Phi(s)$.

Hence, the solution of our problem has the form

$$t(x, \tau) = \int_0^\tau f(\tau - \vartheta)[H(a/\pi\vartheta)^{1/2} \exp[-x^2/4a\vartheta]$$
$$- aH^2 \exp[Hx + aH^2\vartheta] \operatorname{erfc}(\{x/2(a\vartheta)^{1/2}\} + H(a\vartheta)^{1/2})]\,d\vartheta. \tag{7.7.8}$$

c. ***Analysis of the Solution.*** If $f(\tau) = t_a =$ const (the ambient temperature is a constant value), then solution (7.7.8) will become solution (6.1.11):

$$\theta = \frac{t(x, \tau)}{t_a} = \operatorname{erfc}\frac{x}{2\,(a\tau)^{1/2}} - \exp[Hx + aH^2\tau] \operatorname{erfc}\left[\frac{x}{2\,(a\tau)^{1/2}} + H(a\tau)^{1/2}\right]. \tag{7.7.9}$$

Let $t_a = t_m\sqrt{\tau}$, then solution (7.7.8) will acquire the form

$$\frac{t(x, \tau)}{t_m} = \sqrt{\tau}\exp\left[-\frac{x^2}{4a\tau}\right] - \frac{x\sqrt{\pi}}{2\sqrt{a}} \operatorname{erfc}\frac{x}{2\,(a\tau)^{1/2}}$$
$$- \frac{1}{2H}\left(\frac{\pi}{a}\right)^{1/2}\left[\operatorname{erfc}\frac{x}{2\,(a\tau)^{1/2}}\right.$$
$$\left. - \exp(aH^2\tau + Hx) \operatorname{erfc}\left(\frac{x}{2\,(a\tau)^{1/2}} + H(a\tau)^{1/2}\right)\right]. \tag{7.7.10}$$

At $H \to \infty$, $t(0, \tau) = t_m\sqrt{\tau}$, and from solution (7.7.10) we obtain

$$t(x, \tau) = t_m\sqrt{\tau}\exp\left[-\frac{x^2}{4a\tau}\right] - xt_m\frac{\sqrt{\pi}}{2\sqrt{a}} \operatorname{erfc}\left(\frac{x}{2\,(a\tau)^{1/2}}\right). \tag{7.7.11}$$

In general, when $t_a = f(\tau)$ at $H \to \infty$, the temperature of the end of the rod will change in the same way as the ambient temperature, i.e.,

$$t(0, \tau) = f(\tau). \tag{7.7.12}$$

Then we shall have

$$t(x, \tau) = \frac{2}{\sqrt{\pi}} \int_{x/2(a\tau)^{1/2}}^{\infty} f\left(\tau - \frac{x^2}{4a\vartheta^2}\right) \exp[-\vartheta^2]\, d\vartheta$$

$$= \frac{x}{2\,(a\pi)^{1/2}} \int_0^{\tau} \frac{f(\tau - \eta)}{\eta^{3/2}} \exp\left[-\frac{x^2}{4a\eta}\right] d\eta. \qquad (7.7.13)$$

If the surface temperature is constant, then from (7.7.13) we obtain

$$\frac{t(x, \tau)}{t_a} = \operatorname{erfc} \frac{x}{2\,(a\tau)^{1/2}}. \qquad (7.7.14)$$

7.8 Generalized Solution. Duhamel's Theorem

We shall now generalize the problem of heating a body in a medium, the temperature of which is a function of time. Using the theorem of multiplication of transforms, we may prove the well-known Duhamel theorem. For better understanding, first consider the solution of the problem for an infinite plate.

If the ambient temperature is constant and equal to unity ($t_a = 1$), then the solution for the transform for $t_0 = 0$ will have the form

$$T_1(x, s) = \frac{1}{s}\left(\frac{\cosh(s/a)^{1/2}x}{\cosh(s/a)^{1/2}R + \dfrac{1}{H}(s/a)^{1/2}\sinh(s/a)^{1/2}R}\right) = \frac{1}{s}\Phi(s), \qquad (7.8.1)$$

where $1/s$ is the transform of the constant $t_a = 1$.

If $t_a = f(\tau)$ and the transform of $f(\tau)$ is equal to $F(s)$, i.e., $L[f(\tau)] = F(s)$, then the solution for the plate will have the form

$$T(x, s) = F(s)\Phi(s). \qquad (7.8.2)$$

If we multiply and divide expression (7.8.2) by s, then solution (7.8.2) may be represented as the product of two transforms $sF(s)$ and $T_1(x, s)$ as

$$T(x, s) = sF(s)\frac{\Phi(s)}{s} = sF(s)T_1(x, s). \qquad (7.8.3)$$

The original function of the second transform $T_1(x, s)$ is known—this is the solution of the problem for the constant ambient temperature ($t_a = 1$):

$$L^{-1}[T_1(x, s)] = t_1(x, \tau)$$

$$= 1 - \sum_{n=1}^{\infty} A_n \cos \mu_n(x/R) \exp[-\mu_n^2\, \mathrm{Fo}]. \qquad (7.8.4)$$

The original function of the first transform $sF(s)$ may be found in the following way:

$$L[f(\tau)] = F(s), \qquad L[f'(\tau)] = sF(s) - f(0), \tag{7.8.5}$$

$$sF(s) = L[f'(\tau)] + f(0). \tag{7.8.6}$$

If $f(0) = 0$ then the original of the transform $sF(s)$ is $f'(\tau)$. Then applying the theorem of multiplication of transforms, we obtain

$$t(x, \tau) = \int_0^\tau f'(\tau - \vartheta)\, t_1(x, \vartheta)\, d\vartheta. \tag{7.8.7}$$

If $f(0) = \text{const}$, we shall make the following preliminary transformation

$$T(x, s) = sF(s)T_1(x, s) = L[f'(\tau)]T_1(x, s) + f(0)T_1(x, s).$$

Applying the inverse Laplace transformation and the theorem of multiplication of transforms, we obtain:

$$t(x, \tau) = \int_0^\tau f'(\tau - \vartheta)\, t_1(x, \vartheta)\, d\vartheta + f(0)\, t_1(x, \tau). \tag{7.8.8}$$

Relation (7.8.8) is a formulation of the Duhamel theorem for a one-dimensional problem.

For our specific example, we obtain

$$t(x, \tau) = f(0)\left[1 - \sum_{n=1}^{\infty} A_n \cos \mu_n (x/R) \exp[-\mu_n^2 \mathrm{Fo}]\right]$$
$$+ \int_0^\tau f'(\tau - \vartheta)\left[1 - \sum_{n=1}^{\infty} A_n \cos \mu_n (x/R) \exp[-\mu_n^2 (a\vartheta/R^2)]\right] d\vartheta. \tag{7.8.9}$$

Relation (7.8.9) is the solution for an infinite plate when the ambient temperature is a function of time.

Let us generalize our result for a body of any geometry. Let the body be heated in a medium the temperature of which is the function of time $t_a = f(\tau)$. The temperature field for any time is to be found

$$\frac{\partial t(x, y, z, \tau)}{\partial \tau} = a \nabla^2 t(x, y, z, \tau), \tag{7.8.10}$$

$$t(x, y, z, 0) = 0, \tag{7.8.11}$$

$$-(\nabla t)_s + H[f(\tau) - t_s] = 0, \tag{7.9.12}$$

where the subscript s means the surface of the body. Applying the Laplace transform we obtain

$$sT(x, y, z, s) = a\, \nabla^2 T(x, y, z, s)\,, \tag{7.8.13}$$

$$-(\nabla T)_s + HF(s) - HT_s = 0\,. \tag{7.8.14}$$

Let $u(x, y, z, \tau)$ be the solution of our problem when the ambient temperature is equal to unity, i.e., $f(\tau) = 1$, then

$$L[u(x, y, z, \tau)] = U(x, y, z, s)\,. \tag{7.8.15}$$

Since the transform 1 is equal to $1/s$, then the solution of the problem for $t_a = f(\tau)$ for the transform $T(x, y, z, s)$ will be equal to

$$T(x, y, z, s) = sF(s)U(x, y, z, s)\,. \tag{7.8.16}$$

Substituting for the expression $sF(s)$, we obtain

$$sF(s) = L[f'(\tau)] + f(0)\,. \tag{7.8.17}$$

Then we may write

$$\begin{aligned} &T(x, y, z, s) \\ &= \{L[f'(\tau)] + f(0)\}\, U(x, y, z, s) \\ &= f(0)L[u(x, y, z, \tau)] + L[f'(\tau)]\, L[u(x, y, z, \tau)]\,. \end{aligned} \tag{7.8.18}$$

Let us use the inverted Laplace transform and the theorem of multiplication of transforms:

$$t(x, y, z, \tau) = f(0)\, u(x, y, z, \tau) + \int_0^\tau f'(\tau - \vartheta)\, u(x, y, z, \vartheta)\, d\vartheta\,. \tag{7.8.19}$$

The given relation (7.8.19) may be rewritten as

$$t(x, y, z, \tau) = (\partial/\partial\tau) \int_0^\tau f(\tau - \vartheta)\, u(x, y, z, \vartheta)\, d\vartheta\,, \tag{7.8.20}$$

i.e., the well-known Duhamel theorem is obtained: "*If* $f(\tau)$ *and the derivative* $f'(\tau)$ *are partially-continuous at* $\tau > 0$, *then the function* $t(x, y, z, \tau)$ *determined by relation* (7.8.20) *is the solution of boundary value problem* (7.8.10) *with boundary conditions* (7.8.11) *and* (7.8.12)".

On the basis of the theorem of multiplication of transforms we may write relation (7.8.20) as follows:

$$t(x, y, z, \tau) = (\partial/\partial\tau) \int_0^\tau f(\vartheta)\, u(x, y, z, \tau - \vartheta)\, d\vartheta\,. \tag{7.8.21}$$

Taking use of the Duhamel theorem we might solve the problems considered in Sections 7.1–7.7 proceeding from the solution for the constant ambient temperature, i.e., all the problems in which the ambient temperature changes with time. But this classical method for such problems has the following drawbacks: (1) it is necessary to initially solve a supplementary problem with constant boundary conditions, thus the solution of the given problem takes considerable time; (2) the solution is obtained in the form of a series which need further elaboration; and (3) in many cases an effective solution is not obtained as it represents some integral of which the final solution is difficult. Therefore, to solve the boundary problems with the so-called varying boundary conditions, we used the Laplace transformation method which has a number of advantages as compared with the classical methods.

Example for illustration. In Section 7.1, a case was considered in which the ambient temperature was a linear function of time, i.e., $t_a(\tau) = t_0 + b\tau$. We shall repeat the solution of the given problem by the Duhamel method. To simplify the derivation, we shall use relation (7.8.9) and put $t_0 = 0$.

$$t(x, \tau) = \int_0^\tau b\left[1 - \sum_{n=1}^{\infty} A_n \cos \mu_n (x/R) \exp\left[-\mu_n{}^2 \frac{a\vartheta}{R^2}\right]\right] d\vartheta. \quad (7.8.22)$$

Relation (7.8.22) is the solution of our problem (see Section 7.1); it represents an integral which is, however, difficult to evaluate as it stands, so that the following transformations are required. From (7.8.22) we obatin

$$t(x, \tau) = b\tau - \frac{bR^2}{a} \sum_{n=1}^{\infty} \frac{A_n}{\mu_n{}^2} \cos \mu_n \frac{x}{R} + \frac{bR^2}{a} \sum_{n=1}^{\infty} \frac{A_n}{\mu_n{}^2} \cos \mu_n \frac{x}{R} \exp\left[-\mu_n{}^2 \frac{a\tau}{R^2}\right]. \quad (7.8.23)$$

This solution differs from (7.1.8) since it is more complicated. The following special proof is needed:

$$\frac{bR^2}{a} \sum_{n=1}^{\infty} \frac{A_n}{\mu_n{}^2} \cos \mu_n \frac{x}{R} = \frac{b}{2a} \left[\left(1 + \frac{2}{\mathrm{Bi}}\right) R^2 - x^2\right].$$

A more general problem with the given complex boundary condition and variable ambient temperature may be reduced to two problems. We have

$$\frac{\partial t}{\partial \tau} = a \nabla^2 t \quad (7.8.24)$$

$$t(0) = \varphi(x, y, z) \quad (7.8.25)$$

$$-(\nabla t)_s + H[t_a(\tau) - t_s] = 0. \tag{7.8.26}$$

Let us set $t = u + \vartheta$; then we shall have

$$\partial\vartheta/\partial\tau = a\nabla^2\vartheta, \tag{7.8.27}$$

$$\vartheta(0) = 0, \tag{7.8.28}$$

$$-(\nabla\vartheta)_s + H[\vartheta_a(\tau) - \vartheta_s] = 0, \tag{7.8.29}$$

$$\partial u/\partial\tau = a\,\nabla^2 u, \tag{7.8.30}$$

$$u(0) = \varphi(x, y, z), \tag{7.8.31}$$

$$u_s = 0. \tag{7.8.32}$$

Thus we shall obtain two problems.

7.9 Hollow Cylinder

a. Statement of the Problem. *Consider an infinite hollow cylinder with the assigned initial temperature distribution $f(r)$. Heat transfer with the external medium takes place according to the Newton law. The temperature of the medium corresponds to the prescribed functions of time $t_{a1} = \varphi_1(\tau)$, $t_{a2} = \varphi_2(\tau)$*

$$t(r, 0) = f(r) \tag{7.9.1}$$

$$\partial t(R_1, \tau)/\partial r + (\alpha_1/\lambda)[\varphi_1(\tau) - t(R_1, \tau)] = 0. \tag{7.9.2}$$

$$-\partial t(R_2, \tau)/\partial r + (\alpha_2/\lambda)[\varphi_2(\tau) - r(R_2, \tau)] = 0 \tag{7.9.3}$$

b. Solution of the Problem. To solve the problem we use the finite integral Hankel transform formula

$$T_H(\mu_n, \tau) = \int_{R_1}^{R_2} rt(r, \tau)\, U_0(\mu_n\,(r/R_1))\, dr, \tag{7.9.4}$$

where the kernel of the transform is

$$U_0\{\mu_n(r/R_1)\} = [Y_0(\mu_n) + (\mu_n/\mathrm{Bi}_1)Y_1(\mu_n)]J_0(\mu_n(r/R_1)) - [J_0(\mu_n) + (\mu_n/\mathrm{Bi}_1)J_1(\mu_n)]Y_0(\mu_n(r/R_1)). \tag{7.9.5}$$

μ_n are the roots of the characteristic equation

$$\frac{U_0(k\mu_n)}{U_1(k\mu_n)} = \frac{\mu_n}{(\alpha_2/\alpha_1)\,\mathrm{Bi}_1}, \tag{7.9.6}$$

$$U_1(k\mu_n) = \left[Y_0(\mu_n) + \frac{\mu_n}{\mathrm{Bi}_1} Y_1(\mu_n) \right] J_1(k\mu_n)$$

$$- \left[J_0(\mu_n) + \frac{\mu_n}{\mathrm{Bi}_1} J_1(\mu_n) \right] Y_1(k\mu_n) ; \qquad (7.9.7)$$

$\mathrm{Bi}_1 = \alpha_1 R_1/\lambda$, $\mathrm{Bi}_2 = \alpha_2 R_2/\lambda$ are the Biot criteria, $k = R_2/R_1$.

The inversion formula has the form

$$t(r, \tau) = \frac{\pi^2 \mathrm{Bi}_1^2}{2R_1^2} \sum_{n=1}^{\infty} \mu_n^2 T_H(\mu_n, \tau)\, U_0(\mu_n (r/R_1))$$
$$\times [(\alpha_2/\alpha_1)\, J_0(k\mu_n) - (\mu_n/\mathrm{Bi}_1)\, J_1(k\mu_n)]^2$$
$$\times \{[J_0(\mu_n) + \mu_n \mathrm{Bi}_1^{-1} J_1(\mu_n)]^2 [\mu_n^2 + (\alpha_2^2/\alpha_1^2)\, \mathrm{Bi}_1^2]$$
$$- [(\alpha_2^2/\alpha_1^2)\, J_0(k\mu_n) - J_1(k\mu_n)\, (\mu_n/\mathrm{Bi}_1)]^2 (\mu_n^2 + \mathrm{Bi}_1^2)\}^{-1}. \qquad (7.9.8)$$

To solve the problem, every term of the differential heat conduction equation should be multiplied by the kernel of the symmetrical transform $rU_0(\mu_n(r/R_1))$ and integrated between R_1 and R_2.

The eigenfunction $U_0(\mu_n(r/R_1))$ is the solution for the zeroth-order Bessel equation under uniform boundary conditions of the first kind. In the integration boundary of conditions (7.9.2) and (7.9.3), the characteristic equation (7.9.6), as well as

$$U_0(\mu_n) = -2/\pi\mathrm{Bi}_1; \qquad U_1(\mu_n) = 2/\pi\mu_n \qquad (7.9.9)$$

were taken into account.

Then the differential equation will be of the form

$$\frac{dT_H(\mu_n, \tau)}{d\tau} + \frac{a\mu_n^2}{R_1^2} T_H(\mu_n, \tau) - a\, \mathrm{Bi}_2\, U_0(k\mu_n)\, \varphi_2(\tau)$$
$$+ \frac{2a}{\pi} \varphi_1(\tau) = 0. \qquad (7.9.10)$$

The transform of the function $f(r)$ will be designated as

$$T_H(\mu_n, 0) = f(\mu_n) = \int_{R_1}^{R_2} rf(r)\, U_0(\mu_n (r/R_1))\, dr. \qquad (7.9.11)$$

The solution of the ordinary differential equation (7.9.10) accounting for initial condition (7.9.11) will have the form

$$T_H(\mu_n, \tau) = f(\mu_n)\exp[-\mu_n^2 \mathrm{Fo}_1] + a\,\mathrm{Bi}_2\, U_0(k\mu_n)\int_0^\tau \varphi_2(\vartheta)$$
$$\times \exp\left[-\frac{\mu_n^2 a(\tau-\vartheta)}{R_1^2}\right] d\vartheta - \frac{2a}{\pi}\int_0^\tau \varphi_1(\tau)$$
$$\times \exp\left[-\frac{\mu_n^2 a(\tau-\vartheta)}{R_1^2}\right] d\vartheta, \tag{7.9.12}$$

where $\mathrm{Fo}_1 = a\tau/R_1^2$. Using inversion formula (7.9.8), we have

$$t(r, \tau) = \frac{2}{R_1^2}$$
$$\times \sum_{n=1}^{\infty} \frac{\mu_n^2\, U_0((\mu_n\,(r/R_1))\exp[-\mu_n^2 \mathrm{Fo}_1]}{(\alpha_2^2/\alpha_1^2)k^2 U_0^2(k\mu_n)[\mu_n^2 + (\alpha_2^2/\alpha_1^2)\mathrm{Bi}_1^2] - (4/\pi^2 \mathrm{Bi}_1^2)(\mu_n^2 + \mathrm{Bi}_1^2)}$$
$$\times \left\{ \int_{R_1}^{R_2} rf(r)\, U_0(\mu_n\,(r/R_1))\, dr + a\,\mathrm{Bi}_2\, U_0(k\mu_n)\int_0^\tau \varphi_2(\vartheta)\right.$$
$$\left.\times \exp[\mu_n^2(a\vartheta/R_1^2)]\, d\vartheta - (2a/\pi)\int_0^\tau \varphi_1(\vartheta)\exp[\mu_n^2(a\vartheta/R_1^2)]\, d\vartheta\right\}. \tag{7.9.13}$$

7.10 Parallelepiped. Ambient Temperature as a Linear Function of Time

With the help of the Duhamel theorem, we may obtain the solution for the plate of the finite dimensions ($2R_1 \times 2R_2 \times 2R_3$). To do this, we use the solution for the parallelepiped exposed to a constant ambient temperature, $t_a = \text{const}$, which is given in Chapter 6, Section 9. If we use relation (7.8.20), then upon integration we obtain

$$t(x, y, z, \tau) = t_0 + b\tau - \frac{bR^2}{a}\sum_{n=1}^{\infty}\sum_{m=1}^{\infty}\sum_{k=1}^{\infty} \frac{A_{n,1}\, A_{m,2}\, A_{k,3}}{\mu_{n,1} K_1^2 + \mu_{m,2} K_2^2 + \mu_{k,3} K_3^2}$$
$$\times \cos\mu_{n,1}(x/R_1)\cos\mu_{m,2}(y/R_2)\cos\mu_{k,3}(z/R_3)$$
$$\times \{1 - \exp[-(\mu_{n,1}^2 K_1^2 + \mu_{m,2}^2 K_2^2 + \mu_{k,3}^2 K_3^2)\,\mathrm{Fo}]\}, \tag{7.10.1}$$

where

$$K_i = \frac{R}{R_i} \quad (i = 1, 2, 3), \qquad \frac{1}{R^2} = \frac{1}{R_1^2} + \frac{1}{R_2^2} + \frac{1}{R_3^2}, \qquad \mathrm{Fo} = \frac{a\tau}{R^2}.$$

$A_{n,1}, A_{m,2}, A_{k,3}$ are the coefficients determined by the given value of the Biot criterion.

If the number $\mathrm{Bi} = \infty$, then $\mu_{n,1} = (2n-1)\frac{1}{2}\pi$, $\mu_{m,2} = (2m-1)\frac{1}{2}\pi$, and $\mu_{k,3} = (2k-1)\frac{1}{2}\pi$. Solution (7.10.1) is more general.

CHAPTER

8

TEMPERATURE FIELD WITH CONTINUOUS HEAT SOURCES

In many heat transfer processes, there are heat sources inside a body. They may be positive (for example, heating a body by electric current, evolution of heat during condensation) or negative (moisture evaporation in a moist body when being heated). Only some of the most typical examples, frequently encountered in thermal engineering, will be considered in this chapter.

The problems may be divided into two kinds: (1) problems with constant or variable sources acting during the entire heat transfer process (continuous heat sources); and (2) problems with point sources acting during an infinitesimal period (for example, at the initial moment a body receives a thermal impulse). The latter case includes problems of heat transfer in conductors in which a short circuit occurs and the internal heat source is practically instantaneous.

It is difficult to solve problems where the heat sources within a body change both with time and the coordinates. Therefore only those problems will be considered where the change occurs either in time or along the coordinates of the body.

8.1 Semi-Infinite Body

a. Statement of the Problem. *Consider a thin semi-infinite rod at temperature t_0 with the side surface thermally insulated. At the initial moment a noninsulated end acquires the temperature $t_a > t_0$ which remains constant* *during the whole heat transfer process (boundary condition of the first kind). There is a heat source inside the rod of specific strength* w (kcal/m^3 hr). *The temperature distribution along the length of the rod and specific heat rate at any moment are to be found.*

The problem stated may be written mathematically as follows

$$\frac{\partial t(x, \tau)}{\partial \tau} = a \frac{\partial^2 t(x, \tau)}{\partial x^2} + \frac{w}{c\gamma} \quad (\tau > 0;\ 0 < x < \infty), \tag{8.1.1}$$

$$t(x, 0) = t_0, \tag{8.1.2}$$

$$\frac{\partial t(\infty, \tau)}{\partial x} = 0, \tag{8.1.3}$$

$$t(0, \tau) = t_a = \text{const.} \tag{8.1.4}$$

b. Solution of the Problem for w ***= const.*** We apply the Laplace transform and obtain

$$T''(x, s) - \frac{s}{a} T(x, s) + \frac{t_0}{a} + \frac{w}{sac\gamma} = 0. \tag{8.1.5}$$

Equation (8.1.5) may be written as

$$T''(x, s) - \frac{s}{a} \left[T(x, s) - \frac{t_0}{s} - \frac{w}{s^2 c\gamma}\right] = 0. \tag{8.1.6}$$

The general solution of Eq. (8.1.6) may be written in two forms:

$$\begin{aligned} T(x, s) - \frac{t_0}{s} - \frac{w}{s^2 c\gamma} &= A \cosh\left(\frac{s}{a}\right)^{1/2} x + B \sinh\left(\frac{s}{a}\right)^{1/2} x \\ &= A_1 \exp\left[\left(\frac{s}{a}\right)^{1/2} x\right] + B_1 \exp\left[-\left(\frac{s}{a}\right)^{1/2} x\right]. \end{aligned} \tag{8.1.7}$$

Boundary conditions (8.1.3) and (8.1.4) for the transform may be written as

$$T'(\infty, s) = 0, \tag{8.1.8}$$

$$T(0, s) = t_a/s. \tag{8.1.9}$$

It follows from condition (8.1.8) that $A_1 = 0$. The constant B_1 is found from boundary condition (8.1.9):

$$\frac{t_a}{s} - \frac{t_0}{s} - \frac{w}{s^2 c\gamma} = B_1. \tag{8.1.10}$$

Then solution (8.1.7) will acquire the form

$$T(x, s) - \frac{t_0}{s} = \frac{w}{s^2 c\gamma} + \frac{(t_a - t_0)}{s} \exp\left[-\left(\frac{s}{a}\right)^{1/2} x\right] - \frac{w}{s^2 c\gamma} \exp\left[-\left(\frac{s}{a}\right)^{1/2} x\right]. \tag{8.1.11}$$

It is seen from expression (8.1.11) that it has three terms each of which is a tabulated transform with the corresponding inverse transform (see relations (2), (50), and (53) of the table of transforms in the Appendix 5).

The final solution of our problem will have the form

$$\theta = \frac{t(x, \tau) - t_0}{t_a - t_0}$$

$$= \operatorname{erfc} \frac{x}{2\,(a\tau)^{1/2}} + \frac{w}{c\gamma(t_a - t_0)}\,\tau - \frac{w}{c\gamma(t_a - t_0)}\,4\tau\ \mathrm{i}^2 \operatorname{erfc} \frac{x}{2\,(a\tau)^{1/2}}. \tag{8.1.12}$$

The value $w\tau/c\gamma(t_a - t_0)$ is a dimensionless value; it represents the ratio of the amount of heat evolved by a heat source for the given period of time per unit volume to the amount of heat which must be imparted to the unit volume to heat it from the initial temperature to the ambient temperature. This value may be represented as the product of the Fourier number and a new number which is called the Pomerantsev criterion

$$\frac{w\tau}{c\gamma(t_a - t_0)} = \mathrm{Fo}_x \frac{wx^2}{\lambda(t_a - t_0)} = \mathrm{Fo}_x\, \mathrm{Po}_x\,, \tag{8.1.13}$$

where $\mathrm{Fo}_x = a\tau/x^2$ is the Fourier number for the coordinate x and $\mathrm{Po}_x = wx^2/\lambda(t_a - t_0)$ is the Pomerantsev criterion. The physical significance of the Pomerantsev criterion is that it shows the ratio of the amount of heat evolved by a source per unit time in the volume x (the volume of a parallelepiped with a base of 1 m^2 and a height x) to the maximum possible amount of heat transferred by conduction through unit area per unit time at the distance x from the end of the rod (under the assumption that the temperature at the given point is equal to the initial temperature, and the temperature distribution occurs by the linear law).

If the heat source is absent ($\mathrm{Po}_x = 0$), then from solution (8.1.12) the ordinary solution is obtained for a semi-infinite body under the boundary condition of the first kind.

The heat flux q will be found by relation

$$q = -\lambda \frac{\partial t(0, \tau)}{\partial x}. \tag{8.1.14}$$

From relation (8.1.11) we have

$$L[q(\tau)] = \frac{\lambda(t_a - t_0)}{(as)^{1/2}} - \frac{\lambda w}{c\gamma s(as)^{1/2}}.$$

Using the table of transforms we find

$$q = (t_a - t_0)\left(\frac{\lambda c\gamma}{\pi\tau}\right)^{1/2} - 2w\left(\frac{a\tau}{\pi}\right)^{1/2}. \tag{8.1.15}$$

It is seen from relation (8.1.15) that at small values of time the body is heated mainly by conduction through the noninsulated end of the rod. For large values of time, the heating takes place at the expense of the heat source, while from the exposed end, heat loss into the surrounding medium occurs (the second term of relation (8.1.15) becomes greater than the first one).

c. Solution of the problem for $w = we_0^{-kx}$. The heat source is an exponential function of the coordinate, $w = w_0\, e^{-kx}$, where w_0 is the maximum specific strength of the source, k is the constant. The initial and boundary conditions remain the same. To simplify calculation we set $t_0 = 0$.

Using the Laplace transform, we obtain

$$T''(x, s) - (s/a)T(x, s) + (w_0/sc\gamma a)\, e^{-kx} = 0. \tag{8.1.16}$$

This nonhomogeneous equation may be easily solved since its general solution is known and one particular solution is readily obtained in the form of Ae^{-kx}. However, if we intend to solve more complex problems and to follow the general method adopted as based on wider application of the integral transform method, the Fourier sine transformation will be applied to Eq. (8.1.16). Denoting

$$T_s(s, p) = (2/\pi)^{1/2} \int_0^\infty T(s, x) \sin px\, dx\,, \tag{8.1.17}$$

and using the boundary condition of Eq. (8.1.16)

$$T(s, 0) = \frac{t_a}{s}, \qquad T'(s, \infty) = 0\,, \tag{8.1.18}$$

we obtain the solution of (8.1.16) in the form

$$T_s(s, p) = \left(\frac{2}{\pi}\right)^{1/2} \frac{p}{p^2 + (s/a)} \left\{\frac{t_a}{s} + \frac{w_0}{\lambda s}\, \frac{1}{p^2 + k^2}\right\}. \tag{8.1.19}$$

Using the tables of definite integrals and applying the inverse Fourier sine transformation, we obtain

$$T(s, x) = (2/\pi)^{1/2} \int_0^\infty T_s(s, p) \sin px \, dx$$

$$= \frac{t_a}{s} \exp\left[-\left(\frac{s}{a}\right)^{1/2} x\right] + \frac{w_0}{\lambda s} \frac{\exp[-kx] - \exp[-(s/a)^{1/2}x]}{k^2 - (s/a)}. \quad (8.1.20)$$

Using the table of transforms (Appendix 5), we obtain

$$\theta = \frac{t(x, \tau)}{t_a} = \operatorname{erfc} \frac{x}{2(a\tau)^{1/2}} - \frac{w_0}{\lambda k^2 t_a} \left[\operatorname{erfc} \frac{x}{2(a\tau)^{1/2}} - \exp[-kx] - \frac{1}{2} \exp[k^2 a\tau - kx] \operatorname{erfc}\left(k(a\tau)^{1/2} - \frac{x}{2(a\tau)^{1/2}}\right) + \frac{1}{2} \exp[k^2 a\tau + kx] \operatorname{erfc}\left(k(a\tau)^{1/2} + \frac{x}{2(a\tau)^{1/2}}\right)\right]. \quad (8.1.21)$$

d. Solution of the problem for $w = \dot{w}_0\tau^{n/2}$. The specific strength of a heat source is some function of time of the form $\dot{w}_0\tau^{n/2}$ where $\dot{w}_0$ is constant and n is an exponent equal to -1.0 or to any other positive value, i.e., $n > -2$.

The differential equation for the transform will be written as

$$T'(\dot{x}, s) - \frac{s}{a} T(x, s) + \frac{t_0}{a} + \frac{\dot{w}_0 \Gamma(1 + \frac{1}{2}n)}{ac\gamma s^{1+\frac{1}{2}n}} = 0, \quad (8.1.22)$$

where

$$L[\tau^{\frac{1}{2}n}] = \frac{\Gamma(1 + \frac{1}{2}n)}{s^{1+\frac{1}{2}n}}$$

and $\Gamma(m)$ is the gamma function.

The solution of Eq. (8.1.22) under the given boundary conditions has the form

$$T(x, s) - \frac{t_0}{s} = \frac{t_a - t_0}{s} \exp\left[-\left(\frac{s}{a}\right)^{1/2} x\right] + \frac{\dot{w}_0 \Gamma(1 + \frac{1}{2}n)}{c\gamma s^{2+\frac{1}{2}n}} \left(1 - \exp\left[-\left(\frac{s}{a}\right)^{1/2} x\right]\right). \quad (8.1.23)$$

Using the table of transforms we find

$$\theta = \frac{t(x, \tau) - t_0}{t_a - t_0} = \operatorname{erfc} \frac{x}{2(a\tau)^{1/2}} + \frac{\dot{w}_0 \tau^{1+\frac{1}{2}n}}{c\gamma(t_a - t_0)(1 + \frac{1}{2}n)} \times \left[1 - \Gamma\left(2 + \frac{1}{2}n\right) 2^{n+2} \, \mathrm{i}^{n+2} \operatorname{erfc} \frac{x}{2(a\tau)^{1/2}}\right]. \quad (8.1.24)$$

Let us complicate the problem by replacing a boundary condition of the first kind by a boundary condition of the third kind, i.e.,

$$\frac{\partial t(0,\tau)}{\partial x} + H[t_a - t(0,\tau)] = 0. \tag{8.1.25}$$

Then the solution for the transform will have the form

$$T(x,s) - (t_0/s) = \frac{t_a - t_0}{s(1 + (1/H)(s/a)^{1/2})} \exp[-(s/a)^{1/2}x] + \frac{\dot{w}_0 \Gamma(1+\frac{1}{2}n)}{c\gamma s^{2+\frac{1}{2}n}} \left[1 - \frac{1}{1+(1/H)(s/a)^{1/2}} \exp[-(s/a)^{1/2}x)\right]. \tag{8.1.26}$$

The solution for the original function (i.e., the inverse transform) will be

$$\begin{aligned}\theta = {} & \operatorname{erfc}\frac{x}{2(a\tau)^{1/2}} - \exp[Hx + H^2 a\tau]\operatorname{erfc}\left(\frac{x}{2(a\tau)^{1/2}} + H(a\tau)^{1/2}\right) \\ & + \frac{\dot{w}_0 \tau^{1+\frac{1}{2}n}}{c\gamma(t_a - t_0)(1+\frac{1}{2}n)} + \frac{\dot{w}_0 \Gamma(1+\frac{1}{2}n)}{c\gamma(t_a - t_0)\, a^{\frac{1}{2}n}(-H)^{n+2}} \\ & \times \left[\exp[Hx + H^2 a\tau]\operatorname{erfc}\left(\frac{x}{2(a\tau)^{1/2}} + H(a\tau)^{1/2}\right)\right. \\ & \left. - \sum_{m=0}^{n+2} (-2H(a\tau)^{1/2})^m \, i^m \operatorname{erfc}\frac{x}{2(a\tau)^{1/2}}\right].\end{aligned} \tag{8.1.27}$$

If a heat source is absent ($\dot{w}_0 = 0$) then as a special case, we obtain the solution for a semi-infinite rod with its noninsulated end exchanging heat with the surrounding medium according to Newton's law of cooling.

8.2 Infinite Plate

First we shall consider a simpler problem (w = const) with boundary conditions of the third kind, and then a more general one ($w = w(\tau)$ and $q(\tau)$).

a. Statement of the Problem. *Consider an infinite plate with the thickness $2R$ at the temperature t_0. At the initial moment it is placed into a medium with the temperature $t_a > t_0$. Heat transfer with the surrounding medium takes place according to the Newton law (boundary condition of the third kind). Inside a plate there is a heat source the specific strength of which is equal to w. The temperature distribution along the thickness of the plate and the specific heat rate at any time are to be found.*

We have

$$\frac{\partial t(x,\tau)}{\partial \tau} = a\frac{\partial^2 t(x,\tau)}{\partial x^2} + \frac{w}{c\gamma} \quad (\tau > 0;\ -R < x < +R), \tag{8.2.1}$$

$$t(x, 0) = t_0, \tag{8.2.2}$$

$$\frac{\partial t(0,\tau)}{\partial x} = 0, \tag{8.2.3}$$

$$-\frac{\partial t(R,\tau)}{\partial x} + H[t_a - t(R,\tau)] = 0. \tag{8.2.4}$$

The origin of coordinates is in the center of the plate.

b. Solution of the Problem for $w = const$. The solution for the transform $T(x, s)$ under condition (8.2.2) was given in the previous section. For the symmetric problem, this solution may be written as

$$T(x, s) - (t_0/s) = (w/c\gamma s^2) + A\cosh(s/a)^{1/2}x. \tag{8.2.5}$$

The constant A is found from boundary condition (8.2.4) which for the transform $T(x, s)$ will have the form

$$-T'(R, s) + H[(t_a/s) - T(R, s)] = 0. \tag{8.2.6}$$

Having combined solution (8.2.5) with boundary condition (8.2.6), we may determine the constant A. Then solution (8.2.5) will acquire the form

$$T(x, s) - (t_0/s) = \frac{w}{s^2 c\gamma} + \frac{(t_a - t_0)\cosh(s/a)^{1/2}x}{s\left[\cosh(s/a)^{1/2}R + \dfrac{1}{H}(s/a)^{1/2}\sinh(s/a)^{1/2}R\right]}$$
$$- \frac{(w/c\gamma)\cosh(s/a)^{1/2}x}{s^2[\cosh(s/a)^{1/2}R + (1/H)(s/a)^{1/2}\sinh(s/a)^{1/2}R]}. \tag{8.2.7}$$

It is seen from solution (8.2.7) that it has three terms. The first term $w/s^2c\gamma$ has the inverse transform $(w/c\gamma)\,\tau$, the second term is solution (6.3.25) for the transform of the problem (the solution of the problem without heat sources), and the third is solution (7.1.7) for the transform of the problem in which $w/c\gamma$ is substituted for b. Therefore, the solution of our problem may be obtained from solutions of the previous problems, i.e., from solutions (6.3.29) and (7.1.8), to which the value $(w/c\gamma)\,\tau$ is added.

The solution of our problem in dimensionless quantities will have the form

$$\theta = \frac{t(x,\tau) - t_0}{t_a - t_0}$$
$$= 1 + \frac{1}{2}\,\mathrm{Po}\left(1 - \frac{x^2}{R^2} + \frac{2}{\mathrm{Bi}}\right) - \sum_{n=1}^{\infty}\left(1 + \frac{\mathrm{Po}}{\mu_n^2}\right) A_n \cos \mu_n \frac{x}{R} \exp[-\mu_n^2 \mathrm{Fo}], \tag{8.2.8}$$

where Po is the Pomerantsev criterion

$$\mathrm{Po} = \frac{wR^2}{\lambda(t_a - t_0)}. \tag{8.2.9}$$

If we set $\mathrm{Po} = 0$ (the absence of a heat source), then solution (8.2.8) becomes solution (6.3.29).

In a stationary state ($\mathrm{Fo} = \infty$), the temperature distribution along the thickness of the plate is

$$t(x,\tau) = t_a + (w/2\lambda)(R^2 - x^2 + \{2R^2/\mathrm{Bi}\}) = T_\infty(x,\tau). \tag{8.2.10}$$

Solution (8.2.8) with (8.2.10) may be written

$$\theta_\infty - \theta = \frac{t_\infty(x,\tau) - t(x,\tau)}{t_a - t_0}$$
$$= \sum_{n=1}^{n} (1 + \{\mathrm{Po}/\mu_n^2\}) A_n \cos \mu_n (x/R) \exp[-\mu_n^2 \mathrm{Fo}]. \tag{8.2.10a}$$

The mean temperature of the plate is equal to

$$\bar\theta = \frac{\bar t(\tau) - t_0}{t_a - t_0}$$
$$= 1 + \tfrac{1}{3}\,\mathrm{Po}(1 + \{3/\mathrm{Bi}\}) - \sum_{n=1}^{\infty} (1 + \{\mathrm{Po}/\mu_n^2\}) B_n \exp[-\mu_n^2 \mathrm{Fo}]. \tag{8.2.11}$$

The values of the constants A_n, B_n and of the characteristic numbers μ_n are found using Eq. (6.3.30), Chapter 6.

The specific heat rate is found by the conventional formula from the values of the mean temperature.

c. Solution for $w = w_0 e^{-k\tau}$. The specific strength of a heat source changes by an exponential law $w_0 e^{-k\tau}$, where w_0 is the maximum specific strength of the source, corresponding to the initial moment of time and k is the constant numerically equal to the maximum relative rate of change of the dimensionless specific source strength, i.e.,

$$-\left[\frac{d(w/w_0)}{d\tau}\right]_{\max} = k. \tag{8.2.12}$$

Since the transform $L[w] = w_0/(s + k)$, then the solution for the transform will have the form

$$T(x, s) - (t_0/s) = \frac{(t_a - t_0)\cosh(s/a)^{1/2}x}{s\left[\cosh(s/a)^{1/2}R + \frac{1}{H}(s/a)^{1/2}\sin(s/a)^{1/2}R\right]}$$

$$+ \left\{\frac{w_0}{s(s + k)c\gamma}\right\}$$

$$\times \left\{1 - \frac{\cosh(s/a)^{1/2}x}{\left[\cosh(s/a)^{1/2}R + \frac{1}{H}(s/a)^{1/2}\sinh(s/a)^{1/2}R\right]}\right\}. \qquad (8.2.13)$$

To invert the transform we shall use Eqs. (6.3.25) and (6.3.29), Chapter 6, and Eqs. (7.4.6) and (7.4.9), Chapter 7. Then, summing up these solutions we shall obtain the following final solution of our problem in the criterial form

$$\theta = \frac{t(x, \tau) - t_0}{t_a - t_0}$$

$$= 1 - \frac{\text{Po}}{\text{Pd}}\left[1 - \frac{\cos(\text{Pd})^{1/2}(x/R)}{\cos(\text{Pd})^{1/2} - (1/\text{Bi})(\text{Pd})^{1/2}\sin(\text{Pd})^{1/2}}\right]\exp[-\text{Pd Fo}]$$

$$- \sum_{n=1}^{\infty}\left(1 - \frac{\text{Po}}{\text{Pd} - \mu_n^2}\right)A_n \cos\mu_n \frac{x}{R}\exp[-\mu_n^2 \text{Fo}], \qquad (8.2.14)$$

where $\text{Po} = w_0R^2/\lambda(t_a - t_0)$ is the Pomerantsev criterion, $\text{Pd} = (k/a)R^2$ is the Predvoditelev criterion which is defined as the maximum rate of change of the dimensionless specific strength of the heat source by the dimensionless time, i.e.,

$$-\left[\frac{d(w/w_0)}{d\,\text{Fo}}\right]_{\max} = (k/a)R^2 = \text{Pd}.$$

d. Solution for $w = \dot{w}_0\tau^{n/2}$. Solution for the transform at $\text{Bi} = \infty$ (the boundary condition of the first kind) may be written as

$$T(x, s) - \frac{t_0}{s} = \frac{(t_a - t_0)\cosh(s/a)^{1/2}x}{s\cosh(s/a)^{1/2}R} + \frac{\dot{w}_0\Gamma(1+\frac{1}{2}n)}{c\gamma s^{2+\frac{1}{2}n}}\left[1 - \frac{\cosh(s/a)^{1/2}x}{\cosh(s/a)^{1/2}R}\right]. \qquad (8.2.15)$$

To invert the transform, we expand $(1/\cosh(s/a)^{1/2}R)$ into a series as in Chapter 4, Section 3. Then using relation (64) of the table of transforms (see Appendix 5), we obtain the solution of the problem in the following form:

$$\theta = \frac{t(x, \tau) - t_0}{t_a - t_0}$$

$$= \sum_{m=1}^{\infty} (-1)^{m+1}\left[\operatorname{erfc} \frac{(2m-1) - (x/R)}{2(\mathrm{Fo})^{1/2}} + \operatorname{erfc} \frac{(2m-1) + (x/R)}{2(\mathrm{Fo})^{1/2}}\right]$$

$$+ \frac{\dot{w}_0 \tau^{\frac{1}{2}n+1}}{c\gamma(t_a - t_0)(1 + \frac{1}{2}n)}\left\{1 - \Gamma(2 + \tfrac{1}{2} n)\, 2^{n+2} \sum_{m=1}^{\infty} (-1)^{m+1}\right.$$

$$\left.\times \left[\mathrm{i}^{n+2} \operatorname{erfc} \frac{(2m-1) - (x/R)}{2(\mathrm{Fo})^{1/2}} + \mathrm{i}^{n+2} \operatorname{erfc} \frac{(2m-1) + (x/R)}{2(\mathrm{Fo})^{1/2}}\right]\right\}. \quad (8.2.16)$$

If the heat source is absent ($\dot{w}_0 = 0$) then from solution (8.2.16) we obtain solution (6.3.27).

Solutions (8.2.14) and (8.2.16) are simplified if we set $t_a = t_0$; this corresponds to the case when the surfaces of the plate are held at the same temperature as the plate initial temperature; heating takes placeonly at the expense of the heat source.

e. Solution of the Problem with a Variable Heat Source w(x,τ) with the Second Kind of Boundary Condition q(τ). Consider a more general problem with the heat source as a function of the coordinate x and time τ. The initial condition is taken in a more general form

$$t(x, 0) = f(x). \quad (8.2.2a)$$

The boundary condition of the second kind is taken as

$$-\lambda \frac{\partial t(R, \tau)}{\partial x} + q(\tau) = 0. \quad (8.2.4a)$$

The solution may be found by the integral Fourier transformation. Using the cosine Fourier transformation

$$T_c(n, \tau) = \int_0^R t(x, \tau) \cos (n\pi x/R)\, dx, \quad (8.2.17)$$

and by the formula for inversion of the transform $T_c(n, \tau)$ of $t(x, \tau)$

$$t(x, \tau) = (1/R)\, T_c(0, \tau) + \frac{2}{R} \sum_{n=1}^{\infty} T_c(n, \tau) \cos (n\pi x/R). \quad (8.2.18)$$

Multiplying both sides of differential equation (8.2.1) by $\cos(n\pi x/R)$, integrating from 0 to R and accounting for boundary conditions (8.2.3) and (8.2.4a) we obtain[1]

$$\frac{dT_c(n, \tau)}{d\tau} + \frac{an^2\pi^2}{R^2} T_c(n, \tau) = (-1)^n \frac{aq(\tau)}{\lambda} + \frac{1}{c\gamma} w_c(n, \tau), \quad (8.2.19)$$

where

$$w_c(n, \tau) = \int_0^R w(x, \tau) \cos(n\pi x/R)\, dx. \quad (8.2.20)$$

The solution of this equation will be

$$T_c(n, \tau) = \exp\left(-\frac{a\, n^2\pi^2\tau}{R^2}\right) \left[C(n) + (-1)^n \frac{a}{\lambda} \int_0^\tau q(\vartheta) \exp\left[\frac{a\, n^2\pi^2\vartheta}{R^2}\right] d\vartheta \right.$$
$$\left. + \frac{1}{c\gamma} \int_0^\tau w_c(n, \vartheta) \exp\left[\frac{a\, n^2\pi^2}{R^2} \vartheta\right] d\vartheta \right]. \quad (8.2.21)$$

To determine $C(n)$ we use the initial condition (8.2.2a)

$$\begin{aligned} T_c(n, 0) &= C(n) \\ &= \int_0^R t(x, 0) \cos\frac{n\pi x}{R}\, dx \\ &= \int_0^R f(x) \cos(n\pi x/R)\, dx. \end{aligned} \quad (8.2.22)$$

Then

$$T_c(n, \tau) = (-1)^n \frac{a}{\lambda} \int_0^\tau q(\vartheta) \exp[-(an^2\pi^2/R^2)(\tau - \vartheta)]\, d\vartheta$$
$$+ \frac{1}{c\gamma} \int_0^\tau w_c(n, \tau) \exp\left[-\frac{a\, n^2\pi^2}{R^2}(\tau - \vartheta)\right] d\vartheta$$
$$+ \exp\left[-\frac{a\, n^2\pi^2\tau}{R^2}\right] \int_0^R f(x) \cos\frac{n\pi x}{R}\, dx. \quad (8.2.23)$$

For convenience of inversion according to (8.2.18) we rewrite the solution for the inverse transform (8.2.21) as

$$T_c(n, \tau) = T_c(0, \tau) + T_{cn}(n, \tau), \quad (8.2.24)$$

in the second summand $n = 1, 2, 3, \cdots$.

[1] See Chapter 5, formulas (5.2.31)–(5.2.33).

We have

$$T_c(n, \tau) = \int_0^R f(x)\,dx + a/\lambda \int_0^\tau q(\vartheta)\,d\vartheta + \frac{1}{c\gamma}\int_0^\tau w_c(0, \vartheta)\,d\vartheta$$
$$+ \{\exp[-(a\,n^2\pi^2\tau/R^2)] \int_0^R f(x)\cos(n\pi\tau/R)\,dx$$
$$+ (-1)^n (a/\lambda) \int_0^\tau q(\vartheta)\exp[-(a\,n^2\pi^2/R^2)(\tau-\vartheta)]\,d\vartheta$$
$$- (1/c\gamma) \int_0^\tau w_c(n, \vartheta)\exp[-(a\,n^2\pi^2/R^2)(\tau-\vartheta)]\,d\vartheta\}, \quad (8.2.25)$$

where

$$w_c(0, \tau) = \int_0^R w(x, \tau)\,dx. \quad (8.2.26)$$

Inversion of the transform $T_c(n, \tau)$ is carried out using formula (8.2.18):

$$t(x, \tau) = \frac{1}{R}\left\{\int_0^R f(x)\,dx + \frac{a}{\lambda}\int_0^\tau q(\vartheta)\,d\vartheta + \frac{1}{c\gamma}\int_0^\tau w(0, \vartheta)\,d\vartheta\right\}$$
$$+ \frac{2}{R}\sum_{n=1}^{\infty} \cos\frac{n\pi x}{R}\exp\left(-\frac{a\,n^2\pi^2\tau}{R^2}\right)\int_0^R f(x)\cos\frac{n\pi x}{R}\,dx$$
$$+ \frac{2a}{\lambda R}\sum_{n=1}^{\infty}(-1)^n\cos\frac{n\pi x}{R}\int_0^\tau q(\vartheta)\exp\left[-\frac{a\,n^2\pi^2}{R^2}(\tau-\vartheta)\right]d\vartheta$$
$$+ \frac{2}{Rc\gamma}\sum_{n=1}^{\infty}\cos\frac{n\pi x}{R}\int_0^\tau w_c(n, \vartheta)\exp\left[-\frac{a\,n^2\pi^2}{R^2}(\tau-\vartheta)\right]d\vartheta. \quad (8.2.27)$$

Solution (8.2.27) is a general solution of the problem stated.

The solution in generalized variables may be written

$$\theta((x/R), \mathrm{Fo}) = \int_0^1 f(x/R)d(x/R) + \int_0^{\mathrm{Fo}} \mathrm{Ki}(\mathrm{Fo}^*)\,d\mathrm{Fo}^* + \int_0^1 \mathrm{Po}((x/R), \mathrm{Fo}^*)d(\mathrm{Fo})^*$$
$$+2\sum_{n=1}^{\infty}\cos\mu_n\frac{x}{R}\exp[-\mu_n^2\,\mathrm{Fo}]\int_0^1 f(x/R)\cos\mu_n\frac{x}{R}\,d(x/R)$$
$$+2\sum_{n=1}^{\infty}\cos\mu_n(x/R)\int_0^{\mathrm{Fo}}(-1)^n\,\mathrm{Ki}(\mathrm{Fo}^*) + \int_0^1 \mathrm{Po}\{(x/R), \mathrm{Fo}^*\}$$
$$\times\cos\mu_n(x/R)\,d(x/R)\exp[-\mu_n^2(\mathrm{Fo}-\mathrm{Fo}^*)]\,d\mathrm{Fo}^*, \quad (8.2.28)$$

where $\mu_n = n\pi$, $\mathrm{Ki}(\mathrm{Fo}) = q(\tau)R/\lambda t_a$ is the Kirpichev number, $\mathrm{Po} = w(x, \tau)R^2/\lambda t_a$ is the Pomerantsev number or the generalized variable, t_a is the fixed value of a temperature, e.g., a medium temperature, and $\theta = t(x, \tau)/t_a$ is the dimensionless temperature.

With the uniform initial temperature distribution

$$t(0, x) = t_0 \tag{8.2.2}$$

and constant heat flux at a plate surface

$$q(\tau) = q_c = \text{const}, \tag{8.2.4b}$$

solution (8.2.28) assumes the form

$$\theta\left(\frac{x}{R}, \text{Fo}\right)$$

$$= \text{Ki}\left[\text{Fo} - \frac{1}{6}(1-3(x^2/R^2)) - \sum_{n=1}^{\infty}(-1)^n \frac{2}{\mu_n^2} \cos \mu_n \frac{x}{R} \exp[-\mu_n^2 \text{Fo}]\right]$$

$$+ \int_0^{\text{Fo}} d\text{Fo}^* \int_0^1 \text{Po}\left(\frac{x}{R}, \text{Fo}^*\right) d(x/R) + 2 \sum_{n=1}^{\infty} \cos \mu_n \frac{x}{R} \exp[-\mu_n^2 \text{Fo}]$$

$$\times \int_0^{\text{Fo}} \exp[\mu_n^2 \text{Fo}^*]\, d\text{Fo}^* \int_0^1 \text{Po}\{(x/R), \text{Fo}^*\} \cos \mu_n(x/R)\, d(x/R). \tag{8.2.29}$$

In addition, the temperature is taken relative to the initial temperature of the body ($t_0 = 0$). The first term in solution (8.2.29) in square brackets is the solution of the problem without heat sources (Po = 0). The remaining terms include the effect of the heat source upon the temperature distribution.

From solution (8.2.29) it is possible to obtain some particular solutions:

(1) A constant heat source [Po $((x/R), \text{Fo}) = \text{Po}_c = \text{const}$]:

$$\theta((x/R), \text{Fo}) = \text{Po}_c \text{Fo} - \varphi((x/R), \text{Fo}), \tag{8.2.30}$$

where

$$\varphi((x/R), \text{Fo}) = \text{Ki}\left[\text{Fo} - \frac{1}{6}(1 - 3(x^2/R^2)\right]$$

$$- \sum_{n=1}^{\infty} (-1)^n (2/\mu_n^2) \cos \mu_n(x/R) \exp[-\mu_n^2 \text{Fo}] \tag{8.2.31}$$

is the solution of the problem without a heat source.

(2) The heat source is a linear function of the coordinate [Po $((x/R), \text{Fo}) = \text{Po}_c(1 - (x/R))$]:

$$\theta((x/R), \text{Fo}) = \text{Po}_c\left\{\frac{1}{2}\text{Fo} + \frac{4}{\pi^2}\sum_{n=1}^{\infty}\frac{\cos(2n-1)\pi\, x/R}{(2n-1)^4}\right.$$

$$\left.\times \cos \mu_n(x/R)[1 - \exp[-\mu_n^2 \text{Fo}]]\right\} + \varphi((x/R), \text{Fo}). \tag{8.2.32}$$

(3) The heat source is a parabolic function of the coordinate $[\mathrm{Po}\,(x/R, \mathrm{Fo}) = \mathrm{Po}_c(1 - (x^2/R^2))]$:

$$\theta((x/R), \mathrm{Fo}) = \mathrm{Po}_c\left\{\frac{2}{3}\,\mathrm{Fo} - \sum_{n=1}^{\infty} \frac{4}{\mu_n^2} (-1)^n \cos \mu_n \frac{x}{R} \times [1 - \exp[-\mu_n^2\,\mathrm{Fo}]]\right\} + \varphi((x/R), \mathrm{Fo}). \quad (8.2.33)$$

(4) The heat source is an exponential function of the coordinate $[\mathrm{Po}\,(x/R, \mathrm{Fo}) = \mathrm{Po}_c \exp[-b(x/R)]]$:

$$\theta((x/R), \mathrm{Fo}) = \mathrm{Po}_c\left\{\frac{1}{b}(1 - \exp[-b])\,\mathrm{Fo} + \sum_{n=1}^{\infty} [1 - (-1)^n \exp[-b]] \frac{2b}{\mu_n^2(b^2 + \mu_n^2)} \times \cos \mu_n \frac{x}{R}\,[1 - \exp(-\mu_n^2\,\mathrm{Fo})]\right\} - \varphi\left(\frac{x}{R}, \mathrm{Fo}\right). \quad (8.2.34)$$

(5) The heat source is a linear function of time $[\mathrm{Po}\,(x/R, \mathrm{Fo}) = \mathrm{Po}_c(1 + \mathrm{Pd}'\,\mathrm{Fo})]$:

$$\theta((x/R), \mathrm{Fo}) = \mathrm{Po}_c\,\mathrm{Fo}(1 + \tfrac{1}{2}\,\mathrm{Pd}'\,\mathrm{Fo}) - \varphi((x/R), \mathrm{Fo}), \quad (8.2.35)$$

where $\mathrm{Pd}' = (k/a)R^2$ is the Predvoditelev criterion equal to the maximum rate of change in the relative specific power of a heat source by the Fourier number

$$\mathrm{Pd}' = (k/a)\,R^2 = -\left(\frac{d(w/w_0)}{d\mathrm{Fo}}\right)_{\max}, \quad (8.2.36)$$

where k is the constant which is numerically equal to the maximum rate of the change in the dimensionless specific power of a heat source and w_0 is the specific power of a heat source at time $\tau = 0$.

(6) The heat source is an exponential function of time $[\mathrm{Po}\,(x/R, \mathrm{Fo}) = \mathrm{Po}_c \exp[-\mathrm{Pd}'\,\mathrm{Fo}]]$:

$$\theta((x/R), \mathrm{Fo}) = (\mathrm{Po}_c/\mathrm{Pd}')[1 - \exp[-\mathrm{Pd}'\,\mathrm{Fo}]] + \varphi((x/R), \mathrm{Fo}). \quad (8.2.37)$$

(7) The heat source is a periodic function of time $[\mathrm{Po}\,((x/R), \mathrm{Fo}) = \mathrm{Po}_c \cos \mathrm{Pd}'\,\mathrm{Fo}]$:

$$\theta((x/R), \mathrm{Fo}) = (\mathrm{Po}_c/\mathrm{Pd}') \sin \mathrm{Pd}'\,\mathrm{Fo} + \varphi((x/R), \mathrm{Fo}). \quad (8.2.38)$$

(8) The heat source depends upon time in the nth power [Po $((x/R)$, Fo) $= \mathrm{Po}_c(\mathrm{Pd}'\ \mathrm{Fo})^n$]:

$$\theta((x/R), \mathrm{Fo}) = \frac{(\mathrm{Pd}'\ \mathrm{Fo})^n}{n+1} \mathrm{Po}_c\,\mathrm{Fo} + \varphi((x/R), \mathrm{Fo}). \qquad (8.2.39)$$

8.3 Sphere (Symmetrical Problem)

a. Statement of the Problem. The problem is similar to the previous one and mathematically is written

$$\frac{\partial[rt(r,\tau)]}{\partial\tau} = a\frac{\partial^2[rt(r,\tau)]}{\partial r^2} + \frac{wr}{c\gamma} \quad (\tau > 0;\ 0 < r < R), \qquad (8.3.1)$$

$$t(r, 0) = t_0, \qquad (8.3.2)$$

$$\frac{\partial t(0,\tau)}{\partial r} = 0; \quad t(0,\tau) \neq \infty, \qquad (8.3.3)$$

$$-\frac{\partial t(R,\tau)}{\partial r} + H[t_a - t(R,\tau)] = 0. \qquad (8.3.4)$$

b. Solution of the Problem for w = const. The solution for the transform $T(r, s)$ is obtained in a similar way. On the basis of conditions (8.3.2) and (8.3.3) the solution has the form

$$T(r, s) - \frac{t_0}{s} = \frac{w}{s^2 c\gamma} + \frac{B \sinh(s/a)^{1/2} r}{r}. \qquad (8.3.5)$$

The constant B is determined from boundary condition (8.3.4) which is transformed. Thus the solution for the transform acquires the form

$$T(r, s) - \frac{t_0}{s} = \left(\frac{t_a - t_0}{s} - \frac{w}{s^2 c\gamma}\right) \frac{\mathrm{Bi}\, R \sinh(s/a)^{1/2} r}{[(\mathrm{Bi}-1)\sinh(s/a)^{1/2}R + (s/a)^{1/2}R\cosh(s/a)^{1/2}R]r} + \frac{w}{s^2 c\gamma}. \qquad (8.3.6)$$

Analyzing solution (8.3.6), we see that it is the algebraic sum of solutions of the problems considered.

Thus, the solution of our problem may be obtained if from solution (6.5.27) we subtract solution (7.2.12), initially having replaced the value b

with $w/c\gamma$ and adding the value $(w/c\gamma)\tau$ which is the original function of the transform $w/s^2c\gamma$.

Hence the solution of our problem will have the form

$$\theta = \frac{t(r,\tau) - t_0}{t_a - t_0}$$

$$= 1 + \frac{1}{6}\,\mathrm{Po}\left(1 + \frac{2}{\mathrm{Bi}} - \frac{r^2}{R^2}\right)$$

$$- \sum_{n=1}^{\infty}\left(1 + \frac{\mathrm{Po}}{\mu_n{}^2}\right) A_n \frac{R \sin \mu_n(r/R)}{r\mu_n} \exp[-\,\mu_n{}^2\,\mathrm{Fo}], \qquad (8.3.7)$$

where $\mathrm{Po} = wR^2/\lambda(t_a - t_0)$ is the Pomerantsev number, A_n are the initial thermal amplitudes determined by the corresponding relations (see Chapter 6, Eqs. (6.5.28) and (6.5.29)).

In a stationary state we shall have a parabolic law of temperature distribution.

The mean temperature of the sphere required for calculation of the specific heat rate is equal to

$$\bar{\theta} = \frac{\bar{t}(\tau) - t_0}{t_a - t_0}$$

$$= 1 + \frac{1}{15}\,\mathrm{Po}(1 + (5/\mathrm{Bi})) - \sum_{n=1}^{\infty}(1 + (\mathrm{Po}/\mu_n{}^2))B_n \exp[-\,\mu_n{}^2\,\mathrm{Fo}]. \qquad (8.3.8)$$

c. Solution for $w = w_0e^{-k\tau}$**.** Applying a similar method of calculation we shall obtain the solution in the form

$$\theta = 1 - \frac{\mathrm{Po}}{\mathrm{Pd}}\left[1 - \frac{R\,\mathrm{Bi}\sin\{(\mathrm{Pd})^{1/2}\ r/R\}}{r[(\mathrm{Bi}-1)\sin(\mathrm{Pd})^{1/2} + (\mathrm{Pd})^{1/2}\cos(\mathrm{Pd})^{1/2}]}\right]\exp(-\,\mathrm{Pd\,Fo}]$$

$$- \sum_{n=1}^{\infty}\left(1 - \frac{\mathrm{Po}}{\mathrm{Pd} - \mu_n{}^2}\right) A_n \frac{R\sin\mu_n(r/R)}{r\mu_n}\exp[-\,\mu_n{}^2\,\mathrm{Fo}], \qquad (8.3.9)$$

where Pd is the Predvoditelev criterion; in the present case it is equal to $\mathrm{Pd} = (k/a)R^2$, Po is the Pomerantsev criterion, $\mathrm{Po} = w_0R^2/\lambda(t_a - t_0)$.

d. Solution for $w = \dot{w}_0\tau^{n/2}$**.** The specific strength of the heat source is some power function of time; $w = \dot{w}_0\tau^{n/2}$, where $n = -1, 0, 1, 2, \cdots$. We solve the problem by considering it the boundary condition of the first kind ($\mathrm{Bi} = \infty$), i.e., $t(R,\tau) = t_a$. Using the method given above we obtain the solution in the form

$$\theta = \sum_{m=1}^{\infty} \frac{R}{r}\left[\operatorname{erfc}\frac{(2m-1)-(r/R)}{2(\mathrm{Fo})^{1/2}} - \operatorname{erfc}\frac{(2m-1)+(r/R)}{2(\mathrm{Fo})^{1/2}}\right]$$

$$+ \frac{\dot{w}_0 \tau^{1+\frac{1}{2}n}}{c\gamma(t_a - t_0)(1+\frac{1}{2}n)}\left\{1 - \Gamma\left(2+\frac{1}{2}n\right) 2^{n+2}\right.$$

$$\left.\times \sum_{m=1}^{\infty} \frac{R}{r}\left[\mathrm{i}^{n+2}\operatorname{erfc}\frac{(2m-1)-(r/R)}{2(\mathrm{Fo})^{1/2}} - \mathrm{i}^{n+2}\operatorname{erfc}\frac{(2m-1)+(r/R)}{2(\mathrm{Fo})^{1/2}}\right]\right\}. \tag{8.3.10}$$

With no heat source ($\dot{w}_0 = 0$), solution (8.3.10) turns into the corresponding solution Eq. (4.4.34), Chapter 4.

e. The Statement of the Problem with a Variable Heat Source and the Second-Kind Boundary Conditions

$$t(r, 0) = f(r), \tag{8.3.11}$$

$$-\lambda[\partial t(R, \tau)/\partial \tau] + q(\tau) = 0, \tag{8.3.12}$$

$$\partial t(0, \tau)/\partial r = 0; \quad t(0, \tau) \neq \infty. \tag{8.3.13}$$

The heat source is a function of the coordinate and time $w(r, \tau)$.

The solution of the problem for $w(r, \tau)$. To solve the problem stated we use the finite cosine Fourier transformation

$$T_s(p, \tau) = \int_0^R r t(r, \tau) \frac{\sin p\tau}{p} dr, \tag{8.3.14}$$

where p is the root of the characteristic equation

$$\sin pR - pR\cos pR = 0. \tag{8.3.15}$$

Inversion of the transform $T_s(p, \tau)$ of the function $t(r, \tau)$ is performed using the formula

$$t(r, \tau) = \frac{3}{R^3} T_s(0, \tau) + \frac{2}{R}\sum_{n=1}^{\infty} \frac{p_n}{\sin^2 p_n R} \frac{\sin p_n r}{r} T_s(p_n, \tau). \tag{8.3.16}$$

Multiplying all the terms of Eq. (8.3.1) by $r(\sin pr/p)$ and integrating by r from 0 to R with account for boundary conditions (8.3.12) and (8.3.13) we obtain[2]

[2] See Chapter 5, formulas (5.3.24)–(5.3.26).

$$\frac{dT_s(p, \tau)}{d\tau} + ap^2 T_s(p, \tau) = \frac{a}{\lambda} R \frac{\sin pR}{p} q(\tau) + \frac{1}{c\gamma} w_s(p, \tau), \tag{8.3.17}$$

where

$$w_s(p, \tau) = \int_0^R r \frac{\sin pr}{p} w(r, \tau)\, dr$$

is the transform for the function of the internal heat source.

The solution of equation (8.3.17) will be

$$T_s(p, \tau) = \exp[-ap^2\tau]\Big\{C(p) + \frac{a}{\lambda} R \frac{\sin pR}{p} \times \int_0^\tau q(\vartheta) \exp[ap^2\vartheta]\, d\vartheta + \frac{1}{c\gamma} \int_0^\tau w_s(p, \vartheta) \exp[ap^2\vartheta)\, d\vartheta\Big\}. \tag{8.3.18}$$

To define $C(p)$ we use the initial condition

$$C(p) = \int_0^R rf(r) \frac{\sin pr}{p}\, dr. \tag{8.3.19}$$

For the convenience of invertion, we initially find $T_s(0, \tau)$

$$T_s(0, \tau) = \int_0^R r^2 f(r)\, dr + \frac{a}{\lambda} R \int_0^\tau q(\vartheta)\, d\vartheta + \frac{1}{c\gamma} \int_0^\tau \int_0^R r^2 w(r, \vartheta)\, d\vartheta\, dr. \tag{8.3.20}$$

Substituting the value of $T_s(0, \tau)$ and $T_s(p, \tau)$ into formula (8.3.16) and using formula (8.3.19) we obtain the solution of our problem

$$\begin{aligned} t(r, \tau) = {} & \frac{3}{R^3} \int_0^R r^2 f(r)\, dr + \frac{3a}{\lambda R^2} \int_0^\tau q(\vartheta)\, d\vartheta \\ & + \frac{3}{c\gamma R^3} \int_0^\tau \int_0^R r^2 w(r, \vartheta)\, d\vartheta\, dr \\ & + \sum_{n=1}^\infty \frac{p_n}{\sin^2 p_n R} \frac{\sin p_n r}{r} \exp[-ap_n^2\tau] \frac{2}{R} \int_0^R rf(r) \frac{\sin p_n r}{p_n}\, dr \\ & + \frac{a}{\lambda} \sum_{n=1}^\infty \frac{Rp_n \sin p_n R \sin p_n r}{p_n \sin^2 p_n R r} \exp[-ap_n^2\tau] \frac{2}{R} \int_0^\tau q(\vartheta) \\ & \times \exp[ap_n^2\vartheta]\, d\vartheta + \frac{1}{c\gamma} \sum_{n=1}^\infty \frac{p_n R}{\sin^2 p_n R} \frac{\sin p_n r}{r} \\ & \times \exp[-ap_n^2\tau] \frac{2}{R} \int_0^\tau \int_0^R w(r, \vartheta) \frac{r \sin p_n r}{p_n R} \exp[ap_n^2\vartheta]\, d\vartheta\, d\tau. \end{aligned} \tag{8.3.21}$$

We designate $\mu_n = p_n R$, $\mathrm{Fo} = a\tau/R^2$, $\mathrm{Ki(Fo)} = q(\tau)R/\lambda t_a$, $\mathrm{Po} = w(r, \tau)R^2/\lambda t_a$. Then, the solution of Eq. (8.3.21) in generalized variables may be written as

$$\begin{aligned}
\theta\left(\frac{r}{R}, \mathrm{Fo}\right)
&= 3\left\{ \int_0^1 \left(\frac{r}{R}\right)^2 f\left(\frac{r}{R}\right) d\left(\frac{r}{R}\right) + \int_0^{\mathrm{Fo}} \left[\mathrm{Ki}\,(\mathrm{Fo}^*)\right.\right. \\
&\quad \left.\left. + \int_0^1 \frac{r^2}{R^2} \mathrm{Po}\left(\frac{r}{R}, \mathrm{Fo}^*\right) d\left(\frac{r}{R}\right)\right] d\mathrm{Fo}^*\right\} + 2\sum_{n=1}^{\infty} \frac{R\mu_n \sin \mu_n(r/R)}{r \sin^2 \mu_n} \\
&\quad \times \exp[-\mu_n^2\,\mathrm{Fo}] \int_0^1 \frac{r}{R} f\left(\frac{r}{R}\right) \frac{R \sin \mu_n(r/R)}{r}\, d(r/R) \\
&\quad + \sum_{n=1}^{\infty} \frac{2R \sin \mu_n r/R}{r \sin^2 \mu_n} \int_0^{\mathrm{Fo}} \left[\sin \mu_n\, \mathrm{Ki}(\mathrm{Fo}^*)\right. \\
&\quad \left. + \int_0^1 \frac{r}{R} \mathrm{Po}\left(\frac{r}{R}, \mathrm{Fo}^*\right) \sin \mu_n \frac{r}{R}\, d\left(\frac{r}{R}\right)\right] \\
&\quad \times \exp[-\mu_n^2\,(\mathrm{Fo} - \mathrm{Fo}^*)]\, d\mathrm{Fo}^*.
\end{aligned} \tag{8.3.22}$$

For a uniform initial temperature distribution

$$t(r, 0) = t_0 = 0. \tag{8.3.23}$$

When the temperature is read from the initial temperature of a body and at a const.nt value of the Kirpichev number

$$\mathrm{Ki(Fo)} = \frac{q_c R}{\lambda t_a} = \mathrm{Ki} = \mathrm{const.} \tag{8.3.24}$$

Solution (8.3.22) assumes the form

$$\begin{aligned}
\theta\left(\frac{r}{R}, \mathrm{Fo}\right) &= \varphi\left(\frac{r}{R}, \mathrm{Fo}\right) + 3 \int_0^{\mathrm{Fo}} d\mathrm{Fo}^* \int_0^1 \left(\frac{r}{R}\right)^2 \mathrm{Po}\left(\frac{r}{R}, \mathrm{Fo}^*\right) d\left(\frac{r}{R}\right) \\
&\quad + \sum_{n=1}^{\infty} \frac{2R \sin \mu_n r/R}{r \sin^2 \mu_n} \exp[-\mu_n^2\,\mathrm{Fo}] \\
&\quad \times \int_0^{\mathrm{Fo}} \exp(\mu_n^2\,\mathrm{Fo}^*)\, d\mathrm{Fo}^* \\
&\quad \times \int_0^1 \left(\frac{r}{R}\right) \mathrm{Po}\left(\frac{r}{R}, \mathrm{Fo}^*\right) \sin \mu_n \frac{r}{R}\, d\left(\frac{r}{R}\right),
\end{aligned} \tag{8.3.25}$$

where

$$\varphi\{(r/R), \text{Fo}\} = \text{Ki}\Big[3\,\text{Fo} - \frac{1}{10}\{3 - 5(r^2/R^2)\} - \sum_{n=1}^{\infty} \frac{2R \sin \mu_n r/R}{\mu_n^2 r \sin \mu_n} \exp[-\mu_n^2\,\text{Fo}] \tag{8.3.26}$$

is the solution of the problem without the internal heat source.

From solution (8.3.25) it is possible to obtain a number of particular solutions:

(1) The constant heat source [Po$(r/R, \text{Fo}) = \text{Po}_c = \text{const}$]:

$$\theta\{(r/R), \text{Fo}\} = \text{Po}_c\,\text{Fo} + \varphi\{(r/R), \text{Fo}\}. \tag{8.3.27}$$

(2) The heat source is a linear function of the coordinate [Po$\{(r/R), \text{Fo}\} = \text{Po}_c\{1 - (r/R)\}$]:

$$\theta\{(r/R), \text{Fo}\} = \text{Po}_c\Big\{\frac{1}{4}\,\text{Fo} + 2\sum_{n=1}^{\infty} \frac{2 - (2 + \mu_n^2)\cos \mu_n}{\mu_n^5 \sin^2 \mu_n} \times \frac{R \sin \mu_n(r/R)}{r}\,[1 - \exp[-\mu_n^2\,\text{Fo}]]\Big\} + \varphi\{(r/R), \text{Fo}\}. \tag{8.3.28}$$

(3) The heat source is a parabolic function of the coordinate [Po$\{(r/R), \text{Fo}\} = \text{Po}_c\{1 - (r^2/R^2)\}$]:

$$\theta\{(r/R), \text{Fo}\} = \text{Po}_c\Big\{\frac{2}{5}\,\text{Fo} - \sum_{n=1}^{\infty} \frac{2}{\mu_n^4 \sin \mu_n} \frac{R \sin \mu_n\, r/R}{r} \times [1 - \exp[-\mu_n^2\,\text{Fo}]]\Big\} + \varphi\{(r/R), \text{Fo}\}. \tag{8.3.29}$$

(4) The heat source is an exponential function of the coordinate [Po$\{r/R, \text{Fo}\} = \text{Po}_c \exp[-br/R]$]:

$$\theta\{(r/R), \text{Fo}\} = \text{Po}_c\Big\{\sum_{n=1}^{\infty} \frac{2b[2\mu_n - e^{-b}(2 + b^2 + 2b + \mu_n^2)\sin \mu_n]}{\mu_n^2 \sin^2 \mu_n(\mu_n^2 + b^2)^2} \times \frac{R \sin \mu_n(r/R)}{r}\,[1 - \exp[-\mu_n^2\,\text{Fo}]] + 3\,\text{Fo}\,\frac{2 - e^{-b}(b^2 + 2b + 2)}{b^3}\Big\} + \varphi\{(r/R), \text{Fo}\}. \tag{8.3.30}$$

(5) The heat source is a linear function of time [Po$\{(r/R), \text{Fo}\} = \text{Po}_c(1 + \text{Pd}'\,\text{Fo})$]:

$$\theta\{(r/R), \text{Fo}\} = \text{Po}_c\,\text{Fo}(1 + \tfrac{1}{2}\,\text{Pd}'\,\text{Fo}) + \varphi\{(r/R), \text{Fo}\}. \tag{8.3.31}$$

(6) The heat source is an exponential function of time [$\mathrm{Po}\{(r/R), \mathrm{Fo}\} = \mathrm{Po}_c \exp[-\mathrm{Pd}' \mathrm{Fo}]$]:

$$\theta\{(r/R), \mathrm{Fo}\} = \frac{\mathrm{Po}_c}{\mathrm{Pd}'}[1 - \exp[-\mathrm{Pd}' \mathrm{Fo}]] + \varphi\{(r/R), \mathrm{Fo}\}. \quad (8.3.32)$$

(7) The heat source is a periodic function of time [$\mathrm{Po}\{(r/R), \mathrm{Fo}\} = \mathrm{Po}_c \cos \mathrm{Pd}' \mathrm{Fo}$]:

$$\theta\{(r/R), \mathrm{Fo}\} = \frac{\mathrm{Po}_c}{\mathrm{Pd}'} \sin \mathrm{Pd}' \mathrm{Fo} + \varphi\{(r/R), \mathrm{Fo}\}. \quad (8.3.33)$$

(8) The heat source depends upon time in the nth power [$\mathrm{Po}\{(r/R), \mathrm{Fo}\} = \mathrm{Po}_c(\mathrm{Pd}' \mathrm{Fo})^n$]:

$$\theta\{(r/R), \mathrm{Fo}\} = \frac{(\mathrm{Pd}' \mathrm{Fo})^n}{n+1} \mathrm{Po}_c \mathrm{Fo} + \varphi\{(r/R), \mathrm{Fo}\}. \quad (8.3.34)$$

8.4 Infinite Cylinder

a. Statement of the Problem. A similar problem for an infinite cylinder is written mathematically as

$$\frac{\partial t(r, \tau)}{\partial \tau} = a\left(\frac{\partial^2 t(r, \tau)}{\partial r^2} + \frac{1}{r}\frac{\partial t(r, \tau)}{\partial r}\right) + \frac{w}{c\gamma} \quad (\tau > 0;\ 0 < r < R). \quad (8.4.1)$$

The boundary conditions are the same as for a sphere.

b. Solution of the Problem for w = const. The solution for the transform is obtained in the following form

$$T(r, s) - \frac{t_0}{s} = \frac{w}{s^2 c\gamma} + \left(\frac{t_a - t_0}{s} - \frac{w}{s^2 c\gamma}\right) \times \frac{I_0((s/a)^{1/2} r)}{I_0((s/a)^{1/2} R) + (1/H)(s/a)^{1/2} I_1((s/a)^{1/2} R)}. \quad (8.4.2)$$

Comparing solution (8.4.2) with corresponding solutions (6.6.21) and (7.3.2) we find

$$\theta = 1 + \frac{1}{4}\mathrm{Po}\left(1 + \frac{2}{\mathrm{Bi}} - \frac{r^2}{R^2}\right) - \sum_{n=1}^{\infty}\left(1 + \frac{\mathrm{Po}}{\mu_n^2}\right) A_n J_0\{\mu_n(r/R)\} \exp[-\mu_n^2 \mathrm{Fo}], \quad (8.4.3)$$

where $\mathrm{Po} = wR^2/\lambda(t_a - t_0)$ is the Pomerantsev criterion.

It is seen from solution (8.4.3) that in a stationary state the temperature distribution is given by the law of a parabola.

The mean temperature of the cylinder is equal to

$$\bar{\theta} = 1 + \frac{1}{8}\,\mathrm{Po}(1 + \{4/\mathrm{Bi}\}) - \sum_{n=1}^{\infty} (1 + \{\mathrm{Po}/\mu_n^{2}\})B_n \exp[-\,\mu_n^{2}\,\mathrm{Fo}]. \tag{8.4.4}$$

The constants A_n and B_n and the roots μ_n of the characteristic equation are determined from the corresponding relations for the infinite cylinder (see Chapter 6, Section 6).

c. Solution of the Problem for $w = w_0 e^{-k\tau}$. In a similar way we find the solution in the form

$$\theta = 1 - \frac{\mathrm{Po}}{\mathrm{Pd}}\left[1 - \frac{J_0\{(\mathrm{Pd})^{1/2}(r/R)\}}{J_0(\mathrm{Pd})^{1/2} - (1/\mathrm{Bi})(\mathrm{Pd})^{1/2}J_1(\mathrm{Pd})^{1/2}}\right]\exp[-\,\mathrm{Pd}\,\mathrm{Fo}]$$
$$- \sum_{n=1}^{\infty}\left(1 - \frac{\mathrm{Po}}{\mathrm{Pd} - \mu_n^{2}}\right) A_n J_0\left\{\mu_n \frac{r}{R}\right\}\exp[-\,\mu_n^{2}\,\mathrm{Fo}]. \tag{8.4.5}$$

The coefficients A_n are determined by the corresponding relation for the cylinder.

d. Variable Heat Source $w(r, \tau)$: The Second Kind of Boundary Condition. From the mathematical viewpoint the problem is written as

$$r(r, 0) = f(r)\,, \tag{8.4.6}$$

$$-\,\lambda\{\partial t(R, \tau)/\partial r\} + q(\tau) = 0\,, \tag{8.4.7}$$

$$\partial t(0, \tau)/\partial r = 0\,, \quad t(0, \tau) \neq \infty. \tag{8.4.8}$$

e. Solution of the Problem. For the solutions of this problem we use the finite Hankel integral transform

$$T_H(p, \tau) = \int_0^R rt(r, \tau)\, J_0(pr)\, dr\,, \tag{4.8.9}$$

where p is the root of the characteristic equation

$$J_0'(pR) = 0. \tag{8.4.10}$$

Inversion of $T_H\,(p, \tau)$ of the function $t(r, \tau)$ is carried out using the formula

$$t(r, \tau) = (2/R^2)T_H(0, \tau) + (2/R^2)\sum_{n=1}^{\infty} T_H(p_n, \tau)\{J_0(p_n, r)/J_0^2(p_n R)\}. \tag{8.4.11}$$

Applying transform (8.4.9) to differential equation (8.4.1), taking into account boundary conditions (8.4.7)–(8.4.8), we have[3]

$$(dT_H(p, \tau)/d\tau) + ap^2T_H(p, \tau)$$
$$= aR(q(\tau)/\lambda)\, J_0(pR) + (1/c\gamma)\, w_H(p, \tau)\,, \tag{8.4.12}$$

where

$$w_H(p, \tau) = \int_0^R w(r, \tau) r J_0(pr)\, dr\,. \tag{8.4.13}$$

The solution of ordinary differential equation (8.4.12) has the form:

$$T_H(p, \tau) = \Big[C(p) + (aR/\lambda)\, J_0(p, R) \int_0^\tau q(\vartheta) \exp[ap^2\vartheta]\, d\vartheta$$
$$+ \frac{1}{c\gamma} \int_0^\tau w_H(p, \vartheta) \exp[ap^2\vartheta]\Big] \exp[-\, ap^2\tau]\,. \tag{8.4.14}$$

To determine the constant $C(p)$ we use initial condition (8.4.6). From solution (8.4.14) it follows that at $\tau \to 0$

$$T_H(p, 0) = C(p)\,. \tag{8.4.15}$$

Moreover, proceeding from the transform $T_H(p, 0)$ we have

$$T_H(p, 0) = \int_0^R t(r, 0) r J_0(pr)\, dr$$
$$= \int_0^R f(r) r J_0(pr)\, dr\,. \tag{8.4.16}$$

Consequently,

$$C(p) = \int_0^R f(r) r J_0(pr)\, dr\,. \tag{8.4.17}$$

If expression (8.4.17) is substituted into solution (8.4.14) instead of $C(p)$, the solution of the problem for $T_H(p, \tau)$ is obtained.

Before inverting the transform of $T(r, \tau)$ according to formula (8.4.11) we determine $T_H(0, \tau)$ from solution (8.4.14):

$$T_H(0, \tau) = \int_0^R f(r) r\, dr + (aR/\lambda) \int_0^\tau q(\vartheta)\, d\vartheta + (1/c\gamma) \int_0^\tau w_H(0, \vartheta)\, d\vartheta\,, \tag{8.4.18}$$

where

$$w_H(0, \tau) = \int_0^R w(r, \tau) r\, dr\,. \tag{8.4.19}$$

[3] See Chapter 5, formula (5.4.14).

Substituting the value of $T_H(0, \tau)$ and $T_H(p, \tau)$ into formula (8.4.11) we obtain the solution

$$t(r, \tau) = \frac{2}{R^2} \int_0^R f(r) r \, dr + \frac{2a}{\lambda R} \int_0^\tau q(\vartheta) \, d\vartheta + \frac{2}{c\gamma R^2}$$
$$\times \int_0^\tau w_H(0, \vartheta) \, d\vartheta + \left\{ \sum_{n=1}^{\infty} \frac{J_0(p_n r)}{J_0^2(p_n R)} \exp[-a p_n^2 \tau] \frac{2}{R^2} \right.$$
$$\times \int_0^R f(r) r J_0(p_n r) \, dr + \frac{aR}{\lambda} \sum_{n=1}^{\infty} \frac{J_0(p_n R) J_0(p_n r)}{J_0^2(p_n R)}$$
$$\times \exp[-a p_n^2 \tau] \frac{2}{R^2} \int_0^\tau q(\vartheta) \exp[a p_n^2 \vartheta] \, d\vartheta + \frac{1}{c\gamma} \sum_{n=1}^{\infty} \frac{J_0(p_n r)}{J_0^2(p_n R)}$$
$$\left. \times \exp[-a p_n^2 \tau] \frac{2}{R^2} \int_0^\tau \int_0^R w(r, \vartheta) r J_0(p_n r) \exp[a p_n^2 \vartheta] \, d\vartheta \, dr \right\}. \tag{8.4.20}$$

If we designate $\mathrm{Fo} = a\tau/R^2$, $\mu_n = p_n R$, $a\vartheta/R^2 = \mathrm{Fo}^*$, then solution (8.4.20) may be written in generalized variables as

$$\theta\left(\frac{r}{R}, \mathrm{Fo}\right) = 2\left\{ \int_0^1 \frac{r}{R} f\left(\frac{r}{R}\right) d\left(\frac{r}{R}\right) + \int_0^{\mathrm{Fo}} \left[\mathrm{Ki}\,(\mathrm{Fo}^*) \right. \right.$$
$$\left. \left. + \int_0^1 \frac{r}{R} \mathrm{Po}\left(\frac{r}{R}, \mathrm{Fo}\right) d\left(\frac{r}{R}\right) \right] d\mathrm{Fo}^* \right\} + 2 \sum_{n=1}^{\infty} \frac{J_0\{\mu_n(r/R)\}}{J_0^2(\mu_n)}$$
$$\times \exp[-\mu_n^2 \mathrm{Fo}] \int_0^1 \frac{r}{R} f\left(\frac{r}{R}\right) J_0\left(\mu_n \frac{r}{R}\right) d\left(\frac{r}{R}\right)$$
$$+ 2 \sum_{n=1}^{\infty} \frac{J_0\{\mu_n(r/R)\}}{J_0^2(\mu_n)} \int_0^{\mathrm{Fo}} \left[J_0(\mu_n) \mathrm{Ki}(\mathrm{Fo}^*) \right.$$
$$\left. + \int_0^1 \frac{r}{R} \mathrm{Po}\left(\frac{r}{R}, \mathrm{Fo}^*\right) J_0\left(\mu_n \frac{r}{R}\right) d\left(\frac{r}{R}\right) \right]$$
$$\times \exp[-\mu_n^2(\mathrm{Fo} - \mathrm{Fo}^*)] \, d\mathrm{Fo}^*, \tag{8.4.21}$$

where

$$\mathrm{Ki}(\mathrm{Fo}) = \frac{q(\tau) R}{\lambda t_a}, \qquad \mathrm{Po}\left(\frac{r}{R}, \mathrm{Fo}\right) = \frac{w(r, \tau) R^2}{\lambda t_a}, \tag{8.4.22}$$

and μ_n are the roots of the characteristic equation $J_0'(\mu) = 0$.

If the temperature distribution at the initial time moment is uniform

$$t(r, 0) = t_0 = \mathrm{const}, \tag{8.4.23}$$

and t_0 is assumed zero without loss of generality, then with a constant heat flux at the body surface ($q(\tau)$ = const, Ki(Fo) = const), solution (8.4.21) may be written as

$$\begin{aligned}\theta\left(\frac{r}{R}, \mathrm{Fo}\right) &= \frac{t(r,\tau)}{t_a}\\ &= \varphi\left(\frac{r}{R}, \mathrm{Fo}\right) + 2\int_0^{\mathrm{Fo}} d\mathrm{Fo}^* \int_0^1 \frac{r}{R}\,\mathrm{Po}\left(\frac{r}{R}, \mathrm{Fo}^*\right) d\left(\frac{r}{R}\right)\\ &\quad + \sum_{n=1}^{\infty} \frac{2J_0(\mu_n r/R)}{J_0^2(\mu_n)} \exp[-\mu_n^2\,\mathrm{Fo}] \int_0^{\mathrm{Fo}} d\mathrm{Fo}^* \exp[\mu_n^2\,\mathrm{Fo}^*]\\ &\quad \times \int_0^1 \frac{r}{R}\,\mathrm{Po}\left(\frac{r}{R}, \mathrm{Fo}^*\right) J_0\left(\mu_n \frac{r}{R}\right) d\left(\frac{r}{R}\right), \end{aligned} \tag{8.4.24}$$

where

$$\begin{aligned}\varphi((r/R), \mathrm{Fo}) = \mathrm{Ki}\Big[2\,\mathrm{Fo} - \frac{1}{4}\{1 - 2(r^2/R^2)\}\\ - \sum_{n=1}^{\infty} \frac{2}{\mu_n^2 J_0(\mu_n)} J_0\{\mu_n(r/R)\} \exp[-\mu_n^2\,\mathrm{Fo}]\Big].\end{aligned} \tag{8.4.25}$$

The temperature was read from the initial temperature of a body ($t_0 = 0$).

If a heat source is absent (Po = 0), solution (8.4.24)

$$\theta\{(r/R), \mathrm{Fo}\} = \varphi\{(r/R), \mathrm{Fo}\}$$

becomes identical to solution (5.4.5).

We shall now find the solution for some particular problems under condition (8.4.23) and a constant value of Ki.

(1) Constant heat source [$\mathrm{Po}\{(r/R), \mathrm{Fo}\} = \mathrm{Po}_c$ = const]:

$$\theta\{(r/R), \mathrm{Fo}\} = \mathrm{Po}_c\,\mathrm{Fo} + \varphi\{(r/R), \mathrm{Fo}\}. \tag{8.4.26}$$

(2) Heat source is a parabolic function of the radial coordinate [$\mathrm{Po}\{(r/R), \mathrm{Fo}\} = \mathrm{Po}_c\{1 - (r^2/R^2)\}$]:

$$\begin{aligned}\theta\{(r/R), \mathrm{Fo}\} = \mathrm{Po}_c\Big\{\frac{1}{3}\,\mathrm{Fo} + \sum_{n=1}^{\infty} \frac{4[2J_1'(\mu_n) - J_0(\mu_n)]}{\mu_n^3 J_0^2(\mu_n)}\\ \times J_0\{\mu_n(r/R)\}[1 - \exp[-\mu_n^2\,\mathrm{Fo}]]\Big\} + \varphi\{(r/R), \mathrm{Fo}\}.\end{aligned} \tag{8.4.27}$$

(3) Heat source is a linear function of time [$\mathrm{Po}\{(r/R), \mathrm{Fo}\} = \mathrm{Po}_c(1 + \mathrm{Pd}'\,\mathrm{Fo})$]:

$$\theta\{(r/R), \mathrm{Fo}\} = \mathrm{Po}_c\,\mathrm{Fo}(1 + \tfrac{1}{2}\,\mathrm{Pd}'\,\mathrm{Fo}) + \varphi\{(r/R), \mathrm{Fo}\} \tag{8.4.28}$$

where $\mathrm{Pd}' = kR^2/d$ is the Predvoditelev number equal to the maximum relative rate of the change in the specific power of a heat source

$$\mathrm{Pd}' = \left[\frac{d(w/w_0)}{d\,\mathrm{Fo}}\right]_{\max}$$

(4) The heat source is an exponential function of time $[\mathrm{Po}\{(r/R), \mathrm{Fo}\} = \mathrm{Po}_c \exp[-\mathrm{Pd}'\,\mathrm{Fo}]]$:

$$\theta\{(r/R), \mathrm{Fo}\} = (\mathrm{Po}_c/\mathrm{Pd}')[1 - \exp[-\mathrm{Pd}'\,\mathrm{Fo}]] + \varphi\{(r/R), \mathrm{Fo}\}. \qquad (8.4.29)$$

(5) The heat source is a periodic function of time $[\mathrm{Po}\{(r/R), \mathrm{Fo}\} = \mathrm{Po}_c \cos[\mathrm{Pd}'\,\mathrm{Fo}]]$:

$$\theta\{(r/R), \mathrm{Fo}\} = (\mathrm{Po}_c/\mathrm{Pd}') \sin(\mathrm{Pd}'\,\mathrm{Fo}) + \varphi\{(r/R), \mathrm{Fo}\}. \qquad (8.4.30)$$

(6) The heat source depends upon time in the nth power $[\mathrm{Po}\{(r/R), \mathrm{Fo}\} = \mathrm{Po}_c(\mathrm{Pd}'\,\mathrm{Fo})^n]$:

$$\theta\{(r/R), \mathrm{Fo}\} = \frac{(\mathrm{Pd}'\,\mathrm{Fo})^n}{\mathrm{n} + 1}\,\mathrm{Po}_c\,\mathrm{Fo} + \varphi\{(r/R), \mathrm{Fo}\}. \qquad (8.4.31)$$

In all the cases the heat rate is determined by the mean temperature $\bar{\theta}(\mathrm{Fo})$ and the mean value of $\bar{\varphi}(\mathrm{Fo})$ is defined by relation (5.4.8).

CHAPTER

9

TEMPERATURE FIELD WITH PULSE-TYPE HEAT SOURCES

Introduction

In some thermal processes, a body is heated as a result of an instantaneous heat source of constant strength (thermal impulse). Such problems include the problem of heating a cable in which a short circuit occurs; as a result, the cable is given an instantaneous thermal impulse. A number of methods for determining thermal properties are based on the laws of an unsteady-state temperature field resulting from an instantaneous heat flux.

Before considering particular problems we shall dwell on the properties of the solution of the differential heat conduction equation for an infinite body in the presence of a pulse-type heat source.

The differential heat conduction equation

$$\frac{\partial t(x, y, z, \tau)}{\partial \tau} = a \nabla^2 t(x, y, z, \tau) \tag{9.1}$$

is satisfied by the following solution

$$t(x, y, z, \tau) = \frac{b_1}{(2(\pi a \tau)^{1/2})^3} \exp\left[-\frac{(x - x_1)^2 + (y - y_1)^2 + (z - z_1)^2}{4a\tau}\right]. \tag{9.2}$$

It is seen from solution (9.2) that the temperature of the body tends to zero when $\tau \to 0$ at all points with the exception of one point (x_1, y_1, z_1) where it becomes infinitely large.

If solution (9.2) is integrated over the whole volume between $-\infty$ and $+\infty$, then we obtain

$$\int_{-\infty}^{\infty}\int_{-\infty}^{\infty}\int_{-\infty}^{\infty} t(x, y, z, \tau)\, dx\, dy\, dz = b_1$$

since

$$\frac{1}{2(\pi a\tau)^{1/2}} \int_{-\infty}^{\infty} \exp\left[-\frac{(x-x_1)^2}{4a\tau}\right] dx$$
$$= \frac{1}{2(\pi a\tau)^{1/2}} \int_{-\infty}^{\infty} \exp\left[-\frac{(y-y_1)^2}{4a\tau}\right] dy$$
$$= \frac{1}{2(\pi a\tau)^{1/2}} \int_{-\infty}^{\infty} \exp\left[-\frac{(z-z_1)^2}{4a\tau}\right] dz = 1.$$

Hence, expression (9.2) is the solution of the problem of the temperature distribution in an infinite body at any moment of time caused by the action of an instantaneous heat source with the strength b_1 at the point x_1, y_1, z_1 at the moment of time $\tau = 0$, since when an amount of heat $Q_1 = c\gamma b_1$ (kcal) is evolved the temperature distribution will be determined by relation (9.2). Hence

$$q_1 = Q_1/c\gamma \quad (\text{deg m}^3).$$

The function

$$G(x, y, z, \tau, x_1, y_1, z_1) = \frac{b_1}{(2(\pi a\tau)^{1/2})^3} \exp\left[-\frac{(x-x_1)^2 + (y-y_1)^2 + (z-z_1)^2}{4a\tau}\right]$$

is called the function of the temperature effect of an instantaneous heat source. This function has the property of symmetry

$$G(x, y, z, \tau, x_1, y_1, z_1) = G(x_1, y_1, z_1, \tau, x, y, z),$$

which is the expression of the reciprocity principle: The action at the point x, y, z of the source being at the point x_1, y_1, z_1 is equal to the action at the point x_1, y_1, z_1 of the same source placed at the point x, y, z. However, there is no such symmetry with respect to the variable τ which shows irreversibility of thermal processes in time.

Let us determine the form of the function G for other cases. The expression

$$t(x, y, \tau) = \frac{b_2}{4\pi a\tau} \exp\left[-\frac{(x-x_1)^2 + (y-y_1)^2}{4a\tau}\right] \tag{9.3}$$

satisfies the differential equation

$$\frac{\partial t(x, y, \tau)}{\partial \tau} = a \left(\frac{\partial^2 t(x, y, \tau)}{\partial x^2} + \frac{\partial^2 t(x, y, \tau)}{\partial y^2} \right)$$

and is the solution of the problem of the temperature distribution in an infinite body with a two-dimensional heat flux caused by the action of an instantaneous line heat source at the location x_1, y_1 (i.e., crossing the point (x_1, y_1) parallel to the axis z) at the moment of time $\tau = 0$ since

$$\int_{-\infty}^{\infty} \int_{-\infty}^{\infty} t(x, y, \tau)\, dx\, dy = b_2 \text{ (deg m}^2\text{)},$$

where $b_2 = Q_2/c\gamma$ and Q_2 is the strength of the heat source per unit length in kcal/m. Solution (9.3) may be obtained from solution (9.2) assuming that z_1 is distributed from $-\infty$ to $+\infty$, i.e., replacing the point source by a linear one. The expression

$$t(x, \tau) = \frac{b_3}{2(\pi a \tau)^{1/2}} \exp\left[-\frac{(x - x_1)^2}{4a\tau} \right] \tag{9.4}$$

satisfies the equation

$$\frac{\partial t(x, \tau)}{\partial \tau} = a \frac{\partial^2 t(x, \tau)}{\partial x^2}$$

and is precisely the solution of the problem of the temperature distribution in an infinite body with one-dimensional heat flow caused by the action of an instantaneous plane (along the plane x_1) heat source of the strength Q_3 per unit area at the time moment $\tau = 0$, since

$$\int_{-\infty}^{\infty} t(x, \tau)\, dx = b_3,$$

where $b_3 = Q_3/c\gamma$ (deg m), Q_3 is the amount of heat in kcal/m^2 evolved. In a similar way we may show that the expression[1]

$$t(r, \tau) = \frac{b_2}{4\pi a \tau} \exp\left[-\frac{r^2 + r_1^2}{4a\tau} \right] I_0\left(\frac{r r_1}{2a\tau} \right) \tag{9.5}$$

satisfies the equation

[1] Solution (9.5) is obtained from the expression

$$\frac{\vartheta r_1\, dr_1\, d\vartheta}{4\pi a \tau} \exp\left[-\frac{r^2 + r_1^2 - 2 r r_1 \cos \vartheta}{4a\tau} \right]$$

by integration with respect to the variable ϑ between 0 and 2π and by replacing $2\pi r_1 \vartheta\, dr_1$ by b_2.

$$\frac{\partial t(r, \tau)}{\partial \tau} = a\left(\frac{\partial^2 t(r, \tau)}{\partial r^2} + \frac{1}{r}\frac{\partial t(r, \tau)}{\partial r}\right)$$

and is the solution of the problem of temperature distribution in an infinite body caused by the action of an instantaneous heat source at the surface of the cylinder $r = r_1$ at the moment of time $\tau = 0$ with the strength Q_2 per unit length. We have

$$2\pi \int_0^\infty rt(r, \tau)\, dr = b_2 \quad [\text{deg m}^2],$$

where $Q_2 = c\gamma b_2$ kcal/m since

$$\int_0^\infty \exp[-\alpha^2 u^2] J_0(\beta u) u\, du = \frac{1}{2\alpha^2}\exp\left[-\frac{\beta^2}{4\alpha^2}\right].$$

Finally, the expression[2]

$$t(r, \tau) = \frac{b_1}{8\pi r r_1(\pi a\tau)^{1/2}}\left\{\exp\left[-\frac{(r-r_1)^2}{4a\tau}\right] - \exp\left[-\frac{(r+r_1)^2}{4a\tau}\right]\right\} \tag{9.6}$$

satisfies the differential equation

$$\frac{\partial t(r, \tau)}{\partial \tau} = a\left(\frac{\partial^2 t(r, \tau)}{\partial r^2} + \frac{2}{r}\frac{\partial t(r, \tau)}{\partial r}\right)$$

and is the solution of the problem of the temperature distribution in an infinite body at any moment of time caused by the action of a heat source with the strength $Q_1 = b_1 c\gamma$ kcal instantaneously distributed over the spherical surface $r = r_1$ at the moment of time $\tau = 0$, since

$$4\pi \int_0^\infty r^2 t(r, \tau)\, dr = b_1 = Q_1/c\gamma \quad (\text{deg m}^3).$$

Relation (9.2) is an analog of Green's function which is widely applied in potential theory. Green's function is determined inside a closed surface as some potential, which becomes zero at the surface, and tends to infinity at the point $P(x_1, y_1, z_1)$ inside the surface.

[2] Solution (9.6) is obtained from relation

$$t = \frac{1}{(2(\pi a\tau)^{1/2})^3}\,\vartheta r_1^2\, dr_1 \sin\vartheta_1\, d\vartheta_1\, d\varphi_1 \exp\left[-\frac{r^2 + r_1^2 - 2rr_1\cos\vartheta_1}{4a\tau}\right]$$

by integrating over the surface of the sphere of the radius r_1 and by replacing $4\pi r_1^2\vartheta\, dr_1$ by b_1.

Relation (9.2) should, therefore, be regarded as the function that determines the temperature at the point (x, y, z) at the time moment τ, caused by the action of an instantaneous heat source, located at the point (x_1, y_1, z_1) at the time moment $\tau = 0$. And we suppose that the initial temperature of the body is equal to zero while the boundary surface is held at zero temperature.

Using relations (9.2)–(9.6) we may solve a number of specific problems on determination of temperature field inside a body in the presence of an instantaneous heat source with corresponding boundary conditions. The given problem should be subdivided into two problems, one of which should satisfy the above given conditions of the Green's function, and the other the given initial and boundary conditions.

9.1 Semi-Infinite Body

a. Statement of the Problem. *Consider a semi-infinite thin rod, the side surface of which is thermally insulated. At the initial moment of time* ($\tau = 0$) *an instantaneous heat source* Q_3 *(kcal/m²) acts in the cross-section of the rod at the distance* x_1 *from its end.*

Between the noninsulated end of the rod and the surrounding medium ($t_a=0$) *heat transfer occurs according to the Newton law. The temperature distribution along the length of the rod at any time is to be found.* The initial temperature of the rod is assumed to be zero. We have

$$\frac{\partial t(x, \tau)}{\partial \tau} = a \frac{\partial^2 t(x, \tau)}{\partial x^2} \quad (\tau > 0;\ 0 < x < \infty), \tag{9.1.1}$$

$$t(x, 0) = 0, \tag{9.1.2}$$

$$t(\infty, \tau) = 0, \tag{9.1.3}$$

$$-\frac{\partial t(0, \tau)}{\partial x} + Ht(0, \tau) = 0. \tag{9.1.4}$$

Thus, the present problem is the problem of cooling a rod which at the initial moment experienced an instantaneous heat impulse.

b. Solution of the Problem. For an infinite body with one-dimensional heat flow in the presence of an instantaneous heat source the solution of differential equation (9.1.1) has the form

$$u(x, \tau) = \frac{b}{2(\pi a \tau)^{1/2}} \exp\left[-\frac{(x - x_1)^2}{4a\tau}\right], \tag{9.1.5}$$

where $bc\gamma = Q_3$ is the amount of heat evolved by an instantaneous source per unit area of the rod cross-section (kcal/m²). Hence the coefficient b has the dimension deg m.

It is seen from solution (9.1.5) that at $\tau \to 0$, $u(x, \tau)$ tends to zero at all x with the exception of $x = x_1$, where it tends to infinity.

Let us divide our problem into two, introducing new variables

$$t(x, \tau) = u(x, \tau) + v(x, \tau). \tag{9.1.6}$$

The variable $v(x, \tau)$ satisfies differential equation (9.1.1):

$$\frac{\partial v(x, \tau)}{\partial \tau} = a \frac{\partial^2 v(x, \tau)}{\partial x^2}. \tag{9.1.7}$$

The initial condition for the variable $v(x, \tau)$ will be

$$v(x, 0) = 0. \tag{9.1.8}$$

For the solution of equation (9.1.7), we shall use the Laplace transform method. We have

$$V''(x, s) - (s/a)V(x, s) = 0, \tag{9.1.9}$$

where $V(x, s) = L[v(x, \tau)]$ is the transform of the function $v(x, \tau)$.

In general form, the solution of Eq. (9.1.9) may be written

$$\begin{aligned} V(x, s) &= A \cosh(s/a)^{1/2}x + B \sinh(s/a)^{1/2}x \\ &= A_1 \exp[(s/a)^{1/2}x] + B_1 \exp[-(s/a)^{1/2}x]. \end{aligned} \tag{9.1.10}$$

Relation (9.1.6) for the transform may be written

$$T(x, s) = U(x, s) + V(x, s), \tag{9.1.11}$$

where $U(x, s)$ is the transform of the function $u(x, \tau)$, i.e.,

$$\begin{aligned} U(x, s) &= L\left[\frac{b}{2(\pi a \tau)^{1/2}} \exp\left[-\frac{(x - x_1)^2}{4a\tau}\right]\right] \\ &= \frac{b}{2(as)^{1/2}} \exp\left[-\left(\frac{s}{a}\right)^{1/2} |x - x_1|\right] \end{aligned} \tag{9.1.12}$$

(see relation (51) of the Table of Transforms in Appendix 5). A necessary condition for the existence of the transform of the function

$$L\left[\frac{1}{(\pi\tau)^{1/2}} \exp\left[-\frac{k^2}{4\tau}\right]\right] = \frac{1}{\sqrt{s}} \exp[-k\sqrt{s}\,],$$

is the fulfillment of the inequality $k \geq 0$. Therefore in transform (9.1.12) we take an absolute value of the difference $(x - x_1)$ and designate it by the symbol $|x - x_1|$. Clearly the condition of the symmetry of the function and of its transform with respect to x_1 is not broken.

With these important restrictions, we shall turn to the solution of our problem. We have

$$T(x, s) = \{b/2(as)^{1/2}\} \exp[-(s/a)^{1/2} |x - x_1|] + A_1 \exp[(s/a)^{1/2}x] + B_1 \exp[-(s/a)^{1/2}x]. \quad (9.1.13)$$

It follows from condition (9.1.3) that $A_1 = 0$. The constant B_1 is determined from boundary condition (9.1.4) which for the transform $T(x, s)$ has the form

$$-T'(0, s) + HT(0, s) = 0. \quad (9.1.14)$$

Hence we have

$$-(s/a)^{1/2}\{b/2(as)^{1/2}\} \exp[-(s/a)^{1/2}x_1] + (s/a)^{1/2}B_1 + \{Hb/2(as)^{1/2}\} \exp[-(s/a)^{1/2}x_1] + B_1 H = 0.$$

The constant B_1 will then be equal to

$$B_1 = \frac{b}{2(as)^{1/2}} \left(\frac{(1/H)(s/a)^{1/2} - 1}{(1/H)(s/a)^{1/2} + 1}\right) \exp\left[-\left(\frac{s}{a}\right)^{1/2} x_1\right]. \quad (9.1.15)$$

Thus the solution for the transform may be written

$$\begin{aligned} T(x, s) &= \frac{b}{2(as)^{1/2}} \left[\exp[-(s/a)^{1/2} |x - x_1|] + \frac{(1/H)(s/a)^{1/2} - 1}{(1/H)(s/a)^{1/2} + 1} \exp[-(s/a)^{1/2} |x + x_1|]\right] \\ &= \frac{b}{2(as)^{1/2}} \exp[-(s/a)^{1/2} |x - x_1|] + \frac{b}{2(as)^{1/2}} \exp[-(s/a)^{1/2}(x + x_1)] \\ &\quad - \frac{b}{(as)^{1/2}[1 + (1/H)(s/a)^{1/2}]} \exp[-(s/a)^{1/2}(x + x_1)]. \end{aligned} \quad (9.1.16)$$

Using the Table of Transforms we find the solution for the original function $t(x, \tau)$:

$$t(x, \tau) = \frac{b}{2(\pi a\tau)^{1/2}} \left\{\exp\left[-\frac{(x - x_1)^2}{4a\tau}\right] + \exp\left[-\frac{(x + x_1)^2}{4a\tau}\right]\right. \left. - bH \exp[H(x + x_1) + aH^2\tau] \operatorname{erfc}\left(\frac{x + x_1}{2(a\tau)^{1/2}} + H(a\tau)^{1/2}\right)\right\}. \quad (9.1.17)$$

The last term in Solution (9.1.17) may be written in the form

$$-\frac{bH}{(\pi a\tau)^{1/2}}\int_0^\infty \exp\left[-H\xi-\frac{(x+x_1+\xi)^2}{4a\tau}\right]d\xi. \qquad (9.1.18)$$

If $H\to\infty$, which corresponds to imposing a constant temperature equal to zero at the end of the rod in the process of cooling (the boundary condition of the first kind), then relation (9.1.18) is equal to zero. In this case an ordinary solution may be obtained from solution (9.1.17) for cooling a semi-infinite rod if the given initial temperature is in the form of some function of x, i.e., $t(x, 0)=f(x)$. We assume

$$db = f(x_1)\,dx_1, \qquad (9.1.19)$$

$$b = \int_0^\infty f(x_1)\,dx_1. \qquad (9.1.20)$$

Substituting this relation into solution (9.1.17) we find

$$t(x,\tau)=\frac{1}{2(\pi a\tau)^{1/2}}\times\int_0^\infty f(x_1)\left\{\exp\left[-\frac{(x-x_1)^2}{4a\tau}\right]+\exp\left[-\frac{(x+x_1)^2}{4a\tau}\right]\right\}dx_1,$$

i.e., we obtain the solution similar to (4.2.4).

9.2 Infinite Plate

a. Statement of the Problem. *Consider an infinite plate at the temperature $t_0=0$. At the initial moment ($\tau=0$) instantaneous symmetrical heat sources act at $x=\pm x_1$ $(-R<x+R)$ of the strength Q_3 per unit area (heat sources act along the planes $+x_1$ and $-x_1$). Heat is transferred from the two surfaces of the plate ($+R$ and $-R$) to the surrounding medium according to the Newton law of cooling (the boundary condition of the third kind). The temperature distribution over the thickness of the plate at any moment of time is to be found.*

We have

$$t(x, 0)=0, \qquad (9.2.1)$$

$$\partial t(0,\tau)/\partial x=0. \qquad (9.2.2)$$

The origin of the coordinates is in the middle of the plate (the problem is symmetrical)

$$\frac{\partial t(R, \tau)}{\partial x} + Ht(R, \tau) = 0. \tag{9.2.3}$$

To simplify calculations, the ambient temperature is taken to be equal to zero ($t_a = t_0 = 0$).

b. Solution of the Problem. Applying a similar method to that used in the last example to solve the problem, we suppose

$$t(x, \tau) = u(x, \tau) + v(x, \tau),$$

where

$$u(x, \tau) = \frac{b}{2(\pi a\tau)^{1/2}} \left\{ \exp\left[- \frac{(x - x_1)^2}{4a\tau} \right] + \exp\left[- \frac{(x + x_1)^2}{4a\tau} \right] \right\} \tag{9.2.4}$$

is the solution of the problem of cooling an infinite body when instantaneous heat sources act along the planes $\pm x_1$ (Fig. 9.1).

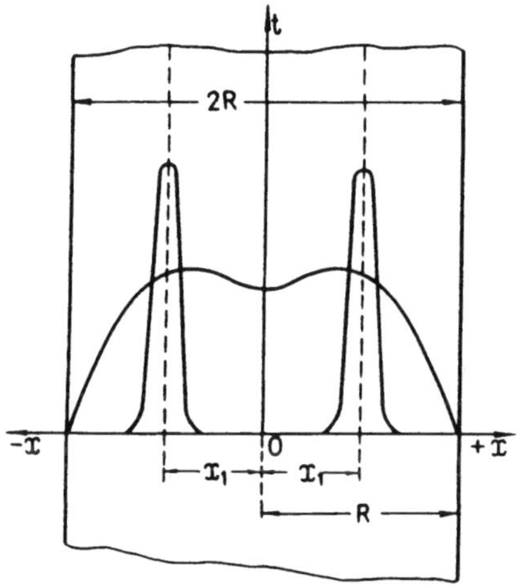

Fig. 9.1. Temperature distribution in an infinite plate in the case of instantaneous heat sources.

The variable $v(x, \tau)$ satisfies the differential heat conduction equation, the solution of which for the transform $V(x, s)$ has been given before (see Solution (9.1.10)).

The expression for the transform may be written

$$\begin{aligned} T(x, s) &= U(x, s) + V(x, s) \\ &= \{b/2(as)^{1/2}\}[\exp[-(s/a)^{1/2} \mid x - x_1 \mid] + \exp[-(s/a)^{1/2}(x + x_1)]] \\ &\quad + A \cosh(s/a)^{1/2}x + B \sinh(s/a)^{1/2}x, \end{aligned} \tag{9.2.5}$$

since the transform $U(x, s) = L[u(x, \tau)]$ is determined by relation (9.1.12).

According to relation (9.2.4) the function $u(x, \tau)$ is an even function with respect to x_1, as is its transform. Therefore, it follows from condition (9.2.2) that

$$\partial v(0, \tau)/\partial x = 0, \qquad V'(0, s) = 0, \tag{9.2.6}$$

hence, $B = 0$.

The constant A is found from boundary condition (9.2.3) which for the transform $T(x, s)$ will be written as

$$T'(R, s) + HT(R, s) = 0.$$

Hence

$$\begin{aligned} &- \{b/2(as)^{1/2}\}(s/a)^{1/2}[\exp[-(s/a)^{1/2}(R - x_1)] + \exp[-(s/a)^{1/2}(R + x_1)]] \\ &\quad + (s/a)^{1/2}A \sinh(s/a)^{1/2}R + \{Hb/2(as)^{1/2}\}[\exp[-(s/a)^{1/2}(R - x_1)] \\ &\quad + \exp[-(s/a)^{1/2}(R + x_1)]] + HA\cosh(s/a)^{1/2}R = 0. \end{aligned} \tag{9.2.7}$$

Having determined the constant A from equality (9.2.7) and substituting its value into solution (9.2.5) we shall have

$$\begin{aligned} T(x, s) = &\frac{b}{2(as)^{1/2}\left[\cosh(s/a)^{1/2}R + \dfrac{1}{H}(s/a)^{1/2}\sinh(s/a)^{1/2}R\right]} \\ &\times \{(\cosh(s/a)^{1/2}R + (1/H)(s/a)^{1/2}\sinh(s/a)^{1/2}R) \\ &\times [\exp[-(s/a)^{1/2}(x - x_1)] + \exp[-(s/a)^{1/2}(x + x_1)]] \\ &+ ((1/H)(s/a)^{1/2} - 1)\cosh(s/a)^{1/2}x \\ &\times [\exp[-(s/a)^{1/2}(R - x_1)] + \exp[-(s/a)^{1/2}(R + x_1)]]\}. \end{aligned} \tag{9.2.8}$$

Replacing the exponential functions by the hyperbolic ones, taking note of the relation $e^{-z} = \cosh z - \sinh z$, we may show that solution (9.2.8) satisfies the expansion theorem. The roots of the characteristic equation are well known (see solution (6.3.25)); they are determined from the corresponding equations (6.3.26) and (6.3.27).

After the necessary manipulations, the solution of our problem will be obtained in the form

$$t(x, \tau) = \frac{2b}{R}\sum_{n=1}^{\infty}\frac{\mu_n}{\mu_n + \sin\mu_n\cos\mu_n}\cos\mu_n\frac{x_1}{R}\cos\mu_n\frac{x}{R}\exp\left[-\mu_n^2\frac{a\tau}{R^2}\right] \tag{9.2.9}$$

where μ_n are the roots of characteristic equation (6.3.27). If the instantaneous heat source is in the middle of the plate ($x_1 = 0$), then $\cos \mu_n(x_1/R) = 1$.

From solution (9.2.9) we may obtain solution (6.3.15) of the problem of cooling an infinite plate in a medium with zero temperature if the given initial plate temperature distribution is in the form of some function $f(x)$. If we put $db = f(x_1)\, dx_1$, then we may write

$$2b = \int_{-R}^{R} f(x_1)\, dx_1.$$

Substitution of this expression into solution (9.2.9) yields

$$t(x, \tau) = \sum_{n=1}^{\infty} \frac{\mu_n}{\mu_n + \sin \mu_n \cos \mu_n} \cos \mu_n \frac{x}{R} \times \frac{2}{R} \int_0^R f(x) \cos \mu_n \frac{x}{R}\, dx \exp[-\mu_n^2 \text{ Fo}], \qquad (9.2.10)$$

i.e., we obtain solution (9.2.10) which is similar to (6.3.15).

Solution (9.2.9) may be written in the form more convenient for calculation since the constant coefficients A_n are tabulated

$$A_n = \frac{2 \sin \mu_n}{\mu_n + \sin \mu_n \cos \mu_n} = (-1)^{n+1} \frac{2\text{Bi}(\text{Bi}^2 + \mu_n^2)^{1/2}}{\mu_n(\text{Bi}^2 + \text{Bi} + \mu_n^2)},$$

i.e.,

$$t(x, \tau) = \sum_{n=1}^{\infty} \frac{bA_n\mu_n}{R \sin \mu_n} \cos \mu_n \frac{x_1}{R} \cos \mu_n \frac{x}{R} \exp[-\mu_n^2 \text{ Fo}]. \qquad (9.2.11)$$

The mean temperature $\bar{t}(\tau)$ required for the determination of the amount of heat lost by a plate in the process of cooling will be equal to

$$\bar{t}(\tau) = \sum_{n=1}^{\infty} \frac{b\mu_n \cos \mu_n(x_1/R)}{R \sin \mu_n} B_n \exp[-\mu_n^2 \text{ Fo}]. \qquad (9.2.12)$$

If $\text{Bi} = \infty$, then $A_n = (-1)^{n+1}(2/\mu_n)$, $\mu_n = (2n - 1)\frac{1}{2}\pi$, and solution (9.2.11) is simplified to

$$t(x, \tau) = \frac{2b}{R} \sum_{n=1}^{\infty} (-1)^{n+1} \cos \frac{(2n-1)\pi x_1}{R} \cos \frac{(2n-1)\pi x}{R} \times \exp[-(2n-1)^2\pi^2\tfrac{1}{4} \text{ Fo}]. \qquad (9.2.13)$$

If the problem is nonsymmetrical ($0 < x < l$, where $l = 2R$ is the thickness of the plate) and there is one plane source at $x = x_1$, then solution of a similar problem for the case Bi $= \infty$ (boundary condition of the first kind) is of the form

$$T(x, s) = \frac{b[\cosh(s/a)^{1/2}(l + x - x_1) - \cosh(s/a)^{1/2}(l - x - x_1)]}{2(as)^{1/2} \sinh(s/a)^{1/2} l}. \quad (9.2.14)$$

Hence we find the solution for the inverse transform, i.e., the solution to our problem.

$$t(x, \tau) = \frac{2b}{l} \sum_{n=1}^{\infty} \sin \frac{n\pi x_1}{l} \sin \frac{n\pi x}{l} \exp[-n^2\pi^2 \mathrm{Fo}]. \quad (9.2.15)$$

9.3 Sphere (Symmetrical Problem)

a. Statement of the Problem. *Consider a spherical body at a temperature equal to zero. At the initial moment, an instantaneous heat source, of strength Q_1 kcal, acts along a spherical surface $r = r_1$. Heat transfer between the surface of the sphere and the surrounding medium occurs according to the Newton law (boundary condition of the third kind). The temperature distribution and the average temperature at any moment of time are to be found.*

We have

$$\frac{\partial[rt(r, \tau)]}{\partial\tau} = a \frac{\partial^2[rt(r, \tau)]}{\partial r^2} \quad (\tau > 0;\ 0 < r < R), \quad (9.3.1)$$

$$t(r, 0) = 0, \quad (9.3.2)$$

$$\frac{\partial t(0, \tau)}{\partial r} = 0, \quad t(0, \tau) \neq \infty, \quad \text{at } \tau > 0, \quad (9.3.3)$$

$$\frac{\partial t(R, \tau)}{\partial r} + Ht(R, \tau) = 0, \quad (9.3.4)$$

$$t(R, \tau) = 0, \quad \text{at Bi} = \infty \ (H = \infty). \quad (9.3.5)$$

b. Solution of the Problem. To shorten the calculation we shall first solve the problem for Bi $= \infty$ (the boundary condition of the first kind) and then give the solution of the problem for finite values of Bi.

We set

$$t(r, \tau) = u(r, \tau) + v(r, \tau), \quad (9.3.6)$$

where $u(r, \tau)$ is the solution of the problem of cooling an infinite body which at the initial moment of time experiences an instantaneous heat impulse $Q_1 = bc\gamma$ (kcal) instantaneously distributed over the spherical surface $r = r_1$. According to relation (9.6) the solution has the form

$$u(r, \tau) = \frac{b}{8\pi r r_1 (\pi a \tau)^{1/2}} \left\{ \exp\left[- \frac{(r - r_1)^2}{4a\tau} \right] - \exp\left[- \frac{(r + r_1)^2}{4a\tau} \right] \right\}. \quad (9.3.7)$$

The transform of this function will be

$$U(r, s) = L[u(x, \tau)] = \frac{b}{8\pi r r_1 (as)^{1/2}} \times \{\exp[-(s/a)^{1/2} |r - r_1|] - \exp[-(s/a)^{1/2}(r + r_1)]\}, \quad (9.3.8)$$

with the same restrictions on $(r - r_1)$ as applied to $(x - x_1)$ (see paragraph preceding Eq. (9.1.13)).

The function $v(r, \tau)$ satisfies Eq. (9.3.1) since $t(r, \tau)$ and $u(r, \tau)$ are its solutions. Hence, the transform of the function $v(r, \tau)$ should satisfy the equation

$$[rV(r, s)]'' - (s/a)rV(r, s) = 0, \quad (9.3.9)$$

where $V(r, s) = L[v(r, \tau)]$.

The solution of this equation under condition (9.3.3) has the form

$$V(r, s) = B\,[\sinh(s/a)^{1/2} r]/r. \quad (9.3.10)$$

Then the solution for the transform $T(r, s)$ may be written as

$$T(r, s) = \frac{b}{8\pi r r_1 (sa)^{1/2}} [\exp[-(s/a)^{1/2} |r - r_1|] - \exp[-(s/a)^{1/2} |r + r_1|]] + \frac{B \sinh(s/a)^{1/2} r}{r}. \quad (9.3.11)$$

The constant B is determined from boundary condition (9.3.5) which for the transform may be written as $T(R, s) = 0$, i.e.,

$$\frac{b}{8\pi r r_1 (as)^{1/2}} [\exp[-(s/a)^{1/2}(R - r_1)] - \exp[-(s/a)^{1/2}(R + r_1)]] + \frac{B \sinh(s/a)^{1/2} R}{R} = 0. \quad (9.3.12)$$

Determining the constant B from equality (9.3.12) and substituting the expression obtained into the solution (9.3.11), we obtain after a minor transformation

$$T(r, s) = \frac{b[\sinh(s/a)^{1/2} r][\sinh(s/a)^{1/2}(R - r_1)]}{4\pi r r_1 (as)^{1/2} \sinh(s/a)^{1/2} R} \quad (0 < r < r_1). \quad (9.3.13)$$

If $R > r > r_1$, then in solution (9.3.13), r and r_1 should be interchanged.

Solution (9.3.13) satisfies all the conditions of the expansion theorem and, therefore, the transition from the transform to the inverse transform is carried out in the usual way,

$$t(r, \tau) = \frac{b}{2\pi R} \sum_{n=1}^{\infty} \frac{1}{rr_1} \sin \frac{n\pi r_1}{R} \sin \frac{n\pi r}{R} \exp[-n^2\pi^2 \text{ Fo}]. \quad (9.3.14)$$

If the Biot criterion is a finite value, then boundary condition (9.3.4) should be used in determining the constant B. Making similar transformation, the solution for finite Bi may be written as

$$t(r, \tau) = \frac{b}{2\pi R} \sum_{n=1}^{\infty} \frac{\mu_n}{\mu_n - \sin \mu_n \cos \mu_n} \times \frac{\sin \mu_n(r_1/R)}{r_1} \frac{\sin \mu_n(r/R)}{r} \exp[-\mu_n^2 \text{ Fo}], \quad (9.3.15)$$

where μ_n are the roots of the characteristic equation (6.5.12).

Since an instantaneous heat source $Q_1 = c\gamma b$ is distributed over the whole spherical surface $4\pi r_1^2$, then the value db will be equal to

$$db = 4\pi r_1^2 f(r_1)\, dr_1, \quad (9.3.16)$$

where $f(r_1)$ is some function of temperature distribution at the initial moment for the problem of cooling a sphere in a medium the temperature of which is equal to zero (see Chapter 6, Section 5).

The value b is then equal to

$$b = 4\pi \int_0^R r_1^2 f(r_1)\, dr_1. \quad (9.3.17)$$

Substitution of this expression into solution (9.3.15) yields

$$t(r, \tau) = \sum_{n=1}^{\infty} \frac{2\mu_n}{\mu_n - \sin \mu_n \cos \mu_n} \frac{\sin \mu_n(r/R)}{r} \frac{1}{R} \times \int_0^R r_1 f(r_1) \sin \mu_n(r_1/R)\, dr_1 \exp[-\mu_n^2 \text{ Fo}], \quad (9.3.18)$$

i.e., a solution is obtained similar to solution (6.5.19).

Since the initial thermal amplitudes A_n determined by relation (6.5.29) are tabulated, we shall write solution (9.3.15) in the form

$$t(r, \tau) = \frac{b}{4\pi R} \sum_{n=1}^{\infty} \frac{[(\text{Bi} - 1)^2 + \mu_n^2]^{1/2} A_n}{\text{Bi}} \frac{\sin \mu_n(r_1/R)}{r_1} \times \frac{\sin \mu_n(r/R)}{r} \exp[-\mu_n^2 \text{ Fo}]. \quad (9.3.19)$$

If the heat source is in the center of the sphere ($r_1 = 0$), then solution (9.3.15) will acquire the form

$$t(r, \tau) = \frac{b}{2\pi R^2 r} \sum_{n=1}^{\infty} \frac{\mu_n^2}{\mu_n - \sin\mu_n \cos\mu_n} \sin\mu_n \frac{r}{R} \exp[-\mu_n^2 \mathrm{Fo}]. \quad (9.3.20)$$

The mean temperature $\bar{t}(r, \tau)$ corresponding to the general solution (9.3.19) is

$$\bar{t}(r, \tau) = \frac{b}{4\pi R r_1 \mathrm{Bi}} \sum_{n=1}^{\infty} [(\mathrm{Bi} - 1)^2 + \mu_n^2]^{1/2} \times \mu_n B_n \exp[-\mu_n^2 \mathrm{Fo}] \sin\mu_n(r_1/R), \quad (9.3.21)$$

where B_n are constant coefficients determined from relation (6.5.49).

9.4 Infinite Cylinder

a. Statement of the Problem. *Consider an infinite cylinder. At the initial moment of time there acts an instantaneous heat source of strength Q_2 (kcal/m) per unit length of the cylindrical surface $r = r_1$. Heat transfer occurs between the surface of the cylinder and the surrounding medium according to the Newton law. The temperature distribution and the mean temperature at any moment of time are to be found.*

We have

$$\frac{\partial t(r, \tau)}{\partial \tau} = a\left(\frac{\partial^2 t(r, \tau)}{\partial r^2} + \frac{1}{r}\frac{\partial t(r, \tau)}{\partial r}\right) \quad (\tau > 0;\ 0 < r < R), \quad (9.4.1)$$

$$t(r, 0) = 0, \quad (9.4.2)$$

$$\frac{\partial t(0, \tau)}{\partial r} = 0, \qquad t(0, \tau) \neq \infty \quad \text{at } \tau > 0, \quad (9.4.3)$$

$$\frac{\partial t(R, \tau)}{\partial r} + Ht(R, \tau) = 0. \quad (9.4.4)$$

To simplify the calculations, the initial temperature of the cylinder and the ambient temperature are assumed to be zero.

If $\mathrm{Bi} = \infty$, then

$$t(R, \tau) = 0 \quad (9.4.5)$$

(boundary condition of the first kind).

b. Solution of the Problem. Let us first solve the problem under the condition of the first kind (Bi $= \infty$). We set

$$t(r, \tau) = u(r, \tau) + v(r, \tau), \tag{9.4.6}$$

where $u(r, \tau)$ is the solution of Eq. (9.4.1) in the presence of an instantaneous heat source $Q_2 = bc\gamma$ (kcal/m) acting on the cylindrical surface $r = r_1$.

According to relation (9.5)

$$u(r, \tau) = \frac{b}{4\pi a\tau} \exp\left[-\frac{r^2 + r_1^2}{4a\tau}\right] I_0\left(\frac{rr_1}{2a\tau}\right). \tag{9.4.7}$$

The transform of this function is found from the table of transforms

$$\begin{aligned} U(r, s) &= \frac{b}{2\pi a} I_0((s/a)^{1/2}r_1)\, K_0((s/a)^{1/2}r), \quad \text{if } r > r_1, \\ U(r, s) &= \frac{b}{2\pi a} I_0((s/a)^{1/2}r)\, K_0((s/a)^{1/2}r_1), \quad \text{if } r < r_1. \end{aligned} \tag{9.4.8}$$

The function $u(r, \tau)$ satisfies differential equation (9.4.1). The solution of equation for the transform $U(r, s)$ under conditions (9.4.2) and (9.4.3) may be written as

$$U(r, s) = AI_0((s/a)^{1/2}r). \tag{9.4.9}$$

Then solution for the transform $T(r, s)$ is of the form

$$T(r, s) = \frac{b}{2\pi a} I_0((s/a)^{1/2}r_1)\, K_0((s/a)^{1/2}r) + AI_0((s/a)^{1/2}r) \quad \text{at } r > r_1. \tag{9.4.10}$$

The constant A is found from boundary condition (9.4.5), i.e., from condition $T(R, s) = 0$. Then solution (9.4.10) will have the form

$$\begin{aligned} T(r, s) = \frac{bI_0((s/a)^{1/2}r_1)}{2\pi a I_0((s/a)^{1/2}R)} &[I_0((s/a)^{1/2}R)\, K_0((s/a)^{1/2}r) \\ &- I_0((s/a)^{1/2}r)\, K_0((s/a)^{1/2}R)], \quad \text{at } r > r_1 \end{aligned} \tag{9.4.11}$$

(if $r < r_1$ in solution (9.4.11), r and r_1 should be interchanged).

Applying the expansion theorem we find the solution for the problem (i.e., the inverse transform)

$$t(r, \tau) = \frac{b}{\pi R^2} \sum_{n=1}^{\infty} \frac{1}{J_1^2(\mu_n)} J_0\left(\mu_n \frac{r}{R}\right) J_0\left(\mu_n \frac{r_1}{R}\right) \exp[-\mu_n^2 \text{ Fo}], \tag{9.4.12}$$

where μ_n are the roots of the Bessel function of the first kind and zeroth order, i.e., they are determined from the characteristic equation

$$J_0(\mu) = 0.$$

Solution (9.4.12) is valid for $r > r_1$ and $r < r_1$, since it is symmetrical with respect to r and r_1.

If heat is exchanged according to the Newton law between the surface of the cylinder and the surrounding medium ($t_a = 0$), then the solution will have the form

$$t(r, \tau) = \frac{b}{\pi R^2} \sum_{n=1}^{\infty} \frac{\mu_n^2}{(\mathrm{Bi}^2 + \mu_n^2) J_0^2(\mu_n)} \times J_0(\mu_n(r_1/R))\, J_0(\mu_n(r/R)) \exp[-\mu_n^2 \,\mathrm{Fo}] \quad (r \lessgtr r_1), \tag{9.4.13}$$

where μ_n are the roots of the corresponding characteristic equation, being the function of the Biot criterion.

From solution (9.4.13) we may obtain the solution for an infinite cylinder, if at the initial moment of time its temperature is some function of r, i.e., $t(r, 0) = f(r)$.

We set

$$db = 2\pi r_1\, f(r_1)\, dr_1;$$

then

$$b = 2\pi \int_0^R r_1 f(r_1)\, dr_1.$$

If the expression obtained is substituted into solution (9.4.13), then

$$t(r, \tau) = \sum_{n=1}^{\infty} \frac{\mu_n^2}{(\mathrm{Bi}^2 + \mu_n^2) J_0^2(\mu_n)} J_0\left(\mu_n \frac{r}{R}\right) \frac{2}{R^2} \times \int_0^R r f(r) J_0(\mu_n(r/R))\, dr \exp[-\mu_n^2 \,\mathrm{Fo}]. \tag{9.4.14}$$

Solution (9.4.14) is equivalent to solution (6.6.15), since

$$\frac{\mu_n^2}{(\mathrm{Bi}^2 + \mu_n^2) J_0^2(\mu_n)} = \frac{1}{[J_0^2(\mu_n) + J_1^2(\mu_n)]}.$$

Solution (9.4.13) may be written in another way using the tabulated coefficient A_n (see relation (6.6.27)):

$$t(r, \tau) = \frac{b}{2\pi R^2} \sum_{n=1}^{\infty} \frac{A_n \mu_n}{J_1(\mu_n)} J_0(\mu_n(r_1/R)) J_0(\mu_n(r/R)) \exp[-\mu_n^2 \,\mathrm{Fo}]. \tag{9.4.15}$$

The corresponding mean temperature $\bar{t}(\tau)$ is equal to

$$\bar{t}(\tau) = \frac{b}{2\pi R^2} \sum_{n=1}^{\infty} \frac{\mu_n}{J_1(\mu_n)} J_0(\mu_n(r_1/R))B_n \exp[-\mu_n^2 \,\mathrm{Fo}], \qquad (9.4.16)$$

where B_n are the constant coefficients determined from relation (6.6.34).

We conclude this section by emphasizing that the method of heat sources not only makes it possible to solve problems with instantaneous heat sources but also problems of cooling or heating a body having an initial temperature distribution as a function of the coordinates. The same results are obtained if these problems are solved by the Fourier–Hankel integral transforms.

9.5 Regular Thermal Regime

G. M. Kondratiev was apparently the first to introduce the concept of a regular regime of cooling or heating of the first kind where the change in a temperature at any point of a body with time is described by a simple exponent (see Chapter 6, Section 10). The derivative of the logarithm of the excess temperature with respect to time will be a constant quantity and is called the rate of heating or cooling of a body ($m = \text{const}$):

$$-\frac{\partial[\ln(t_a - t)]}{\partial\tau} = \frac{1}{t_a - t}\frac{\partial t}{\partial\tau} = m = \text{const}. \qquad (9.5.1)$$

G. M. Kondratiev and his co-workers distinguished the regular regime of the second kind, under which the temperature at any point of a body is the linear function of time

$$\partial t/\partial\tau = b = \text{const}, \qquad (9.5.2)$$

and the temperature distribution within a body is described by some function of coordinates (for one-dimensional symmetrical problems the parabola represents such a function).

Such a regime of heating of a body occurs when the temperature of the ambient medium is a linear time function (third-kind boundary conditions)

$$t_a = t_0 + b\tau, \qquad (9.5.3)$$

or, for the case of a constant heat flux near the body surface,

$$q_s(\tau) = q_c = \text{const}. \qquad (9.5.4)$$

Regular heating regimes arise after a definite time interval determined by the inequality $\mathrm{Fo} > \mathrm{Fo}_1$. The analysis of the solutions (see Chapters 5, 6, and 8) shows that *regular regimes of the first and second kinds have a common property which is characterized by the ratio of the specific heat flux at any point of a body q to the heat flux on its surface q_s which is constant in time*

$$q/q_s = f(x, y, z). \tag{9.5.5}$$

For the regular regime of the first kind (t_a = const), we have (see Chapter 6):

(a) Infinite plate

$$\frac{q(x, \tau)}{q_s} = \frac{\sin \mu_1(x/R)}{\sin \mu_1}, \tag{9.5.6}$$

(b) Sphere

$$\frac{q(r, \tau)}{q_s} = \frac{r \cos \mu_1(r/R) - \sin \mu_1(r/R)}{r^2(R \cos \mu_1 - \sin \mu_1)}, \tag{9.5.7}$$

(c) Infinite cylinder

$$\frac{q(r, \tau)}{q_s} = \frac{J_1(\mu_1 r/R)}{J_1(\mu_1)}. \tag{9.5.8}$$

For the regular regime of the second kind ($\partial t/\partial \tau$ = const) the ratio of heat fluxes is equal to a dimensionless coordinate

$$q(x, \tau)/q_s = x/R; \qquad q(r, \tau)/q_s = r/R. \tag{9.5.9}$$

These relations are also valid for the second-kind boundary condition in the very general case when the heat flux is prescribed as a certain time function. In one-dimensional heat conduction problems, relation (9.5.5). as a general characteristic of the regular regimes of the first and second kinds, remains valid in the presence of a constant heat source. Consequently, *"regularization" of the kinetics of heating occurs not only with respect to the temperature fields but also with respect to heat fluxes.* It is therefore not obligatory to distinguish regular regimes of heating of the first and second kinds.

In developing the principle of regularization it is possible to assume the following relation as a common property of a thermal regular regime:

$$-d\bar{t}/(t_a - \bar{t})\, d\tau = m = \text{const}, \tag{9.5.10}$$

where $\bar{t}$ is the mean temperature of a body with respect to the volume. Consequently, *the rate of heating $d\bar{t}/d\tau$ is directly proportional to a difference between a medium temperature and the mean temperature of a body.*

$$-\,d\bar{t}/d\tau = m(t_a - \bar{t}), \tag{9.5.11}$$

where m is the proportionality factor called as the rate of heating.

We illustrate this by the following examples.

(1) *Constant temperature of a medium* (t_a = const). From solutions (6.3.44), (6.5.48), and (6.6.33), it is possible to write

$$\frac{1}{(t_a - \bar{t})}\frac{d\bar{t}}{d\tau} = -\frac{a}{R^2}\,\frac{\sum_{n=1}^{\infty} \mu_n^2 B_n \exp[-\mu_n^2 \mathrm{Fo}]}{\sum_{n=1}^{\infty} B_n \exp[-\mu_n^2 \mathrm{Fo}]}. \tag{9.5.12}$$

For Fo > Fo_1 it is possible to restrict oneself with the first term of the series, in which case relation (9.5.12) will not depend upon time

$$-\frac{1}{(t_a - \bar{t})}\frac{d\bar{t}}{d\tau} = \frac{a}{R^2}\mu_1^2 = m = \text{const.} \tag{9.5.13}$$

The rate of heating depends upon Bi, the coefficient a and the characteristic dimension of the body R.

(2) *The medium temperature as the linear function of time* [$(\partial t_a/\partial\tau) = b =$ = const]. From solutions (7.1.18), (7.2.18) and (7.3.8) it is possible to write

$$-\frac{1}{(t_a - \bar{t})}\frac{d\bar{t}}{d\tau} = \frac{a}{R^2}\,\frac{\mathrm{Pd}\left[1 + \sum_{n=1}^{\infty} B_n \exp[-\mu_n^2 \mathrm{Fo}]\right]}{\mathrm{Pd}\left[\frac{1}{\Gamma}\left(1 + \frac{\Pi}{\mathrm{Bi}}\right) + \sum_{n=1}^{\infty} \frac{1}{\mu_n^2} B_n \exp[-\mu_n^2 \mathrm{Fo}]\right]}, \tag{9.5.14}$$

where Γ and Π are the constants which take the following values for a plate, cylinder, and sphere respectively:

$$\Gamma = 3, \quad \Pi = 3; \qquad \Gamma = 8, \quad \Pi = 4; \qquad \Gamma = 15, \quad \Pi = 5.$$

If Fo > Fo_1, the series may be neglected, i.e.,

$$-\frac{1}{(t_a - \bar{t})}\frac{d\bar{t}}{d\tau} = \frac{a}{R^2}\,\frac{\Gamma}{(1 + \Pi/\mathrm{Bi})} = \text{const.} \tag{9.5.15}$$

Consequently, the rate of heating determined according to relation (9.5.15) is a constant value dependent on Bi, a, and R as in the first case.

(3) *The medium temperature as a periodic function of time* ($t_a = t_0 + t_m \cos \omega\tau$). From solution (7.6.45) for a periodically steady-state we obtain

$$-\frac{1}{(t_a - \bar{t})} \frac{d\bar{t}}{d\tau} = \text{Pd} \frac{a}{R^2} \cot \bar{M} = \text{const.} \tag{9.5.16}$$

Consequently, the rate of heating of a body depends upon Pd, the coefficient a and the characteristic dimension of the body R since the quantity $\bar{M}$ is a function of Pd.

(4) *Presence of internal continuous heat sources.* When heating bodies without internal heat sources, the final temperature distribution is uniform ($t_a = \text{const}$); in the presence of internal heat sources, the limiting heat state of a body is in nonequilibrium and the temperature field is non-uniform. We designate the nonequilibrium temperature of a steady state through t_∞, i.e.,

$$t_\infty = \lim_{\tau \to \infty} t. \tag{9.5.17}$$

Then from solutions (8.2.11), (8.3.8), and (8.4.4) we may write

$$(t_\infty - \bar{t})/(t_a - t_0) = \sum_{n=1}^{\infty} (1 + (1/\mu_n^2)\,\text{Po})B_n \exp[-\mu_n^2 \,\text{Fo}]. \tag{9.5.18}$$

We obtain the analogous relation for the rate of heating under the conditions of a regular regime

$$-\frac{1}{(t_\infty - \bar{t})} \frac{d\bar{t}}{d\tau} = \frac{a}{R^2} \mu_1^2 = m = \text{const.} \tag{9.5.19}$$

Consequently, the main feature of a thermal regular regime remains the same, only in this case the mean excessive temperature is understood to be $(t_\infty - \bar{t})$.

(5) *Presence of instantaneous heat sources.* From solutions (9.2.12), (9.3.21), and (9.4.16) (for $\text{Fo} > \text{Fo}_1$ we may restrict ourselves with the first term of the whole series), it is possible to write

$$-\frac{1}{\bar{t}} \frac{d\bar{t}}{d\tau} = \frac{a}{R^2} \mu_1^2 = m = \text{const,} \tag{9.5.20}$$

i.e., relation (9.5.10) is valid.

Thus, *under the conditions of a thermal regular regime, the rate of heating a body is directly proportional to the difference between the medium temperature in a steady state* t_∞ *and the mean volumetric temperature of a body* $\bar{t}$

$$- d\bar{t}/d\tau = m(t_\infty - \bar{t}). \tag{9.5.21}$$

The proportionality factor m (*rate of heating*) *is a function of a characteristic dimension of a body, the thermal diffusivity coefficient, and the criterion* Bi *and* Pd. *The specific feature of regularity of kinetics of heating is determined by relation* (9.5.21) *which is also valid in the presence of heat sources.*

CHAPTER

10

BOUNDARY CONDITIONS OF THE FOURTH KIND

Chapters 5 and 6 deal with unsteady-state heat conduction problems in which heat transfer between a body surface and ambient medium mainly occurs by radiation or by convection. If in the problems of steady convective heat transfer, the boundary conditions of the third kind are used, then for those on unsteady convective heat transfer it is necessary to use boundary conditions of the fourth kind. For example, for the case of a flat thin plate in a flow according to the boundary-layer theory the differential heat transfer equation for a liquid may be approximated as

$$\frac{\partial t_1(x, y, \tau)}{\partial \tau} + w(y, x)\frac{\partial t_1(x, y, \tau)}{\partial x} = a_1\frac{\partial^2 t_1(x, y, \tau)}{\partial y^2}$$
$$R \le y \le \infty, \quad 0 \le x \le l, \tag{10.1}$$

where $w(y, x)$ is the velocity of a liquid flow, and l is the plate length in the direction of the liquid motion x. Here we have assumed that convection in the y direction is negligible.

The differential heat conduction equation for the plate will be

$$\frac{\partial t_2(x, y, \tau)}{\partial \tau} = a_2\left(\frac{\partial^2 t_2(x, y, \tau)}{\partial x^2} + \frac{\partial^2 t_2(x, y, \tau)}{\partial y^2}\right);$$
$$0 \le x \le l, \quad -R \le y \le +R, \tag{10.2}$$

where $2R$ is the plate thickness. Subscript 2 refers to the plate (solid) and subscript 1 to the liquid.

Boundary and initial conditions are as follows

$$t_1(x, y, 0) = t_a = \text{const}, \qquad t_2(x, y, 0) = f(x, y)$$
$$t_1(x, \infty, \tau) = t_a, \qquad t_1(0, y, \tau) = t_a. \tag{10.3}$$

$$t_1(x, R, \tau) = t_2(x, R, \tau), \tag{10.4}$$

$$\lambda_1 \frac{\partial t_1(x, R, \tau)}{\partial y} = \lambda_2 \frac{\partial t_2(x, R, \tau)}{\partial y}, \tag{10.5}$$

$$\frac{\partial t_2(x, 0, \tau)}{\partial y} = 0. \tag{10.6}$$

If we assume that heat transfer occurs only through the surface $y = \pm R$, it may be written

$$\frac{\partial t_1(l, y, \tau)}{\partial x} = \frac{\partial t_2(0, y, \tau)}{\partial x} = 0. \tag{10.7}$$

The solution of the system of differential equations (10.1) and (10.2) causes certain difficulties, because the velocity profile $w(y, x)$ is determined by solving the differential hydrodynamics equation (the Navier–Stokes equation). However, as a first approximation it is possible to consider the second term of equation (10.1) as a variable heat source

$$w(y, x) \frac{\partial t_1(x, y, \tau)}{\partial x} = w^*(x, y, \tau). \tag{10.8}$$

Consequently, the solution of the system of equations (10.1) and (10.2) is reduced to that of differential heat conduction equations with heat sources with boundary conditions of the fourth kind.

If the heat conduction coefficient for the body is considerably larger than that for the liquid and the plate length is small, then the problem may be simplified. We designate the mean temperature along x through $\bar{t}(y, \tau)$ as

$$\bar{t}(y, \tau) = (1/l) \int_0^l t(x, y, \tau)\, dx. \tag{10.9}$$

Consequently, the system of differential heat conduction equations is obtained as

$$\frac{\partial \bar{t}_1(y, \tau)}{\partial \tau} = a_1 \frac{\partial^2 \bar{t}_1(y, \tau)}{\partial y^2} - \bar{w}^*(y, \tau), \tag{10.10}$$

$$\frac{\partial \bar{t}_2(y, \tau)}{\partial \tau} = a_2 \frac{\partial^2 \bar{t}_2(y, \tau)}{\partial y^2}, \tag{10.11}$$

where $\bar{w}^*(y, \tau)$ is the heat source averaged along the plate surface.

The solution of the system of differential equations (10.10) and (10.11) under the fourth-kind boundary conditions will be considered. In this chapter, we shall first consider the problems without heat sources, and then the solution of these problems with heat sources.

Problems of heating or cooling a system of contacting bodies (laminated media) when heat transfer between them occurs according to the heat conduction law are related to those of unsteady-state heat conduction.

10.1 System of Two Bodies (Two Semi-Infinite Rods)

a. Statement of the Problem. *The simplest possible system consisting of two semi-infinite rods is taken. The problem is formulated as follows. Two semi-infinite rods with different initial temperatures are considered. The lateral surfaces of both rods are thermally insulated. At the initial moment, the insulated ends of the rods are brought into contact. The temperature distribution is to be determined at any moment provided that thermal coefficients of the rods are different.*

We have

$$\frac{\partial t_1(x, \tau)}{\partial \tau} = a_1 \frac{\partial^2 t_1(x, \tau)}{\partial x^2} \quad (\tau > 0;\ x > 0), \tag{10.1.1}$$

$$\frac{\partial t_2(x, \tau)}{\partial \tau} = a_2 \frac{\partial^2 t_2(x, \tau)}{\partial x^2} \quad (\tau > 0;\ x < 0). \tag{10.1.2}$$

The coordinate origin is in the point of contact of the ends (Fig. 10.1). Boundary conditions are written

$$t_1(x, 0) = f_1(x), \qquad t_2(x, 0) = f_2(x), \tag{10.1.3}$$

$$\frac{\partial t_1(+\infty, \tau)}{\partial x} = \frac{\partial t_2(-\infty, \tau)}{\partial x} = 0, \tag{10.1.4}$$

$$t_1(+0, \tau) = t_2(-0, \tau), \tag{10.1.5}$$

$$\frac{\partial t_1(0, \tau)}{\partial x} = -\frac{\lambda_2}{\lambda_1} \frac{\partial t_2(0, \tau)}{\partial x}. \tag{10.1.6}$$

First the simpler problem will be considered:

$$t_1(x, 0) = t_{01} = \text{const}; \qquad t_2(x, 0) = 0 \tag{10.1.3a}$$

(for initial temperature of the body, subscript 2 designates a reference temperature).

b. Solution of the Problem. The solution of Eqs. (10.1.1) and (10.1.2) for $T_1(x, s)$ and $T_2(x, s)$ with conditions (10.1.3a) and (10.1.4) are of the form

$$T_1(x, s) - (t_{01}/s) = B_1 \exp[-(s/a_1)^{1/2}x] \quad (x > 0), \tag{10.1.7}$$

$$T_2(x, s) = B_2 \exp[-(s/a_2)^{1/2}|x|] \quad (x < 0). \tag{10.1.8}$$

The constants B_1 and B_2 are found from boundary conditions (10.1.5) and (10.1.6) which will be written for the transform as

$$T_1(0, s) = T_2(0, s), \qquad T'(0, s) = -(\lambda_2/\lambda_1)T_2'(0, s). \tag{10.1.9}$$

Then solutions (10.1.7) and (10.1.8) may be written

$$T_1(x, s) = \frac{t_0}{s} - \frac{t_0}{(1 + K_\varepsilon)s} \exp\left[-\left(\frac{s}{a_1}\right)^{1/2} x\right] \quad (x > 0). \tag{10.1.10}$$

$$T_2(x, s) = \frac{K_\varepsilon t_0}{(1 + K_\varepsilon)s} \exp\left[-\left(\frac{s}{a_2}\right)^{1/2} |x|\right] \quad (x < 0), \tag{10.1.11}$$

where

$$K_\varepsilon = \frac{\lambda_1}{\lambda_2}\left(\frac{a_2}{a_1}\right)^{1/2} = \left(\frac{\lambda_1 c_1 \gamma_1}{\lambda_2 c_2 \gamma_2}\right)^{1/2} = \frac{\varepsilon_1}{\varepsilon_2}, \qquad K_\varepsilon = \frac{K_\lambda}{(K_a)^{1/2}};$$

where K_ε is the parameter characterizing the thermal activity of the first rod relative to that of the second; it is equal to the ratio of thermal activity coefficients; K_λ is the parameter characterizing the relative thermal conductivity of a body ($K_\lambda = \lambda_1/\lambda_2$); K_a is the parameter characterizing the thermal inertia of one body relative to that of the other ($K_a = a_1/a_2$). The transforms of (10.1.10) and (10.1.11) are taken directly from the tables. Thus the inversion may be written directly

$$\theta_1 = \frac{t_1(x, \tau)}{t_0} = \frac{K_\varepsilon}{1 + K_\varepsilon}\left(1 + \frac{1}{K_\varepsilon} \operatorname{erf} \frac{x}{2(a_1\tau)^{1/2}}\right) \quad (x > 0), \tag{10.1.12}$$

$$\theta_2 = \frac{t_2(x, \tau)}{t_0} = \frac{K_\varepsilon}{1 + K_\varepsilon} \operatorname{erfc} \frac{|x|}{2(a_2\tau)^{1/2}} \quad (x < 0). \tag{10.1.13}$$

If the initial temperature of the second body is equal to t_{02} and that of the first t_{01}, the solution is of the form

$$\theta_1 = \frac{t_1(x, \tau) - t_{01}}{t_{01} - t_{02}} = \frac{K_\varepsilon}{1 + K_\varepsilon}\left(1 + \frac{1}{K_\varepsilon} \operatorname{erf} \frac{x}{2(a_1\tau)^{1/2}}\right) \tag{10.1.12a}$$

$$\theta_2 = \frac{t_2(x, \tau) - t_{02}}{t_{01} - t_{02}} = \frac{K_\varepsilon}{1 + K_\varepsilon} \operatorname{erfc} \frac{|x|}{2(a_2\tau)^{1/2}}. \tag{10.1.13a}$$

It follows from the solution analysis that at $\tau \to \infty$ *in a steady state the relative temperature of both rods will be the same and equal to*

$$\theta(x, \infty) = K_\varepsilon/(1 + K_\varepsilon).$$

If the thermal activities of the rods are equal ($K_\varepsilon = 1$), *the relative temperature in a steady state will be* $\theta = \frac{1}{2}$. *On the surface of contact this temperature will set in at the moment the rods are brought into contact and will remain constant during the whole heat transfer process* (see Fig. 10.1) *since*

$$\theta(0, \tau) = \theta(x, \infty) = K_\varepsilon/(1 + K_\varepsilon) = \text{const.} \tag{10.1.14}$$

If the thermal activity of one body is considerably lower than that of the other ($\varepsilon_1 \gg \varepsilon_2$), then $K_\varepsilon \gg 1$. In this case $\theta(0, \tau)$ will be of the highest possible value and equal to $\theta(0, \tau) = 1$. If the relative thermal activity of a body is low ($K_\varepsilon \to 0$), $\theta(0, \tau)$ is zero. Consequently, $\theta(0, \tau)$ changes from zero (minimum thermal activity) to unity (maximum thermal activity).

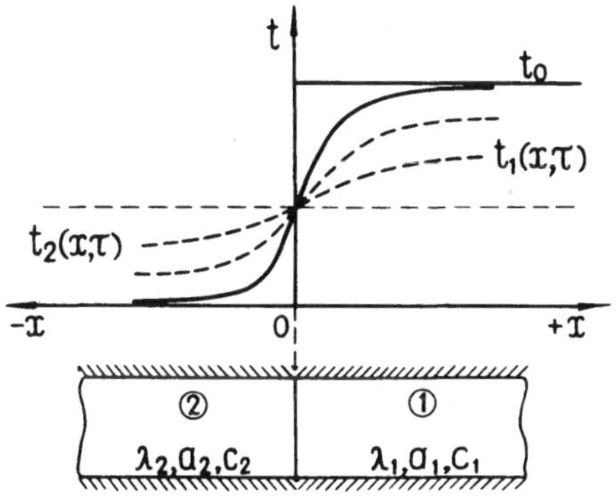

Fig. 10.1. Temperature field of two semi-infinite rods in contact.

Thus the value $\theta(0, \tau)$ is a measure of the decrease of the relative temperature of a semi-infinite body when it is in thermal contact with another semi-infinite body.

Therefore the value

$$1 - \theta(0, \tau) = 1/(1 + K_\varepsilon) = X \tag{10.1.15}$$

may be referred to as *cooling effect*. If $\theta(0, \tau) = 1$ ($X = 0$), then the cooling effect will be zero; vice versa, when $\theta(0, \tau) = 0$ ($X = 1$), the cooling effect will be of the highest value.

The solution of the problem with more general initial conditions (10.1.3) may be obtained [109,115]. We assume that

$$\lambda_1 \frac{\partial t_1(0,\tau)}{\partial x} = -\lambda_2 \frac{\partial t_2(0,\tau)}{\partial x} = j(\tau) \tag{10.1.16}$$

is the unknown function of a heat flux at the boundary of contact of bodies. Applying the infinite cosine Fourier transform to equation (10.1.1) we obtain

$$[t(p,\tau)]_c = (2/\pi)^{1/2} \int_0^\infty t(x,\tau)\cos px\,dx. \tag{10.1.17}$$

Then taking into account conditions (10.1.5) and (10.1.6), we obtain

$$dT_1(p,\tau)_c/d\tau = -p^2 a_1 T_1(p,\tau)_c - (a_1/\lambda_1)(2/\pi)^{1/2} j(\tau) \tag{10.1.18}$$

The solution of Eq. (10.1.18), accounting for initial condition (10.1.3) transformed according to (10.1.17), will be of the form

$$t_1(x,\tau) = \frac{1}{2(\pi a_1 \tau)^{1/2}} \int_0^\infty f_1(\xi)\left\{\exp\left[-\frac{(x+\xi)^2}{4a_1\tau}\right] + \exp\left[-\frac{(x-\xi)^2}{4a_1\tau}\right]\right\} d\xi$$
$$-\frac{1}{\lambda_1}\left(\frac{a_1}{\pi}\right)^{1/2} \int_0^\tau \frac{j(\tau)}{(\tau-\tau^*)^{1/2}} \exp\left[-\frac{x^2}{4a_1(\tau-\tau^*)}\right] d\tau^*. \tag{10.1.19}$$

For the determination of $t_2(x,\tau)$, we transform the region Ω $(-\infty < x \le 0)$ into Ω' $(0 \le z \le \infty)$ by replacing $x = -z$. The solution for $t_2(-z,\tau)$ may be obtained in a similar way as

$$t_2(-z,\tau) = \frac{1}{2(\pi a_2 \tau)^{1/2}} \int_0^\infty f_2(-\xi)\left\{\exp\left[-\frac{(z+\xi)^2}{4a_2\tau}\right] + \exp\left[-\frac{(z-\xi)^2}{4a_2\tau}\right]\right\} d\xi$$
$$+\frac{1}{\lambda_2}\left(\frac{a_2}{\pi}\right)^{1/2} \int_0^\tau \frac{j(\tau)}{(\tau-\tau^*)^{1/2}} \exp\left[-\frac{z^2}{4a_2(\tau-\tau^*)}\right] d\tau^*. \tag{10.1.20}$$

To determine $j(\tau)$, boundary condition (10.1.16) is used. We have

$$\int_0^\tau \frac{j(\tau^*)\,d\tau^*}{(\tau-\tau^*)^{1/2}} = F(\tau),$$

where

$$F(\tau) = \frac{\lambda_1\lambda_2}{[\lambda_1(a_2)^{1/2} + \lambda_2(a_1]^{1/2}\sqrt{\tau}} \left[\frac{1}{(a_1)^{1/2}} \int_0^\infty f_1(\xi)\exp\left[-\frac{\xi^2}{4a_1\tau}\right] d\xi\right.$$
$$\left.-\frac{1}{(a_2)^{1/2}} \int_0^\infty f_2(-\xi)\exp\left[-\frac{\xi^2}{4a_1\tau}\right] d\xi\right]. \tag{10.1.21}$$

Equation (10.1.21) represents the integral Abel equation with respect to the function $j(\tau)$. Its solution has the form

$$j(\tau) = \frac{1}{\pi}\frac{d}{d\tau}\int_0^\tau \frac{F(\tau^*)}{(\tau - \tau^*)^{1/2}}\,d\tau^*. \qquad (10.1.22)$$

Note that in solutions (10.1.19) and (10.1.20), the presence of terms $\exp[-(x \pm \xi)^2/4a\tau]$ leads to convergence of improper integrals for a wide class of functions $f_1(x)$ and $f_2(x)$. Particularly, it may be shown that

$$\lim_{x\to 0+}\frac{\partial t_1}{\partial x} = \frac{j(\tau)}{\lambda_1}, \qquad \lim_{x\to 0-}\frac{\partial t_2}{\partial x} = \frac{j(\tau)}{\lambda_2}.$$

If we assume

$$f_1(x) = t_{01} = \text{const}, \; f_2(x) = t_{02} = \text{const}, \qquad (10.1.23)$$

upon transformation we obtain

$$F(\tau) = \frac{(t_{01} - t_{02})\lambda_1\lambda_2\sqrt{\pi}}{\lambda_1(a_2)^{1/2} + \lambda_2(a_1)^{1/2}} = \text{const}. \qquad (10.1.24)$$

If this value for $F(\tau)$ is substituted into formula (10.1.22), we have

$$j(\tau) = \frac{\varepsilon_1\varepsilon_2}{\varepsilon_1 + \varepsilon_2}\,\frac{(t_{01} - t_{02})}{(\pi\tau)^{1/2}}. \qquad (10.1.25)$$

Solutions (10.1.19) and (10.1.20) take the form

$$\theta_1 = \frac{t_1(x, \tau) - t_{01}}{t_{01} - t_{02}}$$
$$= -\frac{1}{\pi(1 + K_\varepsilon)}\int_0^\tau \exp\left[-\frac{x^2}{4a_1(\tau - \tau^*)}\right]\frac{d\tau^*}{(\tau^*(\tau - \tau^*))^{1/2}}. \qquad (10.1.26)$$

$$\theta_2 = \frac{t_2(x, \tau) - t_{02}}{t_{01} - t_{02}}$$
$$= -\frac{K_\varepsilon}{\pi(1 + K_\varepsilon)}\int_0^\tau \exp\left[-\frac{x^2}{4a_2(\tau - \tau^*)}\right]\frac{d\tau^*}{(\tau^*(\tau - \tau^*))^{1/2}}. \qquad (10.1.27)$$

In solutions (10.1.26) and (10.1.27), integration occurs with respect to time. If integration is carried out with respect to the coordinate, then solutions (10.1.26) and (10.1.27) will be identical to (10.1.12a) and (10.1.13a).

c. Solution of the Case with a Heat Source $w_1(0, \tau) = q_0 = \text{const}$. The problem statement will be changed. At the initial moment, let the temperature of both rods be the same and equal to

$$t_1(x, 0) = t_2(x, 0) = t_0 = \text{const.} \tag{10.1.28}$$

Just at the moment when the ends are brought into contact, a constant heat source acts at the interface with the output q_0 per unit area of contact (kcal/m^2 hr).

The solutions in this case will be of the form

$$\theta_1 = \frac{t_1(x, \tau) - t_0}{t_0} = \frac{2q_0}{\lambda_1 t_0}(a_1\tau)^{1/2}\frac{K_\varepsilon}{1 + K_\varepsilon}\ \text{ierfc}\ \frac{x}{2(a_1\tau)^{1/2}}, \tag{10.1.29}$$

$$\theta_2 = \frac{t_2(x, \tau) - t_0}{t_0} = \frac{2q_0}{\lambda_2 t_0}(a_2\tau)^{1/2}\frac{K_\varepsilon}{1 + K_\varepsilon}\ \text{ierfc}\ \frac{|x|}{2(a_2\tau)^{1/2}}. \tag{10.1.30}$$

The following notations will be introduced

$$\text{Fo}_1 = \frac{a_1\tau}{x^2}, \qquad \text{Fo}_2 = \frac{a_2\tau}{x^2}, \qquad \text{Ki}_1 = \frac{q_0 x}{\lambda_1 t_0}, \qquad \text{Ki}_2 = \frac{q_0 x}{\lambda_2 t_0}.$$

The solutions (19.1.29) and (10.1.30) may be rewritten

$$\theta_1 = \frac{2\ \text{Ki}_1 K_\varepsilon}{1 + K_\varepsilon}\ \text{Fo}_1^{1/2}\ \text{i erfc}\ \frac{1}{2(\text{Fo}_1)^{1/2}} \tag{10.1.31}$$

$$\theta_2 = \frac{2\ \text{Ki}_2}{1 + K_\varepsilon}\ \text{Fo}_2^{1/2}\ \text{i erfc}\ \frac{1}{2(\text{Fo}_2)^{1/2}}. \tag{10.1.32}$$

10.2 System of Two Bodies (Finite and Semi-Infinite Rods)

a. Statement of the Problem. *A finite rod is brought into contact with a semi-infinite rod having different thermal coefficients. The lateral surfaces of the rods are thermally insulated. At the initial moment, the free end of the finite rod is heated instantaneously to the temperature t_a which is maintained constant during the entire heating process (boundary condition of the first kind). The temperature distribution along the length of the rods is to be found.*

Two cases will be considered: (a) when the initial temperature of the rods is zero, and (b) when the initial temperatures of the rods are different.

We have (see Fig. 10.2)

$$\frac{\partial t_1(x, \tau)}{\partial \tau} = a_1 \frac{\partial^2 t_1(x, \tau)}{\partial x^2} \qquad (\tau > 0;\ 0 < x < R), \tag{10.2.1}$$

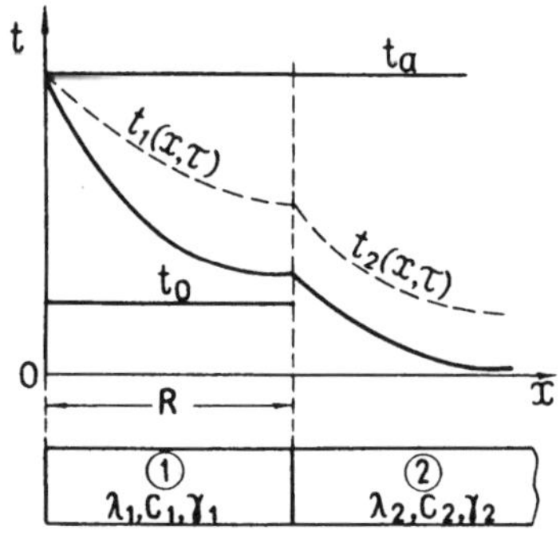

Fig. 10.2. System of two bodies (semi-infinite and infinite rods).

$$\frac{\partial t_2(x, \tau)}{\partial \tau} = a_2 \frac{\partial^2 t_2(x, \tau)}{\partial x^2} \quad (\tau > 0;\ R < x < \infty), \tag{10.2.2}$$

$$t_1(x, 0) = t_2(x, 0) = 0, \tag{10.2.3}$$

$$t_1(R, \tau) = t_2(R, \tau), \tag{10.2.4}$$

$$\frac{\lambda_1}{\lambda_2} \frac{\partial t_1(R, \tau)}{\partial x} = \frac{\partial t_2(R, \tau)}{\partial x}, \tag{10.2.5}$$

$$t_1(0, \tau) = t_a = \text{const}, \tag{10.2.6}$$

$$t_2(\infty, \tau) = 0. \tag{10.2.7}$$

b. Solution of the Problem when $t_1(x, 0) = t_2(x, 0) = 0$. The solution of Eqs. (10.2.1) and (10.2.2) with conditions (10.2.3) and (10.2.7) for the transformed functions $t(r, \tau)$, will be of the form

$$T_1(x, s) = A_1 \exp[(s/a_1)^{1/2}x] + B_1 \exp[-(s/a)^{1/2}x], \tag{10.2.8}$$

$$T_2(x, s) = C_1 \exp[-(s/a)^{1/2}x]. \tag{10.2.9}$$

The constants A_1, B_1, and C_1 are found from boundary conditions (10.2.4)–(10.2.6) which for the transform may be written

$$T_1(R, s) = T_2(R, s), \qquad K_\lambda T_1'(R, s) = T_2'(R, s) \tag{10.2.10}$$

$$T_1(0, s) = t_a/s. \tag{10.2.11}$$

Upon determining the constants, solutions (10.2.8) and (10.2.9) will be

$$T_1(x, s) = (t_a/s)\exp[-(s/a_1)^{1/2}x] - \left\{\frac{ht_a}{s}\right\}$$
$$\times \left\{\exp[-(s/a_1)^{1/2}R]\left[\frac{\exp[(s/a_1)^{1/2}x] - \exp[-(s/a_1)^{1/2}x]}{\exp[(s/a_1)^{1/2}R] - h\exp[-(s/a_1)^{1/2}R]}\right]\right\}. \tag{10.2.12}$$

$$T_2(x, s) = \left\{(t_a/s)\exp[-(s/a_1)^{1/2}R] - \left[\frac{\exp[(s/a_1)^{1/2}R] - \exp[-(s/a_1)^{1/2}R]}{\exp[(s/a_1)^{1/2}R) - h\exp[-(s/a_1)^{1/2}R]}\right]\right.$$
$$\left.\times \frac{ht_a}{s}\exp[-(s/a_1)^{1/2}R]\right\}\exp[-(s/a_2)^{1/2}(x - R)], \tag{10.2.13}$$

where

$$h = \frac{1 - K_\varepsilon}{1 + K_\varepsilon}, \qquad K_\varepsilon = \frac{\varepsilon_1}{\varepsilon_2} = \left(\frac{\lambda_1 c_1 \gamma_1}{\lambda_2 c_2 \gamma_2}\right)^{1/2}.$$

Since $|h| < 1$, we may use the transformation

$$\frac{1}{\exp((s/a_1)^{1/2}R) - h\exp[-(s/a_1)^{1/2}R]} = \frac{\exp[-(s/a_1)^{1/2}R]}{1 - h\exp[-2(s/a_1)^{1/2}R]}$$
$$= \sum_{n=1}^{\infty} h^{n-1}\exp[-(2n-1)(s/a_1)^{1/2}R]. \tag{10.2.14}$$

on the basis of the expansion

$$1/(1 - x) = 1 + x + x^2 + \cdots, \quad \text{if } |x| < 1.$$

$K_\varepsilon > 0$ and $|h| < 1$; therefore, series (10.2.14) converges rapidly.

Solutions (10.2.12) and (10.2.13) may be written

$$\frac{T_1(x, s)}{t_a} = (1/s)\exp[-(s/a_1)^{1/2}x] - (h/s)\sum_{n=1}^{\infty} h^{n-1}$$
$$\times \{\exp[-(2nR - x)(s/a_1)^{1/2}] - \exp[-(2nR + x)(s/a_1)^{1/2}]\}, \tag{10.2.15}$$

$$T_2(x, s)/t_a = [(1-h)/s]\sum_{n=1}^{\infty} h^{n-1}\exp\{-[(x-R) + (2n-1)R(a_2/a_1)^{1/2}](s/a_2)^{1/2}\}. \tag{10.2.16}$$

We obtain the inversion

$$\theta_1 = \frac{t_1(x, \tau)}{t_a} = \operatorname{erfc}\frac{x}{2(a_1\tau)^{1/2}} - h\sum_{n=1}^{\infty} h^{n-1}\left[\operatorname{erfc}\frac{2nR - x}{2(a_1\tau)^{1/2}} - \operatorname{erfc}\frac{2nR + x}{2(a_1\tau)^{1/2}}\right], \tag{10.2.17}$$

$$\theta_2 = \frac{t_2(x, \tau)}{t_a} = \frac{2\,K_\varepsilon}{1 + K_\varepsilon} \sum_{n=1}^{\infty} h^{n-1} \operatorname{erfc}\left[\frac{x - R + (2n-1)K_a^{-1/2}R}{2(a_2\tau)^{1/2}}\right], \quad (10.2.18)$$

where $K_a = a_1/a_2$, $K_a^{-1/2} = (a_2/a_1)^{1/2}$.

If the thermal properties of the rods are the same ($K_\varepsilon = 1$, $h = 0$), then from Eqs. (10.2.17) and (10.2.18) we obtain the solution for a single semi-infinite rod

$$\theta = \operatorname{erfc} \frac{x}{2(a\tau)^{1/2}}.$$

c. Solution of the Problem when $t_1(x, 0) = t_0$, $t_2(x, 0) = 0$. The second case in which the initial temperatures are different will be considered. Boundary conditions remain the same. Using the same calculation method, we obtain similar solutions as above differing only by additional terms containing the factor $[1/(1 + K_\varepsilon)]t_0$ for the finite rod and t_0 for the semi-infinite rod.

The final form for the original will be

$$\begin{aligned}\theta_1 &= \frac{t_1(x, \tau) - t_0}{t_a - t_0} \\ &= \operatorname{erfc} \frac{x}{2(a_1\tau)^{1/2}} \mp h \sum_{n=1}^{\infty} h^{n-1} \operatorname{erfc} \frac{2nR \mp x}{2(a_1\tau)^{1/2}} \\ &\quad \mp \frac{t_0}{(1 + K_\varepsilon)(t_a - t_0)} \sum_{n=1}^{\infty} h^{n-1} \operatorname{erfc} \frac{(2n-1)R \mp x}{2(a_1\tau)^{1/2}}, \end{aligned} \quad (10.2.19)$$

$$\begin{aligned}\theta_2 &= \frac{t_2(x, \tau)}{t_a - t_0} \\ &= \frac{2\,K_\varepsilon}{1 + K_\varepsilon} \sum_{n=1}^{\infty} h^{n-1} \operatorname{erfc}\left[\frac{x - R + (2n-1)K_a^{-1/2}R}{2(a_2\tau)^{1/2}}\right] \\ &\quad + \frac{t_0 K_\varepsilon}{(t_a - t_0)(1 + K_\varepsilon)} \operatorname{erfc} \frac{x - R}{2(a_2\tau)^{1/2}} + \frac{2K_\varepsilon}{(1 + K_\varepsilon)^2(t_a - t_0)} \\ &\quad \times \sum_{n=1}^{\infty} h^{n-1} \operatorname{erfc}\left[\frac{x - R + 2nK_a^{-1/2}R}{2(a_2\tau)^{1/2}}\right]. \end{aligned} \quad (10.2.20)$$

Here the notation $\mp\, \Phi(\mp\, z) = -\,\Phi(-\,z) + \Phi(+\,z)$ is introduced.

If $t_0 = 0$ is assumed, then solutions (10.2.19) and (10.2.20) become solutions (10.2.17) and (10.2.18).

The present problem may also be considered as that of heating a finite plate, one surface of which is in contact with an infinite solid medium (boundary condition of the fourth kind), and the opposite surface is maintained at the constant temperature.

d. The Solution of the Problem with a Variable Surface Temperature. *The temperature of the free surface of the rod is a linear function of time*

$$t(0, \tau) = t_0 + b\tau. \tag{10.2.21}$$

The solution of a system of equations under conditions (10.2.3)–(10.2.5) and (10.2.21) has the form

$$\begin{aligned} t_1(x, \tau) = 4b\tau\Bigg\{&\left(\mathrm{i}^2\,\mathrm{erfc}\,\frac{x}{2(a_1\tau)^{1/2}}\right) + \sum_{n=2}^{\infty} h^{n-1}\,\mathrm{i}^2\,\mathrm{erfc}\,\frac{2(n-1)R + x}{2(a_1\tau)^{1/2}} \\ &- h\left[\mathrm{i}^2\,\mathrm{erfc}\,\frac{2R - x}{2(a_1\tau)^{1/2}} + \sum_{n=2}^{\infty} h^{n-1}\,\mathrm{i}^2\,\mathrm{erfc}\,\frac{2nR - x}{2(a_1\tau)^{1/2}}\right]\Bigg\}. \end{aligned} \tag{10.2.22}$$

$$\begin{aligned} t_2(x, \tau) = 4b\tau\Bigg\{&(1 - h)\,\mathrm{i}^2\,\mathrm{erfc}\left[\left(\frac{R}{2(a_1\tau)^{1/2}} - \frac{R - x}{2(a_2\tau)^{1/2}}\right)\right. \\ &+ (1 - h)\sum_{n=2}^{\infty} h^{n-1}\,\mathrm{i}^2\,\mathrm{erfc}\left.\left[\frac{(2n-1)R}{2(a_1\tau)^{1/2}} - \frac{R - x}{2(a_2\tau)^{1/2}}\right]\right]\Bigg\}. \end{aligned} \tag{10.2.23}$$

If the initial temperature of rods is equal to $t_0 = \text{const}$, the solution in generalized variables may be written as

$$\begin{aligned} \theta_1 &= \frac{t_1(x, \tau) - t_0}{t_0} \\ &= \mathrm{Pd}_1\,\mathrm{Fo}_1\Bigg\{\mathrm{i}^2\,\mathrm{erfc}\,\frac{x/R}{2\,(\mathrm{Fo}_1)^{1/2}} + \sum_{n=2}^{\infty} h^{n-1}\,\mathrm{i}^2\,\mathrm{erfc}\,\frac{(2n-1) + (x/R)}{2\,(\mathrm{Fo}_1)^{1/2}} \\ &\quad - h\left[\mathrm{i}^2\,\mathrm{erfc}\,\frac{2 - (x/R)}{2(\mathrm{Fo}_1)^{1/2}} + \sum_{n-2}^{\infty} h^{n-1}\,\mathrm{i}^2\,\mathrm{erfc}\,\frac{2n - (x/R)}{2(\mathrm{Fo}_1)^{1/2}}\right]\Bigg\}. \end{aligned} \tag{10.2.24}$$

$$\begin{aligned} \theta_2 &= \frac{t_2(x, \tau) - t_0}{t_0} \\ &= \mathrm{Pd}_1\,\mathrm{Fo}_1\Bigg\{(1 - h)\,\mathrm{i}^2\,\mathrm{erfc}\left(\frac{1}{2\,(\mathrm{Fo}_1)^{1/2}} - \frac{1 - (x/R)}{2(\mathrm{Fo}_2)^{1/2}}\right) \\ &\quad + (1 - h)\sum_{n=2}^{\infty} h^{n-1}\,\mathrm{i}^2\,\mathrm{erfc}\left(\frac{2n - 1}{2\,(\mathrm{Fo}_1)^{1/2}} - \frac{1 - (x/R)}{2(\mathrm{Fo}_2)^{1/2}}\right)\Bigg\}. \end{aligned} \tag{10.2.25}$$

e. Solution of Problem under the Condition $t_1(0, \tau) = t_m \sin \omega\tau$. The solution of our problem will be of the form

$$\begin{aligned} &\frac{t_1(x, \tau)}{t_m} \\ &= \frac{1}{M}\left[\exp\left[-\left(\tfrac{1}{2}\,\mathrm{Pd}_1\right)^{1/2}\frac{x}{R}\right]\sin\left(\mathrm{Pd}_1\,\mathrm{Fo}_1 - \left(\tfrac{1}{2}\,\mathrm{Pd}_1\right)^{1/2}\frac{x}{R}\right)\right. \end{aligned}$$

$$- h \exp\left[-(\tfrac{1}{2}\,\mathrm{Pd}_1)^{1/2}\left(2 - \frac{x}{R}\right)\right] \sin\left[\mathrm{Pd}_1\,\mathrm{Fo}_1 + (\tfrac{1}{2}\,\mathrm{Pd}_1)^{1/2}\left(2 - \frac{x}{R}\right)\right]$$

$$- h\left\{\exp\left[-(\tfrac{1}{2}\,\mathrm{Pd}_1)^{1/2}\left(4 - \frac{x}{R}\right)\right] \sin\left[\mathrm{Pd}_1\,\mathrm{Fo}_1 + (\tfrac{1}{2}\,\mathrm{Pd}_1)^{1/2} x/R\right]\right.$$

$$\left. + \exp\left[-(\tfrac{1}{2}\,\mathrm{Pd}_1)^{1/2}\left(2 - \frac{x}{R}\right)\right]\sin\left[\mathrm{Pd}_1\,\mathrm{Fo}_1 - (\tfrac{1}{2}\,\mathrm{Pd}_1)^{1/2}\left(2 - \frac{x}{R}\right)\right]\right\}$$

$$+ \frac{\mathrm{Pd}^*}{\pi}\left[\int_0^\infty \exp(-\mathrm{Pd}^*\,\mathrm{Fo}_1)\sin\left[\frac{x}{R}(\mathrm{Pd}^*)^{1/2} - \arctan\frac{-h\sin 2(\mathrm{Pd}^*)^{1/2}}{1 - h\cos 2(\mathrm{Pd}^*)^{1/2}}\right]\right.$$

$$\times \frac{d\mathrm{Pd}^*}{(\mathrm{Pd}^*)^2 + \mathrm{Pd}_1^{\,2}} - h\int_0^\infty \exp[-\,\mathrm{Pd}^*\,\mathrm{Fo}_1]\sin\left[-\frac{x}{R}(\mathrm{Pd}^*)^{1/2}\right.$$

$$\left.\left. - \arctan\frac{-\sin 2(\mathrm{Pd}^*)^{1/2}}{\cos 2(\mathrm{Pd}^*)^{1/2} - h}\right]\frac{d\mathrm{Pd}^*}{(\mathrm{Pd}^*)^2 + \mathrm{Pd}_1^{\,2}}\right]; \qquad (10.2.26)$$

$$\frac{t_2(x, \tau)}{t_m}$$

$$= \frac{1-h}{M}\left\{\exp\left[-\,(\tfrac{1}{2}\,\mathrm{Pd}_1)^{1/2} - (\tfrac{1}{2}\,\mathrm{Pd}_2)^{1/2}\left(\frac{x}{R} - 1\right)\right]\sin\left[\mathrm{Pd}_1\,\mathrm{Fo}_1\right.\right.$$

$$\left. - (\tfrac{1}{2}\,\mathrm{Pd}_1)^{1/2} - (\tfrac{1}{2}\,\mathrm{Pd}_2)^{1/2}\left(\frac{x}{R} - 1\right)\right] - h\exp\left[-\,3(\tfrac{1}{2}\,\mathrm{Pd}_1)^{1/2}\right.$$

$$\left.\left. - (\tfrac{1}{2}\,\mathrm{Pd}_2)^{1/2}\left(\frac{x}{R} - 1\right)\right]\sin\left[\mathrm{Pd}_1\,\mathrm{Fo}_1 + (\tfrac{1}{2}\,\mathrm{Pd}_1)^{1/2} - (\tfrac{1}{2}\,\mathrm{Pd}_2)^{1/2}\left(\frac{x}{R} - 1\right)\right]\right\}$$

$$+ (1-h)\,\mathrm{Pd}_1\frac{1}{\pi}\int_0^\infty \exp[-\mathrm{Pd}^*\,\mathrm{Fo}_1]\sin\left[\left(\frac{x}{R} - 1\right)(\mathrm{Pd}^*)^{1/2}K_a^{1/2}\right.$$

$$\left. + \arctan\frac{\tan(\mathrm{Pd}^*)^{1/2}}{K_\varepsilon}\right]\frac{d\mathrm{Pd}^*}{(\mathrm{Pd}^*)^2 + \mathrm{Pd}_1^{\,2}}, \qquad (10.2.27)$$

where

$$M = 1 - 2h\exp[-\,2(\tfrac{1}{2}\,\mathrm{Pd}_1)^{1/2}]\cos 2(\tfrac{1}{2}\,\mathrm{Pd}_1)^{1/2} + h^2\exp[-\,4(\tfrac{1}{2}\,\mathrm{Pd}_1)^{1/2}];$$

$\mathrm{Pd}_k = (\omega/a_k)l^2$; $k = 1,2$ $\mathrm{Pd}^* = (\omega^*/a_1)l^2$ is the Predvoditelev criterion.

10.3 System of Two Bodies (Two Infinite Plates)

a. Statement of the Problem. *Two infinite plates with the thickness R_1 and R_2 and with different thermal coefficients are in contact. Their initial temperatures are both equal to zero. At the initial moment, one free surface is instantaneously heated to the temperature t_a which is maintained constant* *during the entire heating process. The temperature of the opposite surface is maintained equal to the initial one. The temperature distribution through the thickness of the system consisting of two plates is to be found.*

The origin of the coordinates is taken in the plane of contact (Fig. 10.3).

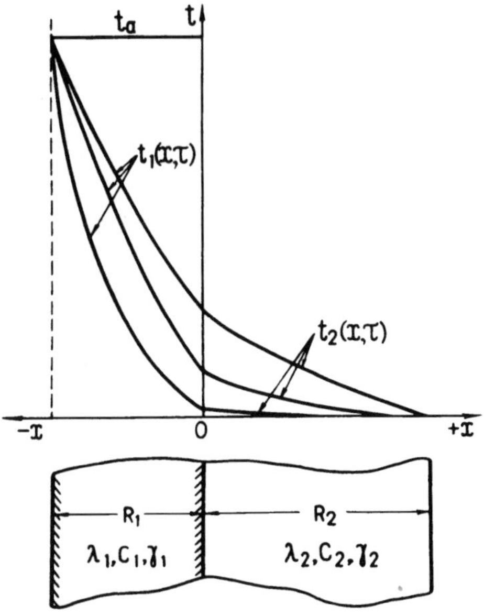

Fig. 10.3. System of two infinite plates.

The boundary conditions will be written as

$$t_1(x, 0) = t_2(x, 0) = 0\,, \tag{10.3.1}$$

$$t_1(-R_1, \tau) = t_a\,, \qquad t_2(R_2, \tau) = 0\,,$$

$$t_1(0, \tau) = t_2(0, \tau)\,, \tag{10.3.2}$$

$$\frac{\lambda_1}{\lambda_2}\frac{\partial t_1(0, \tau)}{\partial x} = \frac{\partial t_2(0, \tau)}{\partial x}. \tag{10.3.3}$$

b. Solution of the Problem. The transformed solutions are found by the conventional method using boundary conditions (10.3.1)–(10.3.3). The solutions will be of the form

$$\frac{T_1(x, s)}{t_a} = \frac{K_\varepsilon \cosh q_1 x \sinh q_2 R_2 - \sinh q_1 x \cosh q_2 R_2}{s[K_\varepsilon \cosh q_1 R_1 \sinh q_2 R_2 + \sinh q_1 R_1 \cosh q_2 R_2]} \tag{10.3.4}$$

$$\frac{T_2(x, s)}{t_a} = \frac{K_\varepsilon \sinh q_2(R_2 - x)}{s[K_\varepsilon \cosh q_1 R_1 \sinh q_2 R_2 + \sinh q_1 R_1 \cosh q_2 R_2]} \tag{10.3.5}$$

where

$$K_\varepsilon = \varepsilon_1/\varepsilon_2, \qquad q_1 = (s/a_1)^{1/2}, \qquad q_2 = (s/a_2)^{1/2}.$$

Using the expansion theorem (solutions (10.3.4) and (10.3.5) satisfy all the conditions of the theorem), and assuming the denominator of relation (10.3.4) and (10.3.5) to be zero, we obtain the characteristic equation

$$\sin\mu \cos K_a^{1/2}K_{R_1}\mu + K_\varepsilon \cos\mu \sin K_a^{1/2}K_{R_1}\mu = 0. \tag{10.3.6}$$

It should be noted that the roots of Eq. (10.3.6) are also the roots of

$$\cot K_a^{1/2}K_{R_1}\mu + K_\varepsilon \cot\mu = 0, \tag{10.3.6a}$$

and the general roots (if they exist) of

$$\sin\mu = 0; \qquad \sin K_a^{1/2}K_{R_1}\mu = 0. \tag{10.3.6b}$$

The system of equations (10.3.6b) has a solution when $K_a^{1/2}K_{R_1}$ is a rational number. Then the solution of the problem is

$$\begin{aligned}\theta_1 &= \frac{t_1(x,\tau)}{t_a}\\ &= \frac{K_\lambda - \dfrac{x}{R_2}}{K_\lambda + \dfrac{R_1}{R_2}} - \sum_{n=1}^{\infty} \frac{2\{K_\varepsilon \cos\mu_n(x/R_1)\sin K_a^{1/2}\mu_n K_{R_1} - \sin\mu_n(x/R_1)\cos K_a^{1/2}\mu_n K_{R_1}\}}{\mu_n[(K_\varepsilon + K_a^{1/2}K_{R_1})\sin\mu_n \sin K_a^{1/2}\mu_n K_{R_1} - (1 + K_\varepsilon K_a^{1/2}K_{R_1})\cos\mu_n \cos K_a^{1/2}\mu_n K_{R_1}]}\\ &\quad\times \exp[-\mu_n^2\,\mathrm{Fo}_1],\end{aligned} \tag{10.3.7}$$

$$\begin{aligned}\theta_2 &= \frac{t_2(x,\tau)}{t_a}\\ &= \frac{R_2 - x}{R_2 + R_1K_\lambda^{-1}} - \sum_{n=1}^{\infty} \frac{2K_\varepsilon \sin K_a^{1/2}\left(\dfrac{R_2 - x}{R_1}\right)\mu_n}{\mu_n[(K_\varepsilon + K_a^{1/2}K_{R_1})\sin\mu_n \sin K_a^{1/2}K_{R_1}\mu_n - (1 + K_\varepsilon K_a^{1/2}K_{R_1})\cos\mu_n \cos K_a^{1/2}K_{R_1}\mu_n]}\\ &\quad\times \exp[-K_aK_{R_1}^2\mu_n^2\,\mathrm{Fo}_2],\end{aligned} \tag{10.3.8}$$

where

$$K_\varepsilon = \varepsilon_1/\varepsilon_2, \qquad K_\lambda = \lambda_1/\lambda_2, \qquad K_a = a_1/a_2, \qquad K_{R_1} = R_1/R_2,$$
$$\mathrm{Fo}_1 = a_1\tau/R_1^2, \qquad \mathrm{Fo}_2 = a_2\tau/R_2^2.$$

If the coefficient $K_a^{1/2}K_{R_1}$ is a proper fraction, i.e.,

$$K_a^{1/2}K_{R_1} = \beta/b,$$

then $K_a^{1/2}K_{R_1}\mu_n = (\beta/b)\mu_n = n\pi$. Consequently, $\mu_n = bn\pi/\beta = bm\pi$. Then we should add two more expressions to solutions (10.3.7) and (10.3.8), i.e.,

$$-\frac{2}{\pi(\beta K_\varepsilon + b)}\sum_{m=1}^{\infty}\frac{\cos mb\pi}{m}\sin\frac{mb\pi x}{R_1}\exp[-m^2b^2\pi^2\,\mathrm{Fo}_1], \tag{10.3.9}$$

$$-\frac{2K_\varepsilon}{\pi(\beta K_\varepsilon + b)}\sum_{m=1}^{\infty}\frac{\cos mb\pi}{m}\sin\frac{\beta m\pi x}{R_1}\exp[-b^2m^2\pi^2K_aK_{R_1}\,\mathrm{Fo}_2], \tag{10.3.10}$$

respectively.

If the coefficient $K_a^{1/2}K_{R_1}$ is an irrational number, series (10.3.9) and (10.3.10) do not appear in the solution, and the roots of characteristic equation (10.3.6) still exist.[1] Using Eq. (10.3.6), we can simplify solutions (10.3.7) and (10.3.8) to

$$\theta_1 = \frac{K_\lambda R_2 - x_2}{K_\lambda R_2 + R_1} - \sum_{n=1}^{\infty}\frac{2K_\varepsilon(\sin^2 K_a^{1/2}K_{R_1}\mu_n \sin\mu_n\left\{\dfrac{(R_1+x)}{R_1}\right\}}{\mu_n[K_\varepsilon\sin^2 K_a^{1/2}K_{R_1}\mu_n + K_a^{1/2}K_{R_1}\sin^2\mu_n]}\exp[-\mu_n{}^2\,\mathrm{Fo}_1]. \tag{10.3.11}$$

$$\theta_2 = \frac{(R_2 - x)K_\lambda}{K_\lambda R_2 + R_1} - \sum_{n=1}^{\infty}\frac{2K_\varepsilon[\sin\mu_n \sin K_a^{1/2}K_{R_1}\mu_n \sin K_a^{1/2}(K_{R_1} - \{x/R_1\})\mu_n]}{\mu_n[K_\varepsilon\sin^2 K_a^{1/2}K_{R_1}\mu_n + K_a^{1/2}K_{R_1}\sin^2\mu_n]} + \exp[\mu_n{}^2K_aK_{R_1}^2\,\mathrm{Fo}_2]. \tag{10.3.12}$$

In a steady state ($\mathrm{Fo} = \infty$), the temperature distribution follows the linear law. The relative temperature on the left-hand surface of the plate ($x = -R_1$) is unity and on the opposite surface ($x = R_2$) it is zero. On the contact surface ($x = 0$) the relative temperature is

$$\theta(0, \infty) = \frac{K_\lambda K_{R_1}}{K_\lambda K_{R_1} + 1}. \tag{10.3.13}$$

If the thermal coefficients of the plates are the same, $K_\lambda = 1$ at $R_1 = R_2 (K_{R_1} = 1)$, then $\theta(0, \infty) = \frac{1}{2}$.

c. Solution with Boundary Conditions of the Second Kind. In contrast to the above problem one surface of a system is thermally insulated, and

[1] Additional roots of the transcendental characteristic equation are also possible in other problems (to be discussed). These additional roots are found in a similar way; accordingly, such cases will not be dealt with here.

a constant heat flux is supplied to the second surface

$$t_1(x, 0) = t_2(x, 0) = t_0 = \text{const.} \tag{10.3.14}$$

$$\partial t_1(R_2, \tau)/\partial x = 0 \tag{10.3.15}$$

$$t_1(0, \tau) = t_2(0, \tau) \tag{10.3.16}$$

$$\frac{\lambda_1}{\lambda_2}\frac{\partial t_1(0, \tau)}{\partial x} = \frac{\partial t_2(0, \tau)}{\partial x} \tag{10.3.17}$$

$$\frac{\partial t_1(-R_1, \tau)}{\partial x} - \frac{q}{\lambda_2} = 0. \tag{10.3.18}$$

The solution for the transform will be

$$\begin{aligned} T_1(x, s) - (t_0/s) = \frac{q}{s}\,&[K_\lambda \cosh(s/a_1)^{1/2}x \cosh(s/a_2)^{1/2}R_2 \\ &- K_a \sinh(s/a_1)^{1/2}x \cosh(s/a_2)^{1/2}R_2]\{[K_\lambda \cosh(s/a_1)^{1/2}R_2 \\ &\times \sinh(s/a_1)^{1/2}R_1 + K_a \sinh(s/a_2)^{1/2}R_2 \cosh(s/a_1)^{1/2}R_1] \\ &\times \lambda_1(s/a_1)\}^{-1}. \end{aligned} \tag{10.3.19}$$

$$T_2(x, s) - (t_0/s) = \frac{qK_\lambda \cosh(s/a_2)^{1/2}(R_2 - x)}{s\{K_\lambda \cosh(s/a_2)^{1/2}R_2 \sinh(s/a_1)^{1/2}R_1 + K_a \sinh(s/a_2)^{1/2}R_2 \cosh(s/a_1)^{1/2}R_1\}\,\lambda_1(s/a_1)^{1/2}}. \tag{10.3.20}$$

We now use the expansion theorem. From the denominator (expression in brackets in formulas (10.3.19) and (10.3.20)), we obtain the characteristic equation[2]

$$K_\lambda \tan \mu + K_a \tan \mu K_a^{1/2} K_{R_2} = 0. \tag{10.3.21}$$

The solution of this problem may be written

$$\begin{aligned} \theta_1 &= \frac{t_1(x, \tau)}{t_0} \\ &= 1 + [\mathrm{Ki}_1/(K_\lambda + K_a K_{R_1})]\{K_\lambda \mathrm{Fo}_1 + \tfrac{1}{2}[K_\lambda K_a K_{R_2}^2 + K_\lambda (x/R_1)^2 \\ &\quad - 2K_a K_{R_2}(x/R_1) - \tfrac{1}{2}[K_\lambda^2 K_a K_{R_2}^2 + K_\lambda K_a K_{R_2} + \tfrac{1}{3}K_\lambda K_a^2 K_{R_2}^3 + \tfrac{1}{3}K_\lambda^2] \\ &\quad \times [K_\lambda + K_a K_{R_2}]^{-1}\} + 2\mathrm{Ki}_1 \sum_{n=1}^{\infty} (1/\mu_n^2 \varphi_n)(\cos \mu_n (x/R_1) \cos \mu_n K_a^{1/2} K_{R_2} \\ &\quad + (K_a/K_\lambda) \sin \mu_n (x/R_1) \sin \mu_n K_a^{1/2} K_{R_2}) \exp[-\mu_n^2 \mathrm{Fo}_1], \end{aligned} \tag{10.3.22}$$

[2] Here, the appearance of additional roots is possible because of the division of the denominator of expressions (10.3.19)–(10.3.20) by $\cos \mu \cos \mu K_a^{1/2} K_{R_2}$. Consideration of this special case is similar to that given in the previous chapter.

$$\theta_2 = 1 + \frac{\mathrm{Ki}_1 K_\lambda}{K_\lambda + K_a K_{R_2}} \{\mathrm{Fo}_1 + \tfrac{1}{2}K_a(K_{R_2} - (x/R_1))^2 - \tfrac{1}{2}[K_\lambda K_a K_{R_2}^2 + K_a K_{R_2}$$
$$+ \tfrac{1}{3}K_a{}^2 K_{R_2}^3 + \tfrac{1}{3}K_\lambda][K_\lambda + K_a K_{R_2}]^{-1}\} + 2\mathrm{Ki}_1 \sum_{n=1}^{\infty} (1/\varphi_n \mu_n{}^2)$$
$$\times \cos \mu_n K_a^{1/2}(K_{R_2} - (x/R_1)) \exp[-\mu_n{}^2 \mathrm{Fo}_1], \qquad (10.3.23)$$

where

$$\varphi_n = (1 + (1/K_\lambda)K_a^{3/2}K_{R_2}) \cos \mu_n \cos \mu_n K_a^{1/2}K_{R_2} - (K_{R_2}K_a^{1/2} + (1/K_\lambda)K_a)$$
$$\times \sin \mu_n \sin \mu_n K_a^{1/2}K_{R_2}, \qquad (10.3.24)$$

where

$$K_\lambda = \lambda_1/\lambda_2; \qquad K_a = a_1/a_2; \qquad K_{R_2} = R_2/R_1; \qquad \mathrm{Fo}_1 = a_1\tau/R_1{}^2;$$
$$\mathrm{Ki}_1 = q_1 R_2/\lambda_1 t_0.$$

If heat transfer between a surface ($x = -R_1$) of the system of plates occurs according to the Newton law $q = \alpha[t_a - t(-R_1, \tau)]$, then the temperature at this surface ($x/R_1 = -1$) and at the boundary of their contact ($x/R_1 = 0$) is found from the diagrams depicted in Figs. 10.4 and 10.5. These are

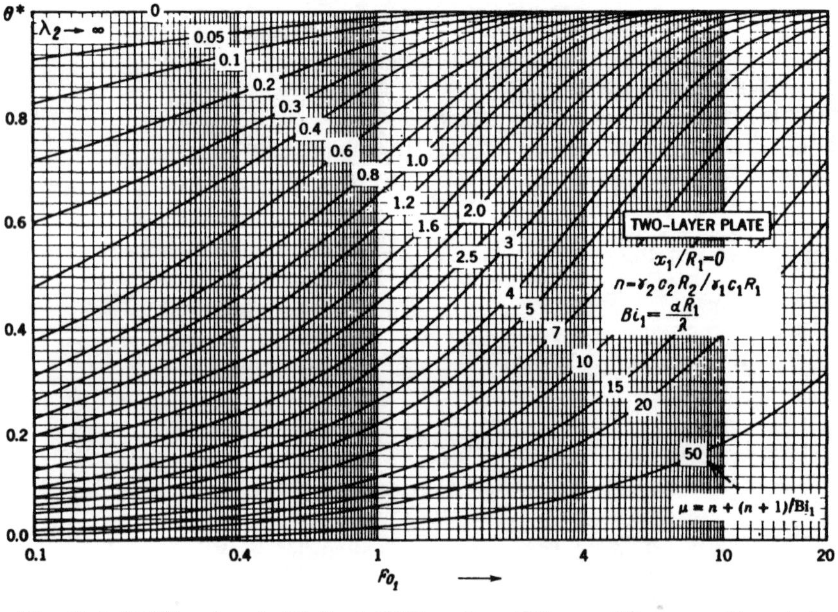

Fig. 10.4. θ^* $\{\theta^* = 1 - \{\mu\,\mathrm{Bi}_1/(n + 1)\}[1 - (t - t_0)/(t_a - t_0)]\}$ versus Fo_1 at the plane of contact of the plates ($x = 0$) for various values of μ [102].

obtained for the case when the heat conduction coefficient of the second plate is infinitely large ($\lambda_2 \to \infty$) and for different values of μ [$\mu = n + (n + 1)\,\mathrm{Bi}_1$] where $n = c_2\gamma_2 R_2/c_1\gamma_1 R_1$, and $\mathrm{Bi}_1 = \alpha R_1/\lambda_1$.

In Fig. 10.4 the quantity

$$\theta^* = 1 - \frac{\mu\,\mathrm{Bi}_1}{(n+1)}\left[1 - \frac{t(0, \tau) - t_0}{t_a - t_0}\right]$$

is the ordinate, and the Fourier number $\mathrm{Fo}_1 = a_1\tau/R_1^2$ is the abscissa.

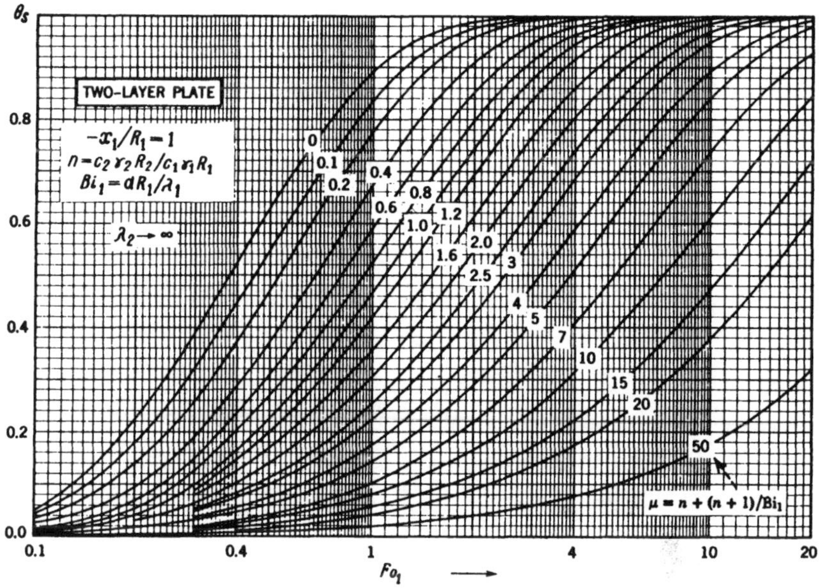

Fig. 10.5. Generalized variable θ_s [$\theta_s = (t_s - t_0)/(t_a - t_0)$] versus Fo_1 for plate surface ($x = -R_1$) for various values of μ [102].

In Fig. 10.5 the quantity

$$\theta_s = \frac{t(-R_1, \tau) - t_0}{t_a - t_0}$$

is the ordinate, and the Fourier number Fo_1 is the abscissa.

10.4 System of Two Spherical Bodies (Sphere inside Sphere)

a. Statement of the Problem. *At the initial moment a sphere with the temperature t_0 is placed inside a hollow sphere with the same initial temperature. Both are in perfect thermal contact. The temperature of the external* *surface of the outside sphere is maintained equal to zero during the whole process of heating. The temperature distribution at any moment in the system "sphere inside a sphere"* (Fig. 10.6) *is to be found.*

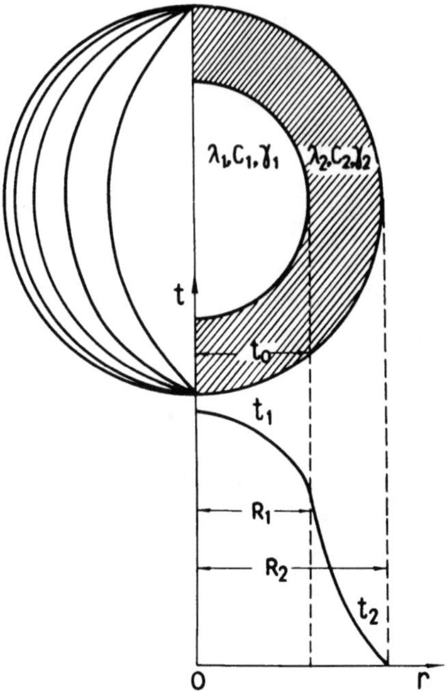

Fig. 10.6. System of two bodies (sphere inside sphere).

We have

$$\frac{\partial[rt_1(r, \tau)]}{\partial\tau} = a_1 \frac{\partial^2[rt_1(r, \tau)]}{\partial r_2} \quad (\tau > 0;\ 0 < r < R_1), \tag{10.4.1}$$

$$\frac{\partial[rt_2(r, \tau)]}{\partial\tau} = a_2 \frac{\partial\ [rt_2(r, \tau)]}{\partial r^2} \quad (\tau > 0;\ R_1 < r < R_2), \tag{10.4.2}$$

$$t_1(r, 0) = t_0, \quad t_2(r, 0) = t_0, \tag{10.4.3}$$

$$t_1(R_1, \tau) = t_2(R_1, \tau), \quad K_\lambda \frac{\partial t_1(R_1, \tau)}{\partial r} = \frac{\partial t_2(R_1, \tau)}{\partial r}, \tag{10.4.4}$$

$$t_2(R_2, \tau) = 0, \quad t_1(0, \tau) \neq \infty. \tag{10.4.5}$$

b. Solution of the Problem. The solution is found by the operational method with conditions (10.4.3)–(10.4.5). The solution for the transform is to be obtained.

$$\frac{T_1(r, s)}{t_0} = \frac{1}{s} - \frac{R_1 R_2 q_2 \sinh q_1 r}{rs[\psi_2(R_2) \sinh q_1 R_1 + K_\lambda \psi_1(R_1) \sinh q_2(R_2 - R_1)]}, \tag{10.4.6}$$

$$\frac{T_2(r, s)}{t_0} = \frac{1}{s} - \frac{R_2[\psi_2(r) \sinh q_1 R_1 + K_\lambda \psi_1(R_1) \sinh q_2(r - R_1)]}{rs[\psi_2(R_2) \sinh q_1 R_1 + K_\lambda \psi_1(R_1) \sinh q_2(R_2 - R_1)]}, \tag{10.4.7}$$

where

$$\psi_1(r) = q_1 R_1 \cosh q_1 r - \sinh q_1 r,$$

$$\psi_2(r) = q_2 R_1 \cosh q_2(r - R_1) + \sinh q_2(r - R_1),$$

$$q_1 = (s/a_1)^{1/2}, \qquad q_2 = (s/a_2)^{1/2}.$$

Assuming the denominator to be zero, we find the roots $s = 0$, and s_n which is an infinite set of the roots, determined from the characteristic equation

$$[K_a^{1/2}\mu \cot K_a^{1/2}(K_{R_2} - 1)\mu + 1] + K_\lambda[\mu \cot \mu - 1] = 0, \tag{10.4.8}$$

where

$$K_a = a_1/a_2, \qquad K_{R_2} = R_2/R_1.$$

Beside these roots, the additional roots of the equations will be found from the equation

$$\sin \mu = 0, \qquad \sin K_a^{1/2}(K_{R_2} - 1)\mu = 0.$$

If $K_a^{1/2}(K_{R_2} - 1)$ is an irrational number, the last equations have no roots and the solutions will be

$$\begin{aligned} \theta_1 &= \frac{t_1(x, \tau)}{t_0} \\ &= \frac{2R_2}{r} \sum_{n=1}^{\infty} \frac{1}{\varphi(\mu_n)} \sin \mu_n \frac{r}{R_1} \sin \mu_n K_\iota^{1/2}(K_{R_2} - 1) \\ &\quad \times \exp[-\mu_n^2 \mathrm{Fo}_1], \end{aligned} \tag{10.4.9}$$

$$\theta_2 = \frac{2R_2}{r} \sum_{n=1}^{\infty} \frac{\sin^2 \mu_n}{\varphi(\mu_n)} \sin K_a^{1/2} \left(K_{R_2} - \frac{r}{R_1}\right) \mu_n \exp[-K_a K_{R_2}^2 \mu_n^2 \mathrm{Fo}_2], \tag{10.4.10}$$

where

$$\begin{aligned} \varphi(\mu_n) &= K_\varepsilon \mu_n \sin^2 K_a^{1/2} \mu_n (K_{R_2} - 1) + K_a^{1/2}(K_{R_2} - 1)\mu_n \sin^2 \mu_n \\ &\quad + \frac{1 - K_a^{1/2} K_\varepsilon}{K_a^{1/2} \mu_n} \sin^2 \mu_n \sin^2 K_a^{1/2}(K_{R_2} - 1)\mu_n, \end{aligned}$$

$$\mathrm{Fo}_1 = a\tau/R_1^2, \qquad \mathrm{Fo}_2 = a_2\tau/R_2^2. \tag{10.4.11}$$

If the quantity $K_a^{1/2}(K_{R_2} - 1)$ is a ratio of integers β/b, then $(\beta/b)\mu = n\pi$ or $\mu = bm\pi$. The following expression should be added to the first solution

$$-\frac{2R_2}{r\pi(\beta K_\varepsilon + b)} \sum_{m=1}^{\infty} (-1)^{(b+\beta)m} \frac{1}{m} \sin \frac{b\pi mr}{R_1} \exp[-m^2 b^2 \pi^2 \,\mathrm{Fo}_1] \tag{10.4.12}$$

and to the second solution the expression

$$\frac{2R_2 K_\varepsilon}{r\pi(\beta K_\varepsilon + b)} \sum_{m=1}^{\infty} \frac{1}{m} \left[\frac{\sin[m\beta\pi(K_{R_2} - r/R_1)]}{(K_\varepsilon - 1)}\right] \exp[-m^2 b^2 \pi^2 \,\mathrm{Fo}_2\, K_a K_{R_2}^2 . \tag{10.4.13}$$

10.5 System of Two Cylindrical Bodies

a. Statement of the Problem. *There is an infinite cylinder surrounded by a thin shell. At the initial time instant the system of two cylinders which have the same initial temperature t_0 is placed into a medium with the temperature $t_a < t_0$. Heat transfer between the shell and medium occurs according to Newton's law of cooling. Find a temperature distribution for the system of two cylindrical bodies* (Fig. 10.7).

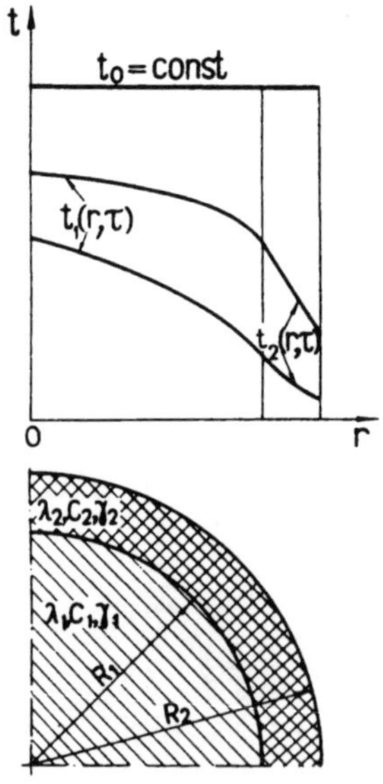

Fig. 10.7. System of two cylindrical bodies.

To the first approximation the thin shell may be considered flat. The problem is formulated as follows

$$\frac{\partial t_1}{\partial \tau} = a_1 \left(\frac{\partial^2 t_1}{\partial r^2} + \frac{1}{r}\frac{\partial t_1}{\partial r}\right) \quad (r > 0;\ 0 \leq r \leq R_1), \tag{10.5.1}$$

$$\frac{\partial t_2}{\partial \tau} = a_2 \frac{\partial^2 t_2}{\partial r^2} \quad (r > 0;\ R_1 \leq r \leq R_2), \tag{10.5.2}$$

$$t_1(r, 0) = t_2(r, 0) = t_0 = \text{const}, \tag{10.5.3}$$

$$t_1(R_1, \tau) = t_2(R_1, \tau); \quad \lambda_1 \frac{\partial t_1(R_1, \tau)}{\partial r} = \lambda_2 \frac{\partial t_2(R_1, \tau)}{\partial r} \tag{10.5.4}$$

$$t_1(0, \tau) < \infty; \quad -\lambda_2 \frac{\partial t_2(R_2, \tau)}{\partial r} + \alpha[t_a - t_2(R_2, \tau)] = 0. \tag{10.5.5}$$

b. Solution of the Problem. Application of the Laplace transformation yields

$$\frac{t_1(r, \tau) - t_0}{t_a - t_0} = 1 - \sum_{n=1}^{\infty} A_n J_0(\mu_n r/R_1) \exp[-\mu_n^2 \text{ Fo}], \tag{10.5.6}$$

$$\frac{t_2(r, \tau) - t_0}{t_a - t_0} = 1 - \sum_{n=1}^{\infty} A_n \{J_0(\mu_n) \cos[\mu_n K_a^{1/2}(r/R_1 - 1)] - K_\varepsilon J_1(\mu_n) \sin[\mu_n K_a^{1/2}(r/R_1 - 1)]\} \exp[-\mu_n^2 \text{ Fo}] \tag{10.5.7}$$

where μ_n are roots of the characteristic equation

$$J_0(\mu)[\text{Bi} \cos K_a^{1/2}(K_{R_2} - 1)\mu - K_a^{1/2} K_{R_2}\mu \sin K_a^{1/2}(K_{R_2} - 1)\mu] - K_\varepsilon J_1(\mu)[\text{Bi} \sin K_a^{1/2}(K_{R_2} - 1)\mu + K_a^{1/2} K_{R_2}\mu \cos K_a^{1/2}(K_{R_2} - 1)\mu] = 0.$$

$$A_n = \frac{2\text{Bi } K_\varepsilon [K_a^{1/2}(K_{R_2} - 1)\mu_n + \text{Bi} \tan K_a^{1/2}(K_{R_2} - 1)\mu_n]}{\mu_n J_0(\mu_n) \sin K_a^{1/2}(K_{R_2} - 1)\mu_n} \times \Bigg\{[K_\varepsilon^2 K_a(K_{R_2} - 1)^2\mu_n^2 + \text{Bi}^2] \cot K_a^{1/2}(K_{R_2} - 1)\mu_n + \frac{2K_\varepsilon K_a^{1/2}(K_{R_2} - 1)}{\sin 2K_a^{1/2}(K_{R_2} - 1)\mu_n}[\text{Bi}^2 + K_a(K_{R_2} - 1)^2\mu_n^2] + [K_a(K_{R_2} - 1)^2\mu_n^2 + 2K_\varepsilon K_a^{1/2}(K_{R_2} - 1)\text{ Bi} + K_\varepsilon^2 \text{ Bi}^2] \tan K_a^{1/2}(K_{R_2} - 1)\mu_n + K_\varepsilon K_a(K_{R_2} - 1)^2\mu_n^2 + 2K_\varepsilon^2 K_a^{1/2}(K_{R_2} - 1)\mu_n \text{ Bi} - 2K_a^{1/2}(K_{R_2} - 1)\mu_n \text{ Bi} - \frac{K_\varepsilon \text{ Bi}^2}{\mu_n}\Bigg]\Bigg\}^{-1} \tag{10.5.8}$$

In the limiting case where $\mathrm{Bi} \to \infty$, the second statement (10.5.5) is replaced by the condition

$$t_2(R_2, \tau) = t_a. \tag{10.5.9}$$

Here the form of equation (10.5.6) remains as before, and equation (10.5.7) becomes of the form

$$\frac{t_2(r, \tau) - t_0}{t_a - t_0}$$

$$= 1 - \sum_{n=1}^{\infty} \frac{2 \sin[K_a^{1/2}(K_{R_2} - r/R_1)\mu_n] \exp(-\mu_n^2 \mathrm{Fo}]}{\mu_n\left[\dfrac{K_\varepsilon^2 - 1}{K_\varepsilon} \sin^2 K_a^{1/2} (K_{R_2} - 1)\mu_n - \dfrac{1}{2\mu_n} \sin 2K_a^{1/2} (K_{R_2} - 1)\mu_n + b\right]} \tag{10.5.10}$$

where $b = K^{1/2}(K_{R_2} - 1) + 1/K_\varepsilon$ (for further details, see Luikov and Mikhailov [73]).

10.6 Infinite Plate

a. Statement of the Problem. *Consider an infinite plate with temperature t_0. At the initial moment, it is placed into a medium with the temperature $t_a < t_0$. The plate is cooled by heat conduction. The temperature distribution at any moment is to be found.*

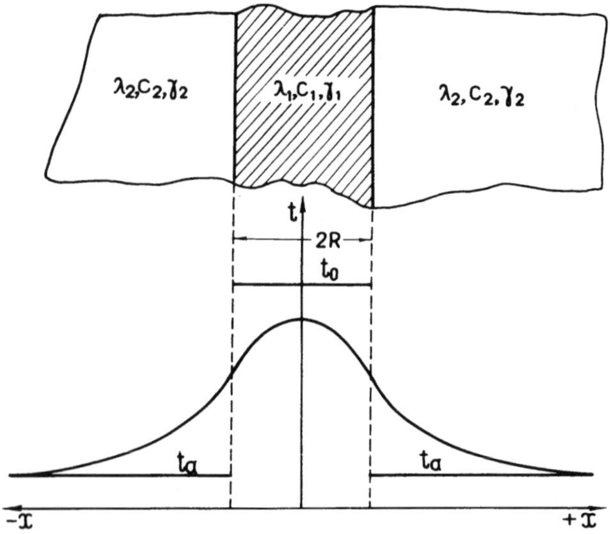

Fig. 10.8. Cooling a plate in an infinite medium.

We place the origin of the coordinates in the center of the plate (Fig. 10.8); subscript 1 refers to the plate and subscript 2 to the medium

$$\frac{\partial t_1(x,\tau)}{\partial \tau} = a_1 \frac{\partial^2 t_1(x,\tau)}{\partial x^2} \quad (\tau > 0;\ -R < x < +R), \tag{10.6.1}$$

$$\frac{\partial t_2(x,\tau)}{\partial \tau} = a_2 \frac{\partial^2 t_2(x,\tau)}{\partial x^2} \quad (\tau > 0;\ R < x < \infty), \tag{10.6.2}$$

$$t_1(x,0) = t_0, \qquad t_2(x,0) = t_a, \tag{10.6.3}$$

$$\pm K_\lambda \frac{\partial t_1(\pm R,\tau)}{\partial x} = \pm \frac{\partial t_2(\pm R,\tau)}{\partial x}, \tag{10.6.4}$$

$$t_1(\pm R,\tau) = t_2(\pm R,\tau), \tag{10.6.5}$$

$$\frac{\partial t_1(0,\tau)}{\partial x} = 0, \qquad \frac{\partial t_2(\pm\infty,\tau)}{\partial x} = 0. \tag{10.6.6}$$

b. Solution of the Problem. The transformed solution of Eqs. (10.6.1) and (10.6.2), taking into account Eqs. (10.6.3) and (10.6.6), may be written

$$T_1(x,s) - (t_0/s) = A\cosh(s/a_1)^{1/2}x = A_1[\exp[(s/a_1)^{1/2}x] + \exp[-(s/a_1)^{1/2}x]], \tag{10.6.7}$$

$$T_2(x,s) - (t_a/s) = B_1\exp[-(s/a_2)^{1/2}x]. \tag{10.6.8}$$

The constants A_1 and B_1 are determined from boundary conditions (10.6.4) and (10.6.5); then solutions (10.6.7) and (10.6.8) will be of the form

$$T_1(x,s) = \frac{t_0}{s} - \frac{(t_0 - t_a)}{s}\left(\frac{1}{1+K_\varepsilon}\right)\frac{\exp[(s/a_1)^{1/2}x] + \exp[-(s/a_1)^{1/2}x]}{\exp[(s/a_1)^{1/2}R] + h\exp[-(s/a_1)^{1/2}R]}, \tag{10.6.9}$$

$$T_2(x,s) - \frac{t_a}{s} = \frac{(t_0 - t_a)}{s}\left(\frac{K_\varepsilon}{1+K_\varepsilon}\right)\frac{\exp[(s/a_1)^{1/2}R] - \exp[-(s/a_1)^{1/2}R]}{\exp[(s/a_1)^{1/2}R] + h\exp[-(s/a_1)^{1/2}R]} \times \exp[-(s/a_2)^{1/2}(x-R)], \tag{10.6.10}$$

where $K_\varepsilon = \varepsilon_1/\varepsilon_2$ is the criterion of the thermal activity of the plate; $h = (1 - K_\varepsilon)/(1 + K_\varepsilon)$ is a dimensionless number, the absolute value of which is always less than unity ($|h| < 1$).

After the following manipulations

$$\frac{1}{\exp[(s/a_1)^{1/2}R] + h\exp[-(s/a_1)^{1/2}R]} = \frac{\exp[-(s/a_1)^{1/2}R]}{1 + h\exp[-2(s/a_1)^{1/2}R]}$$

$$= \sum_{n=1}^{\infty} (-h)^{n-1}\exp[-(2n-1)(s/a_1)^{1/2}R] \tag{10.6.11}$$

the series obtained converges rapidly for $|h| < 1$.

Thus, the solutions may be rewritten as

$$T_1(x, s) = \frac{t_0}{s} - \frac{(t_0 - t_a)}{s(1 + K_\varepsilon)} \sum_{n=1}^{\infty} (-h)^{n-1} \times \{\exp[-(s/a_1)^{1/2}[(2n-1)R - x]] + \exp[-(s/a_1)^{1/2}[(2n - 1(R + x]]\}, \quad (10.6.12)$$

$$T_2(x, s) - \frac{t_a}{s} = \frac{(t_0 - t_a)K_\varepsilon}{s(1 + K_\varepsilon)} \sum_{n=1}^{\infty} (-h)^{n-1} [\exp[-(2n-2)R(s/a_1)^{1/2}] - \exp[-2nR(s/a_1)^{1/2}]] \exp[-(s/a_2)^{1/2}(x - R)]$$

$$= \frac{(t_0 - t_a)K_\varepsilon}{s(1 + K_\varepsilon)} \Big\{\exp[-(s/a_2)^{1/2}(x - R)] - (1+h) \sum_{m=1}^{\infty} (-h)^{m-1} \times \exp[-(2mR(a_2/a_1)^{1/2} + x - R)(s/a_2)^{1/2}]\Big\}. \quad (10.6.13)$$

Using the table of transforms, we obtain the solution in the following form

$$\theta_1 = \frac{t_1(x, \tau) - t_a}{t_0 - t_a} = 1 - \frac{1}{1 + K_\varepsilon} \sum_{n=1}^{\infty} (-h)^{n-1} \Big\{\operatorname{erfc} \frac{(2n-1)R - x}{2(a_1\tau)^{1/2}} + \operatorname{erfc} \frac{(2n-1)R + x}{2(a_1\tau)^{1/2}}\Big\}. \quad (10.6.14)$$

$$\theta_2 = \frac{t_2(x, \tau) - t_a}{t_0 - t_a} = \frac{K_\varepsilon}{1+K_\varepsilon} \operatorname{erfc} \frac{x - R}{2(a_2\tau)^{1/2}} - \frac{K_\varepsilon(1+h)}{1 + K_\varepsilon} \sum_{n=1}^{\infty} (-h)^{n-1} \operatorname{erfc} \frac{x - R + 2nR(a_2/a_1)^{1/2}}{2(a_2\tau)^{1/2}}. \quad (10.6.15)$$

Solutions (10.6.14) and (10.6.15) satisfy the appropriate equations, the initial conditions (since $\sum_{n=1}^{\infty}(-h)^{n-1} = 1/(1 + h) = (1 + K_\varepsilon)/2)$ and boundary conditions.

Let $K_\varepsilon \to 0$ $(h \to 1)$ then the thermal activity of the plate is infinitesimal compared to that of the medium; the process of heat transfer from the infinite plate is accompanied by a rapid fall of the plate surface temperature to that of the medium. Further, if $K_\varepsilon = 0$ $(h = 1)$ is assumed, then $\theta_2(x, \tau) = 0$, and the solution (10.6.14) will become of the following form

$$\theta_1 = 1 - \sum_{n=1}^{\infty} (-1)^{n-1} \left[\operatorname{erfc} \frac{(2n-1)R - x}{2(a_1\tau)^{1/2}} + \operatorname{erfc} \frac{(2n-1)R + x}{2(a_1\tau)^{1/2}}\right]. \quad (10.6.16)$$

Equation (10.6.16) is the solution of the problem of cooling an infinite plate when the temperature on its surface instantaneously falls to that of the medium remaining constant during the entire cooling process (boundary condition of the first kind). This solution was obtained in Chapter 4, Section 3.

The temperature on the surface of contact in the case of finite values of K_ε is

$$\theta_s = \frac{K_\varepsilon}{1+K_\varepsilon} - \frac{2K_\varepsilon}{(1+K_\varepsilon)^2}\sum_{n=1}^{\infty}(-h)^{n-1}\operatorname{erfc}\frac{nR}{(a_1\tau)^{1/2}}. \tag{10.6.17}$$

Thus, $\theta(R, \tau)$ continuously decreases; its maximum value corresponds to the initial moment (at $\tau \to 0$ the function $\operatorname{erfc}\{nR/(a_1\tau)^{1/2}\}$ approaches zero), i.e.,

$$\theta_{s,0} = K_\varepsilon/(1+K_\varepsilon). \tag{10.6.18}$$

This is precisely the same relative temperature which sets in on the contact surface of two semi-infinite bodies (see Section 10.1). The quantity

$$1 - [K_\varepsilon/(1+K_\varepsilon)] = 1 - \theta(R, 0) = X \tag{10.6.19}$$

is referred to as a cooling effect. *The rate of change of $\theta_{s,0}$ depends on the thermal diffusivity and the thickness of the plate.* With small values of $\mathrm{Fo}_1 = a_1\tau/R^2$, $\theta_{s,0}$ changes slightly.

Analysis of solution (10.6.15) shows that the temperature at any point of the medium first increases, then reaches its maximum, and then diminishes. The highest possible temperature on the contact surface ($x = R$) sets in instantaneously, and the time required for reaching the maximum temperature increases with x ($x > R$).

When the values of $\mathrm{Fo}_2 = a_2\tau/R^2$ are small, we need use only the first term of the whole series of solution (10.6.15); it is then possible to write the approximate equality

$$\theta_2 \approx \frac{K_\varepsilon}{1+K_\varepsilon}\operatorname{erfc}\frac{(x/R)-1}{2(\mathrm{Fo}_2)^{1/2}} - \frac{2K_\varepsilon}{(1+K_\varepsilon)^2}\operatorname{erfc}\frac{(x/R)-1+2K_a^{-1/2}}{2(\mathrm{Fo}_2)^{1/2}}. \tag{10.6.20}$$

If the derivative $\partial\theta_2/\partial\,\mathrm{Fo}_2$ is assumed to be zero, then an equation is obtained which can be used for determining $(\mathrm{Fo}_2)_{\max}$ which is the value of the Fourier number corresponding to the maximum temperature

$$(\mathrm{Fo}_2)_{\max} = \frac{1+\left(\dfrac{x}{R}-1\right)K_a^{1/2}}{K_a \ln\left\{\dfrac{2}{1+K_\varepsilon}\left[1+\dfrac{2}{K_a^{1/2}\left(\dfrac{x}{R}-1\right)}\right]\right\}}. \tag{10.6.21}$$

Relation (10.6.21) shows that the relative time required for the maximum temperature to set in increases with the relative coordinate; for $x = R$, $(\mathrm{Fo}_2)_{\max} = 0$.

c. The Same Problem as (a) but with Heat Source w = const. The problem is complicated by the introduction of a positive heat source with the strength w kcal/m³ hr. The differential equation for the medium remains the same as in the previous case, and for a plate it is of the form

$$\frac{\partial t_1(x,\tau)}{\partial \tau} = a_1 \frac{\partial^2 t_1(x,\tau)}{\partial x^2} + \frac{w}{c\gamma}. \tag{10.6.22}$$

Boundary conditions remain the same (conditions (10.6.3)–(10.6.6)).

The transformed solutions are obtained in the form (a method for the solution of such problems is described in detail in the previous chapter)

$$\begin{aligned} T_1(x,s) &= \frac{t_0}{s} + \frac{w}{s^2 c\gamma} - \left(\frac{t_0 - t_a}{s} + \frac{w}{s^2 c\gamma}\right) \frac{1}{1+K_\varepsilon} \sum_{n=1}^{\infty} (-h)^{n-1} \\ &\times \{\exp[-((2n-1)R - x)(s/a_1)^{1/2}] \\ &+ \exp[-(s/a_1)^{1/2}((2n-1)R + x)]\}, \end{aligned} \tag{10.6.23}$$

$$\begin{aligned} T_2(x,s) - \frac{t_a}{s} &= \left(\frac{t_0 - t_a}{s} + \frac{w}{s^2 c\gamma}\right) \left\{\frac{K}{1+K_\varepsilon} \exp[-(s/a_2)^{1/2}(x - R)] \right. \\ &\left. - \frac{K_\varepsilon(1+h)}{1+K_\varepsilon} \sum_{n=1}^{\infty} (-h)^{n-1} \exp[-(s/a_2)^{1/2}(x - R + 2nR)K_a^{-1/2}]\right\} \end{aligned} \tag{10.6.24}$$

For the inversion of the transform, we use relations (55) and (56) of Appendix 5.

$$\begin{aligned} \theta_1 &= \frac{t_1(x,\tau) - t_a}{t_0 - t_a} \\ &= 1 + \mathrm{Po}\,\mathrm{Fo}_1 - \frac{1}{1+K_\varepsilon} \sum_{n=1}^{\infty} (-h)^{n-1} \\ &\times \left\{\operatorname{erfc} \frac{(2n-1) \mp (x/R)}{2(\mathrm{Fo}_1)^{1/2}} + 4\mathrm{Po}\,\mathrm{Fo}_1\, \mathrm{i}^2 \operatorname{erfc} \frac{(2n-1) \mp (x/R)}{2(\mathrm{Fo}_1)^{1/2}}\right\}, \end{aligned} \tag{10.6.25}$$

$$\begin{aligned} \theta_2 &= \frac{t_2(x,\tau) - t_a}{t_0 - t_a} \\ &= \frac{K_\varepsilon}{1+K_\varepsilon} \left(\operatorname{erfc} \frac{(x/R) - 1}{2(\mathrm{Fo}_2)^{1/2}} + 4\mathrm{Po}\,\mathrm{Fo}_2\, \mathrm{i}^2 \operatorname{erfc} \frac{(x/R) - 1}{2(\mathrm{Fo}_2)^{1/2}}\right) - \left[\frac{K_\varepsilon(1+h)}{1+K_\varepsilon}\right] \\ &\times \sum_{n=1}^{\infty} (-h)^{n-1} \left[\operatorname{erfc} \frac{2nK_a^{-1/2} + (x/R) - 1}{2(\mathrm{Fo}_2)^{1/2}} \right. \\ &\left. + 4\mathrm{Po}\,\mathrm{Fo}_2\, \mathrm{i}^2 \operatorname{erfc} \frac{2nK_a^{-1/2} + (x/R) - 1}{2(\mathrm{Fo}_2)^{1/2}}\right], \end{aligned} \tag{10.6.26}$$

where Po is the Pomerantsev criterion

$$\mathrm{Po} = \frac{wR^2}{\lambda(t_0 - t_a)} = \frac{wR^2}{c\gamma(t_0 - t_a)a}. \tag{10.6.27}$$

Here the notation $-\Phi(\mp z) = -\Phi(-z) - \Phi(+z)$ is introduced.

d. The Same Problem as (a) but with an Instantaneous Heat Source. The statement of the problem is following. An infinite plate placed in an infinite medium receives at the initial moment a heating pulse from an instantaneous source located in the center of the plate ($x_1 = 0$). The strength of the instantaneous heat source is $Q = bc\gamma$ (kcal/m^2). The initial temperature of the plate and the medium is zero.

Differential equations for the plate and the medium as well as boundary conditions are the same as in the previous case [Eqs. (10.6.1) and (10.6.2) and conditions (10.6.4)–(10.6.6)]. The initial condition is

$$t_1(x, 0) = t_2(x, 0) = 0.$$

The problem is solved by the method which is discussed in detail in Chapter 9, Section 2. A solution for the transform $T(x, s)$ is obtained in the form

$$\begin{aligned} T_1(x, s) &= \frac{b}{(a_1 s)^{1/2}} \exp[-(s/a_1)^{1/2}x] \\ &\quad - \frac{bh}{(a_1 s)^{1/2}} \frac{\exp[-(s/a_1)^{1/2}(R - x)] + \exp[-(s/a_1)^{1/2}(R + x)]}{\exp[(s/a_1)^{1/2}R] + h\exp[-(s/a_1)^{1/2}R]} \\ &= \frac{b}{(a_1 s)^{1/2}} \Big(\exp[-(s/a_1)^{1/2}x] - h\sum_{n=1}^{\infty}(-h)^{n-1} \\ &\quad \times \{\exp[-(s/a_1)^{1/2}(2nR - x)] + \exp[-(s/a_1)^{1/2}(2nR + x)]\}\Big), \end{aligned} \tag{10.6.28}$$

$$\begin{aligned} T_2(x, s) &= \frac{b(1 - h)}{(a_1 s)^{1/2}[\exp\{(s/a_1)^{1/2}R\} + h\exp[(s/a_1)^{1/2}R]]} \\ &\quad \times \exp[-(s/a_2)^{1/2}(x - R)] = \frac{b(1 - h)}{(a_1 s)^{1/2}} \sum_{n=1}^{\infty}(-h)^{n-1} \\ &\quad \times \exp\{-(s/a_2)^{1/2}[x - R + (2n - 1)(a_2/a_1)^{1/2}R]\}. \end{aligned} \tag{10.6.29}$$

Solutions (10.6.28) and (10.6.29) are taken from the table of transforms and the inversion is fulfilled directly:

$$\begin{aligned} t_1(x, \tau) &= \frac{b}{R(\pi\,\mathrm{Fo}_1)^{1/2}} \Big\{\exp\Big[-\frac{x^2}{4R^2\,\mathrm{Fo}_1}\Big] - h\sum_{n=1}^{\infty}(-h)^{n-1} \\ &\quad \times \Big[\exp\Big[\frac{[2n - (x/R)]^2}{4\mathrm{Fo}_1}\Big] + \exp\Big[-\frac{[2n + (x/R)]^2}{4\mathrm{Fo}_1}\Big]\Big]\Big\}, \end{aligned} \tag{10.6.30}$$

$$t_2(x, \tau) = \frac{2bK_\varepsilon}{K_a^{1/2}(1 + K_\varepsilon)R(\pi \, \mathrm{Fo}_2)^{1/2}} \sum_{n=1}^{\infty} (-h)^{n-1}$$
$$\times \exp\left[-\frac{[(x/R) - 1 + (2n-1)K_a^{-1/2}]^2}{4\mathrm{Fo}_2}\right]. \qquad (10.6.31)$$

The analysis of solutions (10.6.30) and (10.6.31) shows that when the values of Fo are small, we may use a single term of the whole series and then the calculation formulas become simple.

The temperature $t_2(x, \tau)$ at any point of the infinite medium first increases, reaches a certain maximum value, and then decreases. If the derivative $\partial t_2(x, \tau)/\partial \tau$ is assumed to be zero, then for an approximate solution (in case when only the first term may be taken of the whole series (10.6.31) which is valid for rather low values of Fo) we obtain the relation

$$(\mathrm{Fo}_2)_{\max} = \tfrac{1}{2}[(x/R) - 1 + K_a^{-1/2}]^2, \qquad (10.6.32)$$

i.e., the plot of $(\mathrm{Fo}_2)_{\max}^{1/2}$ versus $(x - R)/R$ is a straight line which intercepts the ordinate at $(2K_a)^{-1/2}$. Thus, the shorter the relative coordinate $(x - R)/R$ the less the time required for the maximum temperature to set in.

10.7 Sphere (Symmetrical Problem)

a. Statement of the Problem. *A spherical body at the temperature t_0 is considered. At the initial moment the sphere is placed into an infinite medium with temperature $t_a < t_0$. The sphere is cooled by heat conduction. The temperature distribution at any moment is to be found.*

We have

$$\frac{\partial[rt_1(r, \tau)]}{\partial \tau} = a_1 \frac{\partial^2[rt_1(r, \tau)]}{\partial r^2} \quad (\tau > 0;\ 0 < r < R), \qquad (10.7.1)$$

$$\frac{\partial[rt_2(r, \tau)]}{\partial \tau} = a_2 \frac{\partial^2[rt_2(r, \tau)]}{\partial r^2} \quad (\tau > 0;\ R < r < \infty), \qquad (10.7.2)$$

$$t_1(r, 0) = t_0, \qquad t_2(r, 0) = t_a, \qquad (10.7.3)$$

$$-K_\lambda \frac{\partial t_1(R, \tau)}{\partial r} = -\frac{\partial t_2(R, \tau)}{\partial r}, \qquad (10.7.4)$$

$$t_1(R, \tau) = t_2(R, \tau), \qquad (10.7.5)$$

$$\frac{\partial t_1(0, \tau)}{\partial r} = \frac{\partial t_2(\infty, \tau)}{\partial r} = 0, \quad t_1(0, \tau) \neq \infty, \qquad (10.7.6)$$

(subscript 1 refers to the sphere; subscript 2 to the medium).

b. Solution of the Problem. The transformed solution can be written

$$(t_0/s) - T_1(r, s) = (A/r)\sinh(s/a)^{1/2}r\,, \tag{10.7.7}$$

$$T_2(r, s) - (t_a/s) = (B_1/r)\exp[-(s/a_2)^{1/2}r]\,, \tag{10.7.8}$$

where

$$B_1 = ((t_0 - t_a)R/s)\exp[(s/a_2)^{1/2}R] - A\sinh(s/a_1)^{1/2}R\exp[(s/a_2)^{1/2}R]\,, \tag{10.7.9}$$

$$A = \frac{(t_0 - t_a)R}{s\Phi(s)}(1 + (s/a_2)^{1/2}R)\,, \tag{10.7.10}$$

$$\Phi(s) = K_\lambda(s/a_1)^{1/2}R\cosh(s/a_1)R + (1 - K_\lambda + (s/a_2)^{1/2}R)\sinh(s/a_1)^{1/2}R. \tag{10.7.11}$$

An approximate solution will be found. For this purpose we shall write approximate expressions for $T_1(r, s)$ and $T_2(r, s)$, which are valid for small values of Fo or large values of $(s/a)^{1/2}R$. We have

$$(t_0/s) - T_1(r, s) \approx [A_1(t_0 - t_a)/rs][\exp[(s/a_1)^{1/2}r] - \exp[-(s/a_1)^{1/2}r]]\,, \tag{10.7.12}$$

$$\begin{aligned} T_2(r, s) - (t_a/s) \approx {} & [(t_0 - t_a)/s]\{(R/r) - (A_1/r)(\exp[(s/a_1)^{1/2}R] \\ & - \exp[-(s/a_1)^{1/2}R])\}\exp[-(s/a_2)^{1/2}(r - R)]\,, \end{aligned} \tag{10.7.13}$$

where

$$A_1 = \frac{R(K_\lambda(s/a_1)^{1/2}R + K_\varepsilon)}{K_\lambda(s/a_1)^{1/2}R(1 + K_\varepsilon) + K_\varepsilon(1 - K_\lambda)}\exp[-(s/a_1)^{1/2}R]. \tag{10.7.14}$$

The approximate solutions obtained are tabulated transforms. Hence, for the inversion, we may write

$$\begin{aligned} \theta_1 = {} & \frac{t_0 - t_1(r, \tau)}{t_0 - t_a} \\ & - \frac{R}{r}\left(\frac{1}{1 - K_\lambda}\right)\left\{\frac{-K_\lambda - K_\varepsilon}{K_\varepsilon + 1}\exp\left[\Pi_1{}^2\,\mathrm{Fo}_1 - \Pi_1\left(1 + \frac{r}{R}\right)\right]\right. \\ & \left.\times\,\mathrm{erfc}\left(\frac{1 \mp (r/R)}{2(\mathrm{Fo}_1)^{1/2}} - \Pi_1(\mathrm{Fo}_1)^{1/2}\right) - \mathrm{erf}\,\frac{1 \mp (r/R)}{2(\mathrm{Fo}_1)^{1/2}}\right\}, \end{aligned} \tag{10.7.15}$$

$$\begin{aligned} \theta_2 = {} & \frac{t_2(r, \tau) - t_a}{t_0 - t_a} \\ = {} & \frac{R}{r}\left\{\mathrm{erfc}\,\frac{(r/R) - 1}{2(\mathrm{Fo}_2)^{1/2}} + \frac{1}{K_\lambda - 1}\,\mathrm{erfc}\,\frac{(r/R) - 1}{2(\mathrm{Fo}_2)^{1/2}}\right. \end{aligned}$$

$$- \frac{K_\lambda + K}{(K_\lambda - 1)(K_\varepsilon + 1)} \exp[\Pi_2^2 \,\mathrm{Fo}_2 - \Pi_2((r/R) - 1)]$$

$$\times \operatorname{erfc}\left(\frac{(r/R) - 1}{2(\mathrm{Fo}_2)^{1/2}} - \Pi_2(\mathrm{Fo}_2)^{1/2}\right) - \frac{1}{K_\lambda - 1} \operatorname{erfc} \frac{(r/R) - 1 + 2K_a^{-1/2}}{2(\mathrm{Fo}_2)^{1/2}}$$

$$+ \frac{K_\lambda + K_\varepsilon}{(K_\lambda - 1)(K_\varepsilon + 1)} \exp[\Pi_2^2 \,\mathrm{Fo}_2 - \Pi_2((r/R) - 1 + 2K_a^{-1/2})]$$

$$\times \operatorname{erfc}\left(\frac{(r/R) - 1 + 2K_a^{-1/2}}{2(\mathrm{Fo}_2)^{1/2}} - \Pi_2(\mathrm{Fo}_2)^{1/2}\right)\Bigg\}, \qquad (10.7.16)$$

where

$$\Pi_1 = \frac{K_\lambda - 1}{K_\lambda + (K_\lambda/K_\varepsilon)}, \qquad K_\lambda = \frac{\lambda_1}{\lambda_2}, \qquad K_\varepsilon = \frac{\varepsilon_1}{\varepsilon_2}, \qquad \Pi_2 = \frac{K_\lambda - 1}{1 + K_\varepsilon}, \qquad (10.7.17)$$

and the notation $\Phi(\mp z) = \Phi(-z) - \Phi(+z)$ is introduced. These expressions are only valid for small values of the Fourier number.

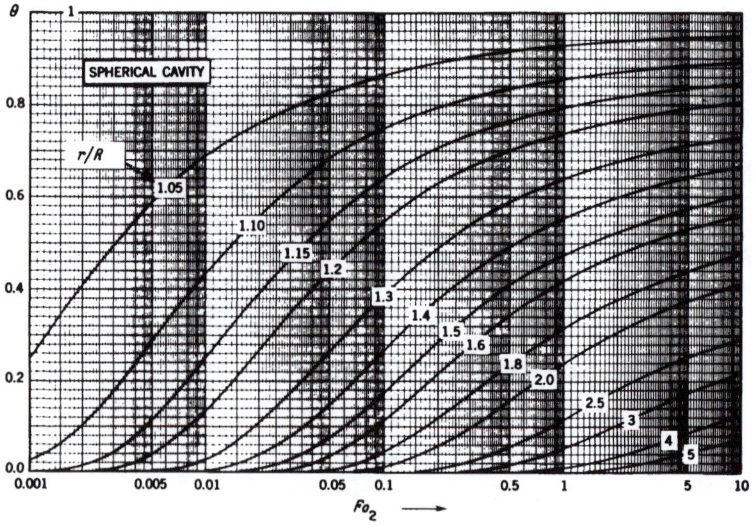

Fig. 10.9. θ [$\theta = (t_2 - t_0)/(t_a - t_0)$ versus Fo_2 for various values of $R^* = r/R$ from 1 to 5 [102].

Figure 10.9 comprises diagrams of the relative excess temperature at different points of the medium for the case when the temperature in the plane of contact is maintained constant during the entire process of heating $t_1(R, \tau) = t_2(R, \tau) = t_a = \text{const}$ when $t_a > t_0$.

10.8 Infinite Cylinder

a. Statement of the Problem. *An infinite cylinder of radius R at the temperature t_0 is considered. At the initial moment it is placed into an infinite medium with a temperature $t_a < t_0$. Heat transfer between the cylindrical surface and the medium follows the heat conduction law. The temperature distribution inside the cylinder and over the medium at any moment is to be obtained.*

We have

$$\frac{\partial t_1(r, \tau)}{\partial \tau} = a_1 \left(\frac{\partial^2 t_1(r, \tau)}{\partial r^2} + \frac{1}{r} \frac{\partial t_1(r, \tau)}{\partial r} \right) \quad (\tau > 0;\ 0 < r < R), \tag{10.8.1}$$

$$\frac{\partial t_2(r, \tau)}{\partial \tau} = a_2 \left(\frac{\partial^2 t_2(r, \tau)}{\partial r^2} + \frac{1}{r} \frac{\partial t_2(r, \tau)}{\partial r^2} \right) \quad (\tau > 0;\ R < r < \infty), \tag{10.8.2}$$

$$t_1(r, 0) = t_0, \qquad t_2(r, 0) = t_a, \tag{10.8.3}$$

$$\frac{\partial t_1(0, \tau)}{\partial r} = 0, \qquad t_1(0, \tau) \neq \infty, \tag{10.8.4}$$

$$\frac{\partial t_2(\infty, \tau)}{\partial r} = 0, \tag{10.8.5}$$

$$t_1(R, \tau) = t_2(R, \tau), \qquad -K_\lambda \frac{\partial t_1(R, \tau)}{\partial r} = -\frac{\partial t_2(R, \tau)}{\partial r}. \tag{10.8.6}$$

b. Solution of the Problem. The transformed solution with conditions (10.8.3)–(10.8.5) are obtained as

$$T_1(x, s) - (t_a/s) = \frac{t_0 - t_a}{s} \left[1 - \frac{K_a^{1/2}}{\Phi(s)} K_1((s/a_2)^{1/2}R) I_0((s/a_1)^{1/2}r) \right], \tag{10.8.7}$$

$$T_2(x, s) - (t_a/s) = \frac{(t_0 - t_a)K_\lambda}{s\Phi(s)} I_1((s/a_1)^{1/2}R) K_0((s/a_2)^{1/2}r), \tag{10.8.8}$$

where $K_1(z)$ and $K_0(z)$ are modified Bessel functions of the second kind and the first and the zeroth order respectively:

$$\Phi(s) = K_a^{1/2} I_0((s/a_1)^{1/2}R) K_1((s/a_2)^{1/2}R) + K_\lambda I_1((s/a_1)^{1/2}R) K_0((s/a_2)^{1/2}R). \tag{10.8.9}$$

Using the inverse Laplace transformation and transforming the integral in the complex plane into the integral of the substantial variable gives us

$$\theta_1 = \frac{t_1(r, \tau) - t_a}{t_0 - t_a}$$

$$= \frac{4}{\pi^2} K_\lambda \int_0^\infty \frac{J_0\left(\mu \dfrac{r}{R}\right) J_1(\mu)}{\mu^2[\varphi^2(\mu) + \psi^2(\mu)]} \exp[-\mu^2 \,\mathrm{Fo}_1]\, d\mu, \tag{10.8.10}$$

$$\theta_2 = \frac{t_2(r, \tau) - t_a}{t_0 - t_a}$$

$$= \frac{2K_\lambda}{\pi \lambda_2 \sqrt{a_2}} \int_0^\infty \frac{J_1(\mu)\left[J_0\left(K_a^{1/2}\mu \dfrac{r}{R}\right) \varphi(\mu) - Y_0\left(K_a^{1/2}\mu \dfrac{r}{R}\right) \psi(\mu)\right]}{\mu[\varphi^2(\mu) + \psi^2(\mu)]}$$

$$\times \exp[-\mu^2 \,\mathrm{Fo}_1]\, d\mu, \tag{10.8.11}$$

where

$$\varphi(\mu) = K_\lambda J_1(\mu) Y_0(K_a^{1/2}\mu) - K_a^{1/2} J_0(\mu) Y_1(K_a^{1/2}\mu),$$
$$\psi(\mu) = K_\lambda J_1(\mu) J_0(K_a^{1/2}\mu) - K_a^{1/2} J_0(\mu) J_1(K_a^{1/2}\mu). \tag{10.8.12}$$

Figures 10.10 and 10.11 represent the diagrams for the change in the relative excess temperature of the medium at different points and the Fourier number ($\mathrm{Fo}_2 = a_2\tau/R^2$) for the case when the temperature of the boundary of contact is maintained constant during the entire process of heating, i.e., $t_1(R, \tau) = t_2(R, \tau) = t_a = \mathrm{const.}$

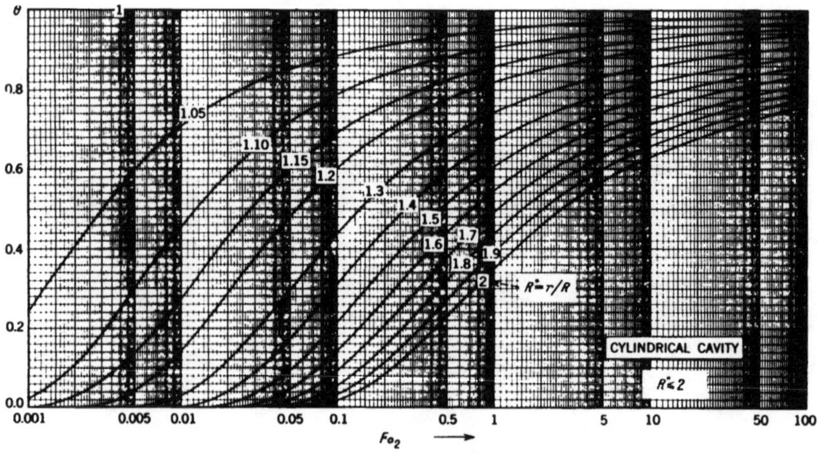

Fig. 10.10. θ [$\theta = (t_2 - t_0)/(t_a - t_0)$] versus Fo_2 for $R^* \leq 2$ [102].

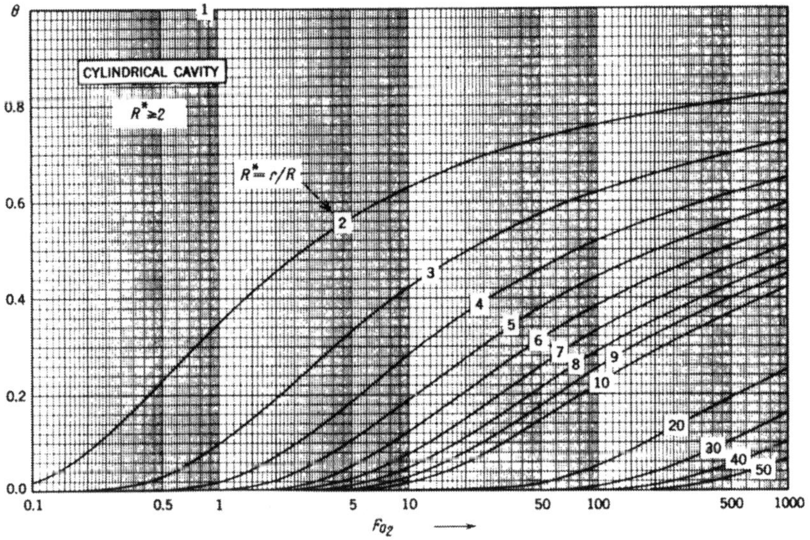

Fig. 10.11. θ $[\theta = (t_2 - t_0)/(t_a - t_0)]$ versus Fo_2 for $R^* \geq 2$ [102].

c. The Same Problem as (a) but with a Heat Source w = const. The problem is complicated by the introduction of a positive heat source with the strength w (kcal/m² hr).

The differential equation for the medium remains the same and for the cylinder it will be of the form

$$\frac{\partial t_1(r, \tau)}{\partial \tau} = a_1 \left(\frac{\partial^2 t_1(r, \tau)}{\partial r^2} + \frac{1}{r} \frac{\partial t_1(r, \tau)}{\partial r} \right) + \frac{w}{c_1 \gamma_1}. \qquad (10.8.13)$$

Boundary conditions will be the same [Eqs. (10.8.4)–(10.8.6)] and the initial temperature of the medium and the cylinder is assumed to be zero:

$$t_1(r, 0) = t_2(r, 0) = t_0 = t_a = 0. \qquad (10.8.14)$$

Then the solution for the transform will be of the form

$$T_1(r, s) = \frac{a_1 w}{\lambda_1 s^2} - \frac{w a_1}{\lambda_1 s^2 \Phi(s)} K_1((s/a_2)^{1/2} R) I_0((s/a_1)^{1/2} r), \qquad (10.8.15)$$

$$T_2(r, s) = \frac{w a_1 I_1((s/a_1)^{1/2} R)}{\lambda_2 s^2 \Phi(s)} K_0((s/a_2)^{1/2} r). \qquad (10.8.16)$$

The quantity $\Phi(s)$ is determined from formula (10.8.9).

The solution for the inversion may be written

$$t_1(r, \tau) = \frac{4wR^2}{\lambda_1 \pi^2} \int_0^\infty (1 - \exp[-\,\mathrm{Fo}_1\, \mu^2]) \frac{J_0\left(\frac{\mu^2}{R}\right) J_1(\mu)\, d\mu}{\mu^4[\varphi^2(\mu) + \psi^2(\mu)]}, \tag{10.8.17}$$

$$t_2(r, \tau) = \frac{2wR^2}{\lambda_2{}^2 \pi \sqrt{a_2}} \int_0^\infty (1 - \exp[-\,\mathrm{Fo}_1\, \mu^2]) \frac{J_1(\mu)}{\mu^3[\varphi^2(\mu) + \psi^2(\mu)]}$$
$$\times [J_0(K_a^{1/2}\mu(r/R))\varphi(\mu) - Y_0\{K_a^{1/2}\mu(r/R)\}\psi(\mu)]\, d\mu, \tag{10.8.18}$$

where $\varphi(\mu)$ and $\psi(\mu)$ are defined by formulas (10.8.12).

10.9 Heat Transfer between a Body and a Liquid Flow

In these concluding sections of this chapter we shall consider the problem of a semi-infinite body in a fluid flow, i.e., that of unsteady-state convective heat transfer.

a. Statement of the Problem. *Consider a liquid flowing through a rectangular tube. The liquid is an incompressibile fluid in steady laminar motion. In the lower wall of the tube there is a hole, into which a finite rod is introduced, whose open surface is a part of the tube surface (see* Fig. 10.12).

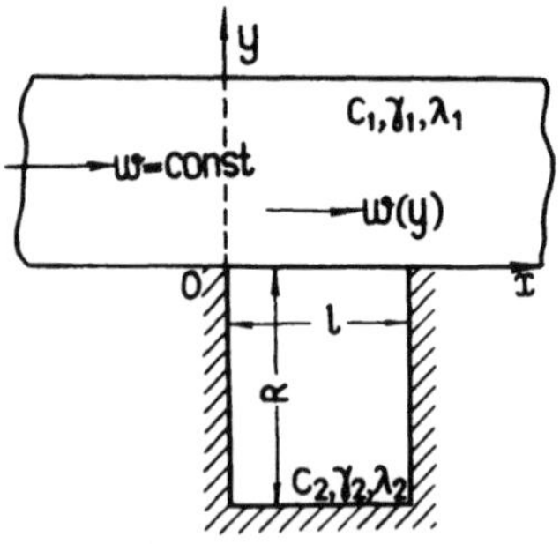

Fig. 10.12. Fluid flow around a semi-infinite body.

All the remaining surfaces of the rod are thermally insulated. The rod was preliminarily heated to a constant temperature $t_2(0) = t_0 =$ const. *At the initial time instant, the rod is introduced into the hydrodynamic tube and cooled by a liquid flow, whose temperature is* $t_a =$ const. *The specific heat flux in unsteady-state heat transfer is to be found.*

The temperature of the flowing liquid is taken as the reference temperature. The rod length in the x direction is l and the thickness is R (see

Fig. 10.12). The resulting transfer between the rod and liquid flow is described by the system of differential equations (10.1) and (10.2) given in the introduction to this chapter.

Boundary and initial conditions may be written as

$$t_1(x, y, 0) = t_a = 0; \qquad t_2(x, y, 0) = t_{20} = \text{const}, \tag{10.9.1}$$

$$t_1(x, \infty, \tau) = t_a = 0; \qquad t_1(0, y, \tau) = t_a = 0, \tag{10.9.2}$$

$$t_1(x, 0, \tau) = t_2(x, 0, \tau), \tag{10.9.3}$$

$$-\lambda_1 \frac{\partial t_1(x, 0, \tau)}{\partial y} = -\lambda_2 \frac{\partial t_2(x, 0, \tau)}{\partial y}. \tag{10.9.4}$$

If the rod length is small or when $l/w \ll R^2/a_2$, formula (1.9.11) for averaging a temperature along x may be used. Then we shall have

$$\frac{\partial \bar{t}_1(y, \tau)}{\partial \tau} + w(y)\frac{t_1(l, y, \tau)}{l} = a_1 \frac{\partial^2 \bar{t}_1(y, \tau)}{\partial y^2}, \tag{10.9.5}$$

$$\frac{\partial \bar{t}_2(y, \tau)}{\partial \tau} = a_2 \frac{\partial^2 \bar{t}_2(y, \tau)}{\partial y^2}. \tag{10.9.6}$$

The temperature $t_1(x, y, \tau)$ is not a monotonically decreasing function of x for the quantity $t_1(l, y, \tau)$ may be expressed through $\bar{t}_1(y, \tau)$ with the help of the theorem on a mean value

$$\bar{t}_1(y, \tau) = t_1(l, y, \tau)(1 - (l^*/l)) \tag{10.9.7}$$

where l^* is some parameter.

We introduce some value of the effective velocity w_l according to the ratio

$$w_l = w(y)(1 - (l^*/l)) \tag{10.9.8}$$

and take it as a constant quantity as in the case of solution of heat transfer problems by the Krischer method [67]. Equation (10.9.5) may be written

$$\frac{\partial \bar{t}_1(y, \tau)}{\partial \tau} + \frac{w_l}{l}\bar{t}_1(y, \tau) = a_1 \frac{\partial^2 \bar{t}_1(y, \tau)}{\partial y^2}. \tag{10.9.5a}$$

The solution of Eqs. (10.9.5a) and (10.9.6), averaged with respect to x under boundary conditions (10.9.1)–(10.9.4), will be of the form

$$t_1(\xi, \eta) = \frac{1}{K_\lambda \sqrt{\pi}} \int_0^\xi \exp\left[-\xi^* - \frac{\eta^2}{4\xi^*}\right] \frac{\psi(\xi - \xi^*)}{\sqrt{\xi^*}}\, d\xi^*, \tag{10.9.9}$$

where

$$\xi = w_l \tau / l = (D/l)\,\mathrm{Pe}\,\mathrm{Fo}_1, \tag{10.9.10}$$

$$\eta = (w_l / a_2 l)^{1/2} y = [(\mathrm{Pe})^{1/2}/(Dl)^{1/2}]y, \tag{10.9.11}$$

$\mathrm{Pe} = w_l D/a_1$ is the Peclet number; $\mathrm{Fo}_1 = a_1\tau/R^2$ is the Fourier number, and D is the equivalent tube diameter, $K_\lambda = \lambda_1/\lambda_2$, $\psi(\xi)$ is the dimensionless heat flux near the body surface determined by the ratio

$$-K_\lambda \frac{\partial t_1(0, \xi)}{\partial \eta} = -\frac{\partial t_2(0, \xi)}{\partial \eta} = \psi(\xi). \tag{10.9.12}$$

A similar ratio may be also written for $t_2(\xi, \eta)$.

The following expression for a heat flux $\psi(\xi)$ is obtained

$$\psi(\xi) = 2K_\lambda \sum_n^{1+[N/\pi]} (1 - \mu_n^2)^{1/2} \exp[-\mu_n^2 \xi]\{NK_\varepsilon(1 - \mu_n^2)^{1/2}$$
$$+ \frac{N}{K_\varepsilon} \mu_n^2/(1 - \mu_n^2)^{1/2} + 1/(1 - \mu_n^2)\}^{-1} + \frac{K_\lambda}{\pi} e^{-\xi}$$
$$\times \int_0^\infty \frac{\sqrt{u}\,\exp[-\xi u]\,du}{1 + u + K_\varepsilon^2 u \cot^2(N(1 + u)^{1/2})}, \tag{10.9.13}$$

where μ_n are the roots of the characteristic equation

$$\mu \tan \mu N = K_\varepsilon(1 - \mu^2)^{1/2}. \tag{10.9.14}$$

Intermediate manipulations are omitted since they may be found (Luikov [67]) The numerical value of the characteristic numbers of μ_n is given in Table 10.1.

For small values of ξ ($\xi = w_l\tau/l \ll 1$) the last term plays an essential role in formula (10.9.13). $\psi(\xi)$ decreases proportionally to $1/\sqrt{\xi}$ with an increase in ξ and the terms entering into the sum become main terms.

In the limit at great values of $\xi (\xi > 1/\mu_1^2)$ only one term is left from the whole sum. The approximate ratio is obtained

$$\psi(\xi) \approx 2K_\lambda(1 - \mu_0^2)^{1/2} \exp[-\mu_0^2 \xi][NK_\varepsilon(1 - \mu_0^2)^{1/2}$$
$$+ \frac{N\mu_0^2}{K_\varepsilon(1 - \mu_0^2)^{1/2}} + \{1/(1 - \mu_0^2)\}]^{-1}. \tag{10.9.15}$$

The value of μ_0 is determined from Table 10.1. In a number of cases it is possible to obtain simple formulas for calculating μ_0. For example, with small K_ε, or $K_\varepsilon N \ll 1$

$$\mu_0^2 = N/K_\varepsilon. \tag{10.9.16}$$

If N is small and K_ε is an arbitrary quantity, then

$$\mu_0{}^2 = \tfrac{1}{2}(K_\varepsilon/N)^2[(1 + 4(N/K_\varepsilon)^2)^{1/2} - 1]. \tag{10.9.17}$$

We now find the specific heat flux as

$$q(\tau) = -\lambda_1 \frac{\partial \bar{t}_1(0, \tau)}{\partial y} = \lambda_2 t_0 \left(\frac{\text{Pe}}{Dl}\right)^{1/2} \psi\left(\frac{w_l \tau}{l}\right). \tag{10.9.18}$$

If formula (10.9.16) is used for μ_0, then from ratio (10.9.15) we have

$$q(\tau) = \lambda_1 t_0 \left(\frac{\text{Pe}}{Dl}\right)^{1/2} \exp\left[-\lambda_1 \left(\frac{\text{Pe}}{Dl}\right)^{1/2} \frac{\tau}{c_2\gamma_2 R}\right]. \tag{10.9.19}$$

Consequently, the specific heat flux decreases with time according to an exponential law.

From solutions (10.9.9) and (10.9.13), it follows that with unsteady convective heat transfer, the Nusselt number

$$\text{Nu} = \frac{q(\tau) l}{[\bar{t}(0, \tau) - t_a]\lambda_1} \tag{10.9.20}$$

will be a variable quantity dependent upon time and the parametric numbers K_λ, K_a, K_l as well as upon the Peclet number. It is therefore advisable that with unsteady heat transfer the generalized variable $\text{Nu}'(\tau)$ should be introduced

$$\text{Nu}'(\tau) = \frac{l q(\tau)}{\lambda_1(t_0 - t_a)}, \tag{10.9.21}$$

where some constant characteristic temperature equal to the initial temperature of the body is introduced instead of the variable temperature of the body surface $\bar{t}(0, \tau)$.

In this case $\text{Nu}'(\tau)$ will be directly proportional to $\psi(\xi)$. By analogy with the Stanton number, it is possible to introduce a number equal to the ratio of the specific heat flux from the body to the liquid related to the unit temperature difference $\Delta t = 1 (t_0 - t_a = 1)$, to the enthalpy heat flux transferred by the liquid flow and recorded with respect to

$$P(\tau) = \frac{q(\tau)}{c_1\gamma_1 w(t_0 - t_a)}. \tag{10.9.22}$$

In addition, it is possible to introduce an integral generalized variable characterizing unsteady convective heat transfer

TABLE 10.1. ROOTS OF THE CHARACTERISTIC EQUATION

$$\mu \tan \mu \; N = K_\varepsilon (1 - \mu^2)^{1/2}$$

N	K_ε	μ_0	μ_1	μ_2	μ_3	μ_4	μ_5	μ_6	μ_7	μ_8	μ_9	μ_{10}	μ_{11}	μ_{12}	μ_{13}	μ_{14}
50	0.01	0.0128	0.0659	0.1272	0.1896	0.2521	0.3148	0.3775	0.4402	0.5030	0.5658	0.6286	0.6913	0.7541	0.8169	0.8797
	0.02	0.0172	0.0685	0.1285	0.1906	0.2529	0.3154	0.3780	0.4406	0.5033	0.5661	0.6288	0.6915	0.7543	0.8171	0.8798
	0.03	0.0197	0.0708	0.1301	0.1916	0.2536	0.3159	0.3785	0.4410	0.5037	0.5664	0.6291	0.6918	0.7545	0.8172	0.8800
	0.04	0.0215	0.0729	0.1315	0.1925	0.2544	0.3166	0.3789	0.4415	0.5040	0.5666	0.6293	0.6920	0.7547	0.8174	0.8801
	0.05	0.0227	0.0746	0.1328	0.1934	0.2551	0.3177	0.3794	0.4418	0.5044	0.5669	0.6295	0.6922	0.7548	0.8175	0.8802
	0.10	0.0263	0.0806	0.1381	0.1977	0.2584	0.3199	0.3818	0.4438	0.5061	0.5684	0.6308	0.6932	0.7557	0.8182	0.8807
40	0.01	0.0149	0.0816	0.1560	0.2366	0.3129	0.3933	0.4717	0.5502	0.6286	0.7071	0.7856	0.8641	0.9426		
	0.02	0.0198	0.0844	0.1573	0.2377	0.3132	0.3939	0.4722	0.5505	0.6289	0.7074	0.7858	0.8642	0.9427		
	0.03	0.0230	0.0868	0.1576	0.2387	0.3139	0.3944	0.4727	0.5509	0.6294	0.7076	0.7860	0.8644	0.9427		
	0.04	0.0252	0.0891	0.1577	0.2397	0.3145	0.3950	0.4731	0.5513	0.6296	0.7078	0.7862	0.7645	0.9428		
	0.05	0.0263	0.0910	0.1579	0.2406	0.3153	0.3956	0.4736	0.5517	0.6299	0.7081	0.7864	0.8647	0.9429		
	0.10	0.0316	0.0983	0.1702	0.2450	0.3213	0.3984	0.4758	0.5595	0.6312	0.7093	0.7873	0.8654	0.9434		
30	0.01	0.0175	0.1080	0.2110	0.3152	0.4196	0.5242	0.6288	0.7334	0.8380	0.9426					
	0.02	0.0235	0.1106	0.2125	0.3161	0.4203	0.5247	0.6292	0.7337	0.8382	0.9427					
	0.03	0.0276	0.1133	0.2140	0.3171	0.4210	0.5252	0.6295	0.7340	0.8384	0.9429					
	0.04	0.0306	0.1158	0.2154	0.3181	0.4217	0.5258	0.6300	0.7343	0.8386	0.9429					
	0.05	0.0330	0.1180	0.2169	0.3190	0.4225	0.5263	0.6304	0.7346	0.8389	0.9431					
	0.10	0.0397	0.1268	0.2231	0.3236	0.4259	0.5289	0.6324	0.7381	0.8399	0.9437					
20	0.01	0.0216	0.1602	0.3157	0.4722	0.6289	0.7858	0.9427								
	0.02	0.0292	0.1631	0.3172	0.4730	0.6295	0.7862	0.9428								
	0.03	0.0352	0.1659	0.3187	0.4740	0.6302	0.7865	0.9430								
	0.04	0.0393	0.1686	0.3201	0.4749	0.6308	0.7869	0.9432								
	0.05	0.0429	0.1712	0.3215	0.4758	0.6314	0.7873	0.9433								
	0.10	0.0538	0.1818	0.3281	0.4802	0.6343	0.7893	0.9442								

TABLE 10.1. *(continued)*

N	K_ε	μ_0	μ_1	μ_2	μ_3	μ_4	μ_5	μ_6	μ_7	μ_8	μ_9	μ_{10}	μ_{11}	μ_{12}	μ_{13}	μ_{14}
10	0.01	0.0311	0.3172	0.6296	0.9428											
	0.02	0.0433	0.3201	0.6308	0.9432											
	0.03	0.0522	0.3229	0.6320	0.9435											
	0.04	0.0593	0.3257	0.6332	0.9439											
	0.05	0.0653	0.3284	0.6344	0.9442											
	0.10	0.0860	0.3411	0.6403	0.9459											
5	0.01	0.0448	0.6308													
	0.02	0.0632	0.6332													
	0.03	0.0774	0.6356													
	0.04	0.0893	0.6379													
	0.05	0.0998	0.6403													
	0.10	0.1302	0.6515													
1	0.01	0.0987														
	0.02	0.1394														
	0.03	0.1720														
	0.04	0.1968														
	0.05	0.2195														
	0.10	0.3029														

$$Q^*(\tau) = \int_0^\tau q(\tau)\, d\tau / c_2\gamma_2 t_0 R_v,$$

where $R_v = V/S$ is the ratio of the body volume to the surface area through which heat transfer occurs.

The function $Q^*(\tau)$ ranges from 0 to 1, it shows what fraction of the whole amount of heat has been already transferred.

10.10 Symmetrical System of Bodies Consisting of Three Infinite Plates

a. Statement of the Problem. *Consider a plate $2R$ thick ($2R = 2l_1$) which is in contact with two plates, each l_2 thick. Ther thermal properties of the extreme plates are identical but they differ from the properties of the middle plate* (see Fig. 10.13). *It is necessary to find the temperature field of the system of three contacting plates.* We have (see Fig. 10.13)

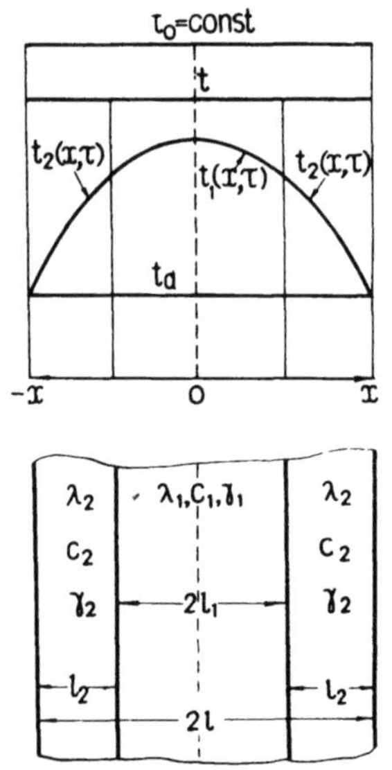

Fig. 10.13. System of three bodies in contact.

$$t_1(x, 0) = t_2(x, 0) = t_0 = \text{const}, \tag{10.10.1}$$

$$t_1(l_1, \tau) = t_2(l_1, \tau); \qquad \partial t_1(0, \tau)/\partial x = 0, \tag{10.10.2}$$

$$\lambda_1 \frac{\partial t_1(l_1, \tau)}{\partial x} = \lambda_2 \frac{\partial t_2(l_1, \tau)}{\partial x}, \tag{10.10.3}$$

$$\lambda_2 \frac{\partial t_2(l, \tau)}{\partial x} + \alpha[t_2(l, \tau) - t_a] = 0. \tag{10.10.4}$$

b. Solution of the Problem. Using the Laplace transform we obtain the solution in the form

$$\theta_1 = \frac{t_0 - t(x, \tau)}{t_0 - t_a}$$

$$= 1 - \sum_{n=1}^{\infty} \frac{2}{\mu_n \psi_n} \cos(\mu_n K_a^{-1/2} x/l_2) \exp[-\mu_n^2 K_l^2 K_a^{-1} \mathrm{Fo}_1], \tag{10.10.5}$$

$$\theta_2 = \frac{t_0 - t(x, \tau)}{t_0 - t_a}$$

$$= 1 - \sum_{n=1}^{\infty} \frac{2}{\mu_n \psi_n} \left[\cos \mu_n \frac{x - l_1}{l_2} \cos \mu_n K_a^{-1/2} K_l \right.$$

$$\left. - K_\varepsilon \sin \mu_n \frac{x - l_1}{l_2} \sin \mu_n K_a^{-1/2} K_l \right] \exp[-\mu_n^2 K_a^{-1} K_l^2 \mathrm{Fo}_1], \tag{10.10.6}$$

where

$$\psi_n = \left[\left(1 + K_\varepsilon K_l K_a^{-1/2} + \frac{1 + K_l}{\mathrm{Bi}}\right) \sin \mu_n + \mu_n \frac{1 + K_l}{\mathrm{Bi}} (1 + \right.$$

$$\left. + K_\varepsilon K_l K_a^{-1/2}) \cos \mu_n \right] \cos \mu_n K_l K_a^{-1/2} + \left[\left(1 + K_\varepsilon^{-1} K_l K_a^{-1/2} + \right.\right.$$

$$\left. + \frac{1 + K_l}{\mathrm{Bi}}\right) \cos \mu_n - \mu_n \frac{1 + K_l}{\mathrm{Bi}} (1 + K_\varepsilon^{-1} K_l K_a^{-1/2}) \sin \mu_n \right]$$

$$\times K_\varepsilon \sin \mu_n K_l K_a^{-1/2}, \tag{10.10.7}$$

μ_n are the roots of the characteristic equation

$$K \frac{\mu}{\mathrm{Bi}} (1 + K_l) \tan(\mu K_l K_a^{-1/2}) = 1 - \frac{\mu}{\mathrm{Bi}} (1 + K_l) \tan \mu$$

$$- K_\varepsilon \tan \mu \tan(\mu K_l K_a^{-1/2}), \tag{10.10.8}$$

$$\mathrm{Bi} = \alpha l/\lambda; \qquad K_l = l_1/l_2; \qquad \mathrm{Fo}_1 = a_1 \tau / l_1^2; \qquad K_a = a_1/a_2. \tag{10.10.9}$$

If $\mathrm{Bi} \to \infty$, boundary condition (10.10.4) takes the form

$$t_2(l, \tau) = t_a = \mathrm{const}. \tag{10.10.4a}$$

In this case the general solution remains but

$$\psi_n = (1 + K_\varepsilon K_l K_a^{-1/2}) \sin \mu_n \cos \mu_n K_l K_a^{-1/2} + K_\varepsilon (1 + K_\varepsilon^{-1} K_l K_a^{-1/2}) \cos \mu_n \sin \mu_n K_l K_a^{-1/2}, \qquad (10.10.10)$$

where μ_n are determined from the equation

$$K_\varepsilon \tan \mu \tan(\mu K_l K_a^{-1/2}) = 1. \qquad (10.10.11)$$

The general case of the system of three plates with a heat source $w_i(x, \tau)$ where $i = 1, 2, 3$, is discussed by Luikov and Mikhailov [73]. Here the general solution is presented for a multi-layer plate.

CHAPTER

11

TEMPERATURE FIELD OF BODY WITH CHANGING STATE OF AGGREGATION

A number of heat transfer processes are associated with changes in the state of aggregation or in the physical-chemical nature of the material. Here the thermodynamic coefficients of a body change discontinuously and the phase transitions may involve heat of melting (sorption and evaporation) or of chemical reactions. Such problems are of importance in metallurgy, thermal structure, and other applied sciences. As an example of this class of problem, and since it is the simplest, we shall deal in this section with the problem of freezing wet ground.

11.1 Freezing of Wet Ground

a. Statement of the Problem. *A moist ground is characterized by a certain assigned temperature distribution* $f(x)$. *At the initial moment, on the ground surface a certain temperature instantaneously sets in,* $t(0, \tau) = \varphi(\tau)$, *which is always lower than the freezing temperature* t_f. *This results in formation of a frozen layer of a variable thickness* $\xi = f(\tau)$. *The temperature of its lower moving boundary is always that of freezing. On this boundary a transition from one state of aggregation to another takes place which requires a heat of transformation* ϱ (kcal/kg). *Thus, the upper boundary* ($x = \xi$) *of the wet region has a constant freezing temperature, and the lower boundary* ($x = l$), *assumed to be at a great depth, has some constant temperature. Often the lower boundary of the freezing region is assumed to lie at an infinite depth* ($l = \infty$). *The thermal coefficients of the freezing and the wet regions* *are different.* Heat transfer through the ground is assumed to be transferred by heat conduction alone.

Thus the problem may be mathematically formulated as follows (see

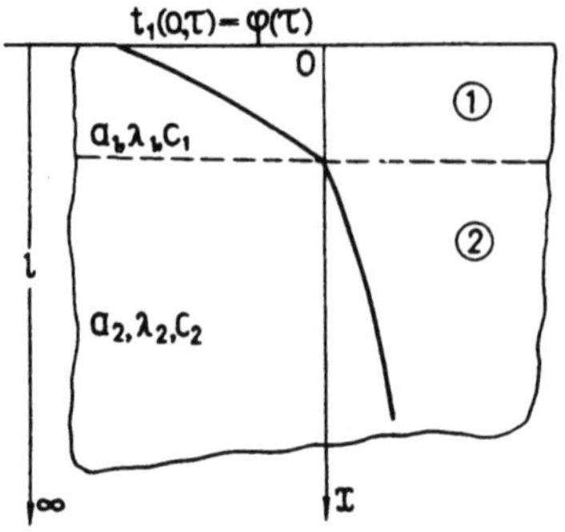

Fig. 11.1. Freezing moist ground.

Fig. 11.1; subscript 1 refers to the freezing region; subscript 2 to the unfrozen or wet region)

$$\frac{\partial t_1(x, \tau)}{\partial \tau} = a_1 \frac{\partial^2 t_1(x, \tau)}{\partial x^2} \quad (\tau > 0;\ 0 < x < \xi), \tag{11.1.1}$$

$$\frac{\partial t_2(x, \tau)}{\partial \tau} = a_2 \frac{\partial^2 t_2(x, \tau)}{\partial x^2} \quad (\tau > 0;\ \xi < x < \infty), \tag{11.1.2}$$

$$t_2(x, 0) = f(x) \quad [\text{since } \xi(0) = 0], \tag{11.1.3}$$

$$t_1(0, \tau) = \varphi(\tau), \tag{11.1.4}$$

$$t_1(\xi, \tau) = t_2(\xi, \tau) = t_f = \text{const}, \tag{11.1.5}$$

$$\partial t_2(\infty, \tau)/\partial x = 0. \tag{11.1.6}$$

On the interface

$$\lambda_1 \frac{\partial t_1(\xi, \tau)}{\partial x} - \lambda_2 \frac{\partial t_2(\xi, \tau)}{\partial x} = \varrho W \gamma_2 \frac{d\xi}{d\tau}, \tag{11.1.7}$$

where W is the moisture content of the ground (kg/kg; the weight of moisture in a unit weight of absolutely dry ground),[1] γ_2 is the density of the ground (kg/m^3). In the frozen ground there are two regions (that of the frozen and the moist ground), the temperature changes in which are

[1] The moisture is assumed to freeze at the temperature t_f, i.e., it is free moisture.

governed by heat conduction equations (11.1.1) and (11.1.2) coupled with boundary conditions (11.1.5) and (11.1.6). Thus, the problem of ground freezing may be formulated as that of the conjugation of two temperature fields with the special condition of a moving boundary.

The problem is simplified to some extent if the temperature of the melting region is assumed to be uniform throughout, to remain the same during the whole heat transfer process, and to be equal to $t_0 = t_2$. Physically, this means that the wet region contains a liquid in which the temperature is constant because of perfect convection. In this case, the problem of ground freezing reduces to that of ice formation in stagnant water.

The condition (11.1.7) becomes of the form

$$\lambda_1 \frac{\partial t_1(\xi, \tau)}{\partial x} = \varrho \gamma_2 \frac{d\xi}{d\tau}. \tag{11.1.8}$$

The main difficulty to the solution is the fact that it becomes one of the class of nonlinear problems by condition (11.1.7) or (11.1.8), i.e., of those with nonlinear boundary conditions. Its solution is discussed in more than 50 original works reported during the last century. The opinion is widely held that in its simplest form, the problem of ground freezing was primarily solved by the Austrian mathematician, Stefan. However, this same problem was treated in 1831 by Lame and Clapeyron, Members of the Russian Academy of Sciences [70]. For the solution, they assumed $t_2(x, 0) = t_0 = \text{const}$ (the initial condition) and $t_1(0, \tau) = 0$ (the surface temperature is a reference one).

b. Solution of Lame and Clapeyron. These workers solved the problem with the simplified boundary condition (11.1.8) assuming the temperature of the water to be at the freezing point, i.e.,

$$t_1(\xi, \tau) = t_2(x, 0) = t_0 = t_f = \text{const}. \tag{11.1.9}$$

Assuming $t_1(x, \tau) = \vartheta(u)$, where $u = x/\sqrt{\tau}$, and substituting it into Eq. (11.1.1) gives

$$a_1 \frac{d^2\vartheta}{du^2} + \frac{u}{2} \frac{d\vartheta}{du} = 0. \tag{11.1.10}$$

We denote

$$d\vartheta/du = z. \tag{11.1.11}$$

Then Eq. (11.1.10) may be written as

$$\frac{dz}{du} + \frac{1}{2a_1} uz = 0. \tag{11.1.12}$$

The solution of Eq. (11.1.12) is of the form

$$z = A \exp[-u^2/4a_1], \tag{11.1.13}$$

where A is the integration constant.

Then, from Eq. (11.1.11)

$$t_1(x, \tau) = A \int_0^{x/\sqrt{\tau}} \exp[-u^2/4a_1]\,du + B$$

is obtained. It follows from the condition $t_1(0, \tau) = 0$ that $B = 0$.

Thus

$$t_1(x, \tau) = A \int_0^{x/\sqrt{\tau}} \exp[-u^2/4a_1]\,du. \tag{11.1.14}$$

Using boundary condition (11.1.8) and assuming $\xi = \beta\sqrt{\tau}$ yields

$$A(\lambda_1/\varrho\gamma) \exp[-\beta^2/4a_1] = \tfrac{1}{2}\beta. \tag{11.1.15}$$

It follows from the condition on the interface ($x = \xi$) that

$$t_f = t_0 = A \int_0^{\beta} \exp[-u^2/4a_1]\,du. \tag{11.1.16}$$

Then from relations (11.1.15) and (11.1.16), the constants A and β may be found. Thus, the final solution of the problem will be of the form

$$\frac{t_1(x, \tau)}{t_0} = \frac{\operatorname{erf}\{x/2(a_1\tau)^{1/2}\}}{\operatorname{erf}\{\beta/2\sqrt{a_1}\}} \tag{11.1.17}$$

$$\frac{\lambda_1 t_0 \exp[-\beta^2/4a_1]}{\sqrt{a_1}\operatorname{erf}[\beta/2\sqrt{a_1}]} = \varrho\gamma_2 \tfrac{1}{2}\sqrt{\pi}\,\beta. \tag{11.1.18}$$

c. Stefan's Solution. The problem is solved with the boundary conditions (11.1.3)–(11.1.7) and with the assumption $f(x) = t_0$, $\varphi(\tau) = t_a = \text{const}$ (Fig. 11.2), i.e.,

$$t_2(x, 0) = f(x) = t_0, \tag{11.1.19}$$

$$t_1(0, \tau) = t_a. \tag{11.1.20}$$

If the value $\xi = f(\tau)$ is sufficiently large, the problem to be solved will be similar to that of cooling a body system (compound semi-infinite rod); the temperature on the contact line of the bodies is maintained constant

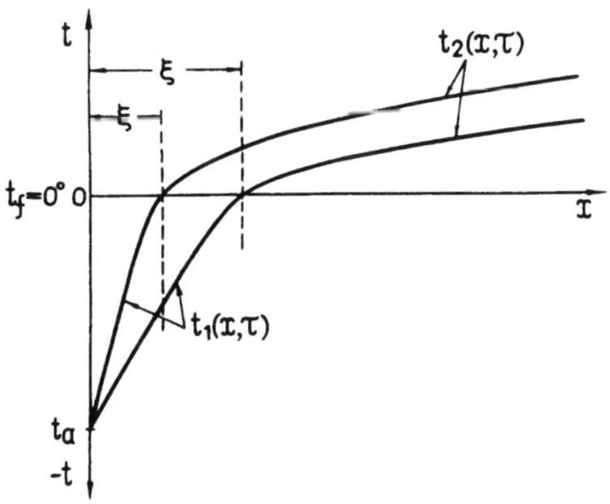

Fig. 11.2. Temperature distribution along the ground depth during freezing process.

and there is a negative heat source at the contact surface. Then the solutions of differential equations (11.1.1) and (11.1.2) will be found of the form

$$t_1(x, \tau) = A_1 + B_1 \operatorname{erf}\{x/2(a_1\tau)^{1/2}\}, \tag{11.1.21}$$

$$t_2(x, \tau) = A_2 + B_2 \operatorname{erf}\{x/2(a_2\tau)^{1/2}\}. \tag{11.1.22}$$

The constants A_1 and A_2 are found from boundary conditions (11.1.19) and (11.1.20) to be

$$A_1 = t_a, \qquad A_2 = t_0 - B_2.$$

Hence

$$t_1(x, \tau) = t_a + B_1 \operatorname{erf}\{x/2(a_1\tau)^{1/2}\} \tag{11.1.23}$$

$$t_2(x, \tau) = t_0 - B_2(1 - \operatorname{erf}\{x/2(a_2\tau)^{1/2}\}. \tag{11.1.24}$$

It follows from condition (11.1.5) that

$$t_a + B_1 \operatorname{erf}\{\xi/2(a_1\tau)^{1/2}\} = t_0 - B_2 \operatorname{erfc}\{\xi/2(a_2\tau)^{1/2}\} = t_f = \text{const}.$$

Since B_1 and B_2 are constants irrespective of the value of τ, the quantity $\xi/\sqrt{\tau}$ should obviously be constant, i.e.,

$$\xi = \beta\sqrt{\tau}, \tag{11.1.25}$$

where β is the proportionality factor characterizing the propagation velocity of the freezing region.

Thus we have

$$B_1 = \frac{t_f - t_a}{\operatorname{erf}\{\beta/2\sqrt{a_1}\}}, \qquad B_2 = \frac{t_0 - t_f}{\operatorname{erfc}\{\beta/2\sqrt{a_2}\}}.$$

Hence, the solution of our problem is of the form

$$t_1(x, \tau) = t_a + (t_f - t_a)[\mathrm{erf}\{x/2(a_1\tau)^{1/2}\}/\mathrm{erf}\{\beta/2\sqrt{a_1}\}], \quad (11.1.26)$$

$$t_2(x, \tau) = t_0 - \frac{(t_0 - t_f)}{\mathrm{erfc}\{\beta/2\sqrt{a_2}\}} \,\mathrm{erfc}\, \frac{x}{2(a_2\tau)^{1/2}}. \quad (11.1.27)$$

The coefficient β is determined from boundary condition (11.1.7), i.e., from the characteristic equation

$$\frac{\lambda_1(t_f - t_a)}{\sqrt{a_1}\,\mathrm{erf}\{\beta/2\sqrt{a_1}\}} \exp\left[-\frac{\beta^2}{4a_1}\right] + \frac{\lambda_2(t_0 - t_f)}{\sqrt{a_2}\,\mathrm{erfc}\{\beta/2\sqrt{a_2}\}} \exp\left[-\frac{\beta^2}{4a_2}\right]$$
$$= \frac{\varrho W\gamma_2\sqrt{\pi}}{2}\beta. \quad (11.1.28)$$

d. Analysis of the Solution. We assume that $t_0 = t_f$. This corresponds to the formation of ice in stagnant water. If the functions $\exp z^2$ and $\mathrm{erf}\, z$ are expanded in series and only the first term is taken of the series, we get

$$\beta = [(2\lambda_1/\varrho\gamma_2)(t_f - t_a)]^{1/2}, \quad (11.1.29)$$

i.e., the coefficient β is independent of the ice heat capacity and depends only on the coefficient λ_1, density γ_2, heat of melting and temperature difference $(t_f - t_a)$.

If in the series expansion of $\exp z^2$ two terms are taken, and in the expansion of $\mathrm{erf}\, z$ the first term is taken, then

$$\beta = \frac{[(2\lambda_1/\varrho\gamma_2)(t_f - t_a)]^{1/2}}{[1 + \{c_1\gamma_1(t_f - t_a)/2\varrho\gamma_2\}]^{1/2}}. \quad (11.1.30)$$

Thus in the second approximation, the coefficient β also depends on the heat capacity of ice.

In general, Eq. (11.1.28) for determining the coefficient β may be written in a criterial form as

$$\frac{\exp[-K_\beta{}^2]}{\mathrm{erf}\, K_\beta} + K_\varepsilon\left[\frac{t_0 - t_f}{t_f - t_a}\right]\frac{\exp[-K_a{}^{-1}K_\beta{}^2]}{\mathrm{erfc}\, K_a{}^{-1/2}K_\beta} = \sqrt{\pi}\,\mathrm{Ko}\, K_\beta, \quad (11.1.31)$$

where $K_\beta = \beta/2\sqrt{a_1}$ is a criterion relating the propagation velocity of the frozen region through the ground and the range of ground cooling, or the relative ability of the ground to freeze; $K_\varepsilon = \varepsilon_2/\varepsilon_1 = (\lambda_2 c_2\gamma_2/\lambda_1 c_1\gamma_1)^{1/2}$ is the parameter relating the thermal activity of the moist ground to that of the frozen region; $K_a = a_2/a_1$ is the ratio of thermal diffusitivites of the wet ground to the frozen one; and $\mathrm{Ko} = \varrho W\gamma_2/c_1\gamma_1(t_f - t_a)$ is the Kossovich criterion. The Kossovich criterion is the ratio of the amount of heat released in freezing a unit volume of the ground to that required for cooling the same unit volume of frozen ground from the freezing temperature to that of the medium.

The characteristic equation (11.1.31) in a criterial form may be solved graphically for the criterion K_β (Fig. 11.3). If the left-hand side of Eq.

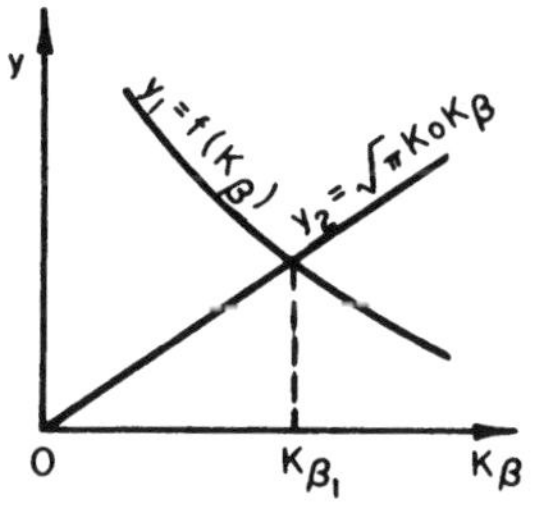

Fig. 11.3. Graphical method for determining K_β from the characteristic equation (11.1.31).

(11.1.31) is denoted by y_1, and the right-hand side by y_2, the value K_β is determined by the intersection of the curve y_1 with the straight line y_2. The slope of the straight line y_2 is $\sqrt{\pi}$ Ko. Thus, the larger the Kossovich criterion, the smaller the value of K_β; this physically represents a decrease in the velocity of propagation of the frozen region because of the increase in the portion of heat absorbed compared to that of cooling the frozen region.

The criterion K_β also depends on the relative thermal activity of the wet ground K_ε, the relative thermal diffusivity K_a, and the temperature parameter K_t.

$$K_t = (t_0 - t_f)/(t_f - t_a). \qquad (11.1.32)$$

The temperature parameter K_t relates the initial temperature of the ground to that of the frozen surface where the reference temperature is taken to be the freezing temperature.

Solutions (11.1.26) and (11.1.27) in criterial form may be written as

$$\theta_1 = \frac{t_1(x, \tau) - t_a}{t_f - t_a} = \frac{\operatorname{erf}\{1/2(\mathrm{Fo}_x)^{1/2}\}}{\operatorname{erf} K_\beta}, \qquad (11.1.33)$$

$$\theta_2 = \frac{t_0 - t_2(x, \tau)}{t_0 - t_f} = \frac{\operatorname{erfc}\{1/2(\mathrm{Fo}_x K_a)^{1/2}\}}{\operatorname{erfc} K_\beta K_a^{-1/2}}, \qquad (11.1.34)$$

$$\xi/x = 2K_\beta(\mathrm{Fo}_x)^{1/2}. \tag{11.1.35}$$

where $\mathrm{Fo}_x = a_1\tau/x^2$ is the Fourier number for the definite coordinate x.

If we assume $\xi = \text{const}$ and $d\xi/d\tau = 0$ (no heat absorption) our problem becomes similar to that in Chapter 10, Section 2 (cooling the system of two bodies: a finite rod with the length ξ in contact with a semi-infinite rod).

The solution of such a problem may be written (see Chapter 10, Section 2)

$$\theta_1 = \frac{t_1(x,\tau) - t_a}{t_0 - t_a}$$

$$= \operatorname{erf}\frac{x}{2(a_1\tau)^{1/2}} - h\sum_{n=1}^{\infty} h^{n-1}\left(\operatorname{erfc}\frac{2n\xi - x}{2(a_1\tau)^{1/2}} - \operatorname{erfc}\frac{2n\xi + x}{2(a_1\tau)^{1/2}}\right), \tag{11.1.36}$$

$$\theta_2 = \frac{t_0 - t_2(x,\tau)}{t_0 - t_a}$$

$$= \frac{2K_\varepsilon'}{1 + K_\varepsilon'}\sum_{n=1}^{\infty} h^{n-1}\operatorname{erfc}\left(\frac{x - \xi + (2n-1)(a_2/a_1)^{1/2}\xi}{2(a_2\tau)^{1/2}}\right), \tag{11.1.37}$$

where

$$K_\varepsilon' = \frac{\varepsilon_1}{\varepsilon_2}, \quad h = (1 - K_\varepsilon')/(1 + K_\varepsilon').$$

Solutions (11.1.36) and (11.1.37) show that for large values of ξ or small values of time, approximate relations may be written as

$$\begin{aligned} \theta_1 &\approx \operatorname{erf}\{x/2(a_1\tau)^{1/2}\}, \\ \theta_2 &\approx \frac{2K_\varepsilon'}{1 + K_\varepsilon'}\operatorname{erfc}\left(\frac{x - \xi + (a_2/a_1)^{1/2}\xi}{2(a_2\tau)^{1/2}}\right). \end{aligned} \tag{11.1.38}$$

Hence at small values of Fo, heat propagates in a rod with the finite length ξ in a way similar to heat distribution in a semi-infinite body. It may be shown that solutions (11.1.21) and (11.1.22) are approximate for a plate of finite size and valid for small values of Fo or for high values of ξ, i.e., for large values of K_β.

In these very simple examples, the temperature on the ground surface $t(0, \tau) = t_a$ was assumed to be constant in time and always below the freezing point. If the temperature $t(0, \tau)$ can alternate above and below the freezing point t_f, then alternating melted and frozen layers may form, leading to a further complication of the problem. In this case, even the simplest problems can be solved only approximately rather than analytically in a closed form.

Some of the more complex heat conduction problems involving changes in the state of aggregation have only been solved as a result of the development of approximate methods. Among methods for the solution of such problems, two of the most important should be mentioned: the methods developed by L.S. Leibenzon which enable us to obtain simple solutions for a large number of problems of practical importance and the method of hydraulic analogies by Lukianov allowing solutions of practically important but complex problems (including two-dimensional effects) with the aid of the hydraulic integrator.

11.2 Approximate Solutions of Problems of Solidification of a Semi-Infinite Body, an Infinite Plate, a Sphere, and an Infinite Cylinder

In 1931, Leibenzon [62] proposed an approximate method for the solution of problems on freezing; this is usually referred to as the first method of Leibenzon. In 1939, he developed another method for the solution of such problems [63]. Both methods have found widespread industrial acceptance and were experimentally verified. The first method will be considered here in detail.

The method of approximate solution involves selecting the functions $t_1(x, \tau)$ and $t_2(x, \tau)$ in such a way that they satisfy initial and boundary conditions. These temperature functions are substituted into the condition of conjugation on the interface and the differential equation obtained is solved with respect to ξ.

a. Solidification of a Semi-Infinite Body (Freezing of Ground). First, the problem of ground freezing will be considered. Boundary conditions are the same as in the Stefan problem.

The temperature distribution in the frozen ground is assumed linear:

$$t_1(x, \tau) = t_a + [(t_f - t_a)/\xi]x , \tag{11.2.1}$$

(this corresponds to the steady state) and the Gauss law is applied for the wet portion of the ground:

$$t_2(x, \tau) = t_f + (t_0 - t_f)\,\mathrm{erf}\{(x - \xi)/2(a_2\tau)^{1/2}\}. \tag{11.2.2}$$

The chosen functions $t_1(x, \tau)$ and $t_2(x, \tau)$ satisfy the initial and boundary conditions.

The temperature gradient on the interface will be

$$\frac{\partial t_1(\xi, \tau)}{\partial x} = \frac{t_f - t_a}{\xi}, \qquad \frac{\partial t_2(\xi, \tau)}{\partial x} = \frac{t_0 - t_f}{(\pi a_2 \tau)^{1/2}}.$$

Then, from the boundary condition (11.1.7)

$$\frac{\lambda_1(t_f - t_a)}{\xi} - \frac{\lambda_2(t_0 - t_f)}{(\pi a_2 \tau)^{1/2}} = \varrho \gamma_2 W \frac{d\xi}{d\tau}. \tag{11.2.3}$$

The solution of Eq. (11.2.3) with respect to ξ and with the initial condition $\xi(0) = 0$ is $\xi = \beta\sqrt{\tau}$. The constant β is determined from the characteristic equation

$$\frac{\lambda_1(t_f - t_a)}{\beta} - \frac{\lambda_2(t_0 - t_f)}{(\pi a_2)^{1/2}} = \tfrac{1}{2}\varrho \gamma_2 W \beta, \tag{11.2.4}$$

which in a criterial form is

$$(1/2K_\beta) - (1/\sqrt{\pi})K_\varepsilon K_t = \text{Ko}\, K_\beta. \tag{11.2.5}$$

Thus the number K_β depends solely on the criterion Ko, K_ε and K_t. Equation (11.2.5) does not contains a K_a criterion. This is explained by the assumed linear temperature distribution in the frozen region of the ground which corresponds to a steady state.

b. Solidification of a Cylinder. An infinite cylinder of fluid at a temperature t_0 will be considered. At the initial moment, the temperature of the surface falls instantaneously to some temperature $t_a < t_f$ which is maintained constant during the whole cooling process. Beginning from the cylinder surface, a frozen liquid layer is formed with the thickness $R - \eta = \xi$ where η is the distance from the cylinder axis to the freezing boundary (Fig. 11.4). To simplify the problem, the fluid temperature is assumed to be uniform throughout and equal to the freezing temperature, i.e.,

$$t_2(x, \tau) = t_0 = t_f = \text{const}. \tag{11.2.6}$$

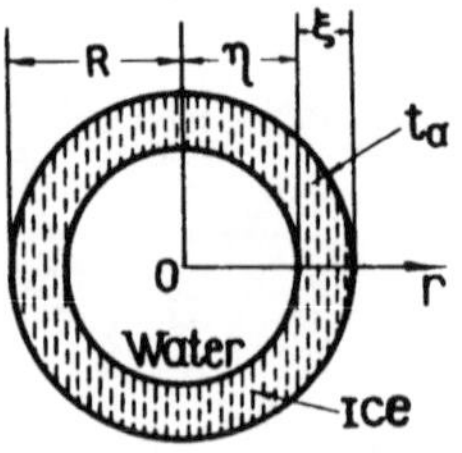

Fig. 11.4. Solidification of an infinite cylinder.

Then the problem is mathematically formulated as

$$\frac{\partial t_1(r,\tau)}{\partial \tau} - a_1\left(\frac{\partial^2 t_1(r,\tau)}{\partial r^2} + \frac{1}{r}\,\frac{\partial t_1(r,\tau)}{\partial r}\right)$$

$$(\tau > 0;\ R > r > \eta;\quad \eta = f(\tau)), \tag{11.2.7}$$

$$t(\eta,\tau) = t_f = \text{const}, \tag{11.2.8}$$

$$t(R,\tau) = t_a = \text{const}, \tag{11.2.9}$$

$$\lambda_1 \frac{\partial t_1(\eta,\tau)}{\partial r} = \varrho\gamma_2 \frac{d\eta}{d\tau}. \tag{11.2.10}$$

The temperature distribution in the frozen region (the ice region) is assumed to follow the temperature distribution law for a hollow cylinder in a steady state, i.e.,

$$t_1(r,\tau) = -\frac{(t_f - t_a)\ln r + t_a \ln\eta - t_f \ln R}{\ln(R/\eta)}. \tag{11.2.11}$$

Solution (11.2.11) satisfies the boundary conditions (11.2.8) and (11.2.9) Substituting solution (11.2.11) into boundary condition (11.2.10)

$$-\lambda_1 \frac{t_f - t_a}{\eta \ln \dfrac{R}{\eta}} = \varrho\gamma_2 \frac{d\eta}{d\tau} \tag{11.2.12}$$

and integrating gives us

$$\frac{\eta^2}{2}\ln\frac{R}{\eta} - \tfrac{1}{4}(R^2 - \eta^2) = -\frac{\lambda_1(t_f - t_a)}{\varrho\gamma_2}\tau, \tag{11.2.13}$$

which allows us to obtain the relation between η and the time τ.

The time of complete pipeline freezing (when $\tau = \tau_{\max}$, $\eta = 0$) is

$$\tau_{\max} = \frac{\varrho\gamma_2 R^2}{4\lambda_1(t_f - t_a)} = \frac{R^2}{4a_1}\,\text{Ko}. \tag{11.2.14}$$

*c. **Solidification of a Sphere.*** The problem is similar to the previous one and was solved by Kovner [58]. Initial and boundary conditions remain the same.

The temperature distribution in a hollow ice sphere is assumed to obey the law of a steady-state temperature distribution, i.e.,

$$t_1(r,\tau) = \frac{t_f - t_a}{R - \eta} R(\{\eta/r\} - 1) + t_f. \tag{11.2.15}$$

Solution (11.2.15) satisfies initial and boundary conditions.

Substitution of equation (11.2.15) into boundary condition (11.2.10) gives, after integrating the equation for determining τ,

$$\tau = \frac{\varrho\gamma_2(\eta - R)^2(2\eta + R)}{6\lambda_1(t_f - t_a)R}. \tag{11.2.16}$$

The time for complex solidification of a sphere $\tau_{\max}$ is

$$\tau_{\max} = \frac{\varrho\gamma_2 R^2}{6\lambda_1(t_f - t_a)} = \frac{R^2}{6a_1}\,\mathrm{Ko}. \tag{11.2.17}$$

The relations obtained (11.2.13)–(11.2.17) are approximate; experiments by Pomerantsev [94] showed their accuracy to be about 15 to 20%. To improve the accuracy, Pomerantsev introduced into (11.2.14) and (11.2.17) a correction factor accounting for the effect of the water heat capacity, the value of which is

$$\varDelta = 1 + \frac{c_2\gamma_2}{\varrho\gamma_1}(t_0 - t_f), \tag{11.2.18}$$

where t_0 is the initial water temperature (when $t_0 = t_f$, $\varDelta = 1$).

In the equations for $\tau_{\max}$, the appropriate expression in the right-hand side should be multiplied by $\varDelta$ in order to obtain the relations of Pomerantsev for the time required for complete solidification of a sphere or a cylinder.

In the engineering literature, a considerable number of works is published on the approximate solution of these problems. In most works, the problem of the ground solidification is considerably simplified (usually the ground heat capacity is assumed to be zero); solutions are to be found for the cases of solidifying a half-plane, cylinder, etc. Leibenzon's method, in which the expressions corresponding to a steady state are taken for the functions t_1 and t_2, represents such an approach.

d. Solidification of an Infinite Plate. In the second method of Leibenzon, the condition of conjugation is replaced by an integral condition of solidification. This condition is a direct result of a thermal balance equation.

We shall compose a heat balance equation for solidification. First, the enthalpy of the whole body is estimated, by means of the enthalpy of the body in a liquid state at the temperature t_f.

The enthalpy of the liquid phase is

$$Q_f = \int_{(V_f)} c_2\gamma_2(t - t_f)\,dv, \tag{11.2.19}$$

where V_f is the liquid phase volume changing in time.

The enthalpy of the solid phase is equal to

$$Q_s - \varrho W \gamma_2 V_s + \int_{(V_s)} c_1 \gamma_1 (t - t_f)\, dv\,, \qquad (11.2.20)$$

where V_s is the variable volume of the solid phase. With the aid of the equality

$$(d/d\tau)(Q_s + Q_f + Q_{in}) = 0\,, \qquad (11.2.21)$$

where Q_{in} is the amount of heat transferred from the fixed volume $V = V_s + V_f = \text{const}$. Taking the limit of the integral terms, we can easily obtain condition (11.1.7) for a small volume near the interphase, i.e., the ordinary condition of solidification, which is of the form

$$\lambda_1 \frac{\partial t_1(\xi, \tau)}{\partial x} - \lambda_2 \frac{\partial t_2(\xi, \tau)}{\partial x} = \varrho W \gamma_2 \frac{d\xi}{d\tau}. \qquad (11.2.22)$$

Using this calculation method, Leibenzon obtained the following formula for complete solidification ($\tau = \tau_{\max}$) of an infinite plate with thickness $2R$, the opposite surfaces of which are cooled and maintained at the temperature t_a which is lower than the freezing temperature

$$\tau_{\max} = \frac{R^2(\varrho\gamma_2 + \mu)}{2\lambda_1(t_f - t_a)}\,, \qquad (11.2.23)$$

where

$$\mu = \tfrac{4}{9} c_2 \gamma_2 (t_0 - t_f) + \tfrac{1}{2} c_1 \gamma_1 (t_f - t_a). \qquad (11.2.24)$$

Similar formulas were obtained by Leibenzon for a cylinder and a sphere; they differ from Eqs. (11.2.14) and (11.2.17) by additional terms accounting for the heat content effect of the liquid and ice in the solidification process.[2]

The problem of ground solidification is that of coupling two temperature fields with a special condition on the interface. A different approximate method for the solution of this problem is possible only when the temperature field in the frozen portion of the ground is considered and the effect of the wet region is accounted for by introducing a heat flow below the solidification front. This method is applied for the calculation of seasonal ground freezing. Practice has shown that the effect of the lower (wet region) temperature field on the upper (frozen) one is small, and therefore an approximate value of the heat flow may be used for calculation which

[2] It should be noted that a large change of γ in a phase transition may give rise to shift and strain in both phases. In this case, the problem is complicated since convective terms appear in the equations.

is in addition averaged for the whole period of freezing. In the fundamental work by Lukianov [74] a calculation formula is proposed for determining the freezing depth which was used by Golovko [70] for compilation of a nomogram. This formula is sufficiently simple and gives good agreement between the predicted values and empirical data.

11.3 Metal Solidification with the Heat Conduction Coefficient and Heat Capacity as Functions of Temperature

Propagation velocities of the crystallization front in a metal ingot or of the freezing ground front are usually calculated with constant values of all thermal properties of the material. The accuracy of the results obtained may be estimated only from the more general solutions. Such a solution was obtained by Lyubov [76] for a linear relation between the coefficients and the temperature. The series method used by Lyubov is sufficiently simple and may be applied not only to transfer problems with moving boundaries but also to problems with more general boundary conditions; the method may also be used in the case when the relation between the coefficients and the temperature is nonlinear.

No heating of the material undergoing the phase change is assumed, and the material temperature is further assumed to depend solely on the location of the phase transformation surface and time. In this case, the equation describing the temperature in the region under phase transformation (say, in a solid metal crust) is of the form

$$\gamma(c_0 + \nu t)(\partial t/\partial\tau) = (\partial/\partial x)[(\lambda_0 + \varkappa t)(\partial t/\partial x)], \qquad (11.3.1)$$

$$t(x, 0) = t_{ph}, \text{ at } x > 0, \qquad t(0, \tau) = t_0, \qquad (11.3.2)$$

where t_0 is the metal surface temperature (°C), and t_{ph} is the temperature corresponding to a phase transformation; specifically for ingot solidification, this is the temperature of metal crystallization.

At the phase transformation front, $x = y(\tau)$ where $y(\tau)$ is the coordinate of the phase transformation front

$$t[y(\tau), \tau] = t_{ph}. \qquad (11.3.3)$$

From the condition of the heat balance, the law of change of phase transformation front is

$$\frac{\partial y}{\partial\tau} = \frac{\lambda_0 + \varkappa t_{ph}}{\varrho\gamma}\left(\frac{\partial t}{\partial x}\right)_{x=y(\tau)}, \qquad (11.3.4)$$

where ϱ is the phase transformation heat.

If we denote

$$\bar{\lambda} = \lambda_0 + \varkappa \frac{t_{ph} + t_0}{2}; \qquad \bar{c} = c_0 + \nu \frac{t_{ph} + t_0}{2}; \qquad \bar{a} = \frac{\bar{\lambda}}{\gamma \bar{c}}; \tag{11.3.5}$$

$$\alpha_1 = \frac{\varkappa(t_{ph} - t_0)}{2\bar{\lambda}}; \qquad \alpha_2 = \frac{\nu(t_{ph} - t_0)}{2\bar{c}}; \qquad j = \frac{2\varrho}{\bar{c}(t_{ph} - t_0)}; \tag{11.3.6}$$

$$\theta = \frac{2}{t_{ph} - t_0}\left(t - \frac{t_0 + t_{ph}}{2}\right); \tag{11.3.7}$$

then Eq. (11.3.1) combined with conditions (11.3.2)–(11.3.4) can be written

$$(1 + \alpha_2\theta)(\partial\theta/\partial\tau) = \bar{a}(\partial/\partial x)[(1 + \alpha_1\theta)(\partial\theta/\partial x)]; \tag{11.3.8}$$

$$\theta(0, \tau) = -1; \quad \theta[y(\tau), \tau] = 1; \quad dy/d\tau = (\bar{a}/j)(1 + \alpha_1)(\partial\theta/\partial x)_{x=y(\tau)}. \tag{11.3.9}$$

The substitutions

$$\xi = x/(2\bar{a}\tau)^{1/2} \qquad \text{and} \qquad y(\tau) = \beta(2\bar{a}\tau)^{1/2} \tag{11.3.10}$$

allow us to transform the last equation into the ordinary differential equation

$$(d/d\xi)[(1 + \alpha_1\theta)(d\theta/d\xi)] + \xi(1 + \alpha_2\theta)(d\theta/d\xi) = 0, \tag{11.3.11}$$

with the conditions

$$\theta(0) = -1, \qquad \theta(\beta) = 1, \tag{11.3.12}$$

$$\beta = [(1 + \alpha_1)/j](d\theta/d\xi)_{\xi=\beta}. \tag{11.3.13}$$

If in equations (11.3.11) and (11.3.13) we assume $\alpha_1 = \alpha_2 = 0$, the problem will be reduced to that with constant coefficients when the value of these coefficients is taken as an arithmetic mean of the values of t_{ph} and t_0. The solution will be of the form

$$\theta = 2\,\frac{\operatorname{erf}(\xi/\sqrt{2})}{\operatorname{erf}(\bar{\beta}/\sqrt{2})} - 1. \tag{11.3.14}$$

Condition (11.3.13) when $\alpha_1 = \alpha_2 = 0$ reduces to the equation

$$\operatorname{erf}(\bar{\beta}/\sqrt{2})\exp[-\tfrac{1}{2}(\bar{\beta})^2]\bar{\beta}j = 2\sqrt{2}/\sqrt{\pi}, \tag{11.3.15}$$

which defines $\bar{\beta}$. The solution of equation (11.3.11) is found in the series form

$$\theta(\xi) = \sum_{n=0}^{\infty} \left(\frac{d^n\theta}{d\xi^n}\right)_{\xi=\beta} \frac{(\xi - \beta)^n}{n!}. \tag{11.3.16}$$

Values of $d^n\theta/d\xi^n$ when $\xi = \beta$ and $n > 1$ will be found by successive differentiation of equation (11.3.11), and when $n = 1$, from condition (11.3.13).

$$\left(\frac{d\theta}{d\xi}\right)_{\xi=\beta} = \frac{\beta}{1+\alpha_1} j; \tag{11.7.17}$$

$$\left(\frac{d^2\theta}{d\xi^2}\right)_{\xi=\beta} = -\beta^2\left[\frac{1+\alpha_2}{(1+\alpha_1)^2} j + \frac{\alpha_1}{(1+\alpha_1)^3} j^2\right]; \tag{11.3.18}$$

$$\left(\frac{d^3\theta}{d\xi^3}\right)_{\xi=\beta} = -\beta^2 \frac{1+\alpha_2}{(1+\alpha_1)^2} j + \beta^3\left[\frac{(1+\alpha_2)^2}{(1+\alpha_1)^3} j + \frac{4\alpha_1 + 3\alpha_1\alpha_2 - \alpha_2}{(1+\alpha_1)^4} j^2 + \frac{3\alpha_1^2}{(1+\alpha_1)^5} j^3\right]; \tag{11.3.19}$$

$$\left(\frac{d^4\theta}{d\xi^4}\right)_{\xi=\beta} = \beta^2\left[\frac{3(1+\alpha_2)^2}{(2+\alpha_1)^3} j + \frac{6\alpha_1 + 4\alpha_1\alpha_2 - 2\alpha_2}{(1+\alpha_1)^4} j^2\right] - \beta^4\left[\frac{(1+\alpha_2)^2}{(1+\alpha_1)^4} j + \frac{(1+\alpha_2)(11\alpha_1 + 7\alpha_1\alpha_2 - 4\alpha_2)}{(1+\alpha_1)^5} j^2 + \frac{\alpha_1(25\alpha_1 + 18\alpha_1\alpha_2 - 7\alpha_2)}{(1+\alpha_1)^6} j^3 + \frac{15\alpha_1^3}{(1+\alpha_1)^7} j^4\right]; \tag{11.3.20}$$

and so on. From the first condition of (11.3.12)

$$-1 = \sum_{n=0}^{\infty} (-1)^n \left(\frac{d^n\theta}{d\xi^n}\right)_{\xi=\beta} \frac{\beta^n}{n!}. \tag{11.3.21}$$

Substituting into equation (11.3.21) the expressions obtained for $(d^n\theta/d\xi^n)_{\xi=\beta}$ and after some manipulations, we obtain a series for determining β:

$$\frac{2}{j}(1+\alpha_1) = \beta^2 + \beta^4\left[\frac{1+\alpha_2}{3(1+\alpha_1)} + \frac{\alpha_1}{2(1+\alpha_1)^2} j\right] + \beta^6\left[\frac{(1+\alpha_2)^2}{15(1+\alpha_1)^2} + \frac{5\alpha_1 + 4\alpha_1\alpha_2 - \alpha_2}{12(1+\alpha_1)^3} j + \frac{\alpha_1^2}{2(1+\alpha_1)^4} j^2\right] + \beta^8\left[\frac{(1+\alpha_2)^3}{105(1+\alpha_1)^3} + \frac{(1+\alpha_2)(31\alpha_1+22\alpha_1\alpha_2-9\alpha_2)}{180(1+\alpha_1)^4} j + \frac{\alpha_1(19\alpha_1+15\alpha_1\alpha_2-4\alpha_2)}{30(1+\alpha_1)^5} j^2 + \frac{5\alpha_1^3}{8(1+\alpha_1)^6} j^3\right] + \cdots. \tag{11.3.22}$$

It may be shown that when $\alpha_1 = \alpha_2 = 0$, series (11.3.22) become a series which may be obtained from (11.3.13) [76]. When β is known, values of $\theta(\xi)$ may be obtained for various magnitudes of ξ from expression (11.3.16) and the resulting plot of $\theta(\xi)$ versus ξ allows us to obtain temperature changes in time for various coordinate points. The propagation velocity of the phase transformation front is found from the formula

$$dy/d\tau = \beta\sqrt{\bar{a}}/\sqrt{2\tau}\,. \tag{11.3.23}$$

Among other methods of solution of the similar equations, the integral method [39] has been widely used recently.

CHAPTER

12

TWO-DIMENSIONAL TEMPERATURE FIELD: PARTICULAR PROBLEMS

Solution of problems of unsteady heat conduction, when the temperature is a function of time and two coordinates, is extremely difficult. Some specific problems may be solved by the methods discussed in this book. Particularly, in Chapter 6, the problems of heating a finite cylinder and three-dimensional plate are considered with the condition of symmetry of the temperature field with respect to the body center (symmetrical problems). These solutions are obtained as generalization of solutions for an infinite cylinder and infinite plate.

In the present chapter, some particular problems of a two-dimensional temperature field are considered wherein the solution may be obtained by the integral transform methods.

12.1 Semi-Infinite Plate

a. Statement of the Problem. *A semi-infinite plate with the thickness l is considered; its temperature is uniform throughout and equal to* 0°C. *At the initial moment, one of the bounding surfaces instantaneously acquires the temperature of the medium t_a which remains constant during the entire heating process. The other bounding surfaces are maintained at the initial temperature* (Fig. 12.1). *The temperature distribution inside the plate is to be found.* We have:

$$\frac{\partial t(x, y, \tau)}{\partial \tau} = a\left(\frac{\partial^2 t(x, y, \tau)}{\partial x^2} + \frac{\partial^2 t(x, y, \tau)}{\partial y^2}\right) \tag{12.1.1}$$

$$(\tau > 0;\ 0 < x < l;\ 0 < y < \infty),$$

$$\begin{aligned} t(x, y, 0) = 0, \quad t(0, y, \tau) = 0, \\ t'(x, \infty, \tau) = 0; \quad t(l, y, \tau) = t_a, \quad t(x, 0, \tau) = 0. \end{aligned} \tag{12.1.2}$$

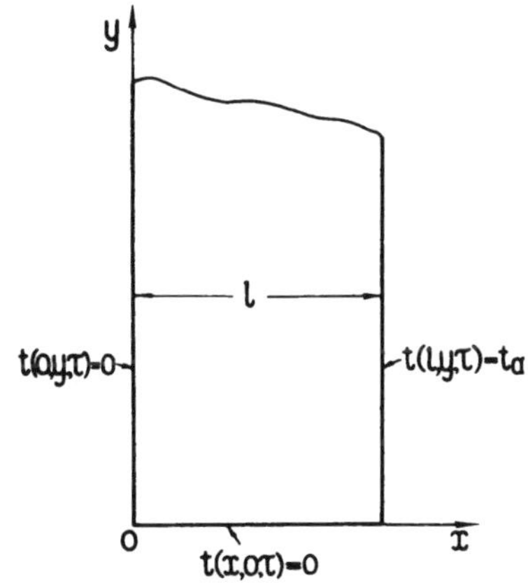

Fig. 12.1. Semi-infinite plate.

b. Solution of the Problem. The Laplace transformation with respect to the time coordinate and the finite Fourier sine transformation with respect to the space coordinate x will be applied.

The transform and inverse transform for the finite Fourier transformations are written respectively

$$F_S(n) = \int_0^\pi f(\xi) \sin n\xi \, d\xi, \qquad f(\xi) = \frac{2}{\pi} \sum_{n=1}^{\infty} \sin n\xi F_S(n),$$

i.e., the variable ξ ranges from 0 to π. To facilitate computation, a new variable $\xi = \pi(x/l)$ is introduced and the dimensionless temperature is designated by θ in accordance with the relation

$$\theta = (t/t_a) - (x/l). \tag{12.1.3}$$

Then the boundary-value problems (12.1.1) and (12.1.2) are written

$$\frac{\partial\theta}{\partial\tau} = a\left(\frac{\pi}{l}\right)^2 \frac{\partial^2\theta}{\partial\xi^2} + a\,\frac{\partial^2\theta}{\partial y^2}. \tag{12.1.4}$$

The initial and boundary conditions become of the form

$$\theta(0, \xi, y) = -(1/\pi)\xi, \tag{12.1.5}$$

$$\theta(\tau, 0, y) = 0, \ \theta(\tau, \pi, y) = 0, \ \partial\theta(\tau, \xi, \infty)/\partial y = 0, \tag{12.1.6}$$

$$\theta(\tau, \xi, 0) = -(1/\pi)\xi. \tag{12.1.7}$$

Successive application of the finite Fourier sine transformation and the Laplace transformation to (12.1.4), and taking into account boundary conditions (12.1.5)–(12.1.7), yields for the function θ_{LS}

$$\frac{d^2\theta_{LS}(y)}{dy^2} = \left[\frac{s}{a} + \left(\frac{\pi}{l} n\right)^2\right]\theta_{LS}(y) - \frac{(-1)^n}{an}, \tag{12.1.8}$$

with the conditions

$$\theta_{LS}(0) = \frac{(-1)^n}{ns}, \qquad \frac{\partial\theta_{LS}(\infty)}{\partial y} = 0. \tag{12.1.9}$$

The solution of Eq. (12.1.8) with condition (12.1.9) becomes of the form

$$\theta_{LS}(y) = \frac{(-1)^n}{n[s + \{(\pi/l)n\}^2 a]} + \frac{(-1)^n (\pi/l)^2 an}{s[s + \{(\pi/l)n\}^2 a]} \times \exp\left[-\left(\frac{s}{a} + \left(\frac{\pi}{l} n\right)^2\right)^{1/2} y\right]. \tag{12.1.10}$$

For inversion of the Laplace transform, we use the relation

$$L^{-1}\left\{\frac{1}{s + \{(\pi/l)\, n\}^2 a}\right\} = \exp[-\{(\pi/l)n\}^2 a\tau].$$

The second term of Eq. (12.1.10) is inverted by the Fourier sine inversion formula.

We have

$$\frac{1}{s\left(\dfrac{s}{a} + \{n^2\pi^2/l^2\}\right)} \exp\left[-y\left(\frac{s}{a} + \{n^2\pi^2/l^2\}\right)^{1/2}\right]$$

$$= \frac{l^2}{n^2\pi^2}\left(\frac{1}{s} - \frac{1}{s + (an^2\pi^2/l^2)}\right) \exp\left[-y\left(\frac{s}{a} + \{n^2\pi^2/l^2\}\right)^{1/2}\right]. \tag{12.1.11}$$

Further,

$$L^{-1}\left[\frac{1}{s + \{an^2\pi^2/l^2\}} \exp\left[-y\left(\frac{s}{a} + \frac{n^2\pi^2}{l^2}\right)^{1/2}\right]\right]$$

$$= \exp[-n^2\pi^2(a\tau/l^2)] \operatorname{erfc} \frac{y}{2(a\tau)^{1/2}};$$

$$L^{-1}\left[\frac{1}{s} \exp\left[-y\left(\frac{s}{a} + \frac{n^2\pi^2}{l^2}\right)^{1/2}\right]\right]$$

$$= \frac{1}{2}\left\{\exp[n\pi(y/l)] \operatorname{erfc}\left[\frac{y}{2(a\tau)^{1/2}} + \frac{n\pi}{l}(a\tau)^{1/2}\right] + \exp[-n\pi(y/l)] \operatorname{erfc}\left[\frac{y}{2(a\tau)^{1/2}} - \frac{n\pi}{l}(a\tau)^{1/2}\right]\right\}. \tag{12.1.12}$$

The inversion of the Fourier sine transform should now be carried out. Returning to the initial variables, we find the final form of solution of this problem to bc

$$\theta = \frac{t(x, y, \tau)}{t_a}$$
$$= \frac{x}{l} + \frac{2}{\pi} \sum_{n=1}^{\infty} \frac{(-1)^n}{n} \sin n\pi \frac{x}{l} \exp[-n^2\pi^2(a\tau/l^2)]$$
$$+ \frac{2}{\pi} \sum_{n=1}^{\infty} \frac{(-1)^n}{n} \sin n\pi \frac{x}{l} \frac{1}{2} \left\{ \exp[n\pi(y/l)] \operatorname{erfc}\left[\frac{y}{2(a\tau)^{1/2}} + \frac{n\pi}{l}(a\tau)^{1/2}\right] \right.$$
$$+ \exp[-n\pi(y/l)] \operatorname{erfc}\left(\frac{y}{2(a\tau)^{1/2}} - \frac{n\pi}{l}(a\tau)^{1/2}\right)$$
$$\left. - 2\exp[-n^2\pi^2(a\tau/l^2)] \operatorname{erfc} \frac{y}{2(a\tau)^{1/2}} \right\}. \tag{12.1.13}$$

Thus, by means of two successive integral Laplace transformations, the solution of a two-dimensional problem on heat propagation in a semi-infinite plate is obtained.

12.2 Two-Dimensional Plate

a. Statement of the Problem. *Consider a two-dimensional problem for a rectangle, the sides of which have a temperature distribution which is controlled to change with time in an assigned way:*

$$\frac{\partial t}{\partial \tau} = a\left(\frac{\partial^2 t}{\partial x^2} + \frac{\partial^2 t}{\partial y^2}\right) \quad (\tau > 0;\ 0 \le x \le h;\ 0 \le y \le d), \tag{12.2.1}$$

with the initial condition

$$t(x, y, 0) = f(x, y), \tag{12.2.2}$$

and boundary conditions

$$t(0, y, \tau) = \varphi(y, \tau); \qquad t(h, y, \tau) = 0, \tag{12.2.3}$$

$$t(x, 0, \tau) = 0; \qquad t(x, d, \tau) = 0. \tag{12.2.4}$$

b. Solution of the Problem. This may be obtained by combining the integral Fourier sine transformation

$$T_F = T_F(m, n, \tau)$$
$$= \int_0^h \int_0^d t(x, y, \tau) \sin \frac{m\pi x}{h} \sin \frac{n\pi y}{d}\, dx\, dy, \tag{12.2.5}$$

with the inversion formula

$$t(x, y, \tau) = \frac{4}{hd} \sum_{m=1}^{\infty} \sum_{n=1}^{\infty} T_F(m, n, \tau) \sin \frac{m\pi x}{h} \sin \frac{n\pi y}{d}. \qquad (12.2.6)$$

Application of transformation (12.2.5) to Eq. (12.2.1) and using boundary conditions (12.2.3) and (12.3.4) gives us

$$\frac{dT_F}{d\tau} + a\pi^2 \left(\frac{m^2}{h^2} + \frac{n^2}{d^2}\right) T_F - \frac{am\pi}{h} \int_0^d \varphi(y, \tau) \sin \frac{n\pi y}{d}\, dy = 0,$$

with the condition

$$T_F|_{\tau=0} = \int_0^h \int_0^d f(x, y) \sin(\pi mx/h) \sin(\pi ny/d)\, dx\, dy.$$

The solution of the last equation is of the form

$$T_F = \exp[-a\sigma\tau] \left[\int_0^h \int_0^d f(x, y) \sin(\pi mx/h) \sin(\pi ny/d)\, dx\, dy \right.$$
$$\left. + (am\pi/h) \int_0^\tau \int_0^d \exp[a\sigma\tau']\, \varphi(y, \tau') \sin(\pi ny/d)\, dy\, d\tau' \right].$$

Finally, from inversion formula (12.2.6), we obtain

$$t(x, y, \tau) = \frac{4}{hd} \sum_{m=1}^{\infty} \sum_{n=1}^{\infty} \exp[-a\sigma\tau] \sin(\pi mx/h) \sin(\pi ny/d)$$
$$\times \left\{ \int_0^\tau \int_0^d \exp[a\sigma\tau']\, \varphi(y', \tau') \sin(\pi ny'/d)\, dy'\, d\tau' \right.$$
$$\left. + \int_0^h \int_0^d f(x', y') \sin(\pi mx'/h) \sin(\pi ny'/d)\, dx'\, dy' \right\}, \qquad (12.2.7)$$

where

$$\sigma = \pi^2 \left(\frac{m^2}{h^2} + \frac{n^2}{d^2}\right). \qquad (12.2.8)$$

The solutions presented by Sneddon [109] and Grinberg [41] are particular cases of solution (12.2.7).

For a semi-infinite plate, $d = \infty$; in this case, it may be shown that for the following conditions

$$t(x, y, 0) = 0, \qquad t(h, y, \tau) = t_a;$$
$$t(0, y, \tau) = t(x, 0, \tau) = t(x, d, \tau) = 0, \qquad (12.2.9)$$

solution (12.2.7) may be rewritten as

$$\frac{t(x, y, \tau)}{t_a} = \frac{x}{h} + \frac{2}{\pi} \sum_{m=1}^{\infty} \frac{(-1)^m}{m} \sin \frac{m\pi x}{h} \exp\left[-\frac{m^2\pi^2}{h^2} a\tau\right]$$

$$+ \frac{2}{\pi} \sum_{m=1}^{\infty} \frac{(-1)^m}{m} \sin \frac{m\pi x}{h} \left\{\exp\left[\frac{m\pi y}{h}\right] \operatorname{erfc}\left(\frac{y}{2(a\tau)^{1/2}} + \frac{m\pi}{h}(a\tau)^{1/2}\right)\right.$$

$$+ \exp\left(-m\pi\frac{y}{h}\right) \operatorname{erfc}\left(\frac{y}{2(a\tau)^{1/2}} - \frac{m\pi}{h}(a\tau)^{1/2}\right)$$

$$\left. - 2\exp\left[-\frac{m^2\pi^2}{h^2} a\tau\right] \operatorname{erfc} \frac{y}{2\,(a\tau)^{1/2}}\right\}, \tag{12.2.10}$$

which is the same as solution (12.1.13).

12.3 Semi-Infinite Cylinder

a. Statement of the Problem. *A semi-infinite cylinder of diameter $2R$ is at the temperature t_0. At the initial moment, its surface experiences a constant temperature t_a which is maintained constant during the whole heating process. The temperature of the base remains constant and equal to the initital temperature. The temperature distribution inside the cylinder is to be found.* We have

$$\frac{\partial t(r, z, \tau)}{\partial \tau} = a\left(\frac{\partial^2 t(r, z, \tau)}{\partial r^2} + \frac{1}{r}\frac{\partial t(r, z, \tau)}{\partial r} + \frac{\partial^2 t(r, z, \tau)}{\partial z^2}\right)$$

$$(\tau > 0;\ 0 < r < R;\ 0 < z < \infty), \tag{12.3.1}$$

$$t(r, z, 0) = t_0, \qquad t(r, 0, \tau) = t_0, \qquad t(R, z, \tau) = t_a, \tag{12.3.2}$$

$$t(0, z, \tau) \neq \infty, \qquad \partial t(0, z, \tau)/\partial r = 0, \qquad \partial t(r, \infty, \tau)/\partial z = 0. \tag{12.3.3}$$

b. Solution of the Problem. The transformed solution of Eq. (12.3.1) with conditions (12.3.2)–(12.3.3) is of the form

$$T(r, z, s) - \frac{t_0}{s} = \frac{(t_a - t_0) I_0\left(\left(\frac{s}{a}\right)^{1/2} r\right)}{s I_0\left(\left(\frac{s}{a}\right)^{1/2} R\right)}$$

$$- \sum_{n=1}^{\infty} \frac{2\mu_n J_0\left(\frac{\mu_n r}{R}\right)(t_a - t_0)}{R^2 s\left(\frac{s}{a} + \frac{\mu_n^2}{R^2}\right) J_1(\mu_n)} \exp\left[-z\left(\frac{s}{a} + \frac{\mu_n^2}{R^2}\right)^{1/2}\right], \tag{12.3.4}$$

where μ_n are the roots of the characteristic equation $J_0(\mu) = 0$.

With the aid of the transform table it is found:

$$\theta = \frac{t(r, z, \tau) - t_0}{t_a - t_0} = 1 - \sum_{n=1}^{\infty} \frac{J_0\{\mu_n(r/R)\}}{\mu_n J_1(\mu_n)} \left\{2 \exp[-\mu_n^2 \text{ Fo}] \operatorname{erf} \frac{z}{2(a\tau)^{1/2}} \right.$$

$$+ \exp\left[\mu_n \frac{z}{R}\right] \operatorname{erfc}\left(\frac{z}{2(a\tau)^{1/2}} + \mu_n(\text{Fo})^{1/2}\right)$$

$$\left. + \exp\left[-\mu_n \frac{z}{R}\right] \operatorname{erfc}\left(\frac{z}{2(a\tau)^{1/2}} - \mu_n(\text{Fo})^{1/2}\right)\right\}, \tag{12.3.5}$$

where $\text{Fo} = a\tau/R^2$ is the Fourier number.

A related problem will be now considered. At the initial moment the base of the cylinder instantaneously acquires a temperature t_a which is maintained constant. Heat from the cylinder surface is transferred by convection into the surrounding medium with the medium temperature equal to the initial cylinder temperature t_0.

The differential equation and the initial condition remain the same. The other boundary condition may be written as

$$t(r, 0, \tau) = t_a, \qquad \frac{\partial t(R, z, \tau)}{\partial r} + H[t(R, z, \tau) - t_0] = 0, \tag{12.3.6}$$

$$t(0, z, \tau) \neq \infty, \qquad \frac{\partial t(0, \infty, \tau)}{\partial z} = 0. \tag{12.3.7}$$

The solution of the problem is obtained in a similar way and the result is

$$\theta = \frac{t(r, z, \tau) - t_0}{t_a - t_0} = \sum_{n=1}^{\infty} \frac{2 \text{ Bi } J_0\{\mu_n(r/R)\}}{J_0(\mu_n)(\text{Bi}^2 + \mu_n^2)} \exp\left[-\mu_n \frac{z}{R}\right]$$

$$+ \sum_{n=1}^{\infty} \frac{\text{Bi } J_0\{\mu_n(r/R)\}}{J_0(\mu_n)(\text{Bi}^2 + \mu_n^2)} \left[\exp\left[\mu_n \frac{z}{R}\right] \operatorname{erfc}\left(\mu_n(\text{Fo})^{1/2} + \frac{z}{2(a\tau)^{1/2}}\right)\right.$$

$$\left. - \exp\left[-\mu_n \frac{z}{R}\right] \operatorname{erfc}\left(\mu_n(\text{Fo})^{1/2} - \frac{z}{2(a\tau)^{1/2}}\right)\right]. \tag{12.3.8}$$

where μ_n are the roots of the characteristic equation

$$\frac{J_0(\mu)}{J_1(\mu)} = \frac{1}{\text{Bi}} \mu. \tag{12.3.9}$$

The same method may be used for the solution of three-dimensional problems in stationary temperature field.

Thus, the Laplace method may be used for the solution of a large number of problems; some approximate solutions result (e.g., special formulas for large and small values of Fo), which represent the main advantage of the operational method.

12.4 Heat Transfer in Cylindrical Regions

The differential heat conduction equation for a circular three-dimensional cylindrical region in cylindrical coordinates r, z, φ, τ, is of the form:

$$\frac{\partial t}{\partial \tau} = a\left(\frac{\partial^2 t}{\partial r^2} + \frac{1}{r}\frac{\partial t}{\partial r} + \frac{\partial^2 t}{\partial z^2} + \frac{1}{r^2}\frac{\partial^2 t}{\partial \varphi^2}\right) + \frac{1}{c\gamma} w(r, z, \varphi, \tau). \tag{12.4.1}$$

If there is symmetry with respect to the axis z, then the operator $\partial/\partial\varphi$ is identically equal to zero, and we obtain

$$\frac{\partial t}{\partial \tau} = a\left(\frac{\partial^2 t}{\partial r^2} + \frac{1}{r}\frac{\partial t}{\partial r} + \frac{\partial^2 t}{\partial z^2}\right) + \frac{1}{c\gamma} w(r, z, \tau). \tag{12.4.2}$$

In addition, if the cylinder of interest is sufficiently long and the initial and boundary conditions are such that parallel cylinder sections normal to its axis have the same transfer potential distribution, then the operator d/dz is also identically equal to zero.

a. Axisymmetrical Cylinder. *Here the solution of* Eq. (12.4.2) *will be considered with boundary conditions of the first, second, and the third kind.*

The solution of such a problem is necessary for the study of various engineering problems, e.g., for simulation of thermal processes in a reactor core, for studying heat transfer from fuel elements of a reactor, heat transfer between a pipe-line and the ground, mass transfer involving chemical transformations, mass transfer through a porous medium, etc.

(1) *First of all we shall consider the solution of an equation for a finite cylinder, l in length, with boundary conditions of the second kind on its surface.* It is assumed that one of the cylindrical bases ($z = 0$) is thermally insulated, while the potential of the other is a function of time and the radial coordinate, i.e., the solution of Eq. (12.4.2) is to be found with the boundary conditions

$$t(0, z, \tau) \neq \infty; \quad \frac{\partial t(0, z, \tau)}{\partial r} = 0; \quad \frac{\partial t(R, z, \tau)}{\partial r} = \frac{1}{\lambda} q(z, \tau);$$

$$\frac{\partial t(r, 0, \tau)}{\partial z} = 0; \quad t(r, l, \tau) = \varphi(r, \tau); \quad t(r, z, 0) = f(r, z). \tag{12.4.3}$$

The solution of this problem may be obtained by various methods, in particular, by the combined usage of finite integral Fourier and Hankel transformations.

For simplicity in subsequent calculations, Eq. (12.4.2) and conditions (12.4.3) are written in a dimensionless form as

$$\frac{\partial \theta}{\partial \mathrm{Fo}} = \frac{1}{X}\frac{\partial}{\partial X}\left(X\frac{\partial \theta}{\partial X}\right) + b^2\frac{\partial^2 \theta}{\partial Z^2} + \mathrm{Po}(X, Z, \mathrm{Fo}); \quad (12.4.4)$$

$$\theta(X, Z, 0) = F(X, Z); \quad (12.4.5)$$

$$\frac{\partial \theta(1, Z, \mathrm{Fo})}{\partial X} = \mathrm{Ki}(Z, \mathrm{Fo}); \quad (12.4.6)$$

$$\frac{\partial \theta(X, 0, \mathrm{Fo})}{\partial Z} = 0; \qquad \theta(X, \pi, \mathrm{Fo}) = \Phi(X, \mathrm{Fo}) \quad (12.4.7)$$

where $\theta = (t - t^*)/t^*$ is a dimensionless temperature (t^* is some initial temperature value, fixed for a definite point of the cylinder), $X = r/R$, $Z = \pi z/l$ are dimensionless coordinates, $F(X, Z) = (f(r, z) - t^*)/t^*$; $\Phi(X, \mathrm{Fo}) = (\varphi(r, \tau) - t^*)/t^*$, $b = \pi R/l$, $\mathrm{Po}(X, Z, \mathrm{Fo}) = [R^2/\lambda t^*]\, w(r, z, \tau)$ is the Pomerantsev criterion, and $\mathrm{Ki}(Z, \mathrm{Fo}) = (R/\lambda t^*)q(z, \tau)$ is the Kirpichev criterion.

We shall use the finite integral Hankel transformation with respect to the variable X to obtain

$$\{\theta(X, Z, \mathrm{Fo})\}_H = \{\theta\}_H = \int_0^1 X J_0(\mu X)\theta(X, Z, \mathrm{Fo})\, dX \quad (12.4.8)$$

with the inversion formula

$$\theta(X, Z, \mathrm{Fo}) = 2 \sum_{\mu} \frac{J_0(\mu X)}{J_0^2(\mu)} \{\theta(X, Z, \mathrm{Fo})\}_H \quad (12.4.9)$$

(where μ is the positive root of the characteristic equation $J_1(\mu) = 0$), and finite Fourier cosine transformation with respect to the variable Z

$$\{\theta(X, Z, \mathrm{Fo})\}_{HC} = \{\theta\}_{HC} = \int_0^{\pi} \{\theta\}_H \cos(n + \tfrac{1}{2})Z\, dZ, \quad (12.4.10)$$

with the inversion formula

$$\{\theta(X, Z, \mathrm{Fo})\}_H = \frac{2}{\pi} \sum_{n=0}^{\infty} \{\theta\}_{HC} \cos(n + \tfrac{1}{2})Z. \quad (12.4.11)$$

The above transformations allow us to obtain the solution of the problme in the following form:

$$\theta(X, Z, \mathrm{Fo}) = \frac{4}{\pi} \sum_{m=1}^{\infty} \frac{J_0(\mu_m X)}{J_0^2(\mu_m)} \Big\{ \sum_{n=0}^{\infty} \cos(n + \tfrac{1}{2})Z$$
$$\times \Big[\exp[-[\mu_m^2 + b^2(n + \tfrac{1}{2})^2]\,\mathrm{Fo}]\{F(X, Z)\}_{HC}$$
$$+ \int_0^{\mathrm{Fo}} \exp[-[\mu_m^2 + b^2(n + \tfrac{1}{2})^2](\mathrm{Fo} - \mathrm{Fo}^*)]$$
$$\times \Big(J_0(\mu_m)\{\mathrm{Ki}(Z, \mathrm{Fo}^*)\}_C + (-1)^n b^2(n + \tfrac{1}{2})\{\Phi(X, \mathrm{Fo}^*)\}_H$$
$$+ \{\mathrm{Po}(X, Z, \mathrm{Fo}^*)\}_{HC}\Big)\, d\mathrm{Fo}^*\Big]\Big\}. \qquad (12.4.12)$$

If it is assumed in (12.4.12) that $f(r, z) = \varphi(r, \tau) = t_0$ and $t^* = t_0$, we shall find

$$\theta = \frac{t - t_0}{t_0}$$
$$= \frac{4}{\pi} \sum_{m=1}^{\infty} \frac{J_0(\mu_m X)}{J_0^2(\mu_m)} \Big\{ \sum_{n=0}^{\infty} \cos \mu_n Z \int_0^{\mathrm{Fo}} \exp[-(\mu_m^2 + b^2\mu_n^2)(\mathrm{Fo} - \mathrm{Fo}^*)]$$
$$\times \Big[J_0(\mu_m) \int_0^{\pi} \mathrm{Ki}(Z, \mathrm{Fo}^*) \cos \mu_n Z\, dZ$$
$$+ \int_0^{\pi} \cos \mu_n Z \Big[\int_0^1 X J_0(\mu_m X)\, \mathrm{Po}(X, Z, \mathrm{Fo}^*)\, dX\Big] dZ\Big] d\mathrm{Fo}^*\Big\}, \qquad (12.4.12a)$$

where $\mu_n = (2n + 1)/2$ $(n = 0, 1, 2, \ldots)$.

(2) *The solution of Eq.* (12.4.4) *will be now considered with boundary conditions of the third kind on the cylindrical surface.* Let the boundary conditions have the form

$$\frac{\partial t(R, z, \tau)}{\partial r} + \frac{\alpha}{\lambda} [t(R, z, \tau) - t_a] = 0,$$
$$\frac{\partial t(x, 0, \tau)}{\partial z} = 0, \qquad t(x, l, \tau) = t_a, \qquad t(x, z, 0) = t_0$$

or a nondimensional form

$$\theta(X, Z, 0) = \theta_0, \qquad (12.4.13)$$

$$\frac{\partial \theta(1, Z, \mathrm{Fo})}{\partial X} + \mathrm{Bi}\, \theta(1, Z, \mathrm{Fo}) = 0, \qquad (12.4.14)$$

$$\frac{\partial \theta(X, 0, \mathrm{Fo})}{\partial Z} = 0; \qquad \theta(X, \pi, \mathrm{Fo}) = 0. \qquad (12.4.15)$$

In equation (12.4.4) and conditions (12.4.13)–(12.4.15) in contradiction to the previous problem the dimensionless potential θ is of the form:

$$\theta(X, z, \text{Fo}) = (t - t_a)/t_0.$$

In addition,

$$\theta_0 = (t_0 - t_a)/t_0; \qquad \text{Bi} = \alpha R/\lambda.$$

Successive application of transformations (12.4.8) and (12.4.10) to Eq. (12.4.4) and conditions (12.4.13)–(12.4.15) yields an ordinary nonuniform transformed differential equation. Solution of this equation and inversion of (12.4.11) and (12.4.9) after some manipulations gives us the final solution

$$\theta(X, Z, \text{Fo}) = \frac{2\text{Bi}}{\pi}\theta_0 \int_Z^\pi \vartheta_1\left(\frac{Z^*}{2\pi}, i\frac{b^2}{\pi}\text{Fo}\right) dZ^*$$
$$\times \sum_{m=1}^{\infty} \frac{\exp[-\mu_m{}^2 \text{Fo}]}{\mu_m{}^2 + \text{Bi}^2} \frac{J_0(\mu_m X)}{J_0(\mu_m)}$$
$$+ \frac{4}{\pi} \sum_{m=1}^{\infty} \frac{\mu_m{}^2}{\mu_m{}^2 + \text{Bi}^2} \frac{J_0(\mu_m X)}{J_0{}^2(\mu_m)} \left\{ \sum_{n=0}^{\infty} \cos \mu_n Z \int_0^{\text{Fo}} \{\text{Po}(X, Z, \text{Fo}^*\}_{HC} \right.$$
$$\left. \times \exp[-(\mu_m{}^2 + b^2\mu_n{}^2)(\text{Fo} - \text{Fo}^*)]\, d\text{Fo}^* \right\}, \tag{12.4.16}$$

where

$$\{\text{Po}(X, Z, \text{Fo})\}_{HC} = \int_0^\pi \left[\int_0^1 X J_0(\mu_m X)\, \text{Po}(X, Z, \text{Fo})\, dX \right] \cos \mu_n Z\, dZ,$$

and μ_m are positive roots of the characteristic equation

$$J_0(\mu)/J_0{}'(\mu) = -(1/\text{Bi})\mu; \qquad \mu_n = \tfrac{1}{2}(2n + 1) \quad (n = 0, 1, 2, 3, \ldots).$$

Expansion of the theta-function ϑ_1 is introduced into (12.4.16)

$$\vartheta_1\left(\frac{Z}{2\pi}, i\frac{b^2}{\pi}\text{Fo}\right) = 2\sum_{n=0}^{\infty} (-1)^n \exp[-b^2\mu_n{}^2 \text{Fo}] \sin \mu_n Z.$$

Some particular cases of solution (12.4.16) will be considered next. For some heat transfer processes of high rate, a heat source (the Pomerantsev criterion) may be approximated by the following expression

$$\text{Po}(X, Z, \text{Fo}) = \text{Po}_1[1 - \exp[-\beta\, \text{Fo}]] J_0(\mu_1 X) \sin Z, \tag{12.4.17}$$

where μ_1 is the smallest positive root of the Bessel function of the zeroth order

$$J_0(\mu_1) = 0, \quad \mu_1 \approx 2.405.$$

From relation (12.4.16) taking into account (12.4.17) and the Poisson summation formula for the condition $\mu_m^2 > \beta$ we obtain [89]

$$\theta(X, Z, \mathrm{Fo}) = \frac{2\mathrm{Bi}}{\pi}\theta_0 \int_Z^{\pi} \vartheta_1\left(\frac{Z^*}{2\pi}, i\frac{b^2}{\pi}\mathrm{Fo}\right) dZ^*$$

$$\times \sum_{m=1}^{\infty} \frac{\exp[-\mu_m^2 \mathrm{Fo}]}{\mu_m^2 + \mathrm{Bi}^2} \frac{J_0(\mu_m X)}{J_0(\mu_m)} + 2\mathrm{Po}_1\, \mu_1 J_1(\mu_1)$$

$$\times \sum_{m=1}^{\infty} \frac{\mu_m^2}{(\mu_m^2 + \mathrm{Bi}^2)(\mu_m^2 - \mu_0^2)} \frac{J_0(\mu_m X)}{J_0(\mu_m)} \left\{\frac{1}{\mu_m^2 + b^2}\right.$$

$$\times \left[\frac{b}{\mu_m} \frac{\cosh(\mu_m/b)(\pi - Z)}{\sinh(\mu_m/b)\pi} - 2\frac{b}{\mu_m} \frac{\cosh(\mu_m/b)(2\pi - Z)}{\sinh 2(\mu_m/b)\pi} - \sin Z\right]$$

$$- \frac{\exp[-\beta\,\mathrm{Fo}]}{\mu_m^2 + b^2 - \beta} \left[\frac{b}{(\mu_m^2 - \beta)^{1/2}} \frac{\cosh\{(\mu_m^2 - \beta)^{1/2}/b\}(\pi - Z)}{\sinh\{(\mu_m^2 - \beta)^{1/2}/b\}\pi}\right.$$

$$\left. - 2\frac{b}{(\mu_m^2 - \beta)^{1/2}} \frac{\cosh\{(\mu_m^2 - \beta)^{1/2}/b\}(2\pi - Z)}{\sinh 2\{(\mu_m^2 - \beta)^{1/2}/b\}\pi} - \sin Z\right] + \frac{2}{\pi}\frac{\beta}{b^4}$$

$$\left. \times \exp[-\mu_m^2 \mathrm{Fo}] \sum_{n=0}^{\infty} \frac{\exp[-b^2\mu_n^2 \mathrm{Fo}] \cos \mu_n Z}{(\mu_n^2 - 1)(\mu_n^2 + \{\mu_m^2/b\})\{\mu_n^2 + (\mu_m^2 - \beta)/b^2\}}\right\}. \tag{12.4.18}$$

If $\mu_m^2 < \beta$, then in (12.4.18) $(\mu_m^2 - \beta)^{1/2}$ should be replaced by $i(\beta - \mu_m^2)^{1/2}$ throughout.

Some other solutions for boundary conditions simpler than (12.4.13)–(12.4.15) will be shown.

If the transfer through the cylinder bases may be neglected compared to that through its surface, then (12.4.15) is of the form

$$\frac{\partial\theta(X, 0, \mathrm{Fo})}{\partial Z} = \frac{\partial\theta(X, \pi, \mathrm{Fo})}{\partial Z} = 0, \tag{12.4.19}$$

and the solution similar to (12.4.16) may be found by the formula

$$\theta(X, Z, \mathrm{Fo}) = 2\mathrm{Bi}\,\theta_0 \sum_{m=1}^{\infty} \frac{\exp[-\mu_m^2 \mathrm{Fo}]}{\mu_m^2 + \mathrm{Bi}^2} \frac{J_0(\mu_m X)}{J_0(\mu_m)}$$

$$+ \frac{2}{\pi} \sum_{m=1}^{\infty} \frac{\mu_m^2}{\mu_m^2 + \mathrm{Bi}^2} \frac{J_0(\mu_m)X}{J_0^2(\mu_m)} \exp[-\mu_m^2 \mathrm{Fo}] \left[\int_0^{\mathrm{Fo}} \exp[\mu_m^2 \mathrm{Fo}^*]\right.$$

$$\times \left\{\int_0^{\pi} \left(\int_0^1 XJ_0(\mu_m X)\,\mathrm{Po}(X, Z, \mathrm{Fo}^*)\, dX\right) dZ\right\} d\mathrm{Fo}^*$$

$$+ \sum_{n=1}^{\infty} \exp[-b^2 n^2 \mathrm{Fo}] \cos nZ \int_0^{\mathrm{Fo}} \exp[(\mu_m^2 + b^2 n^2)\mathrm{Fo}^*]$$

$$\left. \times \left\{\int_0^{\pi} \left(\int_0^1 XJ_0(\mu_m X)\,\mathrm{Po}(X, Z, \mathrm{Fo}^*)\, dX\right) \cos nZ\, dZ\right\} d\mathrm{Fo}^*\right]. \tag{12.4.20}$$

If the dimensionless temperature on the bases is zero $\theta(X, 0, \mathrm{Fo}) = \theta(X, \pi, \mathrm{Fo}) = 0$, the problem solution can be obtained using the finite Fourier sine transformation

$$\theta(X, Z, \mathrm{Fo}) = \frac{4\mathrm{Bi}}{\pi}\theta_0 \int_0^Z \vartheta_2\left(\frac{\mathrm{Fo}^*}{\pi}, i\frac{4b^2}{\pi}\mathrm{Fo}\right) d\mathrm{Fo}^* \sum_{m=1}^{\infty} \frac{\exp[-\mu_m^2\,\mathrm{Fo}]}{\mu_m^2 + \mathrm{Bi}^2}$$
$$\times \frac{J_0(\mu_m X)}{J_0(\mu_m)} + \frac{4}{\pi}\sum_{m=1}^{\infty}\frac{\mu_m^2}{\mu_m^2 + \mathrm{Bi}^2}\,\frac{J_0(\mu_m X)}{J_0^2(\mu_m)}\exp[-\mu_m^2\,\mathrm{Fo}]$$
$$\times \left[\sum_{n=1}^{\infty}\exp[-b^2n^2\,\mathrm{Fo}]\sin(nZ)\int_0^{\mathrm{Fo}}\exp[(\mu_m^2 + b^2n^2)\,\mathrm{Fo}^*]\right.$$
$$\left.\times\left\{\int_0^{\pi}\left(\int_0^1 XJ_0(\mu_m X)\,\mathrm{Po}(X, Z, \mathrm{Fo}^*)\,dX\right)\sin nZ\,dZ\right\} d\mathrm{Fo}^*\right]. \tag{12.4.21}$$

Some other solutions for the cylinder with source of the kind (12.4.17) are presented in the work of Perelman [89]. Particular solutions of Eq. (12.4.2) or (12.4.4) with no source are given in monographs [10, 70]. It should be noted here that solutions for the case without sources may be obtained by superposition of particular solutions. For example, the solution for a finite cylinder is a superposition of solutions for the infinite one-dimensional cylinder and the infinite plate.

Methods for obtaining nonstationary temperature fields of a hollow axisymmetrical cylinder are the same as those for a solid cylinder. In this case, however, other variations of the Hankel finite integral transformation should be used instead of the (12.4.8). If, for example, boundary conditions of the third kind are prescribed, then on the internal and external cylinder surfaces, the following transformation should be used:

$$\{t(r, z, \tau)_H = \int_{R_1}^{R_2} t(r, z, \tau) r U_0(\mu_m r/R_1)\,dr, \tag{12.4.22}$$

where

$$U_0(\mu_m r/R_1) = \left[J_0(\mu_m) + \frac{1}{\mathrm{Bi}_1}\mu_m J_1(\mu_m)\right] Y_0(\mu_m r/R_1)$$
$$- \left[Y_0(\mu_m) + \frac{1}{\mathrm{Bi}_1}Y_1(\mu_m)\right] J_0(\mu_m r/R_1),$$

and μ_m are the roots of the equation

$$\frac{U_0(\varkappa\mu_m)}{U_1(\varkappa\mu_m)} = \frac{\varkappa\mu_m}{\mathrm{Bi}_2} \quad (m = 1, 2, \ldots),$$

$$U_1(\mu_m r/R_1) = \left[J_0(\mu_m) + \frac{\mu_m}{\mathrm{Bi}_1} J_1(\mu_m)\right] Y_1(\mu_m r/R_1) - \left[Y_0(\mu_m) + \frac{1}{\mathrm{Bi}_1} \mu_m Y_1(\mu_m)\right] Y_1(\mu_m r/R_1),$$

$$\mathrm{Bi}_1 = \frac{\alpha_1 R_1}{\lambda}; \qquad \mathrm{Bi}_2 = \frac{\alpha_2 R_2}{\lambda}; \qquad \varkappa = \frac{R_2}{R_1}. \tag{12.4.23}$$

The inversion of (12.4.22) is

$$t(r, z, \tau) = \sum_{m=1}^{\infty} 2\{t\}_H U_0(\mu_m r/R_1) R_1^{-2} \left\{ \varkappa^2 U_0^2(\varkappa\mu_m)\left[1 + \left(\frac{\mathrm{Bi}_2}{\varkappa\mu_m}\right)^2\right] - \frac{4}{\pi^2 \mathrm{Bi}_1^2}\left[1 + \left(\frac{\mathrm{Bi}_1}{\mu_m}\right)^2\right]\right\}^{-1}. \tag{12.4.24}$$

Here, we will present only the final solution for the problem with the boundary conditions

$$t(r, z, 0) = t_0; \tag{12.4.25}$$

$$\frac{\partial t(R_1, z, \tau)}{\partial r} - \frac{\alpha_1}{\lambda} t(R_1, z, \tau) = 0; \quad \frac{\partial t(R_2, z, \tau)}{\partial r} + \frac{\alpha_2}{\lambda} t(R_2, z, \tau) = 0; \tag{12.4.26}$$

$$\frac{\partial t(r, 0, \tau)}{\partial z} = 0; \qquad \frac{\partial t(r, l, \tau)}{\partial z} + \frac{\alpha}{\lambda} t(r, l, \tau) = 0$$

$$(R_1 \le r \le R_2, \; -l \le z \le l). \tag{12.4.27}$$

It may be written as [91]

$$\theta = \frac{t - t_0}{t_a - t_0} = 1 - \sum_{m=1}^{\infty}\sum_{n=1}^{\infty} A_m A_n U_0(\mu_m r/R_1) \cos \mu_n(z/l) \exp[-(\mu_m^2 + b^2\mu_n^2)\,\mathrm{Fo}_1], \tag{12.4.28}$$

$$A_m = \frac{2\,[\mathrm{Bi}_2\, U_0(\varkappa\mu_m) + 2/\pi]}{\varkappa^2 U_0^2(\varkappa\mu_m)[\mu_m^2 + (\mathrm{Bi}_2/\varkappa)^2] - (4/\pi^2\, \mathrm{Bi}_1^2)[\mu_m^2 + \mathrm{Bi}_1^2]}; \tag{12.4.29}$$

$$A_n = (-1)^{n+1} \frac{2\,\mathrm{Bi}(\mathrm{Bi}^2 + \mu_n^2)^{1/2}}{\mu_n^2(\mathrm{Bi}^2 + \mathrm{Bi} + \mu_n^2)}, \qquad \mathrm{Bi} = \frac{\alpha l}{\lambda}. \tag{12.4.30}$$

With the same values of transfer coefficients on the external and internal surfaces ($\alpha_1 = \alpha_2 = \alpha$), the coefficient A_m becomes of the form

$$A_m = \frac{2\,\mathrm{Bi}_1}{(\mu_m^2 + \mathrm{Bi}_1^2)[\varkappa U_0(\varkappa\mu_m) - (2/\pi\, \mathrm{Bi}_1)]}. \tag{12.4.31}$$

The infinite sum in solution (12.4.28) converges rapidly, so for practical calculations only one or two terms of the series are often sufficient. For convenience for the case $\alpha_1 = \alpha_2$, Plyat [91] presented plots for the first two roots μ_m of Eq. (12.4.23) and coefficients A_m as well as functions $U_0(\varkappa\,\mu_m)$ for the values of the Biot criterion $\mathrm{Bi}_1 = \alpha R_1/\lambda = 1$–$10$ and for the ratio $\varkappa = R_2/R_1 = 1.5$–$4.0$. These are shown in Figs. 12.2–12.4.

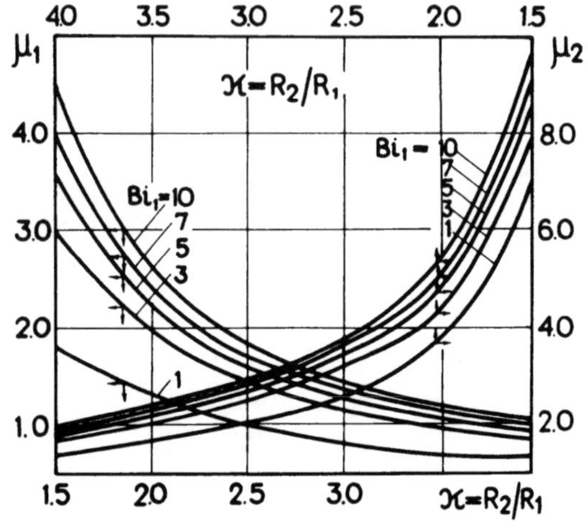

Fig. 12.2. Roots μ_1 and μ_2 of the characteristic equation for various Bi versus $\varkappa = R_2/R_1$ (infinite cylinder).

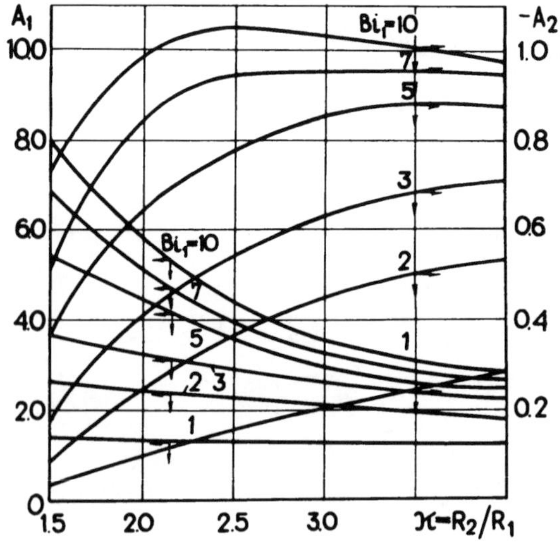

Fig. 12.3. Expansion coefficients A_1 and A_2 for various Bi_1 versus $\varkappa = R_2/R_1$ (infinite cylinder).

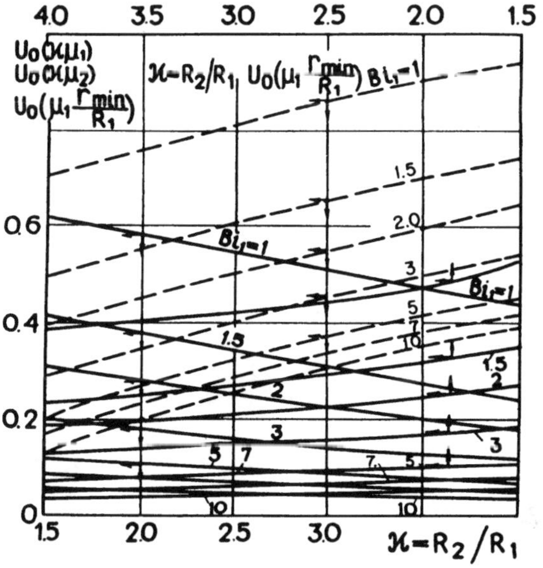

Fig. 12.4. Functions $U_0(\varkappa\mu_1)$ (solid lines), $U_0(\varkappa\mu_2)$ (broken lines) and $U_0(\mu_1\, r_{\min}/R_1)$ for various Bi_1 versus $\varkappa = R_2/R_1$ (infinite cylinder).

Other solutions for a hollow axisymmetrical cylinder for both constant and variable ambient temperature can be found in the above mentioned work by Plyat [91], as well as in works of Danilova [18], Plyatsko [92].

b. Axiasymmetric Cylinder. We conclude the chapter with a discussion of the general solution method of the axiasymmetric differential equation (12.4.1). According to Gavrilova and Prudnikov [35], Eq. (12.4.1) with general boundary conditions will be

$$t(r, \varphi, z, 0) = f(r, \varphi, z)\,, \tag{12.4.32}$$

$$\frac{\partial t(r, \varphi, 0, \tau)}{\partial z} = \psi(r, \varphi, \tau)\,, \tag{12.4.33}$$

$$t(r, \varphi, l, \tau) = \chi(r, \varphi, \tau)\,,$$

$$\lambda\,\frac{\partial t(R, \varphi, z, \tau)}{\partial r} + t(R, \varphi, z, \tau) = 0\,. \tag{12.4.34}$$

The solution will be found in the form of the Fourier series

$$t = \frac{t_0}{2} + \sum_{m=1}^{\infty} [t_m^{(1)} \sin m\varphi + t_m^{(2)} \cos m\varphi]\,. \tag{12.4.35}$$

Suitable representations of the functions w, f, ψ and χ reduce the problem of interest to the solution of the differential equation

$$\frac{\partial t_m^{(i)}}{\partial \tau} = a\left(\frac{\partial^2 t_m^{(i)}}{\partial r^2} + \frac{1}{r}\frac{\partial t_m^{(i)}}{\partial r} - \frac{m^2}{r^2} u_m^{(i)} + \frac{\partial^2 t_m^{(i)}}{\partial z^2}\right) + \frac{1}{c\gamma} w_m^{(i)}, \tag{12.4.36}$$

with boundary conditions

$$t_m^{(i)}(r, z, 0) = f_m^{(i)}(r, z), \tag{12.4.37}$$

$$\frac{\partial t_m^{(i)}(r, 0, \tau)}{\partial z} = \psi_m^{(i)}(r, \tau), \tag{12.4.38}$$

$$t_m^{(i)}(r, l, \tau) = \chi_m^{(i)}(r, \tau),$$

$$\lambda \frac{\partial t_m^{(i)}(R, z, \tau)}{\partial r} + t_m^{(i)}(R, z, \tau) = 0 \quad (i = 1, 2). \tag{12.4.39}$$

Further solution of the problem may be carried out in different ways. One way is by using finite Hankel transformations particularly suited to the problem (as Elistratova [29]); in this case, the solution is obtained in the series form with respect to eigenfunctions of the corresponding Sturm–Liouville problem. Another method is by using the Dini–Bessel series expansion of the functions entering (12.4.37)–(12.4.39):

$$t_m^{(i)} = \sum_{n=1}^{\infty} t_{mn}^{(i)} J_m(\mu_{mn} r/R), \tag{12.4.40}$$

where

$$t_{mn}^{(i)}(z, \tau) = \frac{2}{\pi R^2 [J_m'(\mu_{mn})]^2} \int_0^R t_m^{(i)} r J_m(\mu_{mn} r/R) dr, \tag{12.4.41}$$

and μ_{mn} are the roots of the equation

$$\frac{\gamma x J_m'(x) + R J_m(x)}{x^m} = 0. \tag{12.4.42}$$

Equation (12.4.36) and conditions (12.4.37)–(12.4.39) are then reduced to a new problem

$$\frac{\partial t_{mn}^{(i)}}{\partial \tau} = a\left[\frac{\partial^2 t_{mn}^{(i)}}{\partial z^2} - \left(\frac{\mu_{mn}}{R}\right)^2 t_{mn}^{(i)}\right] + \frac{1}{c\gamma} w_{mn}^{(i)}, \tag{12.4.43}$$

$$t_{mn}^{(i)}(z, 0) = f_{mn}^{(i)}(z),$$

$$\frac{\partial t_{mn}^{(i)}(0, \tau)}{\partial z} = \psi_{mn}^{(i)}(\tau); \quad t_{mn}^{(i)}(l, \tau) = \chi_{mn}^{(i)}(\tau). \tag{12.4.44}$$

The solution of the new problem is not very difficult. The methods are similar to those used for the solution of ordinary transfer equations with boundary conditions of the second kind and the solution may be carried out by means of combined application of the Fourier and Laplace transformations. As the result, we obtain

$$
\begin{aligned}
t_{mn}^{(i)}(z, \tau) = \exp\left[-a\left(\frac{\mu_{mn}}{R}\right)^2\right]\Bigg\{\frac{1}{2l}\int_0^l \left[\vartheta_2\left(\frac{z-\xi}{2l}, \frac{a\tau}{l^2}\right)+\vartheta_2\left(\frac{z+\xi}{2l}, \frac{a\tau}{l^2}\right)\right]\Bigg\} \\
\times f_{mn}^{(i)}(\xi)\, d\xi + \frac{a}{l}\left\{\int_0^\tau \frac{\partial}{\partial z}\vartheta_2\left[\frac{l-z}{2l}, \frac{a(\tau-\tau^*)}{l^2}\right]\chi_{mn}^{(i)}(\tau^*)\, d\tau^*\right. \\
\left. - \int_0^\tau \vartheta_2\left[\frac{z}{2l}, \frac{a(\tau-\tau^*)}{l^2}\right]\psi_{mn}^{(i)}(\tau^*)\, d\tau^*\right\} \\
+ \frac{1}{2l}\int_0^l\int_0^\tau \left\{\vartheta_2\left[\frac{z-\xi}{2l}, \frac{a(\tau-\tau^*)}{l^2}\right] + \vartheta_2\left[\frac{z+\xi}{2l}, \frac{a(\tau-\tau^*)}{l^2}\right]\right\} \\
\times \frac{1}{c\gamma} w_{mn}^{(i)}(\xi, \tau^*)\, d\xi\, d\tau^*\Bigg\}, \qquad (12.4.45)
\end{aligned}
$$

where

$$
\vartheta_2(x, \tau) = 2\sum_{n=0}^{\infty} \exp[-\pi^2(n+\tfrac{1}{2})^2\tau]\cos\pi(2n+1)x. \qquad (12.4.46)
$$

A number of particular problems for an axiasymmetric cylinder is discussed in the well-known monographs by Carslaw and Jaeger [8] and we shall not therefore consider them in detail here.

Consideration of more complex problems of two-dimensional and three-dimensional temperature fields is not the aim in this book. The interested readers may be referred to original works and special monographs [8, 57, 109, 122].

CHAPTER

13

HEAT CONDUCTION WITH VARIABLE TRANSFER COEFFICIENTS

Generally speaking, the values of the various transfer coefficients and thermodynamic properties of the material or medium may differ from point to point. Sometimes they change drastically with varying transfer potentials. A large number of scientific and technological problems may be solved with greater accuracy with the aid of correction factors, taking into account the variable nature of the coefficients. Present-day applications of high-rate transfer processes on a large scale in many branches of modern technology require us to take property variations into account.

It should also be mentioned that suitable substitutions reduce many problems on convective diffusion, heat conduction, fluid dynamics of viscous liquid, etc., to differential equations of the heat conduction type with variable coefficients. This provides additional motivation for the accumulation and generalization of solutions of the nonuniform and nonlinear heat conduction equations: it further suggests that we should continue to develop improved methods of dealing with such nonlinear equations.

As a heat transfer process occurs, the material changes its structural properties to some extent. With random or insignificant changes of body properties along the coordinate, it is permissible, when studying transfer phenomena, to assume thermodynamic properties and transfer coefficients constant and equal to their mean effective values. In a number of cases, however, the nonuniformity of physical properties become so significant and their change along the coordinate so regular, that it is impossible to neglect this nonuniformity. Then, instead of solving the transfer differential equations with constant coefficients, we must solve an equation where some or all of the coefficients are functions of the coordinates.

The solution of the resulting differential transfer equations with variable coefficients involves great difficulties. Exact analytical solutions are currently available for a limited range of problems. Often we are forced to restrict ourselves to various approximations or numerical methods. In this connection, the main problem to be solved by the analytical heat transfer theory is the development of methods for the solution of differential transfer equation systems with variable coefficients.

13.1 Semi-Infinite Body, Heat Conductivity, and Heat Capacity as Power Functions of Coordinates

In this section, solution methods for the following one-dimensional differential heat conduction equation will be discussed:

$$c\gamma \frac{\partial t}{\partial \tau} = \frac{\partial}{\partial x}\left(\lambda \frac{\partial t}{\partial x}\right). \tag{13.1.1}$$

Solutions of Eq. (13.1.1.) will be considered with the boundary conditions

$$t(x, 0) = 0 \quad (0 < x < \infty), \tag{13.1.2}$$

$$t(\infty, \tau) = 0, \qquad t(0, \tau) = t_a. \tag{13.1.3}$$

(a) Let $c\gamma = \text{const}$; the thermal conductivity λ is assumed to depend on the coordinate as

$$\lambda = \lambda_0 x^n. \tag{13.1.4}$$

Applying the Laplace transformation to Eq. (13.1.1) and taking into account (13.1.2) and (13.1.4) gives us

$$x^n T''(x, s) + n x^{n-1} T'(x, s) - (s/a_0) T(x, s) = 0, \tag{13.1.5}$$

where $a_0 = \lambda_0 / c\gamma$.

Substitution of $x = k\xi^m$ reduces Eq. (13.1.5) to

$$m^2 k^2 x^{2m+n-2}(d^2T/d\xi^2) + m(m + n - 1)k x^{m+n-2}(dT/d\xi) - (s/a_0)T = 0. \tag{13.1.6}$$

In the substitution, k and m were chosen arbitrarily. We now make them conform to the conditions

$$2m + n - 2 = 0, \qquad k = 1.$$

Equation (13.1.6) is then transformed into one of the Bessel type as

$$\frac{d^2T}{d\xi^2} + \frac{n}{2-n}\,\frac{1}{\xi}\,\frac{dT}{d\xi} - \frac{4}{(2-n)^2}\,\frac{s}{a_0}\,T = 0.$$

Its particular solutions are

$$\xi^\nu I_\nu\left(\xi\,\frac{2}{2-n}\left(\frac{s}{a_0}\right)^{1/2}\right) \quad \text{and} \quad \xi^\nu K_\nu\left(\xi\,\frac{2}{2-n}\left(\frac{s}{a_0}\right)^{1/2}\right),$$

where I_ν and K_ν are the Bessel functions of the imaginary νth-order argument of the first and second kind, $\nu = (1-n)/(2-n)$, where n is any number except two. Using the relation $\xi = x^{(2-n)/2}$, we return to the initial variable x:

$$x^{\frac{1}{2}(1-n)} I_\nu\left(x^{1-\frac{1}{2}n}\,\frac{2}{2-n}\left(\frac{s}{a_0}\right)^{1/2}\right) \text{ and } x^{\frac{1}{2}(1-n)} K_\nu\left(x^{1-\frac{1}{2}n}\,\frac{2}{2-n}\left(\frac{s}{a_0}\right)^{1/2}\right). \tag{13.1.7}$$

In the majority of cases n is less than unity ($0 < n < 1$).

The first solution does not conform to the first condition (13.1.3) and the solution of Eq. (13.1.5) is therefore taken in the form

$$T(x, s) = Ax^{\frac{1}{2}(1-n)} K_\nu\left(x^{1-\frac{1}{2}n}\,\frac{2}{2-n}\left(\frac{s}{a_0}\right)^{1/2}\right),$$

where A is a constant with respect to x determined from the second boundary condition (13.1.3); when $x \to 0$, $T(x, s) \to t_a/s$, and when $z \to 0$, $z^\nu K_\nu(z) \to 2^{\nu-1}\Gamma(\nu)$ where $\nu > 0$ and $\Gamma(\nu)$ is the gamma function. The constant A is therefore defined as

$$A = \frac{2t_a}{s\Gamma(\nu)(2-n)^\nu}\left(\frac{s}{a_0}\right)^{\nu/2}.$$

Hence, the transformed solution will be of the form

$$T(x, s) = t_a\,\frac{2x^{\frac{1}{2}(1-n)}}{s(2-n)^\nu\Gamma(\nu)}\left(\frac{s}{a_0}\right)^{\nu/2} K_\nu\left(\frac{2x^{1-\frac{1}{2}n}}{2-n}\left(\frac{s}{a_0}\right)^{1/2}\right). \tag{13.1.8}$$

Using the table of transforms, we obtain from (13.1.8) the solution for the original function as

$$\theta - \frac{t(x, \tau)}{t_a} = \frac{1}{\Gamma(\nu)} \int_x^\infty e^{-u} u^{\nu-1}\, du = \frac{\Gamma(\nu, X)}{\Gamma(\nu)},$$

where

$$X = x^{2-n}/(2-n)^2 a_0 \tau .$$

$\Gamma(\nu, X)$ is a gamma function.

Problems with zero initial conditions and for a finite body ($0 \leq x \leq R$) may be solved in a similar way. For the case $n = 2$, the solution of Eq. (13.1.5) will be

$$x^{[-1 \pm (1+4s/a_0)^{1/2}]/2} .$$

(b) If the heat capacities and thermal conductivities depend on the coordinate raised to the power n, then

$$c = c_0 x^n , \qquad \lambda = \lambda_0 x^n , \qquad \gamma = \text{const} , \tag{13.1.9}$$

then, applying the Laplace transformation and accounting for the initial condition (13.1.2) and the variable coefficients (13.1.9) gives us

$$T''(x, s) + (n/x)T'(x, s) - (s/a_0)T(x, s) = 0 ,$$

where

$$a_0 = \lambda_0 / c_0 \gamma .$$

The solutions of this equation are

$$x^\nu I_\nu((s/a_0)^{1/2} x) \quad \text{and} \quad x^\nu K_\nu((s/a_0)^{1/2} x) , \tag{13.1.10}$$

where

$$\nu = \tfrac{1}{2}(1 - n) .$$

Taking into account the first boundary condition (13.1.3) as we did in the previous problem, we obtain

$$T(x, s) = A x^\nu K_\nu((s/a_0)^{1/2} x) . \tag{13.1.11}$$

Combining (13.1.11) with the second boundary condition (13.1.3), we find

$$A = \frac{t_a}{s 2^{\nu-1} \Gamma(\nu)} \left(\frac{s}{a_0} \right)^{\nu/2} .$$

Hence the solution of Eq. (13.1.10) will be

$$T(x, s) = t_a \frac{x^\nu}{s 2^{\nu-1} \Gamma(\nu)} \left(\frac{s}{a_0}\right)^{\nu/2} K_\nu\left(\left(\frac{s}{a_0}\right)^{1/2} x\right).$$

The inversion of this solution is

$$\theta = \frac{t(x, \tau)}{t_a} = \frac{\sqrt{a_0}}{x} \frac{\Gamma(\nu, X)}{\Gamma(\nu)},$$

where

$$X = \frac{x^2}{4a_0\tau}.$$

(c) If heat capacity and thermal conductivities obey the power laws

$$c = c_0 x^m, \qquad \lambda = \lambda_0 x^n, \qquad \gamma = \text{const} \quad (m > -1),$$

then application of the Laplace transformation yields the transformed solution

$$T''(x, s) + \frac{n}{x} T'(x, s) - \frac{s}{a_0} \frac{1}{x^{n-m}} T(x, s) = 0.$$

The solution of this equation is similar to that of problem (a):

$$x^{-\frac{1}{2}(n-1)} I_\nu \left[\frac{2x^{\frac{1}{2}(m-n+2)}}{m-n+2} \left(\frac{s}{a_0}\right)^{1/2}\right]; \qquad x^{-\frac{1}{2}(n-1)} K_\nu \left[\frac{2x^{\frac{1}{2}(m-n+2)}}{m-n+2} \left(\frac{s}{a_0}\right)^{1/2}\right],$$

where

$$\nu = \frac{1-n}{m-n+2}.$$

It should be noted that the solutions for cylindrical and spherical shapes with the coefficients c, λ, and γ, which are power functions of the coordinates, are similar to the problems just discussed. For example, the transformed solution for a sphere is of the form

$$T'' + \frac{n+2}{r} T' - \frac{s}{a_0} \frac{1}{r^{n-m}} T = 0,$$

i.e., in this case the constants n and m are two units greater than before.

The physical meaning of the condition $n < 1$ lies in the absence of infinitely large thermal resistance of the infinitesimal section at $x = 0$. Otherwise, it would be impossible to satisfy boundary conditions at $x = 0$. Quite similarly, for $m \leq -1$, it is impossible to satisfy boundary conditions at $x = 0$ because of the infinitely large heat capacity of this section.

Particular cases of the problem previously considered are those in which the coefficients change with the coordinate following the linear law. Suitable substitution reduces them to a specific form of the solutions just discussed. Thus, in the case $c\gamma = \text{const}$, $\lambda = \lambda_0(1 + \alpha x)$ the new variable $\xi = 1 + \alpha x$ is introduced. In this case, we obtain the equation

$$\frac{d^2T}{d\xi^2} + \frac{1}{\xi}\frac{dT}{d\xi} - \frac{s}{a_0\xi}T = 0\,,$$

which is identical with Eq. (13.1.5), when $n = 1$. Its solutions will be

$$I_0\left[2\frac{(1+\alpha x)^{1/2}}{\alpha}\left(\frac{s}{a_0}\right)^{1/2}\right] \quad \text{and} \quad K_0\left[2\frac{(1+\alpha x)^{1/2}}{\alpha}\left(\frac{s}{a_0}\right)^{1/2}\right].$$

In the same way, we may show that in the case $c = c_0(1 + \alpha x)$, $\lambda = \lambda_0(1 + \alpha x)$, $\gamma = \text{const}$, the solutions of the transformed equation are

$$I_0\left[\frac{1+\alpha x}{\alpha}\left(\frac{s}{a_0}\right)^{1/2}\right] \quad \text{and} \quad K_0\left[\frac{1+\alpha x}{\alpha}\left(\frac{s}{a_0}\right)^{1/2}\right].$$

In a more general case, when

$$c = c_0(1 + \alpha x)^m \quad \text{and} \quad \lambda = \lambda_0(1 + \alpha x)^n\,,$$

introduction of the new variables

$$\xi = (1+lx)^{(m-n+2)/2}, \quad \eta = (l^2\lambda/4c)(m-n+2)^2\tau, \quad \zeta = (1-n)/(m-n+2),$$

reduces Eq. (13.1.1) to the equation

$$\gamma\xi^{1-2\zeta}(\partial t/\partial\eta) = \frac{\partial}{\partial\xi}\left(\xi^{1-2\zeta}(\partial t/\partial\xi)\right).$$

The remaining procedure is the same as for the first problem. A number of specific problems of this type is given in Chudnovsky's monograph [15]. Particularly, in this work as well as in that of Korenev [56] and Luikov and Mikhailov [73], solutions for a number of problems on thermal waves are presented. These problems are also discussed in detail in works by Korenev [56] and other authors.

13.2 Finite Plate. Thermal Conductivity as an Exponential Function of the Coordinate

The solution of equation (13.1.1) *with boundary conditions of the first kind will be considered when the thermal conductivity is an exponential function of the coordinate:*

$$\lambda = \lambda_0 \exp[-\varkappa x], \qquad c\gamma = \text{const} \quad (0 < x \le R;\ \varkappa > 0). \tag{13.2.1}$$

Let

$$t(x, 0) = 0, \tag{13.2.2}$$

$$t(0, \tau) = \varphi(\tau), \qquad t(R, \tau) = \psi(\tau). \tag{13.2.3}$$

Applying the Laplace transformation to Eq. (13.1.1) and conditions (13.2.1)–(13.2.3) we find

$$T''(x, s) - \varkappa T'(x, s) - (s/a_0)\exp[-\varkappa x]T(x, s) = 0, \tag{13.2.4}$$

$$T(0, s) = \Phi, \qquad T(R, s) = \Psi. \tag{13.2.5}$$

The substitution $\xi = (2/\varkappa)(s/a_0)^{1/2}\exp(\varkappa x/2)$ transforms (13.2.4) into the equation

$$T''(\xi, s) - \frac{1}{\xi}T'(\xi, s) - T(\xi, s) = 0,$$

whose solution is

$$T(\xi, s) = A\xi I_1(\xi) + B\xi K_1(\xi). \tag{13.2.6}$$

For the coordinate representation x, Eq. (13.2.6) will be written

$$\begin{aligned} T(x, s) = {} & A\frac{2}{\varkappa}\left(\frac{s}{a_0}\right)^{1/2}\exp\left[\frac{\varkappa x}{2}\right]I_1\left[\frac{2}{\varkappa}\left(\frac{s}{a_0}\right)^{1/2}\exp\left[\frac{\varkappa x}{2}\right]\right] \\ & + B\frac{2}{\varkappa}\left(\frac{s}{a_0}\right)^{1/2}\exp\left[\frac{\varkappa x}{2}\right]K_1\left[\frac{2}{\varkappa}\left(\frac{s}{a_0}\right)^{1/2}\exp\left[\frac{\varkappa x}{2}\right]\right]. \end{aligned} \tag{13.2.7}$$

Constants A and B will be found using boundary conditions (13.2.5). After some manipulations we obtain

$$\begin{aligned} T(x, s) = {} & \left\{\left[\frac{2\Psi}{\varkappa\sqrt{a_0}}K_1\left(\frac{2}{\varkappa}\left(\frac{s}{a_0}\right)^{1/2}\right) - \frac{2\Phi}{\varkappa\sqrt{a_0}}\exp\left[\frac{\varkappa R}{2}\right]\right.\right. \\ & \times K_1\left\{\frac{2}{\varkappa}\exp\left[\frac{\varkappa R}{2}\right]\left(\frac{s}{a_0}\right)^{1/2}\right\}\bigg]\exp\left[\frac{\varkappa x}{2}\right]I_1\left\{\frac{2}{\varkappa}\exp\left[\frac{\varkappa x}{2}\right]\left(\frac{s}{a_0}\right)^{1/2}\right\} \\ & + \left[\frac{2\Phi}{\varkappa\sqrt{a_0}}\exp\left[\frac{\varkappa R}{2}\right]I_1\left\{\frac{2}{\varkappa}\exp\left[\frac{\varkappa R}{2}\right]\left(\frac{s}{a_0}\right)^{1/2}\right\}\right. \\ & \left.\left. - \frac{2\Psi}{\varkappa\sqrt{a_0}}I_1\left\{\frac{2}{\varkappa}\left(\frac{s}{a_0}\right)^{1/2}\right\}\right]\exp\left[\frac{\varkappa x}{2}\right]K_1\left\{\frac{2}{\varkappa}\exp\left[\frac{\varkappa x}{2}\right]\left(\frac{s}{a_0}\right)^{1/2}\right\}\right\} \end{aligned}$$

$$\times \left\{\left[K_1\left\{\frac{2}{\varkappa}\left(\frac{s}{a_0}\right)^{1/2}\right\} I_1\left\{\frac{2}{\varkappa}\exp\left[\frac{\varkappa R}{2}\right]\left(\frac{s}{a_0}\right)^{1/2}\right\} - I_1\left\{\frac{2}{\varkappa}\left(\frac{s}{a_0}\right)^{1/2}\right\}\right.\right.$$

$$\left.\left.\times K_1\left\{\frac{2}{\varkappa}\exp\left[\frac{\varkappa R}{2}\right]\left(\frac{s}{a_0}\right)^{1/2}\right\}\right]\frac{2}{\varkappa\sqrt{a_0}}\exp\left[\frac{\varkappa R}{2}\right]\right\}^{-1}. \tag{13.2.8}$$

Expression (13.2.8) is the solution of the problem considered in transformation form. The inversion (13.2.8) may be obtained from the inversion formula

$$t(x, \tau) = \frac{1}{2\pi i}\int_{\sigma-i\infty}^{\sigma+i\infty} T(x, s)\exp[s\tau]\, ds, \tag{13.2.9}$$

where the integration is carried out along any straight line parallel to the imaginary axis of the plane of the complex variable s passing to the right of all the singularities of the integrand.

The contour integral (13.2.9) is to be evaluated subject to the conditions [Eq. (13.2.5)]

$$\Phi = 0 \qquad \text{and} \qquad \Psi = t_a/s, \tag{13.2.10}$$

which correspond to the boundary conditions

$$t(0, \tau) = 0; \qquad t(R, \tau) = t_a. \tag{13.2.11}$$

After a few manipulations and taking into account (13.2.10), we can rewrite Eq. (13.2.9) in the form

$$t(x, \tau) = \frac{t_a}{2\pi i}\int_{\sigma-i\infty}^{\sigma+i\infty}\frac{\exp\left[\frac{\varkappa}{2}(x-R)\right]\left[K_1\left(\frac{2}{\varkappa}\left(\frac{s}{a_0}\right)^{1/2}\right)I_1\left(\frac{2}{\varkappa}\left(\frac{s}{a_0}\right)^{1/2}\exp[\varkappa x/2]\right)\right]}{K_1\left(\frac{2}{\varkappa}\left(\frac{s}{a_0}\right)^{1/2}\right)I_1\left(\frac{2}{\varkappa}\left(\frac{s}{a_0}\right)^{1/2}\exp[\varkappa R/2]\right)}$$

$$-\frac{I_1\left(\frac{2}{\varkappa}\left(\frac{s}{a_0}\right)^{1/2}\right)K_1\left(\frac{2}{\varkappa}\left(\frac{s}{a_0}\right)^{1/2}\exp\left[\frac{\varkappa x}{2}\right]\right)\Big]}{I_1\left(\frac{2}{\varkappa}\left(\frac{s}{a_0}\right)^{1/2}\right)K_1\left(\frac{2}{\varkappa}\left(\frac{s}{a_0}\right)^{1/2}\exp[\varkappa R/2]\right)}\;\frac{\exp[s\tau]}{s}\,ds. \tag{13.2.12}$$

Calculation of this integral yields

$$\theta = \frac{t(x, \tau)}{t_a} = \frac{\exp[\varkappa x] - 1}{\exp[\varkappa R] - 1} - \pi\exp[\tfrac{1}{2}\varkappa(x - R)]$$

$$\times \sum_{n=1}^{\infty}\frac{J_1(\mu_n)J_1[\mu_n\exp[\frac{1}{2}\varkappa R]]}{J_1^{\,2}(\mu_n) - J_1^{\,2}[\mu_n\exp[\frac{1}{2}\varkappa R]]}\{J_1[\mu_n\exp[\tfrac{1}{2}\varkappa x]]$$

$$\times\; Y_1(\mu_n) - J_1(\mu_n)Y_1[\mu_n\exp[\tfrac{1}{2}\varkappa x]]\}\exp[-\tfrac{1}{4}a_0\varkappa^2\mu_n^{\,2}\tau], \tag{13.2.13}$$

where μ_n are the roots of the characteristic equation

$$\frac{J_1[\mu \exp[\frac{1}{2}\varkappa R]]}{J_1(\mu)} = \frac{Y_1[\mu \exp[\frac{1}{2}\varkappa R]]}{Y_1(\mu)}. \tag{13.2.14}$$

In the steady state ($\tau \to \infty$), solution (13.2.13) becomes of the form

$$t(x, \infty) = t_a \frac{\exp[\varkappa x] - 1}{\exp[\varkappa R] - 1}. \tag{13.2.15}$$

Several solutions for other problems with the exponential relation between the transfer coefficients and the coordinate, particularly for thermal waves, have been collected [15, 56].

13.3 Nonstationary Temperature Fields in Nonlinear Temperature Processes

In high-rate processes, the heat transfer potential can significantly change during short time intervals. This situation is encountered in problems concerned with diffusion processes, gas filtration in a porous medium, thermal explosions, chemical transformations, etc. A valid description of transfer phenomena involving a wide range of temperature must take into account the variations of the transport coefficients with temperature. Under these conditions, mass and heat flows become nonlinear, and determining the transfer potential fields involves the solution of the nonlinear differential equation

$$c(t)\gamma(t)(\partial t/\partial \tau) = \operatorname{div}[\lambda(t) \operatorname{grad} t]. \tag{13.3.1}$$

The solution of Eq. (13.3.1) with the appropriate boundary conditions gives rise to greater difficulties than those of the previously discussed problems in which the coefficients depended on the coordinates; for its solution, various approximate methods are widely used. A complete review of the state of the art is given by Friedman [33] and Crank [17] to whose works the interested reader is referred.

To solve the problems of a nonlinear transfer, a number of methods is currently used. The linearization method based on approximating the nonlinear coefficient involves matching a special relation for the coefficient which linearizes Eq. (13.3.1) [42]. The method of various substitutions demands the introduction of new variables allowing us to reduce nonlinear partial equation (13.3.1) to an ordinary nonlinear total equation, the solution of which is a simpler problem. There are some other methods for the solution of a nonlinear transfer equation (see, e.g., Luikov and Mikhailov [73]). Some methods of procedure used for the solution of a nonlinear transfer problem will be discussed next.

a. Linearization of the Nonlinear Differential Transfer Equation

Usually, the heat conduction coefficient is plotted as the slightly sloping curve $\lambda = \lambda(t)$ which may be correlated with sufficient accuracy by a linear or exponential relation. Charny [13] proposed for this case two linearization methods of the transfer equation for the condition $c\gamma = \text{const}$.

Equation (13.3.1) may be rewritten in the form

$$c\gamma(\partial t/\partial\lambda)(\partial\lambda/\partial\tau) = \operatorname{div}[\lambda(t)(\partial t/\partial\lambda)\operatorname{grad}\lambda]. \tag{13.3.2}$$

If we assume in Eq. (13.3.2) that

$$\lambda\frac{\partial t}{\partial\lambda} = A = \text{const} \qquad \text{and} \qquad \frac{\partial t}{\partial\lambda} = B = \text{const}\,, \tag{13.3.3}$$

then a linear equation with constant coefficients will be obtained. The first condition (13.3.3) corresponds to the exponential form of the curve $\lambda(t)$

$$A = (t_1 - t_0)/\{\ln(\lambda_1/\lambda_0)\}$$

and the second to its linear approximation

$$B = (t_1 - t_0)/(\lambda_1 - \lambda_0).$$

It is easy to see that conditions (13.3.3) are contradictory excluding the case $\lambda = \text{const}$. It may be therefore supposed that either is rigorously fulfilled, but in the latter the quantity A or B should be replaced by its mean value in the temperature range considered.

Another linearization method involves introduction of the new function

$$G = \int_{t_0}^{t} \lambda(t')\,dt'\,,$$

which transforms Eq. (13.3.1) into

$$c\gamma\,\frac{\partial t}{\partial G}\,\frac{\partial G}{\partial\tau} = \nabla^2 G. \tag{13.3.4}$$

If we assume in Equation (13.3.4) that

$$\partial t/\partial G = \text{const}\,,$$

then a linear equation is also obtained. This approximation suggests that a certain section of the curve $G = G(t)$ is replaced by an appropriately plotted chord.

As proved by Storm [111], it is possible to linearize a one-dimensional transfer equation for the case where the rate of heat or mass transfer through the boundary surface ($x = 0$) is a known function of time $\varphi(\tau)$, and the thermal conductivity is described by the relation

$$\lambda^{-1/2} = \lambda_0^{-1/2}[1 + a(t - t_0)]. \tag{13.3.5}$$

In this case, relation (13.3.5) is the approximation for problems with an exponential relation between coefficient λ and the temperature

$$\lambda = \lambda_0 \exp[-2a(t_0 - t)],$$

where $a = \text{const}$ and λ_0 corresponds to the value of the conductivity of the potential t_0. The substitutions

$$\xi = \int_0^x \lambda^{-1/2}\, dx', \qquad \eta = \int_{t_0}^t (\lambda(t'))^{1/2}\, dt',$$

transform one-dimensional equation (13.3.1) into the linear equation

$$c\gamma \frac{\partial \omega}{\partial \tau} = \frac{\partial^2 \omega}{\partial \xi^2} - \frac{a}{\lambda_0^{1/2}} \varphi(\tau) \frac{\partial \omega}{\partial \xi},$$

where

$$\log \omega = -(a/\lambda_0^{1/2})\eta.$$

Storm applied this method to the solution of the problem on a nonstationary temperature distribution in a semi-infinite medium, the transfer rate across the surface $x = 0$ being constant.

We conclude the section by discussing one more method for linearizing the nonlinear transfer equation developed by Wiedenburg [131].

The one-dimensional equation (13.3.1) will be considered provided that the relation between the heat conduction coefficient and the temperature is linear $\lambda = \lambda_0(1 + \alpha t)$. In this case, Eq. (13.3.1) may be written

$$\frac{\partial t}{\partial \tau} = \frac{\partial}{\partial x}\left[a_0(1 + \alpha t)\frac{\partial t}{\partial x}\right] \tag{13.3.6}$$

or, equivalently

$$\frac{\partial t}{\partial \tau} = a_0 \frac{\partial^2 t}{\partial x^2} + \frac{a_0 \alpha}{2} \frac{\partial^2 t^2}{\partial x^2}, \tag{13.3.6a}$$

where $a_0 = \lambda_0/c\gamma = \text{const}$.

The boundary conditions will be written in the form

$$t(x, 0) - t_0 \quad (0 < x < +\infty), \qquad t(0, \tau) = 0, \qquad \partial t(\infty, \tau)/\partial x = 0.$$

When the value of the thermal diffusivity $a(t) = a_0$ ($\alpha = 0$) is constant, the solution of this equation is known and determined by the relation

$$\{t\}_{\alpha=0} = t_0[1 - \operatorname{erfc}(x/2(a_0\tau)^{1/2})]. \qquad (13.3.7)$$

To linearize Eq. (13.3.6), Wiedenburg suggested that we chose, for the term of the equation containing α, a value of the temperature which would conform to the appropriate expression if $\alpha = 0$ [i.e., to solution (13.3.7)]. This substitution makes the nonlinear equation (13.3.6) a linear differential equation with a source dependent on the space coordinate and on time:

$$\frac{\partial t}{\partial \tau} = a_0 \frac{\partial^2 t}{\partial x^2} + a_0 \frac{\alpha t_0^2}{2} \frac{\partial^2}{\partial x^2}\left[\operatorname{erf}\left(\frac{x}{2(a_0\tau)^{1/2}}\right)\right]^2. \qquad (13.3.8)$$

If a new variable, $\xi = x/2(a_0\tau)^{1/2}$, the argument of the Gauss function, is introduced into Eq. (13.3.8), we obtain an ordinary differential equation

$$\frac{d^2 t}{d\xi^2} + 2\xi \frac{dt}{d\xi} = -\frac{\alpha t_0^2}{2} \frac{d^2}{d\xi^2} [\operatorname{erf} \xi]^2,$$

with the conditions

$$t = 0 \quad \text{at} \quad \xi = 0 \qquad \text{and} \qquad t = t_0 \quad \text{at} \quad \xi = +\infty. \qquad (13.3.9)$$

The solution of the latter is to be found in the form

$$t = \varphi_1(\xi) + \varphi_2(\xi)\,\Phi(\xi),$$

wherein for abbreviation, we denote $\Phi(\xi) = \operatorname{erfc}(\xi)$. After the substitution, it becomes apparent that the functions φ_1 and φ_2 are determined by equations

$$\frac{d\varphi_1}{d\xi} = -\Phi \frac{d\varphi_2}{d\xi} \qquad \text{and} \qquad \frac{d\varphi_2}{d\xi}\frac{d\Phi}{d\xi} = -\frac{\alpha t_0^2}{2} \frac{d^2(1-\Phi)^2}{d\xi^2}. \qquad (13.3.10)$$

The solution of Eq. (13.3.10) yields

$$\begin{aligned}\varphi_1(\xi) = {} & A + \alpha t_0^2 \xi^2 \Phi + \alpha t_0^2 \tfrac{1}{2}(1 - \Phi) - (\alpha t_0^2/\sqrt{\pi})\xi \exp[-\xi^2] - \alpha t_0^2 \xi^2 \Phi^2 \\ & - \alpha t_0^2 \tfrac{1}{2}(1 - \Phi^2) + 2\alpha t_0^2 (\Phi/\sqrt{\pi})\xi \exp[-\xi^2] \\ & + \frac{\alpha t_0^2}{\pi}[1 - \exp[-2\xi^2]] + \tfrac{1}{2}(\alpha t_0^2)\Phi^2,\end{aligned}$$

$$\varphi_2(\xi) = B - \alpha t_0^2\xi^2 + \alpha t_0^2\xi^2\Phi + \alpha t_0^2\,\tfrac{1}{2}(1-\Phi) - (\alpha t_0^2/\sqrt{\pi})\xi \times \exp[-\xi^2] - \alpha t_0^2\Phi^2.$$

Hence

$$t = A + B\Phi - (\alpha t_0^2/\sqrt{\pi})\,[1-\Phi]\xi\exp[-\xi^2] + (\alpha t_0^2/\pi) \times [1-\exp[-2\xi^2]] - \tfrac{1}{2}(\alpha t_0^2)\Phi^2,$$

where A and B are constants with respect to ξ. Combining the solution obtained with the boundary conditions (13.3.9), we find

$$A = t_0 - (\alpha t_0^2/\pi) \qquad \text{and} \qquad B = \tfrac{1}{2}(\alpha t_0^2) + (\alpha t_0^2/\pi) - t_0.$$

In a similar way, the solution of Eq. (13.3.6) may be obtained with the conditions

$$t(x, 0) = t_1, \quad 0 < x < +\infty,$$

$$t(x, 0) = t_2, \quad -\infty < x < 0, \quad t_2 > t_1,$$

$$\partial t(\pm\infty, \tau)/\partial x = 0.$$

Using the solution with $\alpha = 0$

$$\{t\}_{\alpha=0} = t_1 + \tfrac{1}{2}(t_2 - t_1)\Phi,$$

we shall find

$$t = A + B\Phi + \frac{\alpha t_1(t_2-t_1)}{2\sqrt{\pi}}\,\xi\exp[-\xi^2] + \frac{\alpha(t_2-t_1)^2}{4\sqrt{\pi}}\,\Phi\xi\exp[-\xi^2] + \frac{\alpha(t_2-t_1)^2}{4\pi}\,[1-\exp[-2\xi^2]] - \frac{\alpha(t_2-t_1)^2}{8}\,(1-\Phi+\Phi^2) - \frac{\alpha t_1(t_2-t_1)}{4}\,(1-\Phi),$$

where

$$A = t_1 + \frac{\alpha(t_2-t_1)^2}{4}\left(\frac{1}{2}-\frac{1}{\pi}\right) + \frac{\alpha t_1(t_2-t_1)}{4},$$

$$B = \frac{t_2-t_1}{2} + \frac{\alpha(t_2-t_1)^2}{8} - \frac{\alpha t_1(t_2-t_1)}{4}.$$

It should be noted that transformation of t alone does not give exact linearization with $a \neq \text{const}$; exact linearization in this case is only obtained by Storm's method. This linearization was achieved by introduction of transformation of the coordinates which depend on t (through λ). Other linearization methods are presented in the work of Friedman [33].

b. Transformations Used for Solution of Nonlinear Differential Transfer Equations

In general, the solution of nonlinear differential equations, whether exact or approximate, is simplified considerably if they are transformed into ordinary nonlinear differential equations. The efficiency of some transformations of this kind was shown when the problem of linearizing a nonlinear equation was first discussed, which we shall now pursue in more detail.

(1) Consider the one-dimensional differential heat-conduction equation

$$c(t)\gamma(t)\frac{\partial t}{\partial \tau} = \frac{\partial}{\partial x}\left[\lambda(t)\frac{\partial t(x, \tau)}{\partial x}\right]. \tag{13.3.11}$$

To determine necessary substitutions, the method of the generalized variables theory can be used.

It follows from dimensional considerations, that this equation remains invariable if the length scale is changed by factor k and simultaneously the time scale is changed by factor k^2. After such changes of scales end conditions for uniform initial and boundary conditions

$$t(x, 0) = t_0, \tag{13.3.12}$$

$$t(0, \tau) = t_a, \tag{13.3.13}$$

also remain invariable, i.e.,

$$t(x, \tau) \equiv t(kx, k^2\tau)$$

with any values of x, τ, and k.

Assuming $k = 1/2\sqrt{\tau}$ yields

$$t(x, \tau) = t\{(x/2\sqrt{\tau}), \tfrac{1}{4}\} = t^*f(\xi)$$

where t^* is a certain fixed value of t,

$$\xi = x/2\sqrt{\tau}. \tag{13.3.14}$$

Thus t depends on the argument ξ alone. We now attempt to find the solution by transforming the differential equation into terms of τ and ξ.

We have

$$\frac{\partial t}{\partial \tau} = -\frac{xt^*}{4\sqrt{\tau^3}}\frac{\partial f}{\partial \xi} = -t^*\frac{\xi}{2\tau}\frac{df}{d\xi}, \qquad \frac{\partial t}{\partial x} = \frac{t^*}{2\sqrt{\tau}}\frac{df}{d\xi};$$

$$\frac{\partial}{\partial x}\left(\lambda\frac{\partial t}{\partial x}\right) = \frac{\partial}{\partial x}\left(\lambda\frac{t^*}{2\sqrt{\tau}}\frac{df}{d\xi}\right) = \frac{t^*}{4\tau}\frac{d}{d\xi}\left(\lambda\frac{df}{d\xi}\right).$$

Substituting the relations obtained into Eq. (13.3.11), we find the total equation for f to be

$$-2c\gamma\xi\frac{df}{d\xi} = \frac{d}{d\xi}\left[\lambda(f)\frac{df}{d\xi}\right]. \qquad (13.3.15)$$

Boundary conditions (13.3.12) and (13.3.13) will then become of the form

$$f(\infty) = t_0, \qquad f(0) = t_a.$$

If finding the function f is difficult or impossible analytically, it may be obtained by numerical integration.

The substitution $\xi = x/2\sqrt{\tau}$ is sometimes referred to as the Boltzmann transformation as it was first used by Boltzmann [4] in 1894 for the solution of Eq. (13.3.11). Substitution (13.3.14) may be applied to Eq. (13.3.11) when transfer phenomena take place in an infinite or semi-infinite medium. In this case, the independent variable must be specified so that the boundary condition depends solely on the variable ξ. In particular, substitution (13.3.14) cannot be used for a finite body, R in thickness, and with the boundary condition

$$t(0, \tau) = t_a, \qquad t(R, \tau) = t_a,$$

since upon transformation of (13.3.14), the argument in the second condition becomes different from zero and depends on both ξ and τ.

As a particular example of using the Boltzmann substitution, we shall solve Eq. (13.3.11) for a semi-infinite body ($\tau > 0$, $0 < x < \infty$) with the boundary conditions

$$t(x, 0) = t_0, \qquad t(0, \tau) = t_a, \quad \partial t(\infty, \tau)/\partial x = 0. \qquad (13.3.16)$$

Let $k = 1/\sqrt{\tau}$; in this case $\xi = x/\sqrt{\tau}$. Then the heat conduction equation will be

$$-\frac{\xi}{2}c\gamma\frac{\partial t}{\partial \xi} = \frac{\partial}{\partial \xi}\left(\lambda\frac{\partial t}{\partial \xi}\right). \qquad (13.3.17)$$

We denote $\partial t/\partial\xi = t'$. Then Eq. (13.3.17) may be written

$$-\frac{\xi}{2}c\gamma\frac{d\xi}{\lambda} = \frac{d(\lambda t')}{\lambda t'}.$$

Hence we shall have

$$\ln(\lambda t') = -c\gamma\int_0^\xi (\xi\, d\xi/2\lambda) + \ln A$$

or

$$\lambda t' = A\exp\left[-c\gamma\int_0^\xi (1/2\lambda)\xi\, d\xi\right].$$

Further integration yields

$$t = A\int_0^\xi \exp\left[-c\gamma\int_0^\xi (\xi/2\lambda)\, d\xi\right](d\xi/\lambda) + B.$$

The constants A and B are determined from boundary conditions (13.3.16). The final solution is of the form

$$\theta = \frac{t_a - t(x,\tau)}{t_a - t_0} = \frac{\int_0^\xi \exp\left[-c\gamma\int_0^{\xi'} (1/2\lambda)\xi\, d\xi\right](d\xi'/\lambda)}{\int_0^\infty \exp\left[-c\gamma\int_0^{\xi'} (\xi/2\lambda)\, d\xi\right](d\xi'/\lambda)}. \quad (13.3.18)$$

Since its right-hand side contains t, relation (13.3.18) is a nonlinear integral equation and not an explicit expression for the solutions. One possible method for such a calculation of a nonstationary temperature field θ is the so-called iteration method. The essence of the iteration method is in the fact that, if the first approximate solution of the equation

$$\theta = \varphi(t)$$

equal to θ_0 is known, then substituting it into the initial equation yields the second approximation $\theta_1 = \varphi(\theta_0)$. The substitution of θ_1 gives the third approximation θ_2, and so on. This operation repeated several times allows us to obtain the solution with any prescribed accuracy.

For simplification, we assume that $c\gamma = \text{const}$. Then (13.3.18) may be rewritten as

$$\theta = \frac{t_a - t(x,\tau)}{t_a - t_0} = \frac{\int_0^\eta \exp\left[-\int_0^\eta (2\eta/\lambda^*)\, d\eta\right](d\eta/\lambda^*)}{\int_0^\infty \exp\left[-\int_0^\eta (2\eta/\lambda^*)\, d\eta\right](d\eta/\lambda^*)}, \quad (13.3.19)$$

where

$$\lambda^* = \lambda/\lambda_0 = \lambda^*(\theta). \tag{13.3.20}$$

Here $\eta = x/2(a_0\tau)^{1/2}$, $a_0 = \lambda_0/c\gamma$ where λ_0 corresponds to the initial value of the temperature t_0.

With a constant value of thermal conductivity $\lambda = \lambda_0 = \text{const}$, relation (13.3.19) becomes the solution

$$\theta = \text{erf}(\eta), \tag{13.3.21}$$

which we use as the first approximation for the calculation.

Using (13.3.21) for equal intervals of change in η, we shall find the approximate values of θ_0. It should be remembered that even when $\eta = 3.0$, with the accuracy to the fifth decimal place, erf $\eta = 1$ and consequently, a stationary potential distribution sets in through the material.

For each of the obtained θ_0, the values of $\lambda_0{}^*$ are estimated by formula (13.3.20). Substituting these values into (13.3.19) allows us to determine new values of θ_1 for the chosen η using any convenient numerical integration formula if the integration is not exact. The calculation procedure is then repeated to obtain the approximations θ_2, θ_3, and so on, till the difference between the successive approximations is of the desired accuracy. This corresponds to approximate integration by the trapezoidal method. Application of other methods is similar.

(2) For numerical calculations of the nonlinear transfer equation (13.3.11) on electronic computers or analog models, Eq. (13.3.11) may be conveniently handled in another form. In this case, the substitution

$$G = \int_{t_0}^{t} \{\lambda(t')/\lambda_0(t_0)\}\, dt' \tag{13.3.22}$$

is advisable.

Substitution (13.3.22) reduces Eq. (13.3.11) to the equation

$$\partial G/\partial \tau = a(G)\, \nabla^2 G, \tag{13.3.23}$$

where $a(G) = \lambda(G)/c(G)\gamma(G)$. It should be noted here that Eq. (13.3.23) is of considerable significance in fluid dynamics. Its solution with appropriate boundary conditions is obtained as the relation between the function G, the coordinates, and time. The function and the temperature t are related by (13.3.22), which may be expressed either in an explicit form (when integration of the expression is possible) or by a plot (when numerical integration is necessary).

For a number of materials or fluids, the relation between the integral thermal conductivity and the volume heat capacity with temperature found by formula (13.3.22) is well approximated by the exponential law

$$\int_0^t (\lambda(t')/\lambda_0)\, dt' = \alpha t^n; \qquad \int_0^t (\gamma c(t')/\gamma_0 c_0)\, dt' = \beta t^m, \quad (13.3.24)$$

wherein λ_0, γ_0, c_0, α, n, β, and m are constants. For example, integral thermal conductivities and volume heat capacities for pure aluminum and iron in a wide range of temperature are described by relations (13.3.24) and have the following constants of the values:

aluminum: $\alpha = 1.57$; $n = 0.905$; $\beta = 0.871$; $m = 1.03$;

iron: $\alpha = 2.04$; $n = 0.84$; $\beta = 0.561$; $m = 1.125$.

Transformation (13.3.22) may also be extended to a variable heat capacity coefficient. Then the transformation of transfer equation (13.3.11) will be of the form

$$\frac{\partial}{\partial \tau} \int_0^t \frac{\gamma c(t')}{\gamma_0 c_0}\, dt' = a_0 \frac{\partial^2}{\partial x^2} \int_0^t \frac{\lambda(t')}{\lambda_0}\, dt'.$$

Friedman [32, 33] solved this equation with conditions

$$t(x, 0) = t(\infty, \tau) = 0, \qquad t(0, \tau) = t_1,$$

using relation (13.3.24). For the approximation of the second order, the solution will be of the form

$$\begin{aligned}[t(\xi)/t_1]^n = {} & \operatorname{erfc} \xi - 4s\{\tfrac{1}{2}(\xi^2 - \tfrac{1}{2})(\operatorname{erfc} \xi)^2 - (1/2\sqrt{\pi}) \exp[-\xi^2][2\xi \operatorname{erfc} \xi \\ & - (1/\sqrt{\pi}) \exp[-\xi^2]] + \tfrac{1}{2}\xi^2 \operatorname{erfc} \xi + (1 + \operatorname{erfc} \xi)[(1/\sqrt{\pi}) \\ & \times \exp[-\xi^2] - \xi \operatorname{erfc} \xi + \tfrac{1}{4}\{(1 + 2\xi^2) \operatorname{erfc} \xi - (2/\sqrt{\pi})\xi \\ & \times \exp[-\xi^2]\}] - \{\tfrac{1}{4} + (1/2\pi)\} \operatorname{erfc} \xi\}, \end{aligned} \quad (13.3.25)$$

where

$$\xi = \frac{x}{2(k t_1^n \tau)^{1/2}}, \qquad s = 1 - \frac{m}{n}, \qquad k = \frac{\lambda_0}{\gamma c_0} \frac{\alpha}{\beta} \frac{n}{m}.$$

Maximum divergence of the result obtained by formula (13.3.25) from that obtained by the subsequent third approximation was not greater than 3% (5°C).

Similar solutions were also obtained for a linear change of the coefficients λ and c with the temperature t for both a semi-infinite and finite medium [10, 32]. Comparison of the results with exact solutions shows good agreement in all cases.

(3) Consider now a number of special substitutions used for the solution of nonlinear heat conduction equations when the relations between the thermal conductivities and the temperature are linear or exponential. If we assume that $c\gamma = \text{const}$, Eq. (13.3.11) may be rewritten

$$\frac{\partial t}{\partial \tau} = \frac{\partial}{\partial x}\left[a(t)\,\frac{\partial t}{\partial x}\right], \tag{13.3.26}$$

where

$$a(t) = (1/c\gamma)\lambda(t).$$

(a) An infinite medium will be considered. Let the temperature satisfy the initial conditions

$$t(x, 0) = t_1, \quad \text{at } x < 0, \tag{13.3.27}$$

$$t(x, 0) = t_2, \quad \text{at } x > 0, \tag{13.3.28}$$

and the thermal diffusivity $a(t)$ be a linear function of the temperature

$$a = \bar{a}[1 + \tfrac{1}{2}\,\alpha(t_1 + t_2) - \alpha t].$$

Here $\bar{a}$ corresponds to a value of the coefficient a with average value of the temperature $t = \frac{1}{2}(t_1 + t_2)$. The substitution

$$\psi = \left[\frac{1 + \frac{1}{2}\alpha(t_1 + t_2) - \alpha t}{1 + \frac{1}{2}\alpha(t_1 - t_2)}\right]^2 \qquad \xi = \frac{x}{2[1 + \frac{1}{2}\alpha(t_1 - t_2)]^{1/2}(\bar{a}\tau)^{1/2}}$$

transforms Eq. (13.3.26) into

$$\frac{d^2\psi}{d\xi^2} = -\frac{2\xi}{\sqrt{\psi}}\,\frac{d\psi}{d\xi}, \tag{13.3.29}$$

with the boundary conditions

$$\psi = 1 \quad \text{at } \xi = +\infty, \qquad \psi = b^2 \quad \text{at } \xi = -\infty$$

where

$$b = \frac{1 - \frac{1}{2}\alpha(t_1 - t_2)}{1 + \frac{1}{2}\alpha(t_1 - t_2)}.$$

The numerical solution of Eq. (13.3.29) was obtained by Stokes [110].

The same case of a linear relation for the temperature diffusivity

$$a = a_0(t/t_0)$$

was considered by Polubarinova–Kochina [93]. Indeed, for a semi-infinite medium with the conditions

$$t(0, \tau) = t_0, \qquad t(x, 0) = 0,$$

the substitution

$$\xi = x/2(a_0\tau)^{1/2}$$

transforms Eq. (13.3.26) into

$$2\xi \frac{d\theta}{d\xi} + \left(\frac{d\theta}{d\xi}\right)^2 + \theta \frac{d^2\theta}{d\xi^2} = 0. \tag{13.3.30}$$

Here $\theta = t/t_0$. The equation may be further simplified by the substitution $\nu = \theta^2$. Then (13.3.30) becomes

$$\frac{d^2\nu}{d\xi^2} + \frac{2\xi}{\sqrt{\nu}} \frac{d\nu}{d\xi} = 0,$$

with boundary conditions

$$\nu = 1 \quad \text{at } \xi = 0, \qquad \nu = 0 \quad \text{at } \xi = \infty.$$

(b) Consider a semi-infinite medium with the condition

$$t(x, 0) = t_0, \qquad t(0, \tau) = t_s \tag{13.3.31}$$

where the thermal diffusivity a follows the exponential law

$$a = a_s \exp[\beta(t - t_s)]$$

and where the subscript s characterizes the value of the parameter on the surface. Then, after transformation we obtain

$$\frac{d}{d\xi}\left(e^{\zeta} \frac{d\zeta}{d\xi}\right) + 2\xi \frac{d\zeta}{d\xi} = 0,$$

where

$$\xi = \frac{x}{2(a_s\tau)^{1/2}}, \qquad \zeta = \beta(t - t_s). \tag{13.3.32}$$

Because of conditions (13.3.31) and (13.3.32) integration in this case should be carried out beginning from $\zeta = 0$ and $\xi = 0$. A more detailed procedure of numerical calculation and its results are presented by Crank [17]. In the same monograph, other transformations are given which are used when the relation between the thermal diffusivity and t is exponential.

(4) We now consider the Newton–Kantorovich iteration method. Methods of solution of general nonlinear problems are numerous and we shall only discuss very briefly the Newton–Kantorovich iteration method, referring the interested reader to the original literature for details [52].

We have an approximate solution of the nonlinear equation

$$u(x, y, \cdots, \tau). \tag{13.3.33}$$

The following approximation in the form $u + \alpha\, \delta u$ is substituted into the equation (boundary, initial, and other conditions) and expanded in series with respect to α, with only the linear terms being conserved. Then a new equation is obtained with respect to $\alpha\, \delta u$ where the coefficients only depend on independent variables.

Upon the solution of this equation, we shall take $(u + \alpha\, \delta u)$ as the initial approximation and repeat this process. Very rapid convergence in particular cases is a peculiarity of Newton's method, so that no more than two or three iterations are usually used. The error estimation in this case is of the form $\sim q^{2^k}$ with some $| q | < 1$ rather than $q_1{}^k$ with usual iterations. Consequently, convergence is very rapid. In this case, estimation of q involves the second derivatives of the coefficients in the equation and boundary conditions with respect to the unknown quantity and the estimation for the general solution of linear problems to intermediate approximations. Accuracy is achieved by a new, more exact problem which is stated in each step.

In the case when the second derivatives of the coefficients do not exist (or they are very large), Newton's method loses its advantages.

It should be noted that in the solution of general linear problems obtained in each step of Newton's method, different approximate methods should be used. Therefore, application of the method to unsteady-state problems is not always reasonable since approximate step-by-step solution allows corrections of nonlinearity in each step. Newton's method is the most promising one for the solution of boundary value problems, e.g., those of steady-state and quasistationary nonlinear transfer, including problems in multidimensional regions especially with solution by electronic digital computers.

As an example, we shall write the procedure of application of Newton's method to Eq. (13.3.15) with $c\gamma = \text{const}$. We have a certain (initial or previous) approximation f_{k-1}. Let

$$f_k = f_{k-1} + \alpha\, \delta f_k. \tag{13.3.34}$$

Substitution of f_k into the equation and conservation of the terms in power α^0 and α in the Taylor series expansion give the equation for correction as

$$2c\gamma\xi \frac{d(\alpha\delta f_k)}{d\xi} + \frac{d}{d\xi}\left[\lambda(f_{k-1}(\xi))\frac{d(\alpha\delta f_k)}{d\xi}\right] + \frac{d}{d\xi}\left(\alpha\delta f_k \frac{d\lambda(f_{k-1}(\xi))}{d\xi}\right)$$
$$= -\left[2c\gamma\xi \frac{df_{k-1}}{d\xi} + \frac{d}{d\xi}\lambda(f_{k-1}(\xi))\frac{df_{k-1}}{d\xi}\right], \tag{13.3.35}$$

with end conditions

$$(\alpha\,\delta f_k)_\infty = t_0 - f_{k-1}(\infty), \qquad (\alpha\,\delta f_k)_0 = t_a - f_{k-1}(0). \tag{13.3.36}$$

Further, upon the solution of problem (13.3.35)–(13.3.36), the same procedure should be carried out for transition from f_k to $f_{k+1} = f_k + \alpha\,\delta f_{k+1}$, and so on. The detailed description is presented by Kantorovich and Akimov [52].

c. *Some Solutions of the Nonlinear Heat Conduction Equation*

By studying solutions of the nonlinear differential heat conduction equations, the effects of the nonlinearity of the transfer coefficients on the temperature distribution will be shown.

(1) *Semi-infinite medium*

$$a(t) = a_0/(1 - \lambda t) \qquad (a_0,\ \lambda \text{ are constants}). \tag{13.3.37}$$

Boundary conditions are of the form

$$t(x, 0) = 0; \qquad t(0, \tau) = t_0. \tag{13.3.38}$$

With the introduction of new variables

$$\theta = t/t_0, \qquad A(\theta) = a(t)/a_0, \qquad \xi = x/2(a_0\tau)^{1/2}, \tag{13.3.39}$$

Eq. (13.3.26) reduces to the ordinary differential equation

$$-2\xi \frac{d\theta}{d\xi} = \frac{d}{d\xi}\left[A(\theta)\frac{d\theta}{d\xi}\right]$$

and boundary conditions (13.3.38) and (13.3.39) become

$$\theta = 0 \quad \text{with } \xi = \infty, \qquad \theta = 1 \quad \text{with } \xi = 0.$$

In addition

$$A(\theta) = 1/(1 - \alpha\theta) \quad (\alpha = \lambda t_0). \tag{13.3.40}$$

Introduction of the new variable

$$g = \int_0^\theta A(\theta')\, d\theta' \Big/ \int_0^1 A(\theta)\, d\theta \tag{13.3.41}$$

into (13.3.40) gives as

$$\ln(1 - \alpha\theta)/\ln(1 - \alpha) = g. \tag{13.3.42}$$

The new variable (13.3.41) transforms the previously obtained differential equation and boundary conditions into the equation

$$- 2\xi\, e^{-\beta g}(dg/d\xi) = d^2g/d\xi^2 \tag{13.3.43}$$

with the conditions

$$g = 0 \quad \text{at } \xi = \infty, \tag{13.3.44}$$

$$g = 1 \quad \text{at } \xi = 0. \tag{13.3.45}$$

$\beta = - \ln(1 - \alpha)$ relates β and α. For the present solution, λ and α are assumed positive and $0 < \alpha < 1$. Hence, $\beta > 0$.

Then we introduce some substitutions defined by the relations

$$dg/d\xi = - \varphi, \qquad \exp[- \beta g]/\beta = q. \tag{13.3.46}$$

In this case, Eq. (13.3.43) will become of the form

$$q(d^2\varphi/dq^2) = - 2/\beta\varphi.$$

Integration of the last equation yields

$$\ln q + C_1 = - \int_0^{\varphi/\sqrt{q}} (C_2 + \tfrac{1}{4}z^2 - (4/\beta)\ln z)^{-1/2}\, dz, \tag{13.3.47}$$

where C_1 and C_2 are the integration constants.

It follows from (13.3.44) and (13.3.46) that $\varphi \to 0$ as $q \to 1/\beta$. This condition allows us to determine C_1 in Eq. (13.3.47). If we account for the resulting expression for C_1, we obtain Eq. (13.3.47) in the form

$$\ln q\beta = - \int_0^{\varphi/\sqrt{q}} (C_2 + \tfrac{1}{4}z^2 - (4/\beta)\ln z)^{-1/2}\, dz. \tag{13.3.48}$$

From condition (13.3.45), it follows that

$$d\varphi/dq = 0; \qquad q = e^{-\beta}/\beta.$$

Substituting the above conditions into (13.3.48) and making some manipulation give us

$$C_2 = (4/\beta)\ln\varepsilon; \qquad \varepsilon = (\varphi/\sqrt{q})_{q=\exp(-\beta)/\beta}$$

and we can write Eq. (13.3.48) after accounting for C_2 as

$$\ln r = -\int_0^u (u_1^2 - \mu\ln u_1^2)^{-1/2}\,du_1, \tag{13.3.49}$$

where

$$r = (q\beta)^{1/2}; \qquad u = \varphi/(\varepsilon\sqrt{q}); \qquad \mu = 8/(\beta\varepsilon^2)$$

are new variables.

This equation is used for obtaining μ, and consequently, the unknown parameter ε as the function of the known parameter β defined by the relation

$$\beta = 2\int_0^1 (u^2 - \mu\ln u^2)^{-1/2}\,du. \tag{13.3.50}$$

Relation (13.3.55) is obtained from Eq. (13.3.43) accounting for the condition $\varphi/\sqrt{q} = \varepsilon$ when $q = e^{-\beta}/\beta$. Integral (13.3.50) relates parameters β and μ. To determine μ for a prescribed β, the numerical solution of (13.3.55) is necessary, e.g., by approximate integration.

The expression for ξ relating it to u and r is obtained from equation (13.3.43) and the subsequent conditions

$$\xi = -\frac{1}{(2\mu)^{1/2}}\left(\frac{u}{r} + \frac{1}{r^2}\frac{du}{dr}\right).$$

Eliminating r from this equation, we obtain with the aid of (13.3.43)

$$\xi = \frac{1}{(2\mu)^{1/2}}\left[(u^2 - \mu\ln u^2)^{1/2} - u\right]\exp\left[\int_0^u (u_1^2 - \mu\ln u_1^2)^{-1/2}\,du_1\right]. \tag{13.3.51}$$

On the other hand, a combination of the second relation (13.3.46), Eq. (13.3.49), and $r = (q\beta)^{1/2}$ yields the expression

$$g = \frac{2}{\beta}\int_0^u (u_1^2 - \mu\ln u_1^2)^{-1/2}\,du_1.$$

Accounting for relation (13.3.41), we obtain the final expression for θ as

$$\theta = \frac{1}{1-\exp[-\beta]}\left[1 - \exp\left\{-2\int_0^u (u_1^2 - \mu\ln u_1^2)^{-1/2}\,du_1\right\}\right]. \tag{13.3.52}$$

It should be noted that the quantity μ in the above equation is related to the assigned parameter β by (13.3.50).

The above solution was obtained by Fujita [34] who showed that the solution for a constant value of the coefficient ($\alpha = \beta = \lambda = 0$) becomes the well-known probability function. The solution method for assigned values of μ is as follows:

(i) β is calculated by numerical integration of Eq. (13.3.50) and hence α is obtained from the relation $\beta = -\ln(1 - \alpha)$.

(ii) ξ and θ are found by numerical integration of (13.3.51) and (13.3.52), respectively.

The relation between α and $\ln \mu$ is presented by (13.3.50) and $\beta = -\ln(1 - \alpha)$. It is plotted in Fig. 13.1. If the value of α is found, this plot allows direct determination of $\ln \mu$. Calculation results are shown in Fig. 13.2.

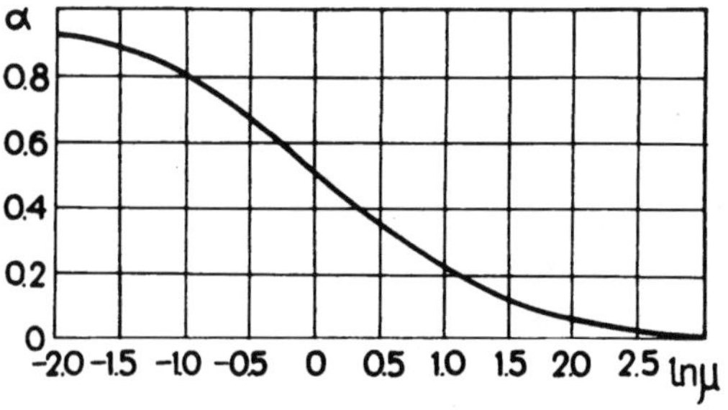

Fig. 13.1. Plot of α versus $\ln \mu$.

(2) *Semi-infinite medium*

$$a(t) = a_0/(1 - \lambda t)^2.$$

Here, the boundary conditions remain the same. The substitution

$$\theta = t/t_0, \qquad \xi = x/2(a_0\tau)^{1/2}, \qquad \alpha = t_0\lambda$$

allows us to obtain the ordinary differential equation with the variables ξ and θ as

$$-2\xi \frac{d\theta}{d\xi} = \frac{d}{d\xi}\left[\frac{1}{(1-\alpha\theta)^2}\frac{d\theta}{d\xi}\right], \tag{13.3.53}$$

with the boundary conditions

$$\theta = 0, \quad \text{at } \xi = \infty, \tag{13.3.54}$$

$$\theta = 1, \quad \text{at } \xi = 0. \tag{13.3.55}$$

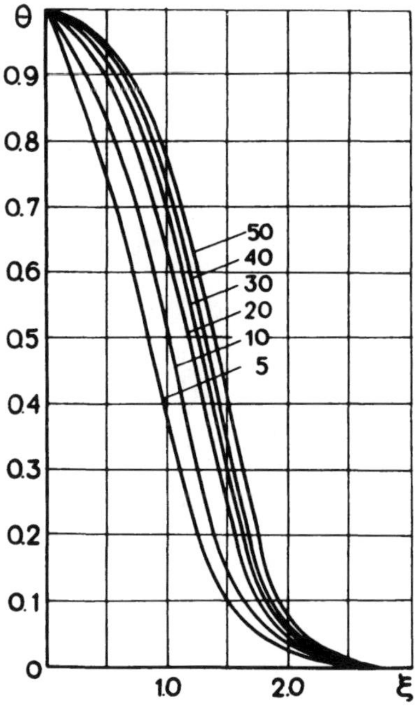

Fig. 13.2. Dimensionless temperature $\theta = t/t_0$ versus $\xi = x/(4a_0\tau)^{1/2}$ for $a(\theta) = a_0/(1 - \alpha\theta)$ (the numbers on the curves represent $1/(1 - \alpha)$ and constitute the ratio $a(\theta)$ when $\theta = \theta_0$ to a_0 when $\theta = 0$).

The solution of Eq. (13.3.53) with boundary conditions (13.3.54)–(13.3.55) gives a relation between θ and ξ.

The final result obtained by Fujita will be presented here; detailed calculation and the analysis of the obtained results are given by Fujita [34]. The solution may be presented in the following form:

$$\theta = \frac{\psi(u, \beta)}{\alpha[1 - \alpha + \psi(u, \beta)]}, \tag{13.3.56}$$

$$\xi = \frac{\beta}{1 - \alpha}\{[1 - \alpha + \psi(u, \beta)]u - \exp[\beta^2(1 - u^2)]\}, \tag{13.3.57}$$

where $\psi(u, \beta)$ is determined from the relation

$$\psi(u, \beta) = \sqrt{\pi}\,\beta[1 - \mathrm{erf}(\beta u)]\exp[\beta^2]$$

and β is the constant obtained as a function of the parameter from the equation

$$\psi(1, \beta) = \alpha \quad (0 < \alpha < 1).$$

Equations (13.3.56) and (13.3.57) give the solution with the parameter u changing within the range $1 \leq u \leq \infty$. The solution of the problem is shown in Fig. 13.3.

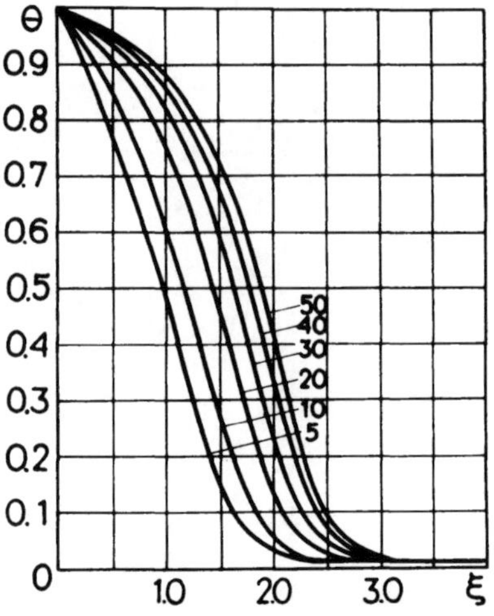

Fig. 13.3. Temperature θ plotted versus ξ for $a(\theta) = a_0/(1 - \alpha\theta)^2$ (the numbers on the curves represent $1/(1-\alpha)^2$ and constitute the ratio of $a(\theta)$ at $\theta = \theta_0$ to a_0 at $\theta = 0$).

(3) *Semi-infinite medium*

$$a(t) = \frac{a_0}{1 + 2\varkappa t + \nu t^2} \quad (\varkappa \text{ and } \nu \text{ are constants}).$$

Boundary conditions again remain the same.

The assumed relation between the thermal diffusivity and the temperature includes the previous problems as particular cases. Moreover, this relation allows consideration of problems for the conditions when the thermal diffusivity has a minimum or maximum with changing temperature. Solutions for this problem have also been obtained by Fujita.

The substitutions

$$\theta = t/t_0, \qquad \xi = x/2(a_0\tau)^{1/2}, \qquad \alpha = \varkappa t_0, \qquad \beta = \nu t_0^2$$

transform Eq. (13.3.1) into the ordinary differential equation

$$-2\xi \frac{d\theta}{d\xi} = \frac{d}{d\xi}\left[\frac{1}{1 + 2\alpha\theta + \beta\theta^2} \frac{d\theta}{d\xi}\right]$$

with the boundary conditions

$$\theta = 0 \quad \text{at } \xi = \infty, \qquad \theta = 1 \quad \text{at } \xi = 0.$$

Following Fujita, we shall consider the case when $a(t)$ passes through a maximum value in the range between $t = 0$ and $t = t_0$. The latter corresponds to the condition when the function $f(\theta)$ defined by the expression

$$f(\theta) = 1 + 2\alpha\theta + \beta\theta^2$$

has its maximum between $\theta = 0$ and $\theta = 1$. It is easy to show that this condition is determined by the inequalities

$$0 < -\alpha/\beta < 1, \tag{13.3.58}$$

$$\beta > 0. \tag{13.3.59}$$

It follows from (13.3.58) and (13.3.59) that

$$\alpha < 0. \tag{13.3.60}$$

Physically, any maximum $a(t)$ in the region of the temperature change should be always finite; the latter demands that the approximate minimum $f(\theta)$ be positive which is equivalent to the condition

$$0 < \alpha^2/\beta < 1. \tag{13.3.61}$$

Relations (13.3.58)–(13.3.59) correspond to the conditions when the coefficient a passes through the maximum.

For the convenience, we denote $\alpha = -\gamma$ and the above equalities can be written

$$0 < \gamma/\beta < 1, \qquad 0 < \gamma^2/\beta < 1.$$

A finite solution of the problem may be given in the following form:

$$\beta\theta = (\beta - \gamma^2)^{1/2} \tan[F(u, \varepsilon) - \tan^{-1}k] + \gamma, \quad 0 < \theta < \theta_1, \tag{13.3.62}$$

$$\beta\theta = (\beta - \gamma^2)^{1/2} \tan[\tan^{-1} m - F(u, \varepsilon) - F(u_m, \varepsilon)] + \gamma, \quad \theta_1 < \theta < 1, \tag{13.3.63}$$

where

$$\beta\theta_1 = (\beta - \gamma^2)^{1/2} \tan[F(1, \varepsilon) - \tan^{-1} k] + \gamma.$$

Besides,

$$\xi = \left[\frac{\beta}{\varepsilon(\beta - \gamma^2)(1 + z^2)}\right]^{1/2} [zu + (1 - u^2 - \varepsilon \ln u)^{1/2}], \quad -k < z < z_1, \tag{13.3.64}$$

$$\xi = \left[\frac{\beta}{\varepsilon(\beta - \gamma^2)(1 + z^2)}\right]^{1/2} [zu - (1 - u^2 - \varepsilon \ln u)^{1/2}],$$

$$z_1 < z < m, \tag{13.3.65}$$

where

$$\tan^{-1} z_1 = F(1, \varepsilon) - \tan^{-1} k,$$

$$\tan^{-1} z = F(u, \varepsilon) - tan^{-1} k, \quad -k < z < z_1,$$

and

$$\tan^{-1} z = \tan^{-1} m - F(u, \varepsilon) + F(u_m, \varepsilon), \quad z_1 < z < m.$$

Auxiliary variables in these equations are determined from the relations

$$k = \frac{\gamma}{(\beta - \gamma^2)^{1/2}}, \qquad m = \frac{\beta}{(\beta - \gamma^2)^{1/2}} \left(1 - \frac{\gamma}{\beta}\right), \tag{13.3.66}$$

where $\gamma = -\alpha$ and β are constant quantities entering the expressions for the thermal diffusivities. In addition, ε and u_m are related to k and m by the expressions

$$\varepsilon \ln u_m + (m^2 + 1)u_m^2 = 1 \tag{13.3.67}$$

and

$$\tan^{-1} k + \tan^{-1} m = 2F(1, \varepsilon) - F(u_m, \varepsilon), \tag{13.3.68}$$

where

$$F(u, \varepsilon) = \int_0^u (1 - u_1^2 - \varepsilon \ln u_1)^{-1/2} \, du_1.$$

Estimation of (13.3.62) and (13.3.63) with respect to u allows us to notice that in relation (13.3.62) u changes from 0 to 1, and in Eq. (13.3.63), it is limited by $u_m < u < 1$.

13.4 Boundary-Value Problems for the Heat Conduction Equation with the Coefficients Dependent upon the Coordinate

The above solutions may be generalized and obtained from a more general solution. First, consider the results of Barenblatt [133] and Barenblatt and Levitan [134] for the equation of the form:

$$c(x)\gamma(x) \frac{\partial t(x, \tau)}{\partial \tau} = \frac{\partial}{\partial x} \left(K(x) \frac{\partial t(x, \tau)}{\partial x}\right). \tag{13.4.1}$$

(1) Following Barenblatt [133], we introduce a new variable

$$z = \int_0^x dx/K(x). \tag{13.4.2}$$

Then Eq. (13.4.1) assumes the form[1]

$$\frac{\partial t(z, \tau)}{\partial \tau} = \frac{1}{q(z)} \frac{\partial^2 t(z, \tau)}{\partial z^2}, \tag{13.4.3}$$

where

$$q(z) = c(x)\gamma(x)K(x). \tag{13.4.4}$$

It is assumed that the function $q(z)$ satisfies the following conditions within the range $0 \leq z < \infty$.

(a) The function $q(z)$ does not take on negative values and may become zero only at $z = 0$.

(b) The function $q(z)$ has a continuous second derivative everywhere, with the possible exception at $z = 0$.

(c) Near the origin of the coordinates, the function $q(z)$ has the form

$$q(z) \underset{z \to 0}{\simeq} z^s(1 + p(z)),$$

where $p(0) = 0$, $s \geq 0$.

(d) The integral

$$\int_0^z [q(z)]^{1/2}\, dz$$

diverges with $z \to \infty$.

(e) The modulus

$$\frac{5}{16} \frac{q'^2(z)}{q^{5/2}(z)} - \frac{q''(z)}{4q^{3/2}(z)}$$

decreases rather quickly since the integral

$$\int_0^\infty \left| \frac{5}{16} \frac{q'^2(z)}{q^{5/2}(z)} - \frac{q''(z)}{4q^{3/2}(z)} \right| dz$$

converges.

[1] With new variables $dy = (q(z))^{1/2}\, dz$, Eq. (13.4.3) assumes the form of the heat conduction equation for a rod with variable heat capacity and conductivity, whereas the constant thermal diffusivity is equal to unity. Following Barenblatt [133], such a rod will be called an "equivalent" rod.

(2) For constructing the solutions of the boundary-value problems of Eq. (13.4.3) with the function $q(z)$ satisfying boundary conditions, we shall summarize the results for the problems on the eigenvalues of the equation

$$(d^2u/dz^2) + \lambda^{s+2}q(z)u = 0, \tag{13.4.5}$$

which is obtained from Eq. (13.4.3) if it is assumed

$$t(z, \tau) = u(z, \lambda)\exp[-\lambda^{s+2}\tau]. \tag{13.4.6}$$

The proof of the results and the details considering the Sturm–Liouville problem for Eq. (13.4.5) on a semi-infinite interval may be found in Barenblatt [133] as well as in other works [135, 136].

Consider the solution of Eq. (13.4.5) satisfying the conditions

$$u_\alpha(0, \lambda) = \sin\alpha, \qquad \frac{du_\alpha(0, \lambda)}{dz} = \cos\alpha \quad (0 \le \alpha \le \pi/2). \tag{13.4.7}$$

Introducing the new dependent and independent variables according to the relations

$$v(\xi, \lambda) = \xi^{-s/4}q^{1/4}(z)u(z, \lambda)$$

and

$$\xi = \left(\frac{1}{2}(s+2)\int_0^z (q(z))^{1/2}\,dz\right)^{2/(s+2)}, \tag{13.4.8}$$

we transform Eq. (13.1.5) to

$$\frac{d^2v}{d\xi^2} + \lambda^{s+2}\xi^s v = R(\xi)v, \tag{13.4.9}$$

where

$$R(\xi) = \frac{s}{4}\left(\frac{s}{4}+1\right)\frac{1}{\xi^2} - \frac{5}{16}\,\frac{\xi^s q'(z)}{q^3(z)} + \frac{1}{4}\,\frac{\xi^s q''(z)}{q^2(z)}. \tag{13.4.10}$$

If the right-hand side of Eq. (13.4.9) is known and condition (13.4.7) is used, then we obtain by the method of variation of the arbitrary constant

$$\begin{aligned} v(\xi, \lambda) = {} & aA(\xi)(\lambda\xi)^{1/2}J_{-1/(s+2)}[(2/(s+2))(\lambda\xi)^{(s+2)/2}] \\ & + bB(\xi)(\lambda\xi)^{1/2}J_{1/(s+2)}[(2/(s+2))(\lambda\xi)^{(s+2)/2}], \end{aligned} \tag{13.4.11}$$

where

$$A(\xi) = \sin\alpha - (b/\lambda)\int_0^\xi (\lambda\eta)^{1/2}J_{1/(s+2)}[(2/(s+2))(\lambda\eta)^{(s+2)/2}]R(\eta)v(\eta)\,d\eta, \tag{13.4.12}$$

$$B(\xi) = -(\cos\alpha/\lambda) + (a/\lambda)\int_0^{\xi} (\lambda\eta)^{1/2} J_{-1/(s+2)}[(2/(s+2))(\lambda\eta)^{(s+2)/2}] \times R(\eta)v(\eta)\, d\eta\,,$$

and

$$a = \frac{\Gamma\{(s+1)/(s+2)\}}{(s+2)^{1/(s+2)}}, \qquad b = \frac{\Gamma\{1/(s+2)\}}{(s+2)^{(s+1)/(s+2)}}. \tag{13.4.13}$$

(3) We now present the asymptotic expression for the function $v(\xi, \lambda)$ from formula (13.4.11) at $z \to \infty$. Using the property (e) of the function $q(z)$ and the known properties of the Bessel functions, it is possible to show that in the expression for $A(\xi)$, and $B(\xi)$, the integrals of (13.4.12) converge at $\xi \to \infty$, hence on the basis of the property (d) of the function $q(z)$ and the asymptotic behavior of the Bessel functions (see Appendix 1) we obtain

$$v(\xi, \lambda) \simeq \left(\frac{s+2}{\pi}\right)^{1/2} \frac{1}{(\lambda\xi)^{s/4}} \times \left\{M(\lambda)\cos\left(\lambda^{(s+2)/2}\int_0^z (q(z))^{1/2}\, dz\right) + N(\lambda)\sin\left(\lambda^{(s+2)/2}\int_0^z (q(z))^{1/2} dz\right)\right\}, \tag{13.4.14}$$

where

$$M(\lambda) = aA(\infty)\sin\left(\frac{s+4}{4(s+2)}\,\pi\right) + bB(\infty)\cos\left(\frac{s+4}{4(s+2)}\,\pi\right),$$
$$N(\lambda) = aA(\infty)\cos\left(\frac{s+4}{4(s+2)}\,\pi\right) + bB(\infty)\sin\left(\frac{s+4}{4(s+2)}\,\pi\right). \tag{13.4.15}$$

Passing to the initial function $u_\alpha(z, \lambda)$ we obtain the asymptotic representation for it at $z \to \infty$ as

$$u_\alpha(z, \lambda) \simeq [(s+2)/\pi]^{1/2}\lambda^{-s/4}q^{-1/4}(z) \times \left\{M(\lambda)\cos(\lambda^{(s+2)/2}\int_0^z (q(z))^{1/2}\, dz) + N(\lambda)\sin(\lambda^{(s+2)/2}\int_0^z (q(z))^{1/2} dz)\right\}. \tag{13.4.16}$$

From formula (13.4.11) and (13.4.12) it is also easy to obtain the asymptotic behavior for the eigenvalue function $u_\alpha(z, \lambda)$ at $\lambda \to \infty$ in the following form (at $\alpha \neq 0$):

$$u_\alpha(z, \lambda) \underset{\lambda\to\infty}{\simeq} [(s+2)/\pi]^{1/2}q^{-1/4}(z)\lambda^{-s/4} \times \left\{P(\lambda)\cos(\lambda^{(s+2)/2}\int_0^z (q(z))^{1/2}\, dz) + Q(\lambda)\sin(\lambda^{(s+2)/2}\int_0^z (q(z))^{1/2}\, dz)\right\}, \tag{13.4.17}$$

where

$$P(\lambda) = a \sin \alpha \sin(s+4)/4(s+2), \quad Q(\lambda) = a \sin \alpha \cos(s+4)/4(s+2). \tag{13.4.18}$$

For the derivative $du_\alpha(z, \lambda)/dz$ at $\lambda \to \infty$ we have ($\alpha \neq 0$)

$$\frac{du_\alpha(z, \lambda)}{dz} \simeq ((s+2)/\pi)^{1/2} q^{1/4}(z) \lambda^{(s+4)/4} \Big\{ - P(\lambda) \sin(\lambda^{(s+2)/2} \int_0^z (q(z))^{1/2}\, dz) + Q(\lambda) \cos(\lambda^{(s+2)/2} \int_0^z (q(z))^{1/2}\, dz) \Big\}. \tag{13.4.19}$$

Similarily, at $\alpha = 0$ and $\lambda \to \infty$ we have for $u_0(z, \lambda)$ and $du_0(z, \lambda)/dz$

$$u_0(z, \lambda) \simeq - ((s+2)/\pi)^{1/2} q^{-1/4}(z) \lambda^{-(s+4)/4} \Big\{ K(\lambda) \cos(\lambda^{(s+2)/2} \int_0^z (q(z))^{1/2}\, dz) + L(\lambda) \sin(\lambda^{(s+2)/2} \int_0^z (q(z))^{1/2}\, dz) \Big\}, \tag{13.4.20}$$

$$\frac{du_0(z, \lambda)}{dz} \simeq ((s+2)/\pi)^{1/2} q^{1/4}(z) \lambda^{s/4} \Big\{ - K(\lambda) \sin(\lambda^{(s+2)/2} \int_0^z (q(z))^{1/2}\, dz) + L(\lambda) \cos(\lambda^{(s+2)/2} \int_0^z (q(z))^{1/2}\, dz) \Big\}, \tag{13.4.21}$$

where

$$K(\lambda) = b \cos \frac{s+4}{4(s+2)} \pi, \qquad L(\lambda) = b \sin \frac{s+4}{4(s+2)} \pi. \tag{13.4.22}$$

(4) In the previous section, some information of the properties of the eigenfunctions of Eq. (13.4.5) are given. The problem on the possibility of of the expansion of the functions into the eigenfunctions of Eq. (13.4.5) over the semi-infinite range $0 \leq z < \infty$ may be studied when considering the Sturm–Liouville problem for Eq. (13.4.5) over a finite range $0 \leq z \leq l$ with the boundary conditions

$$u(0, \lambda) \cos \alpha - u'(0, \lambda) \sin \alpha = 0 \quad (0 \leq \alpha \leq \pi/2), \tag{13.4.23}$$

$$u(l, \lambda) \cos \beta - u'(l, \lambda) \sin \beta = 0 \quad (0 \leq \beta \leq \pi/2). \tag{13.4.24}$$

The above solutions of $u_\alpha(z, \lambda)$ already satisfy condition (13.1.23) and the eigenvalues of λ_n are determined from condition (13.1.24).

It is known that for any continuous function $f(z)$ over the range $(0, l)$ the following equality is valid:

$$\int_0^l f^2(z) q(z)\, dz = \int_0^\infty F^2(\lambda)\, d\varrho_l(\lambda), \tag{13.4.25}$$

where

$$F(\lambda) = \int_0^l q(z)f(z)u_a(z, \lambda)\,dz\,, \tag{13.4.26}$$

$$\varrho_l(\lambda) = \sum_{0\le\lambda_n\le\lambda} \left(\int_0^l q(z)u_a^2(z, \lambda_n)\,dz\right)^{-1} \tag{13.4.27}$$

It is possible to show [134, 136] that a monotonically nondecreasing function $\varrho(\lambda)$ exists which is the limit of the function $\varrho_l(\lambda)$ [see Eq. (13.4.27)] with $l \to \infty$. Further, in formulas (13.4.25) and (13.4.26), it is possible to pass to the limit $l \to \infty$ if $\int_0^\infty f^2(z)q(z)\,dz$ exists. Thus, relations (13.4.25) and (13.4.26) are also valid for the case of a semi-infinite interval ($l = \infty$).

From equality (13.4.25) at $l = \infty$, it is easy to show that

$$f(z) = \int_0^\infty F(\lambda)u_a(z, \lambda)\,d\varrho(\lambda)\,, \tag{13.4.28}$$

provided that the last integral uniformly converges in any finite interval.

To complete the discussion of the Sturm–Liouville problem for a semi-infinite interval, we present the method of construction and the expression for the function $\varrho(\lambda)$. Using asymptotic expression (13.4.16) for the function, it is not difficult to show that at $l \to \infty$

$$\int_0^l q(z)u_a^2(z, \lambda)\,dz \simeq \{(s+2)/\pi\}\lambda^{-s/2}\,\tfrac{1}{2}(M^2(\lambda) + N^2(\lambda)) \int_0^l (q(z))^{1/2}\,dz\,, \tag{13.4.29}$$

where $M(\lambda)$ and $N(\lambda)$ are determined according to formulas (13.4:15). From the known properties of the eigenvalues of λ_n, from the determination of (13.4.27), and the estimation of (13.4.29) it follows that

$$\varrho_l(\lambda+\Delta) - \varrho_l(\lambda) = \sum_{\lambda\le\lambda_n\le\lambda+\Delta} \frac{1}{\int_0^l q(z)u_a^2(z, \lambda_n)\,dz}$$

$$\underset{l\to\infty}{\simeq} \int_\lambda^{\lambda+\Delta} \frac{\lambda^s\,d\lambda}{M^2(\lambda) + N^2(\lambda)}\,, \tag{13.4.30}$$

where in the last equality (asymptotic) the sum is transformed into the integral.

Hence, we finally have for the function

$$\varrho(\lambda+\Delta) - \varrho(\lambda) = \int_\lambda^{\lambda+\Delta} \frac{\lambda^s\,d\lambda}{M^2(\lambda) + N^2(\lambda)} \tag{13.4.31}$$

and

$$d\varrho(\lambda) = \frac{\lambda^s\,d\lambda}{M^2(\lambda) + N^2(\lambda)}\,. \tag{13.4.32}$$

We also present the asymptotic estimations for the function $\varrho(\lambda)$ at $\lambda \to \infty$. If in condition (13.4.23), $\alpha \neq 0$, then from formulas (13.4.15) and (13.4.14) we have

$$M^2(\lambda) + N^2(\lambda) \simeq a^2 \sin^2 \alpha ,$$

i.e.,

$$d\varrho(\lambda) \simeq \lambda^s \, d\lambda / a^2 \sin^2 \alpha \tag{13.4.33}$$

or

$$\varrho(\lambda) \simeq \lambda^{s+1}/(s+1)a^2 \sin^2 \alpha . \tag{13.4.34}$$

Similarly, at $\alpha = 0$ we have

$$\varrho(\lambda) \simeq \lambda^{s+3}/(s+3)b^2 . \tag{13.4.35}$$

From these results (in particular, from formula (13.5.32), it directly follows that the spectrum of the operator $(1/q(z))(d^2/dz^2)$ (see equation (13.4.5)), under conditions (13.4.1)–(13.4.5) imposed upon the function $q(z)$, is continuous, i.e., the eigenvalues of λ cover the whole semi-infinite straight line $(0, \infty)$. Regarding the spectrum continuity, property (d) plays a special role for the function $q(z)$ since the integral $\int_0^\infty (q(z)^{1/2}) \, dz$ serves as the length of an "equivalent" rod. Therefore, when satisfying property (d) when the "equivalent" rod is infinite, boundary condition (13.4.24) appears to be unimportant. In case of the convergence of the integral $\int_0^\infty (q(z))^{1/2} \, dz$, we would have the problems of the finite interval for an ordinary heat conduction equation, the spectrum of the eigenvalues would be discrete, and condition (13.4.24) would also be important after the transition $l \to \infty$.

(5) Before studying the boundary-value problems for heat conduction equation (13.4.3), we present the generalization of the Poisson formula confirming that

$$\lim_{\theta \to \tau} \frac{1}{2\sqrt{\pi}} \int_a^b \varphi(x') \frac{\exp[-(x-x')^2/4(\tau-\theta)]}{(\tau-\theta)^{1/2}} \, dx'$$
$$= \begin{cases} 0, & \text{if } x \text{ lies outside } [a, b], \\ \frac{1}{2}[\varphi(a+0)], & \text{if } x = a, \\ \frac{1}{2}[\varphi(b-0)], & \text{if } x = b, \\ \frac{1}{2}[\varphi(x+0) + \varphi(x-0)], & \text{if } x \text{ lies inside } (a, b). \end{cases}$$

The generalized Poisson formula [134] confirms that if $\varphi(x)$ is a bounded function, piecewise continuous with respect to x, and continuous with respect to t, then $(0 \leq a < b \leq \infty)$

$$\lim_{\theta\to\tau} \int_a^b f(x', \tau)q(x')\, dx' \int_0^\infty \exp[-\lambda^{s+2}(\tau-\theta)]u_\alpha(x, \lambda)u_\alpha(x'\lambda)\, d\varrho(\lambda)$$

$$= \begin{cases} 0, \text{ if } x \text{ lies outside } [a, b], \\ \frac{1}{2}f(a+0, \tau), \quad \text{at } x = a, \\ \frac{1}{2}f(b-0, \tau), \quad \text{at } x = b, \\ \frac{1}{2}[f(x+0, \tau) + f(x-0, \tau)] \quad \text{if } x \text{ lies inside } (a, b). \end{cases} \qquad (13.4.36)$$

(6) We shall now discuss the main subject of the present section, i.e., boundary-value problems of Eq. (13.4.3).

(a) Let

$$t(0, \tau)\cos\alpha - (\partial t(0, \tau)/\partial z)\sin\alpha = 0 \quad (0 \le \alpha \le \pi/2), \qquad (13.4.37)$$

$$t(z, 0) = f(z). \qquad (13.4.38)$$

It is a direct result of the generalized Poisson formula (13.4.36) that the solution of this boundary-value problem is of the form

$$t(z, \tau) = \int_0^\infty f(\zeta)q(\zeta)\, d\zeta \int_0^\infty \exp[-\lambda^{s+2}\tau]u_\alpha(z, \lambda)u_\alpha(\zeta, \lambda)\, d\varrho(\lambda), \qquad (13.4.39)$$

where $u_\alpha(z, \lambda)$ are eigenfunctions of the boundary-value problem for the ordinary equation (13.4.5) with the condition (13.4.23).

(b) With the original initial condition (13.4.38), let the nonhomogeneous boundary condition of the first kind be prescribed as

$$t(0, \tau) = \varphi(\tau). \qquad (13.4.40)$$

$t(z, \tau)$ will be sought as the sum $t(z, \tau) = U(z, \tau) + V(z, \tau)$ where U and V satisfy Eq. (13.4.3) and the conditions

$$U(0, \tau) = \varphi(\tau), \qquad U(z, 0) = 0, \qquad (13.4.41)$$

$$V(0, \tau) = 0, \qquad V(z, 0) = f(z). \qquad (13.4.42)$$

Consideration of the boundary-value problem (13.4.37)–(13.4.38) allows us to conclude that the function $V(z, \tau)$ is the particular case when $\alpha = 0$ in condition (13.4.37). Thus the function $V(z, \tau)$ is defined by the formula similar to (13.4.39):

$$V(z, \tau) = \int_0^\infty f(\zeta)q(\zeta)\, d\zeta \int_0^\infty \exp[-\lambda^{s+2}\tau]u_0(z, \lambda)u_0(\zeta, \lambda)\, d\varrho_0(\lambda), \qquad (13.4.43)$$

where $u_0(z, \lambda)$ is an eigenfunction of Eq. (13.4.5) such that $u_0(0, \lambda) = 0$,

$\partial u_0(0, \lambda)/\partial z = 1$, and $\varrho_0(\lambda)$ is the function $u_0(z, \lambda)$ corresponding to $\varrho(\lambda)$ (see (13.4.27)).

We need only to determine the function $U(z, \tau)$. It will be shown that it may be determined by the formula

$$U(z, \tau) = \int_0^\tau \varphi(\theta)\, d\theta \int_0^\infty \exp[-\lambda^{s+2}(\tau-\theta)] u_0(z, \lambda)\, d\varrho_0(\lambda). \quad (13.4.44)$$

For the proof, it is sufficient to see that first, the function U satisfies Eq. (13.4.3). It may be easily checked by direct substitution of formula (13.4.44) into (13.4.3). Second, it is evident from (13.4.44) that $U(z, \tau) = 0$ when $\tau \to 0$ $(z \neq 0)$, and consequently, the second condition (13.4.41) is fulfilled.

Now it must be shown that the integral in the right-hand side of formula (13.4.44), with $z \to 0$, has the limit $\varphi(\tau)$. With $z \to 0$, formulas (13.4.11) and (13.4.8) yield

$$u_0(z, \lambda) \underset{z\to 0}{\simeq} \frac{b}{\lambda} q^{-1/4}(z)\xi^{s/4}(\lambda\xi)^{1/2} J_{1/(s+2)}\left(\frac{2}{s+2}(\lambda\xi)^{(s+2)/2}\right) + \cdots, \quad (13.4.45)$$

where the elipses denote the terms of higher order with respect to z.

Further, since it is evident from (13.4.45) that only high values of λ contribute to the integral in the right-hand side of Eq. (13.4.44), asymptotic formula (13.4.35) may be used for $d\varrho_0(\lambda)$, i.e., the substitution in (13.4.44) should be made:

$$d\varrho_0(\lambda) \underset{\lambda\to\infty}{\simeq} \left(\frac{\lambda^{s+2}}{b^2} + \cdots\right) d\lambda, \quad (13.4.46)$$

where the elipses in the parentheses denote the terms of asymptotic expansion which increases not more than λ^{s+1} when $\lambda \to \infty$. Thus asymptotic relations (13.4.45) and (13.4.46) are substituted into the integral. Then with $z \to 0$ integral (13.4.44) takes the form

$$(1/b)q^{-1/4}(z)\xi^{s/4} \int_0^\tau \varphi(\theta)\, d(\theta)$$
$$\times \int_0^\infty \exp[-\lambda^{s+2}(\tau-\theta)](\lambda\xi)^{1/2} J_{1/(s+1/2)}[(2/(s+2))(\lambda\xi)^{(s+2)/2}]\lambda^{s+1}\, d\lambda + \cdots. \quad (13.4.47)$$

It was rigorously proved by Barenblatt and Levitan [134] that all the following terms of asymptotic expansion which are denoted in formula (13.4.47) by elipses vanish when $z \to 0$. It must be therefore shown that when $z \to 0$, integral (13.4.47) appears to be equal to $\varphi(\tau)$. Note that when $z \to 0$, the factor $q^{-1/4}(z)\xi^{s/4}$ approaches unity; this follows from property (c) of the function $q(z)$ and the second formula of (13.4.8).

Integration over λ in (13.4.47) may be carried out, yielding Eq. (13.4.44) in the form

$$\frac{\xi}{b(s+2)^{(s+3)/(s+2)}} \int_0^\tau \varphi(\theta) \frac{\exp\{-\xi^{s+2}/(s+2)^2(\tau-\theta)\}}{(\tau-\theta)^{(s+3)/(s+2)}} \, d\theta . \tag{13.4.48}$$

It will be shown that the function

$$F(\tau, \xi) = \begin{cases} \dfrac{\xi}{b(s+2)^{(s+3)/(s+2)}} \dfrac{\exp\{-\xi^{s+2}/(s+2)^2\tau\}}{\tau^{(s+3)/(s+2)}}, & \text{with } \tau > 0, \\ 0, & \text{with } \tau < 0, \end{cases} \tag{13.1.49}$$

has the limit $\delta(\tau - 0)$ at $z \to 0$ ($\xi \to 0$) in a sense of a generalized function. For this, it must be shown [137] that first, with any real τ_1 and τ_2 ($\tau_1 < \tau_2$), the integral $\int_{\tau_1}^{\tau_2} F(\tau, \xi)\, d\tau$ is limited above by a constant, independent of τ_1, τ_2, and ξ. Indeed, since the function $F(\tau, \xi)$ is not negative, then

$$\begin{aligned} \int_{\tau_1}^{\tau_2} F(\tau, \xi)\, d\tau &\leq \int_0^\infty F(\tau, \xi)\, d\tau \\ &= \frac{\xi}{b(s+2)^{(s+3)/(s+2)}} \int_0^\infty \frac{\exp[-\xi^{s+2}/(s+2)^2\tau]}{\tau^{(s+3)/(s+2)}}\, d\tau = 1 . \end{aligned}$$

Second, with any τ_1 and τ_2 different from 0

$$\lim_{\xi \to 0} \int_{\tau_1}^{\tau_2} F(\tau, \xi)\, d\tau = \begin{cases} 0 & \text{with } \tau_1 < \tau_2 < 0, \\ & \text{or} \quad 0 < \tau_1 < \tau_2, \\ 1 & \text{with } \tau_1 < 0 < \tau_2 . \end{cases}$$

This may be easily checked.

Thus, it is proved that formula (13.4.44) is indeed the solution of Eq. (13.4.3) with conditions (13.4.41).

(c) Consider the boundary-value problem of the second kind

$$\partial t(0, \tau)/\partial z = \psi(\tau) \tag{13.4.50}$$

and initial condition (13.4.38). Then $t(z, \tau)$ will be sought in the form $t(z, \tau) = U(z, \tau) + V(z, \tau)$, where U and V now satisfy the conditions

$$\partial U(0, \tau)/\partial z = \psi(\tau), \quad U(z, 0) = 0, \tag{13.4.51}$$

$$\partial V(0, \tau)/\partial z = 0, \quad V(z, 0) = f(z), \tag{13.4.52}$$

as well as Eq. (13.4.3).

The function $V(z, \tau)$ is a particular case of the solution (13.5.43) with $\alpha = \pi/2$. Then $V(z, \tau)$ is of the form

$$V(z, \tau) = \int_0^\infty f(\zeta)q(\zeta)\, d\zeta \int_0^\infty \exp[-\lambda^{s+2}\tau]u_{\pi/2}(z, \lambda)u_{\pi/2}(\zeta, \lambda)\, d\varrho_{\pi/2}(\lambda), \tag{13.4.53}$$

where $u_{\pi/2}(z, \lambda)$ is a solution of Eq. (13.4.5) such that $\partial u_{\pi/2}(0, \lambda)/\partial z = 0$, and $\varrho_{\pi/2}(\lambda)$ is the function $\varrho(\lambda)$ corresponding to $u_{\pi/2}(z, \lambda)$.

Similarly to case (b), it may be shown that the function U is defined by the formula (see (13.4.44)):

$$U(z, \tau) = \int_0^\tau \psi(\theta)\, d\theta \int_0^\infty \exp[-\lambda^{s+2}(\tau - \theta)]u_{\pi/2}(z, \lambda)\, d\varrho_{\pi/2}(\lambda), \tag{13.4.54}$$

which satisfies Eq. (13.4.3) and conditions (13.4.51).

(d) We shall now construct the solution of Eq. (13.4.3) with the non-homogeneous equation

$$t(0, \tau) \cos\alpha - (\partial t(0, \tau)/\partial z) \sin\alpha = \varphi(\tau) \quad (\alpha \neq 0 \text{ or } \alpha \neq \pi/2) \tag{13.4.55}$$

and initial condition (13.4.38). For this, $t(z, \tau)$ will be expressed as the sum $t(z, \tau) = U(z, \tau) + V(z, \tau)$ where U and V satisfy Eq. (13.4.3) and the conditions

$$U(0, \tau) \cos\alpha - (\partial u(0, \tau)/\partial z) \sin\alpha = 0, \qquad U(z, 0) = f(z), \tag{13.4.56}$$

$$\begin{gathered} V(0, \tau) \cos\alpha - (\partial V(0, \tau)/\partial z) \sin\alpha = \varphi(\tau), \\ V(z, 0) = 0, \qquad \partial V(\infty, \tau)/\partial z = 0. \end{gathered} \tag{13.4.57}$$

The function $U(z, \tau)$ is expressed by formula (13.4.39). For determination of the function $V(z, \tau)$, the unknown function

$$\Phi(\tau) = -\partial V(0, \tau)/\partial z \tag{13.4.58}$$

will be introduced. If the function $\Phi(\tau)$ were known, the function $V(z, \tau)$ could be determined by formula (13.4.54), i.e.,

$$V(z, \tau) = \int_0^\tau \Phi(\theta)\, d\theta \int_0^\infty \exp[-\lambda^{s+2}(\tau - \theta)]u_{\pi/2}(z, \lambda)\, d\varrho_{\pi/2}(\lambda). \tag{13.4.59}$$

Assuming in the last formula $z = 0$ and using the fact that $u_{\pi/2}(0, \lambda) = 1$ permits us to obtain

$$V(0, \tau) = \int_0^\tau \Phi(\theta)\, d\theta \int_0^\infty \exp[-\lambda^{s+2}(\tau - \theta)]\, d\varrho_{\pi/2}(\lambda). \tag{13.4.60}$$

Substituting Eq. (13.4.60) in the first condition of (13.4.57) gives for the function the integral Volterra equation of the second kind,

$$\Phi(\tau) + h\int_0^\tau K(\tau-\theta)\Phi(\theta)\,d\theta = \chi(\tau),$$

$$K(\tau) = \int_0^\infty \exp[-\lambda^{s+2}\tau]\,d\varrho_{\pi/2}(\lambda), \tag{13.4.61}$$

where

$$h = \cot\alpha, \qquad \chi(\tau) = \varphi(\tau)/\sin\alpha. \tag{13.4.62}$$

Equation (13.4.61) may be solved in an explicit form by the Laplace transformation for particular kinds of the kernel $K(\tau)$ or by the method of successive approximations. Then, as already mentioned, the function $V(z, \tau)$ will be determined by formula (13.4.54).

(e) Consider the boundary-value problem for the equation with the source

$$q(z)\frac{\partial t}{\partial t} = \frac{\partial^2 t}{\partial z^2} + Q(z, \tau) \tag{13.4.63}$$

with boundary conditions (13.4.38) and (13.4.55).

We express $t(z, \tau)$ in the form

$$t(z, \tau) = U(z, \tau) + V(z, \tau),$$

where U satisfies equation (13.4.3) with boundary conditions (13.4.38) and (13.4.55) and the function V is the solution of Eq. (13.4.63) with the conditions

$$V(0, \tau)\cos\alpha - (\partial V(0, \tau)/\partial z)\sin\alpha = 0, \qquad V(z, 0) = 0. \tag{13.4.64}$$

The function $U(z, \tau)$ coincides with the solution of the problem considered in (d) [Eq. (13.4.61)].

The function $V(z, \tau)$ may be expressed in the form

$$V(z, \tau) = \int_0^\tau d\theta \int_0^\infty Q(\zeta, \theta)\,d\zeta \int_0^\infty \exp[-\lambda^{s+2}(\tau-\theta)]u_\alpha(z, \lambda)u_\alpha(\zeta, \lambda)\,d\varrho(\lambda), \tag{13.4.65}$$

To prove this statement, it should be shown that the function V satisfies Eq. (13.4.63) and conditions (13.4.64).

Estimation of the derivatives

$$\begin{aligned}\partial^2 V/\partial z^2 &= \int_0^\tau d\theta \int_0^\infty Q(\zeta, \theta)\,d\zeta \int_0^\infty \exp[-\lambda^{s+2}(\tau-\theta)](\partial^2 u_\alpha(z, \lambda)/\partial z^2) \\ &\quad\times u_\alpha(\zeta, \lambda)\,d\varrho(\lambda) \\ &= -\int_0^\tau d\theta \int_0^\infty Q(\zeta, \theta)\,d\zeta \\ &\quad\times \int_0^\infty \lambda^{s+2}\exp[-\lambda^{s+2}(\tau-\theta)]u_\alpha(z, \lambda)u_\alpha(\zeta, \lambda)\,d\varrho(\lambda),\end{aligned}$$

$$\partial V/\partial\tau = \lim_{\theta\to\tau} \int_0^\infty Q(\zeta,\theta)\,d\zeta \int_0^\infty \exp[-\lambda^{s+2}(\tau-\theta)]u_a(z,\lambda)$$
$$\times\, u_a(\zeta,\lambda)\,d\varrho(\lambda) - \int_0^\tau d\theta \int_0^\infty Q(\zeta,\theta)\,d\zeta$$
$$\times \int_0^\infty \lambda^{s+2}\exp[-\lambda^{s+2}(\tau-\theta)]u_a(z,\lambda)u_a(\zeta,\lambda)\,d\varrho(\lambda)\,,$$

and assuming that the density of the source $Q(z,\tau)$ satisfies the condition of validity of the generalized Poisson formula [i.e., $Q(z,\tau)$ is bounded, piecewise continuous with respect to z and continuous with respect to τ], yield the first summand in the expression for $\partial t/\partial\tau$ equal to $q^{-1}(z)Q(z,\tau)$. Comparison of the derivatives $\partial V/\partial\tau$ and $\partial^2V/\partial z^2$ shows that V satisfies Eq. (13.4.63). Verification of the fact that Eq. (13.4.64) are satisfied is even more simple.

Consequently, for construction of solutions of the boundary-value problems for Eq. (13.4.3) [or (13.4.63)], knowledge of the solution of ordinary equation (13.4.5) is necessary.

(7) We conclude the chapter by an illustration of the application of the general theory developed for Eq. (13.4.3) to the particular case

$$q(z) = z^s \tag{13.4.66}$$

considered in the previous section.

For example, we shall take the boundary-value problem when

$$t(0,\tau) = \varphi(\tau)\,, \qquad t(z,0) = 0. \tag{13.4.67}$$

The eigenfunction is easily obtained from the solution of Eq. (13.4.5) with $q(z)$ of the form (13.4.66):

$$\varphi_0(z,\lambda) = (b/\lambda)(\lambda z)^{1/2}J_{1/(s+2)}(\{2/(s+2)\}(\lambda z)^{(s+2)/2}). \tag{13.4.68}$$

According to (13.4.15) we have

$$M^2(\lambda) + N^2(\lambda) = [s+2]^{-2(s+1)/(s+2)}\,\Gamma^2\!\left(\frac{s+1}{s+2}\right)\frac{1}{\lambda^2}. \tag{13.4.69}$$

Hence, we determine $d\varrho_0(\lambda)$ using formula (13.4.32). According to (13.4.44) and using the value of the integral, we have

$$t(z,\tau) = \frac{z}{(s+2)^{2/(s+2)}\Gamma[1/(s+2)]}\int_0^\tau \varphi(\tau)\,\frac{\exp[-z^{s+2}/(s+2)^2(\tau-\theta)]}{(\tau-\theta)^{(s+3)/(s+2)}}\,d\theta. \tag{13.4.70}$$

Similarly, with the boundary condition

$$\partial t(0, \tau)/\partial z = -\psi(\tau), \qquad t(z, 0) = 0, \tag{13.4.71}$$

we may obtain

$$\varphi_{\pi/2}(z, \lambda) = a(\lambda z)^{1/2} J_{-1/(s+2)}(\{2/(s+2)\}(\lambda z)^{(s+2)/2}), \tag{13.4.72}$$

$$M^2(\lambda) + N^2(\lambda) = \frac{1}{(s+2)^{2/(s+2)}} \Gamma^2\left(\frac{s+1}{s+2}\right), \tag{13.4.73}$$

and according to (13.4.54)

$$t(z, \tau) = \frac{1}{(s+2)^{s/(s+2)}\Gamma[(s+1)/(s+2)]} \int_0^\tau \psi(\theta) \frac{\exp[-z^{s+2}/(s+2)^2(\tau-\theta)]}{(\tau-\theta)^{(s+1)/(s+2)}}. \tag{13.4.74}$$

CHAPTER

14

FUNDAMENTALS OF THE INTEGRAL TRANSFORMS

The solution of heat conduction equations for some problems (those with variable boundary conditions, systems of nonhomogeneous bodies, etc.) can be very difficult. Solutions are often obtained in the form of integrals or series scarcely suitable for practical application, and further transformations are therefore necessary. In recent years, operational calculus methods have been widely and successfully used yielding not only the rigorous solution but also a number of approximate ones with considerable accuracy. Operational calculus methods also permit us to successfully solve problems in which the desired function is discontinuous; this is a great advantage, since such problems are encountered in thermal physics rather frequently.

Operational methods have been used for a very long time as a supplementary mathematical tool to facilitate solution. Further development of the methods has shown that they are useful mathematically and possess certain advantages in comparison with the classical methods when they are applied to the solution of partial differential equations.

Symbolic or operational calculus as an independent mathematical method was originally developed by M. Vashchenko-Zakharchenko, Professor of the Kiev University [124]. In this monograph, Vashchenko-Zakharchenko presented a systematic development of operational calculus and derived basic relations and their application to the solution of differential equations with constant and variable coefficients.

He was the first to derive the expansion theorem, including the case of multiple roots, which is usually ascribed to Heaviside, and to consider the case of multiple roots.

Vashchenko-Zakharchenko's expansion theorems are formulated in the following way: if $f(D)$ is an integral rational function of the differentiation operator D, then

$$f^{-1}(D)x = \frac{1}{f'(\alpha_1)}(D-\alpha_1)^{-1}x + \frac{1}{f'(\alpha_2)}(D-\alpha_2)^{-1}x + \cdots + \frac{1}{f'(\alpha_n)}(D-\alpha_n)^{-1}x \tag{14.1}$$

where $\alpha_1, \alpha_2, \alpha_3, \cdots, \alpha_n$ are the roots of the equation $f(x) = 0$ (the case of simple roots).

In the same monograph, the expansion theorem for multiple roots is also deduced. Thus, Vashchenko-Zakharchenko may be rightfully considered the originator of operational calculus.

At the end of the last century, Heaviside applied the operational calculus to the solution of some engineering problems. He introduced the operator p the function of which is defined by

$$p^{\alpha}(\sum_n B_n\{x^n/n!\}) = \sum_n B_n\{x^{n-\alpha}/(n-\alpha)!\}. \tag{14.2}$$

In this definition, the quantity under the factorial sign (we denote it by m) satisfies the functional equation

$$f(m) = mf(m-1), \qquad f(0) = 1. \tag{14.3}$$

In the case where m is a fraction $(0 < m < 1)$

$$m! = \Gamma(m+1), \tag{14.4}$$

where $\Gamma(m+1)$ is a gamma function. If m is a positive integer

$$\Gamma(m+1) = m! = 1 \cdot 2 \cdot 3 \cdots m.$$

For example,

$$p^{-1/2}(1) = p^{-1/2}\left(\frac{x^0}{0!}\right) = \frac{x^{1/2}}{(\frac{1}{2})!} = \frac{x^{1/2}}{\Gamma(\frac{3}{2})} = 2\left(\frac{x}{\pi}\right)^{1/2}.$$

If α is an integer (positive or negative), the effect of operator p is the same as D. Heaviside presented no mathematical arguments for this method.

It should be noted that the idea of fractional differentiation is associated with Letnikov who in his work [64a] (published in 1868 considerably before Heaviside) applied the fractional differentiation method to the solution of differential equations.[1]

The operational method of solving a differential equation consists of the fact that in the equation

$$\sum_{m,n} a_{mn} \frac{\partial^{m+n} u(x, \tau)}{\partial \tau^m \, \partial x^n} = 0, \tag{14.5}$$

differentiation with respect to time is replaced by operator p, and the equation acquires the form

$$\sum_{m,n} a_{mn} p^m \frac{d^n u(x, \tau)}{dx^n} = 0. \tag{14.6}$$

This equation is considered an ordinary differential equation with respect to the variable x with the parameter p.

Strict mathematical support of the operational calculus was presented much later, when the relation between the functional Laplace transformation $\int_0^\infty f(\tau) e^{-p\tau} \, d\tau$ and the operational calculus was found. When the Laplace transformation is applied, the differentiation operator is replaced by multiplication procedure involving a certain complex quantity. Strict mathematical arguments were presented by Efros and Danilevsky [28] in their monograph. They established a number of new relations and rules for operational calculus. The theorem proposed by Efros is an effective mean for obtaining initial functions from their transforms. Ditkin contributed substantially to operational calculus, when in his fundamental work [21a] he gave a justification of operational calculus on the basis of modern mathematical concepts. Thus, operational calculus is currently a very well developed mathematical method.

Operational methods are successfully used because in the majority of cases, they are the most direct methods, considerably reducing calculation procedure. In many cases when the solution by classical methods is extremely difficult, a great many problems may be rapidly and effectively solved which makes them, therefore, of exceptional value to the engineer and physicist.

Determination of the temperature field of a solid in heat conduction problems involves the solution of a differential equation with various boundary conditions. The solutions of such problems should be in a form suitable for practical application. We shall consider the most general and simple methods of the Laplace transformation, i.e., the functional Laplace transformation will be applied:

[1] This problem was developed further by Sonin and Nekrasov [103b].

$$F(s) = \int_0^\infty f(\tau)e^{-s\tau}\,d\tau. \tag{14.7}$$

In many works concerned with the solution of electrical engineering problems by the operational calculus methods, the functional Laplace–Carson transformation is used

$$\Phi(p) = p\int_0^\infty f(\tau)e^{-p\tau}\,d\tau. \tag{14.8}$$

The transform of one and the same function is of a different form depending on the integral transformation used. The Laplace transform $F(s)$ corresponds to the Laplace–Carson transform $\Phi(p)/p$. This should be remembered when using transform tables presented in various monographs on operational calculus.

Bearing in mind that the present book is intended mainly for engineers and students, the author has attempted to present the functional Laplace transformation method in the simplest form omitting some details, general discussions, and generalizations.

14.1 Definitions

The Laplace transformation method consists of the fact that it is not the function itself (original) that is under consideration, but the transform function (transform). This transformation is fulfilled by multiplying a certain exponential function and integrating within certain limits.

Let the function of interest $y = f(\tau)$ be a piecewise continuous function of the independent variable τ. A piecewise continuous function is a single-valued function having in a finite interval $(0 \le \tau \le \theta)$ and a finite number of discontinuities at the points $\tau_1, \tau_2, \cdots, \tau_k$. In each interval (τ_{i-1}, τ_i), the function $f(\tau)$ is continuous and tends to a finite limit when approaching the boundary. The function $y = f(\tau)$ is referred to as an original.

The Laplace transformation for the function $y = f(\tau)$ consists in multiplying it by $e^{-s\tau}$ and integrating it over the limits between 0 and ∞.

$$\int_0^\infty f(\tau)e^{-s\tau}\,d\tau = F(s), \tag{14.1.1}$$

where $s = \xi + i\eta$ is some complex quantiy. Integration results in some function $F(s)$ which is called a transformed function. Thus the Laplace transformation is an integral transformation. This is denoted by $L[f(\tau)]$:

$$L[f(\tau)] = F(s) = \int_0^\infty f(\tau)e^{-s\tau}\,d\tau. \tag{14.1.2}$$

It should be noted here that the transform $F(s)$ exists, if integral (14.1.1) converges.

More detailed discussion of the function $f(\tau)$ will be found in Section 14.9 where the conditions for the existence of the function $F(s)$ are discussed.

To simplify the main relations, we restrict the class of functions considered. The function $f(\tau)$ is assumed to be piecewise continuous and different from zero only when $\tau > 0$. The quantity $f(0)$ is taken as $f(+0) = \lim_{\tau \to +0} f(\tau)$ and $f(-0)$ is zero. Then, out of the class of piecewise continuous functions, we single out a subclass of functions characterized by the fact that the asymptotic value of the function $f(\tau)$ at $\tau \to \infty$ is less than the asymptotic value of the function $e^{\sigma\tau}$ where $\sigma > 0$, i.e., when τ is sufficiently large

$$|f(\tau)| < Me^{\sigma\tau} \quad (\sigma > 0,\ M > 0)$$

or

$$|e^{-s\tau}f(\tau)| < Me^{-(s-\sigma)\tau},$$

where σ is a certain finite positive number.

With the above restrictions imposed on the function $f(\tau)$, integral (14.1.1) is a regular function of s in a half-plane on the right of the straight line σ (see Section 14.9), i.e., the function $F(s)$ has derivatives of all orders in the above region and all its singular points lie in a complex plane to the left of the straight line σ.

In the subsequent sections, we shall denote the original function by small letters, and its transform by capital letters. For example: $y(\tau)$ is the original and $Y(s)$ is its transform

$$L[y(\tau)] = Y(s).$$

We now consider some examples.

Example 1. Let the original function be a constant value

$$f(\tau) = A = \text{const} \quad (\tau > 0).$$

Then

$$L[A] = \int_0^{+\infty} Ae^{-s\tau}\, d\tau = -(A/s)e^{-s\tau}\Big|_0^{\infty} = A/s \quad (\text{if } s > 0). \quad (14.1.3)$$

Example 2. Let $f(\tau) = A\tau$. Then

$$L[A\tau] = \int_0^{\infty} A\tau e^{-s\tau}\, d\tau = A/s^2. \quad (14.1.4)$$

Example 3. Let $f(\tau) = e^{k\tau}$ $(\tau > 0)$. Then

$$L[e^{k\tau}] = \int_0^\infty e^{-(s-k)\tau}\, d\tau = 1/(s-k) \quad (\text{if } s > k). \tag{14.1.5}$$

Hence

$$L[e^{-k\tau}] = 1/(s+k). \tag{14.1.6}$$

Example 4. Let $f(\tau) = 1/\sqrt{\tau} = \tau^{-1/2}$. Then

$$L[\tau^{-1/2}] = \int_0^\infty \tau^{-1/2} e^{-s\tau}\, d\tau .$$

We assume

$$s\tau = x^2, \qquad d\tau = 2x\, dx/s,$$

$$L[\tau^{-1/2}] = (\pi/s)^{1/2} \frac{2}{\sqrt{\pi}} \int_0^\infty \exp[-x^2]\, dx$$

$$= (\pi/s)^{1/2} \quad \text{since} \left(2/\sqrt{\pi} \int_0^\infty \exp[-x^2]\, dx = 1\right). \tag{14.1.7}$$

Example 5. Let $f(\tau) = \tau^k$ where k may not only be an integer, but also a fraction $(k > -1)$. Then

$$L[\tau^k] = \int_0^\infty \tau^k e^{-s\tau}\, d\tau .$$

If we assume $s\tau = z$, then

$$L[\tau^k] = \int_0^\infty z^k e^{-z} (dz/s^{k+1}) = \Pi(k)/s^{k+1}, \tag{14.1.8}$$

where $\Pi(k) = \Gamma(k+1)$ (see Appendix 1).

If k is an integer $(k = n)$, then $\Pi(n) = n!$ and

$$L[\tau^k] = n!/s^{n+1}. \tag{14.1.9}$$

In the same way, transformations of a number of other functions may be found. In Appendix 5, a table is given which presents appropriate transforms for the function $f(\tau)$.

It should be noted that not every function $F(s)$ is a transform. For example, there is no original for the function $F(s) = \tan s$ since the poles of this function are located on the whole substantial axis ξ and not only to the left of the straight line σ (see Sections 14.9 and 14.10). It may be shown however that if $\Phi(s)$ is the transform, then the appropriate original will be the only one which would be a piecewise continuous function.

If the function $f(\tau)$ grows more rapidly than $e^{\sigma\tau}$ then it has no transform. For example, the function $f(\tau) = \exp[\tau^2]$ has no transform since its Laplace integral diverges. However, the discontinuous function $f(\tau) = 1/\sqrt{\tau}$ (which approaches the infinity when $\tau \to 0$) has the transform $F(s) = (\pi/s)^{1/2}$, since the Laplace integral converges. The function $f(\tau)$ may be a step function; e.g.,

$$g_k(\tau) = \begin{cases} 0 & \text{at } 0 < \tau < k, \\ 1 & \text{at } \tau > k. \end{cases}$$

Its transform is

$$L[g_k(\tau)] = \int_0^\infty g_k(\tau)e^{-s\tau}\,d\tau = \int_k^\infty e^{-s\tau}\,d\tau = -\frac{1}{s}\,e^{-s\tau}\Big|_k^\infty = e^{-ks}/s. \tag{14.1.10}$$

14.2 Laplace Transformation Properties

a. Linearity Property. If the Laplace transformation is linear, i.e., if A and B are constant, then from the definition of the Laplace transformation, we may write

$$L[Af(\tau) + Bg(\tau)] = AL[f(\tau)] + BL[g(\tau)] = AF(s) + BG(s), \tag{14.2.1}$$

where $F(s)$ and $G(s)$ are transforms of the function $f(\tau)$ and $g(\tau)$, respectively.

Using this property, we may find transforms for a number of functions.

Example 1. Let $f(\tau) = \sinh k\tau$. Then

$$L[\sinh k\tau) = L[\tfrac{1}{2}\,e^{k\tau} - \tfrac{1}{2}\,e^{-k\tau}] = \tfrac{1}{2}[1/(s-k)] - \tfrac{1}{2}[1/(s+k)] = k/(s^2-k^2) \tag{14.2.2}$$

i.e.,

Example 2. Let $f(\tau) = \cosh k\tau$. Then

$$L[\cosh k\tau] = \tfrac{1}{2}\{L[e^{k\tau} + e^{-k\tau}]\} = \tfrac{1}{2}(1/(s-k) + 1/(s+k)) = s/s^2 - k^2. \tag{14.2.3}$$

b. Transform for the Derivative. Let $L[f(\tau)] = F(s)$. We are to find $L[f'(\tau)]$, where $f'(\tau) = df(\tau)/d\tau$.

We have

$$L[f'(\tau)] = \int_0^\infty f'(\tau)e^{-s\tau}\,d\tau = e^{-s\tau}f(\tau)\Big|_0^\infty + s\int_0^\infty f(\tau)e^{-s\tau}\,d\tau. \tag{14.2.4}$$

If $f(\tau)$ enters the subclass with the above asymptotic property, then $e^{-s\tau}f(\tau) \to 0$ when $\tau \to \infty$, and equals $f(0)$ when $\tau \to 0$, i.e.,

$$L[f'(\tau)] = sF(s) - f(0). \tag{14.2.5}$$

Thus, differentiation of the original function corresponds to multiplying the transformed function by s and subsequently subtracting the constant $f(0)$, i.e., the value s possesses the operator property. Therefore, using the functional Laplace transformation, we replace differentiation of the original function by an algebraic operation with the transform. This is the operational property of the Laplace transformation.

If $f(0) = 0$, then $L[f'(\tau)] = sF(s)$, but the quantity s is not identical to the operator $D = d/d\tau$, since for the constant A we have $L[A] = A/s$ and $DA = 0$.

We are now to find the transform of the derivative of the second order:

$$\begin{aligned} L[f''(\tau)] &= sL[f'(\tau)] - f'(0) = s\{sL[f(\tau)] - f(0)\} - f'(0) \\ &= s^2F(s) - sf(0) - f'(0). \end{aligned}$$

In a similar way, we may find the transform for $f'''(\tau)$:

$$L[f'''(\tau)] = s^3F(s) - s^2f(0) - sf'(0) - f''(0).$$

In general

$$L[f^{(n)}(\tau)] = s^nF(s) - s^{n-1}f(0) - s^{n-2}f'(0) - \cdots - f^{(n-1)}(0). \tag{14.2.6}$$

Thus for the function $f(\tau)$ with the imposed asymptotic property, there is a transform for the derivative $f^{(n)}(\tau)$, provided that its continuous derivatives exist up to $f^{n-1}(\tau)$. Formula (14.2.6) is of great practical importance for obtaining transforms. It allows us to find the transformed functions referred to in Section 14.1.

Example 1. Let $f(\tau) = A\tau$, $f'(\tau) = A$. Then

$$L[f'(\tau)] = sL[f(\tau)] - f(0),$$

i.e.,

$$L[A] = sL[A\tau].$$

It is known that $L[A] = A/s$; hence

$$L[A\tau] = A/s^2, \tag{14.2.7}$$

i.e., we also obtain the relation (14.1.4).

Example 2. Let $f(\tau) = \sin k\tau$, $f'(\tau) = k \cos k\tau$, $f''(\tau) = -k^2 \sin k\tau$. Then

$$L[f''(\tau)] = s^2F(s) - sf(0) - f'(0),$$
$$-k^2L[\sin k\tau] = s^2L[\sin k\tau] - k.$$

Hence

$$L[\sin k\tau] = k/(s^2 + k^2). \tag{14.2.8}$$

Example 3. Let $f(\tau) = \cos k\tau$, $f'(\tau) = -k \sin k\tau$, $f''(\tau) = -k^2 \cos k\tau$. Then

$$L[f(\tau)] = F(s),$$
$$-k^2L[\cos k\tau] = s^2L[\cos k\tau] - s.$$

Hence

$$L[\cos k\tau] = s/(s^2 + k^2). \tag{14.2.9}$$

Example 4. Let $f(\tau) = \tau^{1/2}$, $f'(\tau) = \frac{1}{2}\tau^{-1/2}$. Then

$$L[\tfrac{1}{2}\tau^{-1/2}] = sL[\tau^{1/2}].$$

From formula (14.1.7), we have $L[\tau^{-1/2}] = (\pi/s)^{1/2}$. Thus

$$L[\tau^{1/2}] = (1/2s)(\pi/s)^{1/2}. \tag{14.2.10}$$

Example 5. Let $f(\tau) = \tau^{n+1/2}$, $f'(\tau) = (n + \frac{1}{2})\tau^{n+1/2-1}$,

$$f''(\tau) = (n + \tfrac{1}{2})(n + \tfrac{1}{2} - 1)\tau^{n+1/2-2}, \ \ldots, \ f^{(n)}(\tau) = (n + \tfrac{1}{2})\cdots\tfrac{3}{2}\tau^{1/2},$$

$$f^{(n+1)}(\tau) = (n + \tfrac{1}{2})\cdots\tfrac{3}{2}\cdot\tfrac{1}{2}\tau^{-1/2} = \frac{1\cdot 3\cdot 5\cdots(2n+1)}{2^{n+1}}\tau^{-1/2},$$

$$L[f^{(n+1)}(\tau)] = s^{(n+1)}L[\tau^{n+1/2}] - s^nf(0) - \cdots,$$

$$L[\tau^{n+1/2}] = \frac{1\cdot 3\cdot 5\cdots(2n+1)}{2^{n+1}s^{n+1}}L[\tau^{-1/2}].$$

Hence, using (14.1.7), we obtain

$$L[\tau^{n+1/2}] = \frac{1\cdot 3\cdot 5\cdots(2n+1)}{2^{n+1}}\frac{\pi^{1/2}}{s^{n+3/2}}. \tag{14.2.11}$$

c. *Integration of the Original Function.* We are to find the transform of the function $g(\tau) = \int_0^\tau f(\theta)\,d\theta$, i.e., to determine

$$L[g(\tau)] = L\left[\int_0^\tau f(\theta)\,d\theta\right].$$

If $g(\tau)$ satisfies the imposed conditions, then

$$g'(\tau) = f(\tau),$$

$$L[g'(\tau)] = L[f(\tau)] = sL[g(\tau)] = sL\left[\int_0^\tau f(\theta)\, d\theta\right].$$

Since $L[f(\tau)] = F(s)$, then

$$L\left[\int_0^\tau f(\theta)\, d\theta\right] = (1/s)F(s). \tag{14.2.12}$$

Thus integration of the original function $f(\tau)$ corresponds to dividing the transform $F(s)$ by the quantity s, i.e., this quantity s^{-1} has the integration operator property.

Using the same method, we may show that double integration of the original function corresponds to division of the transform by s^2, i.e.,

$$L\left[\int_0^\tau \int_0^\theta f(\xi)\, d\xi\, d\theta\right] = (1/s^2)F(s). \tag{14.2.13}$$

d. Substitution Theorems. We now consider two relations between the original function and its transform which are referred to as substitution theorems.

The first theorem. Let $F(s)$ be the transform of the function $f(\tau)$. After substituting $a\tau$ for τ where a is a constant, we can write

$$L[f(a\tau)] = \int_0^\infty e^{-s\tau} f(a\tau)\, d\tau = \frac{1}{a}\int_0^\infty e^{-(s/a)\theta} f(\theta)\, d\theta = (1/a)F(s/a), \tag{14.2.14}$$

where $\theta = a\tau$.

The substitution of τ/a for τ gives us

$$L[f(\tau/a)] = \int_0^\infty e^{-s\tau} f(\tau/a)\, d\tau = a\int_0^\infty e^{-ast} f(t)\, dt = aF(as), \tag{14.2.15}$$

where $t = \tau/a$, i.e., the substitution of $a\tau$ for the independent variable τ in the original function corresponds to the substitution of s/a for s in the transform and to division of the transform by a. This theorm is often referred to as the similarity theorem.

Example 1. We have $L[\cos \tau] = s/(s^2 + 1)$. Then

$$L[\cos k\tau] = \frac{1}{k}\,\frac{s/k}{(s/k)^2 + 1} = \frac{s}{s^2 + k^2},$$

i.e., the result is the same as in Example 3 [formula (14.2.9)].

The second theorem. The function $f(\tau)$ satisfies the ordinary conditions. Its transform is $F(s)$. In the transform, the quantity s will be replaced by $s - a$, where a is a constant. Then

$$\begin{aligned} F(s-a) &= \int_0^\infty e^{-(s-a)\tau} f(\tau)\, d\tau \\ &= \int_0^\infty e^{-s\tau} e^{a\tau} f(\tau)\, d\tau = L[e^{a\tau} f(\tau)], \end{aligned} \tag{14.2.16}$$

i.e., the substitution of $(s - a)$ for s in the transforms corresponds to multiplication of the original function by the quantity $e^{a\tau}$. This theorem is often referred to as the displacement theorem.

Example 2. It is known that $m!/s^{m+1} = L[\tau^m]$ $(s > 0,\ m = 1, 2, \cdots)$. Using the substitution theorem, we obtain

$$m!/(s-a)^{m+1} = L[\tau^m e^{a\tau}]. \tag{14.2.17}$$

Example 3. We have $L[\cos k\tau] = s/(s^2 + k^2)$. Then

$$\frac{s+a}{(s+a)^2 + k^2} = L[e^{-a\tau} \cos k\tau]. \tag{14.2.18}$$

Simultaneous usage of both theorems allows us to write

$$F(as - b) = F\{a(s - \{b/a\})\} = L[(1/a)\, e^{(b/a)\tau} f(\tau/a)]. \tag{14.2.19}$$

e. The Lag Theorem. Let the function $f(\tau)$, which is different from zero only when $\tau > 0$, control a certain process (Fig. 14.1). Consider the func-

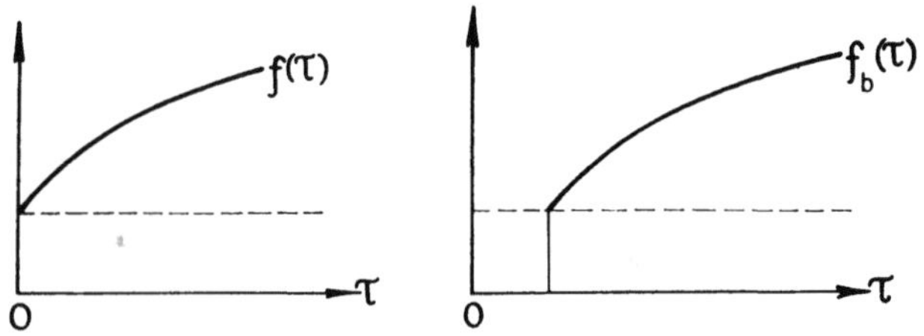

Fig. 14.1. Curves $f(\tau)$ and $f_b(\tau)$.

tion $f_b(\tau)$ controlling a similar process but having the time lag b (see Fig. 14.1):

$$f_b(\tau) = \begin{cases} 0 & \text{at } 0 < \tau < b, \\ f(\tau - b) & \text{at } \tau > b. \end{cases} \tag{14.2.20}$$

We have

$$L[f(\tau)] = \int_0^\infty f(\tau)e^{-s\tau}\,d\tau = F(s). \qquad (14.2.21)$$

We are to find the transform of the function $f_b(\tau)$:

$$\begin{aligned} F_b(s) = L[f_b(\tau)] &= \int_0^\infty f_b(\tau)e^{-s\tau}\,d\tau \\ &= \int_0^b f_b(\tau)e^{-s\tau}\,d\tau + \int_b^\infty f_b(\tau)e^{-s\tau}\,d\tau \\ &= \int_b^\infty f(\tau - b)e^{-s\tau}\,d\tau. \end{aligned} \qquad (14.2.22)$$

Introducing the new variable $\vartheta = \tau - b$ and noting that the new integration limits are from 0 to ∞ gives us

$$F_b(s) = \int_0^\infty f(\vartheta)e^{-s(\vartheta+b)}\,d\vartheta = e^{-sb}\int_0^\infty f(\vartheta)e^{-s\vartheta}\,d\vartheta = e^{-sb}F(s), \qquad (14.2.23)$$

i.e.,

$$L[f_b(\tau)] = e^{-sb}F(s). \qquad (14.2.24)$$

f. Inverse Laplace Transformation. The symbol $L[f(\tau)] = F(s)$ denotes the transformed function $f(\tau)$, i.e., we found the transform from the original function. This construction is referred to as the direct Laplace transformation. In a large number of problems, the function itself has to be obtained from its transform $F(s)$. The symbol $L^{-1}[F(s)]$ is used to denote the inversion of the Laplace transform (i.e., it denotes the original function).

While the direct transformation gives the transform of the function

$$L[f(\tau)] = F(s),$$

the inversion of the transform should give the original function

$$L^{-1}[F(s)] = f(\tau).$$

For example,

$$L^{-1}[1/(s-k)] = e^{k\tau}, \qquad L^{-1}[k/(s^2+k^2)] = \sin k\tau.$$

More strict considerations allow us to conclude that the inverse Laplace transformation is not always the original function and only under certain conditions gives the original. For example, the inverse transform of the function $1/(s-k)$ is $f_1(\tau) = e^{k\tau}$, since the direct transformation of $e^{k\tau}$ gives the transform $1/(s-k)$. But we may obtain another function

$$f_2(\tau) = \begin{cases} e^{k\tau} & \text{when } 0 < \tau < 2 \text{ and } \tau > 2, \\ 1 & \text{when } \tau = 2, \end{cases}$$

which gives the same transform. It should be noted that $f_2(\tau)$ is discontinuous at $\tau = 2$. Any other continuous function for the assigned transform is impossible. Therefore, the transform $F(s)$ cannot have more than one original function $f(\tau)$, which would be continuous for each value of τ. In the majority of problems of mathematical physics, the inverse transformation is single valued.

The inverse transformation is linear, which follows as a direct result of relation (14.2.1), i.e.,

$$L^{-1}[AF(s) + BG(s)] = Af(\tau) + Bg(\tau) = AL^{-1}[F(s)] + BL^{-1}[G(s)]. \tag{14.2.25}$$

14.3 Method of Solution for Simplest Differential Equations

Proceeding from the basic properties of the Laplace transformation, we may solve the simplest ordinary differential equations.

The solution method consists of three steps:

(1) The Laplace transformation is applied to the differential equation and, instead of the original differential equation, an equation for the transform is obtained.

Since the Laplace transformation is an integral transformation and possesses operator properties, an algebraic equation with respect to the transform is obtained instead of the ordinary differential equation for the original.

(2) The algebraic equation obtained is solved with respect to the transform, where s is considered a number. Thus the second step is reduced to determining the solution for the transform.

(3) With the aid of the known relations between the transform $F(s)$ and the original $f(\tau)$, the solution for the inversion is obtained (i.e., the desired original function).

Thus, we first use direct transformation and then inversion. The advantage of the method lies in the fact that we are not required to solve a differential equation for the original function, but rather an algebraic equation for the transform.

We illustrate this by several examples.

Example 1. We have

$$(d^2z/d\tau^2) - k^2z = 0. \tag{14.3.1}$$

Let the desired function $z(\tau)$ be equal to A at $\tau = 0$, i.e., $z(0) = A = \text{const}$ and its derivative $z'(0) = dz(0)/d\tau = D = \text{const}$.

The differential equation may be rewritten

$$z''(\tau) - k^2 z(\tau) = 0.$$

Application of the direct Laplace transformation results in

$$Z(s) = \int_0^\infty e^{-s\tau} z(\tau)\, d\tau = L[z(\tau)],$$

i.e.,

$$L[z''(\tau)] - k^2 L[z(\tau)] = 0.$$

Using formula (14.2.6) gives us

$$s^2 Z(s) - As - D - k^2 Z(s) = 0.$$

The latter is a simple algebraic equation with respect to the transform $Z(s)$ In the solution of the equation, we consider s as a sample number:

$$Z(s) = \frac{As + D}{s^2 - k^2} = A\frac{s}{s^2 - k^2} + D\frac{k}{(s^2 - k^2)k}.$$

As a result, the solution with respect to the transform is obtained.

Using relations (14.2.2) and (14.2.3), we can find the solution with respect to the original function $z(\tau)$, i.e., using the inverse Laplace transformation

$$L^{-1}[Z(s)] = AL^{-1}\left[\frac{s}{s^2 - k^2}\right] + \frac{D}{k}\, L^{-1}\left[\frac{k}{s^2 - k^2}\right],$$

hence

$$z(\tau) = A\cosh k\tau + (D/k)\sinh k\tau = A\cosh k\tau + B\sinh k\tau, \quad (14.3.2)$$

where

$$B = D/k = \text{const}.$$

If $z'(0) = dz(0)/d\tau = D = 0$, the solution of the differential equation will be of the form

$$z(\tau) = A\cosh k\tau. \quad (14.3.3)$$

Example 2. We have

$$z''(\tau) - z'(\tau) - 6z(\tau) = 2.$$

The desired function conforms to the conditions $z(0) = 1$, $z'(0) = 0$.

The direct Laplace transformation will give

$$L[z''(\tau)] - L[z'(\tau)] - 6L[z(\tau)] = L[2],$$

$$s^2 Z(s) - s - sZ(s) + 1 - 6Z(s) = 2/s,$$

and the algebraic equation obtained can be solved for $Z(s)$ as

$$Z(s) = \frac{s^2 - s + 2}{s(s^2 - s - 6)}.$$

Before applying the inverse Laplace transformation, we can rewrite the solution obtained as

$$Z(s) = \frac{A}{s} + \frac{B}{s-3} + \frac{C}{s+2} = -\frac{1}{3}\frac{1}{s} + \frac{8}{15}\frac{1}{s-3} + \frac{4}{5}\frac{1}{s+2},$$

hence

$$L^{-1}[Z(s)] = -\frac{1}{3} L^{-1}\left[\frac{1}{s}\right] + \frac{8}{15} L^{-1}\left[\frac{1}{s-3}\right] + \frac{4}{5} L^{-1}\left[\frac{1}{s+2}\right],$$

$$z(\tau) = -\frac{1}{3} + \frac{8}{15} e^{3\tau} + \frac{4}{5} e^{-2\tau}. \tag{14.3.4}$$

Example 3. We have

$$z'''(\tau) - 2z''(\tau) + 5z'(\tau) = 0,$$

$$z(0) = z'(0) = 0, \qquad z''(0) = 1.$$

With the aid of the same method we find

$$s^3 Z(s) - 1 - 2s^2 Z(s) + 5sZ(s) = 0,$$

$$Z(s) = \frac{1}{s(s^2 - 2s + 5)} = \frac{1}{5}\left(\frac{1}{s} - \frac{s-2}{s^2 - 2s + 5}\right)$$

$$= \frac{1}{5}\frac{1}{s} - \frac{1}{5}\frac{s-1}{(s-1)^2 + 4} + \frac{1}{10}\frac{2}{(s-1)^2 + 4}$$

and the desired solution is obtained for

$$z(\tau) = \frac{1}{5} - \frac{1}{5} e^{\tau} \cos 2\tau + \frac{1}{10} e^{\tau} \sin 2\tau. \tag{14.3.5}$$

For better comprehension of the method of the solution of differential equations and basic relations of the Laplace transformation, the reader is asked to solve the following problems:

(1) $y''(\tau) - k^2 y(\tau) = 0;$

(Answer) $y(\tau) = C_1 e^{k\tau} + C_2 e^{-k\tau}.$

(2) $y''(\tau) - (a + b)y'(\tau) + aby(\tau) = 0;$

(Answer) $y(\tau) = C_1 e^{a\tau} + C_2 e^{b\tau}.$

(3) $y''(\tau) + k^2 y(\tau) = a;$

(Answer) $y(\tau) = C_1 \sin k\tau + C_2 \cos k\tau + (a/k^2).$

(4) $y''(\tau) - 2ay'(\tau) + (a^2 + b^2)y(\tau) = 0;\ y(0) = 0,\ y'(0) = 1;$

(Answer) $y(\tau) = (1/b)e^{a\tau} \sin b\tau.$

(5) $y'''(\tau) + y'(\tau) = e^{2\tau};\ y(0) = y'(0) = y''(0) = 0;$

(Answer) $y(\tau) = -\frac{1}{2} + \frac{1}{10}e^{2\tau} - \frac{1}{5}\sin\tau + \frac{2}{5}\cos\tau.$

(6) $y''(\tau) + y'(\tau) = \tau^2 + 2\tau;\ y(0) = 4;\ y'(0) = -2;$

(Answer) $y(\tau) = \frac{1}{3}\tau^3 + 2e^{-\tau} + 2.$

(7) $y^{(VI)}(\tau) + y'''(\tau) = \cos\tau;\ y(0) = y'(0) = y'''(0) = 0;\ y''(0) = \text{const};$

(Answer) $y(\tau) = -1 + \tau + C\tau^2 + \frac{1}{2}(e^{-\tau} + \cos\tau - \sin\tau).$

(8) $y'(\tau) - z'(\tau) - 2y(\tau) + 2z(\tau) = 1 - 2\tau,\quad y''(\tau) - 2z'(\tau) + y(\tau) = 0$
$(y(0) = z(0) = y'(0) = 0);$

(Answer) $y(\tau) = 2 - 2e^{-\tau} - 2\tau e^{-\tau};\ z(\tau) = 2 - 2e^{-\tau} - 2\tau e^{-\tau} - \tau$

(9) $y^{(IV)}(\tau) + 2y''(\tau) + y(\tau) = 0,\quad y(0) = 0;\quad y'(0) = 1;\quad y''(0) = 2;$
$y'''(0) = -3;$

(Answer) $y(\tau) = \tau(\sin\tau + \cos\tau).$

14.4 Other Properties of the Laplace Transformation

In Section 14.2, we considered the change of the transformed function caused by the differentiation or integration of the original with respect to the variable τ. In this section, we shall discuss the inverse problem, i.e., we shall differentiate and integrate with respect to the parameter s of the transform and determine the corresponding mathematical operation associated with the original function.

a. Differentiation of the Transform. Let $F(s) = L[f(\tau)]$. Taking a number of derivatives of $F(s)$ with respect to s we have

$$F'(s) = dF(s)/ds = \int_0^\infty e^{-s\tau}(-\tau)f(\tau)\,d\tau = L(-\tau f(\tau)),$$

$$F''(s) = \int_0^\infty e^{-s\tau}\tau^2 f(\tau)\,d\tau = L[\tau^2 f(\tau)].$$

In general,

$$F^{(n)}(s) = L[(-\tau)^n f(\tau)]. \tag{14.4.1}$$

Thus n-fold differentiation of the transform corresponds to multiplication of the original by $(-\tau)^n$.

Example. It is known that

$$k/(s^2 + k^2) = L[\sin k\tau].$$

Application of the above rule gives

$$-2ks/(s^2 + k^2)^2 = L[-\tau \sin k\tau]$$

which yields

$$L[\tau \sin k\tau] = 2ks/(s^2 + k^2)^2. \tag{14.4.2}$$

From formula (14.4.2) a new relation between the new original function and its transform may be obtained, if we bear in mind that division of the transform by s corresponds to integration of the original:

$$\int_0^\tau \theta \sin k\theta\,d\theta = (1/k^2)(\sin k\tau - k\tau \cos k\tau).$$

Then we obtain

$$L[\sin k\tau - k\tau \cos k\tau] = 2k^3/(s^2 + k^2)^2. \tag{14.4.3}$$

Using theorem (14.4.1), we may obtain a number of new transformations.

b. Integration of the Transform. Let the function $f(\tau)$ satisfy the usual conditions and let its transform be $F(p)$, i.e.,

$$F(p) = \int_0^\infty e^{-p\tau} f(\tau)\,d\tau.$$

Integrating $F(p)$ from s to b we obtain

$$\int_s^b F(p)\,dp = \int_s^b \int_0^\infty e^{-p\tau} f(\tau)\,d\tau\,dp$$
$$= \int_0^\infty f(\tau) \int_s^b e^{-p\tau}\,dp\,d\tau = \int_0^\infty (f(\tau)/\tau)(e^{-s\tau} - e^{-b\tau})\,d\tau.$$

If the function $f(\tau)$ is one such that the limit $f(\tau)/\tau$ exists at $\tau \to 0$, the integral uniformly converges with respect to b. If $b \to \infty$, then

$$\int_0^{\infty} f(p)\, dp = \int_0^{\infty} (f(\tau)/\tau) e^{-s\tau}\, d\tau$$
$$= L\left[\frac{f(\tau)}{\tau}\right]. \tag{14.4.4}$$

Thus the theorem is obtained that integration of the transform with respect to the parameter s from s to ∞ corresponds to division of the original by τ.

Example 1. As

$$L[\sin k\tau] = k/(s^2 + k^2),$$

then

$$\int_s^{\infty} \frac{k}{p^2 + k^2}\, dp = \frac{\pi}{2} - \arctan \frac{s}{k} = L\left[\frac{\sin k\tau}{\tau}\right]. \tag{14.4.5}$$

If the original of the new function is integrated from 0 to τ, the transform should be divided by s.

The assumption that $k = 1$ gives us

$$\int_0^{\tau} (\sin\theta/\theta)\, d\theta = \mathrm{Si}(\tau),$$

hence

$$L[\mathrm{Si}(\tau)] = (1/s) \arctan s.$$

Example 2. We have

$$L\left[\frac{e^{-a\tau} - e^{-b\tau}}{\tau}\right] = \int_0^{\infty} \left(\frac{1}{p+a} - \frac{1}{p+b}\right) dp = \ln \frac{s+b}{s+a}$$
$$(s > -a \text{ and } s > -b).$$

At $a = 0$ and $b = 1$, we have

$$L\left[\frac{1 - e^{-\tau}}{\tau}\right] = \ln\left(1 + \frac{1}{s}\right). \tag{14.4.6}$$

c. Multiplication of Transforms. Let $F_1(s)$ and $F_2(s)$ be transforms of the appropriate functions $f_1(\tau)$ and $f_2(\tau)$, i.e.,

$$F_1(s) = L[f_1(\tau)], \qquad F_2(s) = L[f_2(\tau)].$$

Then the product of the transforms $F_1(s)$, $F_2(s)$ is

$$F_1(s)F_2(s) = L\left[\int_0^\tau f_1(\theta)f_2(\tau-\theta)\,d\theta\right] = L\left[\int_0^\tau f_1(\tau-\theta)f_2(\theta)\,d\theta\right]. \tag{14.4.7}$$

Relation (14.4.7) is known as the multiplication theorem for transforms or the Borel theorem. This theorem will be demonstrated in Section 14.9.

We now introduce the notation

$$\begin{aligned} f_2(\tau)f_1{}^*(\tau) &= \int_0^\tau f_1(\tau-\theta)f_2(\theta)\,d\theta\,, \\ f_1(\tau)f_2{}^*(\tau) &= \int_0^\tau f_1(\theta)f_2(\tau-\theta)\,d\theta\,, \end{aligned} \tag{14.4.8}$$

where

$$f_1{}^*(\tau)f_2(\tau) = f_1(\tau)f_2{}^*(\tau).$$

Then

$$F_1(s)F_2(s) = L[f_1{}^*(\tau)f_2(\tau)]\,, \qquad L^{-1}[F_1(s)F_2(s)] = f_1{}^*(\tau)f_2(\tau). \tag{14.4.9}$$

Relation (14.4.9) is sometimes formulated using the fact that the product of the transforms corresponds to convolution of the original. This is of tremendous importance in operational calculus.

d. The Efros Theorem. Efros has demonstrated an important theorem of which the Borel theorem is a particular case. If $F(s)$ is the transform of the function $f(\tau)$, i.e.,

$$F(s) = \int_0^\infty f(\tau)e^{-s\tau}\,d\tau\,, \tag{14.4.10}$$

then the original

$$F[\varphi(s)]\Phi(s) = \int_0^\infty f^*(\vartheta)e^{-s}\,d\vartheta \tag{14.4.11}$$

is given by the formula

$$f^*(\vartheta) = \int_0^\infty f(\tau)\psi(\tau,\vartheta)\,d\tau\,, \tag{14.4.12}$$

where $\psi(\tau, \vartheta)$ is the solution of the integral equation

$$e^{-\tau\varphi(s)}\Phi(s) = \int_0^\infty \psi(\tau,\vartheta)e^{-s\vartheta}\,d\vartheta. \tag{14.4.13}$$

The demonstration of this theorem will be presented in Section 14.9.

From the Efros theorem, the following consequence is obtained as a particular case. If $f(\tau)$ is the original of the transform $F(s)$, i.e.,

$$L[f(\tau)] = F(s),$$

then

$$\frac{F(\sqrt{s})}{\sqrt{s}} = L\left[\frac{1}{(\pi\tau)^{1/2}} \int_0^\infty \exp\left[-\frac{u^2}{4\tau}\right] f(\tau)\, d\tau\right]. \tag{14.4.14}$$

Example 1.

$$L^{-1}\left[\frac{1}{s^2}\,\frac{1}{s-k}\right] = \tau^* e^{k\tau} = \int_0^\tau (\tau - \theta) e^{k\theta}\, d\theta$$
$$= \frac{1}{k^2}(e^{k\tau} - k\tau - 1).$$

If

$$f_1(\tau) = f_2(\tau) = f(\tau),$$

then

$$[F(s)]^2 = L[f^*(\tau) f(\tau)].$$

Example 2.

$$L^{-1}\left[\frac{s^2}{(s^2+k^2)^2}\right[= \cos k\tau^* \cos k\tau$$
$$= \int_0^\tau \cos k(\tau - \theta) \cos k\theta\, d\theta$$
$$= (1/2k)[\sin k\tau + k\tau \cos k\tau].$$

Using the transforms of three functions $f_1(\tau)$, $f_2(\tau)$, $f_3(\tau)$,

$$F_1(s)F_2(s)F_3(s) = L[f_1^*(\tau) f_2^*(\tau) f_3(\tau)]. \tag{14.4.15}$$

Using relations (14.4.9) and (14.4.15), it may be again shown that integration of the original corresponds to division of the transfrom by s, i.e.,

$$L^{-1}\left[\frac{1}{s} F(s)\right] = 1^* f(\tau) = \int_0^\tau f(\theta)\, d\theta,$$

$$L^{-1}[(1/s^2)F(s)] = 1^* \int_0^\tau f(\theta)\, d\theta = \int_0^\tau \int_0^\theta f(\xi)\, d\xi\, d\theta,$$

thus, relations (14.2.12) and (14.2.13) are obtained.

Example 3.

$$L^{-1}\left[\frac{1}{s-1}\,\frac{1}{\sqrt{s}}\right] = e^{\tau} * \frac{1}{(\pi\tau)^{1/2}} = e^{\tau}\frac{2}{\sqrt{\pi}}\int_0^{\tau} e^{-\theta}\frac{d\theta}{2\sqrt{\theta}} = e^{\tau}\operatorname{erf}(\tau) \tag{14.4.16}$$

where

$$\operatorname{erf}(z) = \frac{2}{\sqrt{\pi}}\int_0^{z}\exp[-z^2]\,dz.$$

Example 4. The inversion of the transform $F(s) = 1/(\sqrt{s}-1)$ is to be found. First of all, we may write

$$\frac{1}{\sqrt{s}-1} = \frac{\sqrt{s}+1}{s-1} = \frac{1}{s-1} + \frac{1}{\sqrt{s}}\left(1+\frac{1}{s-1}\right);$$

then

$$L^{-1}\left[\frac{1}{\sqrt{s}-1}\right] = \frac{1}{(\pi\tau)^{1/2}} + e^{\tau}[1+\operatorname{erf}\sqrt{\tau}\,]. \tag{14.4.17}$$

e. Transforms of Some Particular Functions. We conclude this section by considering the direct transformations of some particular functions which are usually encountered in heat conduction problems. To find the transform, relation (14.4.9) will be used.

Example 1. Let the function be $f(\tau) = \tau^{-3/2}\exp(-k^2/4\tau)$ where $k > 0$. We find the transform of this function to be

$$\begin{aligned} F(s) = L[f(\tau)] &= \int_0^{\infty} e^{-s\tau}\exp[-k^2/4\tau]\tau^{-3/2}\,d\tau \\ &= (4/k)\int_0^{\infty}\exp\{-[\theta^2+(k^2s/4\theta^2)]\}\,d\theta \\ &= (4/k)e^{-k\sqrt{s}}\int_0^{\infty}\exp[-\{\theta-(b/\theta)\}^2]\,d\theta \end{aligned}$$

where $\theta = k/2\sqrt{\tau}$, $b = (k/2)\sqrt{s}$. The latter integral equals $\sqrt{\pi}/2$ so that

$$F(s) = (2\sqrt{\pi}/k)\exp[-k\sqrt{s}\,].$$

Thus

$$L\left[\frac{k}{2(\pi\tau^3)^{1/2}}\exp\left[-\frac{k^2}{4\tau}\right]\right] = \exp[-k\sqrt{s}\,] \quad (k>0,\ s>0). \tag{14.4.18}$$

Example 2. If the original in Eq. (14.4.18) is multiplied by $(-\tau)$, then the derivative of the transform with respect to s should be taken as

$$L\left[\frac{1}{\sqrt{\pi\tau}}\exp\left[-\frac{k^2}{4\tau}\right]\right] = \frac{1}{\sqrt{s}}\exp[-k\sqrt{s}\,].$$

If $k = 0$, the formula derived previously will be obtained:

$$L\left[\frac{1}{\sqrt{\pi\tau}}\right] = \frac{1}{\sqrt{s}}.$$

Example 3. In relation (14.4.18) the original is integrated and the transform is divided by s:

$$\begin{aligned} L^{-1}\left[\frac{1}{s}\exp[-k\sqrt{s}\,]\right] &= \frac{k}{2\sqrt{\pi}}\int_0^\tau \exp\left[-\frac{k^2}{4\theta}\right]\theta^{-3/2}\,d\theta \\ &= \frac{2}{\sqrt{\pi}}\int_{k/2\sqrt{\tau}}^\infty \exp[-\xi^2]\,d\xi \\ &= \frac{2}{\sqrt{\pi}}\int_0^\infty \exp[-\xi^2]\,d\xi - \frac{2}{\sqrt{\pi}}\int_0^{k/2\sqrt{\tau}} \exp[-\xi^2]\,d\xi \\ &= 1 - \operatorname{erf}\frac{k}{2\sqrt{\tau}}. \end{aligned} \tag{14.4.19}$$

As a result, we obtain one of the most useful relations for heat conduction problems:

$$L^{-1}[(1/s)e^{-k\sqrt{s}}] = 1 - \operatorname{erf}(k/2\sqrt{\tau}\,) = \operatorname{erfc}(k/2\sqrt{\tau}\,). \tag{14.4.20}$$

Example 4. In relation (14.4.19) the function is integrated and the transform is divided by s with the result

$$\begin{aligned} L^{-1}\left[\frac{1}{s\sqrt{s}}\exp[-k\sqrt{s}\,]\right] &= \int_0^\tau \frac{1}{(\pi\tau)^{1/2}}\exp\left[-\frac{k^2}{4\theta}\right]d\theta \\ &= 2\left(\frac{\tau}{\pi}\right)^{1/2}\exp\left[-\frac{k^2}{4\tau}\right] - k\operatorname{erfc}\frac{k}{2\sqrt{\tau}} \\ &= 2\sqrt{\tau}\left[\frac{1}{\sqrt{\pi}}\exp\left[-\frac{k^2}{4\tau}\right] - \frac{k}{2\sqrt{\tau}}\operatorname{erfc}\frac{k}{2\sqrt{\tau}}\right] \\ &= 2\sqrt{\tau}\,\operatorname{i\,erfc}\frac{k}{2\sqrt{\tau}}, \end{aligned} \tag{14.4.21}$$

where

$$\text{i erfc } u = \frac{1}{\sqrt{\pi}} e^{-u^2} - u \text{ erfc } u = \int_u^\infty \text{erfc } \xi \, d\xi. \tag{14.4.22}$$

Hence the function *i* erfc *u* means the integral of the function erfc *u* in the range from *u* to ∞.

In a similar way, it may be shown that

$$\begin{aligned}\frac{1}{s^2}\exp[-k\sqrt{s}] &= \left[\tau + \frac{k^2}{2}\right]\text{erfc}\frac{k}{2\sqrt{\tau}} - k\left(\frac{\tau}{\pi}\right)^{1/2}\exp\left[-\frac{k^2}{4\tau}\right] \\ &= 4\tau\left\{\left(\frac{1}{4} + \frac{k^2}{8\tau}\right)\text{erfc}\frac{k}{2\sqrt{\tau}} - \frac{k}{4}\frac{1}{\sqrt{\pi\tau}}\exp\left[-\frac{k^2}{4\tau}\right]\right\} \\ &= 4\tau \, \text{i}^2 \, \text{erfc}\frac{k}{2\sqrt{\tau}},\end{aligned} \tag{14.4.23}$$

where

$$\begin{aligned}\text{i}^2 \text{ erfc } u &= \int_u^\infty \text{i erfc } \xi \, d\xi = \tfrac{1}{4}[(1 - 2u^2)\text{ erfc } u - (2/\sqrt{\pi})ue^{-u}] \\ &= \tfrac{1}{4}[\text{erfc } u - 2u \text{ i erfc } u].\end{aligned} \tag{14.4.24}$$

This method applied in succession for obtaining new transforms gives the relation

$$\frac{1}{s^{1+n/2}}\exp\left[-\left(\frac{s}{a}\right)^{1/2}x\right] = L\left[(4\tau)^{n/2}\,\text{i}^n\text{ erfc }\frac{x}{2(a\tau)^{1/2}}\right], \tag{14.4.25}$$

where

$$\text{i}^n \text{ erfc } u = \int_u^\infty \text{i}^{n-1}\text{ erfc } \xi \, d\xi \quad (n = 0, 1, 2, 3, \ldots). \tag{14.4.26}$$

Consequently,

$$\text{i}^0 \text{ erfc } u = \text{erfc } u. \tag{14.4.27}$$

A general recurrent formula for i^n erfc *u* is

$$2n\,\text{i}^n\text{ erfc } u = \text{i}^{n-2}\text{ erfc } u - 2u\,\text{i}^{n-1}\text{ erfc } u. \tag{14.4.28}$$

It may be demonstrated that $z = \text{i}^n \text{ erfc } u$ is the solution of the differential equation

$$\frac{d^2z}{du^2} + 2u\frac{dz}{du} - 2nz = 0. \tag{14.4.29}$$

It follows from formula (14.4.28) that

$$\mathrm{i}^n \operatorname{erfc} 0 = \frac{1}{2^n \Gamma(\frac{1}{2}n + 1)} = \frac{1}{2^n \Pi(\frac{1}{2}n)}. \tag{14.4.30}$$

Relations (14.4.25) and (14.4.30) are widely encountered in the solution of heat conduction problems.

All of the relations between the original function and its transform derived in the present chapter are compiled in the table in Appendix 5.

14.5 Solution of the Linear Differential Equation with Constant Coefficients by Operational Methods

We will consider the linear differential equation with constant coefficients

$$A_n \frac{d^n z(\tau)}{d\tau^n} + A_{n-1} \frac{d^{n-1} z(\tau)}{d\tau^{n-1}} + \cdots + A_1 \frac{dz(\tau)}{d\tau} + A_0 z(\tau) = \varphi(\tau), \tag{14.5.1}$$

which may be rewritten as

$$A_n z^{(n)}(\tau) + A_{n-1} z^{(n-1)}(\tau) + \cdots + A_1 z'(\tau) + A_0 z(\tau) = \varphi(\tau). \tag{14.5.2}$$

For the simplicity of derivation, we assume the initial conditions

$$z^{(n)}(0) = z^{(n-1)}(0) = \cdots = z(0) = 0. \tag{14.5.3}$$

More general initial conditions could have been taken (e.g., every derivative of z at $\tau = 0$ is equal to a constant); the method of solution, however, will be the same. We shall use the method discussed in Section 14.3.

Application of the direct Laplace transformation gives an algebraic equation of the first power with respect to one unknown:

$$A_n s^n Z(s) + A_{n-1} s^{n-1} Z(s) + \cdots + A_1 s Z(s) + A_0 Z(s) = \Phi(s) \tag{14.5.4}$$

where $\Phi(s) = L[\varphi(\tau)]$. The solution of this algebraic equation is of the form

$$Z(s) = \Phi(s)/\psi(s), \tag{14.5.5}$$

where $\psi(s) = A_n s^n + A_{n-1} s^{n-1} + A_{n-2} s^{n-2} + \cdots + A_1 s + A_0$ is a polynomial of the nth power with respect to s.

The solution of the considered differential equation is obtained by the inverse Laplace transformation

$$z(\tau) = L^{-1}[\Phi(s)/\psi(s)]. \tag{14.5.6}$$

Inversion of the transform may be carried out rapidly if the transform coincides with one of the above transforms. In this case, the relation should be known which allows us to obtain the original of the function if its transform is of the form $\Phi(s)/\psi(s)$ where $\psi(s)$ is a polynomial of the nth power with respect to s. The original function is usually obtained by means of a contour integral in the complex variable domain (see Section 14.8). In this book, the author has not used this method of inversion, since this would require the reader's familiarity with the fundamentals of the theory of functions of a complex variable.

To obtain the inversion of the form (14.5.6), we make use of the Vashchenko-Zakharchenko method. Thus, the main difficulty in the solution of differential equations with constant coefficients arises when the inverse Laplace transformation is obtained, i.e., in inversion of the transform.

14.6 Expansion Theorems

Expansion theorems were originally derived by Vashchenko-Zakharchenko in the form obtained by means of the Laplace transformation. Heaviside obtained expansion theorems long after Vashchenko-Zakharchenko. Heaviside's theorems are of another form since the transformation in this case was carried out according to Laplace–Carson. The Laplace–Carson method appears to be less convenient than the Laplace method.

Let the transform $F(s)$ of the function $f(\tau)$ be presented as the ratio of two polynomials

$$F(s) = \Phi(s)/\psi(s), \tag{14.6.1}$$

where

$$\psi(s) = A_0 + A_1 s + A_2 s^2 + \cdots + A_n s^n \tag{14.6.2}$$

is the polynomial with respect to s in power n, and

$$\Phi(s) = B_0 + B_1 s + B_2 s^2 + \cdots + B_m s^m \tag{14.6.3}$$

is a polynomial in power m with respect to s and $n > m$. The case $m > n$ is imossible since the original of the function corresponding to the integral positive power s is some discontinuous function of a special kind.

The polynomial $\psi(s)$ of the power n may have n roots $s_1, s_2, s_3, \cdots, s_n$. Two cases will be considered separately.

a. The Case of Simple Roots. Let all the roots $s_1, s_2, \ldots, s_n$ be different, i.e.,

$$\psi(s) = (s - s_1)(s - s_2) \cdots (s - s_n). \tag{14.6.4}$$

The relation (14.6.1) may be written

$$\frac{\Phi(s)}{\psi(s)} = \frac{C_1}{s - s_1} + \frac{C_2}{s - s_2} + \cdots + \frac{C_n}{s - s_n}. \tag{14.6.5}$$

The coefficients $C_1, C_2, \cdots, C_n$ are independent of s and consequently, if they are found, the inversion of series (14.6.5) is not difficult, as it is known that

$$L^{-1}[C_n/(s - s_n)] = C_n\, e^{s_n \tau}.$$

To find coefficients C_n, we multiply both sides of the equality (14.6.5) by $(s - s_1)$ to obtain

$$\frac{\Phi(s)(s - s_1)}{\psi(s)} = C_1 + (s - s_1)\left[\frac{C_2}{s - s_2} + \cdots + \frac{C_n}{s - s_n}\right]. \tag{14.6.6}$$

Let $s \to s_1$. Then all the terms of the right-hand side of Eq. (14.6.6) except the first one will be zero. The left-hand side of the equality becomes indeterminate, i.e., equal to 0/0, since both numerator and denominator are zero. We determine its value by L'Hospital's rule

$$\begin{aligned} C_1 &= \lim_{s \to s_1} \left[\frac{\Phi(s)(s - s_1)}{\psi(s)}\right] \\ &= \lim_{s \to s_1} \left[\frac{\Phi'(s)(s - s_1) + \Phi(s)}{\psi'(s)}\right] = \frac{\Phi(s_1)}{\psi'(s_1)}. \end{aligned} \tag{14.6.7}$$

Thus to obtain the coefficient C_1 in the function $\Phi(s)$, the root s_1 should be substituted for s, the derivative of $\psi(s)$ with respect to s taken, the root s_1 substituted for s in $\psi'(s)$, and finally, $\Phi(s_1)$ divided by $\psi'(s_1)$.

The coefficient C_2 is found in a similar way. For this purpose, both sides of equality (14.6.5) are multiplied by $(s - s_2)$ and $s \to s_2$ is assumed:

$$\frac{\Phi(s)(s - s_2)}{\psi(s)} = C_2 + (s - s_2)\left[\frac{C_1}{s - s_1} + \frac{C_3}{s - s_3} + \cdots + \frac{C_n}{s - s_n}\right]. \tag{14.6.8}$$

Hence

$$C_2 = \lim_{s \to s_2} \frac{\Phi(s)(s - s_2)}{\psi(s)} = \frac{\Phi(s_2)}{\psi'(s_2)}. \tag{14.6.9}$$

Similarly, for the coefficient C_3

$$C_3 = \Phi(s_3)/\psi'(s_3).$$

In general

$$C_n = \Phi(s_n)/\psi'(s_n). \qquad (14.6.10)$$

Inverse Laplace transformation is now applied to series (14.6.5) with the result

$$\begin{aligned} f(\tau) &= L^{-1}[F(s)] = L^{-1}[\Phi(s)/\psi(s)] \\ &= C_1 e^{s_1\tau} + C_2 e^{s_2\tau} + \cdots + C_n e^{s_n\tau} \\ &= \{\Phi(s_1)/\psi'(s_1)\}e^{s_1\tau} + \{\Phi(s_2)/\psi'(s_2)\}e^{s_2\tau} + \cdots + \{\Phi(s_n)/\psi'(s_n)\}e^{s_n\tau} \\ &= \sum_{n=1}^{n} \{\Phi(s_n)/\psi'(s_n)\}\, e^{s_n\tau}. \end{aligned} \qquad (14.6.11)$$

Then the expansion theorem of Vashchenko-Zakharchenko is obtained as

$$f(\tau) = L^{-1}[\Phi(s)/\psi(s)] = \sum_{n=1}^{n} \{\Phi(s_n)/\psi'(s_n)\}e^{s_n\tau}. \qquad (14.6.12)$$

Relation (14.6.12) is different from the Heaviside expansion theorem because of the fact that the operator p enters into the latter instead of s, and the product $p_n\psi'(p_n)$ appears instead of $\psi'(s_n)$. Besides, in the Heaviside formula there is additional term $\Phi(0)/\psi(0)$ which is inconvenient to a certain extent for practical calculations.

b. Generalization of the Theorem. The expansion is also valid for the case when the transform $F(s)$ is a ratio of transcendental functions $\Phi(s)$ and $\psi(s)$. In the theory of complex variable functions, it is demonstrated that such a function expands in a common fraction series of the form (14.6.5). As a result, we obtain relation (14.6.12)

$$L^{-1}[\Phi(s)/\psi(s)] = \sum_{n=1}^{n} \{\Phi(s_n)/\psi'(s_n)\}e^{s_n\tau}. \qquad (14.6.13)$$

Here it is assumed that the function $\psi(s)$ is a generalized polynomial and has the simple roots $s_1, s_2, \cdots, s_n$ only, this function having no zeroth roots (all the roots of $F(s)$ lie on the left of the straight line σ. See Section 14.8). If $\Phi_1(s)$ and $\psi_1(s)$ are not generalized polynomials, but multiplication by $s^k(|k| < 1)$ may reduce them to generalized polynomials, then the expansion theorem may also be used.

Let $\Phi_1(s)s^k = \Phi(s)$, $\psi_1(s)s^k = \psi(s)$, where $|k| < 1$, $s^k \neq 0$. Then

$$\lim_{s\to s_n} \frac{\Phi(s)}{\psi'(s)} = \lim_{s\to s_n} \frac{s^k\Phi_1(s)}{[s^k\psi_1(s)]'} = \lim_{s\to s_n} \left[\frac{s^k\Phi_1(s)}{s^k\psi_1{}'(s) + ks^{k-1}\psi_1(s)}\right].$$

If s_n is the root of the equation $\psi_1(s)$, i.e., $\psi_1(s_n) = 0$, we obtain

$$\lim_{s\to s_n} \frac{\Phi(s)}{\psi'(s)} = \lim \frac{\Phi_1(s)}{\psi_1{}'(s)}. \tag{14.6.14}$$

If for some root $s_k = 0$, the transform $F(s)$ should be presented as the ratio of two converging power series with respect to s of which the exponents should be natural numbers, i.e.,

$$\psi(s) = a_0 + a_1 s + a_2 s^2 + \cdots, \tag{14.6.15}$$

$$\Phi(s) = b_0 + b_1 s + b_2 s^2 + \cdots. \tag{14.6.16}$$

Such power series of $\Phi(s)$ and $\psi(s)$ may be referred to as generalized polynomials, or polynomials of infinitely high power.

It is necessary condition that the ratio $\Phi(s)/\psi(s)$ is not equal to a constant or to the function s^r (where r is any positive integer), since they have no inverse transform.

If the ratio of two integral transcendental functions may be reduced to the ratio of two converging power series with exponent in the form of natural numbers (generalized polynomials), the above conditions are also imposed on them. Thus, the necessary condition of the expansion theorem lies in the fact that the transform $F(s)$ should be presented as the ratio of two converging power series in which the exponents are natural numbers and the constant a_0 should be zero when $b_0 \neq 0$.

c. The Case of Multiple Roots. Let $\psi(s)$ be a polynomial of the nth power ($n > m$) with n roots, of which several are multiple roots ($s_r = s_{r+1} = \cdots = s_m$), i.e.,

$$\psi(s) = (s - s_1)(s - s_2)\cdots(s - s_m)^k\cdots(s - s_n),$$

where k is the multiplicity number of the root s_m (k is an integer which is more than unity). As before, we may write

$$F(s) = \frac{\Phi(s)}{\psi(s)} = \frac{C_1}{s - s_1} + \frac{C_2}{s - s_2} + \cdots + \frac{D}{(s - s_m)^k} + \cdots + \frac{C_n}{(s - s_n)}$$
$$= \frac{D}{(s - s_m)^k} + W(s). \tag{14.6.17}$$

Coefficients $C_1, C_2, \ldots, C_n$ are obtained by the conventional method, i.e., the expansion theorem is applied to the case of simple roots. Thus, the original of the transform $W(s)$ is found by relation (14.6.12). Consider the transform $D/(s - s_m)^k$. We are to determine the coefficient D. For this, both sides of equality (14.6.17) are multiplied by $(s - s_m)^k$, i.e.,

$$\Phi(s)(s - s_m)^k/\psi(s) = D + (s - s_m)^k W(s). \tag{14.6.18}$$

Let $s \to s_m$. Then

$$D = \lim_{s \to s_m} \frac{\Phi(s)(s - s_m)^k}{\psi(s)}.$$

On the other hand, we may write

$$\frac{D}{(s - s_m)^k} = \frac{1}{(k-1)!} L[D\tau^{k-1}e^{s_m\tau}]$$

$$= \frac{1}{(k-1)!} L\left[\frac{d^{k-1}}{ds_m^{k-1}} (De^{s_m\tau})\right],$$

since it is known (see (14.2.17)) that

$$L[\tau^{k-1}e^{s_m\tau}] = (k-1)!/(s - s_m)^k.$$

Thus, we obtain

$$f_2(\tau) = L^{-1}\left[\frac{D}{(s - s_m)^k}\right]$$

$$= \frac{1}{(k-1)!} \lim_{s \to s_m} \left\{\frac{d^{k-1}}{ds^{k-1}} \left[\frac{\Phi(s)(s - s_m)^k}{\psi(s)} e^{s\tau}\right]\right\}. \tag{14.6.19}$$

It should be noted that relation (14.6.19) may be obtained strictly from the integral formula for the inverse Laplace transformation, since it is the well-known Cauchy formula for obtaining the residue of the pole $s = s_m$ of order k (see Section 14.9). If $k = 1$ (a simple root), then relation (14.6.19) becomes relation (14.6.18).

Applying formulas (14.6.12)–(14.6.14) and (14.6.19) to obtain the original of the function, any differential equation with constant coefficients may be solved.

d. A Particular Case of the Expansion Theorem. The transform $F(s)$ is restricted by an additional condition that it be an analytical function of the argument $1/s$ at $1/|s| < \varrho$. Then $F(s)$ may be expanded in a converging power series.

$$F(s) = \frac{D_1}{s} + \frac{D_2}{{s_2}^2} + \cdots = \sum_{n=1}^{\infty} \frac{D_n}{s^n}. \tag{14.6.20}$$

The function $F(s)$ should be one such that the series would not contain a constant D_0, since it has no inverse transform as it is not a Laplace transform of some time function.

Since series (14.6.20) converges uniformly at $|\,1/s\,| < \varrho$, then termwise inverse Laplace transformation leads to

$$f(\tau) = L^{-1}[F(s)] = \sum_{n=1}^{\infty} \{D_n/(n-1)!\}\tau^{n-1}. \tag{14.6.21}$$

Relation (14.6.21) may be obtained as a particular case of the expansion theorem (for multiple roots), if we assume that $\psi(s)$ has an infinite number of zeroth roots with various multiplying numbers (from 1 to n).

Most problems on heat transfer may be solved by using formulas (14.6.13) and (14.6.19) for obtaining the original. Formulas (14.6.13) and (14.6.19) are therefore the main ones for the solution of the differential heat conduction equation.

We shall illustrate the above statements by several examples which were discussed previously.

In Example 1, Section 14.3, the transform was obtained in the form

$$Z(s) = \frac{As + D}{s^2 - k^2} = \frac{\Phi(s)}{\psi(s)},$$

where $\psi(s) = s^2 - k^2 = (s - k)(s + k)$ is a second-order polynomial. It has two simple roots $s_1 = k$, $s_2 = -k$; $\psi'(s) = 2s$.

Application of formula (14.6.13) yields

$$z(\tau) = \frac{Ak + D}{2k} e^{k\tau} + \frac{Ak - D}{2k} e^{-k\tau} = A \cosh k\tau + \frac{D}{k} \sinh k\tau,$$

i.e., the same relation as (1.43.2).

In Example 2, Section 14.3, the following relation was obtained in the solution of the differential equation for the transform

$$Z(s) = \frac{s^2 - s + 2}{s(s^2 - s - 6)} = \frac{\Phi(s)}{\psi(s)},$$

where $\psi(s) = s(s^2 - s - 6) = s(s - 3)(s + 2)$ is a third-order polynomial. It has three roots, $s_1 = 0$, $s_2 = 3$, $s_3 = -2$; $\psi'(s) = 3s^2 - 2s - 6$.

Then

$$f(\tau) = L^{-1}[Z(s)] = \sum_{n=1}^{3} \{\Phi(s_n)/\psi'(s_n)\}e^{s_n\tau}$$
$$= \frac{2}{(-6)} + \frac{(9-3+2)}{(27-6-6)} e^{3\tau} + \frac{(4+2+2)}{(12+4-6)} e^{-2\tau}$$
$$= -\frac{1}{3} + \frac{8}{15} e^{3\tau} + \frac{4}{5} e^{-2\tau}.$$

i.e., the same result is obtained.

Example 1. Invert the transform, i.e., find the original $\cos k\tau$ from the transform.

Formula (14.6.13) yields

$$f(\tau) = L^{-1}\left[\frac{s}{s^2+k^2}\right] = \sum_{n=1}^{2} \frac{\Phi(s_n)}{\psi'(s_n)} e^{s_n\tau},$$

where $\Phi(s) = s$, $\psi(s) = s^2 + k^2 = (s+ik)(s-ik)$, $\psi'(s) = 2s$; $s_1 = ik$, $s_2 = -ik$. Hence

$$f(\tau) = \frac{ik}{2ik} e^{ik\tau} + \frac{(-ik)}{2(-ik)} e^{-ik\tau} = \cos k\tau.$$

Example 2. Invert $L^{-1}[C^3/s(s+C)^3]$.

$$f(\tau) = L^{-1}\left[\frac{\Phi(s)}{\psi(s)}\right] = L^{-1}\left[\frac{C^3}{s(s+C)^3}\right], \qquad \Phi(s) = C^3.$$

Here, $\psi(s) = s(s+C)^3$ is a fourth-order polynomial. Its roots are $s_1 = 0$, $s_2 = -C$ (threefold root).

$$f(\tau) = \frac{\Phi(0)}{\psi'(0)} + \frac{1}{2!} \lim_{s \to -C} \left\{\frac{d^2}{ds^2}\left[\frac{C^3}{s} e^{s\tau}\right]\right\}$$
$$= 1 - (1 + C\tau + \tfrac{1}{2}C^2\tau^2)e^{-C\tau}.$$

Example 3. The functions $y(\tau)$ and $z(\tau)$ are to be found which satisfy the following system of differential equations

$$y''(\tau) - z''(\tau) + z'(\tau) - y(\tau) = e^\tau - 2,$$
$$2y''(\tau) - z''(\tau) - 2y'(\tau) + z(\tau) = -\tau,$$

with the initial conditions

$$y(0) = y'(0) = z(0) = z'(0) = 0.$$

The solution is carried out in the usual way:

$$s^2Y(s) - s^2Z(s) + sZ(s) - Y(s) = [1/(s-1)] - (2/s),$$
$$2s^2Y(s) - s^2Z(s) - 2sY(s) + Z(s) = -1/s^2.$$

These equations are rewritten

$$(s+1)Y(s) - sZ(s) = -(s-2)/s(s-1)^2,$$
$$2sY(s) - (s+1)Z(s) = -1/s^2(s-1).$$

The solution of two algebraic equation with two unknowns yields

$$Y(s) = \frac{1}{s(s-1)^2}, \qquad Z(s) = \frac{2s-1}{s^2(s-1)^2}.$$

To obtain the original, we use the relation (14.6.19).

The polynomial $\psi(s) = s(s-1)^2$ has the roots $s_1 = 0$, $s_2 = 1$ (twofold root). Then

$$y(\tau) = \frac{\Phi(0)}{\psi'(0)} + \lim_{s\to 1}\left[\frac{d}{ds}\left(\frac{1}{s}\, e^{s\tau}\right)\right] = 1 + \tau e^{\tau} - e^{\tau}.$$

Similarly we find

$$z(\tau) = \lim_{s\to 0}\left\{\frac{d}{ds}\left[\frac{2s-1}{(s-1)^2}\, e^{s\tau}\right]\right\} + \lim_{s\to 1}\left\{\frac{d}{ds}\left[\frac{2s-1}{s^2}\, e^{s\tau}\right]\right\} = \tau e^{\tau} - \tau.$$

As examples of the use of the inverse transformation by formulas (14.6.12) and (14.6.19) it is suggested that the reader solve the following problems:

(1) $y''(\tau) + 3y'(\tau) + 2y(\tau) = 4;\ y(0) = 2;\ y'(0) = 0$:
(Answer) $y(\tau) = 2$.

(2) $y''(\tau) - 2y'(\tau) + y(\tau) = 1$:
(Answer) $y(\tau) = (C_1 + C_2\tau)e^{t} + 1$.

(3) $y''(\tau) + y(\tau) = 3\sin\tau;\ y(0) = 1;\ y'(0) = -\frac{3}{2}$:
(Answer) $y(\tau) = (1 - \frac{3}{2}\tau)\cos\tau$.

(4) $y''(\tau) + n^2y(\tau) = a\sin(m\tau + \alpha),\ m \neq n;\ y(0) = y'(0) = 0$:

$$\text{(Answer)}\quad y(\tau) = \frac{a}{n(m^2 - n^2)}\{m\cos\alpha\sin n\tau + n\sin\alpha\cos n\tau - n\sin(m\tau + \alpha)\}.$$

(5) $y''(\tau) - m^2 y(\tau) = ae^{m\tau} + be^{n\tau};\ y(0) = y'(0) = 0$:

(Answer) $y(\tau) = (a/2m^2)(m\tau e^{m\tau} - \sinh(m\tau) + b/2m(m^2 - n^2)$
$\times \{(m-n)e^{-m\tau} + (m+n)e^{m\tau} - 2me^{n\tau}\}$.

(6) $y''(\tau) + y(\tau) = \sin\tau \sin 2\tau$:

(Answer) $y(\tau) = [y(0) - \frac{1}{16}]\cos\tau + \frac{1}{16}\cos 3\tau + [y'(0) + \frac{1}{4}\tau]\sin\tau$

(7) $y'''(\tau) + y(\tau) = 1 + \tau + \frac{1}{2}\tau^2;\ y(0) = y''(0) = 0;\ y'(0) = -1$:

(Answer) $y(\tau) = 1 + \tau + \frac{1}{2}\tau^2 + \frac{2}{3}e^{-\tau}$
$- \frac{2}{3}e^{\tau/2}(\cos\frac{1}{2}\tau\sqrt{3} + \sqrt{3}\sin\frac{1}{2}\tau\sqrt{3})$.

(8) $y''(\tau) + y(\tau) = \tau\cos 2\tau$:

(Answer) $y(\tau) = y(0)\cos\tau + y'(0)\sin\tau - \sin\tau\,\frac{5}{9}$
$- \frac{1}{3}\tau\cos 2\tau + \frac{4}{9}\sin 2\tau$.

14.7 Solution of Some Differential Equations with Variable Coefficients

A linear differential equation whose coefficients are polynomials with respect to τ may be transformed into a linear differential equation for the transformed function with coefficients which will be constant with respect to s. The solution of such a transform equation is simpler than that of the original solution if the polynomial order is lower than that of the equation.

If the coefficients of a differential equation for the original function are first-order polynomials, the differential equation for the transform will be a linear first-order equation which can be solved by conventional methods. The difficulty arises in transition from the solution for the transform to that for the original, i.e., in obtaining the original function from its transform.

As an example, two differential Bessel equations will be considered, the solutions of which were used in problems on heat conduction.

Example 1. The differential Bessel equation of the first order will be considered

$$\frac{d^2z(\tau)}{d\tau^2} + \frac{1}{\tau}\frac{dz(\tau)}{d\tau} - k^2 z(\tau) = 0, \tag{14.7.1}$$

with the boundary conditions $z(0) = 1$, $z'(0) = 0$.

We can rewrite the equation in the form

$$\tau z''(\tau) + z'(\tau) - k^2\tau z(\tau) = 0. \tag{14.7.2}$$

For the transformation, relation (14.4.1) will be used, with the result

$$L[\tau z''(\tau)] = -\frac{d}{ds}[s^2Z(s) - s] = -2sZ(s) - s^2Z'(s) + 1,$$
$$L[\tau z(\tau)] = -\frac{d}{ds}Z(s) = -Z'(s). \tag{14.7.3}$$

Then

$$2sZ(s) + s^2Z'(s) - 1 - sZ(s) + 1 - k^2Z'(s) = 0,$$
$$(s^2 - k^2)Z'(s) + sZ(s) = 0. \tag{14.7.4}$$

A differential first-order equation with constant coefficients can be obtained which is solved by the ordinary method

$$dZ/Z = -s\,ds/(s^2 - k^2), \qquad Z(s) = C/(s^2 - k^2)^{1/2}.$$

where C is the integration constant.

For the inversion, $Z(s)$ is expanded in series, which yields

$$Z(s) = \frac{C}{s}\left(1 - \frac{k^2}{s^2}\right)^{-1/2}$$
$$= \frac{C}{s}\left[1 + \frac{1}{2}\left(\frac{k}{s}\right)^2 + \frac{1}{2^2}\frac{3}{2!}\left(\frac{k}{s}\right)^4 + \cdots\right]$$
$$= C\sum_{n=0}^{\infty}\frac{(2n)!}{(2^n n!)^2}\frac{k^{2n}}{s^{2n+1}}. \tag{14.7.5}$$

We shall use the known relation. Then

$$z(\tau) = C\sum_{n=0}^{\infty}\frac{1}{(2^n n!)^2}(k\tau)^{2n}. \tag{14.7.6}$$

The constant $C = 1$, since $z(0) = 1$ from the condition.

The series obtained for $z(\tau)$ represents a first kind of Bessel function of the zeroth order of the imaginary argument $(k\tau)$, i.e.,

$$z(\tau) = I_0(k\tau). \tag{14.7.7}$$

Thus, we obtain a new transformation

$$L^{-1}[1/(s^2 - k^2)^{1/2}] = I_0(k\tau). \tag{14.7.8}$$

If we assume $k = i$ $(k^2 = -1)$, then $I_0(i\tau) = J_0(\tau)$. Hence

$$L^{-1}[1/(s^2 + 1)^{1/2}] = J_0(\tau). \tag{14.7.9}$$

Example 2. Consider the differential Bessel equation of the νth order

$$\tau^2 z''(\tau) + \tau z'(\tau) + (\tau^2 - \nu^2)z(\tau) = 0, \tag{14.7.10}$$

to which the Laplace transformation is applied (ν is a positive integer).

Using the same method of transformation of the original function in the differential equation, we obtain

$$(s^2 + 1)Z''(s) + 3sZ'(s) + (1 - \nu^2)Z(s) = 0. \tag{14.7.11}$$

The equation obtained is rather complex, except for a particular case when $\nu = 1$.

Introduction of a new variable $y(\tau) = \tau^\nu z(\tau)$ leads to

$$\tau y''(\tau) + (1 - 2\nu)y'(\tau) + \tau y(\tau) = 0. \tag{14.7.12}$$

Since $y(0) = 0$, we have

$$(s^2 + 1)Y'(s) + (1 - 2\nu)sY(s) = 0. \tag{14.7.13}$$

The solution of the above differential equation is of the form

$$Y(s) = \frac{C\nu!}{(2\nu)!} \sum_{n=0}^{\infty} \frac{(-1)^n(2n + 2\nu)!}{2^{2n}n!(\nu + n)!} \frac{1}{s^{2n+2\nu+1}}. \tag{14.7.14}$$

Hence

$$y(\tau) = \frac{C\nu!}{(2\nu)!} \sum_{n=0}^{\infty} \frac{(-1)^n}{2^{2n}n(\nu! + n)!} \tau^{2\nu+2n}. \tag{14.7.15}$$

The constant C will be chosen so that $\nu!C/(2\nu)! = 1/2^\nu$; then the desired function $z(\tau)$ is

$$z(\tau) = \sum_{n=0}^{\infty} \frac{(-1)^n}{n!(\nu + n)!} \left(\frac{\tau}{2}\right)^{\nu+2n} = J_\nu(\tau), \tag{14.7.16}$$

where $J_\nu(\tau)$ is a Bessel function of the first kind of the νth order.

From this example, the new relation for the transform of the function $y(\tau)$ is obtained as

$$L[\tau^\nu J_\nu(\tau)] = \frac{2\nu!}{2^\nu \nu!(s^2 + 1)^{\nu+1/2}}. \tag{14.7.17}$$

14.8 Integral Transformations and Operational Methods

All the operational methods are based on integral transformation of the function and they may therefore be called methods of integral transformation.

In the theory of analytical functions, it is demonstrated that any function $u(z)$ of the complex variable $z = x + iy$, which satisfies particular conditions, may be transformed to another function $v(\zeta)$ of the complex variable $\zeta = \xi + i\eta$ by means of the relation

$$u(z) = \int_C K(z, \zeta)v(\zeta)\, d\zeta. \tag{14.8.1}$$

Here the kernel $K(z, \zeta)$ is the analytical function of every complex variable, and the integration path of C should be chosen each time in an appropriate way.

With the kernel

$$K(z, \zeta) = e^{z\zeta} \quad \text{or} \quad K(z, \zeta) = e^{iz\zeta},$$

the Laplace transformation is obtained. We shall consider this in more detail.

Let some function $F(s)$ of the complex variable $s = \xi + i\eta$ be regular in the range $\alpha < \xi < \beta$ wherein $\int_{-\infty}^{\infty} |F(\xi + i\eta)|\, d\eta$ converges. We assume that in every narrower range $\alpha + \delta \leq \xi \leq \beta - \delta$ ($\delta > 0$ is an arbitrary small constant positive integer) the function uniformly tends to zero with increasing absolute value of the ordinate η. Then for real positive values of x

$$g(x) = (1/2\pi i) \int_{\sigma-i\infty}^{\sigma+i\infty} x^{-s}F(s)\, ds, \tag{14.8.2}$$

$$F(s) = \int_0^\infty x^{s-1}g(x)\, dx. \tag{14.8.3}$$

Formulas (14.8.2) and (14.8.3) are known as the Mellin transform formulas.[2]

Substitution of the real variable $x = e^{\tau}$ yields $g(x) = f(\tau)$ and

$$f(\tau) = (1/2\pi i) \int_{\sigma-i\infty}^{\sigma+i\infty} e^{s\tau}F(s)\, ds, \tag{14.8.4}$$

$$F(s) = \int_{-\infty}^{\infty} e^{-s\tau}f(\tau)\, d\tau. \tag{14.8.5}$$

Integral relation (14.8.5) is the Laplace transform.

[2] These are demonstrated by Courant and Hilbert [15a].

Hence, if integral transform (14.8.5) is assumed as the direct transform of the function $f(\tau)$ of the real variable, relation (14.8.4) is an inversion, i.e., the function $F(s)$ is the transform of $f(\tau)$.

Relations (14.8.4) and (14.8.5) are obtained directly from the Mellin transforms, but they may be demonstrated by a simplified method which clearly shows the expression of the function as the integral of the complex variable. In the subsequent portion this method is presented in a somewhat simplified form.[3]

Consider the integral

$$(1/2\pi i) \int_{\sigma-i\infty}^{\sigma+i\infty} e^{s\tau}\, ds/s. \tag{14.8.6}$$

The integral is taken along the straight line parallel to the imaginary axis and lying on the right of this axis at the distance σ (Fig. 14.2). The integrand modulus at $\tau < 0$ approaches zero rapidly when $s \to \infty$ in the right-hand half-plane in the direction parallel to the imaginary axis. For $\tau > 0$, the integral is 1, since to the left of σ the integrand has only one singular point which is a simple pole $s = 0$.

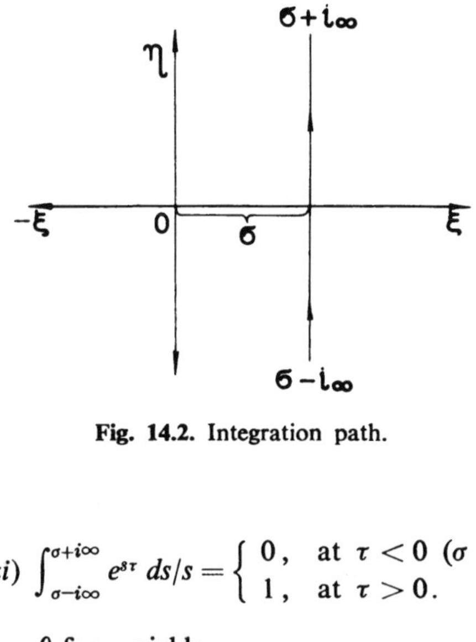

Fig. 14.2. Integration path.

Thus, we have

$$(1/2\pi i) \int_{\sigma-i\infty}^{\sigma+i\infty} e^{s\tau}\, ds/s = \begin{cases} 0, & \text{at } \tau < 0 \ (\sigma > 0) \\ 1, & \text{at } \tau > 0. \end{cases} \tag{14.8.7}$$

Substitution of $\tau - \theta$ for τ yields

$$(1/2\pi i) \int_{\sigma-i\infty}^{\sigma+i\infty} e^{s(\tau-\theta)}\, ds/s = \begin{cases} 0, & \text{at } \tau < \theta, \\ 1, & \text{at } \tau > \theta. \end{cases} \tag{14.8.8}$$

[3] More extensive discussion is presented by Efros and Danilevsky [28].

Substituting θ_1 and $\theta_2(\theta_1 < \theta_2)$ for θ and subtracting the second integral from the first, leads to

$$I = \frac{1}{2\pi i} \int_{\sigma-i\infty}^{\sigma+i\infty} e^{s\tau} \frac{e^{-s\theta_1} - e^{-s\theta_2}}{s} ds = \begin{cases} 0 \text{ at } \tau < \theta_1, \\ 1 \text{ at } \theta_1 < \tau < \theta_2, \\ 0 \text{ at } \tau > \theta_2. \end{cases} \tag{14.8.9}$$

Thus, the integral I is a discontinuous function which is plotted in Fig. 14.3.

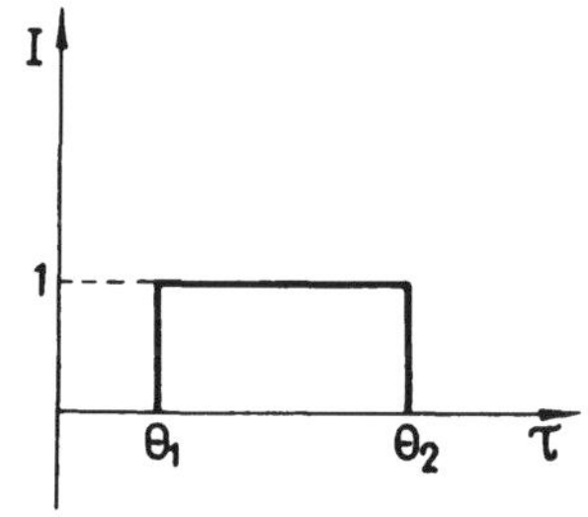

Fig. 14.3. Plot of a single function.

Take an arbitrary function $f(\tau)$ of the independent variable τ from τ_1 to τ_2. This function is plotted in Fig. 14.4. The continuous curve $f(\tau)$ may be replaced by a stepwise line $\varphi_n(\tau)$ the values of which coincide with $f(\tau)$

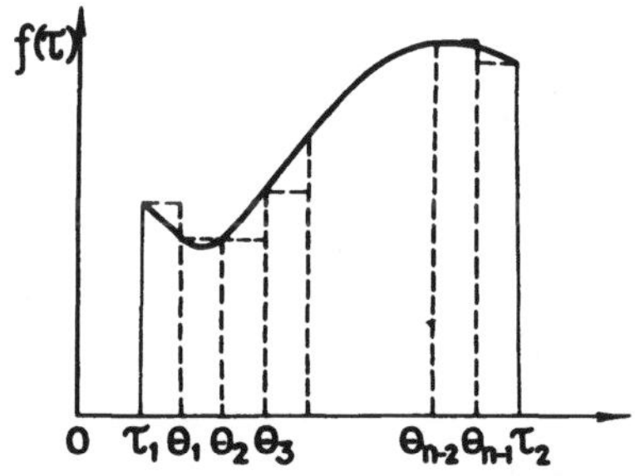

Fig. 14.4. Representation of a function for the Mellin transform derivation.

at n points $\theta_0, \theta_1, \theta_2, \cdots, \theta_{n-1}, \tau_1 = \theta_0 < \theta_1 < \theta_2 < \cdots < \theta_n = \tau_2$. Each step (θ_k, θ_{k+1}) may be expressed by means of the integral I, and the whole stepwise line can be expressed as the sum of integrals I as

$$\begin{aligned} \varphi_n(\tau) &= \sum_{k=0}^{n-1} f(\theta_k)(1/2\pi i) \int_{\sigma-i\infty}^{\sigma+i\infty} \exp[s\tau]\{\exp[-s\theta_k] - \exp[-s\theta_{k+1}]\}\, ds/s \\ &= (1/2\pi i) \int_{\sigma-i\infty}^{\sigma+i\infty} \exp[s\tau] \left(\sum_{k=0}^{n-1} \exp[-s\theta_k] f(\theta_k)\, \Delta\theta_k \right) ds, \end{aligned} \tag{14.8.10}$$

where

$$\begin{aligned} \Delta\theta_k &= \{1 - \exp[-s(\theta_{k+1} - \theta_k)]\}/s \\ &= \theta_{k+1} - \theta_k - (1/2!)(\theta_{k+1} - \theta_k)^2 s + \cdots . \end{aligned} \tag{14.8.11}$$

If n is infinitely increased so that all the differences $(\theta_{k+1} - \theta_k) \to 0$, then $\varphi_n(\tau) \to f(\tau)$ and $\Delta\theta_k$ will be different from $(\theta_{k+1} - \theta_k)$ by an infinitesimal value of the second order. Then

$$\begin{aligned} f(\tau) &= (1/2\pi i) \int_{\sigma-i\infty}^{\sigma+i\infty} e^{s\tau} \lim_{n\to\infty} \left(\sum_{k=0}^{n-1} e^{-s\theta_k} f(\theta_k) \right)(\theta_{k+1} - \theta_k)\, ds \\ &= (1/2\pi i) \int_{\sigma-i\infty}^{\sigma+i\infty} e^{s\tau} \int_{\tau_1}^{\tau_2} e^{-s\tau} f(\tau)\, d\tau\, ds. \end{aligned} \tag{14.8.12}$$

We can change the integration limits of the function $f(\tau)$ with respect to τ from $-\infty$ to $+\infty$ ($\tau_1 = -\infty$, $\tau_2 = \infty$) and denote the integral of $f(\tau)$ with respect to τ by $F(s)$ to obtain

$$f(\tau) = (1/2\pi i) \int_{\sigma-i\infty}^{\sigma+i\infty} e^{s\tau} F(s)\, ds, \tag{14.8.13}$$

$$F(s) = \int_{-\infty}^{+\infty} e^{-s\tau} f(\tau)\, d\tau. \tag{14.8.14}$$

Since in the present problems $\tau \geq 0$, then

$$F(s) = \int_0^{\infty} e^{-s\tau} f(\tau)\, d\tau. \tag{14.8.15}$$

Carson used the modified integral transformation (14.8.15) for justification of the operational calculus

$$W(p) = p \int_0^{\infty} e^{-pt} f(t)\, dt, \tag{14.8.16}$$

where p is a parameter equal to s. Relation (14.8.16) is referred to as the Laplace–Carson integral. Proceeding from relation (14.8.16) we can show that the parameter p possesses properties of the Heaviside operator ($p = d/dt$).

Transform (14.8.16) differs from Laplace transform (14.8.15) only by the function $W(p)$ which enters the transfrom instead of $F(s)$. The function $W(p)$ may also be called the transform of the function $f(\tau)$ according to Laplace–Carson. The relations between $W(p)$ and $f(\tau)$ may be used for obtaining the relations between $F(s)$ and $f(\tau)$, if it is assumed that

$$f(t) = f(\tau) \qquad \text{and} \qquad F(s) = W(p)/p. \tag{14.8.16a}$$

In the majority of works on operational calculus, the Laplace–Carson relation is adopted as an integral transformation. To invert relation (14.8.16) the function $F(s)$ in relation (14.8.13) should be replaced by $W(p)/p$, s by p, and τ by t. Then

$$f(t) = (1/2\pi i) \int_{\sigma-i\infty}^{\sigma+i\infty} e^{pt}(W(p)/p)\, dp. \qquad (14.8.17)$$

Integral relation (14.8.17) is called the Bromwich formula; it is used for the inversion of the transform, if the transformation is carried out following Laplace–Carson.

Thus, all the operational methods are based on integral transformation (14.8.1) and in particular, on Mellin transforms.

It may be noted that the Fourier transform is directly obtainable from the Laplace transform. For this purpose it is sufficient to write the appropriate expression $s = \xi + i\eta$ instead of the complex variable, i.e.,

$$F(\xi + i\eta) = \int_0^\infty e^{-i\eta\tau}[e^{-\xi\tau} f(\tau)]\, d\tau. \qquad (14.8.18)$$

Since in formula (14.8.13) integration is carried out along the straight line $\mathrm{Re}\, s = \sigma$, then substitution of the variable $s = \sigma + i\eta$ yields

$$f(\tau) = (e^{\sigma\tau}/2\pi) \int_{-\infty}^{\infty} F(\sigma + i\eta) e^{i\eta\tau}\, d\eta. \qquad (14.8.18a)$$

Introduction of the new designations $f(\tau)e^{-\sigma\tau} = g(\tau)$, $\{1/(2\pi)^{1/2}\}F(\sigma + i\eta) = G(\eta)$ gives the form of Eqs. (14.8.18) and (14.8.18a) as

$$G(\eta) = 1/(2\pi)^{1/2} \int_0^\infty g(\tau) e^{-i\eta\tau}\, d\tau, \qquad (14.8.18b)$$

$$g(\tau) = 1/(2\pi)^{1/2} \int_{-\infty}^{\infty} G(\eta) e^{i\eta\tau}\, d\eta. \qquad (14.8.18c)$$

Thus, the Laplace transform relating $f(\tau)$ and $F(s)$ is the Fourier transform relating the functions $g(\tau)$ and $G(\eta)$ where σ is an arbitrary real number the value of which is larger than the growth of the function $f(\tau)$. The range of validity of the Fourier transformation is considerably narrower than that of the Laplace transformation, since for the improper integral to converge, the function $g(\tau)$ must satisfy a rather severe condition at infinity, for example, the condition of absolute integrability, i.e., convergence of the integral

$$\int_{-\infty}^{\infty} |\, g(\tau)\, |\, d\tau.$$

Of all the integral transformation methods discussed, we shall only consider in detail the Laplace transformation

$$F(s) = L[f(\tau)] = \int_0^\infty e^{-s\tau} f(\tau)\, d\tau\,, \tag{14.8.19}$$

which is a direct transformation of the original $f(\tau)$ to $F(s)$.

For the inversion of the transform, we use

$$f(\tau) = L^{-1}[F(s)] = (1/2\pi i) \int_{\sigma-i\infty}^{\sigma+i\infty} e^{s\tau} F(s)\, ds. \tag{14.8.20}$$

The Laplace integral (14.8.19) converges absolutely, if the considered functions $f(\tau)$ are piecewise continuous at $\tau > 0$ and equal to zero at $\tau < 0$. Besides, at $\tau \to \infty$, it should be

$$|f(\tau)| < Me^{\sigma_0 \tau}.$$

In this situation, integral (14.8.19) represents the function $F(s)$ of the complex variable s, which is regular in the half-plane Re $s \geq s_1 > \sigma_0$. If s increases infinitely with the modulus and remains in the same half-plane, then

$$\lim_{|s|\to\infty} F(s) = 0.$$

In the inverse transform, the integration is carried out along the straight line $\sigma = \text{const}$, the number σ being arbitrary but greater than σ_0. Integration is carried out along the straight line $\sigma = \text{const}$ in the plane $s = \xi + i\eta$ which is parallel to the imaginary axis and lies in the half-plane

$$\text{Re } s \geq s_1 > \sigma_0.$$

14.9 Inversion of the Transform

To obtain the original from its transform, so far we have used the known relations between the original function and its transform which are obtained by the direct Laplace transformation or by specially derived expression theorems when $F(s)$ represents the ratio of two converging power series with respect to s of which the exponents are natural numbers.

It follows from the previous section that the original function is given by the Laplace transformation

$$f(\tau) = L^{-1}[F(s)] = (1/2\pi i) \int_{\sigma-i\infty}^{\sigma+i\infty} e^{s\tau} F(s)\, ds. \tag{14.9.1}$$

Clearly, the above restrictions must be imposed on $F(s)$, for $L^{-1}[F(s)]$ to exist. The main restriction is that the function $F(s)$ should uniformly approach zero with respect to the argument s when $|s| \to \infty$. Integration is carried out along the straight line σ and the function $F(s)$ should be one such that all the singular points would lie to the left of the integration path.

Since the integrand function is regular to the right of the straight line $\sigma = \text{const}$, the integration path σ may be replaced by another one provided that it would end at $\sigma \pm i\infty$.

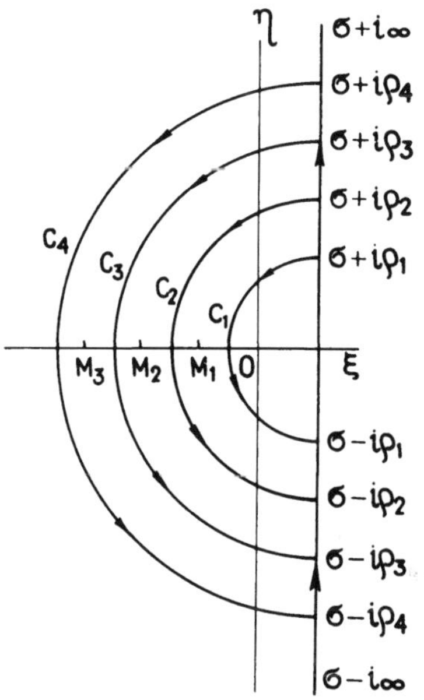

Fig. 14.5. System of integration contours (poles lie on the real axis).

In the majority of problems, the function $F(s)$ is one in which all the poles lie on the negative real axis or on the imaginary one; then in the first case, the integration path may be taken as a semicircle with the center on the straight line $\sigma = \text{const}$ and in the second case as a rectangle (Figs. 14.5 and 14.6).

If all the singular points are inside the contour and the integrand function on the contour is regular and single-valued, the closed contour integral C equals the sum of residues with respect to all the singular points inside the contour

$$(1/2\pi i) \int_{\sigma-i\infty}^{\sigma i+\infty} e^{s\tau} F(s)\, ds = (1/2\pi i) \int_C e^{s\tau} F(s)\, ds = \sum_{n=1}^{k} \operatorname{res}[e^{s\tau} F(s)]. \quad (14.9.2)$$

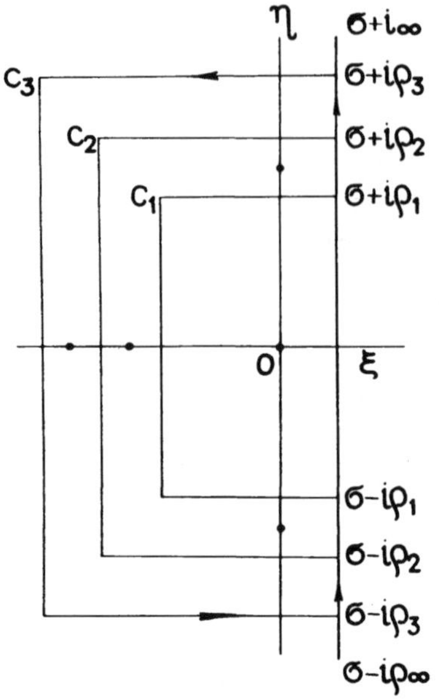

Fig. 14.6. System of integration contours (poles lie on the real and imaginary axes).

Using (14.9.1) and (14.9.2), all the relations discussed in the previous sections may be obtained by inverse transformation.

Example 1. Let $F(s) = \Phi(s)/\psi(s)$, where $\psi(s) = (s - s_1)(s - s_2)\cdots(s - s_n)$ is a polynomial to the nth power with respect to s. Then

$$L^{-1}[F(s)] = (1/2\pi i) \int_C e^{s\tau}\{\Phi(s)/\psi(s)\}\, ds = \sum_1^n \operatorname{res}[e^{s\tau}\{\Phi(s)/\psi(s)\}]. \tag{14.9.3}$$

The polynomial roots $s_1, s_2, \ldots, s_n$ are poles of the function $F(s)$, since $\psi(s)$ is a polynomial with respect to s. With no multiple roots, all the poles are simple ones. Then

$$L^{-1}[F(s)] = \sum_1^n \{\Phi(s_n)/\psi'(s_n)\}e^{s_n\tau}, \tag{14.9.4}$$

i.e., the expansion theorem is obtained (the case of simple roots).

Example 2. If the polynomial $\psi(s)$ has simple multiple poles $s_r = s_{r+1} = s_{r+2} = \cdots = s_m$ with the multiplicity number k and all the remaining roots are simple ($k = 1$), then

$$L^{-1}[F(s)] = \frac{1}{2\pi i} \int_C e^{s\tau} \frac{\Phi(s)}{\psi(s)}\, ds$$
$$= \sum_1^n \operatorname{res}\left[e^{s\tau} \frac{\Phi(s)}{\psi(s)}\right]$$
$$= \sum_1^n \left\{\frac{1}{(k-1)!} \lim_{s\to s_n} \left\{\frac{d^{k-1}}{ds^{k-1}} \left[\frac{\Phi(s)(s-s_k)^k}{\psi(s)}\, e^{s\tau}\right]\right\}\right\}. \quad (14.9.5)$$

If all the roots are simple ($k = 1$), then from (14.9.5) relation (14.9.4) is obtained. If one root s_m includes the multiplicity number k, the residue will be

$$\frac{1}{(k-1)!} \lim_{s\to sm} \left\{\frac{d^{k-1}}{ds^{k-1}} \left[\frac{\Phi(s)(s-s_m)^k}{\psi(s)}\, e^{s\tau}\right]\right\} = L^{-1}\left[\frac{D}{(s-s_m)^k}\right], \quad (14.9.6)$$

i.e., the relation is obtained which is identical to formula (14.6.19) of the expansion theorem (the case of multiple roots).

The restrictions imposed on the expansion theorem result from this. The function $F(s)$ should be single-valued and have as singular points the poles lying to the left of the straight line σ. If these conditions are not fulfilled, then the expansion theorem cannot be applied and in this case relation (14.9.1) should be used.

Instead of the ordinary inversion formula (14.9.1) *for the Laplace transform, the inversion formula may be derived which involves no contour integration, but only differentiation and transition to the limit.*

For derivation of such a formula we shall first determine the operator e^{aD} where D is the differentiation operator

$$e^{aD}f(x) = f(x+a) \quad (14.9.7)$$

irrespective of whether the function is differentiable or not.[4]

Further, we introduce the transform of convolution

$$\psi(x) = \int_{-\infty}^{\infty} G(x-y)\varphi(y)\, dy, \quad (14.9.8)$$

and the two-side Laplace transformation of the kernel of convolution (14.9.8)

$$\int_{-\infty}^{\infty} G(x)e^{-sx}\, dx = 1/E(s). \quad (14.9.9)$$

[4] We have $e^{aD}f(x) = \sum_{k=0}^{\infty} a^k/k!\, f^k(x) = f(x+a)$, i.e., the effect of the differentiation operator e^{aD} is reduced to the shift of the argument of the affected function.

Then

$$[1/E(D)]\varphi(x) = \left\{ \int_{-\infty}^{\infty} G(y)e^{-yD}\, dy \right\} \varphi(x)$$
$$= \int_{-\infty}^{\infty} G(y)\varphi(x-y)\, dy$$
$$= \int_{-\infty}^{\infty} G(x-y)\varphi(y)\, dy = \psi(x), \qquad (14.9.10)$$

i.e., the inversion of convolution (14.9.8) is obtained in the form

$$\varphi(x) = E(D)\psi(x). \qquad (14.9.11)$$

Let the kernel $G(x)$ in formula (14.9.8) be of the form

$$G(x) = \exp[x - e^{-x}]. \qquad (14.9.12)$$

Then

$$\{1/E(s)\} = \int_{-\infty}^{\infty} \exp[-e^{x}] \exp[(1-s)x]\, dx = \Gamma(1-s) \quad (\operatorname{Re} s < 1), \qquad (14.9.13)$$

and inversion formula (14.9.11) is of the form

$$\{1/\Gamma(1-D)\}\psi(x) = \varphi(x). \qquad (14.9.14)$$

Integral function $1/\Gamma(z)$ may be expanded into an infinite product

$$\{1/\Gamma(1-z)\} = \lim_{n\to\infty} \exp[b_n z - \gamma z] \prod_{k=1}^{n} (1 - \{z/k\}) \exp[z/k], \qquad (14.9.15)$$

where γ is the Euler number, and b_n is the succession of material numbers with the property $\lim_{n\to\infty} b_n = 0$. The last condition of the form

$$b_n = \ln n - \sum_{k=1}^{n} \frac{1}{k} + \gamma. \qquad (14.9.16)$$

Accounting for the last equality and substituting the variables in (14.9.8) (with the kernel $G(x)$ defined by (14.9.12))

$$\psi(x) = e^{x}F(e^{x}), \qquad (14.9.17)$$

$$\varphi(y) = f(e^{-y}), \qquad (14.9.18)$$

$$x = \ln(\tau), \qquad (14.9.19)$$

$$y = -\ln(\vartheta), \qquad (14.9.20)$$

we obtain from relation (14.9.15)

$$f(\tau) = \lim_{n\to\infty} \left\{ \frac{(-1)^n}{n!} \left(\frac{n}{\tau}\right)^{n+1} F^{(n)}\left(\frac{n}{\tau}\right) \right\}, \qquad (14.9.21)$$

where $F(s)$ is a transform of the function $f(\tau)$.

We can illustrate the material formula for the Laplace transformation obtained by a simple example. Let

$$F(s) = 1/(s+1), \qquad (14.9.22)$$

then

$$F^{(n)}(s) = (-1)^n n!/(s+1)^{n+1}. \qquad (14.9.23)$$

Consequently,

$$\begin{aligned} f(\tau) &= \lim_{n\to\infty} \left\{ \frac{(-1)^n}{n!} \left(\frac{n}{\tau}\right)^{n+1} \frac{(-1)^n n!}{(1+\{n/\tau\})^{n+1}} \right\} \\ &= \lim_{n\to\infty} \frac{1}{(1+\{\tau/n\})^{n+1}} \\ &= \lim_{n\to\infty} \frac{1}{(1+\{\tau/n\})^n} = 1/e^\tau = e^{-\tau}. \end{aligned} \qquad (14.9.24)$$

We shall take another example. Consider a function with a branch point. For example,

$$F(s) = (\pi/s)^{1/2}, \qquad (14.9.25)$$

$$\cdots$$

$$F^{(n)}(s) = \sqrt{\pi}\, \frac{(-1)^n}{2^n}\, \frac{(2n-1)!!}{s^{n+1/2}}. \qquad (14.9.26)$$

Thus,

$$\begin{aligned} f(\tau) &= \lim_{n\to\infty} \left\{ \frac{(-1)^n}{n!} \left(\frac{n}{\tau}\right)^{n+1} \sqrt{\pi}\, \frac{(-1)^n}{2^n}\, \frac{(2n-1)!!}{\left(\frac{n}{\tau}\right)^{n+1/2}} \right. \\ &= \left(\frac{\pi}{\tau}\right)^{1/2} \lim_{n\to\infty} \left\{ \frac{(2n-1)!!\sqrt{n}}{2^n n!} \right\} \\ &= \frac{1}{\sqrt{\tau}} \lim_{n\to\infty} \left\{ \frac{\Gamma(n+\frac{1}{2})}{\Gamma(n+1)} \sqrt{n} \right\}. \end{aligned} \qquad (14.9.27)$$

We use the known asymptotic expansion for the gamma function (Stirlings formula)

$$\Gamma(z) \simeq (2\pi)^{1/2} z^{z-1/2} e^{-z},$$

which yields

$$f(\tau) = 1/\sqrt{\tau}. \tag{14.9.28}$$

We close the present section by deriving the theorem of transform multiplication and the Efros theorem.

Let the transforms of the functions $f_1(\tau)$ and $f_2(\tau)$ be $F_1(s)$ and $F_2(s)$, respectively, i.e.,

$$F_1(s) = L[f_1(\tau)], \qquad F_2(s) = L[f_2(\tau)]. \tag{14.9.29}$$

We shall find the original $h(\tau)$ which corresponds to the product of the transforms $F_1(s)$, $F_2(s)$:

$$h(\tau) = L^{-1}[F_1(s)F_2(s)]. \tag{14.9.30}$$

Application of integral relation (14.9.1) gives

$$h(\tau) = (1/2\pi i) \int_{\sigma-i\infty}^{\sigma+i\infty} e^{s\tau} F_1(s)F_2(s)\, ds$$

It is known that

$$F_2(s) = \int_0^\infty e^{-s\tau} f_2(\tau)\, d\tau.$$

Hence

$$h(\tau) = (1/2\pi i) \int_0^\infty f_2(\theta)\, d\theta \int_{\sigma-i\infty}^{\sigma+i\infty} F_1(s) e^{s(\tau-\theta)}\, ds. \tag{14.9.31}$$

However

$$(1/2\pi i) \int_{\sigma-i\infty}^{\sigma+i\infty} F_1(s) e^{s(\tau-\theta)}\, ds = f_1(\tau-\theta). \tag{14.9.32}$$

Hence

$$h(\tau) = \int_0^\infty f_2(\theta) f_1(\tau-\theta)\, d\theta.$$

It is known that $f_1(\tau) = 0$, when $\tau < 0$. Thus $f_1(\tau-\theta) = 0$ if $\theta > \tau$ and the upper integration limit ∞ may therefore be replaced by τ. Then

$$h(\tau) = \int_0^\tau f_2(\theta) f_1(\tau-\theta)\, d\theta. \tag{14.9.33}$$

Relation (14.9.32) may be substituted with respect to the function $f_2(\tau-\theta)$ with the result

$$h(\tau) = \int_0^\tau f_1(\theta) f_2(\tau-\theta)\, d\theta. \tag{14.9.34}$$

Thus, we obtain the Borel theorem

$$L^{-1}[F_1(s)F_2(s)] = \int_0^\tau f_1(\theta)f_2(\tau-\theta)\,d\theta = \int_0^\tau f_2(\theta)f_1(\tau-\theta)\,d\theta$$
$$= f_1(\tau)f_2{}^*(\tau) = f_1{}^*(\tau)f_2(\tau). \tag{14.9.35}$$

One more theorem may be derived from the relation. Comparison of (14.9.32) with the integral

$$f_1(\tau) = L^{-1}[F_1(s)] = (1/2\pi i)\int_{\sigma-i\infty}^{\sigma+i\infty} e^{s\tau}F_1(s)\,ds$$

yields

$$L^{-1}[F_1(s)e^{s\theta}] = f_1(\tau-\theta). \tag{14.9.36}$$

Thus multiplication of the transform by $e^{s\theta}$ where $\theta > 0$ corresponds to substitution of $\tau - \theta$ in the original for τ (the lag theorem). Here

$$L^{-1}[F_1(s)e^{s\theta}] = \begin{cases} 0, & \tau < \theta \\ f_1(\tau-\theta), & \tau > \theta \end{cases} \qquad (\theta > 0). \tag{14.9.37}$$

The Efros theorem may be formulated in the following way: If $F(s)$ is the transform of the function $f(\tau)$, i.e.,

$$F(s) = \int_0^\infty f(\tau)e^{-s\tau}\,d\tau, \tag{14.9.38}$$

then the original $f^*(\tau)$ of the transform

$$F[\varphi(s)]\Phi(s) = \int_0^\infty f^*(\vartheta)e^{-s\vartheta}\,d\vartheta \tag{14.9.39}$$

is given by the formula

$$f^*(\vartheta) = \int_0^\infty f(\tau)\psi(\tau,\vartheta)\,d\tau, \tag{14.9.40}$$

where $\psi(\tau, \vartheta)$ is the original of the transform

$$e^{-\tau\varphi(s)}\Phi(s) = \int_0^\infty \psi(\tau,\vartheta)e^{-s\vartheta}\,d\vartheta. \tag{14.9.41}$$

The original function $f(\tau)$ is defined by the relation

$$f(\tau) = (1/2\pi i)\int_{\sigma-i\infty}^{\sigma+i\infty} e^{s\tau}F(s)\,ds. \tag{14.9.42}$$

If $\varphi(s)$ is introduced into (14.9.38) instead of s, we obtain from formulas (14.9.38) and (14.9.39)

$$F[\varphi(s)] = \int_0^\infty f(\tau)e^{-\tau\varphi(s)}\, d\tau \tag{14.9.43}$$

$$f^*(\vartheta) = (1/2\pi i)\int_{\sigma-i\infty}^{\sigma+i\infty} e^{s\vartheta}F[\varphi(s)]\Phi(s)\, ds. \tag{14.9.44}$$

We substitute relation (14.9.43) into relation (14.9.44) and change the integration order to obtain

$$f^*(\vartheta) = (1/2\pi i)\int_0^\infty f(\tau)\int_{\sigma-i\infty}^{\sigma+i\infty} \exp[-\tau\varphi(s) + s\vartheta]\Phi(s)\, ds\, d\tau. \tag{14.9.45}$$

Evidently, here it is necessary that the integral

$$(1/2\pi i)\int_{\sigma-i\infty}^{\sigma+i\infty} \exp[s\vartheta - \tau\varphi(s)]\Phi(s)\, ds = \psi(\tau, \vartheta). \tag{14.9.46}$$

converges.

It follows from relation (14.9.46) that $\psi(\tau, \vartheta)$ is the original of the transform

$$\exp[-\tau\varphi(s)]\Phi(s) = \int_0^\infty e^{-s\vartheta}\psi(\tau, \vartheta)\, d\vartheta. \tag{14.9.47}$$

Relation (14.9.45) may be written

$$f^*(\vartheta) = \int_0^\infty f(\tau)\psi(\tau, \vartheta)\, d\tau \tag{14.9.48}$$

i.e., the theorem is demonstrated.

Relations (14.9.45) and (14.9.48) are of great importance for obtaining the original from the transform, and vice versa.

14.10 Integral Fourier and Hankel Transforms

Inversion formulas of integral Fourier and Hankel transforms may be obtained from the integral Fourier formulas.

The function $f(x)$ with the period $2\pi\lambda$ is assumed to be expressed by the Fourier series

$$f(x) = \tfrac{1}{2}a_0 + \sum_{n=1}^{\infty} (a_n \cos(nx/\lambda) + b_n \sin(nx/\lambda)). \tag{14.10.1}$$

Constant coefficients a_0, a_n, b_n are obtained by multiplication of each term of series (14.10.1) by $\cos(nx/\lambda)$, $\sin(nx/\lambda)$ and by integration with respect to x from $-\pi\lambda$ to $+\pi\lambda$.

Since the trigonometric functions

$$\int_{-\pi\lambda}^{\pi\lambda} \sin(nx/\lambda)\,dx = 0\,, \qquad \int_{-\pi\lambda}^{\pi\lambda} \cos(nx/\lambda)\,dx = 0\,, \tag{14.10.2}$$

$$\int_{-\pi\lambda}^{\pi\lambda} \sin(nx/\lambda)\cos(nx/\lambda)\,dx = 0\,, \quad \int_{-\pi\lambda}^{\pi\lambda} \cos(nx/\lambda)\sin(nx/\lambda)\,dx = 0\,, \tag{14.10.3}$$

$$\begin{aligned} \int_{-\pi\lambda}^{\pi\lambda} \sin(nx/\lambda)\,\sin(mx/\lambda)\,dx &= \begin{cases} 0, & m \neq n, \\ \pi\lambda, & m = n, \end{cases} \\ \int_{-\pi\lambda}^{\pi\lambda} \cos(nx/\lambda)\,\cos(mx/\lambda)\,dx &= \begin{cases} 0, & m \neq n, \\ \pi\lambda, & m = n, \end{cases} \end{aligned} \tag{14.10.4}$$

are orthogonal, we obtain from series (14.10.1)

$$\begin{aligned} \pi\lambda a_0 &= \int_{-\pi\lambda}^{\pi\lambda} f(x')\,dx'\,, \qquad \pi\lambda a_n = \int_{-\pi\lambda}^{\pi\lambda} f(x')\cos(nx'/\lambda)\,dx'\,, \\ \pi\lambda b_n &= \int_{-\pi\lambda}^{\pi\lambda} f(x')\sin(nx'/\lambda)\,dx'\,. \end{aligned} \tag{14.10.5}$$

Thus, expansion in series of the function $f(x)$ may be expressed as

$$f(x) = (1/2\pi\lambda)\int_{-\pi\lambda}^{\pi\lambda} f(x')\,dx' + (1/\pi\lambda)\sum_{n=1}^{\infty}\int_{-\pi\lambda}^{\pi\lambda} f(x')\cos(n(x-x')/\lambda)\,dx'. \tag{14.10.6}$$

Assuming $n/\lambda = \alpha$, $1/\lambda = d\alpha$, and $\lambda \to \infty$, instead of the sum (14.10.6), we have the Fourier integral

$$\pi f(x) = \int_0^{\infty} d\alpha \int_{-\infty}^{+\infty} f(x')\cos\alpha(x-x')\,dx'\,. \tag{14.10.7}$$

The above derivation is somewhat formal. For rigorous arguments and for revealing the conditions of existence of the Fourier integral, the reader is referred to special manuals. It should be however noticed that the function $f(x)$ should satisfy Dirichlet's conditions for any finite range and the integral $\int_{-\infty}^{\infty} f(x)\,dx$ should absolutely converge.

Inversion formulas may be derived from the Fourier integral. This is possible because of the fact that in thermophysical problems of interest, for which solution integral transforms are used, the conditions of validity of inversion formulas are always fulfilled.

Formula (14.10.7) is rewritten as

$$\begin{aligned} \pi f(x) &= \int_0^{\infty}\left[\int_{-\infty}^{\infty} f(x')\cos px'\,dx'\right]\cos px\,dp \\ &\quad + \int_0^{\infty}\left[\int_{-\infty}^{\infty} f(x')\sin px'\,dx'\right]\sin px\,dp\,. \end{aligned} \tag{14.10.8}$$

(a) If the function $f(x)$ is odd, then

$$\int_{-\infty}^{\infty} f(x') \cos px' \, dx' = 0,$$
$$\int_{-\infty}^{\infty} f(x') \sin px' \, dx' = 2 \int_{0}^{\infty} f(x') \sin px' \, dx' = 2f_{Fs}(p), \tag{14.10.9}$$

where $f_s(p)$ is the Fourier sine transform of the function defined by

$$f_{Fs}(p) = \int_{0}^{\infty} f(x) \sin px \, dx. \tag{14.10.10}$$

Thus, the inversion formula for the Fourier sine transformation is of the form

$$f(x) = (2/\pi) \int_{0}^{\infty} f_{Fs}(p) \sin px \, dp. \tag{14.10.11}$$

(b) If the function $f(x)$ is even, then

$$\int_{-\infty}^{\infty} f(x') \sin px' \, dx' = 0,$$
$$\int_{-\infty}^{+\infty} f(x') \cos px' \, dx' = 2 \int_{0}^{\infty} f(x') \cos px' \, dx' = 2f_{Fc}(p), \tag{14.10.12}$$

where $f_{Fc}(p)$ is the Fourier cosine transform defined by the relation

$$f_{Fc}(p) = \int_{0}^{\infty} f(x) \cos px \, dx. \tag{14.10.13}$$

Hence, the inversion formula for the Fourier cosine transform is of the form

$$f(x) = (2/\pi) \int_{0}^{\infty} f_{Fc}(p) \cos px \, dp. \tag{14.10.14}$$

Example 1. Find the transform of the function $f(x) = e^{-x}$. The Fourier sine transform for the function $f(x)$ will be of the form

$$f_{Fs}(p) = \int_{0}^{\infty} e^{-x} \sin px \, dx = 1/(1 + p^2). \tag{14.10.15}$$

The Fourier cosine transformation for the function $f(x)$ is of the form

$$f_{Fc}(p) = \int_{0}^{\infty} e^{-x} \cos px \, dx = 1/(1 + p^2). \tag{14.10.16}$$

Conversely, if the transform of the Fourier sine transformation $f_{Fs}(p) = p/(1 + p^2)$ is known, the inversion of the function will be

$$f(x) = (2/\pi) \int_{0}^{\infty} [p/(1 + p^2)] \sin px \, dp = e^{-x}. \tag{14.10.17}$$

If the Fourier cosine transform $f_{Fc}(p) = 1/(1 + p^2)$ is known, then the inversion will be

$$f(x) = (2/\pi) \int_0^\infty [1/(1 + p^2)] \cos px \, dp = e^{-x}. \qquad (14.10.18)$$

(c) Fourier integral (14.10.7) may be written as

$$\frac{1}{2} \int_{-\infty}^{\infty} d\alpha \int_{-\infty}^{\infty} f(x') \cos \alpha(x - x') \, dx'. \qquad (14.10.19)$$

It is known that

$$\int_{-\infty}^{\infty} d\alpha \int_{-\infty}^{\infty} f(x') \sin \alpha(x - x') \, dx' = 0. \qquad (14.10.20)$$

Then we may write formula (14.10.7), accounting also for (14.10.19), as

$$2\pi f(x) = \int_{-\infty}^{\infty} \exp[i\alpha x] \, d\alpha \int_{-\infty}^{\infty} f(x') \exp[- i\alpha x'] \, dx'. \quad (14.10.21)$$

It should be noted that from formula (14.10.21), the inversion formula for the Laplace transform may be obtained. However, we shall make use of relation (14.10.21) to obtain the inversion formula for a complex Fourier transform defined by

$$f_{Fi}(p) = \int_{-\infty}^{\infty} f(x) e^{ipx} \, dx. \qquad (14.10.22)$$

Denoting $\alpha = - p$, we obtain from (14.10.21) the inversion formula for the complex Fourier transform

$$f(x) = \frac{1}{2\pi} \int_{-\infty}^{\infty} e^{-ipx} f_{Fi}(p) \, dp. \qquad (14.10.23)$$

Example 2. Find the complex Fourier transform of the function $f(x) = e^{-|x|}$.

In accordance with formula (14.10.22)

$$f_{Fi}(p) = \int_0^\infty \exp[- (1 - ip)x] \, dx + \int_{-\infty}^0 \exp[(1 + ip)x] \, dx$$
$$= (1 - ip)^{-1} + (1 + ip)^{-1} = 2/(1 + p^2). \qquad (14.10.24)$$

Conversely, if a complex Fourier transform $f_{Fi}(p) = 2/(1 + p^2)$ is given, the inversion is

$$f(x) = (1/\pi) \int_{-\infty}^{\infty} 1/(1 + p^2) \exp[- ipx] \, dp \qquad (14.10.25)$$

If we assume $x \geq 0$, then the integral (14.10.25) taken along the circle with the center at the coordinate origin and lying in the plane p below the real axis, tends to zero as the circle radius tends to infinity. Thus, integral (14.10.25) may be replaced by that over a closed contour which may be evaluated as the product of $-2\pi i$ by the residue at $p = -i$. The sign is negative because the contour is clockwise.

We have

$$f(x) = \frac{-2i \exp[-ix(-i)]}{[(d/dp)(1+p^2)]_{p=-i}} = \exp[-x]. \qquad (14.10.26)$$

For the case $x \leq 0$, the integration path is closed by a semicircle lying above the real axis. The integral equals the product of $2\pi i$ by the residue at $p = i$:

$$f(x) = e^x. \qquad (14.10.27)$$

Both results are expressed by the same formula (14.10.23).

(d) Inversion formulas (14.10.22)–(14.10.23) give inversion Hankel formulas. We shall extend formulas (14.10.22)–(14.10.23) to the case of two independent variables x and y.

We shall denote

$$f_F(s, t) = \int_{-\infty}^{\infty} \int_{-\infty}^{\infty} f(x, y) \exp[i(sx + ty)]\, dx\, dy. \qquad (14.10.28)$$

Then

$$4\pi^2 f(x, y) = \int_{-\infty}^{\infty} \int_{-\infty}^{\infty} f_F(s, t) \exp[-i(xs + yt)]\, ds\, dt. \qquad (14.10.29)$$

Assuming $x = r\cos\theta$, $y = r\sin\theta$, $s = p\cos\alpha$, $t = p\sin\alpha$ we obtain the following form of (14.10.28)–(14.10.29)

$$f_F(p, \alpha) = \int_0^{\infty} r\, dr \int_0^{2\pi} f(r, \theta) \exp[ipr\cos(\theta - \alpha)]\, d\theta, \qquad (14.10.30)$$

$$4\pi^2 f(r, \theta) = \int_0^{\infty} p\, dp \int_0^{2\pi} f_F(p, \alpha) \exp[-ipr\cos(\theta - \alpha)]\, d\alpha. \qquad (14.10.31)$$

Take the function $f(r, \theta)$ as the function $e^{-in\theta} f(r)$. Then from (14.10.30) we have

$$f_F(p, \alpha) = \int_0^{\infty} f(r) r\, dr \int_0^{2\pi} \exp\{i[-n\theta + pr\cos(\theta - \alpha)]\}\, d\theta. \qquad (14.10.32)$$

Denoting $\varphi = \alpha - \theta - \frac{1}{2}\pi$, we may write the integral with respect to θ as

$$\exp[in(\tfrac{1}{2}\pi - \alpha)] \int_0^{2\pi} \exp[i(n\varphi - pr \sin \varphi)]\, d\varphi$$
$$= 2\pi \exp[in(\tfrac{1}{2}\pi - \alpha)] J_n(pr), \tag{14.10.33}$$

since the integral with respect to φ is the known Bessel integral.

The Hankel transform of the function $f(r)$ is denoted by $f_H(p)$

$$f_H(p) = \int_0^\infty r J_n(pr) f(r)\, dr. \tag{14.10.34}$$

Then instead of (14.10.32) we may write

$$f_F(p, \alpha) = 2\pi \exp[in(\tfrac{1}{2}\pi - \alpha)] f_H(p). \tag{14.10.35}$$

Substituting $e^{-in\theta} f(r)$ for $f(r, \theta)$ and expression for $f_F(p, \alpha)$ from (14.10.35) into (14.10.31) yields

$$2\pi f(r) \exp[-in\theta] = \int_0^\infty p f_H(p)\, dp \times \int_0^{2\pi} \exp\left\{i\left[n\left(\frac{\pi}{2} - \alpha\right) - pr \cos(\theta - \alpha)\right]\right\} d\alpha. \tag{14.10.36}$$

Assuming $\varphi = \theta - \alpha + \frac{1}{2}\pi$, we may express the integral with respect to α as

$$\exp[-in\theta] \int_0^{2\pi} \exp[i(n\varphi - pr \sin \varphi)]\, d\varphi.$$

If we again use the Bessel integral, then this expression may be presented as $2\pi e^{-in\vartheta} J_n(pr)$. Thus, the final inversion formula for the Hankel transform is

$$f(r) = \int_0^\infty p J_n(pr) f_H(p)\, dp. \tag{14.10.37}$$

Example 3. Assume that the function $f(r) = 1/r$. Then the inversion of the function in the integral Hankel transform is of the form

$$f_H(p) = \int_0^\infty J_n(pr)\, dr = 1/p. \tag{14.10.38}$$

Conversely, if we have the transform $f_H(p) = 1/p$, then we obtain form formula (14.10.37)

$$f(r) = \int_0^\infty J_n(pr)\, dp = 1/r. \tag{14.10.39}$$

To conclude this section, we shall consider the domain of validity of the Fourier transform in heat conduction problems.

Fourier and Hankel transforms discussed in the present section may be used when the independent variable ranges from 0 to ∞ (heat conduction of semi-infinite solids). Care has to be taken that integrals entering into inversion formulas converge. The choice of the Fourier sine or cosine transforms is decided by the form of boundary conditions (heat transfer conditions on the interface " body-surroundings ").

This will be considered in detail. The second temperature derivative along the coordinate $\partial^2\vartheta/\partial x^2$ enters into the differential heat conduction equation for one-dimensional problems.

We shall make use of the Fourier sine transform

$$\int_0^\infty \sin px(\partial^2\vartheta/\partial x^2)\,dx = \Big[(\partial\vartheta/\partial x)\sin px\Big]_{x=0}^{x=\infty} - p\int_0^\infty \cos px(\partial\vartheta/\partial x)\,dx. \tag{14.10.40}$$

The first term of the right-hand side of formula (14.10.40) vanishes at $x = 0$, since $\sin px = 0$. If $\partial\vartheta/\partial x \to 0$ at $x \to \infty$, which is usually the case in heat conduction problems, the first term of (14.10.40) will be zero even at $x \to \infty$. Integration by parts of the second term in (14.10.40) gives

$$\int_0^\infty (\partial^2\vartheta/\partial x^2)\sin px\,dx = -p\Big[\vartheta\cos px\Big]_{x=0}^{x=\infty} - p^2\int_0^\infty \vartheta \sin px\,dx. \tag{14.10.41}$$

The second term of the right-hand side of the formula is the product of the transform ϑ (ϑ_{Fs}) by p^2, and the first term is $-p\vartheta_{x=0}$ provided that $\vartheta \to 0$ as $x \to \infty$.

Hence

$$\int_0^\infty (\partial^2\vartheta/\partial x^2)\sin px\,dx = p[\vartheta]_{x=0} - p^2\vartheta_{Fs}, \tag{14.10.42}$$

where $[\vartheta]_{x=0}$ is the value of the function at the boundary i.e., the temperature at the body surface.

In a similar manner we shall find the Fourier cosine transform of

$$\int_0^\infty (\partial^2\vartheta/\partial x^2)\cos px\,dx = -(\partial\vartheta/\partial x)_{x=0} - p^2\vartheta_{Fc}, \tag{14.10.43}$$

where ϑ_{Fc} is the cosine Fourier transform. Here it is assumed that conditions

$$\vartheta_{x=\infty} = (\partial\vartheta/\partial x)_{x=\infty} = 0$$

are fulfilled. The value of the temperature gradient at the body surface $(\partial\vartheta/\partial x)_{x=0}$ should be given in the problem condition.

Thus, the Fourier sine transform should be applied when the temperature at the body surface is given (boundary conditions of the first kind), a cosine transform is appropriate when heat flux is prescribed (boundary conditions of the second kind).

It should be noted that contrary to the Laplace transform, the derivative $\partial\vartheta/\partial x$ or any other derivative of odd order cannot be removed by a sine or cosine transform. Thus integration by parts of the integrals

$$\int_0^\infty (\partial\vartheta/\partial x) \sin px \, dx\,, \qquad \int_0^\infty (\partial\vartheta/\partial x) \cos px \, dx$$

yields the integrals

$$\int_0^\infty \vartheta \cos px \, dx\,, \qquad \int_0^\infty \vartheta \sin px \, dx\,,$$

which cannot be expressed by ϑ_{Fs} and ϑ_{Fc}.

However, application of the Fourier sine and cosine transforms present great advantages in comparison with the Laplace transformation when the initial temperature distribution as the function of coordinates is prescribed.

14.11 Finite Integral Fourier and Hankel Transforms

The above integral transforms are usually called infinite Laplace–Fourier–Hankel transforms since the variable to be removed ranges from 0 to ∞ or from $-\infty$ to $+\infty$. If the time variable in transient heat conduction processes ranges from 0 to ∞ and the Laplace transforms with respect to these variables are always valid, then infinite Fourier and Hankel transforms applied to space coordinates are valid for infinite and semi-infinite solids only.

Recently, Koshlyakov [57], Grinberg [41], Sneddon [109] and Doetch [25] extended Fourier and Hankel transforms to the finite range of variables to be eliminated.

(a) The finite Fourier sine transform is defined by the formula

$$F_s[f(x)] = f_{Fs}(p) = \int_0^\pi f(x) \sin px \, dx\,, \qquad (14.11.1)$$

where p is a positive integer. The fact that π is chosen as the upper integration limit does not restrict the generality of transform (14.11.1) since the integration range may always be reduced to the section $(0, \pi)$ by appropriate substitution.

The inversion formula is derived from relations of the Fourier series theory. It is known that the function $f(x)$ may be expanded into the Fourier sine series of which coefficients a_p are defined by the formula

$$a_p = (2/\pi) \int_0^\pi f(x) \sin px \, dx = (2/\pi) f_{Fs}(p). \tag{14.11.2}$$

Formula (14.11.2) is identical with (14.11.1). The inversion formula is thus of the form

$$f(x) = (2/\pi) \sum_{p=1}^{\infty} f_{Fs}(p) \sin px. \tag{14.11.3}$$

If the function $f(x)$ changes in the range $0 < x < c$, then introduction the variable $z = \pi x/c$ yields

$$\int_0^c f(x) \sin(p\pi x/c) \, dx = (c/\pi) \int_0^\pi f(cz/\pi) \sin pz \, dz = (c/\pi) F_s(cx/\pi). \tag{14.11.4}$$

Example. For $f(x) = x$ $(0 < x < c)$ we have

$$F_{s\pi}[x] = \frac{c}{\pi} F_s\left[\frac{cx}{\pi}\right] = \frac{c^2(-1)^{p+1}}{p} \quad \text{where } p = 1, 2, 3, \cdots. \tag{14.11.5}$$

We are to find the transform of the second derivative of the function

$$\begin{aligned} \int_0^\pi f''(x) \sin px \, dx &= f'(x) \sin px \Big|_0^\pi - p \int_0^\pi f'(x) \cos px \, dx \\ &= - p \cos px f(x) \Big|_0^\pi = - p^2 \int_0^\pi f(x) \sin px \, dx, \end{aligned} \tag{14.11.6}$$

and consequently,

$$F_s[f''(x)] = - p^2 F_s[f(x)] + p[f(0) - (-1)^p f(\pi)]. \tag{14.11.7}$$

We can follow the method of the example $f(x) = x^2$. We have for $f''(x) = 2$.

$$F_s[2] = - p^2 F_s[x^2] - p(-1)^p \pi^2. \tag{14.11.8}$$

However

$$F_s[2] = 2F_s[1] = 2[1 - (-1)^p](1/p),$$

thus,

$$F_s[x^2] = (\pi^2/p)(-1)^{p-1} - (2/p^3)[1 - (-1)^p]. \tag{14.11.9}$$

Similarly, it may be shown that

$$F_s[f^{(4)}(x)] = p^4 f_{Fs}(p) - p^3[f(0) - (-1)^p f(\pi)] + p[f''(0) - (-1)^p f''(\pi)]. \quad (14.11.10)$$

In Table 14.1, transforms are presented of certain functions transformed by the Fourier sine transformation.

(b) The Fourier cosine transform is defined by the relation

$$F_c[f(x)] = f_{Fc}(p) = \int_0^\pi f(x) \cos px \, dx, \quad (14.11.11)$$

where p is a positive integer or zero. If the function $f(x) = 1$, then $f_{Fc}(p) = 0$ $(p = 1, 2, 3, \cdots)$, and $f_{Fc}(0) = \pi$.

If the function $f(x) = x$, then

$$f_{Fc}(p) = -(1/p^2)[(-1)^p - 1]; \quad f_{Fc}(0) = \pi^2/2. \quad (14.11.12)$$

For the Fourier cosine transform, the following relations are valid

$$F_c[f(x) + A] = f_{Fc}(p), \quad p \neq 0 \quad (14.11.13)$$

$$F_c[f(x) + A] = f_{Fc}(0) + \pi A \quad (14.11.14)$$

where A is constant.

The inversion formula is found in a similar way for a finite Fourier cosine transform

$$f(x) = F_c^{-1}[f_{Fc}(p)] = (1/\pi) f_{Fc}(0) + \frac{2}{\pi} \sum_{p=1}^{\infty} f_{Fc}(p) \cos px, \quad (14.11.15)$$

$$f_{Fc}(0) = \int_0^\pi f(x) \, dx. \quad (14.11.16)$$

For the Fourier cosine transform, the following relations are valid:

$$F_c[f''(x)] = -p^2 F_c[f(x)] - f'(0) + (-1)^p f'(\pi), \quad (14.11.17)$$

and

$$\lim_{p \to \infty} f_{Fc}(p) = 0.$$

Similarly to (14.11.10) we have

$$F_c[f^4(x)] = p^4 f_{Fc}(p) + p^2[f'(0) - (-1)^p f'(\pi)] - f'''(0) + (-1)^p f'''(\pi). \quad (14.11.18)$$

Transforms $f_{Fc}(p)$ for some functions $f(x)$ are given in Table 14.2 (taken from reference [15a]).

TABLE 14.1. FINITE SINE TRANSFORMS

$f_{Fs}(p)$	$f(x)$
$f_{Fs}(p) = \int_0^{\pi} f(x) \sin px \, dx;$ $p = 1, 2, 3, \cdots$	$f(x) \quad 0 < x < \pi$
$(-1)^{p+1} f_{Fs}(p)$	$f(\pi - x)$
$1/p$	$(\pi - x)/\pi$
$(1/p)(-1)^{p+1}$	x/π
$(1/p)[1 - (-1)^p]$	1
$(\pi/p^2) \sin pc \quad (0 < c < \pi)$	$\begin{cases} (\pi - c)x; & x \leq c, \\ c(\pi - x); & x \geq c \end{cases}$
$(\pi/p) \cos pc \quad (0 \leq c \leq \pi)$	$\begin{cases} -x; & x < c, \\ \pi - x; & x > c \end{cases}$
$\frac{1}{p^3}(-1)^{p+1}$	$(1/6\pi)x(\pi^2 - x^2)$
$(1/p^3)[1 - (-1)^p]$	$\frac{1}{2}x(\pi - x)$
$(1/p)\pi^2(-1)^{p-1} - (2/p^3)[1 - (-1)^p]$	x^2
$\pi(-1)^p \left(\frac{6}{p^3} - \frac{1}{p}\pi^2\right)$	x^3
$\{p/(p^2 + c^2)\}[1 - (-1)^p e^{c\pi}]$	e^{cx}
$\frac{p}{p^2 + c^2}$	$\frac{1}{\sinh c\pi} \sinh c(\pi - x)$
$\frac{p}{p^2 - k^2} \quad (\lvert k \rvert \neq 0, 1, 2, \cdots)$	$\frac{\sin k(\pi - x)}{\sin k\pi}$
$f_{Fs}(m) = \frac{1}{2}\pi;\ 0(n \neq m)$	$\sin mx \quad (m = 1, 2, 3, \cdots)$
$\frac{p}{p^2 - k^2}[1 - (-1)^p \cos k\pi]$	$\cos kx \quad (\lvert k \rvert \neq 1, 2, 3, \cdots)$
$\frac{p}{p^2 - m^2}[1 - (-1)^{p+m}] \quad (p \neq m)$	$\cos mx \quad (m = 1, 2, 3, \cdots)$

TABLE 14.2. Finite Cosine Transforms

$f_{Fc}(p)$	$f(x)$
$f_{Fc}(p) = \int_0^\pi f(x)\cos px\,dx$ $(p = 0, 1, 2, 3, \cdots)$	$f(x) \quad (0 < x < \pi)$
$(-1)^p f_{Fc}(p)$	$f(\pi - x)$
$f_{Fc}(0) = \pi$, and 0 when $p = 1, 2, 3, \ldots$	1
$(2/p)\sin pc;\quad f_{Fc}(0) = 2c - \pi$	$\begin{cases} 1 & (0 < x < c) \\ -1 & (c < x < \pi) \end{cases}$
$-[1 - (-1)^p]/p^2;\quad f_{Fc}(0) = \dfrac{\pi^2}{2}$	x
$(-1)^p/p^2;\quad f_{Fc}(0) = \dfrac{1}{6}\pi^2$	$\dfrac{x^2}{2\pi}$
$1/p^2;\quad f_{Fc}(0) = 0$	$\dfrac{1}{2\pi}(\pi - x)^2 - \dfrac{\pi}{6}$
$\dfrac{3\pi^2}{p^2}(-1)^p + \dfrac{6}{p^4}[1 - (-1)^p]$ $f_{Fc}(0) = \dfrac{\pi^2}{4}$	x^3
$((-1)^p e^{p\pi} - 1)/(p^2 + c^2)$	$(1/c)e^{cx}$
$\dfrac{1}{p^2 + c^2}$	$\dfrac{\cosh c(\pi - x)}{c \sinh c\pi}$
$1/(p^2 - k^2)[(-1)^p \cos k\pi - 1]$ $(\lvert k \rvert \neq 0, 1, 2, 3, \ldots)$	$(1/k)\sin kx$
$\{1/(p^2 - m^2)\}[(-1)^{p+m} - 1];$ $f_{Fc}(m) = 0 \quad (m = 1, 2, 3, \ldots)$	$(1/m)\sin mx$
$\dfrac{1}{p^2 - k^2} \quad (\lvert k \rvert \neq 0, 1, 2, 3, \ldots)$	$-\dfrac{\cos k(\pi - x)}{k \sin k\pi}$
$0 (n \neq m);\quad f_{Fc}(m) = \pi/2$ $(m = 1, 2, 3, \ldots)$	$\cos mx$

When finite integral Fourier transforms are applied, the second temperature derivative along the body coordinate in the case of a sine transform becomes

$$\int_0^{\pi} (\partial^2\vartheta/\partial x^2) \sin px \, dx = p[\vartheta_0 - (-1)^p\vartheta_\pi] - p^2\vartheta_{Fs} \quad (14.11.19)$$

and the case of a cosine transform

$$\int_0^{\pi} (\partial^2\vartheta/\partial x^2) \cos px \, dx = (-1)^p(\partial\vartheta/\partial x)_\pi - (\partial\vartheta/\partial x)_0 - p^2\vartheta_{Fc}. \quad (14.11.20)$$

Thus, when a sine transform is used, the surface temperature should be known (ϑ_0 and ϑ_π) and with a cosine transform, the temperature gradients on the surface $[(\partial\vartheta/\partial x)_\pi, (\partial\vartheta/\partial x)_0]$ should be known. Boundary conditions of the third kind imply prescription of the law of heat transfer between the surface and the surrounding medium. For example, in the case of one-dimensional symmetrical problems, boundary conditions are of the form

$$\{(\partial\vartheta/\partial x) + H\vartheta\}_{x=R} = 0, \qquad (\partial\vartheta/\partial x)_{x=0} = 0. \quad (14.11.21)$$

The temperature ϑ is read from the medium temperature. Then the integral Fourier cosine transform is defined by the relation

$$f_{Fc}(p) = \int_0^R f(x) \cos px \, dx, \quad (14.11.22)$$

where p is not a positive integer, but is a positive root of the transcendental equation

$$p \tan pR = H. \quad (14.11.23)$$

It may be shown that if p and q are the roots of equation (14.11.23) then[5]

$$\int_0^R \cos px \cos qx \, dx = 0, \quad p \neq q, \quad (14.11.24)$$

$$\int_0^R \cos^2 px \, dx = \frac{R(p^2 + H^2) + H}{2(p^2 + H^2)}, \quad p = q. \quad (14.11.25)$$

The inversion formula for the Fourier series expansion is

$$f(x) = \sum_p a_p \cos px, \quad (14.11.26)$$

[5] Details are found in Chapter 6, Section 3.

where summation is made over all positive roots of (14.11.23) and the coefficients a_p are defined by the relation

$$a_p = \frac{2(p^2 + H^2)}{R(p^2 + H^2) + H} \int_0^R f(x) \cos px \, dx. \tag{14.11.27}$$

Thus, the inversion formula for the Fourier cosine transform with boundary conditions of the third kind is of the form

$$f(x) = \sum_p \frac{2(p^2 + H^2)}{R(p^2 + H^2) + H} f_{Fc}(p) \cos px. \tag{14.11.28}$$

(c) The finite integral Hankel transform is defined by the relation

$$f_H(p) = \int_0^1 rf(r) J_n(pr) \, dr. \tag{14.11.29}$$

where p is the positive root of the equation

$$J_n(p) = 0. \tag{14.11.30}$$

The upper limit is chosen to be unity for convenience, since it is appropriate for solution of problems in generalized variables. The function $f(r)$ may be expressed within the range ($0 \leq r \leq 1$) by the Fourier–Bessel series

$$f(r) = \sum_p a_p J_n(pr), \tag{14.11.31}$$

where the coefficients defined by formula (14.11.32) are related with the transform of $f_H(p)$ by the simple expression

$$a_p = [2/J_{n+1}^2(p)] \int_0^1 rf(r) J_n(pr) \, dr = 2f_H(p)/J_{n+1}^2(p). \tag{14.11.32}$$

Thus, the inversion formula for a finite integral Hankel transform is

$$f(r) = \sum_p [2J_n(pr)/J_{n+1}^2(p)] f_H(p), \tag{14.11.33}$$

where summation is carried out over all positive roots of Eq. (14.11.30). For boundary conditions of the third kind, the finite Hankel transform is defined also by relation (14.11.29) where p is defined by the positive root of the equation

$$pJ_n'(p) + HJ_n(p) = 0. \tag{14.11.34}$$

In this case the inversion formula will be

$$f(r) = \sum_p \frac{2p^2}{H^2 + p^2 - n^2} \frac{J_n(pr)}{J_n^2(p)} f_H(p). \qquad (14.11.35)$$

Summation is carried out over all positive roots of Eq. (14.11.34).

(d) The finite integral Hankel transform for a hollow cylinder when the variable r ranges between R_1 and R_2 is of the form

$$f_H(p) = \int_{R_1}^{R_2} rf(r)B_n(pr)\, dr, \quad R_2 > R_1, \qquad (14.11.36)$$

where

$$B_n(pr) = J_n(pr)Y_n(pR_1) - Y_n(pr)J_n(pR_1) \qquad (14.11.37)$$

$Y_n(pr)$ is the Bessel function of the second kind of the order n, and p is a positive root of the equation

$$J_n(pR_2)\ Y_n(pR_1) = Y_n(pR_2)J_n(pR_1). \qquad (14.11.38)$$

The inversion formula is expressed by

$$f(r) = \frac{\pi^2}{2} \sum_p B_n(pr) \frac{p^2 J_n^2(pR_2)}{J_n^2(pR_1) - J_n^2(pR_2)} f_H(p). \qquad (14.11.39)$$

The above manipulation allows us to determine the group of terms from the differential heat conduction equation

$$f(\vartheta) \equiv \frac{1}{r} \frac{\partial}{dr}\left(r \frac{\partial\vartheta}{\partial r}\right) - \frac{n^2\vartheta}{r}. \qquad (14.11.40)$$

The method for determining kernels of finite integral Fourier and Hankel transforms and appropriate inversion formulas will be considered in Section 14.12.

In the present section, the author's intention was to demonstrate the essence of finite integral transformations and their relation with formulas for Fourier and Bessel series expansion, i.e., their relation with the classical method for solution of unsteady heat conduction problems. Using these ideas, we may extend the integral Laplace transform to the finite range of the variable to be removed. In this case, the finite Laplace transform may be used to eliminate space variables for bodies of finite dimensions. However, when finite integral Fourier and Hankel transforms are used for solution of unsteady heat conduction problems, such a generalized method is not necessary, and the most adequate method is the integral Laplace transformation for elimination of the time variable which ranges from 0 to ∞.

14.12 Kernels of Finite Integral Transforms

When finite integral transforms are used, the main difficulty is finding the kernel of the transform which conforms with the prescribed boundary conditions. The kernel of the finite integral transform is found from the solution of the appropriate Sturm–Liouville problem. This problem is reduced to the solution of the homogeneous differential equation

$$\varrho(x)\frac{\partial t(x,\tau)}{\partial \tau} = \frac{\partial}{\partial x}\left[p(x)\frac{\partial t(x,\tau)}{\partial x} - q(x)t(x,\tau)\right] \quad a \leq x \leq b, \tag{14.12.1}$$

with homogeneous boundary conditions

$$\alpha t(a,\tau) + \beta\left(\frac{\partial t}{\partial x}\right)_{x=a} = 0; \quad \alpha^2 + \beta^2 \neq 0, \tag{14.12.2}$$

$$\gamma t(b,\tau) + \delta\left(\frac{\partial t}{\partial x}\right)_{x=b} = 0; \quad \gamma^2 + \delta^2 \neq 0. \tag{14.12.3}$$

To find a nontrivial solution (which is not identically zero) of Eq. (14.12.1), we shall use the Fourier method of separation of variables. The solution is to be sought as the product of the functions $X(x)$ and $T(\tau)$. It should be noted that the function $X(x)$ depends only on x, and $T(\tau)$ on τ alone:

$$t(x,\tau) = X(x)T(\tau). \tag{14.12.4}$$

Substitution of (14.12.4) into (14.12.3) yields

$$\varrho XT' = T(pX')' - qXT. \tag{14.12.5}$$

Separation of variables in (14.12.5) gives

$$\frac{T'}{T} = \frac{(pX')' - qX}{\varrho X}. \tag{14.12.6}$$

The left-hand side of the equation may be equal to the right-hand side only when the both sides are equal to the same constant, λ.

We then obtain

$$T' + \lambda T = 0, \tag{14.12.7}$$

$$(pX')' + (\lambda\varrho - q)X = 0. \tag{14.12.8}$$

The boundary conditions become

$$\alpha X(a) + \beta X'(a) = 0, \tag{14.12.9}$$

$$\gamma X(b) + \delta X'(b) = 0. \tag{14.12.10}$$

Equations (14.12.7)–(14.12.8) are differential Sturm–Liouville equations. The solution of Eqs. (14.12.7)–(14.12.10) is that of the boundary-value Sturm–Liouville problem. The values of λ for which the solution of the Sturm–Liouville problem is nontrivial are the eigenvalues of the problem, and the solutions $X(x, \lambda)$ corresponding to these values are eigenfunctions.

The boundary-value Sturm–Liouville problem will be further referred to as a regular problem, if the interval $[a, b]$ is finite, and functions $p(x)$, $q(x)$ and $\varrho(x)$ are continuous where $p > 0$ and $q \geq 0$.

Equation (14.12.8) is a linear homogeneous equation of the second order and its general solution is of the form

$$X(x) = C_1 X_1(x, \lambda) + C_2 X_2(x, \lambda), \tag{14.12.11}$$

where C_1 and C_2 are arbitrary constants and $X_1(x, \lambda)$ and $X_2(x, \lambda)$ are partial linearly independent solutions of Eq. (14.12.8), i.e., the ratio of these solutions is not constant

$$[X_1(x, \lambda)/X_2(x, \lambda) \neq \text{const}].$$

Substitution of (14.12.11) into (14.12.9)–(14.12.10) yields

$$\alpha[C_1 X_1(a, \lambda) + C_2 X_2(a, \lambda)] + \beta[C_1 X_1'(a, \lambda) + C_2 X_2'(a, \lambda)] = 0, \tag{14.12.12}$$

$$\gamma[C_1 X_1(b, \lambda) + C_2 X_2(b, \lambda)] + \delta[C_1 X_1'(b, \lambda) + C_2 X_2'(b, \lambda)] = 0 \tag{14.12.13}$$

or

$$\varphi_1(\lambda) C_1 + \varphi_2(\lambda) C_2 = 0, \tag{14.12.14}$$

$$\varphi_3(\lambda) C_1 + \varphi_4(\lambda) C_2 = 0, \tag{14.12.15}$$

where $\varphi_i(\lambda)$ $(i = 1, 2, 3, 4)$ have evident values, e.g.,

$$\varphi_1(x) = \alpha X_1(a, \lambda) + \beta X_1'(a, \lambda).$$

The system of linear homogeneous equations for C_1 and C_2 may be solved when the determinant of the system is zero

$$\begin{vmatrix} \varphi_1(\lambda) & \varphi_2(\lambda) \\ \varphi_3(\lambda) & \varphi_4(\lambda) \end{vmatrix} = 0 \tag{14.12.16}$$

The eigenvalues λ are found from the characteristic equation. It is proved that a regular boundary-value problem has a countable set of eigenvalues $\lambda_1, \lambda_2, \lambda_3, \ldots, \lambda_n$. For the particular boundary conditions the single eigenfunction $X_n(x)$ corresponds to each eigenvalue λ_n. This function is found in the following way. From Eq. (14.12.14)

$$C_2 = \frac{\varphi_1(\lambda_k)}{\varphi_2(\lambda_k)} C_1. \quad (14.12.17)$$

Substituting C_2 into (14.12.11) yields

$$X(x) = C_1\left[X_1(x, \lambda_k) - \frac{\varphi_1(\lambda_k)}{\varphi_2(\lambda_k)} X_2(x, \lambda_k)\right]. \quad (14.12.18)$$

Since the eigenfunctions are found accurate to a constant factor, by omitting the constant C_1 we shall find the eigenfunction

$$X_k(x) = X_1(x, \lambda_k) - \frac{\varphi_1(\lambda_k)}{\varphi_2(\lambda_k)} X_2(x, \lambda_k). \quad (14.12.19)$$

In the functional theory, we use the following theorem: The system of eigenfunctions X_1, X_2, X_3 is orthogonal with the weight $\varrho(x)$ within the range $[a, b]$

$$\int_a^b \varrho(x)X_k(x)X_n(x)\,dx = 0 \quad (14.12.20)$$

for all $k \neq n$.

Now we return to our problem. We shall find the solution of Eq. (14.12.1) with boundary conditions (14.12.2) and (14.12.3) in the form of the eigenfunction expansion X_n

$$t(x, \tau) = \sum_{n=1}^{\infty} T_n(\tau)X_n(\tau). \quad (14.12.21)$$

The function $T_n(\tau)$ depends only on τ and is independent of x; it is constant (with respect to x) coefficient in the expansion in eigenfunctions $X_n(x)$.

The expansion coefficients $T_n(\tau)$ are found in the following way: first, we multiply the both sides of equality (14.12.21) by $\varrho(x)X_k(x)$ and integrate between a and b assuming validity of term-wise integration

$$\int_a^b \varrho(x)t(x, \tau)X_k(x)\,dx = \sum_{k=1}^{\infty} T_n(\tau) \int_a^b \varrho(x)X_k(x)X_n(x)\,dx. \quad (14.12.22)$$

For all $k \neq n$, the integrals in the right-hand side become zero, therefore only one term of the whole series remains at $k = n$

$$\int_a^b \varrho(x)t(x, \tau)X_k(x)\,dx = T_n(\tau) \int_a^b \varrho(x)X_n^2(x)\,dx. \quad (14.12.23)$$

Hence

$$T_n(\tau) = \int_a^b \varrho(x)t(x, \tau)X_n(x)\, dx \bigg/ \int_a^b \varrho(x)X_n^2(x)\, dx. \quad (14.12.24)$$

The numerator depends on τ alone. We denote it by $t_n^*(\tau)$ and obtain the desired finite integral transform

$$t_n^*(\tau) = \int_a^b \varrho(x)t(x, \tau)X_n(x)\, dx. \quad (14.12.25)$$

The appropriate problems of the heat conduction theory are to be solved by this transformation with the kernel $\varrho(x)X_n(x)$. Formula (14.12.24) may be written

$$T_n(\tau) = t_n^*(\tau) \bigg/ \int_a^b \varrho(x)X_n^2(x)\, dx. \quad (14.12.26)$$

Using relation (14.12.26), we find the inversion formula for final integral transformation (14.12.25)

$$t(x, \tau) = \sum_{n=1}^{\infty} \left\{ t_n^*(\tau) \bigg/ \int_a^b \varrho(x)X_n^2(x)\, dx \right\} X_n(x). \quad (14.12.27)$$

We shall illustrate this by an example. We are to find the integral transform and the inversion formula for

$$\frac{\partial t(x, \tau)}{\partial \tau} = a \frac{\partial^2 t(x, \tau)}{\partial x^2}; \quad (14.12.28)$$

$$\frac{\partial t(0, \tau)}{\partial x} = 0, \qquad \frac{\partial t(R, \tau)}{\partial x} + Ht(R, \tau) = 0, \quad (14.12.29)$$

where H is the dimensionless heat transfer factor.

The system of differential Sturm-Liouville equations is of the form

$$aX'' + \lambda X = 0; \qquad X'(0) = 0; \qquad X'(R) + HX(R) = 0. \quad (14.12.30)$$

Solving this problem, we shall find that the eigenvalues λ_k should satisfy the transcendental equation

$$-(\lambda/a)^{1/2} \sin(\lambda/a)^{1/2}R + H \cos(\lambda/a)^{1/2}R = 0. \quad (14.12.31)$$

Using the notation

$$(\lambda/a)^{1/2}R = \mu, \qquad \cot \mu = (1/\mathrm{Bi})\, \mu, \quad (14.12.32)$$

we can readily find from the roots of this characteristic equation the eigenvalues

$$\lambda_n = a\mu_n^2/R^2 \quad \text{where } n = 1, 2, 3, \ldots .$$

It is easy to demonstrate that the eigenfunctions $X_n(x)$ correspond to these λ_n:

$$X_n(x) = \cos(\lambda_n/a)^{1/2}x = \cos \mu_n \, x/R. \tag{14.12.33}$$

Thus with the prescribed initial condition, the present problem may be solved by the finite integral transformation

$$t_n^*(\tau) = \int_0^R t(x, \tau) \cos(\mu_n x/R) \, dx \tag{14.12.34}$$

with the inversion formula

$$t(x, \tau) = \frac{2}{R} \sum_{n=1}^{\infty} \frac{\mu_n}{\mu_n + \sin \mu_n \cos \mu_n} t_n^*(\tau) \cos \frac{\mu_n x}{R}, \tag{14.12.35}$$

since

$$\int_0^R \cos^2 \frac{\mu_n x}{R} = \frac{R}{2} \left[1 + \frac{1}{\mu_n} \sin \mu_n \cos \mu_n\right]. \tag{14.12.36}$$

Table 14.3 furnishes the forms of the finite integral Fourier transforms and the inversion formulas for various boundary conditions.

In the solution of homogeneous problems, the method of finite integral transforms has no advantage over the classical method, although integral transforms are extremely useful for the solution of nonhomogeneous heat conduction problems.

TABLE 14.3. FINITE FOURIER TRANSFORM FOR AN INFINITE PLATE

$t_s(\tau)$	$t_n^*(\tau)$	$t(x, \tau)$
$t(0, \tau) = t(0)$ $t(l, \tau) = t(l)$	$\int_0^l t(x, \tau) \sin \mu_n (x/l)\, dx;$ $\sin \mu = 0$	$\frac{2}{l} \sum_{n=1}^{\infty} t_n^*(\tau) \sin \mu_n (x/l)$
$t(0, \tau) = t(0)$ $t'(l, \tau) = t'(l)$	$\int_0^l t(x, \tau) \sin \mu_n (x/l)\, dx;$ $\cos \mu = 0$	$\frac{2}{l} \sum_{n=1}^{\infty} t_n^*(\tau) \sin \mu_n (x/l)$
$t'(0, \tau) = t'(0)$ $t(l, \tau) = t(l)$	$\int_0^l t(x, \tau) \cos \mu_n (x/l)\, dx;$ $\cos \mu = 0$	$\frac{2}{l} \sum_{n=1}^{\infty} t_n^*(\tau) \cos \mu_n (x/l)$
$t'(0, \tau) = t'(0)$ $t'(l, \tau) = t'(l)$	$\int_0^l t(x, \tau) \cos \mu_n (x/l)\, dx;$ $\sin \mu = 0$	$\frac{2}{l} t(0) + \frac{2}{l} \sum t_n^*(\tau) \cos \mu_n (x/l)$
$t(0, \tau) = t(0)$ $t'(l, \tau) + Ht(l, \tau) = t'(l) + Ht(l)$	$\int_0^l t(x, \tau) \sin \mu_n (x/l)\, dx$ $\mu \cos \mu + Hl \sin \mu = 0$	$\frac{2}{l} \sum_{n=1}^{\infty} \frac{(\mu_n^2 + H^2l^2) t_n^*(\tau) \sin \mu_n (x/l)}{Hl + \mu_n^2 + H^2l^2}$
$t'(0, \tau) = t'(0)$ $t'(0, \tau) + Ht(l, \tau) = t'(l) + Ht(l)$	$\int_0^l t(x, \tau) \cos \mu_n (x/l)\, dx$ $Hl \cos \mu l = \mu \sin \mu = 0$	$\frac{2}{l} \sum_{n=1}^{\infty} \frac{(\mu_n^2 H^2 l^2) t_n^*(\tau) \cos \mu_n (x/l)}{Hl + \mu_n^2 + H^2l^2}$
$t'(0, \tau) - Ht(0, \tau) = t'(0) - Ht(0)$ $t(l, \tau) = t(l)$	$\int_0^l t(x, \tau)[\mu_n \cos \mu_n (x/l) + Hl \sin \mu_n (x/l)](dx/l)$ $\mu \cos \mu + Hl \sin \mu = 0$	$\sum_{n=1}^{\infty} \frac{2t_n^*(\tau)[\mu_n \cos \mu_n (x/l) + Hl \sin \mu_n (x/l)]}{Hl + \mu_n^2 + H^2l^2}$
$t'(0, \tau) - Ht(0, \tau) = t'(0) - Ht(0)$ $t'(l, \tau) = t'(l)$	$\int_0^l t(x, \tau)[\mu_n \cos \mu_n (x/l) + Hl \sin \mu_n (x/l)](dx/l)$ $Hl \cos \mu - \mu \sin \mu = 0$	$\sum_{n=1}^{\infty} \frac{2t_n^*(\tau)[\mu_n \cos \mu_n (x/l) + Hl \sin \mu_n (x/l)]}{Hl + \mu_n^2 + H^2l^2}$
$t'(0, \tau) - H_1 t(0, \tau) = t'(0) - H_1 t(0)$ $t'(l, \tau) + Ht(l, \tau) = t'(l) + Ht(l)$	$\int_0^l t(x, \tau)[\mu_n \cos \mu_n (x/l) + H_1 l \sin \mu_n (x/l)](dx/l)$ $2Hl \mu \cos \mu + (H^2l^2 - \mu^2) \sin \mu = 0$	$\sum_{n=1}^{\infty} \frac{2t_n^*(\tau)[\mu_n \cos \mu_n (x/l) + Hl \sin \mu_n (x/l)]}{2Hl + \mu_n^2 + H^2l^2}$
$t'(0, \tau) - H_1 t(0, \tau) = t'(0) - H_1 t(0)$ $t'(l, \tau) + H_2 t(l, \tau) = t'(l) + H_2 t(l)$	$\int_0^l t(x, \tau)[\mu_n \cos \mu_n (x/l) + H_1 l \sin \mu_n (x/l)](dx/l)$ $(H_1 + H_2) l \mu \cos \mu + (H_1 H_2 l^2 - \mu^2) \sin \mu = 0$	$\sum_{n=1}^{\infty} \frac{2(\mu_n^2 + H_2^2 l^2) t_n^*(\tau)[\mu_n \cos \mu_n (x/l) + H_1 l \sin \mu_n (x/l)]}{(H_1 l + H_2 l)(\mu_n^2 + H_1 H_2 l^2) + (\mu_n^2 + H_1^2 l^2)(\mu_n^2 + H_2^2 l^2)}$

CHAPTER

15

ELEMENTS OF THE THEORY OF ANALYTIC FUNCTIONS AND ITS APPLICATION

The present chapter contains the minimum information necessary to apply the theory of analytic functions to the heat conduction theory of mathematical physics.

The fact that the analytic function is fully determined by the nature and distribution of its singularities makes it possibile to evaluate the behavior of different functions with complex integral concepts. For example, proceeding from the singularities of the Laplace transform, it is possible for us to judge the asymptotic behavior of the original function without calculating the appropriate contour integral.

The Cauchy and residue theorems give wide possibilities for the transformation of integrals and sums, in particular, for the calculation of contour integrals necessary for the inversion of the Laplace transform. However, since, in practice, the majority of integrals are not calculated in their infinite form, the different methods of asymptotic estimates, some of which are also presented here, are of special importance.

The material contained in this chapter is stated as simply as possible. The reader is referred to special manuals for additional information and strict proofs [60, 25].

15.1 Analytic Functions

We shall consider the complex variable $z = x + iy$. As is known, such a variable may be presented by a two-dimensional vector with components x and y along the appropriate coordinate axes. The length of this vector, i.e., the modulus of the complex number, is equal to $(x^2 + y^2)^{1/2}$ and the angle with the positive direction of the axis x is measured counterclockwise and equals arc $\tan(y/x)$.

If the law which makes it possible to determine one or more sets of values of the other complex variable w associated with the variable z is prescribed, then the single-valued or multivalued function $w = f(z)$ is prescribed, respectively. The set of values taken by the independent variable z is called the region of function determination. The formula of the function of the complex variable $w = f(z)$ is equivalent to that of two functions of the real variable $w = u + iv$ where $u = (x, y)$ and $v = v(x, y)$.

Some simple examples of complex-variable functions are[1]

$$w = z^2 = x^2 - y^2 + 2ixy\,, \qquad w = z\bar{z} = x^2 + y^2\,,$$
$$w = e^z = e^x(\cos y + i \sin y)\,, \qquad w = \sqrt{z}\,.$$

From all possible complex-variable functions it appears advisable to single out a rather narrow class of functions, the analytic functions. They may be determined if the derivative of the function $f(z)$ is considered at some point z_0. The derivative is usually determined as the limit of the ratio

$$\{f(z_0 + \Delta z) - f(z_0)\}/\Delta z$$

as $\Delta z \to 0$ if this limit exists. However, by contrast to the real variable function $f(x)$, here the increment $\Delta z = \Delta x + i\,\Delta y$ represents a vector and may tend to zero by very different ways. For example, $\Delta y = 0$ may be assumed, and then we may pass to the limit $\Delta x \to 0$ along the real axis, and so on. If the single-valued function $f(z) = u + iv$ has a derivative at a point z_0 and this derivative is uniquely determined, i.e., it does not depend upon the way of transition to the limit $\Delta z \to 0$, then such a function is called analytic at the point z_0.

Let us determine the conditions, which will help us to establish whether a given function is analytic at some point. For this purpose we write out a detailed definition of a derivative as

$$f'(z) = \lim_{\Delta z \to 0} \left\{\frac{f(z_0 + \Delta z) - f(z_0)}{\Delta z}\right\}$$
$$= \lim_{\Delta z \to 0, \Delta y \to 0} \left\{\left(\frac{\partial u}{\partial x} + i\frac{\partial v}{\partial x}\right)\Delta x + \left(\frac{\partial u}{\partial y} + i\frac{\partial v}{\partial y}\right)\Delta y\right\} \Big/ \{\Delta x + i\,\Delta y\}. \tag{15.1.1}$$

[1] A conjugate-complex quantity will be designated by an overbar.

We rewrite formula (15.1.1) as

$$f'(z) = \lim_{\Delta x\to 0, \Delta y\to 0} \left(\left\{\left\{\frac{\partial u}{\partial x} + i\frac{\partial v}{\partial x}\right\} \Big/ \left\{1 + i\frac{\Delta y}{\Delta x}\right\}\right\}\right.$$
$$\left.\times \left[1 + i\frac{\Delta y}{\Delta x}\left\{\frac{\partial v}{\partial y} - i\frac{\partial u}{\partial y}\right\} \Big/ \left\{\frac{\partial u}{\partial x} + i\frac{\partial v}{\partial x}\right\}\right]\right). \quad (15.1.2)$$

The derivative $f'(z)$, determined by the relation $\Delta y/\Delta x$, is equal to the tangent of the inclination angle of the vector Δz. This no longer holds when the denominator in the upper line of formula (15.1.2) and the factor in square brackets are equal, i.e.,

$$\left\{\frac{\partial v}{\partial y} - i\frac{\partial u}{\partial y}\right\} \Big/ \left\{\frac{\partial u}{\partial x} + i\frac{\partial v}{\partial x}\right\} = 1. \quad (15.1.3)$$

Equating real and imaginary parts of the last equality, we obtain:

$$\frac{\partial u}{\partial x} = \frac{\partial v}{\partial y}, \qquad \frac{\partial u}{\partial y} = -\frac{\partial v}{\partial x}. \quad (15.1.4)$$

The conditions necessary for the function $f(z)$ to be analytic at some point are called the Cauchy–Riemann conditions. These conditions are necessary and sufficient if the derivatives $\partial u/\partial x$, $\partial u/\partial y$, $\partial v/\partial x$, $\partial v/\partial y$ are continous at the point under consideration.

The points at which condition (15.1.4) is not satisfied are called singularities of the function. For example, the function $f(z) = 1/z$ has no derivative at $z = 0$. As can be easily verified for the function $f(z) = (z - 1)^{1/2}$, the point $z = 1$ will be such a point where the appropriate derivative converts into infinity, and the function $f(z) = z\bar{z}$ is not analytic everywhere.

15.2 Contour Integration of Complex Variable Functions

Let the complex variable function $f(z)$ be prescribed on some curve C (See Fig. 15.1). Divide the contour C into a finite number of portions with ends $[z_0, z_1]$, $\cdots$, $[z_k, z_{k+1}]$, $\cdots$, $[z_{n-1}, z_n]$, etc. By analogy with the theory of integrals of functions of a real variable, we compose the sum

$$\sum_{k=0}^{n} f(\zeta_k)(z_{k+1} - z_k) \quad (15.2.1)$$

after designating any arbitrary point on the kth portion through ζ_k. With

$n \to \infty$ and provided that the lengths of all the portions tend to zero, the limit of this sum is called the integral of the function $f(z)$ with respect to C and is designated through

$$\int_C f(z)\, dz. \tag{15.2.2}$$

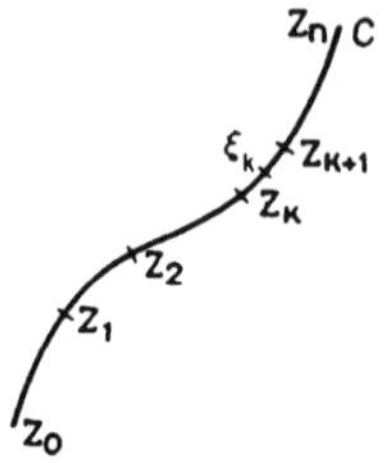

Fig. 15.1. Division of contour C into portions.

If the function $f(z)$ is piecewise continuous and the contour C may be composed of continuously adjacent arcs with a continuously varying tangent (piecewise-smooth contour), then integral (15.2.2) always exists.

A contour integral possesses properties of usual real contour integrals: in particular, a constant multiplier may be factored out of the integral sign, the integral of the sum of the functions is equal to the sum of integrals of summands and the integral sign changes with the direction of traverse of the integral contour.

If the contour C represents a closed curve, the direction is taken as positive if this closed contour is traversed such that the internal region limited by this contour remains to the left.

The region bounded by a closed contour is singly connected if the border of the region consists of a single linking part (See Fig. 15.2a) or more exact if any closed contour within this region may be constricted to a point by means of continuous deformation without breaking the border of the region. The multiple-connected region is presented in Fig. 15.2b; here the

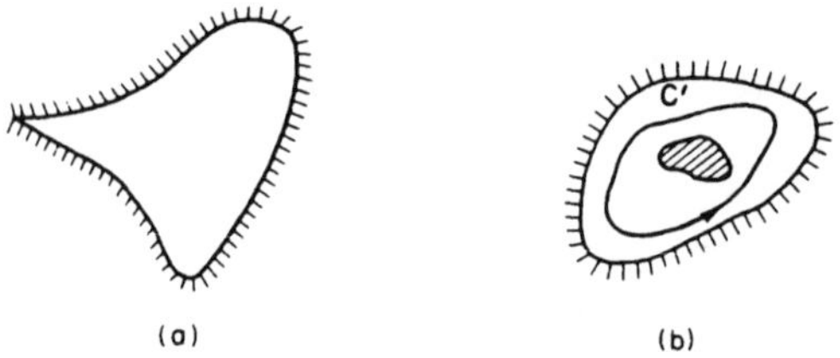

Fig. 15.2. (a) single-linked and (b) two-linked regions.

contour C' inside the region cannot be subjected to the above deformation without breaking the border of the region.

We now prove that if $f(z)$ is an analytic function which is continuous inside the closed contour C and on the contour itself, then

$$\int_C f(z)\, dz = 0.$$

This important confirmation is called the Cauchy theorem.

We shall prove this theorem when the contour C bounds a star-shaped region, and the derivative $f'(z)$ (which does exist since the function $f(z)$ is analytic) is bounded inside and on the contour C.

The region is star-shaped if in it there exists such a point z_0, from which any ray intersects the border of the region only one time.

We assume that point z_0 is the origin. Consider the integral

$$\psi(\lambda) = \lambda \int_C f(\lambda z)\, dz \quad (0 \leq \lambda \leq 1). \tag{15.2.3}$$

We must prove that $\psi(1) = 0$. We differentiate $\psi(\lambda)$ with respect to λ to obtain

$$\psi'(\lambda) = \int_C f(\lambda z)\, dz + \lambda \int_C z f'(\lambda z)\, dz. \tag{15.2.4}$$

Using the boundedness of the derivative $f'(z)$, we integrate the second integral in (15.2.4) by parts which gives us

$$\psi'(\lambda) = \int_C f(\lambda z)\, dz + \lambda \Big[(z f(\lambda z)/\lambda) - (1/\lambda) \int_C f(\lambda z)\, dz \Big], \tag{15.2.5}$$

where for the term $zf(\lambda z)/\lambda$ the difference of its values at finite and initial points of the contour should be taken. For the closed contour C, owing to the single-valuedness of the function $zf(\lambda z)$, this difference converts into zero. Hence it follows that $\psi'(\lambda) = 0$, i.e., the function $\psi(\lambda)$ is also constant; assuming in formula (15.2.3) that $\lambda = 0$ we find that this constant is equal to zero. Thus, $\psi(1) = 0$ which was to be proved.

The Cauchy theorem is not directly applicable to a multilinked region. However, it is also easily extended to this case. Consider the doubly-connected region depicted in Fig. 15.3. The Cauchy theorem becomes applicable to the contour integral consisting of C_1, C_2 and sections C_3 and C_4. The directions of the bypass of different parts of the complete contour obey the general rule given in Fig. 15.3. Integrals with respect to C_3 and C_4 cancel out, and we have

$$\int_{C_1\circlearrowleft} f(z)\,dz + \int_{C_2\circlearrowright} f(z)\,dz = 0 \tag{15.2.6}$$

or

$$\int_{C_1\circlearrowleft} f(z)\,dz - \int_{C_2\circlearrowleft} f(z)\,dz = 0\,, \tag{15.2.6a}$$

i.e., the external contour integral is equal to the internal one if both are counter-clockwise.

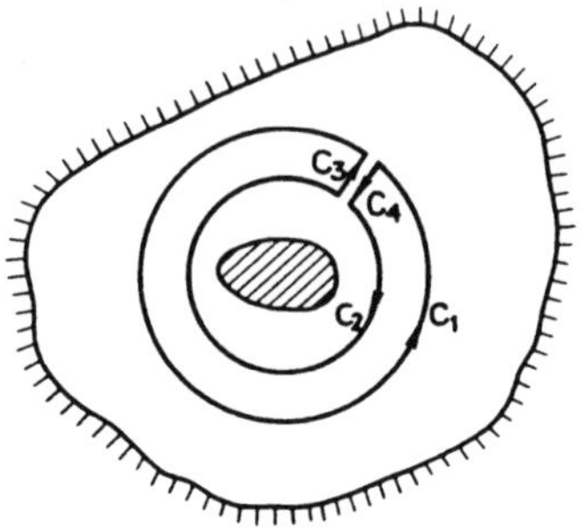

Fig. 15.3. Contour in a multilinked region.

It directly follows from the Cauchy theorem that if $f(z)$ is a function analytic inside some region, then the integral $\int_{z_1}^{z_2} f(z)\,dz$ taken along any contour connecting points z_1 and z_2 (but being completely in the region of the analyticity of the function $f(z)$) depends only upon z_1 and z_2 and does not depend upon the path of integration. For the proof, it is sufficient to compare the integrals by two different contours C_1 and C_2 connecting points z_1 and z_2 (See Fig. 15.4). The Cauchy theorem may be applied to the region

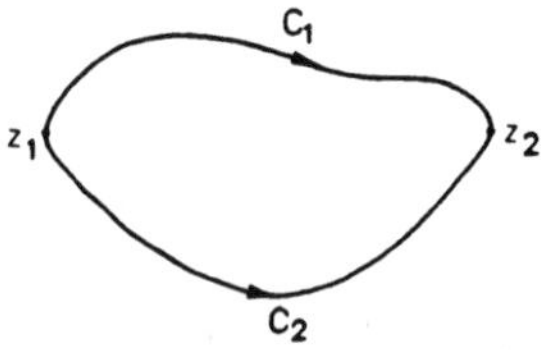

Fig. 15.4. Independence of integral $\int f(z)\,dz$ upon the integration path inside the analyticity region.

bounded by the contour C_2 and the contour C_1 in a reverse direction from z_2 to z_1. Hence it follows directly that

$$\int_{C_1} f(z)\,dz = \int_{C_2} f(z)\,dz \tag{15.2.7}$$

(here C_1 is passed by in the direction shown in Fig. 15.4). Hence it also follows that the value of the integral does not vary with arbitrary contour deformation, if in this deformation only, the contour does not leave the region of analyticity of the integrand function, i.e., it does not intersect either of its singularities. Keeping to this rule, it is possible to continuously deform the path of integration in contour integrals by choosing a contour more suitable for their calculation.

Note that the following theorem [60], which is the reverse of the Cauchy theorem, may be proved: if the function $f(z)$ is continuous and single-valued inside some closed contour C and $\int f(z)\, dz$ equals zero for any closed contour inside C, then $f(z)$ is analytic inside C.

Thus, this theorem gives an integral indication of the function analyticity equivalent to the Cauchy–Riemann conditions.

We now proceed to the derivation of the integral Cauchy formula which is very important for applications.

Consider the integral

$$F(z) = \int_C (f(z')\, dz')/(z' - z) \tag{15.2.8}$$

with respect to some closed contour C, on and inside of which the function $f(z)$ is analytic. The point z is an arbitrary one inside the contour. As we have shown, the contour C may be deformed without changing the value of the integral. If we replace C by a circle with the center at the point z and small radius ϱ and assume $z' = z + \varrho e^{i\varphi}$, we obtain

$$F(z) = i \int_0^{2\pi} f(z + \varrho e^{i\varphi})\, d\varphi \tag{15.2.9}$$

and passing to the limit $\varrho \to 0$ we find

$$F(z) = 2\pi i f(z),$$

i.e.,

$$f(z) = (1/2\pi i) \int_C f(z')/(z' - z)\, dz'. \tag{15.2.10}$$

The Cauchy formula makes it possible to study different properties of analytic functions, and in particular, allows the calculation of the value of the function $f(z)$ at any point inside the contour with respect to its value on the contour. Thus, the values of the analytic function in different parts of a complex plane are not arbitrary but closely related.

With the help of the integral representation of analytic functions presented by the Cauchy formula, it is possible to prove that the derivatives of this function are also analytic in the same region as the function $f(z)$. When performing the conditions under which the Cauchy formula was derived, we have

$$
\begin{aligned}
f'(z) &= \lim_{\Delta z\to 0} \frac{f(z+\Delta z)-f(z)}{\Delta z} \\
&= \frac{1}{2\pi i} \lim_{\Delta z\to 0} \left\{ \frac{1}{\Delta z} \int_C f(z') \left[\frac{1}{z'-z-\Delta z} - \frac{1}{z'-z} \right] dz' \right\} \\
&= \frac{1}{2\pi i} \lim_{\Delta z\to 0} \int_C \frac{f(z')}{(z'-z)(z'-z-\Delta z)}\, dz' \\
&= \frac{1}{2\pi i} \int_C \frac{f(z')}{(z'-z)^2}\, dz'. \qquad (15.2.11)
\end{aligned}
$$

The possibility of the limiting transition under the sign of the integral may be proved from the analyticity (and, consequently, from boundedness) of the function $f(z)$ on the contour C.

Similarly, by calculating the higher derivatives, we obtain a formula for the nth derivative, expressing its value at any point inside the contour with the help of that of the function on the contour as

$$
f^{(n)}(z) = \frac{n!}{2\pi i} \int_C \frac{f(z')}{(z'-z)^{n+1}}\, dz'. \qquad (15.2.12)
$$

15.3 Representation of Analytic Functions by Series

Consider an infinite series, whose terms are functions of the complex variable z prescribed in some region D. This region, at all points of which the series

$$
\sum_{k=1}^{\infty} u_k(z) \qquad (15.3.1)
$$

converges, is called its convergence region.

As in the case of functions of real variables, we introduce the concept of uniform convergence of a series with variable terms (15.3.1). Uniform convergence of series (15.3.1) in some region denotes that the series converges equally well at all the points of this region. To be precise, series (15.3.1) is called uniformly converging in some region D, if at any prescribed positive value of ε there exists such a positive number N, which is the same for all values of z in the region D, that

$$\left| \sum_{k=n+1}^{n+m} u_k(z) \right| < \varepsilon , \qquad (15.3.2)$$

at $n > N$ and arbitrary positive integer m.

For uniform convergence of series (15.3.1), it is enough that for all z in the region D the estimate for the modulus of the terms of series (15.3.1) should be valid

$$| u_k(z) | < m_k$$

at all values of k when m_k are positive numbers forming a converging series. In addition, series (15.3.1) converges absolutely, i.e., in the series consisting of moduli of the series terms that (15.3.1) converges.

For uniformly converging series, it is possible to prove some interesting statements regarding operational procedure with such series and properties of their sum.

First, if the terms of series (15.3.1) $u_k(z)$ $(k = 1, 2, \ldots)$ are continuous functions in some region D, and the series uniformly converges in this region, then the sum of the series will be a continuous function.

Second, if series (15.3.1) consisting of continuous functions converges uniformly on some curve, then this series may be integrated by terms along this curve.

And, finally, if the terms of series (15.3.1) are analytic functions in a closed region D with the contour C, and the series uniformly converges on the contour C, then it converges uniformly within the whole region D, its sum is an analytic function inside the region D, and series (15.3.1) may be differentiated term by term infinitely (the Weierstrass theorem).

Now consider the particular case of series (15.3.1) of the form

$$\sum_{n=0}^{\infty} a_n(z - b)^n , \qquad (15.3.3)$$

which are called power series.

For power series, the Abel theorem [60] may be proved rather simply: if series (15.3.3) converges for some $z = z_0$, then it absolutely converges at any point z satisfying the condition

$$| z - b | < | z_0 - b | ,$$

i.e., being closer to the point b than to z_0. In addition, series (15.3.3) will converge uniformly in any circle with b as the center and the radius less than the distance $| z_0 - b |$.

From the Abel theorem it follows that if series (15.3.3) converges at some point z_1, then it also converges at any point which is farther from the point b than z_1. Consequently, for any power series (15.3.3) there exists such a positive number R that the series converges at $|z - b| < R$ and diverges at $|z - b| > R$. The number R is called the radius of convergence of a power series, and the circle $|z - b| < R$, the circle of convergence of this series. From the Abel theorem it follows that series (15.3.3) will converge uniformly inside its own circle of convergence.[2]

If the convergence radius R is equal to infinity, then the corresponding power series converges along the whole complex plane.

We again emphasize that series (15.3.3) converges uniformly inside its own circle of convergence, that the above theorems can be formulated for a general case of series with variable terms, and that the Weierstrass theorem, in particular, can be applied to it. The sum of power series (15.3.3) is therefore an analytic function inside the convergence circle of this series, and the series may be integrated and differentiated by terms infinitely.

We now show that any function $f(z)$ analytic in some circle $|z - b| < R$ with b as the center may be represented by power series (15.3.3) inside this circle and that such an expansion is unique. The representation of the function by means of such a power series is called the Taylor expansion in series.

To prove the above statement, we build the contour C having the shape of a circle, with b as the center and the radius less than that of the convergence circle R but containing some fixed point z. Then, all the conditions necessary for applying the Cauchy formula will be satisfied for the function $f(z)$ and we can write

$$f(z) = (1/2\pi i) \oint_C \{f(z')/(z' - z)\}\, dz'. \qquad (15.3.4)$$

Part of the integrand in formula (15.3.4) may be presented as

$$\frac{1}{z' - z} = \frac{1}{z' - b}\, 1 \Big/ \left(1 - \frac{z - b}{z' - b}\right) = \sum_{n=0}^{\infty} \frac{(z - b)^n}{(z' - b)^{n+1}}, \qquad (15.3.5)$$

where we used the formula for the sum of an infinite geometrical progression which in its tern is applicable because $|z - b| < |z' - b|$ ($|z' - b|$ is the radius of the circle C and according to the condition, the point z is inside this circle).

For the moduli of the terms of series (15.3.5) we have the estimate

$$\frac{(z - b)^n}{(z' - b)^{n+1}} = \frac{1}{R_C}\, q^n, \qquad (15.3.6)$$

[2] The problem of convergence of the series on the circle $|z - b| = R$ demands special investigation.

where R_C is the radius of the circle C, and $q = (z - b)/(z' - b)$ and $0 \leq q < 1$. Thus, from the theorems formulated at the beginning of the present section for series with variable terms and from equation (15.3.6), it follows that the series will converge uniformly with respect to the variable z' and, consequently, it may be integrated by terms. On using formula (15.3.4) multiplying the left- and right-hand sides of relation (15.3.5) by $(1/2\pi i)f(z')$ and integrating with respect to the contour C, we obtain

$$f(z) = \sum_{n=0}^{\infty} a_n(z - b)^n, \tag{15.3.7}$$

where

$$a_n = \frac{1}{2\pi i} \int_C \frac{f(z')}{(z' - b)^{n+1}} dz', \tag{15.3.8}$$

or according to (15.2.12) for the nth derivative of the function $f(z)$, we have

$$a_n = f^{(n)}(b)/n!, \tag{15.3.9}$$

i.e., finally the value of the function $f(z)$ at any point inside the circle $|z - b| < R$ where $f(z)$ is analytic may be presented by the Taylor series

$$f(z) = \sum_{n=0}^{\infty} \{f^{(n)}(b)/n!\}(z - b)^n. \tag{15.3.10}$$

It is easy to show that expansion (15.3.10) is unique. For this it is sufficient to show that the coefficients a_n in formula (15.3.7) are determined uniquely by formula (15.3.9). Assuming $z = b$ in formula (15.3.7), we have $a_0 = f(b)$; further differentiating series (15.3.7) and upon each differentiation assuming $z = b$, we obtain formula (15.3.9) for the series coefficients.

From this, it follows that the Taylor series at the point b of the function $f(z)$ converges inside the circle with the center at the point b, where $f(z)$ is analytic. Thus, the convergence radius of the Taylor series coincides with the distance from b where the expansion is sought, up to the singularity of $f(z)$ nearest to it, where $f(z)$ is not analytic. For example, the function $1/(1 + z)$ which at a point $z = 0$ expands into the Taylor series of type

$$1/(1 + z) = 1 - z + z^2 - \cdots, \tag{15.3.11}$$

has a singularity $z = -1$ where it and its derivatives are infinite. The convergence radius of series (15.3.10) is therefore equal to unity, i.e., it converges only at $|z| < 1$. This very function $1/(1 + z)$ with expansion at a point $z = 4$ will have the Taylor series

$$1/(1 + z) = \sum_{n=0}^{\infty} \frac{(-1)^n}{5^{n+1}} (z - 4)^n , \tag{15.3.12}$$

which converges and represents the function $1/(1 + z)$ in the circle with the center at $z = 4$ of radius 5 equal to a distance from $z = 4$ up to the nearest (in this case the only one) singularity $z = -1$.

Obviously, the function cannot be expanded into the Taylor series in the vicinity of its singularity. However, it is also possible to obtain some expansion valid near some types of singularities. For this purpose, consider the power series of the form

$$\sum_{n=-\infty}^{\infty} a_n(z - b)^n , \tag{15.3.13}$$

containing not only positive but also negative powers $(z - b)$. Such a series is called the Laurent series. We shall determine the region of its convergence. The part of series (15.3.13) containing the positive powers of $(z - b)$ is an ordinary power series and, consequently, has some circle of convergence $| z - b | < R_1$. To consider the part of series (15.3.13) with negative powers of $(z - b)$, we introduce a new variable $z' = (z - b)^{-1}$. Then, this part of the series converts into an ordinary power series of the form

$$a_{-1}z' + a_{-2}z'^2 + a_{-3}z'^3 + \cdots . \tag{15.3.14}$$

Such a series has some circle of convergence with the radius $1/R_2$ and the point $z' = 0$ as the center. Thus, the region of convergence of series (15.3.14) is determined by the inequality $| z' | < 1/R_2$. Proceeding to the former variable z, we find that series (15.3.14) therefore converges at $| z - b | > R_2$. Therefore series (15.3.13) converges in the region defined by the inequalities

$$| z - b | < R_1 , \qquad | z - b | > R_2 . \tag{15.3.15}$$

These inequalities may be satisfied and, consequently, determine some region only in the case of $R_1 > R_2$. This region is a ring bounded by concentric circles of the radii R_1 and R_2 with the point b as the center. Since in the previous considerations we divided the Laurent series (15.3.13) into two power series, then from the properties of the power series it follows that series (15.3.13) inside its convergence ring converges absolutely and uniformly, its sum is the analytic function in the ring, and the series may be differentiated by terms.

Now we prove that if the function $f(z)$ is analytic inside the range $R_2 < | z - b | < R_1$ (inside the circle $| z - b | = R_2$ and outside the circle

$|z - b| = R_1$ the function $f(z)$ may have singularities), then it may be expanded into the Laurent series inside this ring.

After narrowing the external circle C_{R_1} and widening the internal one C_{R_2} a little (see Fig. 15.5), it is possible to consider the function $f(z)$ analytical on both circles. Then the Cauchy formula may be extended to the internal part of the ring. For an arbitrary point z inside the ring, we have

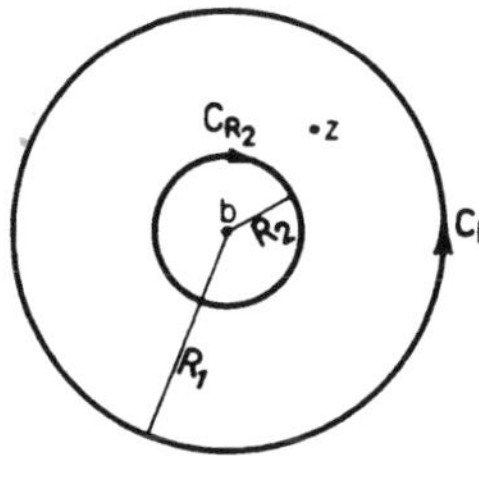

Fig. 15.5. Contour used for deriving the expansion into the Laurent series near a singularity, $z = b$.

$$f(z) = \frac{1}{2\pi i} \int_{C_{R_1}} \frac{f(z')}{z' - z}\, dz' + \frac{1}{2\pi i} \int_{C_{R_2}} \frac{f(z')}{z' - z}\, dz'. \qquad (15.3.16)$$

Applying the considerations which are quite similar to those in the case of the Taylor expansion to each integral in formula (15.3.16), we obtain for the coefficients of series (15.3.13)

$$a_n = (1/2\pi i) \int_{C_{R_1}} [f(z')/(z' - b)^{n+1}]\, dz' \quad (n = 0, 1, 2, \ldots) \qquad (15.3.17)$$

and

$$a_{-n} = -\,(1/2\pi i) \int_{C_{R_2}} f(z')(z' - b)^{n-1}\, dz' \quad (n = 1, 2, \ldots). \qquad (15.3.17')$$

The directions of the traverses of contours C_{R_1} and C_{R_2} are given in Fig. 15.5. It is not difficult to show that under the stated conditions, expansions (15.3.13) of the function $f(z)$ with the coefficients determined according to relations (15.3.17) and (15.3.17′) will be unique.

As an example of the expansion of the functions in the Laurent series, consider the function $e^{1/z}$. A point $z = 0$ will be a singularity of this function. Consequently, excepting the origin of the coordinates, the whole complex plane will be the convergence ring of the Laurent series, which is of the form $\sum_{n=0}^{\infty} 1/n!\, z^n$, and all the coefficients a_{-n} $(n = 1, 2, \ldots)$ will be equal to zero.

15.4 Classification of Analytic Functions by Their Singularities. The Concept of Analytic Continuation

The representation of the functions by the Laurent series makes it possible to classify isolated singularities of the analytic function, i.e., those singularities which are the center of some sufficiently small circle where there are no other singularities of the function. Near such a point, the analytic function may be expanded into the Laurent series with a convergence ring, whose internal circle degenerates into a point. If expansion (15.3.13) contains a finite number of terms with negative powers $(z - b)$,[3] i.e., it is of the form

$$f(z) = \sum_{n=-N}^{\infty} a_n(z - b)^n, \tag{15.4.1}$$

then at the point b, the function has a pole of the Nth order.

If the series of the negative powers in the Laurent expansion is infinite, then the point b is called an essential singularity of the function.

As an example, consider the function

$$e^{1/(z-1)}/(e^z - 1). \tag{15.4.2}$$

At $z = 2\pi in$ ($n = 0, \pm 1, \pm 2, \ldots$), the numerator of the function is analytic and differs from zero, and the denominator becomes zero as $(z - 2\pi in)$ at $z \to 2\pi in$. Thus, at points $z = 0, \pm 2\pi i, \pm 4\pi i$, and so on, the function has poles of the first order (simple poles). Further, at $z = 1$ the denominator is analytic and the numerator has an isolated singularity, where it is represented by the series

$$e^{1/(z-1)} = \sum_{n=0}^{\infty} \frac{1}{n!} \frac{1}{(z - 1)^n},$$

and the whole function (15.4.2) will therefore contain an infinite number of terms with negative powers $(z - 1)$, and function (15.4.2) has an essential singularity at $z = 1$.

Analytic functions with isolated singularities may be classified according to the type and location of their singularities in a complex plane. First of all, show that the function which has no singularities (also at a point $z = \infty$) is reduced to the trivial case when it is constant. This statement, called the Liouville theorem, is formulated more exactly as follows: if $f(z)$

[3] If at the point b the expansion does not contain terms with negative powers $(z - b)$, then it is the Taylor series, and the function is analytical at b.

is the analytic function at all the values of z and $|f(z)| \lesseqgtr M$ where M is some positive number at all z, then $f(z)$ is constant.

The estimate of the derivative $f'(z)$ is determined according to (15.2.12) as

$$f'(z) = (1/2\pi i) \int_C [f(z')/(z' - z)^2]\, dz',$$

where the contour C is chosen in the form of the circle with the radius R and z as the center. We have an estimate

$$\begin{aligned} |f'(z)| &= \left| (1/2\pi i) \int_C [f(z')/(z' - z)^2]\, dz' \right| \\ &= (1/2\pi R) \left| \int_0^{2\pi} e^{-i\varphi} f(z + Re^{i\varphi})\, d\varphi \right| \\ &\leq (1/2\pi R) \int_0^{2\pi} \left| e^{-i\varphi} f(z + Re^{i\varphi}) \right| d\varphi \\ &\leq (1/2\pi R)\, M 2\pi = M/R, \end{aligned}$$

thus

$$|f'(z)| \leq M/R.$$

Since $f(z)$ is analytic and is bounded over the whole plane, then R may be chosen as large as is desired. When $R \to \infty$, we have $f'(z) \equiv 0$ at all z, i.e., $f(z)$ is constant.

Thus, all the nontrivial analytic functions certainly have singularities (probably at infinity).

The considerations similar to those for proving the Liouville theorem may show that the polynomial of power $\leq m$ is the single function which is analytic at all the finite values of z and its rate of increase is not more than $|z|^m$ at $z \to \infty$.

Polynomials are a particular case of integer functions, i.e., of those which are analytic at all finite z. For example, we encounter very often the integer functions e^z, $\sin z$, $\cos z$, $J_\nu(z)/z^\nu$ and many others. All these functions have no singularities at finite z and at a point at infinity they have an essential singularity (it is easy to show by introducing a new variable $z' = 1/z$ and expanding the above functions into the Laurent series. near a point $z' = 0$)

Further, the functions form the class of meromorphic functions where the finite number of poles are the only singularities at any finite part of the complex plane.

It is not difficult to show that a rational function, i.e., that representing the relation of two polynomials, is a single meromorphic function which has no singularities except the poles, including also the point at infinity $z = \infty$.

Some examples of transcendental functions are: tan z, where simple poles lie on a real axis at points $(n + 1/2)\pi$, $n = 0, \pm 1$, $\Gamma(z)$[4] (simple poles are at points $z = -n$, $n = 0, 1, 2, \ldots$) and generally the relation of any two integer functions $\tan z = \sin z/\cos z$, etc., the zeroes of the integer function in the denominator being the poles of the appropriate meromorphic function.

Thus, we have found that, according to the character of singularities and their location in the complex plane, the single-valued analytic functions with isolated singularities may be subdivided into the following classes:

(*a*) the function without any singularities is constant;

(*b*) the functions, the only singularity of which is located in infinity, form the class of the integer functions. The particular case of the integer functions, the only singularity of which at $|z| \to \infty$ represents a pole of the nth order, is reduced to the polynomial of the nth power with respect to z.

(c) The functions which at all finite $|z|$ have no singularities, except poles, are called meromorphic ones. If, moreover, a point at infinity is not an essential singularity, then in this particular case, we have the simplest meromorphic function, i.e., the rational function, which is the ratio of two polynomials.

Besides the above classes of single-valued functions with isolated singularities, i.e., poles and essential singularities, we often meet with multivalued functions in applications which have specific nonisolated singularities called points of branching. When passing along some contour round such a point, the function assumes the value different from the initial one.

For example, consider the function $\sqrt{z}$. When bypassing the point $z = 0$ along the circle with the center at this point and with the radius r, the value of the function $\sqrt{z} = \sqrt{r}\, e^{i\theta/2}$ changes from $\sqrt{r}$ at $\theta = 0$ to $-\sqrt{r}$ at $\theta = 2\pi$. Thus, $z = 0$ is the point of branching of the function $\sqrt{z}$. From the similar considerations, it is seen that $z = \infty$ is also the point of branching of the function $\sqrt{z}$. Generally, the points of branching always exist in pairs.

Multivaluedness of the function may be eliminated if we restrict ourselves to some partial region of its determination. Then, it may be said that in this region, the branch of the function is determined. For example, for $\sqrt{z}$ the set of the values determined by the formula $\sqrt{r}\, e^{i\theta/2}$ at $-\pi < \theta < \pi$ is the branch of the function in the region obtained by the edge along the negative real half-axis from the point $z = 0$ to $z = \infty$, which are the points

[4] The function $\Gamma(z)$ as known may be determined by the integral $\Gamma(z) = \int_0^\infty e^{-t} t^{z-1}\, dt$ at $\operatorname{Re} z > 0$.

of branching of this function. The set of the values prescribed by the formula $-\sqrt{r}\, e^{i\theta/2}$ will be another branch of the function in the same region. The function $\sqrt{z}$ considered has two branches. Similarly, wc obtain thc fact that the function $z^{m/n}$, where m/n is the rational real number, has $n - m$ branches if the fraction m/n does not cancel. Ln z is the most important among the functions with points of branching of an infinite order. The infinite number of its branches is determined by the formula ln z.

$$\ln r + i(\theta + 2\pi n) \quad (-\pi < \theta < \pi).$$

Each $n = 0, 1, 2$ gives the branch ln z. $z = 0$ and $z = \infty$ are the points of branching.

If the function z^a is determined where a is irrational as $e^{a \ln z}$, then knowing the properties of the function ln z, it is not difficult to see that the function z^a will have an infinite set of branches.

Then problem of the possibility of the single analytic continuation of the function is connected with the presence of the points of branching if the initial determination of the function is valid only in the bounded part of a complex plane. For example, the power series

$$1 + z + z^2 + \cdots$$

directly determines the function $1/(1 - z)$ inside the circle $|z| < 1$ only, and the integral

$$\int_0^\infty \exp[-x(1 - z)]\, dx$$

determines the same function but in the whole half-plane Re $z < 1$. There arises the problem of ways of performing the analytic continuation of the function initially prescribed in some part of a complex plane as well as determining under which conditions the process of the analytic continuation of the function will give the one and only result. If the function is, for example, given by a power series with a finite convergence radius, then at least the process of the analytic continuation may be theoretically performed as follows. Let the function $f(z)$ at a point z_0 be prescribed by a power series with the convergence radius R_0 equal to the distance up to the closest singularity of the function $f(z)$ from z_0. Then at any point (designate it through z_1) lying inside the convergence circle $|z - z_0| < R_0$ and different from z_0, it is also possible to calculate all the derivatives and to construct a series (see Fig. 15.6)

$$f(z) = \sum_{n=0}^{\infty} (f^{(n)}(z_1)/n!)(z - z_1)^n,$$

with some convergence radius R_1 equal to the distance from z_1 to the nearest singularity, which generally speaking is different from the previous one. Now, constructing the power series with the center at some point of a circle $|z - z_1| < R_1$, etc., it is possible to obtain the value of the function $f(z)$ in the whole complex plane. However, as is seen from the above consider-

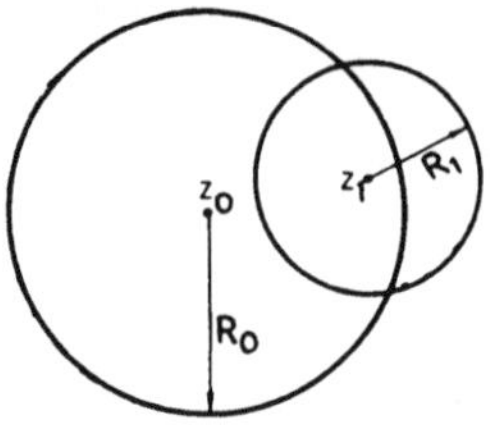

Fig. 15.6. Analytic continuation of the function $f(z)$ by means of a power series.

ations, the process of the analytic continuation might not be applied if singularities densely fill the boundary of the initial region of determination. Then, we may say that the function is not continued beyond the limits of its natural region of existence. Such a function is the function prescribed by the power series of the form

$$f(z) = 1 + z^2 + z^4 + \cdots + z^{2n} + \cdots,$$

which converges inside the circle $|z| < 1$. Since $f(1) = \infty$, then $z = 1$ is a singularity, but for $f(z)$

$$f(z) = z^2 + f(z^2).$$

Then at $z^2 = 1$ (i.e., also at $z = -1$), the function $f(z)$ will have singularities. As

$$f(z) = z^2 + z^4 + f(z^4),$$

then all the points determined by the equation $z^4 = 1$ will be singular too. Repeating all these considerations, we have that all the points determined from the equations

$$z^{2^n} = 1 \quad (n = 1, 2, \ldots)$$

will be singular. There will be infinitely many roots of these equations on any small portion of the circle. The function under consideration cannot, therefore, be continued beyond the limits of a solid line of singularities $|z| = 1$.

The analytic continuation is not obligatorily performed by constructing a sequence of power series. It is also possible to use functional relations, with the help of which we may compare the values of the function in the initial region with those in that region where it was not initially determined. For example, the gamma function may be determined in a half-plane $\operatorname{Re} z > 0$ by the integral

$$\Gamma(z) = \int_0^\infty e^{-t} t^{z-1}\, dt.$$

However, with the help of the known functional relation $z\Gamma(z) = \Gamma(z+1)$, it may be also continued into the half-plane $\operatorname{Re} z < 0$.

Finally, regarding the uniqueness of the analytic continuation, the following confirmation may be proved: if the function $f_1(z)$ determined in the region D_1 (see Fig. 15.7) and the function $f_2(z)$ determined in the region D_2 are obtained by the analytic continuation of the function $f(z)$ prescribed initially in the region D, and if the region D_3, common for D_1 and D_2, overlaps D, then $f_1(z) = f_2(z)$ in D_3.

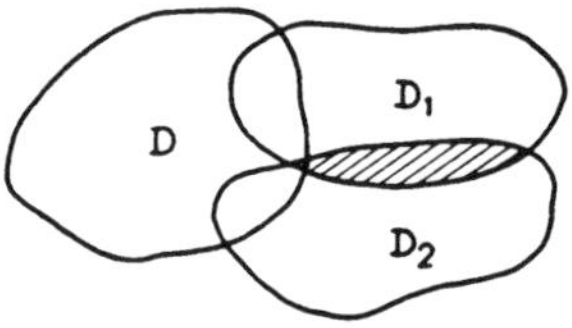

Fig. 15.7. The theorem of uniqueness of analytic continuation.

Thus, in the region D_3, we obtain the fact that $f_1(z)$ coincides with $f_2(z)$, irrespective of the way of the continuation of the function $f(z)$.

The theorem of uniqueness of the analytic continuation will, however, be broken, if between two different ways of the continuation of the function there is a singularity of the function, i.e., its point of branching.

15.5 Residue Theory and Its Application to Calculating Integrals and Summing Up Series

Consider the expansion of the function $f(z)$ into the Laurent series near the isolated singularity b. In a general case, this expansion is of the form (see (15.3.13))

$$f(z) = \sum_{n=-\infty}^{\infty} a_n (z-b)^n. \tag{15.5.1}$$

In this expansion at $(z-b)^{-1}$, the coefficient a_{-1} plays an essential role.

This coefficient a_{-1} is called the residue of the function $f(z)$ at the considered singularity b (at a pole or essential singularity). We shall show that the integral with respect to some contour surrounding b may be expressed through the residue of the function at this point. For this purpose we integrate formula (15.5.1) with respect to some small closed contour C surrounding b assuming that series (15.5.1) uniformly converges on this contour. Therefore,

$$\int_C f(z)\,dz = \sum_{n=-\infty}^{\infty} a_n \int_C (z-b)^n\,dz. \tag{15.5.2}$$

According to the Cauchy theorem, the contour C may be deformed into the circle of a small radius ϱ with b as the center. Then, formula (15.5.2) may be written

$$\int_C f(z)\,dz = i \sum_{n=-\infty}^{\infty} a_n \varrho^{n+1} \int_0^{2\pi} \exp[i(n+1)\varphi]\,d\varphi. \tag{15.5.3}$$

It is not difficult to see that all the integrals $\int_0^{2\pi} \exp[i(n+1)\varphi]\,d\varphi$ are equal to zero, except the one corresponding to $n = -1$ which equals 2π. Thus, we have

$$\int_C f(z)\,dz = 2\pi i a_{-1}. \tag{15.5.4}$$

In a similar way, we can also consider a more general case when the function $f(z)$ is analytic in some region D with the contour C, except where the finite number of points are poles or essential singularities of the function. We encircle each singularity $b_1, b_2, \ldots, b_k$ with a contour, e.g., a small circle C_k (see Fig. 15.8). Then, according to the Cauchy theorem (see formula (15.2.6a)) we have

$$\int_C f(z)\,dz = \sum_{k=1}^{k} \int_{C_k} f(z)\,dz. \tag{15.5.5}$$

As is easy to establish by repeating the considerations used for deriving

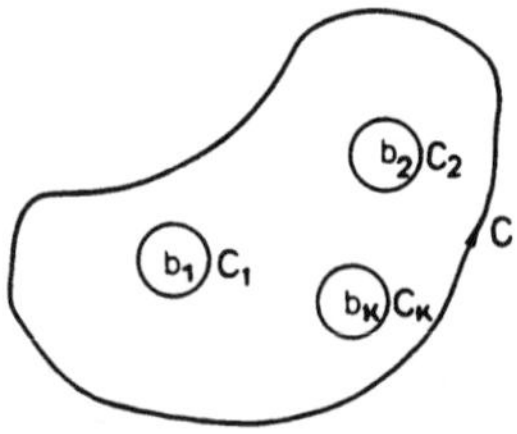

Fig. 15.8. Contour for deriving residue theorem.

formula (15.5.4), that each of the integrals in the right-hand side of formula (15.5.5), with respect to circles C_k, is equal to $2\pi i a_{-1}^{(k)}$ where $a_{-1}^{(k)}$ is the residue at b_k. Then

$$\int_C f(z)\, dz = 2\pi i \sum_{k=1}^{m} a_{-1}^{(k)}. \tag{15.5.6}$$

Now we may finally formulate the residue theorem.

If the function $f(z)$ is analytic everywhere in the region bounded by the closed contour C, except for the finite number of singularities lying inside the contour C, then the value of the integral $\int_C f(z)\, dz$ equals the sum of residues of the integrand $f(z)$ multiplied by $2\pi i$ at all its singularities inside C.

Before passing to the numerous applications of this theorem, it is necessary to notice simple ways to determine residues. Let the function $f(z)$ have a pole of the mth order at b. At this point, the Laurent series is of the form

$$f(z) = \sum_{n=-m}^{\infty} a_n (z-b)^n. \tag{15.5.7}$$

The function $(z-b)^m f(z)$ will be analytic at b and, consequently, it may be presented by the Taylor series at this point. For the expansion of the function $(z-b)^m f(z)$ into the Taylor series, the coefficient a_{-1} will be that at $(z-b)^{m-1}$. Recollecting formula (15.3.9) for the coefficients in the Taylor series, we obtain

$$a_{-1} = \frac{1}{(m-1)!} \left\{ \frac{d^{m-1}}{dz^{m-1}} [(z-b)^m f(z)] \right\}_{z=b}. \tag{15.5.8}$$

Note one more frequently used case of formula (15.5.8) when the function $f(z)$ is of the form $f(z) = \varphi(z)/\psi(z)$ where functions $\varphi(z)$ and $\psi(z)$ are analytic at the point b, and $\psi(z)$ has a simple root at this point. Then, if $\varphi(b) \neq 0$, the function $f(z)$ has a simple pole at the point b. Using general formula (15.5.8) at $m = 1$ and taking into account that $\psi(z) = \psi'(b) \times (z-b) + \cdots$, we obtain

$$a_{-1} = \varphi(b)/\psi'(b). \tag{15.5.9}$$

We now pass to the applications based on the fact that with the help of the residue theorem, the contour integration may be reduced to the determination of residues, which in a number of cases, is calculated by a simple differentiation.

We apply the residue theorem to calculating definite integrals. Consider the integral of the form

$$\int_{-\infty}^{\infty} f(x)\, dx, \tag{15.5.10}$$

where $f(z)$ is the function analytic everywhere in the upper half-plane $\operatorname{Im} z > 0$, except for the finite number of singularities, which, however, do not lie on a real axis $-\infty < x < \infty$. At $|z| \to \infty$, the function $zf(z)$ should tend to zero, i.e., the function $|f(z)|$ should tend to zero at $|z| \to \infty$ quicker than $A/|z|$. To calculate integral (15.5.10), we shall consider the contour integral consisting of an intercept of the real axis and the half-circle of a large radius R with the center at the origin, lying above the real axis (see Fig. 15.9). Then, according to the residue theorem, the integral with respect to this contour is

$$\int_C f(z)\,dz = \int_{-R}^{R} f(x)\,dx + \int_\Gamma f(z)\,dz = 2\pi i \sum a_{-1}, \quad (15.5.11)$$

where the sum of residues of the function $f(z)$ with respect to singularities lying above the real half-axis is designated through $\sum a_{-1}$ and Γ is the half-circle.

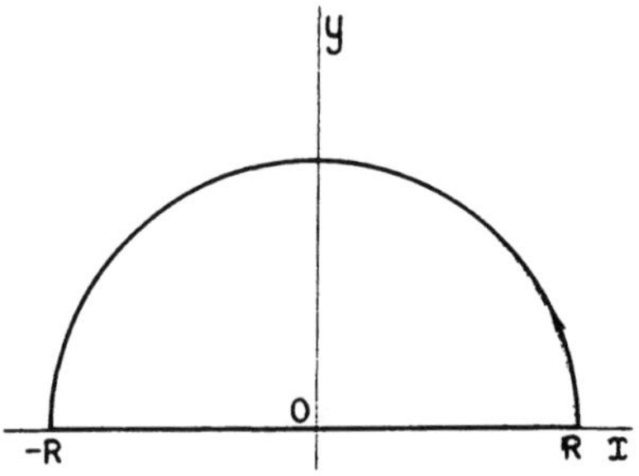

Fig. 15.9. Contour for calculating integrals having infinite limits with the help of the residue theorem.

For estimation of the integral with respect to the half-circle, we have $(z = Re^{i\varphi})$

$$\left| \int_\Gamma f(z)\,dz \right| = \left| \int_0^\pi f(Re^{i\varphi})Re^{i\varphi} i\,d\varphi \right|$$
$$\leq \max |f(Re^{i\varphi})|\; R \int_0^\pi d\varphi = \pi R \max |f(Re^{i\varphi})|,$$

where $\max |f(Re^{i\varphi})|$ is the maximum of a modulus of the function $f(z)$ on the half-circle Γ and where we take into account that $|ie^{i\varphi}| = 1$. Since $R|f(Re^{i\varphi})| \to 0$ at $R \to \infty$ for all $0 \leq \varphi \leq \pi$, the considered integral with respect to the half-circle converts to zero.

Note that with the use of the same method, it is possible to prove a more general statement called the Jordan lemma: if in the upper half-plane and on the real axis the function $f(z)$ satisfies the condition that $f(z)$ uniformly tends to zero at $z \to \infty$, and γ is some positive number, then at $R \to \infty$

$$\int_{\Gamma} f(z)e^{i\varphi z}\, dz \to 0,$$

where the contour Γ is the half-circle in the upper half-plane with the center at the origin of coordinates and the radius R. We use this lemma later.[5] Now, considering integral (15.5.10) we finally obtain from (15.5.11) at $R \to \infty$ and from the estimate of the integral with respect to the half-circle, that

$$\int_{-\infty}^{\infty} f(x)\, dx = 2\pi i \sum \left\{ \begin{array}{l} \text{of residues of } f(z) \text{ with respect to sin-} \\ \text{gularities lying in the upper half-plane} \end{array} \right. \tag{15.5.12}$$

We now consider the examples.

1. The integral

$$\int_{-\infty}^{\infty} dx/(x^2+1)^3 \tag{15.5.13}$$

belongs to (15.5.10) and satisfies, as is easy to check, all the conditions necessary for use of formula (15.5.12). The integrand $f(z) = 1/(z^2+1)^3$.has poles of the third order at points $z = i$ and $z = -i$. Only the pole $z = i$ lies in the upper half-plane with a residue $-(3/16)i$ calculated by formula (15.5.8). According to (15.5.12) we have

$$\int_{-\infty}^{\infty} dx/(x^2+1)^3 = 2\pi i\left(-\frac{3}{16}i\right) = \frac{3}{8}\pi.$$

2. The integral

$$\int_0^{\infty} \cos x\, dx/(x^2+a^2)(x^2+b^2) \qquad (\operatorname{Re} a > \operatorname{Re} b > 0) \tag{15.5.14}$$

may be presented as

$$\frac{1}{2}\int_{-\infty}^{\infty} \frac{\cos x\, dx}{(x^2+a^2)(x^2+b^2)} = \frac{1}{2}\int_{-\infty}^{\infty} \frac{e^{ix}\, dx}{(x^2+a^2)(x^2+b^2)},$$

[5] To determine contour integrals arising at the inversion of the Laplace transform, it is necessary to have the modification of the Jordan lemma which is obtained by replacing the variable $iz = s$, i.e., it confirms that for any function $F(s)$ tending uniformly to zero at $R \to \infty$ on the half-circle Γ' of the radius R lying in the left half-plane, we have at any positive ν

$$\lim_{R\to\infty} \int_{\Gamma'} F(s)e^{s\nu}\, ds = 0.$$

since the integral of the odd function $\sin x/(x^2 + a^2)(x^2 + b^2)$ with respect to a symmetric interval becomes zero.

The last integral has the form of (15.5.10). In particular, it is easy to make sure that the integrand

$$f(z) = e^{iz}/(z^2 + a^2)(z^2 + b^2)$$

tends to zero with an increase in z in the upper half-plane (i.e., when $\operatorname{Im} z > 0$) at a greater increase than any degree of z. Singularities of this function are four poles of the first order at points $z = \pm ia$, $z = \pm ib$. Two of them, $z = ia$ and $z = ib$, lie on the real axis. The residues at these points determined by formula (15.5.9) are equal respectively to

$$-\frac{e^{-a}}{2ia(a^2 - b^2)} \quad \text{and} \quad \frac{e^{-b}}{2ib(a^2 - b^2)}.$$

Hence we have

$$\int_0^\infty \frac{\cos x\, dx}{(x^2 + a^2)(x^2 + b^2)} = \frac{\pi}{2} \frac{1}{a^2 - b^2} \left(\frac{e^{-b}}{b} - \frac{e^{-a}}{a} \right).$$

Further, we consider a new form of integrals

$$\int_0^{2\pi} f(\cos\theta, \sin\theta)\, d\theta\,, \tag{15.5.15}$$

where the function $f(\cos\theta, \sin\theta)$ is the rational function of its own arguments $\cos\theta$ and $\sin\theta$. To determine such integrals, it is necessary to express $\cos\theta$ and $\sin\theta$ through a complex variable $z = e^{i\theta}$ according to the Euler formula, after this integral (15.5.15) transforms into a contour integral with respect to the circle of a single radius. From the residue theorem, this contour integral is expressed as the sum of residues with respect to singularities lying inside the single circle.

As an example, we shall calculate the integral

$$\int_0^{2\pi} \{\sin^2\theta/(a + b\cos\theta)\}\, d\theta \quad (a > b > 0). \tag{15.5.16}$$

Letting $\sin\theta = (z - 1/z)/2i$, $\cos\theta = (z + 1/z)/2$ and $d\theta = -i(dz/z)$ respectively, we find that integral (15.5.16) is equal to the following contour integral with respect to the single circle:

$$\frac{i}{2b} \int_{|z|=1} \frac{(z^2 - 1)^2}{z^2(z^2 + (2a/b)z + 1)}\, dz.$$

The integrand in this integral has singularities: the second-order pole at a point $z = 0$, a simple pole at a point $z = \{(a^2 - b^2)^{1/2} - a\}/b$ and one more simple pole at a point $z = -\{(a^2 - b^2)^{1/2} - a\}/b$. Only the first two points from three singularities (since $a/b > 1$) lie inside the single circle. The The residue at the point $z = 0$ equals $-2a/b$, and at the point $z = \{(a^2 - b^2)^{1/2} - a\}/b$ it is equal, respectively, to $2(\{a^2/b^2\} - 1)^{1/2}$. We have

$$\int_0^{2\pi} \frac{\sin^2\theta}{a + b\cos\theta}\,d\theta = \frac{i}{2b}\int_{|z|=1} \frac{(z^2-1)^2}{z^2(z^2 + 2(a/b)z + 1)}\,dz$$
$$= \frac{i}{2b}\,2\pi i\left(-\frac{2a}{b} + 2\left(\frac{a^2}{b^2} - 1\right)^{1/2}\right)$$
$$= \frac{2\pi}{b^2}\,(a - (a^2 - b^2)^{1/2}).$$

We now pass to the case when the real integral is calculated from the multivalued function. Consider the integral

$$\int_0^\infty x^{\mu-1} f(x)\,dx\,, \qquad (15.5.17)$$

where $f(z)$ is the rational function without poles on a positive part of the real axis, μ is the real number and $x^\mu f(x) \to 0$ at $x \to 0$ and $x \to \infty$. For calculation of integral (15.5.17), we introduce the contour integral

$$\int_C (-z)^{\mu-1} f(z)\,dz\,, \qquad (15.5.18)$$

where C is the closed contour presented in Fig. 15.10. The integrand in expression (15.5.18) is the multivalued function due to the presence of the

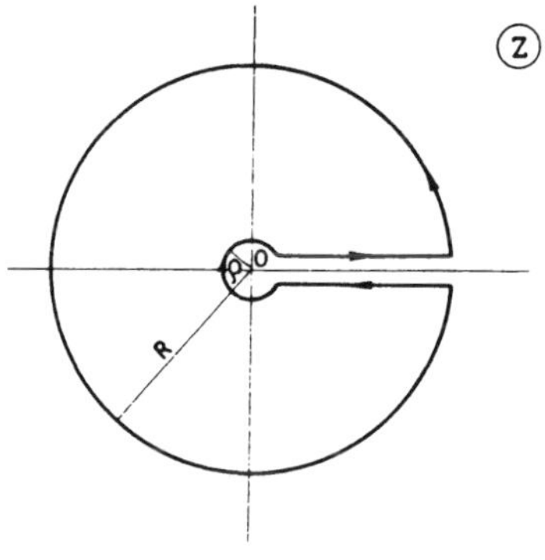

Fig. 15.10. Contour for calculating integrals of type (5.5.18) with a branching point $z = 0$.

multiplier $(-z)^{\mu-1}$ if μ is not an integer. In order to make the function single-valued, in Fig. 15.10, a section along the positive axis connecting points of branching $z = 0$ and $z = \infty$ is made. The complex plane is cut in such a way that the integrand will be already single-valued and there is no other choice but a fixed one for a certain branch of this function. For this purpose, we agree that on the upper edge of the section where z is positive, $-\pi$ will be the argument of a negative quantity $-z$. When bypassing the origin of coordinates counter-clockwise, the argument obtains the increment 2π. Consequently, on the lower edge of the section, bypassing the point $z = 0$, the argument $-z$ will be $-\pi + 2\pi = \pi$. Thus, we have that $-z = ue^{-i\pi}$ on the upper edge of the section and $-z = ue^{i\pi}$ on the lower edge, where u is the modulus of z. For integral (15.5.18), writing it as a sum of integrals with respect to the upper and lower edges of the section and with respect to small and great circles, we have

$$\begin{aligned}\int_C (-z)^{\mu-1}f(z)\,dz &= \int_\varrho^R e^{-i\pi(\mu-1)}u^{\mu-1}f(u)\,du - \int_\varrho^R e^{i\pi(\mu-1)}u^{\mu-1}f(u)\,du \\ &\quad + \int_{C_\varrho} (-z)^{\mu-1}f(z)\,dz + \int_{C_R\circlearrowleft}(-z)^{\mu-1}f(z)\,dz \\ &= 2\pi i \sum \left\{\begin{array}{l}\text{residues of the function } (-z)^{\mu-1}f(z) \\ \text{with respect to the function poles } f(z).\end{array}\right.\end{aligned} \tag{15.5.19}$$

With the help of simple estimates it is not difficult to show that the integral $\int_{C_\varrho\circlearrowleft}(-z)^{\mu-1}f(z)\,dz$ converts to zero at $\varrho \to 0$ due to the condition $z^\mu f(z) \to 0$ at $z \to 0$. The integral $\int_{C_R\circlearrowleft}(-z)^{\mu-1}f(z)\,dz$ vanishes in its turn at $R \to \infty$ since $z^\mu f(z) \to 0$ at $z \to \infty$. Thus, formula (15.5.19) takes the form

$$-e^{-i\pi\mu}\int_0^\infty u^{\mu-1}f(u)\,du + e^{i\pi\mu}\int_0^\infty u^{\mu-1}f(u)\,du = 2\pi i \sum [\text{residues}],$$

or finally (passing to the former notation for a variable of integration)

$$\int_0^\infty x^{\mu-1}f(x)\,dx = (\pi/\sin \pi\mu) \sum [\text{residues}]. \tag{15.5.20}$$

As a simple example, we calculate the integral

$$\int_0^\infty \{x^{\mu-1}/(1 + x^2)\}\,dx \qquad (0 < \mu < 2),$$

which satisfies all the conditions of the applicability of formula (15.5.20). The integrand $f(z) = (-z)^{\mu-1}/(z^2 + 1)$, in addition to the point of branching at $z = 0$, has simple poles at points $z = i$ and $z = -i$ with residues $\frac{1}{2}(\exp[-i\,\frac{1}{2}\pi\mu])$ and $\frac{1}{2}(\exp[i\,\frac{1}{2}\pi\mu])$, respectively.

According to formula (15.5.20), we have

$$\int_0^\infty \frac{x^{\mu-1}}{1+x^2}\,dx = \frac{\pi}{\sin \pi\mu}\left(\frac{\exp[-i\,\frac{1}{2}\pi\mu]}{2} + \frac{\exp[i\,\frac{1}{2}\pi\mu]}{2}\right)$$

$$= \frac{\pi}{\sin \pi\mu}\cos\frac{\pi\mu}{2} = \frac{1}{2}\pi \operatorname{cosec}\frac{1}{2}\pi\mu. \tag{15.5.21}$$

Because of the special importance for the heat conduction theory, with the help of some examples we now illustrate the application of the residue theory for determining a special class of contour integrals due to the inversion of the Laplace transform, i.e., for calculating the function $f(\tau)$ prescribed in the interval $\tau \geq 0$ with respect to its original function $F(s)$. Thus, we shall calculate the contour integral

$$f(\tau) = (1/2\pi i)\int_{a-i\infty}^{a+i\infty} F(s)e^{s\tau}\,ds, \tag{15.5.22}$$

where a is a real number, such that the straight line $\operatorname{Re} s = a$ lies to the right from the singularities of the function $F(s)$.

(1) Let $F(s) = 1/(s+b)$ $(b > 0)$. The only singularity of $F(s)$ is a simple pole at a point $s = -b$. For calculating $f(\tau)$, we consider the integral

$$\int_C F(s)e^{s\tau}\,ds$$

with respect to the contour C given in Fig. 15.11, consisting of a part of an

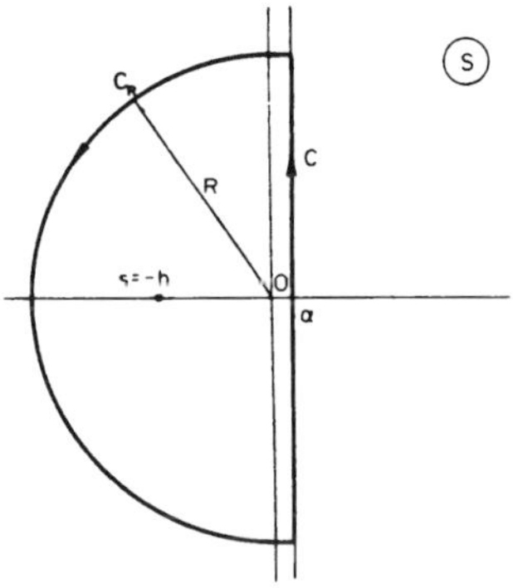

Fig. 15.11. Contour for calculating the inverse Laplace transform of single-valued functions.

imaginary axis and the left half-circle C_R of the radius R. The residue of the integrand $e^{s\tau}/(s+b)$ at a point $s = -b$ is equal to $e^{-a\tau}$. Consequently

$$\int_C \{e^{s\tau}/(s+b)\}\, ds = \int_{a-iR}^{a+iR} \{e^{s\tau}/(s+b)\}\, ds + \int_{C_R} [e^{s\tau}/(s+b)]\, ds$$
$$= 2\pi i e^{-b\tau}. \tag{15.5.23}$$

When R tends to ∞, then the integral with respect to the imaginary axis transforms into desired integral (15.5.22), and that with respect to the half-circle converts zero according to the Jordan lemma. Finally from (15.5.23) it follows that

$$f(\tau) = (1/2\pi i) \int_{a-i\infty}^{a+i\infty} [e^{s\tau}/(s+b)]\, ds = e^{-b\tau}. \tag{15.5.24}$$

(2) $F(s) = 1/\sqrt{s}\,(s - b^2)$. The singularities are a simple pole at $s = b^2$ and a branching point at $s = 0$. Consider the integral

$$\frac{1}{2\pi i} \int_C \frac{e^{s\tau}}{\sqrt{s}\,(s - b^2)}\, ds, \tag{15.5.25}$$

with respect to the contour C given in Fig. 15.12. A complex plane s is cut along a negative part of the real axis in order to single out a single-valued branch of the integrand. Inside the contour C consisting of a part of the imaginary axis, a left half-circle C_R, upper and lower edges of a section, and a small circle C_ϱ with a point $s = 0$ as a center, there is the only one

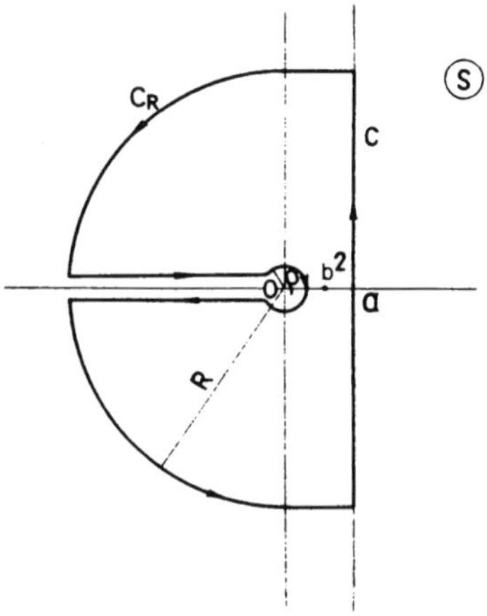

Fig. 15.12. Contour for calculating the inverse Laplace transform of the functions with the points of branching $z = 0$ and $z = \infty$.

singularity, i.e., a pole $s = b^2$ with a residue $\exp[b^2\tau]/b$. Integral (15.5.25) is presented as

$$\frac{1}{2\pi i}\int_C \frac{e^{s\tau}}{\sqrt{s}\,(s-b^2)}\,ds = \frac{e^{b^2\tau}}{b} = \frac{1}{2\pi i}\int_{a-iR}^{a+iR}\frac{e^{s\tau}}{\sqrt{s}\,(s-b^2)}\,ds$$

$$+\frac{1}{2\pi i}\int_{C_R\swarrow}\frac{e^{s\tau}}{\sqrt{s}\,(s-b^2)}\,ds + \frac{1}{2\pi i}\int_{C_\varrho\nwarrow}\frac{e^{s\tau}}{\sqrt{s}\,(s-b^2)}\,ds$$

$$-\frac{1}{2\pi i}\int_\varrho^R \frac{e^{-u\tau}\,du}{\sqrt{u}\,e^{i\pi/2}(u+b^2)} + \frac{1}{2\pi i}\int_\varrho^R\frac{e^{-u\tau}\,du}{\sqrt{u}\,e^{-i\pi/2}(u+b^2)}, \tag{15.5.26}$$

where in the two last integrals, section $s = ue^{i\pi}$ is on the upper edge, and $s = ue^{-i\pi}$ is on the lower edge.

At $R \to \infty$, the integral with respect to C_R according to the Jordan lemma becomes zero. Estimating the integral with respect to C_ϱ at $\varrho \to 0$ assuming $s = \varrho e^{i\varphi}$, we have

$$\lim_{\varrho\to 0}\frac{1}{2\pi i}\int_{C_\varrho\nearrow}\frac{\exp[s\tau]}{\sqrt{s}\,(s-b^2)}\,ds = \lim_{\varrho\to 0}\frac{1}{2\pi i}\int_0^{2\pi}\frac{\exp[\tau\varrho\exp[i\varphi]]i\varrho\exp[i\varphi]\,d\varphi}{\sqrt{\varrho}\,\exp[i\varphi/2](\varrho\exp[i\varphi]-b^2)}$$
$$= 0.$$

Integral (15.5.26) is now written

$$\frac{e^{b^2\tau}}{b} = f(\tau) + \frac{1}{\pi}\int_0^\infty \frac{e^{-u\tau}}{\sqrt{u}\,(u+b^2)}\,du. \tag{15.5.27}$$

The last integral may be expressed as a probability. In order to illustrate this, we first of all introduce a new variable $u = z^2$, and obtain

$$j(\tau) = \frac{2}{\pi}\int_0^\infty \frac{e^{-z^2\tau}}{z^2+b^2}\,dz.$$

As is not difficult to verify directly, the function $j(\tau)$ satisfies the differential equation

$$j'(\tau) - b^2 j(\tau) = -1/(\pi\tau)^{1/2}$$

provided that $j(0) = 1/b$. The solution of this equation is of the form

$$j(\tau) = \frac{e^{b^2\tau}}{b} - \frac{e^{b^2\tau}}{b}\,\frac{2}{\sqrt{\pi}}\int_0^{b\sqrt{\tau}}\exp[-z^2]\,dz.$$

Finally formula (15.5.27) is written as

$$f(\tau) = \{\exp[b^2\tau]/b\} \operatorname{erf}(b\sqrt{\tau}). \tag{15.5.28}$$

(3) $F(s) = \exp[-b\sqrt{s}]/\sqrt{s}$. A singularity is a point of branching at $s = 0$. The contour is the same as in the previous case. Since inside this contour with a section the function $F(s)$ has no singularities, then

$$(1/2\pi i) \int_C \{\exp[-b\sqrt{s} + s\tau]/\sqrt{s}\}\, ds = 0. \tag{15.5.29}$$

It is easy to check that both integrals with respect to the left of the half-circle at $R \to \infty$ and those with respect to a small circle with a point $s = 0$ as a center at $\varrho \to 0$ vanish. Equality (15.5.29) may now be rewritten as a sum of integrals along the imaginary axis equal to $f(\tau)$ together with those with respect to the edges of the sections.

Assuming on the upper edge $s = ue^{i\pi}$ and on the lower one $s = ue^{-i\pi}$

$$f(\tau) + \frac{1}{2\pi i}\int_0^\infty \frac{\exp[-bi\sqrt{u} - u\tau]}{i\sqrt{u}}\, du + \frac{1}{2\pi i}\int_0^\infty \frac{\exp[bi\sqrt{u} - u\tau]}{i\sqrt{u}}\, du = 0,$$

i.e.,

$$f(\tau) = (1/\pi) \int_0^\infty \exp[-u\tau]\{\cos b\sqrt{u}/\sqrt{u}\}\, du. \tag{15.5.30}$$

To calculate this integral, we introduce a new variable of integration $z = \sqrt{u}$. Then

$$\begin{aligned} f(\tau) &= \frac{2}{\pi}\int_0^\infty \exp[-z^2\tau] \cos bz\, dz \\ &= \frac{1}{\pi}\int_{-\infty}^\infty \exp[-z^2\tau] \cos bz\, dz \\ &= \frac{1}{\pi}\int_{-\infty}^\infty \exp[-z^2\tau + ibz]\, dz, \end{aligned}$$

Assuming here $z = (x/\sqrt{\tau}) + (ib/2\tau)$, we obtain

$$f(\tau) = \frac{1}{\pi}\,\frac{\exp[-b^2/4\tau]}{\sqrt{\tau}} \int_{-\infty}^\infty \exp[-x^2]\, dx = \frac{1}{\sqrt{\pi}}\,\frac{\exp[-b^2/4\tau]}{\sqrt{\tau}}. \tag{15.5.31}$$

(4) $F(s) = \ln(s + \beta)/(s + \alpha)$. The function $\ln(s + \beta)/(s + \alpha)$ has points of branching at $s = -\beta$ and $s = -\alpha$. The point at infinity is not the point

of branching. In reality, if a closed contour, bypassing around both points $-\beta$ and $-\alpha$ in a positive direction, is fitted, then $\ln(s+\beta)$ and $\ln(s+\alpha)$ obtain one and the same term $2\pi i$ and the difference $\ln(s+\beta) - \ln(s+\alpha) = F(s)$ does not change. Thus, the function $F(s)$ will be single-valued in the plane with the section connecting the points $s = -\beta$ and $s = -\alpha$.

To calculate the original function $F(s)$, we consider the integral with respect to the contour presented in Fig. 15.13. In the region bounded by

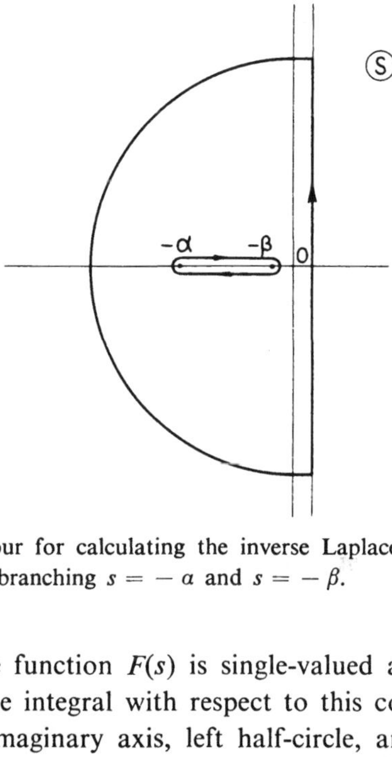

Fig. 15.13. Contour for calculating the inverse Laplace transform of the functions with the points of branching $s = -\alpha$ and $s = -\beta$.

this contour, the function $F(s)$ is single-valued and has no singularities. Consequently, the integral with respect to this contour, consisting of the portion of the imaginary axis, left half-circle, and the upper and lower edges of the section connecting $-\beta$ and $-\alpha$, is equal to zero. Assuming $s = -\beta + ue^{i\pi}$ and $s = -\beta + ue^{-i\pi}$ on the upper and lower edges of the section, respectively, and taking into account that the integral with respect to the half-circle whose radius tends to infinity becomes zero, we have, according to the Jordan lemma

$$f(\tau) + \frac{1}{2\pi i}\int_0^{\alpha-\beta} e^{-\beta\tau-u\tau}\ln\frac{ue^{i\pi}}{\alpha-\beta+ue^{i\pi}}\,du$$

$$-\frac{1}{2\pi i}\int_0^{\alpha-\beta} e^{-\beta\tau-u\tau}\ln\frac{ue^{-i\pi}}{\alpha-\beta+ue^{-i\pi}}\,du = 0$$

or

$$f(\tau) = -\frac{e^{-\beta\tau}}{2\pi i}\int_0^{\alpha-\beta} e^{-u\tau}\ln\left(\frac{\alpha-\beta+ue^{-i\pi}}{\alpha-\beta+ue^{i\pi}}\,e^{2\pi i}\right)du$$
$$= +\frac{e^{-\beta\tau}}{2\pi i}\ln(e^{2\pi i})\frac{e^{(\beta-\alpha)\tau}-1}{\tau}$$
$$= \frac{e^{-\alpha\tau}-e^{-\beta\tau}}{\tau}. \tag{15.5.32}$$

Thus, we shall no longer consider contour integrals, which may be calculated in finite form, i.e., they may be expressed through elementary and special functions. However, the application of the residue theory, as we shall see, is not exhausted by calculating such integrals. This is particularly so as they are not numerous. In particular, the majority of such finite-form contour integrals (15.5.22) inverting the Laplace transform are given in tables and handbooks [113] on operational calculus.

If the solution of some problem is obtained as a contour integral which is not solved in finite form, then the residue theory makes it possible to express this contour integral through ordinary integrals with respect to a real variable. Such an integral representation of the solution frequently makes it possible to visualize its behavior and to facilitate numerical calculations. As an example, consider a heat conduction problem for a body occupying a semispace $0 \le x < \infty$, whose density or heat capacity linearly increases with a removal from a surface $x = 0$ where there occurs heat transfer according to the Newton law. It is necessary to determine a heat flux density through this surface. Thus, the problem is formulated as

$$c\gamma\,\frac{x}{l}\,\frac{\partial t}{\partial \tau} = \lambda\,\frac{\partial^2 t}{\partial x^2}, \tag{15.5.33}$$

$$-\frac{\partial t(0,\tau)}{\partial x} + h[t_a - t(0,\tau)] = 0,$$
$$t(0,x) = 0, \qquad \frac{\partial t(\infty,\tau)}{\partial x} = 0. \tag{15.5.34}$$

Here t_a, c, γ, λ, l, h are constants. It is possible to show [89, 112] that the necessary flow $q(\tau) = -\lambda(\partial t(0,\tau)/\partial x)$ satisfies the following Volterra integral equation of the second kind with a difference core

$$\lambda h t_a - q(\tau) = b\int_0^{\tau}\frac{q(\tau')}{(\tau-\tau')^{2/3}}\,d\tau', \tag{15.5.35}$$

where

$$b = \frac{h}{\Gamma(\frac{2}{3})}\left(\frac{\lambda l}{3c\gamma}\right)^{1/3}.$$

Since in the right-hand side of integral equation (15.5.35) there is the convolution of the functions $q(\tau)$ and $1/\tau^{2/3}$, it is easily solved by the Laplace transform. The solution is of the form

$$q(\tau) = \lambda h t_a \frac{1}{2\pi i} \int_{c-i\infty}^{c+i\infty} \frac{e^{s\tau}\, ds}{s^{2/3}(s^{1/3} + b\Gamma(\frac{1}{3}))} \quad (c > 0). \qquad (15.5.36)$$

This contour integral may be easily transformed into a real integral by integrating the integrand with respect to the contour consisting of the part of the imaginary axis, the section connecting points of branching $s = 0$ and $s = \infty$, and the left half-circle (see Fig. 15.14). Since inside this contour the

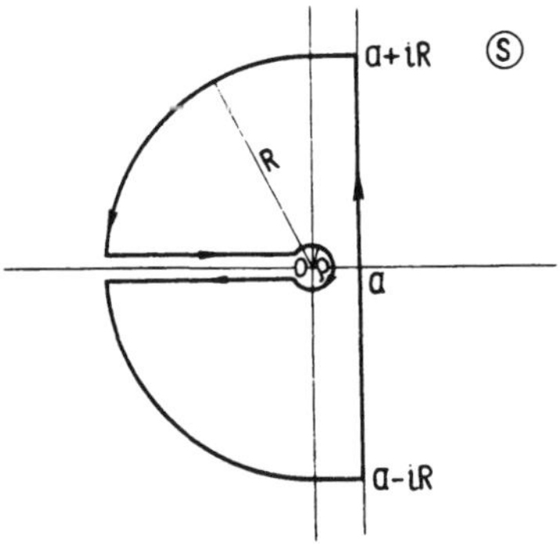

Fig. 15.14. Contour for calculating the integral characterizing the heat flux density (15.5.36).

integrand has no singularities, then the desired integral is equal simply to the integrals along the edges of the section (integrals with respect to the half-circle vanish according to the Jordan lemma when the radius of the semicircle tends to infinity).

We have

$$\frac{q(\tau)}{\lambda h t_a} - \frac{1}{2\pi i} \int_0^\infty \frac{e^{-u\tau}\, du}{u^{2/3}e^{-i2\pi/3}(u^{1/3}e^{-i\pi/3} + b\Gamma(\frac{1}{3}))} + \frac{1}{2\pi i} \int_0^\infty \frac{e^{-u\tau}\, du}{u^{2/3}e^{i2\pi/3}[u^{1/3}e^{i\pi/3} + b\Gamma(\frac{1}{3})]} = 0.$$

Upon simple transformations, we have

$$\frac{q(\tau)}{\lambda h t_a} = \frac{3\sqrt{3}}{2\pi} \int_0^\infty \frac{\exp[-\xi z^3]}{z^2 + z + 1}\, dz, \qquad (15.5.37)$$

where

$$\xi = [b\Gamma(1/3)]^3 \tau.$$

From the solution of (15.5.37) it is immediately seen that $q(\tau)$ is a monotonically decreasing (from $\lambda h t_a$ to 0) function τ. Integral (15.5.37) is integrated by terms, and for large values of ξ, a simple asymptotic estimate is obtained for this integral in the next section.

The methods of the residue theory may also be used for the problems contrasting in certain ways to the above ones. Frequently it appears useful to express some function through a contour integral. In this case, the integral representation of a complex function may appear to be convenient for investigation if the integrand in the contour integral has a simple form and contains elementary functions. Moreover, by conforming the contour according to the Cauchy theorem, it is possible to obtain approximate estimates. In particular, if the function is given by a series, then the representation of the sum of a series by means of a contour integral allows the sum of a series in finite form to be found in some cases.

Let $f(z)$ be a rational function with a finite number of poles at points $z_1, \ldots, z_p$. Let the power of a polynomial in the numerator of the function $f(z)$ be smaller than that in the denominator by not less than two units, so that $f(z)$ decreases at $z \to \infty$ at a rate not less than A/z^2. Consider the integral with respect to the radius circle $(n + \frac{1}{2})$

$$\int_{|z|=n+1/2} f(z) \cot \pi z \, dz \quad (n = 0, 1, \ldots). \tag{15.5.38}$$

The meromorphic function $\cot \pi z$ is limited everywhere, except for the circles with radii with integer values. Therefore, for integral (15.5.38) we have the estimate

$$\left| \int_{|z|=n+1/2} f(z) \cot \pi z \, dz \right| \leq \int_{|z|=n+1/2} \left| f(z) \cot \pi z \right| dz$$
$$\leq \frac{A}{(n + \frac{1}{2})^2} M 2\pi (n + \tfrac{1}{2}), \tag{15.5.39}$$

where M is the maximum of a modulus $\cot \pi z$ in a plane z with remote circles with integer values. From the estimate of (15.5.39) it follows that the integral $\int_{|z|=n+1/2} f(z) \cot \pi z \, dz \to 0$ at $n \to \infty$. This same integral may be calculated with the help of the residue theorem. Simple poles of meromorphic function $\cot \pi z$ at points $z = 0, \pm 1, \pm 2, \ldots, \pm n$ with residues $(1/\pi) f(z)$, and poles of the rational function $f(z)$, will be singularities of the integrand function in a circle $|z| = n + \frac{1}{2}$. We assume that n is sufficiently great so that all the poles $z_1, z_2, \ldots, z_p$ of the function $f(z)$ enter the circle $|z| < n + \frac{1}{2}$. Then, according to the residue theorem we have

$$\int_{|z|=n+1/2} f(z) \cot \pi z \, dz = 2\pi i \Big\{ (1/\pi) \sum_{k=-n}^{n} f(k) + \sum_{1}^{p} \text{residues}[f(z) \cot \pi z] \text{ in poles } f(z) \Big\}.$$

Passing to the limit $n \to \infty$, we obtain

$$\sum_{n=-\infty}^{\infty} f(n) = -\pi \sum_{1}^{p} \begin{cases} \text{residues } [f(z) \cot \pi z] \\ \text{in poles } f(z). \end{cases} \tag{15.5.40}$$

The formula for the series of the form

$$\sum_{n=-\infty}^{\infty} (-1)^n f(n)$$

may be obtained in a similar way where $f(z)$ possesses the former properties. It is necessary only to replace $\cot \pi z$ by $1/(\sin \pi z)$ in integral (15.5.38). As a result we have

$$\sum_{n=-\infty}^{\infty} (-1)^n f(n) = -\pi \sum_{1}^{p} \begin{cases} \text{residues } [f(z)/\sin \pi z] \\ \text{at poles } f(z). \end{cases} \tag{15.5.41}$$

As an illustration of the application of the formula considered, we shall sum the series

$$\sum_{n=1}^{\infty} 1/(n^2 + a^2). \tag{15.5.42}$$

First of all, we write (15.5.42) in the form

$$\sum_{n=1}^{\infty} \frac{1}{n^2 + a^2} = \frac{1}{2} \left(\sum_{n=-\infty}^{\infty} \frac{1}{n^2 + a^2} - \frac{1}{a^2} \right).$$

Further, the rational function $1/(z^2 + a^2)$ has two simple poles at points $z = \pm ia$, and residues of the function $\cot \pi z/(z^2 + a^2)$ at these poles are equal to $\cot \pi iz/2ia = -\coth \pi z/2a$, respectively. Thus, according to (15.5.40) we have

$$\begin{aligned} \sum_{n=1}^{\infty} \frac{1}{n^2 + a^2} &= -\frac{1}{2a^2} + \frac{1}{2} \sum_{n=-\infty}^{\infty} \frac{1}{n^2 + a^2} \\ &= -\frac{1}{2a^2} + \frac{1}{2} \pi \, 2 \frac{\coth \pi a}{2a} \\ &= \frac{\pi}{2a} \coth \pi a - \frac{1}{2a^2}. \end{aligned} \tag{15.5.43}$$

This ends our discussion of the residue theory.

15.6 Some Analytical Properties of Laplace Transforms and Asymptotic Estimates

In the previous chapter, we dealt with the bases of operational calculus and with the use of the properties of the Laplace transforms related to operational calculus. Effectiveness of operational calculus is quite obvious because of its simplicity and the possibility of the use of extensive tables of integral transforms. It is, however, very useful to examine the works on direct studies of contour integrals, to which the inverse Laplace transform is reduced. This is because the form of the solution of the problem expressed as the contour integral is more flexible and allows us to obtain the representation of the methods of solution and different, sufficiently simple approximations for solution with the help of deformation of a contour integral and the study of an integrand behavior in a complex plane.

Without dwelling on the contents of Chapter 14, we shall examine analytic properties of the Laplace transform and the simplest methods of obtaining asymptotic estimations. We must remember that the functions to which the Laplace transform is applied, should be piecewise-continuous, different from zero only at $\tau \geq 0$, and should increase at a rate not greater than an exponential function. The latter denotes that there exist such constants $A > 0$ and $\sigma_0 \geq 0$ so that for all $\tau > 0$

$$|f(\tau)| < Ae^{\sigma_0 \tau}. \tag{15.6.1}$$

The number σ_0 is called the exponent of the growth of the function $f(\tau)$. Under these conditions, as has already been mentioned in Chapter 14, it may be shown that the transform $F(s)$ will exist in a half-plane $\operatorname{Re} s > \sigma_0$ and, moreover, will be an analytic function there. Hence, in its turn, it follows that for any transform $F(s)$ at $\operatorname{Re} s = \sigma \to \infty +$, $F(s) \to 0$. This results immediately from inequality (15.6.1) and the estimate

$$|F(s)| = \left| \int_0^\infty e^{-s\tau} f(\tau)\, d\tau \right| < A \int_0^\infty e^{-(\sigma-\sigma_0)\tau}\, d\tau = A/(\sigma - \sigma_0).$$

The condition $F(s) \to 0$ at $\operatorname{Re} s \to +\infty$ is necessary for the function $F(s)$ to be the Laplace transform, therefore the function s^α $(\alpha \geq 0)$ cannot be the transform of the function in its ordinary meaning.[6] However, the condition $F(s) \to 0$ at $\operatorname{Re} s \to +\infty$ is not sufficient to be a transform, as is seen from the function e^{-s}.[6]

[6] We do not consider the theory of the Laplace transformation of the so-called generalized functions [22, 36], i.e., for example, the known Dirac delta function.

Note that the region where the function $F(s)$ is analytic, is wider as a rule than the half-plane Re $s >|\sigma_0$ where $F(s)$ may be presented by the integral $\int_0^\infty e^{-st} f(\tau)\, d\tau$. We may easily see this from the tables of the transforms (see Appendix 5). For example, the function $F(s) = \Gamma(\nu + 1)/s^{\nu+1}$, being the transform of the function τ^ν, has only one singularity at $s = 0$ which in case of the positive integer $\nu = m$ ($m = 0, 1, \cdots$) will be a pole of the $(m + 1)$th order along the whole plane, except at this pole, the function $m!/s^{m+1}$ is analytic. If ν is not an integer, then $F(s)$ will be a multivalued function with branching at $s = 0$ and $s = \infty$, and the whole plane with a section along the real negative half-axis will be the region of analyticity of its single-valued branch determined by the condition $-\pi < \theta < \pi$ $(s = re^{i\theta})$. For example, the function $F(s) = 1/(s^2 + 1)$ being a transform of $\sin \tau$ only in the half-plane Re $s > 0$, will be analytic over the whole plane except for two simple poles at points $s = \pm i$. Therefore, although different relations for the Laplace transform and, in particuar, all the properties of such a kind derived in the previous chapter are established only for the appropriate half-plane of convergence of the integrals $\int_0^\infty e^{-s\tau} f(\tau)\, d\tau$, these relations may be found by the extension of the analytic continuation to the whole analycity of the appropriate functions.

Proceeding from the analytic properties of the Laplace transform in a complex plane, we now establish two practical important limiting relations.

First, if the function $f(\tau)$ satisfies the inequality

$$|f(\tau)| < M$$

for all $\tau > 0$ where $M > 0$ is some constant, then

$$\lim_{s \to +0} sF(s) = f(\infty) \tag{15.6.2}$$

provided that $f(\infty) = \lim_{\tau \to +\infty} f(\tau)$ exists.

Second, if the function $f(\tau)$ satisfies the inequality

$$|f(\tau)| < Ae^{\sigma_0 \tau}$$

for all $\tau > 0$ where A and σ_0 are positive quantities, then

$$\lim_{s \to +\infty} sF(s) = f(0) \tag{15.6.3}$$

assuming that $f(0) = \lim_{\tau \to +0} f(\tau)$ exists.

In formulas (15.6.2) and (15.6.3), s tends to the appropriate limiting values along the real axis.

To prove limiting relation (15.6.2), we consider the difference

$$sF(s) - f(\infty) = s\left[\int_0^\infty e^{-s\tau} f(\tau)\, d\tau - f(\infty)\int_0^\infty e^{-s\tau}\, d\tau\right] \quad (15.6.4)$$

since, according to the theorem conditions, s may be considered real, and introduction of a new variable of integration $x = s\tau$ into formula (15.6.4) gives

$$\begin{aligned} sF(s) - f(\infty) &= \int_0^\infty [f(x/s) - f(\infty)]e^{-x}\, dx \\ &= \int_0^{x_0} [f(x/s) - f(\infty)]e^{-x}\, dx + \int_{x_0}^\infty [f(x/s) - f(\infty)]e^{-x}\, dx, \end{aligned} \quad (15.6.5)$$

where $x_0 > 0$ is still arbitrary.

First consider the integral

$$J_1 = \int_0^{x_0} [f(x/s) - f(\infty)]e^{-x}\, dx.$$

Since the function $f(\tau)$ is restricted by $(|f]\tau)| < M$ for all $\tau > 0)$, then it is obvious that

$$|J_1| < M(1 - e^{-x_0}).$$

In the last inequality, x_0 may be chosen small enough that

$$|J_1| < Mx_0$$

becomes smaller than any small prescribed quantity designated through $\varepsilon > 0$. Having fixed such x_0, we choose some $s_0 > 0$ small enough that at $x > x_0$ for all $0 < s < s_0$ the difference

$$|f(x/s) - f(\infty)| < \varepsilon.$$

Then, for the integral

$$J_2 = \int_{x_0}^\infty [f(x/s) - f(\infty)]e^{-x}\, dx$$

we have

$$|J_2| < \varepsilon \int_{x_0}^\infty e^{-x}\, dx = \varepsilon e^{-x_0} < \varepsilon.$$

Summing the estimates obtained for $|J_1|$ and $|J_2|$ we have that

$$|sF(s) - f(\infty)| < 2\varepsilon.$$

Thus, it is possible to choose a sufficiently small s so that $|sF(s) - f(\infty)|$ would be smaller than the prescribed small positive number that is proved by relation (15.6.2). Formula (15.6.3) is proved in a similar way.

The limiting relations obtained allow the determination of the value of the function $f(\tau)$ at $\tau = 0$ and $\tau = \infty$ according to the known transform $F(s)$ without calculating the contour integral which serves as an inversion of the Laplace transform if it is known that $f(+0)$ and $f(\infty)$ exists. In heat conduction problems, the existence of these values may be frequently established from the physical considerations. For example, if from the problem conditions the existence of a steady-state temperature field is obvious, then it may be determined with the help of relation (15.6.2) according to the transform of the solution.

We now consider some examples.

(1) In Chapter 4, Section 2, it was shown that the Laplace transform of a semispace temperature $x > 0$ with an initial temperature t_0, on the boundary of which a zero temperature is maintained, is equal to

$$\frac{T(x, s)}{t_0} = \frac{1 - \exp[-(s/a)^{1/2}x]}{s}.$$

Hence, it is possible to check whether the expression for $T(x, s)$ obtained as a result of subtraction satisfies the initial condition. We have

$$\frac{t(x, 0)}{t_0} = \lim_{s\to+\infty} s\frac{T(x, s)}{t_0} = \lim_{s\to+\infty} (1 - \exp[-(s/a)^{1/2}x] = 1.$$

Further, find the steady-state value of the temperature to be

$$t(x, \infty) = \lim_{s\to+0} sT(x, s) = t_0 \lim_{s\to+0} (1 - \exp[-(s/a)^{1/2}x]) = 0.$$

(2) As some less trivial example, consider the problem of a temperature field in two semispaces being in thermal contact (see Chapter 10, Section 1). The right semispace has a temperature t_0 at the initial time moment and the left one has a zero initial temperature. The transforms will be of the form

$$\begin{aligned} T_1(x, s) &= \frac{t_0}{s} - \frac{t_0}{1 + k_\varepsilon}\frac{\exp[-(s/a)^{1/2}x]}{s} \quad (x > 0), \\ T_2(x, s) &= \frac{k_\varepsilon t_0}{(1 + k_\varepsilon)s}\exp[(s/a_2)^{1/2}x] \quad (x < 0), \end{aligned} \qquad (15.6.6)$$

where

$$k_\varepsilon = \left(\frac{\lambda_1 c_1 \gamma_1}{\lambda_2 c_2 \gamma_2}\right)^{1/2}.$$

We have

$$\lim_{s\to+0} sT_1(x, s) = t_0 k_\varepsilon/(1 + k_\varepsilon),$$

$$\lim_{s\to+0} sT_2(x, s) = t_0 k_\varepsilon/(1 + k_\varepsilon).$$

Thus, at $\tau \to \infty$, both semispaces have the same (as it should be in the presence of an equilibrium state) temperature $t_0 k_\varepsilon/(1 + k_\varepsilon)$.

It is also easy to check, with the help of limiting relation (16.5.3), that expression (15.6.6) satisfies the initial conditions of the problem.

(3) Let $F(s) = 1/(s^2 + 1)$, then

$$\lim_{s\to+\infty} sF(s) = \lim_{s\to+\infty} s/(s^2 + 1) = 0$$

and

$$\lim_{s\to+0} sF(s) = \lim_{s\to+0} [s/(s^2 + 1)] = 0.$$

However, only the first of the equalities gives the value of the function $f(\tau) = \sin \tau$ at $\tau = 0$. But $\lim_{s\to+0} sF(s)$ does not give the value of the limit $f(\tau)$ at $\tau \to \infty$ as this limit does not exist.

The same situation may be also encountered in some heat conduction problems, for example, in the problem on a semispace $x > 0$, on the boundary of which the temperature changes in time according to the harmonic law. Under the simplest boundary conditions, it is possible to write

$$\begin{gathered} \partial t(x, \tau)/\partial\tau = a\partial^2 t(x, \tau)/\partial x^2 \quad (0 \le x < \infty), \\ t(x, 0) = 0, \quad t(0, \tau) = t_m \sin \omega\tau, \quad \partial t(\tau, \infty)/\partial x = 0. \end{gathered} \tag{15.6.7}$$

It is easy to calculate that

$$T(x, s) = t_m\{\omega/(s^2 + \omega^2)\} \exp[-(s/a)^{1/2} x]. \tag{15.6.8}$$

Although from (15.6.8) it follows that

$$\lim_{s\to+0} sT(x, s) = 0,$$

this does not denote that

$$\lim_{\tau\to\infty} t(x, \tau) = 0,$$

since the last limit does not exist.

The above limiting relations for the Laplace transform expressed by formulas (15.6.2) and (15.6.3) are a very special case of asymptotic estimations.

We now consider some methods of the asymptotic estimations useful for studying heat conduction problems.

Two functions $f(x)$ and $g(x)$ are called asymptotically equal when their argument x tends to some value of x_0 if the ratio $f(x)/g(x)$ tends to unity at $x \to x_0$.

The asymptotic equality will be written as

$$f(x) \simeq g(x) \quad (x \to x_0).$$

For example

$$x+1 \simeq x \quad , \quad \text{at } x \to \infty, \qquad \sinh x \simeq e^x/2, \quad \text{at } x \to +\infty,$$

$$\frac{x^2+3x+2}{5x^3+\frac{x^2}{2}+1} \simeq \frac{1}{5x}, \quad \text{at } x \to +\infty, \qquad \sin x \simeq x, \quad \text{at } x \to 0,$$

and so on.

The case when the function $g(x)$ is more simple from the point of view of its calculation at $x \to x_0$ as compared to the function $f(x)$, is of practical importance. Frequently, at $x \to \infty$, the descending power series τ may serve as such simple functions

$$f(x) \simeq \sum_{n=0}^{\infty} C_n/x^{\lambda_n} \quad (\lambda_0 < \lambda_1 < \lambda_2 < \cdots). \tag{15.6.9}$$

Series (15.6.9) does not obligatorily converge but in some practical cases for the asymptotic diverging series, the error resulting from the replacement of the function $f(x)$ by a part of series (15.6.9) $\sum_{n=0}^{N} C_n/x^{\lambda_n}$ may be less than the last term in this sum C_N/x^{λ_N}, i.e.,

$$x^{\lambda_n}\left[f(x) - \sum_{n-0}^{N} \frac{C_n}{x^{\lambda_n}}\right] \to 0, \quad \text{at } x \to \infty. \tag{15.6.10}$$

We further note that the asymptotic expansion of the function, if it exists, is determined once. However, one and the same asymptotic series may serve as asymptotic expansion of different functions. For example, two functions $f(x)$ and $f(x) + e^{-x}$ have the same asymptotic expansions at $x \to +\infty$, since $\lim_{x \to +\infty} x^{\lambda_n} e^{-x} = 0$ for any λ_n.

The asymptotic series may be summed up by terms, multiplied, and integrated.

If it is known that an arbitrary function allows the asymptotic expansion, then it may be obtained by differentiating the asymptotic expansion of the function.

The practical importance of the asymptotic estimations is very great. This is explained by the fact that the solution of many nontrivial problems of mathematical physics is very tedious or complex. For example, it may be prescribed by a complex functional series or a contour integral. At the same time, there is frequently no need to know a precise solution of all the values of the parameters and variables of the problem but only at some limiting values. For example, it is sufficient to know the behavior of the solution after a long period of time. In these cases, it is possible to judge the behavior of a complex precise solution according to its asymptotic expansion. Since the solution of nonlinear problems of the heat conduction theory may always be expressed in terms of contour integrals (and in some cases in terms of integrals with respect to a real variable), then it is first necessary to consider the methods of the asymptotic estimations of the integrals of the appropriate types.

First consider real integrals such as

$$F(\sigma) = \int_a^b \varphi(x) e^{\sigma h(x)} dx \tag{15.6.11}$$

where all the quantities are real. At large values of a positive parameter σ, integral (15.6.11) may be estimated by the Laplace method, whose essence is that if the function $h(x)$ has a maximum on the line (a, b), then at large values of σ this maximum will be very prominent, and the vicinity of a maximum point will give the basic contribution to the value of the integral. If the function $h(x)$ has some maxima, then the integration interval in (15.6.11) may be divided into the finite number of intervals, so that the function $h(x)$ takes a maximum value only at one of the endpoints of each interval (for example, on the left end) and does not achieve the maximum value at other points. It is therefore sufficient to restrict oneself to the case in which the function $h(x)$ in integral of (15.6.11) has the only maximum at $x = a$, i.e., $h(x) < h(a)$ at all $a < x \leq b$. Let the function $h(x)$ have a continuous derivative of the second order. At a maximum point $h'(a) = 0$ and $h''(a) < 0$, there exists such a sufficiently small number $\gamma > 0$, that the derivative $h'(x) < 0$ at $a < x < a + \gamma$. At $\sigma \to +\infty$ we have, according to the main principle of the Laplace method,

$$F(\sigma) \simeq \int_a^{a+\gamma} \varphi(x) e^{\sigma h(x)} dx. \tag{15.6.12}$$

We assume also that the function $\varphi(x)$ is continuous and introduce a new variable of integration y instead of x according to the ratio

$$h(a) - h(x) = y^2. \qquad (15.6.13)$$

Then, (15.6.12) may be rewritten

$$F(\sigma) \simeq -2e^{\sigma h(a)} \int_0^Y y \frac{\varphi[x(y)]}{h'[x(y)]} e^{-\sigma y^2} dy, \qquad (15.6.14)$$

where

$$Y = (h(a) - h(a+\gamma))^{1/2} > 0.$$

As already mentioned, only the value of the integrand in the vicinity of a point $x = a$ (i.e., $y = 0$) is essential in Eq. (15.6.14). In (15.6.14), $\varphi(x)$ may be approximately replaced by $\varphi(a)$ under the integration sign, and $y(x)/h'(x)$ by

$$\lim_{x\to a+0} y(x)/h'(x) = \lim_{x\to a+0} \frac{(h(a) - h(x))^{1/2}}{h'(x)}$$

$$= \lim_{x\to a+0} \frac{\left(-\dfrac{h''(a)}{2}(x-a)^2\right)^{1/2}}{h''(a)(x-a)}$$

$$= -\frac{1}{(-2h''(a))^{1/2}}.$$

Taking into account these results, relation (15.6.14) may now be written

$$F(\sigma) \simeq (2/-h''(a))^{1/2}\varphi(a) \exp[\sigma h(a)] \int_0^Y \exp[-\sigma y^2\, dy]. \qquad (15.6.15)$$

At $\sigma \to +\infty$, only the values of y close to 0 contribute to the integral in formula (15.6.15). It is therefore possible to replace the upper limit of integration Y by infinity without error. Finally we obtain an asymptotic estimation of integral (15.6.11) for $\sigma \to +\infty$ provided that $h(x)$ has maximum at $x = a$

$$F(\sigma) \simeq \{\pi/[-2\sigma h''(a)]\}^{1/2} e^{\sigma h(a)} \varphi(a). \qquad (15.6.16)$$

By means of similar considerations, it is also possible to obtain the subsequent terms of the asymptotic expansion $F(\sigma)$. We present a complete asymptotic expansion only for one important particular case of integral (15.6.11) when $h(x) = -x^\alpha$ $(\alpha > 0)$, $a = 0$, and $0 < b \le \infty$. If $\int_0^b |\varphi(x)| \exp[-\sigma_0 x^\alpha]\, dx$ also converges for some σ_0, then the integral

$$F(\sigma) = \int_0^b \varphi(x) \exp[-\sigma x^\alpha]\, dx \qquad (15.6.17)$$

has an asymptotic expansion

$$F(\sigma) \simeq \sum_{n=0}^{\infty} \frac{C_n}{\alpha} \Gamma\left(\frac{n+\beta+1}{\alpha}\right) \sigma^{-(n+\beta+1)/\alpha}, \qquad (15.6.18)$$

where $\beta > -1$ and the coefficients C_n are determined from the expansion $\varphi(x)$ into a power series such that

$$\varphi(x) = \sum_{n=0}^{\infty} C_n x^{n+\beta},$$

which is assumed to be converging at $|x| < R > 0$.

The theorem of the asymptotic expansion of the Laplace transform directly follows from the estimate of (15.6.18) for integral (15.6.17) at $\alpha = 1$, $b = \infty$ and from replacing notation $x \to \tau$, $\sigma \to s$. If the integral

$$F(s) = \int_0^\infty e^{-s\tau} f(\tau)\, d\tau$$

converges anywhere and the original function $f(\tau)$ can be expanded near a point $\tau = 0$ into a converging series of the form

$$f(\tau) = \sum_{n=0}^{\infty} C_n \tau^{\lambda_n} \quad (-1 < \lambda_0 < \lambda_1 < \cdots),$$

then the transform $F(s)$ has an asymptotic expansion at $s \to +\infty$ such that

$$F(s) \simeq \sum_{n=0}^{\infty} C_n (\Gamma(\lambda_n + 1)/s^{\lambda_n + 1}). \qquad (15.6.19)$$

It would be more advisable to use the theorem which would allow us to judge the asymptotic behavior of the original function according to some expansion of the transform. Such a theorem will now be presented. The integral representation for a heat flux density through the semispace surface, whose density or heat capacity linearly increases with a removal from the surface (see (15.5.33)–(15.5.37)) may be another example of the asymptotic estimation obtained for the integral of the form (15.6.17). It is shown that the heat flux is proportional to the integral

$$F(\xi) = \frac{3\sqrt{3}}{2\pi} \int_0^\infty \frac{\exp[-\xi z^3]}{z^2 + z + 1}\, dz. \qquad (15.6.20)$$

At $\xi \to \infty$ (large time values), we obtain from (15.6.18) ($\beta = 0$, $\alpha = 3$, $\varphi(z) = 1/(z^2 + z + 1) \leq 1$ at $0 \leq z < \infty$)

$$F(\xi) \simeq \frac{\sqrt{3}}{2\pi} \sum_{n=0}^{\infty} \Gamma\left(\frac{n+1}{3}\right) \frac{C_n}{\xi^{(n+1)/3}}, \qquad (15.6.21)$$

where C_n are the expansion coefficients

$$1/(z^2 + z + 1) = \sum_{n=0}^{\infty} C_n z^n \quad (C_0 = 1,\ C_1 = -1,\ C_2 = 0,\ C_3 = 1,\ \ldots).$$

The last power series has a convergence radius different from zero (the convergence radius of this series is equal to the distance to the nearest singularity of the function $1/(z^2 + z + 1)$ at $z = 0$, i.e., it is easy to see that $R_{cx} = 1$). From the estimate of (15.6.21) it follows that at $\tau \to \infty$, the heat flux decreases inversely proportional to a cubic root of time.

Last, we consider the estimation of the integrals of form (15.6.11). The asymptotic estimation which is frequently met in analysis of the function.

$$\Gamma(\sigma + 1) = \int_0^\infty e^{-x} x^\sigma \, dx \tag{15.6.22}$$

is of practical interest at $\sigma \to +\infty$.

Starting with transforming (15.6.22) to the integral of form (15.6.11), we then introduce a new variable of integration $x = \sigma z$, and (15.6.22) can be rewritten

$$\begin{aligned}\Gamma(\sigma + 1) &= \sigma^{\sigma+1} \int_0^\infty z^\sigma \exp[-\sigma z] \, dz \\ &= \sigma^{\sigma+1} \int_0^\infty \exp[-\sigma(z - \ln z)] \, dz \\ &= \sigma^{\sigma+1}[J_1(\sigma) + J_2(\sigma)],\end{aligned} \tag{15.6.23}$$

where

$$J_1(\sigma) = \int_0^1 \exp[-\sigma(z - \ln z)] \, dz$$

and

$$J_2(\sigma) = \int_1^\infty \exp[-\sigma(z - \ln z)] \, dz.$$

The integral $J_2(\sigma)$ belongs to that of (15.6.11). We have $\varphi(z) \equiv 1$, $a = 1$, $h(z) = -(z - \ln z)$, $h'(1) = 0$, $h(1) = -1$, and $h''(1) = -1 < 0$. According to formula (15.6.16), we have an estimate for $J_2(\sigma)$ at $\sigma \to +\infty$ as

$$J_2(\sigma) \simeq (\pi/2\sigma)^{1/2} e^{-\sigma}. \tag{15.6.24}$$

We now transform the integral $J_1(\sigma)$ by introducing a new variable of integration $z = 1/u$, then $J_1(\sigma)$ assumes the form

$$J_1(\sigma) = \int_1^\infty \exp[-\sigma\{(1/u) + \ln u\}] \, du/u^2.$$

In such a form, it is obvious that $J_1(\sigma)$ is also the integral of form (15.6.11) where $h(u) = -\{1/u + \ln u\}$, and the maximum $h(u)$ as necessary is achieved on the lower limit at $u = 1$ (point $u = \infty$ where also $h'(\infty) = 0$ is not a maximum). Further, taking into account that for $J_1(\sigma)$, $\varphi(u) = 1/u^2$, $h(1) = -1$, and $h''(1) = -1 < 0$ we obtain from formula (15.6.16)

$$J_1(\sigma) \simeq (\pi/2\sigma)^{1/2}e^{-\sigma} \tag{15.6.25}$$

at $\sigma \to +\infty$.

Substituting estimates (15.6.24) and (15.6.25) into (15.6.23) we obtain the first term of the asymptotic expansion for the gamma function

$$\Gamma(\sigma + 1) \simeq (2\pi)^{1/2}\sigma^{\sigma+1/2}e^{-\sigma} \tag{15.6.26}$$

that represents the known Stirling formula which was used in the previous chapter (when deriving the real inversion of the Laplace transform).

The so-called method of steepest descent of the asymptotic estimations of the contour integrals of the form

$$J(\sigma) = \int_C e^{\sigma h(z)}\varphi(z)\, dz \tag{15.6.27}$$

is closely connected with the Laplace method which serves for obtaining the asymptotic estimations of the integrals with respect to a real variable. The contour integrals of such a form are closely connected with the Laplace transform. At large values of the parameter ($\operatorname{Re}\sigma \to +\infty$), the integrand in (15.6.27) will vary very quickly due to the presence of the multiplier $\exp[i\operatorname{Im}[\sigma h(z)]]$, which varies with a frequency proportional to σ. These variations make it impossible to directly calculate integral (15.6.27). It is therefore natural to try to deform the integration contour without intersecting singularities and, consequently, without changing the value of the integral according to the Cauchy theorem. Thus the variations of the integrand might be reduced to a minimum, especially to those portions which make a greater contribution to the integral. The value of the integral will be obviously determined by that portion of the integration contour C where the modulus $|\exp[\sigma h(z)]| = \exp[\operatorname{Re}[\sigma h(z)]]$ takes the greatest possible values.

To avoid the variations of the integrand, we change the integration path so that

$$\exp[i\operatorname{Im}[\sigma h(z)]] = \text{const} \tag{15.6.28}$$

on the portion where $\operatorname{Re}[\sigma h(z)]$ takes the greatest value.

At this point, z_0, we have

$$h'(z_0) = 0. \tag{15.6.29}$$

This point cannot be a maximum or minimum of the function $\operatorname{Re} h(z)$, because, according to the Cauchy–Riemann conditions (15.1.4), the real (and imaginary) parts of the analytic functions satisfy the Laplace equation ($z = x + iy$):

$$\frac{\partial^2 \operatorname{Re} h(z)}{\partial x^2} + \frac{\partial^2 \operatorname{Re} h(z)}{\partial y^2} = 0$$

and, as is known, such functions cannot have maxima and minima. Consequently, the point z_0 will be a saddle point. The direction of the path of integration at the saddle point should be chosen according to (15.6.28) and is determined by the equation

$$\operatorname{Im}[\sigma h(z)] = \operatorname{Im}[\sigma h(z_0)]. \tag{15.6.30}$$

Contour integral (15.6.30) passing through z_0 may be already estimated according to the Laplace method, and besides large at values of σ, the contribution to the integral will be defined only by the nearest vicinity of z_0. Repeating the above considerations used for estimating integral (15.6.11), we obtain for the first term of asymptotic expansion (15.6.27)

$$J(\sigma) \simeq \left(\frac{2\pi}{\sigma e^{i\pi} h''(z_0)}\right)^{1/2} e^{\sigma h(z_0)} \varphi(z_0). \tag{15.6.31}$$

Consider one example. The function $\exp[(\sigma/2)\{z - (1/z)\}]$ in the vicinity of its singularity $z = 0$ may be expanded into the Laurent series

$$\exp[(\sigma/2)\{z - (1/z)\}] = \sum_{n=-\infty}^{\infty} C_n(\sigma) z^n, \tag{15.6.32}$$

where $C_n(\sigma)$ are the coefficients of the Laurent expansion. These are determined by formulas (15.3.17) and (15.3.17′). In formula (15.3.17′), changing the direction of the by-pass on the contour into the opposite one, replacing the subscript n, where n covers values 1, 2,... by a subscript n where n assumes values -1, -2, and taking into account that any contour covering a point $z = 0$ may be chosen as C_{R_1} (and C_{R_2}), it is possible to write a common expression for the coefficients $C_n(\sigma)$ as

$$C_n(\sigma) = (1/2\pi i) \int_C \exp[(\sigma/2)\{z - (1/z)\}]\, dz/z^{n+1}. \tag{15.6.33}$$

These coefficients may be directly obtained in terms of a power series with respect to σ by multiplying the expansion

$$e^{\sigma z/2} = \sum_{n=0}^{\infty} (1/n!)(\sigma/2)^n z^n$$

and

$$e^{-\sigma/2z} = \sum_{n=0}^{\infty} \frac{(-1)^n}{n!} \left(\frac{\sigma}{2}\right)^n \frac{1}{z^n}.$$

It is then easy to find that at z^m $(m = 0, 1, \ldots)$ the coefficient is equal to

$$C_m(\sigma) = \sum_{n=0}^{\infty} \frac{(-1)^n}{(m+n)!n!} \left(\frac{\sigma}{2}\right)^{m+2n} \tag{15.6.34}$$

and at $1/z^m$

$$C_{-m}(\sigma) = (-1)^m C_m(\sigma) \quad (m = 1, 2, \ldots). \tag{15.6.35}$$

However, formulas (15.6.34) and (15.6.35) coincide with the definition of the Bessel functions of the first kind of the integer order (see Appendix 1). Thus, formula (15.6.33) gives an integral representation of the function $J_n(\sigma)$. As has been already mentioned, in formula (15.6.33) the contour C should contain only a point $z = 0$. We choose the circle of a single radius as C. Then the integral representation of the Bessel function $J_n(\sigma)$ is written as

$$J_n(\sigma) = \frac{1}{2\pi i} \int_{|z|=1} \exp\left[\frac{\sigma}{2}\left(z - \frac{1}{z}\right)\right] \frac{dz}{z^{n+1}}. \tag{15.6.33a}$$

From this integral representation, the asymptotic behavior of the function $J_n(\sigma)$ at $\sigma \to +\infty$ may be determined by the method of steepest descent. At $\varphi(z) = 1/z^{n+1}$, $h(z) = \frac{1}{2}\{z - (1/z)\}$, the saddle points are $z_1 = i = e^{i\pi/2}$ and $z_2 = -i = e^{-i\pi/2}$. The integration path for each of the saddle points is determined by the equations $\operatorname{Im}\{z - (1/z)\} = \pm 2i$. Summing up the contributions of both saddle points we obtain from (15.6.31)

$$J_n(\sigma) \simeq \frac{1}{(2\pi\sigma)^{1/2}} \left(\exp\left[i\left(\sigma - \frac{\pi}{2}n - \frac{\pi}{4}\right)\right] + \exp\left[-i\left(\sigma - \frac{\pi}{2}n - \frac{\pi}{4}\right)\right]\right)$$

$$= \left(\frac{2}{\pi\sigma}\right)^{1/2} \cos\left(\sigma - \frac{\pi}{2}n - \frac{\pi}{4}\right). \tag{15.6.36}$$

It would be most valuable for us to use the theorems which, according to the known transform $F(s)$ proceeding from its analytic properties, i.e., according to the position and nature of singularities, would allow determination of the asymptotic behavior of the original function $f(\tau)$ at $\tau \to \infty$ without calculating the appropriate contour integral.

First, we consider the cases when the function $F(s)$ has points of branching and poles and then formulate the theorem for the general case.

Thus, we let $F(s)$ have a branching point at $s = 0$ and no other singularities in the finite part of a plane s. Further, at $|s| \to \infty$, we let $F(s)$ tend uniformly to zero in a left half-plane. If $F(s)$ can be expanded into a series

$$F(s) = \sum_{n=0}^{\infty} C_n s^{\lambda_n} \quad (\lambda_0 < \lambda_1 < \lambda_2 < \cdots), \tag{15.6.37}$$

then

$$f(\tau) \simeq \sum_{n=0}^{\infty} \frac{C_n}{\Gamma(-\lambda_n)} \frac{1}{\tau^{\lambda_n+1}}. \tag{15.6.38}$$

For proof, we note that the integral

$$(1/2\pi i) \int_C e^{s\tau} F(s)\, ds = 0, \tag{15.6.39}$$

where C is the contour presented in Fig. 15.14. It consists of a straight line passing to the right and parallel to the imaginary axis from the point $a - iR$ to $a + iR$ $(a > 0)$, the left half-circle with the radius R, upper and lower edges of the slot, and the inner circle with the radius ϱ around the point $s = 0$. From the Jordan lemma it, directly follows that the integral with respect to the half-circle at $R \to \infty$ tends to zero. Thus, equality (15.6.39) may be written

$$f(\tau) = -(1/2\pi i) \int_L e^{s\tau} F(s)\, ds, \tag{15.6.40}$$

where L is the contour consisting of the upper and lower edges along the negative half-axis between points $s = 0$ and the point n at infinity on the circle of the radius ϱ around the point $s = 0$. The direction of the bypass is given in Fig. 15.14.

Substituting (15.6.37) for (15.6.40) and assuming that is is possible to change the order where summation and integration are performed (the validity of such a change may not hold true with some additional conditions which are performed as a rule in practice) we have

$$f(\tau) = -\sum_{n=0}^{\infty} C_n (1/2\pi i) \int_L e^{s\tau} s^{\lambda_n}\, ds.$$

Introducing a new variable $s\tau = s'$ $(\tau > 0)$, i.e., transforming the similarity of the plane s (the circle around the point $s = 0$ is transformed into a concentric circle of the radius $\varrho\tau$, etc.) we obtain

$$f(\tau) = -\sum_{n=0}^{\infty} C_n \tau^{-\lambda_n - 1} - (1/2\pi i) \int_L s'^{\lambda_n} e^{s'} \, ds' . \tag{15.6.41}$$

Consider the integral

$$J(z) = (1/2\pi i) \int_{L'} s^{-z} e^{s} \, ds , \tag{15.6.42}$$

where the contour L' coincides with L but it is bypassed in the opposite direction (see Fig. 15.15). If $z < 1$, then the integral with respect to the circle

$$(1/2\pi) \int_{-\pi}^{\pi} \varrho^{1-z} \exp[\varrho \exp[i\varphi] + i(1-z)\varphi] \, d\varphi$$

tends to zero at $\varrho \to 0$. $J(z)$ is expressed through the integrals with respect

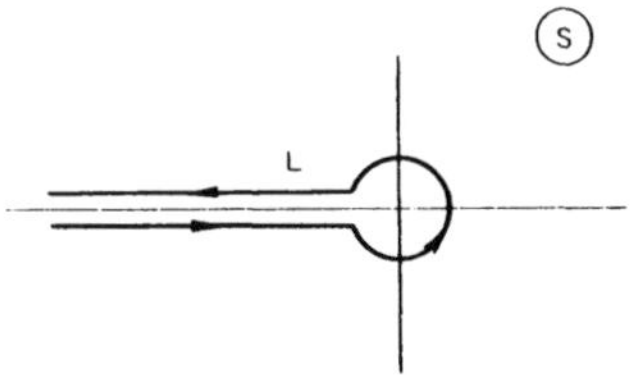

Fig. 15.15. Contour for the integral representation of the γ function.

to the edges (on the upper edge $s = ue^{i\pi}$ and $s^{-z} = x^{-z}e^{-i\pi z}$ and on the lower edge $s = ue^{-i\pi}$ and $s^{-z} = x^{-z}e^{i\pi z}$)

$$\begin{aligned} J(z) &= \frac{e^{\pi i z}}{2\pi i} \int_0^\infty x^{-z} e^{-x} \, dx - \frac{e^{-\pi i z}}{2\pi i} \int_0^\infty x^{-z} e^{-x} \, dx \\ &= \frac{\sin \pi z}{\pi} \int_0^\infty x^{-z} e^{-x} \, dx \\ &= \frac{\sin \pi z}{\pi} \Gamma(1-z) , \end{aligned} \tag{15.6.43}$$

since

$$\Gamma(u) = \int_0^\infty e^{-x} x^{u-1} \, dx$$

at $\operatorname{Re} u > 0$.

The known functional relation

$$\Gamma(z)\Gamma(1-z) = \pi/\sin \pi z \tag{15.6.44}$$

is valid for the gamma function, therefore, formula (15.6.43) may be written as

$$J(z) = 1/\Gamma(z),$$

i.e., for the function $\Gamma(z)$, the integral representation

$$1/\Gamma(z) = (1/2\pi i) \int_{L'} s^{-z} e^{s} \, ds \tag{15.6.45}$$

is valid, which is proved for $z < 1$. However, it may be analytically continued to any z. With the help of integral representation (15.6.41) we obtain expansion (15.6.38).

We also note that if a point of branching is not at a point different from zero but at a point s_0, then according to the displacement theory given in the previous chapter, relation (15.6.38) will be again valid if its right-hand side is multiplied by $e^{s_0 \tau}$.

The theorem proved is called the rule of fractional exponents, since if any of the numbers λ_n is zero or an integer positive number, then the appropriate term in expansion (15.6.38) vanishes as

$$1/\Gamma(-n) = 0, \quad \text{at } n = 0, 1, \ldots.$$

This directly follows from relation (15.6.44) if it is rewritten as

$$1/\Gamma(1 - z) = (1/\pi)\Gamma(z) \sin \pi z$$

and assuming here $z = 1, 2, \ldots$.

Thus, asymptotic expansion (15.6.38) contains only fractional powers of $1/\tau$.

As an illustration, consider the transform

$$F(s) = e^{-\gamma\sqrt{s}}/s \quad (\gamma > 0). \tag{15.6.46}$$

The function $F(s)$ has a point of branching $s = 0$ and satisfies all the conditions, under which the rule of fractional exponents is derived if the branch is chosen for which $\operatorname{Re}\sqrt{s} \geq 0$. Expanding the exponent in (15.6.46) into series we have

$$F(s) = \sum_{n=0}^{\infty} \{(-\gamma)^n/n!\} s^{n/2-1}, \tag{15.6.47}$$

hence, according to (15.6.38)

$$\begin{aligned} f(\tau) &= \sum_{n=0}^{\infty} \frac{(-\gamma)^n}{n!} \frac{1}{\Gamma(1 - \frac{1}{2}n)} \frac{1}{\tau^{n/2}} \\ &= 1 + \sum_{n=1}^{\infty} \frac{(-\gamma)^n}{n!} \frac{1}{\Gamma(1 - \frac{1}{2}n)} \frac{1}{\tau^{n/2}}. \end{aligned}$$

In the sum $\sum_{n=1}^{\infty}$, all the terms with even n vanish. Assuming, therefore, $n = 2m + 1$, we obtain

$$f(\tau) = 1 - \sum_{m=0}^{\infty} \frac{\gamma^{2m+1}}{(2m+1)!\Gamma(\frac{1}{2} - m)} \frac{1}{\tau^{m+1/2}}.$$

From the gamma function theory, it is known that

$$\Gamma(\tfrac{1}{2} - m) = (-1)^m (2^{2m} m!/(2m)!) \sqrt{\pi},$$

therefore

$$f(\tau) = 1 - \frac{2}{\sqrt{\pi}} \sum_{m=0}^{\infty} \frac{(-1)^m}{(2m+1)!\, m!} \left(\frac{\gamma}{2\sqrt{\tau}}\right)^{2m+1}. \qquad (15.6.48)$$

This, as expected, coincides with the expansion of the function erfc $(\gamma/2\sqrt{\tau})$ at small values of $\gamma/2\sqrt{\tau}$.

We now consider the case when poles at points $s_1, s_2, \ldots s_N$ (see Fig. 15.16) are singularities of the transform $F(s)$. Obviously, calculating the integral $(1/2\pi i) \int_C e^{s\tau} F(s)\, ds$ with respect to the contour given in Fig. 15.16,

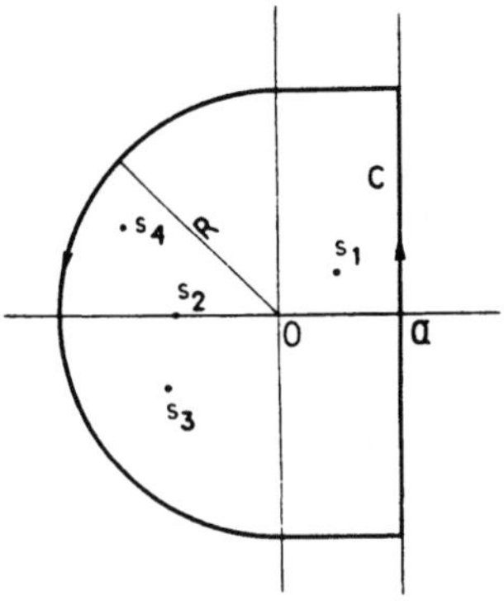

Fig. 15.16. Location of the poles of the functions $F(s)$.

and taking into account that the integral with respect to the left circle at $R \to \infty$ vanishes according to the Jordan lemma, we have ($a > \operatorname{Re} s_k$; $k = 1, 2, \ldots, N$)

$$f(\tau) = (1/2\pi i) \int_{a-i\infty}^{a+i\infty} e^{s\tau} F(s)\, ds$$

$$= \sum_{k=1}^{N} \{\text{residues}[e^{s\tau} F(s)] b s_k\}. \qquad (15.6.49)$$

If the value of τ is large, then it is obviously necessary to leave only the term with the greatest value of Re s_k in the right-hand side of (15.6.49), since due to the presence of the multiplier $e^{s_k\tau}$, it will be large compared the remainder. If there are some such terms, whose Re s_k are equal (and Im s_k are different), then the sum of these terms should be taken. For example, the function

$$F(s) = 1/s^2(s-1)$$

has a twofold pole at $s = 0$ and simple pole at $s = 1$. At $\tau \to \infty$ we have

$$f(\tau) \simeq e^{\tau}$$

then

$$f(\tau) = e^{\tau} - \tau - 1 .$$

Thus, for defining the asymptotic determination of the function at $\tau \to \infty$ if only its transform is known, there is no need to perform contour integration. This asymptotic behavior of the function will be determined by a absolutely correct singularity (singularities) of the transform.

Now the next general theorem is quite obvious. Let $F(s)$ have singularities —poles and points of branching; the function $F(s)$ at $|s| \to \infty$ tends uniformly to zero in a half-plane Re $s < 0$; the number of singularities s_k with the greatest value of Re s_k is finite ($k = 1, 2, \ldots, N$) and if the expansion of $F(s)$ in the vicinity of point $s = s_k$ is given by the series

$$F(s) = \sum_{n=0}^{\infty} C_n^{(k)}(s - s_k)^{\lambda_n^{(k)}} \quad (\lambda_0^{(k)} < \lambda_1^{(k)} < \cdots), \tag{15.6.50}$$

then the asymptotic expansion of $f(\tau)$ will be of the form

$$f(\tau) \simeq \sum_{k=1}^{N} e^{s_k\tau} \sum_{n=0}^{\infty} \frac{C_n^{(k)}}{\Gamma(-\lambda_n^{(k)})\tau^{\lambda_n^{(k)}+1}}, \tag{15.6.51}$$

where $1/\Gamma(-\lambda_n^{(k)}) = 0$ if $\lambda_n^{(k)} = 0, 1, 2, \ldots$.

We shall illustrate the application of this important theorem by some examples.

First, the exact solution of a great number of problems is obtained with the help of the Laplace transformation presented in the previous chapters. We propose that the reader obtain as an exercise the asympmtotic expansion of the solution at $\tau \to \infty$ directly from the transform of the solution, thus determining the most correct singularities of this transform, and prove that the obtained asymptotic expansion of the solution will coincide with the path of the transition $\tau \to \infty$ in the final solution.

Further, we shall consider the problem of a temperature field in a half-space $x \geq 0$ heated by an impulse periodic heat flux (see Fig. 15.17)

$$q(\tau) = q_0 \begin{cases} 1 & \text{at } 0 < \tau < \tau_0, \\ 0 & \text{at } \tau_0 < \tau < 2\tau_0, \end{cases} \tag{15.6.52}$$

$$q(\tau + 2n\tau_0) = q(\tau).$$

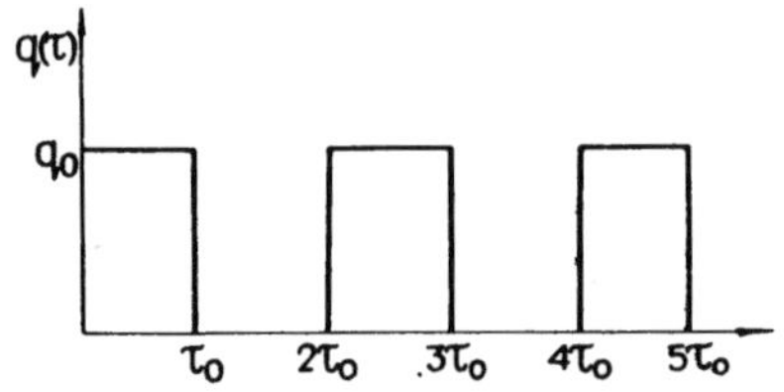

Fig. 15.17. Impulse periodic heat flux q_0.

We formulate the problem

$$\frac{\partial t(x, \tau)}{\partial \tau} = a \frac{\partial^2 t(x, \tau)}{\partial x^2} \quad (0 < x < \infty), \tag{15.6.53}$$

$$t(x, 0) = 0, \qquad \partial t(\infty, \tau)/\partial x = 0, \qquad -\lambda\{\partial t(0, \tau)/\partial x\} = q(\tau).$$

Hence, it is easy to obtain that

$$T(x, s) = (1/\lambda)(a/s)^{1/2} \exp[-(s/a)^{1/2}x] q_L(s). \tag{15.6.54}$$

We now calculate

$$q_L(s) = \int_0^\infty e^{-s\tau} q(\tau)\, d\tau.$$

Let $f(\tau)$ be an arbitrary periodic function with a period τ_i, whose Laplace transform exists. We have

$$f(\tau) = f(\tau + n\tau_1) \quad (n = 1, 2, \ldots),$$

$$\begin{aligned} F(s) &= \int_0^\infty f(\tau) e^{-s\tau}\, d\tau \\ &= \int_0^{\tau_1} f(\tau) e^{-s\tau}\, d\tau = \int_{\tau_1}^{2\tau_1} f(\tau) e^{-s\tau}\, d\tau + \cdots \\ &= \sum_{m=0}^{\infty} e^{-ms\tau_1} \int_0^{\tau_1} f(\tau) e^{-s\tau}\, d\tau \\ &= \left\{ \int_0^{\tau_1} f(\tau) e^{-s\tau}\, d\tau \right\} \Big/ (1 - e^{-s\tau_1}). \end{aligned} \tag{15.6.55}$$

Applying relation (15.6.55) to the heat flux $q(\tau)$ with the period 2τ we obtain

$$q_L(s) = \frac{q_0 \int_0^{\tau_0} e^{-s\tau}\, d\tau}{1 - e^{-2s\tau_0}} = \frac{q_0}{s} \frac{1 - e^{-s\tau_0}}{1 - e^{-2s\tau_0}}.$$

Substituting the result obtained into (15.6.54), we find

$$T(x, s) = \frac{q_0\sqrt{a}}{\lambda} \frac{\exp[-(s/a)^{1/2}x]}{s^{3/2}} \frac{1 - e^{-s\tau_0}}{1 - e^{-2s\tau_0}}. \quad (15.6.56)$$

The law, according to which the temperature of a half-space varies at $\tau \to \infty$, is not obvious. However, it is easily determined with the help of relations (15.6.50) and (15.6.51). The point of branching at $s = 0$, the infinite number of simple poles lying on an imaginary axis at points

$$s_m = \frac{2m + 1}{\tau_0} \pi i \quad (m = 0, \pm 1, \pm 2, \ldots), \quad (15.6.57)$$

are singularities of $T(x, s)$. All the singularities have $\operatorname{Re} s = 0$. Expanding the transform (15.6.56) near the point of branching $s = 0$, we have

$$T(x, s) \approx \frac{q_0\sqrt{a}}{2\lambda} \frac{1}{s^{3/2}} + \cdots. \quad (15.6.58)$$

Hence, according to (15.6.50) and (15.6.51) we find that at $\tau \to \infty$

$$t(x, \tau) \simeq \frac{q_0}{\lambda} \left(\frac{a}{\pi}\tau\right)^{1/2} \quad (15.6.59)$$

The contribution of the poles of (15.6.57) into the asymptotic expansion of the function $t(x, \tau)$ should not be taken into account although $\operatorname{Re} s_m = 0$, as in case of the point of branching. This is explained by the fact that each of these poles gives the term periodically dependent on time as

$$\exp[i(2m + 1)\pi(\tau/\tau_0)]$$

and their sum, also representing the periodic and bounded function τ, will be small at $\tau \to \infty$ compared to the main term of (15.6.59). Thus, the temperature of a half-space, on whose surface the heat flux varying according to law (15.6.52) drops, will increase with $\sim\sqrt{\tau}$ at large values of τ.

Consider one more example. Find the law, according to which the boundary temperature $x = 0$ of a semi-infinite body is related to the heat flux density $q(\tau)$ through this same surface. Applying the Laplace transformation and the convolution theorem it is easy to express the temperature $t(x, \tau)$ through the heat flux density $q(\tau)$. We have, assuming that the initial temperature is equal to zero,

$$t(x, \tau) = \frac{1}{\lambda}\left(\frac{a}{\pi}\right)^{1/2}\int_0^\tau \frac{\exp[-x^2/4a(\tau-\xi)]}{(\tau-\xi)^{1/2}}\, q(\xi)\, d\xi. \tag{15.6.59a}$$

Assuming in the latter equation $x \to 0$, we find the relation between the surface temperature $x = 0$ and the heat flux density through this surface as where the notation $t(0, \tau) = \theta(\tau)$ is introduced.

$$\theta(\tau) = \frac{1}{\lambda}\left(\frac{a}{\pi}\right)^{1/2}\int_0^\tau \frac{q(\xi)}{(\tau-\xi)^{1/2}}\, d\xi \tag{15.6.60}$$

If the heat flux density $q(\tau)$ is prescribed, then the determination of the surface temperature $\theta(\tau)$ is reduced to quadrature. If $\theta(\tau)$ is known and $q(\tau)$ is unknown, then (15.6.60) is an integral equation for $q(\tau)$.

For example, let the surface temperature be prescribed and equal to

$$\theta(\tau) = \theta_0 e^{\gamma\tau}. \tag{15.6.61}$$

Then, solving integral equation (15.6.60) with respect to $q(\tau)$, we obtain with the help of the Laplace transform

$$q_L(s) = \theta_0(\lambda/\sqrt{a}\,)\sqrt{s}\,/(s-\gamma). \tag{15.6.62}$$

The function $\sqrt{s}\,/(s-\gamma)$ has two singularities: a simple pole at $s = \gamma$ and a point of branching $s = 0$. If we are only interested in the behavior of the function $q(\tau)$ at large values of τ, then it may be obtained according to the above theorem. The asymptotic behavior of $q(\tau)$ will be determined by the most correct singularity $q_L(s)$. Consider two cases.

First, if $\gamma = k^2 > 0$, then the pole $s = k^2$ will be a right singularity, according to formulas (15.6.50) and (15.6.51) we have:

$$q_L(s) \approx \theta_0(\lambda/\sqrt{a}\,)[k/(s-k^2)] + \cdots$$

and

$$q(\tau) \simeq \theta_0(\lambda/\sqrt{a}\,)ke^{k^2\tau} + \cdots, \tag{15.6.63}$$

i.e., the heat flux density increases exponentially with time.

If $\gamma = -k^2 < 0$, then the point of branching $s = 0$ will be the most correct singularity of (15.6.62). Expansion (15.6.62) (at $\gamma = -k^2$) in the vicinity of the point $s = 0$ has the form

$$q_L(s) = \theta_0 \frac{\lambda}{\sqrt{a}} \sum_{n=0}^{\infty} \frac{(-1)^n}{k^{2(n+1)}} s^{n+1/2}.$$

Hence, according to formulas (15.6.50) and (15.6.51) we have

$$q(\tau) \simeq \theta_0 \frac{\lambda}{\sqrt{a}} \sum_{n=0}^{\infty} \frac{(-1)^n}{k^{2(n+1)}} \frac{1}{\Gamma(-n-\frac{1}{2})} \frac{1}{\tau^{n+3/2}}. \qquad (15.6.64)$$

It is easy to see that the expansions (15.6.63) and (15.6.64) coincide with the asymptotic expansions which may be determined from the exact inversions of the functions $\sqrt{s}/(s-k^2)$ and $\sqrt{s}/(s+k^2)$ given in Appendix 5.

Finally, we may note once more that the last theorem allows us to immediately judge the behavior of the original function at large values of τ according to the position of singularities of a transform in a complex plane without any calculations. In reality, if among singularities of a transform there are such singularities, for which $\operatorname{Re} s_m > 0$, then the original function will increase with $\tau \to \infty$ exponentially. Conversely, if even for the correct singularity $\operatorname{Re} s_m < 0$, then the original function exponentially decreases with $\tau \to \infty$.

The theorem on the asymptotic expansion of the original function with respect to the known expansion of a transform is especially important in those cases when the latter has a very complex form, and the appropriate contour integral cannot be calculated.

The above methods of asymptotic estimates allow us to obtain the information on the behavior of the solution, which is very valuable if the contour integral (with the knowledge of which the solution of the problem is found) cannot be calculated. However, we note that the proposed methods of the asymptotic estimates are used at a large value of some variable or parameter. If time is such a variable, then it would be interesting, together with the above methods of determination of the asymptotic behavior of a time function at $\tau \to \infty$ with respect to the analytic properties of the Laplace transform, to also have the possibility of studying the behavior of the solution at small time values.

We shall now consider one way.

It is not difficult to see that the solutions of one-dimensional heat conduction problems, with a space coordinate varying within a finite interval under the conditions of the first and second kinds at the ends of this interval, may be expressed through linear combinations, integrals, and derivatives of the next series which play essential role in the heat conduction theory and in other sections of the analysis and which are called the theta function:

$$\begin{aligned}
\theta_1(\tilde{x}, i\tilde{\tau}) &= 2 \sum_{n=0}^{\infty} (-1)^n \exp\left[-\pi\left(n+\frac{1}{2}\right)^2 \tilde{\tau}\right] \sin(2n+1)\pi\tilde{x}, \\
\theta_2(\tilde{x}, i\tilde{\tau}) &= 2 \sum_{n=0}^{\infty} \exp\left[-\pi\left(n+\frac{1}{2}\right)^2 \tilde{\tau}\right] \cos(2n+1)\pi\tilde{x}, \\
\theta_3(\tilde{x}, i\tilde{\tau}) &= 1 + 2 \sum_{n=1}^{\infty} \exp[-\pi n^2 \tilde{\tau}] \cos 2n\pi\tilde{x}, \\
\theta_0(\tilde{x}, i\tilde{\tau}) &= 1 + 2 \sum_{n=1}^{\infty} (-1)^n \exp[-\pi n^2 \tilde{\tau}] \cos 2n\pi\tilde{x} \qquad (\tilde{\tau} > 0).
\end{aligned} \tag{15.6.65}$$

With the help of direct differentiation, it is easy to check that all these series satisfy the heat conduction equation of the form

$$\frac{\partial \theta_i}{\partial \tilde{\tau}} = \frac{1}{4\pi} \frac{\partial^2 \theta_i}{\partial \tilde{x}^2} \quad (i = 1, 2, 3, 0). \tag{15.6.66}$$

Thus, the variables $\tilde{\tau}$ and $\tilde{x}$ are those of the dimensionless time and the coordinate.

The series of form (15.6.65) will converge at large values of τ and vice versa, usually slowly converge at small τ.

It is, however, possible to note a general way of transformation of the function into the quickly converging series of form (15.6.65) for small τ (i.e., transformation of the series themselves and those which follow from them at differentiation and integration).

Consider the series

$$\sum_{n=-\infty}^{\infty} f(2\pi n), \tag{15.6.67}$$

where $f(\tau)$ is a continuous and a continuously differentiable function so that the series

$$\sum_{n=-\infty}^{\infty} f(2\pi n + \varphi) \text{ and } \sum_{n=-\infty}^{\infty} f'(2\pi n + \varphi)$$

converge alsolutely and uniformly for all $0 \leq \varphi < 2\pi$. Consequently, the series $\sum_{n=-\infty}^{\infty} f(2\pi n + \varphi)$ may be expanded into the converging Fourier series for the above series of φ.

Thus, we have

$$\sum_{n=-\infty}^{\infty} f(2\pi n + \varphi)$$
$$= (1/2\pi) \sum_{m=-\infty}^{\infty} \left\{e^{im\varphi} \int_0^{2\pi} e^{-im\omega} \left[\sum_{n=-\infty}^{\infty} f(2\pi n + \omega)\right] d\omega\right\}$$
$$= (1/2\pi) \sum_{m=-\infty}^{\infty} e^{im\varphi}\left\{\sum_{n=-\infty}^{\infty}\left[\int_0^{2\pi} f(2n\pi + \omega)e^{-im\omega}\, d\omega\right]\right\}$$
$$= (1/2\pi) \sum_{m=-\infty}^{\infty} e^{im\varphi}\left\{\sum_{n=-\infty}^{\infty}\left[\int_{2\pi n}^{2\pi(n+1)} f(\omega)e^{-im\omega}\, d\omega\right]\right\}$$
$$= (1/2\pi) \sum_{m=-\infty}^{\infty} e^{im\varphi}\left\{\int_{-\infty}^{\infty} f(\omega)e^{-im\omega}\, d\omega\right\}. \tag{15.6.68}$$

If we assume $\varphi = 0$, then equalities (15.6.68) can be written

$$\sum_{n=-\infty}^{\infty} f(2\pi n) = (1/2\pi) \sum_{m=-\infty}^{\infty} \int_{-\infty}^{\infty} f(\omega)e^{-im\omega}\, d\omega. \tag{15.6.69}$$

The above relation, called the Poisson summation formula, has many important applications, particularly for the transformation of series and their summation if the transformed series in the right-hand side appears to be so simple that its sum is known.

Before applying the Poisson formula for transformation of the theta function, we write it in another form by introducing the function $g(n)$ according to the relation

$$f(2\pi n) = g(n). \tag{15.6.70}$$

Then, formula (15.6.69) assumes the form

$$\sum_{n=-\infty}^{\infty} g(n) = \sum_{m=-\infty}^{\infty} \int_{-\infty}^{\infty} g(\omega) \exp[-\,2\pi im\omega]\, d\omega. \tag{15.6.71}$$

For example, we can apply formula (15.6.71) to the function $\theta_3(\tilde{x}, i\tilde{\tau})$, and for this purpose, transform the appropriate series

$$\theta_3(\tilde{x}, i\tilde{\tau}) = 1 + 2 \sum_{n=1}^{\infty} e^{-\pi n^2 \tilde{\tau}} \cos 2n\pi\tilde{x}$$
$$= \sum_{n=-\infty}^{\infty} \exp[-\,\pi n^2 \tilde{\tau}] \cos 2\pi n\tilde{x}$$
$$= \sum_{=-\infty}^{\infty} \exp[-\,\pi n^2 \tilde{\tau} - 2\pi in\tilde{x}]. \tag{15.6.72}$$

Hence, according to the Poisson summation formula, we have

$$\theta_3(\tilde{x}, i\tilde{\tau}) = \sum_{n=-\infty}^{\infty} \exp[-\pi n^2 \tilde{\tau} - 2\pi i n \tilde{x}]$$

$$= \sum_{m=-\infty}^{\infty} \int_{-\infty}^{\infty} \exp[-\pi\omega^2\tilde{\tau} - 2\pi i\omega(\tilde{x} + m)]\, d\omega. \qquad (15.6.73)$$

The integral under the summation sign has been already calculated above. It appears to be equal to

$$\int_{-\infty}^{\infty} \exp[-\pi\omega^2\tilde{\tau} - 2\pi i\omega(\tilde{x} + m)]\, d\omega = (1/\sqrt{\tilde{\tau}}) \exp[-\pi(\tilde{x} + m)^2/\tilde{\tau}]. \qquad (15.6.74)$$

Substituting the value of the integral into formula (15.6.73), we obtain a new representation of the function $\theta_3(\tilde{x}, i\tilde{\tau})$ in the form of the extremely quickly converging series at small $\tilde{\tau}$ ($\tilde{\tau} > 0$):

$$\theta_3(\tilde{x}, i\tilde{\tau}) = (1/\sqrt{\tilde{\tau}}) \exp[-\pi\tilde{x}^2/\tilde{\tau}] \sum_{m=-\infty}^{\infty} \exp[-(\pi m^2/\tilde{\tau}) - (2\pi m\tilde{x}/\tilde{\tau})]. \qquad (15.6.75)$$

Using the determination of (15.6.72), equality (15.6.75) may be written in the form of the functional relation

$$\theta_3(\tilde{x}, i\tilde{\tau}) = (1/\sqrt{\tilde{\tau}}) \exp[-\pi\tilde{x}^2/\tilde{\tau}]\theta_3\{(\tilde{x}/i\tilde{\tau}), (i/\tilde{\tau})\}. \qquad (15.6.76)$$

The similar relations may be also obtained for the remainder theta function. They are of the form

$$\theta_1(\tilde{x}, i\tilde{\tau}) = (1/\sqrt{\tilde{\tau}}) \exp[-\pi\tilde{x}^2/\tilde{\tau}]\theta_1\{(\tilde{x}/i\tilde{\tau}), (i/\tilde{\tau})\},$$

$$\theta_2(\tilde{x}, i\tilde{\tau}) = \frac{1}{\sqrt{\tilde{\tau}}} \exp[-\pi\tilde{x}^2/\tilde{\tau}]\theta_0\{(\tilde{x}/i\tilde{\tau}), (i/\tilde{\tau})\},$$

$$\theta_0(\tilde{x}, i\tilde{\tau}) = 1/\sqrt{\tilde{\tau}} \exp[-\pi\tilde{x}^2/\tilde{\tau}]\theta_2\{(\tilde{x}/i\tilde{\tau}), (i/\tilde{\tau})\}.$$

For the problems with cylindrical symmetry, the asymptotic expansions of the appropriate series at small τ are obtained by Epstein [30].

APPENDIX

1

SOME REFERENCE FORMULAS

Some relations are now presented which are used for the solution of the heat conduction problem.

1. *Series Expansion of Trigonometric Functions*

$$\cos x = 1 - \frac{x^2}{2!} + \frac{x^4}{4!} - \frac{x^6}{6!} + \cdots, \quad |x| < \infty;$$

$$\sin x = \frac{x}{1!} - \frac{x^3}{3!} + \frac{x^5}{5!} - \frac{x^7}{7!} + \cdots, \quad |x| < \infty;$$

$$\tan x = x + \frac{x^3}{3} + \frac{2x^5}{3\cdot 5} + \frac{17x^7}{3^2\cdot 5\cdot 7} + \cdots, \quad |x| < \frac{\pi}{2};$$

$$\cot x = \frac{1}{x} - \frac{x}{3} - \frac{x^3}{3^2\cdot 5} - \frac{2x^5}{3^3\cdot 5\cdot 7} - \cdots, \quad 0 < x < \pi.$$

2. *Expansion of Hyperbolic Functions in Series*

$$\cosh x = 1 + \frac{x^2}{2!} + \frac{x^4}{4!} + \frac{x^6}{6!} + \cdots, \quad |x| < \infty,$$

$$\sinh x = x + \frac{x^3}{3!} + \frac{x^5}{5!} + \frac{x^7}{7!} + \cdots, \quad |x| < \infty,$$

$$\tanh x = 1 - 2e^{-2x} + 2e^{-4x} - 2e^{-6x} + \ldots, \quad |x| < \pi/2,$$

$$\coth x = 1 + 2e^{-2x} + 2e^{-4x} + 2e^{-6x} + \ldots, \quad |x| < \pi,$$

$$\frac{1}{\sinh x} = 2(e^{-x} + e^{-3x} + e^{-5x} + \cdots), \quad |x| < \pi,$$

$$\frac{1}{\cosh x} = 2(e^{-x} - e^{-3x} + e^{-5x} - e^{-7x} + \cdots), \quad |x| < \pi/2.$$

3. *Relations between Hyperbolic, Trigonometric, and Exponential Functions*

It follows directly from the above expansions that

$$\sinh x = \tfrac{1}{2}(e^{x} - e^{-x}), \quad \cosh x = \tfrac{1}{2}(e^{x} + e^{-x}),$$

$$\sin x = \frac{1}{2i}(e^{ix} - e^{-ix}), \qquad \cos x = \frac{1}{2}(e^{ix} + e^{-ix}),$$

$$e^{ix} = \cos x + i \sin x, \qquad e^{-ix} = \cos x - i \sin x \quad \text{(Euler formulas)},$$

$$\sin ix = i \sinh x, \qquad \cos ix = \cosh x, \qquad \tan ix = i \tanh x,$$

$$\sinh ix = i \sin x, \qquad \cosh ix = \cos x, \qquad \tanh ix = i \tan x,$$

$$\cos(x + iy) = \cos x \cosh y - i \sin x \sinh y,$$

$$\sin(x + iy) = \sin x \cosh y + i \cos x \sinh y.$$

Values of $\sin x$, $\cos x$, $\tan x$, $\sinh x$, $\cosh x$, $\tanh x$, e^{x}, e^{-x} from 0 to 10 are presented in the book by Segal and Semendyaev [102a].

4. *Some Integrals Impossible to Reduce to Elementary Functions*

a. Gauss Error Function

$$\operatorname{erf} x = (2/\sqrt{\pi}) \int_0^{\infty} \exp[-x^2]\, dx,$$

$$\operatorname{erf} \infty = 1, \qquad \operatorname{erf}(-x) = -\operatorname{erf} x,$$

$$\operatorname{erfc} x = 1 - \operatorname{erf} x = (2/\sqrt{\pi}) \int_x^{\infty} \exp[-x^2]\, dx.$$

The series expansion erf x for low values of x will be presented.
For low x, we may be write

$$\operatorname{erf} x = \frac{2}{\sqrt{\pi}} \int_0^{x} dx \sum_{n=0}^{\infty} (-1)^n \frac{x^{2n}}{n!} = \frac{2}{\sqrt{\pi}} \sum_{n=0}^{\infty} (-1)^n \frac{x^{2n+1}}{n!(2n+1)},$$

hence,

$$\frac{1}{2}\sqrt{\pi} \operatorname{erf} x = \frac{x}{1} - \frac{x^3}{1!3} + \frac{x^5}{2!5} - \cdots.$$

For high values of x, applications of the following relation are possible:

$$\frac{1}{2}\sqrt{\pi}\operatorname{erfc} x = \int_x^\infty \exp[-x^2]\,dx = e^{-x^2}/2x - \frac{1}{2}\int_x^\infty (1/x^2)\exp[-x^2]\,dx,$$

hence

$$\frac{1}{2}\sqrt{\pi}\operatorname{erfc} x = \frac{1}{2}e^{-x^2}\left(\frac{1}{x} - \frac{1}{2x^3} + \frac{1\cdot 3}{2^2x^5} - \cdots (-1)^{n-1}\frac{1\cdot 3\cdots(2n-3)}{2^{n-1}x^{2n-1}}\right)$$

$$+ (-1)^n \frac{1\cdot 3\cdots(2n-1)}{2^n}\int_0^\infty \frac{1}{x^{2n}}\exp[-x^2]\,dx.$$

The series obtained does not converge since the ratio of the nth term to the $(n-1)$th term is larger than unity. However, if we take n terms of the series, then the residue (the last term of the series) will be less than the nth term because

$$\int_x^\infty (1/x^{2n})\exp[-x^2]\,dx < e^{-x^2}\int_x^\infty (1/x^{2n})\,dx.$$

Thus erfc x may be expanded into asymptotic series

$$\operatorname{erfc} x \simeq \frac{1}{\sqrt{\pi}}\exp[-x^2]\left(\frac{1}{x} - \frac{1}{2x^3} + \frac{1\cdot 3}{2^2x^5} - \frac{1\cdot 3\cdot 5}{2^3x^7} + \cdots\right).$$

The following relations are valid:

$$\int_0^\infty \exp[-x^2]\,\frac{\sin 2xy}{x}\,dx = \frac{1}{2}\pi\operatorname{erf} y,$$

$$\int_0^\infty \exp[-x^2]\sin 2xy\,dx = \frac{1}{2}\sqrt{\pi}\exp[-y^2]\operatorname{erf} y.$$

In heat conduction problems, the functions erf x and erfc x have to be integrated and differentiated. We introduce the notations

$$d^n \operatorname{erf} x = d^n/dx^n \operatorname{erf} x.$$

Then

$$d \operatorname{erf} x = (2/\sqrt{\pi})\exp[-x^2],$$

$$d^2 \operatorname{erf} x = -(4/\sqrt{\pi})x\exp[-x^2].$$

Derivatives of erf x are usually presented in the same tables for erf x. For integration of the function erfc x, we introduce the notation:

$$\mathrm{i}^n \operatorname{erfc} x = \int_x^\infty \mathrm{i}^{n-1}\operatorname{erfc}\xi\,d\xi.$$

Then

$$\begin{aligned}
i^0 \operatorname{erfc} x &= \operatorname{erfc} x\,, \\
i \operatorname{erfc} x &= \int_x^\infty \operatorname{erfc} \xi \, d\xi \\
&= \frac{1}{\sqrt{\pi}} \exp[-x^2] - x \operatorname{erfc} x\,, \\
i^2 \operatorname{erfc} x &= \int_x^\infty i \operatorname{erfc} \xi \, d\xi \\
&= \frac{1}{4}\left[(1+2x^2)\operatorname{erfc} x - \frac{2}{\sqrt{\pi}} x \exp[-x^2]\right] \\
\cdots &= \tfrac{1}{4}(\operatorname{erfc} x - 2x \, i \operatorname{erfc} x).
\end{aligned}$$

The general recurrence formula is of the form

$$2n\, i^n \operatorname{erfc} x = i^{n-2} \operatorname{erfc} x - 2x i^{n-1} \operatorname{erfc} x.$$

Hence

$$i^n \operatorname{erfc} 0 = \frac{1}{2^n \Pi(\frac{1}{2}n)} = \frac{1}{2^n \Gamma(1+\frac{1}{2}n)}.$$

It may be shown that the function $y = i^n \operatorname{erfc} x$ is the solution of the differential equation

$$y'' + 2xy' - 2ny = 0.$$

The function $i^n \operatorname{erfc} x$ is of great importance in heat conduction problems and it is therefore tabulated for $n = 1, 2, 3, 4, 5$, and 6 in Appendix 6.

The Gauss integral of the complex argument is

$$\begin{aligned}
\operatorname{erf}(\alpha + i\beta) &= \frac{2}{\sqrt{\pi}} \int_0^{\alpha+i\beta} \exp[-\xi^2]\, d\xi \\
&= \frac{2}{\sqrt{\pi}} \int_0^{\alpha} \exp[-\xi^2]\, d\xi + \frac{2}{\sqrt{\pi}} \int_\alpha^{\alpha+i\beta} \exp[-\xi^2]\, d\xi \\
&= \operatorname{erf} \alpha + \frac{2}{\sqrt{\pi}} \exp[-\alpha^2] \int_0^\beta \exp[y^2](\sin 2\alpha y + i \cos 2\alpha y)\, dy.
\end{aligned}$$

For purely imaginary argument we have

$$\operatorname{erf} i\beta = (2i/\sqrt{\pi}) \int_0^\beta \exp[y^2]\, dy.$$

b. Gamma Function or Gauss Pi Function

$$\Pi(k) = \Gamma(k + 1) = \int_0^\infty e^{-x} x^k \, dx ,$$

$$\Pi(k - 1)\, \Pi(- k) = \pi/\sin \pi k ,$$

$$\Pi(k) = k(\Pi)(k - 1).$$

If $k > 0$ and k is an integer ($k = m$), then $\Pi(k) = \Pi(m) = m!$ $= 1\cdot2\cdot3 \cdots n$. Further,

$$\Pi(-\tfrac{1}{2}) = \sqrt{\pi} , \qquad \Pi(+\tfrac{1}{2}) = \sqrt{\pi}/2.$$

5. Bessel Function

The first-kind Bessel function of the νth order may be represented in the series form

$$J_\nu(z) = \sum_{m=0}^{\infty} \frac{(-1)^m}{m!\Gamma(\nu + m + 1)} (\tfrac{1}{2}z)^{\nu+2m}, \quad |\arg z| < \pi ,$$

where ν is a real number, and z may be a complex value.

The function $J_\nu(z)$ is a particular solution of the Bessel differential equation

$$\frac{d^2y}{dz^2} + \frac{1}{z}\frac{dy}{dz} + \left(1 - \frac{\nu^2}{z^2}\right) y = 0.$$

If ν is an integer ($\nu = n$), then the following relation holds:

$$J_n(z) = (-1)^n J_{-n}(z).$$

The second-kind Bessel function of the νth order (for any values of ν) is determined by the relation

$$Y_\nu(z) = \frac{J_\nu(z)\cos \nu\pi - J_{-\nu}(z)}{\sin \nu\pi}, \quad |\arg z| < \pi .$$

If ν is a positive integer ($\nu = n$), then we may write

$$\pi Y_n(z) = 2[\ln(\tfrac{1}{2}z) + C]J_n(z) - \sum_{k=0}^{\infty} (-1)^k \frac{(\tfrac{1}{2}z)^{n+2k}}{k!(n+k)!} \left[\sum_{m=1}^{n+k} m^{-1} + \sum_{m=1}^{k} m^{-1}\right]$$
$$- \sum_{k=0}^{n=1} (\tfrac{1}{2}z)^{-n+2k}\{(n - k - 1)!/k!\}, \quad |\arg z| < \pi$$

where $C = 0.5772$ is the Euler constant.

If $n = 0$, we shall have

$$\begin{aligned}\tfrac{1}{2}\pi Y_0(z) &= [\ln(\tfrac{1}{2}z) + C]J_0(z) + (\tfrac{1}{2}z)^2 - (1 + \tfrac{1}{2})\{(\tfrac{1}{2}z)^4/(2!)^2\} \\ &\quad + (1 + \tfrac{1}{2} + \tfrac{1}{3})\{(\tfrac{1}{2}z)^6/(3!)^2\} - \cdots \\ &= [\ln(\tfrac{1}{2}z) + C]J_0(z) + (z^2/2^2) - (1 + \tfrac{1}{2})\{z^4/2^2\cdot 4^2\} \\ &\quad + (1 + \tfrac{1}{2} + \tfrac{1}{3})(z^6/2^2\cdot 4^2\cdot 6^2) - \cdots .\end{aligned}$$

The modified Bessel equation

$$\frac{d^2y}{dz^2} + \frac{1}{2}\,\frac{dy}{dz} - \left(1 + \frac{\nu^2}{z^2}\right) y = 0$$

has particular solutions defined by the relations

$$I_\nu(z) = \sum_{k=0}^{\infty} \frac{1}{k!\Pi(\nu + k)} (\tfrac{1}{2}z)^{\nu+2k} \qquad |z| < \infty,\ |\arg z| < \pi$$

where $I_\nu(z)$ is the first-kind modified Bessel function of the νth order and $K_\nu(z)$ is the second-kind modified Bessel function of the νth order:

$$K_\nu(z) = \frac{1}{2}\pi\,\frac{I_{-\nu}(z_\nu) - I_\nu(z)}{\sin \nu\pi}.$$

For the particular case $\nu = 0$ we have

$$\begin{aligned}K_0(z) &= -\ [\ln(\tfrac{1}{2}z) + C]I_0(z) + (\tfrac{1}{2}z)^2 + (1 + \tfrac{1}{2})\{(\tfrac{1}{2}z)/(2!)^2\} + \cdots \\ &= -\ [\ln(\tfrac{1}{2}z) + C]I_0(z) + (z^2/2^2) + (1 + \tfrac{1}{2})(z^4/2^2\cdot 4^2) \\ &\quad + (1 + \tfrac{1}{2} + \tfrac{1}{3})(z^6/2^2\cdot 4^2\cdot 6^2) + \cdots .\end{aligned}$$

The relations between $J_\nu(z)$, $Y_\nu(z)$ and $I_\nu(z)$, $K_\nu(z)$ are of the form

$$K_\nu(i^{2k}z) = i^{-2k\nu}K_\nu(z) - i\pi\,\frac{\sin \nu_k\pi}{\sin \nu\pi}\,I_\nu(z),$$

$$Y_\nu(i^{2k}z) = i^{-2k\nu}Y_\nu(z) + 2i \sin k\nu\pi \cot \nu\pi J_\nu(z).$$

Series expansions of $I_\nu(z)$ and $K_\nu(z)$ at high values of $|z|$ are of the form

$$I_\nu(z) = \frac{1}{(2\pi z)^{1/2}}\,e^z\left(1 - \frac{4\nu^2 - 1^2}{1!8z} + \frac{(4\nu^2 - 1^2)(4\nu^2 - 3^2)}{2!(8z)^2} - \cdots\right),$$

$$K_\nu(z) = \left(\frac{\pi}{2z}\right)^{1/2} e^{-z}\left(1 + \frac{4\nu^2 - 1^2}{1!8z} + \frac{(4\nu^2 - 1^2)(4\nu^2 - 3^2)}{2!(8z)^2} + \cdots\right).$$

Relations between the functions and their derivatives are

$$J_0'(z) = -\ J_1(z), \quad Y_0'(z) = -\ Y_1(z), \quad I_0'(z) = I_1(z), \quad K_0'(z) = -\ K_1(z).$$

In a general case

$$zI_\nu'(z) + \nu I_\nu(z) = zI_{\nu-1}(z)\,,$$
$$zI_\nu'(z) - \nu I_\nu(z) = zI_{\nu+1}(z)\,,$$
$$zK_\nu'(z) + \nu K_\nu(z) = -\,zK_{\nu-1}(z)\,,$$
$$zK_\nu'(z) - \nu K_\nu(z) = -\,zK_{\nu+1}(z)\,,$$
$$zJ_\nu'(z) + \nu J_\nu(z) = zJ_{\nu-1}(z)\,,$$
$$zJ_\nu'(z) - \nu J_\nu(z) = -\,zJ_{\nu+1}(z)\,,$$
$$K_{-\nu}(z) = K_\nu(z)\,, \qquad K_{1/2}(z) = (\pi/2z)^{1/2}e^{-z}.$$
$$J_{\nu+1}(z) + J_{\nu-1}(z) = (2\nu/z)\,J_\nu(z)\,,$$
$$J_{\nu+1}(z) + Y_{\nu-1}(z) = (2\nu/z)Y_\nu(z).$$

When z is small

$$J_0(z) \simeq 1 - (z^2/_4) + R, \qquad \text{where } R < z^4/_{64}\,,$$
$$J_1(z) \simeq (z/2) - (z^3/_{16}) + R, \qquad \text{where } R < z^5/_{384}.$$

When z is large

$$J_0(z) \simeq (2/\pi z)^{1/2}\cos(z - \pi/4)\,,$$
$$J_1(z) \simeq (2/\pi z)^{1/2}\cos(z - 3\pi/4)\,,$$
$$Y_0(z) \simeq (2/\pi z)^{1/2}\sin(z - \pi/4)\,,$$
$$Y_1(z) \simeq (2/\pi z)^{1/2}\sin(z - 3\pi/4).$$

APPENDIX

2

THE UNIQUENESS THEOREM

In Chapter 1 it was shown that the solution of the differential heat conduction equation should satisfy not only the equation itself but also the boundary and initial conditions. The question arises as to whether two solutions can exist simultaneously which would satisfy the equation and boundary conditions. In the subsequent section, we shall show that such solutions are impossible. This theorem is referred to as the uniqueness theorem.

Let two solutions $t_1(x, y, z, \tau)$ and $t_2(x, y, z, \tau)$ satisfy the differential equation

$$\partial t/\partial \tau = a\,\nabla^2 t\,, \tag{2.1}$$

and initial and boundary conditions

$$\lim_{\tau\to 0} t = f(x, y, z)\,, \tag{2.2}$$

$$t_s = \varphi(x, y, z, \tau)\,, \tag{2.3}$$

where the suffix s denotes the body surface.

We assume that

$$t_1 - t_2 = u\,; \tag{2.4}$$

then

$$\partial u/\partial \tau = a\,\nabla^2 u\,, \tag{2.5}$$

$$\lim_{\tau\to 0} u = 0\,, \qquad u_s = 0\,. \tag{2.6}$$

Consider the following integral

$$J = \int_V \tfrac{1}{2} u^2 \, dv. \tag{2.7}$$

The integral is taken over the body volume $V(dv = dx\, dy\, dz)$, i.e., it is a triple integral. Then

$$\partial J/\partial \tau = \int_V u(\partial u/\partial \tau)\, \partial v = a \int_V u \nabla^2 u \, dv. \tag{2.8}$$

We apply Green's formula to obtain

$$\int_{(S)} u \frac{\partial u}{\partial n} ds = \int_{(V)} u \nabla^2 u \, dv + \int_{(V)} \left[\left(\frac{\partial u}{\partial x}\right)^2 + \left(\frac{\partial u}{\partial y}\right)^2 + \left(\frac{\partial u}{\partial z}\right)^2 \right] dv,$$

where in the first integral, integration is carried out over the body surface, S. Then we may be write

$$\begin{aligned}\frac{\partial J}{\partial \tau} &= a \int_{(S)} u \frac{\partial u}{\partial n} ds - a \int_{(V)} \left[\left(\frac{\partial u}{\partial x}\right)^2 + \left(\frac{\partial u}{\partial y}\right)^2 + \left(\frac{\partial u}{\partial z}\right)^2 \right] dv \\ &= -a \int_{(V)} \left[\left(\frac{\partial u}{\partial x}\right)^2 + \left(\frac{\partial u}{\partial y}\right)^2 + \left(\frac{\partial u}{\partial z}\right)^2 \right] dv,\end{aligned}$$

since the first integral equals zero according to boundary condition (2.6.), as we integrate over the surface where $u = 0$. Thus

$$\partial J/\partial \tau \leq 0. \tag{2.9}$$

Since $\quad J = 0 \;\text{ at }\; \tau = 0 \;(u = 0 \;\text{ at }\; \tau = 0),$

then

$$J \leq 0. \tag{2.10}$$

But from relation (2.7) it follows that

$$J \geq 0. \tag{2.11}$$

Hence $J = 0$. Thus,

$$u = 0, \qquad t_1 = t_2.$$

Consequently, if some function $t(x, y, z, \tau)$ satisfies the differential equation, both initial and boundary conditions, then it is a unique solution of the problem of interest.

It should be noticed that the solution of the problem may be expressed by various functional relations, but it does not mean the existence of different solutions of the problem; consequently, it does not contradict the uniqueness theorem.

APPENDIX

3

DIFFERENTIAL HEAT CONDUCTION EQUATION IN VARIOUS COORDINATE SYSTEMS

In Chapter 1 the differential heat conduction equation was derived in a Cartesian system as

$$\frac{\partial t}{\partial \tau} = a\,\nabla^2 t = a\left(\frac{\partial^2 t}{\partial x^2} + \frac{\partial^2 t}{\partial y^2} + \frac{\partial^2 t}{\partial z^2}\right). \tag{3.1}$$

Now $\nabla^2 t$ will be expressed in spherical and cylindrical coordinate systems.

We assume

$$x = r \sin\theta \cos\psi, \qquad y = r \sin\theta \sin\psi, \qquad z = r\cos\theta. \tag{3.2}$$

Then

$$\nabla^2 t = \frac{1}{r^2}\left[\frac{\partial}{\partial r}\left(r^2 \frac{\partial t}{\partial r}\right) + \frac{1}{\sin\theta}\frac{\partial}{\partial \theta}\left(\sin\theta \frac{\partial t}{\partial \theta}\right) + \frac{1}{\sin^2\theta}\frac{\partial^2 t}{\partial \psi^2}\right] \tag{3.3}$$

or

$$\nabla^2 t = \frac{\partial^2 t}{\partial r^2} + \frac{2\partial t}{r\,\partial r} + \frac{1}{r^2}\frac{\partial}{\partial \mu}\left[(1-\mu^2)\frac{\partial t}{\partial \mu}\right] + \frac{1}{r^2(1-\mu^2)}\frac{\partial^2 t}{\partial \psi^2}, \tag{3.4}$$

where $\mu = \cos\theta$. Relation (3.4) is the expression of $\nabla^2 t$ in spherical coordinates (r, θ, and ψ).

Assumption

$$x = r\cos\theta, \qquad y = r\sin\theta.$$

yields

$$\nabla^2 t = \frac{1}{r}\left[\frac{\partial}{\partial r}\left(r\,\frac{\partial t}{\partial r}\right)+\frac{\partial}{\partial \theta}\left(\frac{1}{r}\,\frac{\partial t}{\partial \theta}\right)+\frac{\partial}{\partial z}\left(r\,\frac{\partial t}{\partial z}\right)\right], \tag{3.5}$$

or

$$\nabla^2 t = \frac{\partial^2 t}{\partial r^2}+\frac{1}{r}\,\frac{\partial t}{\partial r}+\frac{1}{r^2}\,\frac{\partial^2 t}{\partial \theta^2}+\frac{\partial^2 t}{\partial z^2}. \tag{3.6}$$

Relation (3.6) is the expression of $\nabla^2 t$ in cylindrical coordinates (r, θ). Substitution of expressions (3.4) and (3.6) into (3.1) yields forms of the heat conduction equation in spherical and cylindrical coordinates.

APPENDIX

4

MAIN RULES AND THEOREMS OF THE LAPLACE TRANFORMATION

Item	$f(\tau)$	$F(s) = L[f(\tau)]$
1	$f(\tau) = (1/2\pi i)\int_{\sigma-i\infty}^{\sigma+i\infty} F(s)e^{s\tau}\, d\tau$	$\int_0^\infty f(\tau)e^{-s\tau}\, d\tau = F(s)$
2	$Af(\tau) + Bg(\tau)$	$AF(s) + BG(s)$
3	$f'(\tau)$	$sF(s) - f(+0)$
4	$f^{(n)}(\tau)$	$s^nF(s) - s^{n-1}f(+0) - s^{n-2}f'(+0) - \cdots - f^{n-1}(+0)$
5	$e^{a\tau}f(\tau)$	$F(s-a)$
6	$(1/a)f(\tau/a)$	$F(as)$
7	$(1/a)e^{b\tau}f(\tau/a)$	$F(as-b)$
8	$f(\tau-b)$, if $f(\tau)=0$, at $\tau<0$	$e^{-bs}F(s)$
9	$\int_0^\tau f(\theta)\, d\theta$	$(1/s)F(s)$
10	$\int_0^\tau \int_0^\theta f(\vartheta)\, d\vartheta\, d\theta$	$(1/s^2)F(s)$
11	$\tau f(\tau)$	$-F'(s)$
12	$\tau^n f(\tau)$	$(-1)^nF^{(n)}(s)$
13	$\int_0^\tau f_1(\tau-\theta)f_2(\theta)\, d\theta = f_1{*}f_2$	$F_1(s)F_2(s)$

APPENDIX 4 *(continued)*

Item	$f(\tau)$	$F(s) = L[f(\tau)]$
14	$\{1/(\pi\tau)^{1/2}\} \int_0^\infty f(\vartheta) \exp - \left(\frac{\vartheta^2}{4\tau}\right) d\vartheta$	$(1/\sqrt{s})F(\sqrt{s})$
15	$\int_0^\infty e^{-s\vartheta} f^*(\vartheta)\, d\vartheta$ where $f^*(\vartheta) = \int_0^\infty f(\tau)\psi(\tau, \vartheta)\, d\tau$, $\psi(\tau, \vartheta)$	$F[\varphi(s)]\varphi(s)$ $e^{-\tau\varphi(s)}\varphi(s) = \int_0^\infty e^{-s\vartheta}\psi(\tau, \vartheta)\, d\vartheta$
16	$\sum_{n=1}^{n} \frac{\Phi(s_n)}{\psi'(s_n)} e^{s_n\tau}$	$\Phi(s)/\psi(s)$, where $\psi(s) = (s - s_1) \times (s - s_2) \cdots (s - s_n)$
17	$\sum_{k=1}^{n} \frac{1}{(k-1)!} \lim_{s\to s_n} \left\{ \frac{d^{k-1}}{ds^{k-1}} \times \left[\frac{\Phi(s)(s-s_n)^k}{\psi(s)} e^{s\tau} \right] \right\}$ (the case of multiple roots).	$\Phi(s)/\psi(s)$ where $\psi(s) = (s-s_1)(s-s_2)\cdots \times (s-s_m)^k(s-s_{m+1})\cdots(s-s_n)$

APPENDIX

5

TRANSFORMS OF SOME FUNCTIONS

Item	Transform, $F(s) = L[f(\tau)]$	Original, $f(\tau)$
1	$1/s$	1
2	$1/s^2$	τ
3	$1/s^n \quad (n = 1, 2, 3, \ldots)$	$\tau^{n-1}/(n-1)!$
4	$1/\sqrt{s}$	$1/(\pi\tau)^{1/2}$
5	$s^{-3/2}$	$2\left(\frac{\tau}{\pi}\right)^{1/2}$
6	$s^{-(n+1/2)} \quad (n = 1, 2, 3, \ldots)$	$\frac{2^n \tau^{n-1/2}}{[1\cdot 3\cdot 5\cdots(2n-1)]\sqrt{\pi}}$
7	$\Gamma(m)/s^m \quad (m > 0)$	τ^{m-1}
8	$\frac{\Gamma(m+1)}{s^{m+1}} = \frac{\Pi(m)}{s^{m+1}} \quad (m > -1)$	τ^m
9	$1/(s-a)$	$e^{a\tau}$
10	$1/(s+a)$	$e^{-a\tau}$
11	$1/(s-a)^2$	$\tau e^{a\tau}$
12	$1/(s-a)^n \quad (n = 1, 2, 3, \ldots)$	$[1/(n-1)!]\tau^{n-1}e^{a\tau}$
13	$\Gamma(m)/(s-a)^m \quad (m > 0)$	$\tau^{m-1}e^{a\tau}$
14	$1/(s-a)(s-b)$	$[1/(a-b)](e^{a\tau} - e^{b\tau})$

APPENDIX 5 (*continued*)

Item	Transform, $F(s) = L[f(t)]$	Original, $f(\tau)$
15	$s/(s-a)(s-b)$	$[1/(a-b)]\,(e^{a\tau} - e^{b\tau})$
16	$1/(s-a)(s-b)(s-c)$	$\dfrac{(b-c)e^{a\tau} + (c-a)e^{b\tau} + (a-b)e^{c\tau}}{(a-b)(b-c)(c-a)}$
17	$k/(s^2+k^2)$	$\sin k\tau$
18	$s/(s^2+k^2)$	$\cos k\tau$
19	$k/(s^2-k^2)$	$\sinh k\tau$
20	$s/(s^2-k^2)$	$\cosh k\tau$
21	$k/\{(s+a)^2+k^2\}$	$e^{-a\tau}\sin k\tau$
22	$(s+a)/\{(s+a)^2+k^2\}$	$e^{-a\tau}\cos k\tau$
23	$1/s(s^2+k^2)$	$(1/k^2)(1-\cos k\tau)$
24	$1/s^2(s^2+k^2)$	$(1/k^3)(k\tau - \sin k\tau)$
25	$1/(s^2+k^2)^2$	$(1/2k^3)(\sin k\tau - k\tau\cos k\tau)$
26	$s/(s^2+k^2)^2$	$(\tau/2k)\sin k\tau$
27	$s^2/(s^2+k^2)^2$	$(1/2k)(\sin k\tau + k\tau\cdot\cos k\tau)$
28	$(s^2-k^2)/(s^2+k^2)^2$	$\tau\cos k\tau$
29	$s/(s^2+a^2)(s^2+b^2) \quad (a^2 \neq b^2)$	$(\cos a\tau - \cos b\tau)/(b^2-a^2)$
30	$1/\{(s-a)^2+k^2\}$	$\dfrac{1}{k}\,e^{a\tau}\sin k\tau$
31	$3k^2/(s^3+k^3)$	$e^{-k\tau} - e^{k\tau/2}(\cos \frac{1}{2}k\tau\sqrt{3} - \sqrt{3}\sin\frac{1}{2}k\tau\sqrt{3})$
32	$4k^3/(s^4+4k^4)$	$\sin k\tau\cosh k\tau - \cos k\tau\sinh k\tau$
33	$s/(s^4+4k^4)$	$(1/2k^2)\sin k\tau\sinh k\tau$
34	$1/(s^4-k^4)$	$(1/2k^3)(\sinh k\tau - \sin k\tau)$
35	$s/(s^4-k^4)$	$(1/2k^2)(\cosh k\tau - \cos k\tau)$
36	$s^n/(s^2+k^2)^{n+1}$	$\tau^n \sin k\tau/2^n k n!$
37	$\dfrac{1}{s}\left(\dfrac{s-1}{s}\right)^n$	$l_n(\tau) = \dfrac{1}{n!}\,e^{\tau}\,\dfrac{d^n}{d\tau^n}\,(\tau^n e^{-\tau})$
38	$s/(s-k)^{3/2}$	$\{1/(\pi\tau)^{1/2}\}e^{k\tau}(1+2k\tau)$
39	$(s-a)^{1/2} - (s-b)^{1/2}$	$\{1/2(\pi\tau^3)^{1/2}\}(e^{b\tau} - e^{a\tau})$

APPENDIX 5 *(continued)*

Item	Transform, $F(s) = L[f(\tau)]$	Original, $f(\tau)$
40	$1/(\sqrt{s}+k)$	$\{1/(\pi\tau)^{1/2}\} - ke^{k^2\tau}\,\mathrm{erfc}\,k\sqrt{\tau}$
41	$\sqrt{s}/(s-k^2)$	$1/(\pi\tau)^{1/2} + ke^{k^2\tau}\,\mathrm{erf}\,k\sqrt{\tau}$
42	$\sqrt{s}/(s+k^2)$	$1/(\pi\tau)^{1/2} - (2k/\sqrt{\pi})\exp[-k^2\tau] \times \int_0^{k\sqrt{\tau}} \exp[x^2]\,dx$
43	$1/\sqrt{s}(s-k^2)$	$(1/k)\exp[-k^2\tau]\,\mathrm{erf}\,k\sqrt{\tau}$
44	$1/\sqrt{s}(s^2+k^2)$	$(2/k\sqrt{\pi})\exp[-k^2\tau]\int_0^{k\sqrt{\tau}}\exp[x^2]\,dx$
45	$(b^2-a^2)/(s-a^2)(b+\sqrt{s})$	$\exp[a^2\tau][b - a\,\mathrm{erf}\,a\sqrt{\tau}] - b\exp[b^2\tau]\,\mathrm{erfc}\,b\sqrt{\tau}$
46	$1/\sqrt{s}(\sqrt{s}+k)$	$\exp[k^2\tau]\,\mathrm{erfc}\,k\sqrt{\tau}$
47	$1/(s+k)(s+b)^{1/2}$	$\{1/(b-k)^{1/2}\}e^{-k\tau}\,\mathrm{erf}[(b-k)\tau]^{1/2}$
48	$(b^2-k^2)/\sqrt{s}(s-k^2)\sqrt{s}+b$	$\exp[k^2\tau]\left(\frac{b}{k}\,\mathrm{erf}\,k\sqrt{\tau}-1\right) + \exp[b^2\tau]\,\mathrm{erfc}\,b\sqrt{\tau}$
49	$\exp[-k\sqrt{s}]\quad (k \geq 0)$	$\{k/2(\pi\tau^3)^{1/2}\}\exp[-k^2/4\tau]$
50	$(1/s)\exp[-k\sqrt{s}]\quad (k \geq 0)$	$[1-\mathrm{erf}(k/2\sqrt{\tau})] = \mathrm{erfc}(k/2\sqrt{\tau})$
51	$(1/\sqrt{s})\exp[-k\sqrt{s}]\quad (k \geq 0)$	$(1/\sqrt{\pi\tau})\exp[-k^2/4\tau]$
52	$(1/s\sqrt{s})\exp[-k\sqrt{s}]\quad (k \geq 0)$	$2\sqrt{\tau}\,\mathrm{i\,erfc}(k/2\sqrt{\tau}) = 2(\tau/\pi)^{1/2} \times \exp[-k^2/4\tau] - k\,\mathrm{erfc}(k/2\sqrt{\tau})$
53	$(1/s^2)\exp[-k\sqrt{s}]$	$4\tau\,\mathrm{i}^2\,\mathrm{erfc}(k/2\sqrt{\tau}) = (\tau+\frac{1}{2}k^2) \times \mathrm{erfc}\,\frac{k}{2\sqrt{\tau}} - k(\tau/\pi)^{1/2}\exp[-k^2/4\tau]$
54	$\{1/s(s^n)^{1/2}\}\exp[-k\sqrt{s}]\quad (k \geq 0)$	$(4\tau)^{n/2}\,\mathrm{i}^n\,\mathrm{erfc}(k/2\sqrt{\tau})$
55	$(1/s^n\sqrt{s})\exp[-2(ks)^{1/2}]\quad (k \geq 0)$	$[1/(n-1)!]\int_0^{\tau}(\tau-z)^{n-1} \times \exp[-k/z]dz/(\pi z)^{1/2}$
56	$\dfrac{b\exp[-k\sqrt{s}]}{s(b+\sqrt{s})}\quad (k \geq 0)$	$\mathrm{erfc}(k/2\sqrt{\tau}) - \exp[bk]\exp[b^2\tau] \times \mathrm{erfc}(b\sqrt{\tau} + (k/2\sqrt{\tau}))$

APPENDIX 5 (*continued*)

Item	Transform, $F(s) = L[f(\tau)]$	Original, $f(\tau)$
57	$\exp[-k\sqrt{s}]/\sqrt{s}(b+\sqrt{s})$ $(k \geq 0)$	$\exp[bk]\exp[b^2\tau]\operatorname{erfc}(b\sqrt{\tau}+(k/2\sqrt{\tau})$
58	$1/(1+(s/b)^{1/2})\exp[-k\sqrt{s}]$	$(b/\pi\tau)^{1/2}\exp[-k^2/4\tau]-b\exp[k\sqrt{b}+b\tau]$ $\times \operatorname{erfc}\{(k/2\sqrt{\tau})+(b\tau)^{1/2}\}$
59	$\{1/s\sqrt{s}(\sqrt{s}+b)\}\exp[-k\sqrt{s}]$	$(2/b)(\tau/\pi)^{1/2}\exp[-k^2 4\tau]-[(1+bk)/b^2]$ $\times \operatorname{erfc}(k/2\sqrt{\tau})+(1/b^2)\exp[bk+b^2\tau]$ $\times \operatorname{erfc}\{(k/2\sqrt{\tau})+b\sqrt{\tau}\}$
60	$\{1/(\sqrt{s})^{n+1}(\sqrt{s}+b)\}\exp[-k\sqrt{s}]$	$\{1/(-b)^n\}\exp[bk+b^2\tau]$ $\times \operatorname{erfc}(\{k/2\sqrt{\tau}\}+b\sqrt{\tau})-\dfrac{1}{(-b)^n}$ $\times \sum_{m=0}^{n-1}(-2b\sqrt{\tau})^m \, i^m \operatorname{erfc}(k/2\sqrt{\tau})$
61	$\{1/(\sqrt{s}+b)^2\}\exp[-k\sqrt{s}]$	$-2b(\tau/\pi)^{1/2}\exp[-k^2/4\tau]+(1+bk$ $+2b^2\tau)\exp[bk+b^2\tau]$ $\times \operatorname{erfc}\{(k/2\sqrt{\tau})+b\sqrt{\tau}\}$
62	$\{1/s(\sqrt{s}+b)^2\}\exp[-k\sqrt{s}]$	$(1/b^2)\operatorname{erfc}(k/2\sqrt{\tau})-(2/b)(\tau/\pi)^{1/2}$ $\times \exp[-k^2/4\tau]-(1/b^2)(1-bk-2b^2\tau)$ $\times \exp[bk+b^2\tau]\operatorname{erfc}\{(k/2\sqrt{\tau})+b\sqrt{\tau}\}$
63	$\{1/(s-b)\}\exp[-k\sqrt{s}]$	$\frac{1}{2}\exp[b\tau]\{\exp[-k/\sqrt{b}]$ $\times \operatorname{erfc}[(k/2\sqrt{\tau})-\sqrt{b\tau})]$ $+\exp[k/\sqrt{b}]\operatorname{erfc}[(k/2\sqrt{\tau})+(b\tau)^{1/2}]\}$
64	$\{\Pi(\frac{1}{2}n)/s^{2+1/2n}\}\exp[-k\sqrt{s}]$	$[\tau^{1+\frac{n}{2}}/(1+\frac{1}{2}n)]\,\Pi(1+\frac{1}{2}n)\,2^{n+2}$ $\times i^{n+2}\operatorname{erfc}(k/2\sqrt{\tau})$
65	$(1/s^{3/4})\exp[-k\sqrt{s}]$	$(1/\pi)(k/2\tau)^{1/2}\exp[-k^2/8\tau]K_{1/4}(k^2/8\tau)$
66	$(1/s)\exp[-k(s+b)^{1/2}]$	$\frac{1}{2}[\exp[-k\sqrt{b}]\operatorname{erfc}\{(k/2\sqrt{\tau})-\sqrt{b\tau}\}$ $+\exp[k\sqrt{b}]\operatorname{erfc}\{(k/2\sqrt{\tau})+(b\tau)^{1/2}\}]$
67	$(1/s)(s+2b)^{1/2}\exp[-k(s+2b)^{1/2}]$	$\{1/(\pi\tau)^{1/2}\}[\exp-\{(k^2/4\tau)+2b\tau\}]$ $+\frac{1}{2}(2b)^{1/2}\Big[e^{-k\sqrt{2b}}\operatorname{erfc}\{(k/2\sqrt{\tau})$ $-(2b\tau)^{1/2}\}-e^{k\sqrt{2b}}$ $\times \operatorname{erfc}\left(\dfrac{k}{2\sqrt{\tau}}+(2b\tau)^{1/2}\right)\Big]$

APPENDIX 5 *(continued)*

Item	Transform, $F(s) = L[f(\tau)]$	Original, $f(\tau)$
68	$\{1/s(s+2b)^{1/2}\}\exp[-k(s+2b)^{1/2}]$	$\{1/2(2b)^{1/2}\}\Big[\exp[-k\sqrt{2b}]$ $\times \operatorname{erfc}\left(\dfrac{k}{2\sqrt{\tau}} - (2b\tau)^{1/2}\right)$ $- \exp[k\sqrt{2b}]\operatorname{erfc}\left(\dfrac{k}{\sqrt{2\tau}} + (2b\tau)^{1/2}\right)\Big]$
69	$1/(s^2+k^2)^{1/2}$	$J_0(k\tau)$
70	$1/(s^2-k^2)^{1/2}$	$I_0(k\tau)$
71	$\{(s+2k)^{1/2}/\sqrt{s}\} - 1$	$ke^{-k\tau}[I_0(k\tau) + I_1(k\tau)]$
72	$1/(s+k)^{1/2}(s+b)^{1/2}$	$\exp[-\frac{1}{2}(k+b)\tau]I_0[\frac{1}{2}(k-b)\tau]$
73	$\Gamma(m)/(s+k)^m(s+b)^m \quad (m > 0)$	$\sqrt{\pi}\,\{\tau/(k-b)\}^{m-1/2}\exp[-\frac{1}{2}(k+b)\tau]$ $\times I_{m-1/2}\{\frac{1}{2}(k-b)\tau\}$
74	$1/(s+k)^{1/2}(s+n)^{3/2}$	$\tau\exp[-\frac{1}{2}(k+b)\tau][I_0\{\frac{1}{2}(k-b)\tau\}$ $+ I_1[\frac{1}{2}(k-b)\tau)]$
75	$[(s+2k)^{1/2} - \sqrt{s}]/[(s+2k)^{1/2} + \sqrt{s}]$	$(1/\tau)e^{-k\tau}I_1(k\tau)$
76	$(1-s)^n/s^{n+1/2}$	$(n!/(2n)!(\pi\tau)^{1/2})H_{2n}(\sqrt{\tau})$
77	$(1-s)^n/s^{n+3/2}$	$(n!/\sqrt{\pi}\,(2n+1)!)H_{2n+1}(\sqrt{\tau})$
78	$\dfrac{((s^2+k^2)^{1/2} - s)^\nu}{(s^2+k^2)^{1/2}} \quad (\nu > -1)$	$k^\nu J_\nu(k\tau)$
79	$1/(s^2+k^2)^m \quad (m > 0)$	$(\sqrt{\pi}/\Gamma(m))(\tau/2k)^{m-1/2}J_{m-1/2}(k\tau)$
80	$1/(s^2-k^2)^m \quad (m > 0)$	$(\sqrt{\pi}/\Gamma(m))\left(\dfrac{\tau}{2k}\right)^{m-1/2} I_{(m-1/2)}(k\tau)$
81	$(1/s)e^{-k/s}$	$J_0(2(k\tau)^{1/2})$
82	$(1/\sqrt{s})e^{-k/s}$	$\{1/(\pi\tau)^{1/2}\}\cos 2(k\tau)^{1/2}$
83	$(1/\sqrt{s})e^{k/s}$	$\{1/(\pi\tau)^{1/2}\}\cosh 2(k\tau)^{1/2}$
84	$(1/s\sqrt{s})e^{-k/s}$	$\{1/(\pi k)^{1/2}\}\sin 2(k\tau)^{1/2}$
85	$(1/s\sqrt{s})e^{k/s}$	$\{1/(\pi k)^{1/2}\}\sinh 2(k\tau)^{1/2}$
86	$(1/s^m)e^{-k/s} \quad (m > 0)$	$(\tau/k)^{(m-1)/2}J_{m-1}(2(k\tau)^{1/2})$

APPENDIX 5 (*continued*)

Item	Transform, $F(s) = L[f(\tau)]$	Original, $f(\tau)$
87	$(1/s^m)e^{k/s} \quad (m > 0)$	$(\tau/k)^{(m-1)/2}\ I_{m-1}(2(k\tau)^{1/2})$
88	$\dfrac{\exp[-k(s(s+b))^{1/2}]}{(s(s+b))^{1/2}}$	$\begin{cases} 0, & \text{when } 0 < \tau < k, \\ \exp[-\frac{1}{2}b\tau]I_0(\frac{1}{2}b(\tau^2 - k^2)^{1/2}), & \text{when } \tau > k \end{cases}$
89	$\dfrac{\exp[-k(s^2+b^2)^{1/2}]}{(s^2+b^2)^{1/2}}$	$\begin{cases} 0, & \text{when } 0 < \tau < k, \\ J_0(b(\tau^2 - k^2)^{1/2}), & \tau > k \end{cases}$
90	$\dfrac{\exp[-k(s^2-b^2)^{1/2}]}{(s^2-b^2)^{1/2}}$	$\begin{cases} 0, & \text{when } 0 < \tau < k, \\ I_0(b(\tau^2 - k^2)^{1/2}), & \text{when } \tau > k \end{cases}$
91	$\dfrac{\exp[-k(s^2+b^2)^{1/2} - s)]}{(s^2+b^2)^{1/2}} \quad (k \geq 0)$	$J_0(b(\tau^2 + 2k\tau)^{1/2})$
92	$\dfrac{b^{\nu}\text{-}\exp[-k(s^2+b^2)^{1/2}]}{(s^2+b^2)^{1/2}((s^2+b^2+s)^{1/2})^{\nu}} \quad (\nu > -1)$	$\begin{cases} 0, & (0 < \tau < k), \\ \left(\dfrac{\tau - k}{\tau + k}\right)^{1/2\nu} J_{\nu}(b(\tau^2 - k^2)^{1/2}) & (\tau > k) \end{cases}$
93	$(1/s) \ln s$	$\Gamma'(1) - \ln \tau, [\Gamma'(1) = -0.5772]$
94	$(1/s^k) \ln s$	$\tau^{k-1}\left\{\dfrac{\Gamma'(k)}{[\Gamma(k)]^2} - \dfrac{\ln \tau}{\Gamma(k)}\right\}$
95	$(\ln s/s - k) \quad (k > 0)$	$e^{k\tau}[\ln k - \mathrm{Ei}(-k\tau)];$ $\mathrm{Ei}(-\tau) = -\int_{\tau}^{\infty} e^{-x}(dx/x) \quad (\tau > 0)$
96	$\ln s/(s^2 + 1)$	$\cos \tau\ \mathrm{si}(\tau) - \sin \tau\ \mathrm{ci}(\tau)$
97	$\dfrac{1}{s} \ln(1 + ks) \quad (k > 0)$	$-\mathrm{Ei}(-\tau/k)$
98	$\ln[(s-k)/(s-b)]$	$(1/\tau)(e^{b\tau} - {}^{k\tau})$
99	$\ln[(s^2+k^2)/s^2]$	$(2/\tau)(1 - \cos k\tau)$
100	$\ln[(s^2-k^2)/s^2]$	$(2/\tau)(1 - \cosh k\tau)$
101	$\arctan k/s$	$(1/\tau) \sin k\tau$
102	$(1/s) \arctan(k/s)$	$\mathrm{si}(k\tau)$
103	$\exp[k^2s^2]\ \mathrm{erfc}\ ks \quad (k > 0)$	$(1/k\sqrt{\pi}) \exp[-\tau^2/4k]$
104	$(1/s) \exp[k^2s^2]\ \mathrm{erfc}\ ks \quad (k > 0)$	$\mathrm{erf}\ \tau/2k$

APPENDIX 5 (*continued*)

Item	Transform, $F(s) = L[f(\tau)]$	Original, $f(\tau)$
105	$e^{ks}\operatorname{erfc}(ks)^{1/2} \quad (k > 0)$	$\sqrt{k}/\pi\sqrt{\tau}(\tau + k)$
106	$(1/\sqrt{s})\operatorname{erfc}(ks)^{1/2}$	$\begin{cases} 0, & (0 < \tau < k) \\ (\pi\tau)^{-1/2} & (\tau > k) \end{cases}$
107	$(1/\sqrt{s})e^{ks}\operatorname{erfc}(ks)^{1/2} \quad (k > 0)$	$1/(\pi(\tau + k))^{1/2}$
108	$\operatorname{erfc} k/\sqrt{s}$	$(1/\pi\tau)\sin(2k\sqrt{\tau})$
109	$(1/\sqrt{s})\exp[k^2/s]\operatorname{erfc}(k/\sqrt{s})$	$\{1/(\pi\tau)^{1/2}\}\exp[-2k\sqrt{\tau}]$
110	$K_0(ks)$	$\begin{cases} 0 & (0 < \tau < k), \\ (\tau^2 - k^2)^{-1/2} & (\tau > k) \end{cases}$
111	$K_0(k\sqrt{s})$	$(1/2\tau)\exp[-k^2/4\tau]$
112	$(1/s)e^{ks}K_1((ks)^{1/2})$	$(1/k)(\tau(\tau + 2k))^{1/2}$
113	$(1/\sqrt{s})K_1(k\sqrt{s})$	$(1/k)\exp[-k^2/4\tau]$
114	$(1/\sqrt{s})e^{k/s}K_0(k/s)$	$\{2/(\pi\tau)^{1/2}\}K_0(2(2k\tau)^{1/2})$
115	$(1/k\sqrt{s}s)\tanh k\sqrt{s}$	$1 - \sum_{n=1}^{\infty} \frac{8}{(2n-1)^2\pi^2} \times \exp\left[-\frac{(2n-1)^2\pi^2\tau}{4k^2}\right]$
116	$\begin{matrix} I_\nu((s/a)^{1/2}x_1)K_\nu((s/a)^{1/2}x); & x > x_1 \\ I_\nu((s/a)^{1/2}x)K_\nu((s/a)^{1/2}x_1); & x < x_1 \end{matrix}$	$\frac{1}{2\tau}\exp\left(-\frac{x^2 + x_1^2}{4a\tau}\right) I_\nu\left[\frac{xx_1}{2a\tau}\right] \quad (\nu \geq 0)$
117	$s^{\nu/2}K_\nu(k\sqrt{s})$	$\frac{k^\nu \exp[-k^2/4\tau]}{(2\tau)^{\nu+1}}$
118	$s^{\nu/2-1}K_\nu(k\sqrt{s})$	$(2^{\nu-1}/k^{\nu+1})\Gamma\{\nu, (k^2/4\tau)\}$

APPENDIX

6

VALUES OF FUNCTIONS i^n erfc x

x	i erfc x	i^2 erfc x	i^3 erfc x	i^4 erfc x	i^5 erfc x	i^6 erfc x
0.0	0.5642	0.2500	0.0940	0.0313	0.0094	0.0026
0.01	0.5542	0.2444	0.0916	0.0303	0.0091	0.0025
0.02	0.5444	0.2438	0.0891	0.0294	0.0088	0.0025
0.03	0.5350	0.2335	0.0868	0.0285	0.0085	0.0023
0.04	0.5251	0.2282	0.0845	0.0277	0.0082	0.0023
0.05	0.5156	0.2230	0.0822	0.0268	0.0079	0.0022
0.06	0.5062	0.2179	0.0800	0.0260	0.0077	0.0021
0.07	0.4969	0.2129	0.0779	0.0252	0.0074	0.0020
0.08	0.4878	0.2080	0.0757	0.0245	0.0072	0.0019
0.09	0.4787	0.2031	0.0737	0.0237	0.0069	0.0019
0.10	0.4698	0.1984	0.0717	0.0230	0.0067	0.0018
0.11	0.4610	0.1937	0.0697	0.0223	0.0065	0.0017
0.12	0.4523	0.1892	0.0678	0.0216	0.0063	0.0017
0.13	0.4437	0.1847	0.0659	0.0209	0.0060	0.0016
0.14	0.4352	0.1803	0.0641	0.0203	0.0058	0.0015
0.15	0.4268	0.1760	0.0623	0.0197	0.0056	0.0015
0.16	0.4186	0.1718	0.0606	0.0190	0.0055	0.0014
0.17	0.4104	0.1676	0.0589	0.0184	0.0053	0.0014
0.18	0.4024	0.1635	0.0572	0.0179	0.0051	0.0013
0.19	0.3944	0.1596	0.0566	0.0173	0.0049	0.0013
0.20	0.3866	0.1557	0.0540	0.0167	0.0047	0.0012
0.21	0.3789	0.1518	0.0525	0.0162	0.0046	0.0012
0.22	0.3713	0.1481	0.0510	0.0157	0.0044	0.0011

APPENDIX 6 (*continued*)

x	i erfc x	i^2 erfc x	i^3 erfc x	i^4 erfc x	i^5 erfc x	i^6 erfc x
0.23	0.3638	0.1444	0.0495	0.0152	0.0043	0.0011
0.24	0.3564	0.1408	0.0481	0.0147	0.0041	0.0011
0.25	0.3491	0.1373	0.0467	0.0142	0.0040	0.0010
0.26	0.3419	0.1338	0.0454	0.0138	0.0038	0.0009
0.27	0.3348	0.1304	0.0441	0.0133	0.0037	0.0009
0.28	0.3278	0.1271	0.0428	0.0129	0.0035	0.0009
0.29	0.3210	0.1239	0.0415	0.0125	0.0034	0.0009
0.30	0.3142	0.1207	0.0403	0.0121	0.0033	0.0008
0.31	0.3075	0.1176	0.0391	0.0114	0.0032	0.0008
0.32	0.3010	0.1145	0.0379	0.0113	0.0031	0.0008
0.33	0.2945	0.1116	0.0368	0.0109	0.0030	0.0007
0.34	0.2882	0.1087	0.0357	0.0105	0.0029	0.0007
0.35	0.2819	0.1058	0.0346	0.0102	0.0027	0.0007
0.36	0.2758	0.1030	0.0336	0.0099	0.0027	0.0007
0.37	0.2722	0.0998	0.0330	0.0094	0.0026	0.0006
0.38	0.2637	0.0976	0.0316	0.0092	0.0025	0.0006
0.39	0.2579	0.0950	0.0306	0.0089	0.0024	0.0006
0.40	0.2521	0.0925	0.0297	0.0086	0.0023	0.0006
0.41	0.2465	0.0900	0.0288	0.0083	0.0022	0.0005
0.42	0.2409	0.0875	0.0279	0.0080	0.0021	0.0005
0.43	0.2354	0.0852	0.0270	0.0077	0.0020	0.0005
0.44	0.2300	0.0828	0.0262	0.0075	0.0020	0.0005
0.45	0.2247	0.0806	0.0254	0.0072	0.0019	0.0005
0.46	0.2195	0.0783	0.0246	0.0070	0.0018	0.0004
0.47	0.2144	0.0762	0.0238	0.0067	0.0017	0.0004
0.48	0.2094	0.0740	0.0230	0.0065	0.0017	0.0004
0.49	0.2045	0.0720	0.0223	0.0063	0.0016	0.0004
0.50	0.1996	0.0700	0.0216	0.0060	0.0015	0.0004
0.52	0.1902	0.0661	0.0203	0.0056	0.0014	0.0003
0.54	0.1811	0.0623	0.0190	0.0052	0.0013	0.0003
0.56	0.1724	0.0588	0.0177	0.0059	0.0012	0.0003
0.58	0.1640	0.0555	0.0166	0.0045	0.0011	0.0003
0.60	0.1559	0.0523	0.0155	0.0042	0.0010	0.0002
0.62	0.1482	0.0492	0.0145	0.0039	0.0010	0.0002
0.64	0.1407	0.0463	0.0136	0.0036	0.0009	0.0002
0.66	0.1335	0.0436	0.0127	0.0033	0.0008	0.0002
0.68	0.1267	0.0410	0.0118	0.0031	0.0007	0.0002
0.70	0.1201	0.0382	0.0110	0.0028	0.0007	0.0001

APPENDIX 6 (*Continued*)

x	i erfc x	i^2 erfc x	i^3 erfc x	i^4 erfc x	i^5 erfc x	i^6 erfc x
0.72	0.1138	0.0362	0.0103	0.0027	0.0006	0.0001
0.74	0.1077	0.0340	0.0096	0.0025	0.0006	0.0001
0.76	0.1020	0.0319	0.0089	0.0023	0.0005	0.0001
0.78	0.0965	0.0299	0.0083	0.0021	0.0005	0.0001
0.80	0.0912	0.0280	0.0077	0.0020	0.0005	0.0001
0.82	0.0861	0.0262	0.0072	0.0018	0.0004	0.0001
0.84	0.0813	0.0246	0.0067	0.0017	0.0004	0.0001
0.86	0.0767	0.0230	0.0062	0.0015	0.0003	0.0001
0.88	0.0724	0.0215	0.0057	0.0014	0.0003	0.0001
0.90	0.0682	0.0201	0.0053	0.0013	0.0003	0.0001
0.92	0.0642	0.0187	0.0049	0.0012	0.0003	0.0001
0.94	0.0605	0.0175	0.0046	0.0011	0.0003	0.0000
0.96	0.0569	0.0163	0.0043	0.0010	0.0002	0.0000
0.98	0.0535	0.0152	0.0039	0.0009	0.0002	0.0000
1.0	0.0503	0.0142	0.0036	0.0009	0.0002	0.0000
1.1	0.0365	0.0099	0.0025	0.0006	0.0001	0.0000
1.2	0.0260	0.0068	0.0016	0.0004	0.0001	0.0000
1.3	0.0183	0.0046	0.0011	0.0002	0.0000	0.0000
1.4	0.0127	0.0030	0.0007	0.0001	0.0000	0.0000
1.5	0.0086	0.0020	0.0004	0.0001	0.0000	0.0000
1.6	0.0058	0.0013	0.0003	0.0000	0.0000	0.0000
1.7	0.0038	0.0008	0.0002	0.0000	0.0000	0.0000
1.8	0.0025	0.0005	0.0001	0.0000	0.0000	0.0000
1.9	0.0016	0.0003	0.0001	0.0000	0.0000	0.0000
2.0	0.0010	0.0002	0.0000	0.0000	0.0000	0.0000

References

1. G. A. Akselrud, "The Theory of Extraction from Porous Bodies (Teoriya izvlecheniya veshchestv iz poristykh tel)." Izd. Lvovsk. Politekh. Inst., Lvov (1959).
2. V. P. Bakaleev, Possibility of solution of nonlinear heat conduction problems, *Inzh.-Fiz.Zh.* **4** (10), 119–122 (1961).
3. I. S. Berezin and N. P. Zhidkov, "Calculation Methods." Fizmatgiz, Moscow (1960).
4. L. Boltzmann, *Ann. Physik* **93**, 959 (1894).
5. T. G. Bromwich, Symbolical methods in theory of conduction of heat, *Proc. Cambridge Phil. Soc.* **220**, 441 (1921).
6. D. V. Budrin, Hydrostatic integrator for the solution of differential heat conduction equation accounting for the temperature dependence of thermal properties, *Tr. Ural'sk. Politekh. In-ta, Sb.* **53**, 22 (1955).
7. D. V. Budrin and E. A. Sukhanov, Regularization of temperature fields of simple-shape bodies, *Inzh.-Fiz.Zh.* **2** (1), 79 (1959).
8. H. S. Carslaw and J. C. Jaeger, "Conduction of Heat in Solids." Oxford Univ. Press, London and New York (1959).
9. H. S. Carslaw, "Introduction to the Mathematical Theory of the Conduction of Heat in Solids." Dover, New York (1945).
10. H. S. Carslaw and J. C. Jeager, "Operational Methods in Applied Mathematics." Oxford Univ. Press, London and New York (1941).
11. J. R. Carson, "Electric Circuit Theory and Operational Calculus." New York (1926).
12. H. Charles, Sur l'equation de la chaleur dans la theorie l'heredité, *Bull. Soc. Sci. Liege* (1946-8).
13. I. A. Charny, On linearization methods of nonlinear heat conduction equations, *Izv. Akad. Nauk SSSR, OTN*, No. 6, 828–843 (1951).
14. V. S. Chirkin, "Heat Conduction of Industrial Materials." Mashgiz, Moscow (1962).
15. A. F. Chudnovsky, "Thermal Properties of Disperse Materials." Fizmatgiz, Moscow (1962).

15a.R. Courant and D. Hilbert, "Methoden der Mathematischen" Sprınger-Verlag, Berlin (1931),

19. M. B. Coyle, Solution of transient heat conduction problems by the air analogy, *in* "Heat Transfer Problems" (Translation from English). Gosenergoizdat, Moscow-Leningrad (1959).

17. J. Crank, "The Mathematics of Diffusion." Clarendon Press, Oxford (1956).
18. I. N. Danilova, On a temperature field in a finite hollow cylinder with heat transfer at the boundaries, *Izv. Akad. Nauk SSSR. Otd. Tekh. Nauk. Energetika i Avtomatika*, No. 4, 151–155 (1960).
19. I. N. Danilova, Temperature distribution in an infinite hollow cylinder involved in heat transfer with media of variable temperature, *Izv. Akad. Nauk SSSR, OTN*, No. 12, 148–150 (1958).
20. A. Datsev, Godishnik na Sofiisky Universitet, Sofia, **43**, 113–131 (1946/1947).

21. K. G. Denbigh, "The Thermodynamics of the Steady State." Methuen, London (1951).
21a. V. A. Ditkin, "Operational Calculus." Gostekhizdat, Moscow (1947).
22. V. A. Ditkin and A. P. Prudnikov, "Integral Transformation and Operational Calculus." Fizmatgiz, Moscow (1961).
23. V. A. Ditkin and A. P. Prudnikov, "Operational Calculus of Two Variables and Its Application." Fizmatgiz, Moscow (1959).
24. G. Doetsch, Integration von Differentialgleichungen vermittels der endlichen Fourier Transformation, *Math. Ann.* **112**, 52–68 (1935).
25. G. Doetsch, "Theory und Anwendung der Laplace-Transformation." Berlin (1937).
26. G. N. Dulnev, "Heat Transfer in Radioelectronic Devices." Gosenergoizdat, Moscow (1963).
27. E. R. G. Eckert and R. M. Drake, Jr., "Heat and Mass Transfer." McGraw-Hill, New York (1959).
28. A. M. Efros and A. M. Danilevsky, "Operational Calculus and Contour Integrals (Operatsionnoe ischislenie i konturnye integraly)." UONTI Kharkov (1937).
29. M. V. Elistratova, A finite integral Hankel transform, *Inzh.-Fiz.Zh.* **3** (6), 120–125 (1960).
30. L. F. Epstein, Some infinite sums involving zeroes of $J_0(x)$, *J. Res. Natl. Bur. Std. B68*, No. 1, 17 (1964).
31. M. A. Evgrafov, "Analytical Functions (Analiticheskie funktsii)." Nauka, Moscow (1965).
32. N. E. Friedman, Analog method for study of transient heat flow in solids with temperature dependent thermal properties. *J. Appl. Phys.* **26**, 129–130, 783 (1956).
33. N. E. Friedman, Quasilinear heat flow, *Trans. Am. Soc. Mech. Engr.* **80** (3), 635 (1958).
34. H. Fujita, *Text. Res. J.* **22**, 757 (1952).
35. R. I. Gavrilova and A. P. Prudnikov, A problem of the heat-conduction theory, *Inzg.-Fiz.Zh.* **3** (5), 136-137 (1960).
36. I. M. Gelfand and G. E. Shilov, "Generalized Functions and Manipulations with Them." Vol. 1. Fizmangiz, Moscow (1952).
37. S. K. Godunov and V. S. Ryaben'ky, "Introduction to the Difference Scheme Theory (Vvedenie v teoriyu raznostnukh skhem)." Fizmatgiz, Moscow (1962).
38. E. M. Goldfarb, Simultaneous solution of heat conduction equations for a slab, cylinder, and sphere, *Nauk. Dokl. Vyssh. Shkoly, Met.* No. 3, 129 (1958).
39. T. R. Goodman, The Heat-Balance Integral—Further Considerations and Refinements, *Trans. Am. Soc. Mech. Engr.* **C83** (1), 83–86 (1961).
40. U. Grigull, "Temperaturausgleich in einfachen Körpern." Springer-Verlag, Berlin (1959).
41. G. A. Grinberg, "Selected Problems of the Mathematical Theory of Electrical and Magnetic Effects (Izbrannye voprosy matematicheskoi teorii elektricheskikh i magnitnykh yavlenii)," Izd. Akad. Nauk SSSR, Moscow (1948).
42. H. Gröber, E. Erk, and U. Grigull, "Die Grundgezetze der Wärmübertragung." Springer-Verlag, Berlin (1955).
43. S. R. de Groot and P. Mazur, "Nonequilibrium Thermodynamics." North-Holland Amsterdam (1962).
44. S. R. de Groot, "Thermodynamics of Irreversible Processes." North-Holland, Amsterdam (1956).

45. A. A. Gukhman, Introduction to the similarity theory (Vvedeniye v teoriuy podobiya), *Izd. Vyssh. Shkoly* (1963).
46. L. Hamburger, Introducere in Teoria Propagāru Cāldurii. I. Conducti a prin Solide. Editure Republicii Populare Romîna (1956).
47. O. Heaviside, "Electromagnetic Theory." London (1893).
48. F. G. Hooper, C. H. Miller, and J. F. Keffer, A hydraulic analogy to transient conduction in two-dimensional fields, *in* "International Developments in Heat Transfer", Part I. *Internatl. Heat Transfer Conf.*, 1961, p. 39 (1961).
49. G. P. Ivantsov and B. Ya. Lyubov, Heating of a fixed bed of spheres by a hot gas flow, *Dokl. Akad. Nauk SSSR* **86** (2), 293–296 (1962).
50. E. Janke, F. Emde, and F. Lösch, "Tafeln Hoherer Funktionen. B. G. Feubner. Verlagsgesellschaft." Teubner Verlagsgesellschaft, Stuttgart (1960).
51. H. Jeffreys, "Operational Method in Mathematical Physics." Cambridge Univ. Press, London and New York (1931).
52. A. V. Kantorovich and G. P. Akimov, "Functional Analysis in Rated Spaces (Funktsional'nyi analiz v normirovannykh prostranstvakh)." Fizmatgiz, Moscow (1959).
53. M. V. Kirpichev, "The Similarity Theory" (Teoriya podobiya), Izd. Akad. Nauk SSSR, Moscow (1953).
54. P. K. Konakov, "The Similarity Theory (Teoriya podobiya)." Gosenergoizdat, Moscow (1958).
55. G. M. Kondratiev, "Regular Thermal Regime (Regulyarnyi teplovoi rezhim)." Gostekhizdat, Moscow (1964).
56. V. G. Korenev, "Particular Problems of Elasticity and Heat Conduction Theory Solved in Bessel Functions (Nekotorye zadachi teorii uprugosti i teploprovodnosti, reshaemye v Besselevykh funktsiyakh). Fizmatgiz, Moscow (1960).
57. N. S. Koshlyakov, Basic Differential Equations of Mathematical Physics (Osnovnye differentsial'nye uravneniya matematicheskoi fiziki). Gostekhizdat, Moscow (1936).
58. S. S. Kovner, Conditions of thermal similarity in freezing and melting, *Izv. Akad. Nauk SSSR* (*Ser. Geogr. i Geofiz.*) **3**, 143 (1943).
59. L. A. Kozdoba, "Electrical Simulation of Temperature Fileds in Elements of Vessel Power Plants (Elektromodelirovanie temperaturnykh polei v detalyakh sudovykh energeticheskikh ustanovok)." Izd. Sudostr. Prom-sti, Leningrad (1964).
60. S. S. Kutateladze, "Fundamentals of Heat Transfer." Edward Arnold, London (1963).
61. M. A. Lavrentiev and B. M. Shabat, "Methods of the Theory of Complex Variable Functions (Metody teorii funktsii kompleksnogo peremennogo)." Fizmatgiz, Moscow (1958).
62. L. S. Leibenzon, "Manual on Oil Mechanics (Rukovodstvo po neftepromyslovoi mekhanike)." ONTI, Moscow-Leningrad (1931).
63. L. S. Leibenzon, On dynamic temperature condition for formation of folding on the earth surface and in cooling. *Izd. Akad. Nauk SSSR* (*Ser. Geog. i Geofiz.*), No. 6, 625 (1939).
64. G. Liebman, A new electrical analog method for the solution of transient heat conduction problems, *Trans. Am. Soc. Mech. Engr.* No. 3 (1956).
64a. A. V. Letnikov, "Differential Theory with Arbitrary Indicator." Matem. Sborn., Izd. S.-P. Akad. Nauk, St. Peterburg (1868).
65. A. V. Luikov, A contribution to the regular thermal regime theory, *Dokl. Akad. Nauk BSSR* **1** (2), 52–56 (1957).

66. A. V. Luikov, Application of irreversible thermodynamics methods to investigation of heat and mass transfer, *Internatl. J. Heat Mass Transfer* **1** (1965).
67. A. V. Luikov, Heat and mass transfer in capillary-porous bodies. *Advan. Heat Transfer* **1** (1964).
68. A. V. Luikov, "Heat Conduction and Diffusion (Teploprovodnost' i diffuziya)." Gizlegprom, Moscow (1941).
69. A. V. Luikov, "Heat Conduction in Unsteady Processes (Teploprovodnost' nestatsionarnykh protsessov)." Gosenergoizdat, Moscow (1948).
70. A. V. Luikov, "Heat Conduction Theory (Teoriya teploprovodnosti). Gostekhizdat." Moscow (1952); 2nd Ed. Vyssh. Shkola, Moscow (1967).
71. A. V. Luikov, The application of the Heaviside-Bromwich operational method to the solution of a problem in heat conduction. *Phil. Mag.* **7** (22), 239 (1936).
72. A. V. Luikov and Yu. A. Mikhailov, "Theory of Energy and Mass Transfer." Prentice-Hall, New York (1961); Pergamon Press, New York (1965).
73. A. V. Luikov and Yu. A. Mikhailov, "Theory of Heat and Mass Transfer." Gosenergoizdat, Moscow (1963); Theory of heat and mass transfer, *Israel Progr. Sci. Trans.* Cat. No. 2145 (1967).
74. V. S. Lukianov, Hydraulic instruments for engineering calculations, *Izv. Akad. Nauk SSSR, Otd. Tekh. Nauk* No .2 (1939).
75. A. I. Lurie, "Operational Calculus and Its Application to Mechanical Problems (Operatsionnoye ischisleniye i ego primeneniye k zadacham mekhaniki)." Gostekhizdat, Moscow (1950).
76. B. Ya. Lyubov, Calculation of the solidification rate of a casting accounting for the temperature dependence of metal thermal properties, *Dokl. Akad. Nauk SSSR* **42** (4), 763–766 (1953).
77. Sh. E. Mikeladze, Numerical integration of partial differential equations (Chislennye metody integrirovaniya differential'nykh uravnenii s chastnymi proizvodnymi), *Izd. Akad. Nauk SSSR, Moscow* (1936).
78. Yu. A. Mikhailov, Highly intensive heat and mass transfer in dispersed media, *Internatl. J. Heat Mass Transfer* **1** (1) (1960).
79. M. A. Mikheev, "Heat-Transfer Fundamentals (Osnovy teploperedachi)." Gosenergoizdat, Moscow-Leningrad (1956).
80. E. I. Mikulin, A temperature field of two solids with a slot, *Inzh. Fiz. Zh.* **4** (2) (1961).
81. J. Mikusinski, "Rachunek operatorow." Panst. Wyd. Naukowe, Warsaw (1953).
82. A. V. Minyatov, Heating of an infinite cylinder in a shell, *Zh. Tekh. Fiz.* **30** (6) (1960).
83. A. I. Mochalin, Application of the Dirac γ-function to the solution of parabolic partial differential equations, *in* "Heat and Mass Transfer in Evaporation Processes." Izd. Akad. Nauk SSSR, Moscow (1958).
84. A. D. Moore, Soap-film and sanded mapper techniques, *J. Appl. Mech., Trans. Am. Soc. Mech. Engr.* **72**, 291 (1950).
85. G. Myunts, "Integral Equations," Vol. 1. Gostekhizdat, Moskow (1934).
86. A. T. Nikitin, Temperature distribution in an infinite hollow cylinder involved in heat transfer with media of variable temperature. *Inzh. Fiz. Zh.* **3** (8), 74–77 (1960).
87. D. Yu. Panov, "Handbook on Numerical Solution of Partial Differential Equations (Spravochnik po chislennomu resheniyu differentsial'nykh uravnenii v chastnykh proizvodnykh)," 5th Ed. Gostekhizdat, Moscow (1951).

88. T. L. Perelman, On heat transfer in a pipe laminar flow, *Inzh. Fiz. Zh.* **4** (4) (1961).
89. T. L. Perelman, Temperature field in a hollow finite cylinder with internal sources, *Inzh. Fiz. Zh.* **3** (5) (1960).
90. Sh. N. Plyat, Heat and mass transfer of circular hollow cylinders under boundary conditions of the second kind, *Internatl. J. Heat Mass Transfer* **5** (1/2), 59 (1962).
91. Sh. N. Plyat, The distribution of temperature in abrasive workpieces in the process of thermal machining, *Inzh. Fiz. Zh.* **3** (7), 80–87 (1960).
92. G. V. Plyatsko, The temperature field in a hollow cylinder with variable suspended state, *Inzh. Fiz. Zh.* **2** (10), 65–71 (1959).
93. P. Ya. Polubarinova-Kochina, A particular nonlinear partial equation in the filtration theory, *Dokl. Akad. Nauk SSSR* **13** (6), 623–626 (1948).
94. A. A. Pomerantsev, A. S. Chernyshev, and I. L. Ferberov, The coal burning theory in a rectangular channel, *Izv. Akad. Nauk SSSR OTN* No. 7, 1067 (1948).
95. D. Pöschl, Berechnung der Oberflächentemperaturen geometrisch einfacher Körper bei Abkühlung oder Erwärmung der Isolierstoffe, *Z. Zngew. Math. Mech.* **12**, 280 (1932).
96. I. Prigogine, "Introduction to Thermodynamics of Irreversible Processes." Springfield, Thomas (1955).
97. A. P. Prudnikov, Integral solution of a mixed problem for a parabolic differential equation set, *Dokl. Zkad. Nauk SSSR* **115**, 869–871 (1957).
98. A. P. Prudnikov, Investigation of heat and mass transfer in disperse media, *Inzh. Fiz. Zh.* **1** (4) (1958).
99. V. K. Sauliev, "Integration of Parabolic Type Equations by the Net Method (Integrirovanie uravnenii parabolicheskogo tipa metodami setok)." Fizmatizdat, Moscow (1960).
100. M. Sawada, On the general solution of basic expression for heat transmission, *J. Soc. Mech. Engr., Japan* **35** (183), 695 (1932).
101. P. Schneider, "Conduction Heat Transfer." Eddison-Wesley, New York (1957).
102. P. Schneider, "Temperature Response Charts." Wiley, New York (1963).
102a. B. I. Segal and K. A. Semendyaev, "Five-Valued Mathematical Tables." Izd. Akad. Nauk, Moscow (1950).
103. N. G. Shimko, A finite integral Hankel transform for a hollow cylinder, *Inzh. Fiz. Zh.* **3** (10), 39 (1960).
103a. C. J. Shirtiffe and D. G. Stephenson, "Tabulated Values of Special Hyperbolic Functions." Nat. Res. Council, Ottawa (1961).
103b. N. Y. Sonin and P. A. Nekrasov, Mat. Sbornik, Izd. St.-P. Akad. Nauk, St. Petersburg (1888).
104. M. S. Smirnov, A heat conduction problem for a system of two bodies, *in* "Heat and Mass Transfer in Evaporation Processes," pp. 153–155. Izd. Akad. Nauk SSSR, Moscow (1958).
105. M. S. Smirnov, A Temperature Field in Free-Layer Wall at Fourth-Kind Boundary Conditions, *in* "Heat and Mass Transfer in Capillary-Porous Bodies," 8.17-20. Gosenergoizdat, Moscow-Leningrad (1957).
106. M. S. Smirnov, Investigation of transient heat transfer in bakery ovens, *in* "Heat and Mass Transfer of Food Stuffs." *Trudy Moskovsk. Tekh. In-ta Pishch Prom.* **6**, 78-89 (1956).
107. M. S. Smirnov, On a system of differential equations for highly intensive heat and mass transfer, *Internatl. J. Heat Mass Transfer* **1**, 521 (1962).

108. M. S. Smirnov, The heat-conduction problem for a material of variable thermal properties, *Trudy Mosk. Tekh. In-ta Pishch. Prom.* **2**, 250–253 (1958).
109. I. N. Sneddon, "Fourier Transforms." McGraw Hill, New York (1951).
110. R. H. Stokes. *Trans. Faraday Soc.* **48**, 1229 (1950).
111. M. L. Storm, Heat conduction in simple metals. *J. Appl. Phys.* **22**, 940–951 (1951).
112. W. G. Sutton, *Proc. Roy. Soc. Ser.* **A 182**, 48 (1943).
113. "Tables of Integral Transforms," Vols. 1 and 2. McGraw-Hill, New York (1954).
114. A. G. Temkin, "The Analytical Theory of Transient Drying Heat and Mass Transfer and Reverse Problems of the Analytical Drying Theory (Analiticheskaya teoriya nestatsionarnogo teplo-i massoobmena v protsesse sushki i obratnye zadachi analiticheskoi teorii sushki)." Dissertation. Inst. Heat Mass Transf., Minsk (1964).
115. I. M. Tetelbaum, "Electrical Simulation (Elektricheskoye modelirovaniye)." Fizmatgiz, Moscow (1959).
116. A. I. Tikhonov and A. A. Samarsky, "Equations of Mathematical Physics (Uravneniya matematicheskoi fiziki)." Gostekhizdat, Moscow (1953).
117. A. I. Tikhonov and A. A. Samarsky, "Mathematical Physics Equations (Uravneniya matematicheskoi fiziki)." Gostekhizdat, Moscow (1956).
118. C. G. Tranter, "Integral Transforms in Mathematical Physics." Methuen, London (1951).
119. E. P. Trofimov, The problem on a transient temperature field of an infinite hollow cylinder, *Inzh. Fiz. Zh.* **3** (10), 47–53 (1960).
120. P. V. Tsoi, A boundary-value problem for a parabolic-type differential equations set, *Inzh. Fiz. Zh.* **4** (12), 61–69 (1961).
121. P. V. Tsoi, Transient heat transfer of a system of bodies, *Inzh. Fiz. Zh.* **4** (1), 120–123 (1961).
122. B. Van der Pol and H. Bremmer, "Operational Calculus Based on the Two-Sided Laplace Integral." Cambridge Univ. Press, London and New York (1950).
123. A. P. Vanichev, Approximate method of heat conduction problems with variable constants, *Izv. Akad. Nauk SSSR, OTN* No. 12 (1946).
124. M. E. Vashchenko-Zakharchenko, "Symbolic Calculus and Its Application to Integration of Linear Differential Equations (Simvolicheskoye ischesleniye i ego prilozheniye k integrirovaniyu lineinykh differentsial'nykh uravnenii)." Kiev (1862).
125. P. Vernotte, La nouvelle equation de la chaleur, *J. Tranms. de la chaleur* (1961).
126. B. A. Volynsky and V. E. Buckhman, "Models for Boundary-Value Problem Solutions (Modeli dlya resheniya kraevykh zadach)." Fizmatgiz, Moscow (1960).
127. L. A. Vulis, I. F. Zherebyatiev and A. T. Lukiyanov, The solution of nonlinear heat conduction equations on static electrical integrators, *in* "Heat and Mass Transfer," Vol. 5. Izd. Akad Nauk BSSR, Minsk (1963).
128. P. P. Yushkov, Approximate solutions of the transient heat conduction method by the finite-difference method, *Trudy In-ta Energ. Akad. Nauk BSSR* **6**, (1958).
129. P. P. Yushkov, Bessel functions and their application to heat conduction problems (Funktsii Besselya i ikh prilozheniya k resheniyu zadach teploprovodnosti), *Izd. Akad. Nauk BSSR, Minsk* (1962).
130. P. P. Yushkov, Numerical integration of heat conduction equations in polar networks, *Proc. Leningrad Tech. Refrigeration Inst.* **14** (1956).
131. O. Wiedenburg, Über die Hydrodiffusion, *Ann. Phys.* (1879).

132. R. V. Churchill, "Operational Mathematics." McGraw-Hill, New York (1958).
133. G. I. Barenblatt, A particular method of heat conduction equation solution, *Dokl. Akad. Nauk SSSR* **72**, 667 (1950); On the solution of a heat conduction equation with a nonhomogeneous condition, *Dokl. Akad. Nauk SSSR* **74**, 201 (1950).
134. G. I. Barenblatt and B. M. Levitan, On some boundary-value problems of the turbulent diffusion equation, *Izv. Akad. Nauk SSSR, Ser. Mat.* **16**, 253 (1952); On some generalization of the Poisson formula of the heat conduction theory, *Dokl. Akad. Nauk SSSR* **79**, 917 (1951).
135. E. C. Titchmarsh, "Eigenfunction Expansions Associated with Second-Order Differential Equations." Clarendon Press, Oxford (1946).
136. B. M. Levitan, "Eigenfunction Expansions of Second-Order Differential Equations." Gostekhizdat, Moscow (1950).
137. I. M. Gelfand and G. E. Shilov, "Generalized Functions and Manipulations with Them." Fizmatgiz, Moscow (1959).

AUTHOR INDEX

Numbers in parentheses are reference numbers and indicate that an author's work is referred to, although his name is not cited in the text. Numbers in italics show the page on which the complete reference is listed.

SUBJECT INDEX

N

P

Q

R

S

T

U

V

Printed in Dunstable, United Kingdom

91338441R00393